OXFORD STUDIES IN PROBABILITY

OXFORD STUDIES IN PROBABILITY

1. Foundations of the prediction process
F. B. Knight

2. Poisson approximation
A. D. Barbour, L. Holst, and S. Janson

3. Poisson processes
J. F. C. Kingman

4. Limit theorems of probability theory
V. V. Petrov

5. Random geometric graphs
M. Penrose

6. Functional Gaussian Approximation for Dependent Structures
Florence Merlevède, Magda Peligrad, and Sergey Utev

Functional Gaussian Approximation for Dependent Structures

FLORENCE MERLEVÈDE

Université Paris-Est

MAGDA PELIGRAD

University of Cincinnati

SERGEY UTEV

University of Leicester

June 18, 2018

FLORENCE MERLEVÈDE

Université Paris-Est, LAMA and CNRS UMR 8050, Cité Descartes - 5 boulevard Descartes, Champs-sur-Marne, 77454 Marne-la-Vallée Cedex 2, France, Email: florence.merlevede@u-pem.fr

MAGDA PELIGRAD

Department of Mathematical Sciences, University of Cincinnati, PO Box 210025, Cincinnati, Oh 45221-0025, USA, Email: peligrm@ucmail.uc.edu

SERGEY UTEV

Department of Mathematics, University of Leicester, University Road, Leicester LE1 7RH, United Kingdom, Email: su35@le.ac.uk

OXFORD
UNIVERSITY PRESS

OXFORD
UNIVERSITY PRESS

Great Clarendon Street, Oxford, OX2 6DP,
United Kingdom

Oxford University Press is a department of the University of Oxford.
It furthers the University's objective of excellence in research, scholarship,
and education by publishing worldwide. Oxford is a registered trade mark of
Oxford University Press in the UK and in certain other countries

First Edition published in 2019
Impression: 1

Published in the United States of America by Oxford University Press
198 Madison Avenue, New York, NY 10016, United States of America

British Library Cataloguing in Publication Data
Data available

Library of Congress Control Number: 2018960483

ISBN 978–0–19–882694–1

DOI: 10.1093/oso/9780198826941.001.0001

Printed and bound by
CPI Group (UK) Ltd, Croydon, CR0 4YY

PREFACE

In the recent years there has been a tremendous effort towards a better understanding of the structure and asymptotic behavior of stochastic processes. For dependent processes the basic techniques are, after a suitable blocking argument, to approximate the process either by blocks of independent random variables or by martingales. Both these methods were successfully used for sequences having various mixing structures and led to sharp, unimprovable limit theorems of all kinds. Excellent techniques were developed under various mixing conditions as seen in surveys by Philipp (1986) and Dehling and Philipp (2002), a monograph by Rio (2000) and books by Bradley (2007). The only problems are first that the mixing coefficients can be hard to compute and then the conditions hard to verify, and second that many processes, even simple ones, are not mixing in the traditional sense. To treat a larger class of examples, the new direction in modeling the dependence is to redefine the dependence conditions by substantially reducing the classes of functions used in their definitions or by using innovative martingale-like conditions. In this way many new examples are included in the general structures and many general results can be established.

This book is a systematic effort to present new progress in both methods; the approximation of the process by either blocks of independent random variables or by martingale differences. As a matter of fact the book has double scope. First, to present tools for dealing with dependent structures directed to obtain normal approximations. Second, to apply the normal approximations presented in the book to various examples. The main tools consist of inequalities for dependent sequences of random variables, leading to limit theorems, including the functional central limit theorem and functional moderate deviation principle. The results will point out large classes of dependent random variables which satisfy invariance principles, making possible the statistical study of data coming from stochastic processes both with short and long memory.

The closest related monographs to this book are the influential books by Hall and Heyde (1980) on "Martingale limit theory and its applications" and the monograph by Richard Bradley (2007) on "Introduction to mixing conditions". It is the intention of the book to continue these monographs in the following way. Regarding mixing structures, the main goal is to stress the maximal inequalities and their application to the functional central limit theorem. We shall also generalize the mixing coefficients to include classes of negatively dependent random variables and associated sequences.

Concerning martingales, the goal is to characterize the stochastic processes that can be studied via martingale approximation. We shall develop necessary and sufficient conditions for the validity of a martingale approximation and also simple, sufficient conditions that are easy to verify, leading to a better understanding of the structure of some classes of stochastic processes and their asymptotic behavior. All kinds of limiting theorems can be obtained in this way, including conditional functional central limit theorem and moderate deviation. Motivated by practical applications, particular attention will be given to non-stationary structures and to processes which do not start from the equilibrium. For instance, new results, such as those concerning the study of the asymptotic behavior of partial sums associated with non-stationary dependent sequences satisfying projective conditions, will allow us to describe the asymptotic behavior of processes sampled by a shifted Markov chain or of random walks in random scenery under different kinds of assumptions on the random walk and on the random scenery.

Another direction of the book is to study classes of stochastic processes defined by using a restricted maximal coefficient of correlation based only on increasing functions, linking in this way the notion of mixing to the notions of association and negative dependence. The results will unify and extend these fields, providing also a technique to deal with processes whose evolution involves several independent components with different structures.

Several applications have been carefully selected to exhibit the importance of the theoretical results. They range from random walks in random scenery to determinantal processes. Due to their importance in analyzing new data in economics, linear processes with dependent innovations will also be considered and analyzed.

The book is a research book. The proofs are, most of the time, fully developed, so the book is also suitable as a textbook for an advanced course in probability. It is organized as follows: Part I is devoted to tools such as moment inequalities, maximal moment inequalities, variance behavior, central limit theorem and its functional form, or the moderate deviations principle and its functional form, for several classes of dependent random variables and under various dependent conditions, such as mixingale-type conditions, mixing (in the traditional sense) conditions, weak dependent conditions or weakly negatively dependent conditions. Part II contains several advanced examples for which the stochastic limiting behavior can be studied with the tools developed in Part I.

This text has evolved out of several collaborations among the authors during the last decade. Parts of this book were written during mutual visits. In particular we would like to thank University of Cincinnati and University of Paris-Est Marne la Vallée for facilitating numerous interactions between the co-authors. We would also like to thank George Deligiannidis for fruitful discussions concerning the results on random walks in random scenery. Finally, we acknowledge the financial support of two National Science Foundation grants, NSF DMS 1512936 and NSF DMS 1811373 as well as several Taft Research Center grants at the University of Cincinnati.

CONTENTS

LIST OF NOTATIONS

We set out here some notations used throughout this monograph.

$a_n \ll b_n$	For two sequences (a_n) and (b_n) of reals, this means that there exists a numerical constant C not depending on n such that $a_n \leq Cb_n$ for all n.
$a \wedge b, \min(a, b)$	Means the minimum between two reals a and b.
$a \vee b, \max(a, b)$	Means the maximum between two reals a and b.
x^+, x_+	For a real x the number $\max(0, x)$.
x^-, x_-	For a real x the number $\max(0, -x)$.
$f(x-0)$	For a function f, the left limit of f at point x.
$f(x+0)$	For a function f, the right limit of f at point x.
f^{-1}	The generalized inverse function of f.
$\mathbb{E}(X)$	For an integrable random variable X, the expectation of X.
$\mathrm{Var}(X)$	For a square integrable random variable X, the variance of X.
$\mathrm{cov}(X, Y)$	For two square integrable random variables X and Y, the covariance between X and Y.
$\mathrm{corr}(X, Y)$	For two square integrable random variables X and Y, the correlation between X and Y.
$Q_{\lvert X\rvert}$	For a random variable X, the càdlàg inverse of $t \to \mathbb{P}(\lvert X\rvert > t)$.
$\mathbf{1}_B, \mathbf{1}\{B\}, I(B)$	Indicator function of B.
$B^c, \bar{B}$	Complementary of B.
$\mathcal{A} \vee \mathcal{B}$	The σ-field generated by $\mathcal{A} \cup \mathcal{B}$.
$\mathbb{L}^p, \mathbb{L}^p(\mathbb{P}), \mathbb{L}_p, \mathbb{L}_p(\mathbb{P})$	For $p \geq 1$, the space of random variables X such that $\mathbb{E}(\lvert X\rvert^p) < \infty$.
$\mathbb{L}^p(\mathcal{F})$	The space of random variables that are $\mathcal{F}$-measurable and in $\mathbb{L}^p$.

$\mathbb{L}^\infty$	The space of almost surely bounded random variables.
$\|X\|_p$	For $p \geq 1$, the usual norm on $\mathbb{L}^p$.
$\|X\|_\infty$	The usual norm on $\mathbb{L}^\infty$.
$\mathbb{L}^p(\nu)$	For $p \geq 1$ and probability law ν on $\mathcal{X}$, the space of functions f such that $\int_{\mathcal{X}} \|f\|^p d\nu < \infty$.
$\|Y\|_{\nu,p}$ (resp. $\|Y\|_{\nu,\infty}$)	The $\mathbb{L}^p$-norm (resp. the $\mathbb{L}^\infty$-norm) of the r.v. Y on a probability space $(X, \mathcal{X}, \nu)$.
$\mathbb{E}(X\|\mathcal{A})$, $\mathbb{E}_{\mathcal{A}}(X)$	Conditional expectation of X conditionally to the σ-field $\mathcal{A}$.
$\mathrm{Var}_{\mathcal{A}}(X)$	Conditional variance of X conditionally to the σ-field $\mathcal{A}$.
$\mathbb{E}_i(X)$	Means $\mathbb{E}(X\|\mathcal{F}_i)$ when the underlying σ-field is $\mathcal{F}_i$ and when no confusion is possible.
$\mathbf{P}_i(X)$	Means $\mathbb{E}_i(X) - \mathbb{E}_{i-1}(X)$ (this is a projection operator).
$C([0, 1])$	The space of continuous functions from $[0, 1]$ to $\mathbb{R}$.
$D([0, 1])$	The space of càdlàg functions from $[0, 1]$ to $\mathbb{R}$.
$X_n \to^d X, X_n \Rightarrow X$	Convergence in distribution of X_n to X.
$X =^d Y, X =^D Y$	Means that X and Y have the same law.
$X_n \to^{\mathbb{P}} X$	Convergence in probability of X_n to X.
$(\alpha(n))_{n\geq 0}$, $(\alpha_n)_{n\geq 0}$	The sequence of strong mixing coefficients defined in (5.2).
$(\beta(n))_{n\geq 0}$, $(\beta_n)_{n\geq 0}$	The sequence of absolute regular coefficients defined in (5.2).
$(\phi(n))_{n\geq 0}$, $(\phi_n)_{n\geq 0}$	The sequence of ϕ-mixing coefficients defined in (5.3).
$(\rho(n))_{n\geq 0}$, $(\rho_n)_{n\geq 0}$	The sequence of maximal correlation coefficients defined in (5.3).
$(\alpha_{k,\mathbf{X}}(n))_{n\geq 0}$	For a sequence of real-valued r.v.'s $\mathbf{X} = (X_i)_{i\in Z}$ and a non-negative integer k, the α-dependent coefficients of order k defined in (5.13).
$(\beta_{1,\mathbf{X}}(n))_{n\geq 0}$, $(\beta_{2,\mathbf{X}}(n))_{n\geq 0}$	For a sequence of real-valued r.v.'s $\mathbf{X} = (X_i)_{i\in\mathbb{Z}}$, the β-dependent coefficients defined in (5.14) and (5.15).
$(\phi_{k,\mathbf{X}}(n))_{n\geq 0}$	For a sequence of real-valued r.v.'s $\mathbf{X} = (X_i)_{i\in Z}$ and a non-negative integer k, the ϕ-dependent coefficients of order k defined in (5.16).

α^{-1}	The function defined by $\alpha^{-1}(u) = \sum_{i\geq 0} 1_{u<\alpha_i}$.
$\alpha_{k,\mathbf{X}}^{-1}$	The function defined by $\alpha_{k,\mathbf{X}}^{-1}(u) = \sum_{i\geq 0} 1_{u<\alpha_{k,\mathbf{X}}(i)}$.
$(r_n)_{n\geq 1}$	The sequence of r-negative dependence defined in (9.1).
$(r'_n)_{n\geq 1}, (r^*_n)_{n\geq 1}$	The sequences of linear negative dependence defined in (7.16) and (7.17).
$(\rho^*_n)_{n\geq 1}$	The sequence of interlaced ρ-coefficients defined in (10.2).
A^c or $\bar{A}$	Complement of the set A.
$A \subset B$ or $A \subseteq B$	Means that for each $x \in A$ one has $x \in B$. The possibility of $A = B$ is included.
$A - B$ or $A \setminus B$	Is equal to $A \cap B^c$.
$A \Delta B$	Is the symmetric difference of A and B and is equal to $(A \cap B^c) \cup (B \cap A^c)$.

1

Introduction to Stochastic Processes

1.1 Introduction

The aim of this section is first to introduce (essentially heuristically) the main problems that motivated this textbook and also to give a collection of examples of stochastic processes whose stochastic behavior can be handled with the tools developed throughout the book.

1.1.1 Discussion

A stochastic process $\{S_t\}$ can be thought of as the function of time $t \in T$ with a hidden random parameter $w \in \Omega$, i.e.

$$S_t = S(t, w) ,$$

where the hidden random parameter w is not observable. Given $w = \cdot$ in a sample space Ω, $S_t = S(t, \cdot)$ is a random path (or a random element, i.e. a random function in a certain functional space). Given $t = \cdot$ in a time space T, $S_. = S(\cdot, w)$ is a random variable.

Nowadays, stochastic processes are an essential part of mathematics with many fascinating open problems. On the other hand, stochastic modeling is an important part of data analysis. Fitting a model to data and assessing how good is the fit is still one of the crucial problems in both applied and theoretical statistics.

A pragmatic statistical approach is to replace the original statistic S_t by the Gaussian model, that is to apply the Central Limit Theorem (CLT), i.e. to approximate the original statistic by the Gaussian variable with the same mean and same variance

$$S_t \approx N(\mathbb{E}(S_t), \mathrm{Var}(S_t)), \tag{1.1}$$

or with more general parameters (taking off outliers or extreme values, or in a more subtle way as happens very often for dependent variables)

$$S_t \approx N(a_t, b_t). \tag{1.2}$$

Functional Gaussian Approximation for Dependent Structures. Florence Merlevède, Magda Peligrad and Sergey Utev. Oxford University Press (2019).
DOI: 10.1093/oso/9780198826941.001.0001

In both approximations above $N(a, b)$ stands for a Gaussian random variable with mean a and variance b and, for two stochastic processes, (X_t) and (Y_t), $X_t \approx Y_t$ means that the two stochastic processes have "approximatively" the same law if t is large enough.

In addition, to treat more general statistics such as $\max_{1 \le i \le n} S_i$ or $\int_0^1 f(u, S_u)du$, the paths approximation is more appropriate, which leads what is usually called the functional CLT or the (weak) invariance principle (IP). We refer to Section 1.3.2 for more details concerning the functional form of the CLT. In particular, this type of result leads to

$$(S_{t_1}, \dots, S_{t_n}) \approx (N_1(a_1, b_1), \dots, N_n(a_n, b_n)), \tag{1.3}$$

where $(N_1(a_1, b_1), \dots, N_n(a_n, b_n))$ is a Gaussian vector.

For example, a Bayesian approach in MCMC (Markov Chains Monte Carlo methods) is based on

(i) first, constructing or modeling Markov chains (often reversible) to simulate the random parameter treated as a stationary distribution (Markov Chains part) and
(ii) then computing or estimating the parameter as an average (Monte Carlo part).

It is then traditional to declare that the error or the confidence interval is computed from the Gaussian approximation (1.1) (or (1.2)). It immediately motivates several important questions:

- When does the Gaussian approximation (1.1) (or (1.2)) hold? Or even the paths approximation leading to (1.3)?
- Are (1.1), (1.2) and (1.3) equivalent or does the Gaussian approximation hold together with the convergence of the 1st, 2nd or higher order moments?

To make questions more rigorous, we shall essentially work all along this book with the following setting. For any positive integer n, let $S_n = \xi_1 + \dots + \xi_n$ denote the partial sum associated with a sequence $\xi = (\xi_i)_{i \in \mathbb{Z}}$ of real-valued random variables which is assumed to satisfy the following assumptions:

$$(H) \begin{cases} \text{– (SC) condition: it is } \textit{a strictly stationary} \text{ sequence} \\ \text{– (DC) condition: it satisfies some } \textit{dependence conditions} \\ \text{– (P0) condition: } 0 < \mathbb{E}\left[\xi_1^2\right] < \infty \text{ and } \mathbb{E}\left[\xi_1\right] = 0 \end{cases}$$

Different kind of dependence conditions will be considered in this textbook: Mixingale (or projective)-type conditions as those developed in Chapter 4, mixing conditions or weak dependent conditions with or without rate (see Sections 5.1 and 5.2 for the definitions of traditional mixing coefficients and weak dependence coefficients considered in the book), or weakly negatively dependent conditions (as defined in Chapter 7).

Let us state now several basic problems addressed in this book.

Fix a limiting constant $\sigma > 0$ and consider the following properties:

(P1): $\mathrm{Var}(S_n)/n \to \sigma^2$ (*asymptotic linear behavior of the variance*)

(P2): $S_n/\sqrt{n} \to^d N(0, \sigma^2)$ (the CLT question with standard normalization)

(P3): $\{S_{[nt]}/\sqrt{n}, t \in [0,1]\} \to^d \sigma W$ in $D([0,1])$ equipped with the uniform topology, where W is a standard Brownian motion (the IP question with standard normalization)

The following natural questions then arise. If we assume (H): When does one or all of the properties above hold? More exactly, assuming (SC) and (P0) conditions, what dependence conditions do we need to impose to guarantee that one or all properties hold?

The motivation obviously comes from the classical CLT and IP for the *benchmark model of i.i.d. variables*. In this case, we have:

Fact 1.1 *Let $\xi = (\xi_i)_{i\in\mathbb{Z}}$ be a sequence of independent and identically distributed (i.i.d.) real-valued random variables sequence. Then*

$$(P0) \iff (P1) \iff (P2) \iff (P3). \tag{1.4}$$

Contrary to the i.i.d. case, in general, in the stationary setting only, the equivalences in (1.4) do not hold. To convince ourselves, it suffices to consider the so-called telescoping sums $\xi_i = Y_i - Y_{i-1}$ with $(Y_i)_{i\in\mathbb{Z}}$ being a stationary sequence of r.v.'s that can even be i.i.d.

On the other hand, under (P0), (1.4) holds for many classes of stationary sequences such as important benchmark models like ergodic sequences of martingale differences, m-dependent sequences, or mixing sequences with mild mixing rates (like ϕ-mixing or ρ-mixing sequences). But in general, the properties (P1), (P2) or (P3) do not hold for strongly mixing sequences without extra moment assumptions on ξ_1, whatever is the rate of mixing.

For sequences of i.i.d. centered real-valued r.v.'s with finite variance (i.e. satisfying (P0)), the only non-trivial normalization in CLT is proportional to $\sqrt{n}$. Dependent structures give rise to a variety of normalizations, which raises the CLT question of the partial sum or the IP question of the partial sums process $\{S_{[nt]}, t \in [0,1]\}$ under a more general normalization b_n rather than $\sqrt{n}$ (see for instance Section 4.5 or Chapter 13).

From the limiting behavior point of view it may be natural to refer to sequences satisfying (P1)–(P3) as "CLT weakly dependent" sequences. Some questions then arise as: Let $T : (\xi_i)_{i\in\mathbb{Z}} \to (Y_i)_{i\in\mathbb{Z}}$ be a class of transforms which preserves stationarity. Does T preserve the CLT weak dependence?

In particular, we address this question for linear processes where $(T(\xi_i)_{i\in\mathbb{Z}})_k = \sum_{j\geq 0} a_j \xi_{k-j} := Y_k$ (see Chapter 12).

Note that the transform $\xi_i \to f(\xi_i)$ preserves stationarity, i.e. transfers a stationary sequence to another stationary sequence. This particular transform $\xi_i \to f(\xi_i)$ is sometimes referred to as *an instantaneous function.*

Preservation of asymptotic behavior properties is of course a relatively traditional question, and we note that it is not easy to construct a transform which does not preserve ergodicity and that many transforms preserve the CLT property. However, even if we start with a sequence of i.i.d. r.v.'s, just a few transforms preserve large or even moderate deviations.

The problem of finding weak enough dependence conditions to guarantee that one or all of the properties (P1), (P2) or (P3) hold is divided into several subproblems.

For example, the extension from the CLT to FCLT/IP requires maximal inequalities. Thus an essential part of finding conditions for (P3) to hold is the form of the preservation of maximal inequalities under transforms.

As a matter of fact, maximal inequalities and also moment inequalities, maximal moment inequalities, variance behavior and variance and covariance bounds are interesting by themselves and are some of the *essential tools* in this monograph.

In order to accommodate these and other important features under the roof of a monograph of reasonable size, we concentrate on methods and omit some of the algebraic manipulations if they do not lead to special interests. Part I of this book is devoted to tools such as moment inequalities, maximal moment inequalities, variance behavior, CLT, FCLT or moderate deviations for several classes of dependent random variables. Part II contains several advanced examples such as those developed in the following section.

1.1.2 *Examples*

In this part, we survey several approaches to stochastic modeling with a particular emphasis given to processes providing examples and counterexamples to the problems stated in the previous section.

A) Linear processes

(1) *General linear processes.* A simple example of stochastic process is the *deterministic process,* namely a single function $X_t = g(t)$. A slight modification generates a large class by considering arbitrary deterministic behavior $g_i(t)$ on different disjoint regions $A_i \subseteq \Omega$, e.g.

$$X_t = \sum_{j\in\mathbb{Z}} g_j(t) I_{A_j},$$

where the sequence $(I_{A_j})_j$ is actually negatively dependent (see definition (9.2)). A simple modification leads then to linear processes

$$X_t = \sum_{j\in\mathbb{Z}} g_j(t)\eta_j.$$

For example, any complex-valued stochastic process $\{X_t : t \in [0, 1]\}$ with (random) paths in $\mathbb{L}_2$ admits the Fourier decomposition

$$X_t = \sum_{j=-\infty}^{\infty} e^{i2\pi tj}\eta_j, \tag{1.5}$$

over the orthonormal system $(e^{-i2\pi tj})_{j\in\mathbb{Z}}$ with random coefficients η_j, in general. The sequence of $(\eta_j)_{j\in\mathbb{Z}}$ is usually treated as noise. A *strong white noise* corresponds to a sequence of i.i.d. centered r.v.'s in $\mathbb{L}_2$ with positive variance.

Also, a Brownian motion $\{W_t : t \geq 0\}$ with mean zero and variance parameter σ^2 has the following representation via i.i.d. standard normal variables $(Z_i)_{i\geq 1}$

$$W_t = \frac{\sqrt{2}}{\pi}\sum_{k=1}^{\infty} \frac{\sin((k-1/2)\pi t)}{(k-1/2)} Z_k.$$

(2) A large class of *stationary linear processes* in discrete time with stationary innovation $(\xi_j)_{j\in\mathbb{Z}}$ is defined by

$$X_k = \sum_{j=-\infty}^{\infty} a_{k-j}\xi_j.$$

Linear processes provide a rich class of examples or counterexamples to the problems raised in the previous section and to other various generalizations of limit theorems from i.i.d. to stationary sequences.

For example, notice that the uniform variable U on interval $[0, 1]$ has the following dyadic representation

$$U = \sum_{i=1}^{\infty} 2^{-i}\delta_i,$$

where $(\delta_k)_{k\in\mathbb{Z}}$ is a sequence of i.i.d. r.v.'s with $\mathbb{P}(\delta_k = 0) = \mathbb{P}(\delta_k = 1) = 1/2$ which suggests a Bernoulli shift linear process

$$U_k = \sum_{i=1}^{\infty} 2^{-i}\delta_{i+k}.$$

(3) A simple but powerful subclass of counterexamples to necessary conditions for limit theorems for stationary sequences, as comparing with sequences of i.i.d. r.v.'s, is generated by telescopic sums

$$X_k = \xi_k - \xi_{k-1},$$

where $(\xi_i)_{i\in\mathbb{Z}}$ is a stationary sequence.

B) Functional transform processes

A more general class is the *functional transform process* where the noise $\xi = (\xi_i)_{i\in\mathbb{Z}}$ is a stationary sequence and the randomness is only due to this noise ξ, i.e.

$$X_t = f(t, (\xi_i)_i).$$

(1) The classical example with a noise being a sequence of i.i.d. r.v.'s is the *renewal process* or *counting process* defined by

$$N_t = \max\{n \geq 1 : \xi_1 + \ldots + \xi_n \leq t\}.$$

(2) Let $\mathbb{R}^{\mathbb{Z}}$ be a space of double infinite sequences $x = (x_i)_{i\in\mathbb{Z}}$ and S be the left shift operator, i.e. $S[x](i) = x_{i+1}$, $i \in \mathbb{Z}$. Let $f : \mathbb{R}^{\mathbb{Z}} \to \mathbb{R}$ be a measurable function. The pair (S, f) transforms a stationary sequence $(\xi_i)_i$ into another stationary sequence by

$$X_k = f(S^k[(\xi_i)_{i\in\mathbb{Z}}]).$$

(3) Taking $f(x) = \sum_j a_j x_{-j}$ for the linear functional provides *another representation of linear processes*

$$X_k = \sum_{j=-\infty}^{\infty} a_j S^k[(\xi_i)_i](-j) = \sum_{j=-\infty}^{\infty} a_j \xi_{k-j}.$$

(4) *Adapted transform/sequences* are usually referred to the case when the transform is such that $f(x) = f(\dots, x_{-1}, x_0)$, so that the stationary sequence is then defined by

$$\zeta_k = f(\dots, \xi_{k-2}, \xi_{k-1}, \xi_k).$$

Note that adapted transforms preserve ergodicity.

(5) *Linear adapted sequences (also called moving averages)*. The adapted version of the linear functional is $f(x) = \sum_{j=0}^{\infty} a_j x_{-j}$, which generates linear adapted sequences

$$X_k = \sum_{j=0}^{\infty} a_j \xi_{k-j} = \sum_{j=-\infty}^{k} a_{k-j} \xi_j.$$

(6) *Linear time series (ARMA processes)* represent an important subclass of the linear adapted sequences which admits the following linear regression representation

$$X_k = \theta X_{k-1} + \xi_k$$

or more generally

$$X_k = \theta_1 X_{k-1} + \dots + \theta_d X_{k-d} + \xi_k,$$

where ξ_k is the noise, usually a sequence of i.i.d. r.v.'s or an ergodic and stationary sequence of martingale differences, and the $\theta, \theta_1, \dots, \theta_d$ are real numbers.

(7) *m-dependent adapted sequences* are usually referred to the case when $f(x) = f(x_{-m+1}, \dots, x_0)$ so that the stationary sequence is then defined by

$$X_k = f(\xi_{k-m+1}, \dots, \xi_k).$$

For i.i.d. innovations $(\xi_i)_{i\in\mathbb{Z}}$, it defines an adapted m-dependent stationary sequence in the sense that $\sigma(X_\ell, \ell \le i)$ and $\sigma(X_\ell, \ell \ge j)$ are independent if $j - i \ge m$.

C) Recursive sequences and dynamic systems

(1) *Discrete dynamical systems.* We assume that the information is given recursively by

$$x_{k+1} = T(x_k).$$

Given an initial value $x_0 = x$, the recurrent procedure generates a path/orbit

$$x \to T(x) \to T(T(x)) = T^2(x) \to T(T(T(x))) = T^3(x) \ldots$$

Now, let $x = X_0$ be random and assume that the transform T preserves the distribution of X_0, i.e. $T(X_0) =^d X_0$. Then, the recurrence generates a stationary sequence

$$X_k = T^k(X_0),$$

and any stationary sequence $(X_k)_k$ has infinitely many representations with prescribed joint distributions.

(2) *Markov chains as random recursive schemes or non-linear time series* are defined by

$$X_{k+1} = T(X_k, \xi_{k+1}),$$

where $(\xi_k)_k$ is a sequence of i.i.d. r.v.'s and the initial state X_0 is adapted, i.e. measurable with respect to $\sigma(\xi_j, j \leq 0)$ (that is the σ-field generated by $\xi_j, j \leq 0$).

(3) *Strong noise.* An important example of Markov chains is obviously a sequence of i.i.d. r.v.'s $(X_i)_i$. This is a *benchmark model* to compare results with and generalize from.

(4) *Renewal chain.* The Markov chain given on $\mathbb{Z}_+$ is motivated by the counterexamples to ergodic theorem and can be thought of as a simple random algorithm. Given a sequence $(\xi_i)_{i\in\mathbb{Z}}$ of i.i.d. r.v.'s with values in $\mathbb{Z}_+$, the chain is then defined by

$$\begin{aligned} &\text{for } X_k > 0, \quad X_{k+1} = X_k - 1; \\ &\text{for } X_k = 0, \quad X_{k+1} = \xi_{k+1}. \end{aligned}$$

Starting from a positive value, the chain deterministically goes down all the way until it hits 0 and then it renovates.

D) Additive functions of Markov chains

Homogeneous Markov chains not having equilibrium/stationary distribution and more generally additive functions of Markov chains provide another important class of "pseudo" examples to the main questions raised in the previous section.

(1) Let consider S_n be a *random walk/non-homogeneous Markov chain* with i.i.d. jumps ξ_i and starting at $S_0 = 0$, that is

$$S_n = \sum_{i=1}^{n} \xi_i.$$

To determine a two-sided random walk, we continue it to the left by reflecting it; in other words, for any $n \geq 1, S_{-n} = -\sum_{k=1}^{n} \xi_{-k}$ and then $(S_{-n})_{n\geq 0} =^d (-S_n)_{n\geq 0}$.

(2) *Additive functionals of a random walk.* They are defined by $\Sigma_{k=1}^{n} f(S_k)$ where f is a measurable function and S_k is a random walk. They provide another simple source for dependence. Notice that when f is non-decreasing, it creates a positively associated sequence.

(3) *Self intersections.* Applying the shift transform construction to random walks and the functional which counts intersections $f(x) = \sum_{j=-\infty}^{\infty} I(x_j = x_0)$, we discover the stationary sequence

$$X_k = \sum_{j=-\infty}^{\infty} I(S_{k+j} = S_k),$$

which is well-defined for the transient random walk S_k.

For the recurrent case, $X_k = \infty$, it is convenient to consider its time truncated version

$$X_{k,n} = \sum_{j=1}^{n} I(S_j = S_k).$$

For each fixed n, the random variables $X_{k,n}$ are approximately stationary connected and provide an example of strongly dependent variables and examples/counterexamples to the main questions raised in the previous section.

1.2 Stationary Sequences

1.2.1 *Definitions*

We begin with some general definitions. It is convenient to let

$$X_A = (X_i, i \in A), \text{ for a set } A \subseteq \mathbb{Z}.$$

a) Stationarity assumption and stationary sequences

We say that $X_{\mathbb{Z}} = \{X_i\} = \{X_i : i \in \mathbb{Z}\}$ is a (strictly) stationary sequence if for any integers $n, i_1, \ldots, i_n$ and h

$$(X_{i_1+h}, \ldots, X_{i_n+h}) =^d (X_{i_1}, \ldots, X_{i_n}). \tag{1.6}$$

There are many equivalent ways to define the stationary sequence. A common way is to apply the (left) shift operator $S : \mathbb{R}^{\mathbb{Z}} \to \mathbb{R}^{\mathbb{Z}}$ defined by

$$S(\{x_i\})_i = x_{i+1} \ , \ i \in \mathbb{Z}. \tag{1.7}$$

and then, the stationarity means that the original sequence $\{X_i\}$ has the same distribution as the shifted sequence $S(\{X_i\})$ written as

$$S(\{X_i\}) =^d \{X_i\} \quad \text{or} \quad X_{\mathbb{Z}+1} =^d X_{\mathbb{Z}}.$$

In other words, the transform S is measure preserving, i.e.

$$\mathbb{P}(\{X_i\} \in W) = \mathbb{P}(\{X_i\} \in S^{-1}(W)),$$

and the probability measure $\mathbb{P}$ is referred to as invariant measure, or S–invariant measure.

Stationary sequence on a set Now, consider a set of variables $X_{\mathcal{S}} = \{X_i : i \in \mathcal{S}\}$ given on a subset $\mathcal{S}$ of a commutative group $(G, +)$ with unity 0. The stationarity assumption is almost the same

$$(X_{i_1+h}, \dots , X_{i_n+h}) =^d (X_{i_1}, \dots , X_{i_n}).$$

However with an **extra condition** that both the initial set $i_1, \dots, i_n$ and the shifted index $i_1 + h, \dots, i_n + h$ must belong to $\mathcal{S}$ to make sense.

For $\mathcal{S} = \mathbb{Z}^+$, we derive a **one-sided** stationary sequence, which can be extended to a two-sided stationary sequence. For $\mathcal{S} = \mathbb{R}$, we derive a continuous time stationary process. For the finite set $\mathcal{S} = \{1, \dots , n\}$ we derive a stationary connected vector as a part of a stationary connected triangular array.

b) Filtrations

We say that a sequence $\{\mathcal{G}_k\} = \{\mathcal{G}_k \ ; \ k \in \mathbb{Z}\}$ of σ-fields is a non-decreasing filtration or simply a filtration if $\{\mathcal{G}_k\}$ is non-decreasing, i.e. $\mathcal{G}_k \subseteq \mathcal{G}_j$ for $k \le j$.

Let $\sigma(X_A) = \sigma(X_i, i \in A)$ denote the σ–field generated by X_i with indexes $i \in A$. In particular, let

$$\mathcal{F}_{a,b} = \sigma(X_i, a \le i \le b), \quad \text{and}$$
$$\mathcal{F}_{-\infty,k} = \sigma(X_i, i \le k) \quad \text{called a } \textbf{natural filtration.}$$

Notice that $\{\mathcal{F}_{-\infty,k}\}_{k\in\mathbb{Z}}$ is a non-decreasing sequence of σ-fields and such that X_k is $\mathcal{F}_{-\infty,k}$-measurable written as $X_k \in \mathcal{F}_{-\infty,k}$.

For any filtration $\{\mathcal{G}_k\}$, associate the largest and smallest σ-field generated by it

$$\mathcal{G}^{\infty} = \bigvee_k \mathcal{G}_k = \sigma\Big(\bigcup_k \mathcal{G}_k\Big) \ , \ \mathcal{G}_{-\infty} = \bigcap_k \mathcal{G}_k.$$

Stationary filtration Let $\{\mathcal{G}_k\} = \{\mathcal{G}_k;\ k \in \mathbb{Z}\}$ be a non-decreasing sequence of σ-fields. For all integers m and h, let us construct conditional measure valued processes

$$\mathbb{P}_{m,h} \equiv \mathbb{P}_{m,h}(A) = \mathbb{P}((X_\ell, \ell \geq m) \in A|\mathcal{G}_h).$$

We say that $\{\mathcal{G}_k\}$ is a stationary filtration if the sequence $\{\mathbb{P}_{m+i,h+i}\ ;\ i \in \mathbb{Z}\}$ is stationary.

Equivalently, for all integers $n, m_1, \ldots, m_n, h_1, \ldots, h_n, u$, by taking any measurable functionals $f_j : \mathbb{R}^{\mathbb{N}} \to \mathbb{R}, j = 1, \ldots, u$ and measurable function $g : \mathbb{R}^u \to \mathbb{R}$

$$\begin{aligned}&\mathbb{E}g\left(\mathbb{E}[f_1(X_\ell, \ell \geq m_1)|\mathcal{G}_{h_1}], \ldots, \mathbb{E}[f_u(X_\ell, \ell \geq m_n)|\mathcal{G}_{h_n}]\right)\\ &\quad = \mathbb{E}g\left(\mathbb{E}[f_1(X_\ell, \ell \geq m_1 + i)|\mathcal{G}_{h_1+i}], \ldots, \mathbb{E}[f_u(X_\ell, \ell \geq m_n + i)|\mathcal{G}_{h_n+i}]\right) \text{ for any } i \in \mathbb{Z}.\end{aligned}$$

Adapted filtration The non-decreasing set of σ-fields $\{\mathcal{G}_k\}$ is an adapted filtration for the sequence $\{X_k\}$ if $X_k \in \mathcal{G}_k$ for any k or the sequence $\{X_k\}$ is adapted to the filtration $\{\mathcal{G}_k\}$.

Example: *The natural filtration is adapted and stationary.*

Now, we present several ways to construct stationary sequences with stationary filtrations.

1.2.2 Constructions

a) Construction of a stationary sequence via the Kolmogorov theorem

Two-sided family Let k be a positive integer. For a class of finite subsets $\{i_1, \ldots, i_k\} \subset \mathbb{Z}$, consider a family of multivariate distributions $\mu_{i_1,\ldots,i_k}$ such that the following Kolmogorov compatibility conditions hold:

$$\mu_{i_1,\ldots,i_k}(H_1 \times \cdots \times H_k) = \mu_{i_{\pi(1)},\ldots,i_{\pi(k)}}(H_{\pi(1)} \times \cdots \times H_{\pi(k)}),$$

where π is a permutation of $(1, 2, \ldots, k)$, and

$$\mu_{i_1,\ldots,i_{k-1}}(H_1 \times \cdots \times H_{k-1}) = \mu_{i_1,\ldots,i_{k-1},i_k}(H_1 \times \cdots \times H_{k-1} \times \mathbb{R}).$$

In addition, assume that the following **stationarity condition** holds

$$\mu_{i_1+h,\ldots,i_k+h} = \mu_{i_1,\ldots,i_k},$$

for all integers $k, i_1, \ldots, i_k$ and h.

Then, there exists a stationary sequence $\{X_i\} = \{X_i : i \in \mathbb{Z}\}$ with the prescribed multivariate distributions, i.e. for all integers $i_1, \ldots, i_k$

$$(X_{i_1}, \ldots, X_{i_k}) =^d \mu_{i_1,\ldots,i_k}.$$

The sequence $\{X_i\}$ can be constructed on the probability triple $(\mathbb{R}^{\mathbb{Z}}, \mathcal{B}, P)$ by setting $X_j(x) = x_j$ for any $j \in \mathbb{Z}$. $\mathbb{R}^{\mathbb{Z}}$ has the topology generated by the cylinder open sets:

$$\{x : x_{i_j} \in G_j, j = 1, \dots, n\}$$

for all open subsets $G_j \subset \mathbb{R}$ and integers n, $i_1 \dots, i_n$. $\mathcal{B}$ is its Borel σ-field and P is the resulting probability measure constructed on $\mathcal{B}$ satisfying $P(X_{i_j} \in G_j, j = 1, \dots, n) = \mu_{i_1,\dots,i_n}(G_1 \times \cdots \times G_n)$. The existence of such a probability is ensured by the Kolmogorov existence theorem (see Theorem 36.1 in Billingsley (1995)). The measure-preserving map is the (left) shift operator $S : \mathbb{R}^{\mathbb{Z}} \to \mathbb{R}^{\mathbb{Z}}$ defined by (1.7) which is one-to-one. Both S and S^{-1} are continuous, and hence measurable, so S is referred to as a bijective bi-measurable transformation preserving the probability measure.

b) Reconstruction of the stationary sequence

Let $\mathcal{C}$ be a collection of finite subsets $J \subseteq \mathbb{Z}$ and $F_{\mathcal{C}} = \{F_J \; ; \; J \in \mathcal{C}\}$ be a family of joint distributions which satisfies the Kolmogorov compatibility property and the following stationarity restriction: If $J, J + h \in \mathcal{C}$ then

$$F_J = F_{J+h}.$$

Does there exist a stationary sequence X_t with the prescribed multivariate distributions? And if exists, does it uniquely determine the stationary version?

One-sided family and its extension Notice that when $\mathcal{C} = \mathcal{C}A$ is the class of all subsets in an infinite interval $A \in D = \{a, a+1, \dots\}$ (e.g. on $\mathbb{Z}_+$ only) then it does exist and uniquely determines the stationary version of the extended family of distributions for $J \subset \mathbb{Z}$ by simply shifting set J to interval D and observing that

$$\text{for} \quad h > a + \max_{i \in J} |i| \quad \text{we have} \quad F_{J+h} = F_J.$$

c) Stationary sequence as a dynamical system

Another way to define a stationary sequence is to start with the probability space $(\Omega, \mathcal{A}, \mathbb{P})$, and then define $T : \Omega \mapsto \Omega$ as being a measurable transformation preserving the probability, i.e. for any measurable set U,

$$\mathbb{P}(U) = \mathbb{P}(T^{-1}U).$$

Example: *Let $p \geq 2$ be a positive integer and $T : [0, 1] \mapsto [0, 1]$ be defined by $T(x) = px - [px]$. Then this map preserves the Lebesgue measure on $[0, 1]$. Moreover, Fourier transform technique shows that the Lebesgue measure is the only T-invariant absolute continuous distribution.*

Adapted/non-adapted filtration Here the stationary sequence and the filtration are simultaneously defined by the following procedure. Let X_0 be a random variable. Let $\mathcal{G}_0$ be a

σ-field of $\mathcal{A}$ satisfying $\mathcal{G}_0 \subseteq T^{-1}(\mathcal{G}_0)$. We then define the filtration $(\mathcal{G}_i)_{i\in\mathbb{Z}}$ by $\mathcal{G}_i = T^{-i}(\mathcal{G}_0)$ (referred to as the stationary filtration). Finally, define the stationary sequence $(X_i)_{i\in\mathbb{Z}}$ by $X_i = X_0 \circ T^i$. Then, $\{X_k\}$ is a stationary sequence and $\{\mathcal{G}_k\}$ is a stationary filtration for the sequence. The filtration is adapted if and only if X_0 is $\mathcal{G}_0$–measurable.

d) Stationary sequence as a functional of another stationary sequence

Fix a measurable function $g : \mathbb{R}^\infty \to \mathbb{R}$, an underlying real-valued stationary sequence $\{Y_k\}$ and its natural filtration $\{\mathcal{G}_k\}$. Define

$$X_k = g(Y_j, j \le k).$$

Then, $\{X_k\}$ is a stationary sequence and $\{\mathcal{G}_k\}$ is an adapted stationary filtration of $\{X_k\}$.

1.2.3 Ergodicity

a) Invariant σ-field

Invariance in the algebraic sense We say that A is algebraically T–invariant, if $A = T^{-1}A$. The algebraically T–invariant sets form a σ-field.

Example: *The countable algebraically invariant sets consist of at most countable union of the orbits x, $T(x)$, $T(T(x))$, ...*

Invariance in the probability sense Consider the measure-preserving transformation $T : \Omega \mapsto \Omega$ in the complete probability space $(\Omega, \mathcal{A}, \mathbb{P})$. Then, we say that the set A is measure invariant if

$$\mathbb{P}(A \Delta T^{-1}(A)) = 0.$$

Again the class of all measure invariant sets is a σ–field.

This σ-field will be denoted as $\mathcal{I}$ and referred to as the σ–field of invariant subsets for the measure preserving transformation $T : \Omega \mapsto \Omega$ on the complete probability space $(\Omega, \mathcal{A}, \mathbb{P})$.

For the stationary sequence $\{X_i : i \in \mathbb{Z}\}$ with not prescribed transform, we assume that the sequence is given on the probability space $(\mathbb{R}^{\mathbb{Z}}, \mathcal{B}', \mathbb{P})$, where $\mathcal{B}'$ is the Caratheodory–Lebesgue completion of $\mathcal{B}$ of Section 1.2.2 (i.e. by adding all zero measure sets), and $\mathbb{P}$ is defined in Section 1.2.2. Then the invariant σ–field $\mathcal{I}$ is also invariant with respect to the measure-preserving shift operator S defined by (1.7).

b) A stationary field

For a σ-field $\mathcal{G}$, define an obvious filtration by taking $\mathcal{G}_k = \mathcal{G}$. We say that $\mathcal{G}$ is a stationary σ–field, if the associated filtration is stationary.

Equivalently, for all integers n, m_1, ..., m_n, u by taking any measurable functionals $f_j : \mathbb{R}^{\mathbb{N}} \to \mathbb{R}$, $j = 1, ..., u$ and measurable function $g : \mathbb{R}^u \to \mathbb{R}$

$$\begin{aligned} &\mathbb{E}g\big(\mathbb{E}[f_1(X_\ell, \ell \ge m_1)|\mathcal{G}], \dots, \mathbb{E}[f_u(X_\ell, \ell \ge m_n)|\mathcal{G}]\big) \\ &\quad = \mathbb{E}g\big(\mathbb{E}[f_1(X_\ell, \ell \ge m_1 + i)|\mathcal{G}], \dots, \mathbb{E}[f_u(X_\ell, \ell \ge m_n + i)|\mathcal{G}]\big) \quad \text{for any } i \in \mathbb{Z}. \end{aligned}$$

Examples:

(1) *The invariant σ-field is stationary.*

(2) *Given a stationary filtration $\{\mathcal{G}_k\}$,*

(i) *the σ-field $\mathcal{G}^{\infty} = \bigvee_{k\in\mathbb{Z}} \mathcal{G}_k$ is stationary and*

(ii) *also the σ-field $\mathcal{G}_{-\infty} = \bigcap_{k\in\mathbb{Z}} \mathcal{G}_k$ is stationary.*

(3) *Let $\mathcal{G}$ be a stationary field for the stationary sequence $\{Y_k\}$. For a measurable function $f : \mathbb{R}^{\mathbb{Z}} \to \mathbb{R}$, let $X_k = f(Y_j, j \geq k)$. Then, the σ-field $\mathcal{G}$ is stationary for $\{X_k\}$.*

Stationary and non-stationary fields play an important role in checking projective type conditions.

c) Compatibility of a stationary filtration and a stationary field

Now, assume that we have both an adapted stationary filtration $\{\mathcal{G}_k\}$ and a stationary field $\mathcal{F}$. We say that they are compatible if for any measurable function $f : \mathbb{R}^{\mathbb{Z}} \to \mathbb{R}$ and any integer m,

$$\mathbb{E}[f(X_i, i \geq m)|\mathcal{F}] = \mathbb{E}(\mathbb{E}[f(X_i, i \geq m)|\mathcal{G}_m]|\mathcal{F}).$$

Example: *Assume that $\mathcal{F} \subseteq \mathcal{G}_{-\infty}$. Then they are compatible.*

By adapting Proposition 2.12 in Bradley (2007), we derive

Lemma 1.2 *The natural stationary filtration and the invariant σ-field are compatible. In particular,*

$$\mathbb{E}[S_i(S_j - S_i)|\mathcal{I}] = \mathbb{E}\{S_i\mathbb{E}(S_j - S_i|\mathcal{G}_i)|\mathcal{I}\}, \tag{1.8}$$

where $S_k = X_1 + \ldots + X_k$ and $\mathcal{G}_i = \sigma(X_k, k \leq i)$.

Caution: In general $\mathcal{I} \nsubseteq \mathcal{G}_j$ and thus the tower lemma, namely $\mathbb{E}(\mathbb{E}(X|\mathcal{G})|\mathcal{F}) = \mathbb{E}(X|\mathcal{F})$ a.s. for $\mathcal{F} \subset \mathcal{G}$, does not apply and so there exists a stationary sequence $\{X_k\}$, and a non–stationary filtration $\{\mathcal{G}_k\}$, such that

$$\mathbb{E}[X_1X_2|\mathcal{I}] \neq \mathbb{E}\{X_1\mathbb{E}(X_2|\mathcal{G}_1)|\mathcal{I}\}.$$

d) Ergodicity and Birkhoff–Khintchine theorem

Theorem 1.3 (Birkhoff–Khintchine theorem) *Let $X = \{X_i, i \in \mathbb{Z}\}$ be a stationary sequence constructed on $(\mathbb{R}^{\mathbb{Z}}, \mathcal{B}, P)$ (see Section 1.2.2). Let S be the left shift operator from $\mathbb{R}^{\mathbb{Z}}$ into $\mathbb{R}^{\mathbb{Z}}$. Let $\mathcal{I} = \{B \in \mathcal{B} : S(B) = B\}$ be the set of all invariants subsets by S. For each measurable function f defined on a subset of $\mathbb{R}^{\mathbb{Z}}$, taking values in $\mathbb{R}$ and satisfying $\mathbb{E}(|f(X)|) < \infty$,*

$$R_n(f) = \frac{1}{n}\sum_{i=1}^{n} f \circ S^i(X) \to \mathbb{E}(f(X)|X^{-1}(\mathcal{I})) \text{ a.s. and in } \mathbb{L}^1, \text{ as } n \to \infty. \tag{1.9}$$

If we assume in addition that $\mathbb{E}(|f(X)|^p) < \infty$ for a $p \in [1, \infty]$, then the convergence holds in $\mathbb{L}^p$.

For a proof of this result we refer to the monograph of Rosenblatt (1974). There are many fascinating proofs, including the Garsia technique (to be applied later on, see Garsia (1965)) and the Bishop–Ivanov exponential crossing inequality of the fixed interval $[a, b]$ $a < b$, by the path of S_n/n of a non-negative stationary sequence,

$$\mathbb{P}(\text{number of crossing of infinite path } S_n/n \text{ of } [a, b] > k) \leq cq^k$$

valid for partial sums S_n of a stationary sequence of bounded variables with $q < 1$ depending on the stationary sequence and a, b (see Bishop (1966) and Ivanov (1996)).

We say that the sequence $X = \{X_i, i \in \mathbb{Z}\}$ is **ergodic**, if the limit of $R_n(f)$ is constant for each f. Equivalently, each element of $X^{-1}(\mathcal{I})$ has a measure 0 or 1, and in this case the limit is $\mathbb{E}(f(X))$.

Theorem 1.3 can also be formulated with the help of a measure-preserving transformation. More precisely if $(\Omega, \mathcal{A}, \mathbb{P})$ is a probability space, and $T : \Omega \mapsto \Omega$ a measurable transformation preserving the probability $\mathbb{P}$, then *for any measurable function f taking values in* $\mathbb{R}$ *and such that* $f \in \mathbb{L}^1(\Omega, \mathcal{A}, \mathbb{P})$,

$$\frac{1}{n}\sum_{i=0}^{n-1} f(T^i(x)) \to \mathbb{E}(f|\mathcal{I})(x) \text{ for } \mathbb{P} - a.e.\ x \in \Omega \text{ and in } \mathbb{L}^1, \text{ as } n \to \infty,$$

where $\mathcal{I}$ *is the* σ*-algebra of all the algebraically* T*–invariant sets. In addition, if we assume that* $f \in \mathbb{L}^p(\Omega, \mathcal{A}, \mathbb{P})$ *for a* $p \in [1, \infty]$, *then the convergence holds in* $\mathbb{L}^p$.

1.2.4 Projective Decomposition

Given a filtration $\{\mathcal{F}_k\}$, we define the sequence of projections (or equivalently of projective operators) by

$$\mathbf{P}_j(X) = \mathbb{E}[X|\mathcal{F}_j] - \mathbb{E}[X|\mathcal{F}_{j-1}]$$

and use them to derive the following projective decomposition:

$$X_k = \mathbb{E}[X_k|\mathcal{F}_{-\infty}] + \sum_j \mathbf{P}_j(X_k).$$

It follows that the partial sums admit the decomposition

$$S_n = X_1 + \dots + X_n = \left(\sum_{k=1}^{n} \mathbb{E}[X_k|\mathcal{F}_{-\infty}]\right) + \sum_j \sum_{k=1}^{n} \mathbf{P}_j(X_k).$$

Lemma 1.4 *Assume that* $\mathbb{E}[X_k|\mathcal{F}_{-\infty}] = 0$ *almost surely, for all k in* $\{1, \dots, n\}$. *Then, for all* $k \in \{1, \dots, n\}$,

$$X_k = \sum_j \mathbf{P}_j(X_k) \text{ and } S_n = \sum_j \sum_{k=1}^n \mathbf{P}_j(X_k),$$

i.e. we derive the representation of X_k as a linear-type statistic and S_n as a sum of non-stationary martingale differences. Assume, in addition, that the filtration $\{\mathcal{F}_k\}$ is stationary. Then the infinite-dimensional martingale difference processes $(\{\mathbf{P}_j(X_k)\}_{j\in\mathbb{Z}})$ are stationary over k.

1.2.5 *Spectral Density for Second Order Real-valued Stationary Stochastic Processes*

Definition *A real-valued stochastic process $(X_j)_{j\in\mathbb{Z}}$ is called second order stationary if the variables are square integrable for all $j \in \mathbb{Z}$, $\mathbb{E}(X_j) = \mathbb{E}(X_0) = m$ and for all $j, k \in \mathbb{Z}$, $\text{cov}(X_k, X_j) = \mathbb{E}(X_k - m)(X_j - m) = \mathbb{E}(X_0 - m)(X_{j-k} - m) = c_{j-k}$. Note that $c_{-k} = c_k$.*

By Herglotz's Theorem (Brockwell and Davis, 1991), there exists a non-decreasing function G (the spectral distribution function) on $(-\pi, \pi)$, endowed with Borelian sigma algebra, such that, for all $j \in \mathbb{Z}$,

$$c_j = \int_{-\pi}^{\pi} \mathrm{e}^{\mathrm{i}j\theta} dG(\theta).$$

If $G(\theta)$ is absolutely continuous with respect to Lebesgue measure λ, then its derivative $f(\theta)$ is called the spectral density and

$$c_j = \int_{-\pi}^{\pi} \mathrm{e}^{\mathrm{i}j\theta} f(\theta) d\theta. \tag{1.10}$$

In the following lemma we combine a few facts about spectral densities, covariances, behavior of variances of sums and their relationships. The first two points are well known.

Lemma 1.5 *Let $(X_i)_{i\in\mathbb{Z}}$ be a $\mathbb{L}^2$-stationary sequence of real-valued variables with $\mathbb{E}[X_0] = 0$. Let G denote its spectral distribution function and f its spectral density (if exists).*

(i) *For any positive integer n and any real numbers $a_1, ..., a_n$,*

$$\mathbb{E}\left(\sum_{k=1}^n a_k X_k\right)^2 = \int_{-\pi}^{\pi} \left|\sum_{k=1}^n a_k \mathrm{e}^{\mathrm{i}kt}\right|^2 f(t)dt \le 2\pi \|f\|_\infty \sum_{k=1}^n a_k^2$$
$$\le \left(\mathbb{E}[X_0^2] + 2\sum_{k=1}^\infty |\mathbb{E}(X_0 X_k)|\right) \sum_{k=1}^n a_k^2.$$

(ii) *Assume (B) : $\sum_{k=1}^\infty |\mathbb{E}(X_0 X_k)| < \infty$. Then f is continuous. Moreover, if $\mathbb{E}[X_k X_0] \ge 0$ for all k, then the spectral density is bounded if and only if Relation (B) is satisfied.*

(iii) *Assume that the spectral density f is continuous, and let $d^{(n)} = (d_{n,j})_{j\in\mathbb{Z}}$ be a double array of real numbers with $d_n^2 = \Sigma_{j\in\mathbb{Z}} d_{n,j}^2 < \infty$ that satisfies the condition*

$$\frac{1}{d_n^2}\sum_{j=-\infty}^{\infty} |d_{n,j} - d_{n,j-1}|^2 \to 0, \text{ as } n \to \infty. \tag{1.11}$$

Then,

$$\lim_{n\to\infty} \frac{1}{d_n^2}\mathbb{E}\left(\sum_{j=-\infty}^{\infty} d_{n,j}X_j\right)^2 = 2\pi f(0). \tag{1.12}$$

Proof of (iii): Fix $\varepsilon > 0$. By the Stone–Weierstrass theorem there exists a trigonometric polynomial $P_m(t) = \Sigma_{k=-m}^{m} c_k e^{itk}$ such that $\sup_{t\in[-\pi,\pi]} |f(t)-P_m(t)| \leq \varepsilon$. In particular

$$|f(0) - \Sigma_{k=-m}^{m} c_k| \leq \varepsilon. \tag{1.13}$$

Hence and by (i),

$$\begin{aligned} \frac{1}{d_n^2}\mathbb{E}\left(\sum_{j\in\mathbb{Z}} d_{n,j}X_j\right)^2 &= \frac{1}{d_n^2}\int_{-\pi}^{\pi}\left|\sum_{j\in\mathbb{Z}} d_{n,j}e^{itj}\right|^2 f(t)dt \\ &= O(\varepsilon) + \frac{1}{d_n^2}\int_{-\pi}^{\pi}\left|\sum_{j\in\mathbb{Z}} d_{n,j}e^{itj}\right|^2 P_m(t)dt. \end{aligned} \tag{1.14}$$

With the notation $A_{n,k} := d_n^{-2}\Sigma_{j\in\mathbb{Z}} d_{n,j}d_{n,j+k}$ we have

$$\frac{1}{d_n^2}\int_{-\pi}^{\pi}\left|\sum_{j\in\mathbb{Z}} d_{n,j}e^{itj}\right|^2 P_m(t)dt = 2\pi\sum_{k=-m}^{m} c_k A_{n,k}. \tag{1.15}$$

To end the proof, it suffices to notice that, by the fact below, $\lim_{n\to\infty} 2\pi\sum_{k=-m}^{m} c_k A_{n,k} = 2\pi\sum_{k=-m}^{m} c_k$ and then, by combining this convergence with (1.13), (1.15) and (1.14), the proof of the statement is complete.

Fact 1.6 *Let $d^{(n)} = (d_{n,j})_{j\in\mathbb{Z}}$ be a double array of real numbers with $d_n^2 = \Sigma_{j\in\mathbb{Z}} d_{n,j}^2 < \infty$ and satisfying (1.11). Then, for any $k \in \mathbb{Z}$,*

$$\lim_{n\to\infty} |A_{n,k} - 1| = 0,$$

where $A_{n,k} := d_n^{-2}\Sigma_{j\in\mathbb{Z}} d_{n,j}d_{n,j+k}$.

Indeed note that, for any $k \in \mathbb{Z}$, $A_{n,k} = 1-(2d_n^2)^{-1}\sum_{j\in\mathbb{Z}}(d_{n,j+k}-d_{n,j})^2$. Therefore

$$|A_{n,k} - 1| \le \frac{k^2}{2d_n^2} \sum_{j\in\mathbb{Z}} (d_{n,j+1} - d_{n,j})^2,$$

which implies the desired result. ∎

In view of point (iii) of Lemma 1.5, one can ask ourselves about necessary and sufficient conditions ensuring that f is continuous on $(-\pi, \pi)$. This question was solved by Bradley in different papers. We cite Theorem 6.3 in his survey paper (2005) where comments on previous works are also provided. To state Bradley's result, we need to introduce linear dependence coefficients associated with a $\mathbb{L}^2$-stationary sequence $X = (X_i)_{i\in\mathbb{Z}}$ of centered real-valued random variables. For each positive integer n, define

$$\zeta_n(X) := \sup \left| \mathbb{E}\left(\sum_{i\in Q} X_i \sum_{j\in S} X_j \right) \right| \Big/ \operatorname{Card}(Q \cup S), \tag{1.16}$$

the supremum taken over all pairs of non-empty, finite, disjoint sets Q, $S \in \mathbb{Z}$ satisfying $d(Q, S) = \min_{q\in Q, s\in S} |q - s| \ge n$, and

$$\kappa_n(X) := \sup \left| \operatorname{cov}\left(\sum_{i\in Q} a_i X_i, \sum_{j\in S} a_j X_j \right) \right| \Big/ \left(\left(\sum_{k\in Q} |a_k|^2 \right)^{1/2} \left(\sum_{k\in S} |a_k|^2 \right)^{1/2} \right), \tag{1.17}$$

the supremum taken over all pairs of non-empty, finite, disjoint sets Q, $S \in \mathbb{Z}$ satisfying $d(Q, S) \ge n$ and all choices of real numbers a_k, $k \in Q \cup S$. (In both definitions, the sets Q and S can be interlaced; that is, each set can have elements between the ones of the other set.)

Theorem 1.7 (Theorem 6.3 in Bradley (2005)). *For a given $\mathbb{L}^2$-stationary sequence $X = (X_i)_{i\in\mathbb{Z}}$ of centered real-valued variables, the following three conditions are equivalent:*

(a) *X has a continuous spectral density function on $(-\pi, \pi)$.*

(b) *One has $\zeta_n(X) \to 0$ as $n \to \infty$.*

(c) *One has $\kappa_n(X) \to 0$ as $n \to \infty$.*

Remark 1.8 *It is also common to define the spectral density on the unit circle T in the complex plane C, i.e. $T = \{z : |z| \le 1, z \in C\}$. More precisely, let μ_T be the normalized Lebesgue measure on T (so $\mu_T(T) = 1$) and, for each $t \in T$, denote by θ the element of $(-\pi, \pi)$ such that $t = \exp(i\theta)$. Then, for a given real-valued second order stationary process $(X_j)_{j\in\mathbb{Z}}$, a "spectral density function" (if one exists) is a real, non-negative, Borel, integrable function $h : T \to [0, \infty)$ such that*

$$\forall k \in \mathbb{Z}, \operatorname{cov}(X_0, X_k) = \int_{t\in T} t^k h(t) d\mu_T(t). \tag{1.18}$$

Obviously both definitions coincide since the following relation holds: $(2\pi)^{-1} h(e^{i\theta}) = f(\theta)$.

1.3 Introduction to Convergence of Stochastic Processes

1.3.1 *Useful Facts Concerning Convergence in Distribution*

Theorems below are respectively Theorem 3.1 and Theorem 3.2 in Billingsley (1999). In their statements S will be a metric space whose metric is denoted by $\rho(x, y)$.

Theorem 1.9 *Suppose that* $(\mathbf{X}_n, \mathbf{Y}_n)$ *are random elements of* $S \times S$. *If* $\mathbf{X}_n \to^d \mathbf{X}$ *and* $\rho(\mathbf{X}_n, \mathbf{Y}_n) \to^{\mathbb{P}} 0$ *then* $\mathbf{Y}_n \to^d \mathbf{X}$.

Theorem 1.10 *Suppose that* $(\mathbf{X}_{u,n}, \mathbf{X}_n)$ *are random elements of* $S \times S$. *If* $\mathbf{X}_{u,n} \to_n^d \mathbf{Z}_u \to_u^d \mathbf{X}$ *and*

$$\lim_{u\to\infty} \limsup_{n\to\infty} \mathbb{P}\left(\rho\left(\mathbf{X}_{u,n}, \mathbf{X}_n\right) \geq \epsilon\right) = 0$$

for each $\epsilon > 0$, *then* $\mathbf{X}_n \to_n^d \mathbf{X}$.

Next result is Theorem 3.4 in Billingsley (1999).

Theorem 1.11 *Assume that* $\mathbf{X}_n \to^d \mathbf{X}$. *Then* $\mathbb{E}(|\mathbf{X}|) \leq \liminf_{n\to\infty} \mathbb{E}(|\mathbf{X}_n|)$.

The next result is the so-called convergence of moments Theorem (see Theorems 3.5 and 3.6 in Billingsley (1999)).

Theorem 1.12 *Assume that* $\mathbf{X}_n \to^d \mathbf{X}$. *Let* $p \geq 1$. *Then* $\mathbb{E}|\mathbf{X}_n|^p \to \mathbb{E}|\mathbf{X}|^p$ *if and only if* $|\mathbf{X}_n|^p$ *is uniformly integrable.*

Another important fact about convergence in distribution is the so-called continuous mapping theorem.

Theorem 1.13 *Let* h *be a continuous mapping of* S *into another metric space* S'. *Suppose that* $\mathbf{X}_n \to^d \mathbf{X}$. *Then* $h(\mathbf{X}_n) \to^d h(\mathbf{X})$.

The continuity assumption on h can be weakened by assuming only that h is measurable and $\mathbb{P}(\mathbf{X} \in D_h) = 0$ where D_h is the set of discontinuities of h (see Theorem 2.7 in Billingsley (1999)).

Next theorem is Theorem 14.2 in Billingsley (1995) and shows that, for convergence in distribution, there can be only one possible limit type and essentially only one possible sequence of norming constants.

Theorem 1.14 *Let* $(X_n)_{n\in\mathbb{Z}}$ *and* $(Y_n)_{n\in\mathbb{Z}}$ *be two sequences of real-valued random variables such that*

$$\frac{X_n - v_n}{u_n} \to^d X \text{ and } \frac{X_n - b_n}{a_n} \to^d Y, \text{ as } n \to \infty,$$

where $u_n > 0$, $v_n > 0$ *and* X *and* Y *are non-degenerate random variables. Then there exist* a *and* b, $a > 0$ *such that* $a_n/u_n \to a$, $(b_n - v_n)/u_n \to b$ *and* $\dfrac{X-b}{a} =^d Y$.

1.3.2 Weak Convergence of the Partial Sums Process

Let us consider a triangular array of real-valued random variables $(X_{n,i})_{1\leq i\leq n}$ and a triangular array $(t_{k,n})_{0\leq k\leq n}$ of partitions of the interval $[0, 1]$, more exactly,

$$t_0 = t_{0,n} = 0 < t_{1,n} < \dots < t_{n,n} = 1\ ,\ \max_{k=1,\dots,n}\left[t_{k,n} - t_{k-1,n}\right] \to 0 \quad (\text{as } n \to \infty). \tag{1.19}$$

It is convenient to add $t_{n+1,n} = 1$ and $X_{n,j} = 0$ for $j > n$.

Define the partial sums and the partial sums process by

$$S_m = S_{n,m} = \sum_{j=1}^{m} X_{n,j} \text{ and } W_n = W_{n,X} = \{W_n(t), t \in [0, 1]\}, \tag{1.20}$$

where

$$W_n(t) = S_{n,k} \text{ for } t_{k,n} \leq t < t_{k+1,n}.$$

The process W_n has valued in the space $D([0, 1])$ of càdlàg functions on $[0,1]$ that we shall equip, for our convenience, with the uniform topology defined by $\|f\|_\infty = \sup_{t\in[0,1]} |f(t)|$. We are interested in proving the weak convergence of W_n to a process, let's say W, in $(D([0, 1]), \|\cdot\|_\infty)$. In order for

$$W_n \to^d W \text{in } (D([0, 1]), \|\cdot\|_\infty)\,, \tag{1.21}$$

it is necessary and sufficient to prove that (see Section 7 of Billingsley (1999)):

1. the finite-dimensional distributions of W_n converge to those of W
2. the process $\{W_n(t), t \in [0, 1]\}$ is tight, that is, for any $\varepsilon > 0$,

$$\lim_{m\to\infty} \limsup_{n\to\infty} \mathbb{P}\left(\sup_{|t-s|\leq m^{-1}} |W_n(t) - W_n(s)| > \varepsilon \right) = 0. \tag{1.22}$$

The following two lemmas are useful in establishing (1.22) (see also Theorem 7.4 in Billingsley (1999)).

Lemma 1.15 *(Modulus of continuity bounds.) Given the partition $\{t_{k,n}\}$ and the positive integer m, for $i = 1, \dots, m$, we define*

$$\nu_i = \max\{k \in \{0, \dots, n\} : t_{k,n} \leq i/m\} \text{ and } \nu_0 = 0.$$

Then,

$$\sup_{|t-s|\leq m^{-1}} |W_n(t) - W_n(s)| \leq 3 \max_{1\leq i\leq m} \max_{\nu_{i-1}\leq k\leq \nu_i} |S_{n,k} - S_{n,\nu_{i-1}}|. \tag{1.23}$$

Moreover,

$$\sup_{|t-s|\leq m^{-1}} |W_n(t) - W_n(s)| \geq \max_{1\leq i\leq m} \max_{\nu_{i-1}\leq k\leq \nu_i} |S_{n,k} - S_{n,\nu_{i-1}}| - \max_{1\leq i\leq n} |X_{n,i}|,$$

which shows that the upper bound in (1.23) is optimal.

Lemma 1.16 *Hence to prove (1.22), it is enough to show that, for any $\varepsilon > 0$,*

$$\lim_{m\to\infty} \limsup_{n\to\infty} \sum_{i=1}^{m} \mathbb{P}\left(\max_{\nu_{i-1}\leq k\leq \nu_i} |S_{n,k} - S_{n,\nu_{i-1}}| > \varepsilon \right) = 0. \tag{1.24}$$

Now, let us assume $X_{n,j} = b_n^{-1} X_j$ where $(X_j)_{j\in\mathbb{Z}}$ be a stationary sequence and $b_n \to \infty$ be a sequence of positive constants, then (1.24) is satisfied when

$$\lim_{\lambda\to\infty} \limsup_{n\to\infty} \lambda^2 \mathbb{P}\left(\max_{1\leq k\leq n} |S_k| > \lambda b_n \right) = 0. \tag{1.25}$$

Lemma 1.17 *Let $(X_j)_{j\in\mathbb{Z}}$ be a stationary sequence and $b_n \to \infty$ be a sequence of positive constants. Set $S_n = \sum_{k=1}^{n} X_k$. Then, if $\{\max_{1\leq k\leq n} S_k^2/b_n^2\}_{n\geq 1}$ is an uniformly integrable family, condition (1.25) holds and then the process $\{b_n^{-1} S_{[nt]}, t \in [0,1]\}$ is tight.*

The following result is also useful.

Theorem 1.18 *Let two triangular arrays $(X_{n,i})_{1\leq i\leq n}$ and $(Y_{n,i})_{1\leq i\leq n}$ of real-valued random variables. Assume that for any $\varepsilon > 0$,*

$$\lim_{n\to\infty} \mathbb{P}\left(\max_{1\leq k\leq n} \left| \sum_{i=1}^{k} (X_{n,i} - Y_{n,i}) \right| \geq \varepsilon \right) = 0.$$

Then $W_{n,X} \to^d W$ if and only if $W_{n,Y} \to^d W$ where $W_{n,X}$ (resp. $W_{n,Y}$) is the partial sum process associated with $X = (X_{n,i})_{1\leq i\leq n}$ (resp. with $Y = (Y_{n,i})_{1\leq i\leq n}$).

Given a partition $\{t_{i,n}\}$, define $\nu(t) = \max\{j : t_{j,n} \leq t\}$. Note also that the following process (which is the linear spline of the partial sums process) with values in $C([0,1])$

$$\widetilde{W}_n = \{\widetilde{W}_n(t), t \in [0,1]\} \text{ where } \widetilde{W}_n(t) = W_n(t) + \frac{t - t_{\nu(t),n}}{t_{\nu(t)+1,n} - t_{\nu(t),n}} X_{n,\nu(t)+1},$$

will converge in distribution to W if W_n converges in distribution to W and

$$\max_{1\leq i\leq n} |X_{n,i}| \to^{\mathbb{P}} 0.$$

Remark 1.19 *Notice that the most common construction is via the partition $t_{k,n} = k/n$ so that $W_n(t) = S_{[nt]}$ and $\widetilde{W}_n(t) = S_{[nt]} + (nt - [nt]) X_{n,[nt]+1}$. In this case, the convergence $W_n \to^d W$ is often written in the form:*

$$S_{[nt]} \text{ converges in distribution to } W \text{ in } (D([0,1]), \|\cdot\|_\infty).$$

1.3.3 *Maximal Moment Inequalities*

Let $\{X_i\}$ be a sequence of random variables. Let $h(n)$ be a slowly varying function in the sense of Karamata, i.e. h admits an extension on $\mathbb{R}_+$ and the extended function is slowly varying, meaning that for all $a > 0$, $h(ax)/h(x) \to 1$ as $x \to \infty$.

The following lemma is handy for proving maximal inequalities.

Lemma 1.20 *For $k > m$, define the partial sums*

$$S_{m,k} = \sum_{i=m+1}^{k} X_i.$$

Let $\beta > 0$ and $h(n)$ be a slowly varying function. Assume that for some $p \geq 1$, there exist finite positive constants K and $\alpha > 1/\beta$, such that for all integers $n \geq 1$ and $m \geq 0$,

$$\mathbb{E}\left(|S_{m,m+n}|^p\right) \leq K(n^\beta h(n))^\alpha. \tag{1.26}$$

Then there exists a positive constant $C = C(p, \alpha, \beta)$ such that, for all integers $n \geq 1$ and $m \geq 0$,

$$\mathbb{E}\left(\max_{1\leq i\leq n} |S_{m,m+i}|^p\right) \leq KC(n^\beta h(n))^\alpha. \tag{1.27}$$

Proof The proof follows by using the following maximal inequality: Let r be the non-negative integer such that $2^r - 1 < n \leq 2^r$. Then

$$\| \max_{1\leq i\leq n} |S_{m,m+i}| \|_p \leq \sum_{l=0}^{r} \left(\sum_{k=0}^{2^{r-l}-1} \|S_{m,m+(k+1)2^l} - S_{m,m+k2^l}\|_p^p \right)^{1/p}, \tag{1.28}$$

by setting $S_{m,m} = 0$. This maximal inequality follows by applying the simple chaining argument lemma below and by taking the norm in $\mathbb{L}^p$.

Lemma 1.21 *Let $(a_n)_{n\geq 1}$ be a sequence in a Banach space $\mathcal{B}$ with norm $|\cdot|_{\mathcal{B}}$. Write $s_n = a_1 + \ldots + a_n$ and $s_0 = 0$. Let $p \geq 1$. For every $r \geq 0$, we have*

$$\max_{1\leq n\leq 2^r} |s_n|_{\mathcal{B}} \leq \sum_{k=0}^{r} \left(\sum_{\ell=1}^{2^{r-k}} |s_{2^k\ell} - s_{2^k(\ell-1)}|_{\mathcal{B}}^p \right)^{1/p}. \tag{1.29}$$

To prove the lemma above, we proceed by induction on $r \geq 0$. The result is obvious for $r = 0$. Let $1 \leq n \leq 2^r$. We have $|s_{2n-1}|_{\mathcal{B}} \leq |s_{2n-2}|_{\mathcal{B}} + |a_{2n-1}|_{\mathcal{B}}$. Hence, writing $\tilde{a}_n = a_{2n-1} + a_{2n}$ and $\tilde{s}_n = \sum_{k=1}^{n} \tilde{a}_k = s_{2n}$, we get that

$$\max_{1\le k\le 2^{r+1}} |s_k|_{\mathcal{B}} \le \max_{1\le n\le 2^r} |\tilde{s}_n|_{\mathcal{B}} + \left(\sum_{n=1}^{2^r} |a_{2n-1}|_{\mathcal{B}}^p\right)^{1/p},$$

and the result follows.

To end the proof of Lemma 1.20, we take advantage of the assumption (1.26) in (1.28). This implies that

$$\|\max_{1\le i\le n} |S_{m,m+i}|\|_p \le 2^{r/p} K^{1/p} \sum_{\ell=0}^{r} \left(2^{-\ell} 2^{\alpha\beta\ell} (h(2^\ell))^\alpha\right)^{1/p},$$

which, by the properties of the slowly varying functions and since $\alpha\beta > 1$, implies (1.27). ■

Lemma 1.20 can be useful to prove tightness as shown in the next result.

Corollary 1.22 *Let $\{X_i\}$ be a stationary sequence of random variables and $S_n = X_1 + ... + X_n$ be its partial sums. Let $\beta \in]0, 2]$ and $h(n)$ be a slowly varying function.*

(i) *If $\beta \in]0, 1]$ and we assume that, for some $p > 2/\beta$,*

$$\mathbb{E}(|S_n|^p) \le c(n^\beta h(n))^{p/2},$$

then, the sequence $\left\{\frac{S_{[nt]}}{\sqrt{n^\beta h(n)}}\ ;\ t \in [0, 1]\right\}$ *is tight.*

(ii) *If $\beta \in]1, 2]$ and we assume that*

$$\mathbb{E}\left(|S_n|^2\right) \le cn^\beta h(n),$$

then, the sequence $\left\{\frac{S_{[nt]}}{\sqrt{n^\beta h(n)}}\ ;\ t \in [0, 1]\right\}$ *is tight.*

Proof To prove point (i), we use Lemma 1.20 which implies that there exists a constant κ such that, for any $\lambda > 0$ and any $n \ge 1$,

$$\mathbb{P}\left(\max_{1\le k\le n} |S_k| > \lambda\sqrt{n^\beta h(n)}\right) \le \kappa\lambda^{-p}.$$

Since $p > 2$, this proves that (1.25) holds and the tightness follows by Lemma 1.16.

To prove point (ii), we shall rather prove that (1.24) holds with $\nu_i = \left[\frac{ni}{m}\right]$. Using stationarity and applying Lemma 1.20, we get that there exists a constant κ such that, for any $\varepsilon > 0$, any $m \ge 1$ and any n large enough

$$\sum_{i=1}^{m} \mathbb{P}\left(\max_{[n(i-1)/m]<k\le[ni/m]} |S_k - S_{[ni/m]}| > \varepsilon\sqrt{n^\beta h(n)}\right) \le \kappa\varepsilon^{-2} m^{1-\beta}.$$

Since $\beta > 1$, this proves that (1.24) is satisfied and the tightness follows again by Lemma 1.16. ■

1.3.4 Moderate Deviations

Let us consider a triangular array of real valued random variables $(X_{n,i})_{1\leq i\leq n}$ and its partial sums S_n defined in (1.20). The moderate deviation principle question concerns the study the limiting behavior of

$$\mathbb{P}(\sqrt{a_n}S_n \in A)$$

for Borelian sets A and (a_n) a sequence of positive real numbers. When $a_n = 1$ for any $n \geq 1$, and $(X_{n,i})_{1\leq i\leq n}$ is a triangular array of independent, real-valued random variables centered, in $\mathbb{L}^2$ and such that $\sum_{i=1}^{n} \mathbb{E}(X_{n,i}^2) = 1$, then the central limit theorem holds under the so-called Lindeberg condition, namely: For any $\varepsilon > 0$, $\lim_{n\to\infty} \sum_{i=1}^{n} \mathbb{E}\left(X_{n,i}^2 1_{\{|X_{n,i}|>\varepsilon\}}\right) = 0$. When $a_n \to 0$, the Moderate Deviations Principle (MDP) is to find conditions for the following result.

Definition 1.23 *We say that the MDP holds for S_n with speed $a_n \to 0$ and rate function $I(t)$ if for each A Borelian,*

$$\begin{aligned} -\inf_{t\in A^o} I(t) &\leq \liminf_n a_n \log \mathbb{P}(\sqrt{a_n}S_n \in A) \\ &\leq \limsup_n a_n \log \mathbb{P}(\sqrt{a_n}S_n \in A) \leq -\inf_{t\in \bar{A}} I(t), \end{aligned} \tag{1.30}$$

where A^o denotes the interior of A and $\bar{A}$ the closure of A.

The Moderate Deviations Principle is then an intermediate behavior between the central limit theorem ($a_n = a$) and Large Deviation ($a_n = a/n$). Usually, MDP has a simpler rate function, inherited from the approximated Gaussian process, and holds for a larger class of dependent random variables than the Large Deviation Principle.

The following result is useful to prove such a limiting behavior.

Lemma 1.24 *(Gärtner-Ellis) If for any $t \in \mathbb{R}$, there exists σ^2 such that*

$$\lim_{n\to\infty} a_n \log \mathbb{E}(\exp(tS_n/\sqrt{a_n})) = \sigma^2 t^2/2,$$

then S_n satisfies the MDP with speed a_n and rate function $I(t) = t^2/(2\sigma^2)$.

(See Theorem 2.3.6 in Dembo and Zeitouni (1998)).

We are also interested in giving a more general result concerning the Donsker process W_n associated with the partial sums, as defined in (1.20). More exactly, we say that the family of random variables $\{W_n, n > 0\}$ satisfies the MDP in $D([0,1])$ with speed $a_n \to 0$ and good rate function $I(.)$, if the level sets $\{x, I(x) \leq \alpha\}$ are compact for all $\alpha < \infty$, and for all Borel sets $\Gamma \in \mathcal{B}$

$$-\inf_{t\in\Gamma^0} I(t) \le \liminf_n a_n \log \mathbb{P}(\sqrt{a_n}W_n \in \Gamma)$$
$$\le \limsup_n a_n \log \mathbb{P}(\sqrt{a_n}W_n \in \Gamma) \le -\inf_{t\in\bar{\Gamma}} I(t). \tag{1.31}$$

To prove that $\{W_n, n > 0\}$ satisfies the MDP in $D([0, 1])$, it is sufficient to prove that

1. the finite-dimensional distributions of W_n satisfy the MDP, namely: For m a fixed integer and for any $0 < t_1 < t_2 < \cdots < t_m \le 1$, the m -tuple

$$(W_n(t_1), W_n(t_2), \dots, W_n(t_m))$$

satisfy the MDP in $\mathbb{R}^m$ with speed a_n and a rate function $I_{t_1,\cdots,t_m}(u_1, \cdots, u_m)$.

2. the process $\{W_n(t), t \in [0, 1]\}$ is tight in the sense of moderate deviations, that is, for any $\epsilon > 0$,

$$\lim_{\eta\to 0} \limsup_{n\to\infty} a_n \log \mathbb{P}\left(\sup_{|s-t|\le\eta,\, 0\le s,\, t\le 1} \sqrt{a_n}|W_n(t) - W_n(s)| \ge \epsilon \right) = -\infty. \tag{1.32}$$

We end this section by stating an approximation lemma from Dembo and Zeitouni (1998, Theorem 4.2.13, p 130), called the exponential equivalence lemma, which allows us to derive, in some situations, the moderate deviations principle by approximation.

Lemma 1.25 *Let $\xi_n(.) := \{\xi_n(t), t \in [0, 1]\}$ and $\zeta_n(.) := \{\zeta_n(t), t \in [0, 1]\}$ be two processes in $D([0, 1])$. Assume that for any $\delta > 0$,*

$$\limsup_{n\to\infty} a_n \log \mathbb{P}(\sqrt{a_n} \sup_{t\in[0,1]} |\xi_n(t) - \zeta_n(t)| \ge \delta) = -\infty.$$

Then, if the sequence of processes $\xi_n(.)$ satisfies (1.31) then so does the sequence of processes $\zeta_n(.)$.

PART I
Probabilistic Tools

2

Moment Inequalities and Gaussian Approximation for Martingales

The aim of this chapter is to collect some results concerning martingales (moment inequalities, maximal inequalities, exponential inequalities, central limit theorem and moderate deviations principle and their functional forms) that will be useful tools to analyze the asymptotic behavior of partial sums associated with dependent sequences.

2.1 Definitions and Properties

We recall that $\{\mathcal{F}_k\}_{k\in\mathbb{Z}}$ is a non-decreasing filtration or simply a filtration if each $\mathcal{F}_j$ is a σ–field and $\mathcal{F}_k \subseteq \mathcal{F}_j$ for $k \le j$.

We will say that $\{\mathcal{F}_{n,j}\}_{j\in\mathbb{Z},n\ge1}$ is an array of non-decreasing filtrations or simply an array of filtrations if for each n, $\{\mathcal{F}_{n,j}\}_{j\in\mathbb{Z}}$ is a (non-decreasing) filtration.

Definition 2.1 *Let $(\Omega, \mathcal{A}, \mathbb{P})$ be a probability space, $(\mathcal{F}_n)$ a non-decreasing sequence of σ–fields of $\mathcal{A}$, and $(M_n, n \ge 1)$ a sequence of real random variables in $\mathbb{L}^1(\mathbb{P})$. We say that $\{M_n, \mathcal{F}_n\}$ is a martingale if for any integer $n \ge 1$, M_n is $\mathcal{F}_n$–measurable and*

$$\mathbb{E}\,(M_n|\mathcal{F}_{n-1}) = M_{n-1}.$$

It is convenient to take $M_0 = 0$. If the equality is replaced by $\ge$ (resp. $\le$), we say that $\{M_n, \mathcal{F}_n\}$ is a submartingale (resp. supermartingale).

We say that $\{d_n, \mathcal{F}_n\}$ is a martingale differences sequence if for any integer $n \ge 1$, d_n is in $\mathbb{L}^1(\mathbb{P})$, $\mathcal{F}_n$–measurable and

$$\mathbb{E}[d_n|\mathcal{F}_{n-1}] = 0.$$

It follows that $M_n - M_{n-1}$ is a martingale difference as soon as M_n is a martingale.

Fact 2.2 ***(Jensen inequality)*** *Let φ be a convex function on $\mathbb{R}$ and $(M_n, n \ge 1)$ a sequence of real random variables. Assume that for all $n \ge 1$, $\varphi(M_n)$ is in $\mathbb{L}^1(\mathbb{P})$. Then*

Functional Gaussian Approximation for Dependent Structures. Florence Merlevède, Magda Peligrad and Sergey Utev. Oxford University Press (2019).
DOI: 10.1093/oso/9780198826941.001.0001

1. *If M_n is a martingale then $\varphi(M_n)$ is a submartingale.*
2. *If M_n is a submartingale and φ is additionally non-decreasing then $\varphi(M_n)$ is a submartingale.*

The proofs follow directly by applying the conditional Jensen inequality.

Definition 2.3 *Given a filtration $\{\mathcal{F}_n\}$, a positive integer valued random variable (with the possible value $+\infty$) τ is called a stopping time if for each positive integer n, the event $\{\tau = n\} \in \mathcal{F}_n$ i.e. is $\mathcal{F}_n$–measurable.*

Notice that since the filtration is non-decreasing, it is equivalent to say that for all $n \geq 1$, $\{\tau \leq n\} \in \mathcal{F}_n$.

Fact 2.4 *Assume that $\{M_n, \mathcal{F}_n\}_{n\in\mathbb{N}}$ is a (sub-, super-) martingale and τ is a stopping time with respect to $\{\mathcal{F}_n\}_{n\in\mathbb{N}}$. Then the stopped process $\{M_{n\wedge\tau}, n \geq 0\}$ is a (sub-, super-) martingale with respect to $\{\mathcal{F}_n\}_{n\in\mathbb{N}}$. Consequently $\mathbb{E}(M_{n\wedge\tau}) = \mathbb{E}(M_0)$ if $\{M_n, \mathcal{F}_n\}_{n\in\mathbb{N}}$ is a martingale, $\mathbb{E}(M_{n\wedge\tau}) \geq \mathbb{E}(M_0)$ if $\{M_n, \mathcal{F}_n\}_{n\in\mathbb{N}}$ is a submartingale and $\mathbb{E}(M_{n\wedge\tau}) \leq \mathbb{E}(M_0)$ if $\{M_n, \mathcal{F}_n\}_{n\in\mathbb{N}}$ is a supermartingale.*

Proof To prove the above fact, we set $\widetilde{M}_n = M_{n\wedge\tau}$, and we first notice that

$$|\widetilde{M}_n| \leq \max_{1\leq k\leq n} |M_k|,$$

implying that $\mathbb{E}(|\widetilde{M}_n|) \leq \sum_{k=1}^{n} \mathbb{E}(|M_k|) < \infty$. Notice now that on the set $\{\tau > n\}$, $\widetilde{M}_n = M_n$ and $\widetilde{M}_{n+1} = M_{n+1}$, whereas on the set $\{\tau \leq n\}$, $\widetilde{M}_{n+1} = \widetilde{M}_n$. Therefore

$$\widetilde{M}_{n+1} = \widetilde{M}_n + (M_{n+1} - M_n)I(\tau > n).$$

Since the event $\{\tau > n\}$ is $\mathcal{F}_n$-measurable, we get that

$$\mathbb{E}(\widetilde{M}_{n+1}|\mathcal{F}_n) = \widetilde{M}_n + I(\tau > n)\mathbb{E}(M_{n+1} - M_n|\mathcal{F}_n),$$

giving the first part of the result. The second part of the result is direct by noting that $\mathbb{E}(\widetilde{M}_{n+1}) = \mathbb{E}(\mathbb{E}(\widetilde{M}_{n+1}|\mathcal{F}_n))$ and then $\mathbb{E}(\widetilde{M}_{n+1}) = \mathbb{E}(\widetilde{M}_n)$ if $\{M_n, \mathcal{F}_n\}_{n\in\mathbb{N}}$ is a martingale, $\mathbb{E}(\widetilde{M}_{n+1}) \geq \mathbb{E}(\widetilde{M}_n)$ if $\{M_n, \mathcal{F}_n\}_{n\in\mathbb{N}}$ is a submartingale and $\mathbb{E}(\widetilde{M}_{n+1}) \leq \mathbb{E}(\widetilde{M}_n)$ if $\{M_n, \mathcal{F}_n\}_{n\in\mathbb{N}}$ is a supermartingale. ∎

2.2 Maximal and Moment Inequalities for Martingales

We begin with:

2.2.1 *Doob's Maximal Inequality*

Theorem 2.5 ***(Doob's maximal inequality)*** *Suppose that $(M_n, \mathcal{F}_n)$ is a submartingale. Then for each real λ,*

$$\lambda \mathbb{P}(\max_{1 \le k \le n} M_k \ge \lambda) \le \mathbb{E}\left(M_n I(\max_{1 \le k \le n} M_k \ge \lambda)\right).$$

Proof Define the following stopping time,

$$\tau = \min\{k : M_k \ge \lambda\}.$$

Therefore $\{\max_{1 \le k \le n} M_k \ge \lambda\} = \{\tau \le n\}$. Since on the set $\{\tau \le n\}$, $M_\tau \ge \lambda$, it follows that

$$\lambda I(\tau \le n) \le M_\tau I(\tau \le n) = \sum_{k=1}^{n} M_k I(\tau = k).$$

By the submartingale property of $(M_n, \mathcal{F}_n)$ and since $I(\tau = k)$ is $\mathcal{F}_k$–measurable, it follows that for any $1 \le k \le n$,

$$\mathbb{E}(M_k I(\tau = k)) \le \mathbb{E}(\mathbb{E}(M_n | \mathcal{F}_k) I(\tau = k)) = \mathbb{E}(M_n I(\tau = k)).$$

From the above considerations, we get that

$$\lambda \mathbb{P}(\tau \le n) \le \sum_{k=1}^{n} \mathbb{E}(M_n I(\tau = k)) = \mathbb{E}(M_n I(\tau \le n)) = \mathbb{E}\left(M_n I\left(\max_{1 \le k \le n} M_k \ge \lambda\right)\right). \quad \blacksquare$$

Remark 2.6 *Let X be a real-valued random variable. Then the tail function H_X of X is defined by $H_X(t) = \mathbb{P}(X > t)$. The quantile function Q_X of X is the càdlàg inverse of H_X (note that Q_X is nonincreasing). As a byproduct of Doob's maximal inequality, we can state the following inequality in terms of quantile function: Let $(M_0, M_1, ..., M_n)$ be a submartingale in $\mathbb{L}^1$ such that $M_0 \ge 0$ almost surely. Set $M_n^* = \max(M_0, M_1, \dots, M_n)$. Then*

$$Q_{M_n^*}(u) \le u^{-1} \int_0^u Q_{M_n}(v) dv \quad \text{for any } u \text{ in } [0, 1]. \tag{2.1}$$

This inequality is a reformulation of Lemma 1 in Dubins and Gilat (1978) and a complete proof is given in Rio (2017) after the statement of his lemma 2.3. Note also that Dubins and Gilat (1978) proved the optimality of (2.1).

In terms of moments, the Doob maximal inequality gives the following corollary:

Corollary 2.7 *Suppose that $(M_n, \mathcal{F}_n)$ is a submartingale. Let $p \in]1, \infty[$ and q be its conjugate. Then for each $\lambda \ge 0$,*

$$\mathbb{E}\left(\max_{1 \le k \le n} |M_k|^p I\left(\max_{1 \le k \le n} |M_k| \ge \lambda\right)\right) \le q^p \mathbb{E}\left(|M_n|^p I\left(\max_{1 \le k \le n} |M_k| \ge \lambda\right)\right).$$

Proof We start with the following lemma:

Lemma 2.8 *Let X and Y be two non-negative real-valued random variables such that for any real* $\lambda \geq 0$,

$$\lambda \mathbb{P}(X \geq \lambda) \leq \mathbb{E}\,(YI(X \geq \lambda)). \tag{2.2}$$

Then, for any $p \in]1, \infty[$ *and any real* $\lambda \geq 0$,

$$\mathbb{E}(X^p I(X \geq \lambda)) \leq q^p \mathbb{E}\,(Y^p I(X \geq \lambda))$$

where $q = p/(p-1)$.

To prove the lemma, we first write that for any $p \in]1, \infty[$ and any $\lambda \geq 0$,

$$\mathbb{E}(X^p I(X \geq \lambda)) = p \int_\lambda^\infty x^{p-1} \mathbb{P}(X \geq x) dx + \lambda^p \mathbb{P}(X \geq \lambda).$$

Then, using (2.2), we obtain

$$\begin{aligned}\mathbb{E}(X^p I(X \geq \lambda)) &\leq p \int_\lambda^\infty x^{p-2} \mathbb{E}\,(YI(X \geq x))\, dx + \lambda^{p-1} \mathbb{E}\,(YI(X \geq \lambda)) \\ &\leq q \mathbb{E}\left(YX^{p-1} I(X \geq \lambda)\right).\end{aligned}$$

Therefore Hölder's inequality entails that

$$\mathbb{E}(X^p I(X \geq \lambda)) \leq q\,(\mathbb{E}\,(Y^p I(X \geq \lambda)))^{1/p}\,(\mathbb{E}\,(X^p I(X \geq \lambda)))^{1/q},$$

ending the proof of the lemma.

Corollary 2.7 follows by the fact that since $(M_n, \mathcal{F}_n)$ is a submartingale so is $(|M_n|, \mathcal{F}_n)$. Therefore by Theorem 2.5, condition (2.2) is satisfied with $X = \max_{1 \leq k \leq n} |M_k|$ and $Y = |M_n|$ and Corollary 2.7 follows by an application of the above lemma. ∎

As another consequence of the Doob maximal inequality, we formulate the following result:

Proposition 2.9 *Suppose that* $(M_n, \mathcal{F}_n)$ *is a submartingale. Then for each positive* λ,

$$\lambda \mathbb{P}(\max_{1 \leq k \leq n} |M_k| \geq \lambda) \leq 2\mathbb{E}(|M_n| I(|M_n| \geq \lambda/2).$$

Proof Since $(M_n, \mathcal{F}_n)$ is a submartingale so is $(|M_n|, \mathcal{F}_n)$. Therefore by Theorem 2.5,

$$\lambda \mathbb{P}(\max_{1 \leq k \leq n} |M_k| \geq \lambda) \leq \mathbb{E}\left(|M_n| I(\max_{1 \leq k \leq n} |M_k| \geq \lambda)\right).$$

Then, we truncate at the level $\lambda/2$. We have

$$\mathbb{E}\left(|M_n|I(\max_{1\le k\le n}|M_k|\ge\lambda)\right)=\mathbb{E}\left(|M_n|I(|M_n|<\lambda/2,\max_{1\le k\le n}|M_k|\ge\lambda)\right)$$
$$+\mathbb{E}\left(|M_n|I(|M_n|\ge\lambda/2,\max_{1\le k\le n}|M_k|\ge\lambda)\right)$$
$$\le\frac{\lambda}{2}\mathbb{P}\left(\max_{1\le k\le n}|M_k|\ge\lambda\right)+\mathbb{E}\left(|M_n|I\left(|M_n|\ge\lambda/2\right)\right).$$

So,

$$\lambda\mathbb{P}(\max_{1\le k\le n}|M_k|\ge\lambda)\le\frac{\lambda}{2}\mathbb{P}(\max_{1\le k\le n}|M_k|\ge\lambda)+\mathbb{E}\left(|M_n|I(|M_n|\ge\lambda/2)\right)$$

giving the desired result. ■

Proceeding as in the proof of Corollary 2.7 and using Proposition 2.9, we easily get

Corollary 2.10 *Suppose that $(M_n,\mathcal{F}_n)$ is a submartingale. Let $p\in]1,\infty[$ and q be its conjugate. Then for each $\lambda\ge 0$,*

$$\mathbb{E}(\max_{1\le k\le n}|M_k|^pI(\max_{1\le k\le n}|M_k|\ge\lambda))\le 2^pq\mathbb{E}\left(|M_n|^pI(|M_n|\ge\lambda/2)\right).$$

It follows from the corollary above that if $((|M_n|^p)_{n\ge1}$ is an uniformly integrable family then so also is the family $(\max_{1\le k\le n}|M_k|^p)_{n\ge1}$.

Remark 2.11 *Suppose that $(\mathcal{G}_j)_{j\in\mathbb{Z}}$ is a non-increasing filtration meaning that for each $j\in\mathbb{Z}$, $\mathcal{G}_j$ is a σ-field of $\mathcal{A}$ and for each $k\le j$,*

$$\mathcal{G}_j\subseteq\mathcal{G}_k.$$

Suppose in addition that $(\tilde{d}_j)_{j\in\mathbb{Z}}$ is a sequence of real-valued random variables that are in $\mathbb{L}^1(\mathbb{P})$ and such that for any $j\in\mathbb{Z}$, $\tilde{d}_j$ is $\mathcal{G}_j$-measurable and

$$\mathbb{E}(\tilde{d}_j|\mathcal{G}_{j+1})=0\ \text{a.s.}$$

Such a sequence is usually called a reverse martingale differences sequence. If we define $\widetilde{M}_0=0$ and for any integer k in $[0,n]$

$$\widetilde{M}_k=\sum_{j=1}^{k}\tilde{d}_j,\ \widetilde{M}_{k,n}=\widetilde{M}_n-\widetilde{M}_{n-k}\ \text{and}\ \mathcal{F}_{k,n}=\mathcal{G}_{n-k+1},$$

then $(\mathcal{F}_{k,n})_{0\le k\le n}$ is a filtration and $(\widetilde{M}_{k,n},\mathcal{F}_{k,n})_{0\le k\le n}$ is a martingale. With the notations above, it follows that for any positive real λ,

$$\mathbb{P}(\max_{1\le k\le n} |\widetilde{M}_k| \geq 2\lambda) \leq \mathbb{P}(|\widetilde{M}_n| \geq \lambda) + \mathbb{P}(\max_{1\le k\le n} |\widetilde{M}_{k,n}| \geq \lambda).$$

Therefore, applying Markov's inequality to the first term in the right-hand side and Proposition 2.9 to the second one, we get that for any positive real λ,

$$\lambda\mathbb{P}(\max_{1\le k\le n} |\widetilde{M}_k| \geq 2\lambda) \leq 3\,\mathbb{E}\left(|\widetilde{M}_n| I(|\widetilde{M}_n| \geq \lambda/2)\right).$$

Therefore, proceeding as in the proof of Corollary 2.7, we have, for any $p \in]1,\infty[$ and any positive λ,

$$\mathbb{E}(\max_{1\le k\le n} |\widetilde{M}_k|^p I(\max_{1\le k\le n} |\widetilde{M}_k| \geq 2\lambda)) \leq 3\times 2^{2p-1}\left(\frac{p}{p-1}\right)\mathbb{E}\left(|\widetilde{M}_n|^p I(|\widetilde{M}_n| \geq \lambda/2)\right). \quad (2.3)$$

2.2.2 *Moment Inequalities for Martingales*

We now turn to some useful moment inequalities.

Theorem 2.12 *(Burkholder's inequality (1973, 1988)) Suppose that $(M_n, \mathcal{F}_n)_{n\geq 1}$ is a martingale and $1<p<\infty$. Let $d_1 = M_1$ and $d_i = M_i - M_{i-1}$ for $2 \leq i \leq n$. Then there exist positive constants c_1 and c_2 depending only on p such that*

$$c_1\mathbb{E}\left(\sum_{i=1}^n d_i^2\right)^{p/2} \leq \mathbb{E}|M_n|^p \leq c_2\mathbb{E}\left(\sum_{i=1}^n d_i^2\right)^{p/2}.$$

There are many proofs of this inequality, via symmetrization, via truncation or via suitable stopping times (see e.g. Section 2.4 in Hall and Heyde (1980)).

Note that Theorem 2.12 implies the following Marcinkiewicz–Zygmund type inequalities.

Theorem 2.13 *Suppose that $(M_n, \mathcal{F}_n)_{n\geq 1}$ is a martingale. Let $d_1 = M_1$ and $d_i = M_i - M_{i-1}$ for $2 \leq i \leq n$. Then*

$$||M_n||_p^2 \leq (p-1)\sum_{i=1}^n ||d_i||_p^2, \quad \text{for } p \geq 2, \quad (2.4)$$

$$||M_n||_p^p \leq 2\sum_{i=1}^n ||d_i||_p^p, \quad \text{for } 1 \leq p \leq 2. \quad (2.5)$$

For $p \geq 2$, the constant $p-1$ is derived in Rio (2009). For $1 \leq p \leq 2$, the constant is due to von Bahr and Esseen (1965). See also Pinelis (2015) for the optimal constant for the case $1 \leq p \leq 2$ and Cuny, Dedecker and Jan (2017) for the case when $(d_i)_{i\in\mathbb{Z}}$ is a sequence of real-valued random variables adapted to a non-decreasing sequence of σ-fields (see their Proposition 10).

By applying the Doob inequality, we then derive the maximal version of Marcinkiewicz–Zygmund inequality.

Corollary 2.14 *For $p \geq 1$ there exists a constant c_p such that*

$$\|\max_{1\leq k\leq n}|M_k|\|_p^2 \leq c_p \sum_{i=1}^{n} \|d_i\|_p^2, \quad \textit{for} \ \ p \geq 2, \tag{2.6}$$

$$\|\max_{1\leq k\leq n}|M_k|\|_p^p \leq c_p \sum_{i=1}^{n} \|d_i\|_p^p, \quad \textit{for} \ \ 1 \leq p \leq 2. \tag{2.7}$$

Now, let us state square function deviation inequalities.

Theorem 2.15 *(Burkholder (1979)) Suppose that $(M_n, \mathcal{F}_n)_{n\geq 1}$ is a martingale. Let $d_1 = M_1$ and $d_i = M_i - M_{i-1}$ for $i \geq 2$. Then there exists a positive constant C such that for any $\lambda > 0$,*

$$\lambda \mathbb{P}\left(\left[\sum_{k=1}^{\infty} d_k^2\right]^{1/2} > \lambda\right) \leq C \sup_{n\geq 1} \|M_n\|_1.$$

In his 1979 paper, Burkholder obtained $C = 2$ improving the constant $C = 3$ he obtained in his 1973 paper.

Proof of Theorem 2.15 We will give a proof of Theorem 2.15 with $C = 3$ following the proof of Burkholder (1973). In order to simplify the notation, let

$$D_n = \left[\sum_{k=1}^{n} d_k^2\right]^{1/2} \quad \text{and } D = \left[\sum_{k=1}^{\infty} d_k^2\right]^{1/2}.$$

We start with the following lemma.

Lemma 2.16 *Suppose that $(M_n, \mathcal{F}_n)_{n\geq 1}$ is a $\mathbb{L}^1$-bounded martingale. Let τ be the stopping time defined by*

$$\tau(\omega) = \inf\{k \geq 1 \ : \ |M_k(\omega)| > \lambda\},$$

where $\lambda > 0$. Then

$$\|D_{\tau-1}\|_2^2 + \|M_{\tau-1}\|_2^2 = 2\,\mathbb{E}(M_\tau M_{\tau-1}) \leq 2\lambda \sup_{n\geq 1} \|M_n\|_1.$$

Proof Notice that $|M_{\tau-1}| \leq \lambda$, so $\mathbb{E}(M_\tau M_{\tau-1}) \leq \lambda \|M_\tau\|_1 \leq \lambda \sup_{n\geq 1} \|M_n\|_1$. To prove the left-hand side inequality, we notice that for any $n \geq 1$,

$$D_{n-1}^2 + M_{n-1}^2 = 2M_n M_{n-1} - 2\sum_{k=1}^{n} M_{k-1} d_k.$$

Now, setting $g_n = \sum_{k=1}^{n} M_{k-1} d_k$ and $\nu = \min(n, \tau)$, and taking into account that $I(\tau \geq k)$ as well as M_{k-1} are $\mathcal{F}_{k-1}$-measurable, we have that

$$\mathbb{E}(g_\nu) = \sum_{k=1}^{n} \mathbb{E}\left(M_{k-1} I(\tau \geq k) \mathbb{E}(d_k | \mathcal{F}_{k-1})\right) = 0.$$

Therefore

$$\mathbb{E}(D_{\nu-1}^2) + \mathbb{E}(M_{\nu-1}^2) = 2\mathbb{E}(M_\nu M_{\nu-1}). \tag{2.8}$$

Since $\sup_{n\geq 1} M_{\nu-1}^2 \leq \lambda^2$ and $\sup_{n\geq 1} |M_\nu M_{\nu-1}| \leq \lambda^2 + \lambda |M_\tau|$ are integrable, taking the limit in both sides of (2.8) gives the left-hand side inequality. This ends the proof of the lemma.

We turn now to the proof of the inequality given in Theorem 2.15. Fix λ and define τ as in Lemma 2.16. Since $D_{\tau-1} = D$ on the set $\{\tau = \infty\} = \{\sup_{n\geq 1} |M_n| \leq \lambda\}$, we have, by Lemma 2.16,

$$\begin{aligned}\lambda \mathbb{P}\left(D > \lambda, \sup_{n\geq 1} |M_n| \leq \lambda\right) &\leq \lambda \mathbb{P}\left(D_{\tau-1} > \lambda\right) \\ &\leq \lambda^{-1} \|D_{\tau-1}\|_2^2 \leq 2 \sup_{n\geq 1} \|M_n\|_1.\end{aligned}$$

Hence by Doob's maximal inequality given in Theorem 2.5,

$$\begin{aligned}\lambda \mathbb{P}\left(D > \lambda\right) &\leq \lambda \mathbb{P}\left(\sup_{n\geq 1} |M_n| > \lambda\right) + \mathbb{P}\left(D > \lambda, \sup_{n\geq 1} |M_n| \leq \lambda\right) \\ &\leq \sup_{n\geq 1} \|M_n\|_1 + 2 \sup_{n\geq 1} \|M_n\|_1.\end{aligned}$$

■

We now state the Rosenthal's inequality, which follows by careful truncation arguments (see e.g. Section 2.4 in Hall and Heyde (1980) for a proof).

Theorem 2.17 *Suppose that $(M_n, \mathcal{F}_n)_{n\geq 1}$ is a martingale and $2 \leq p < \infty$. Let $d_1 = M_1$ and $d_i = M_i - M_{i-1}$ for $2 \leq i \leq n$. Then there exist constants c_1 and c_2 depending only on p such that*

$$\begin{aligned}&c_1 \left\{\mathbb{E}\left[\left(\sum_{i=1}^{n} \mathbb{E}(d_i^2 | \mathcal{F}_{i-1})\right)^{p/2}\right] + \sum_{i=1}^{n} \mathbb{E}|d_i|^p\right\} \\ &\leq \mathbb{E}|M_n|^p \leq c_2 \left\{\mathbb{E}\left[\left(\sum_{i=1}^{n} \mathbb{E}(d_i^2 | \mathcal{F}_{i-1})\right)^{p/2}\right] + \sum_{i=1}^{n} \mathbb{E}|d_i|^p\right\}.\end{aligned}$$

2.2.3 *Exponential Inequalities for Martingales*

This section contains exponential inequalities for martingales that are useful for generalization to dependent sequences. Refinements of these inequalities can be found in Chapter 3 of the recent monograph by Bercu, Delyon and Rio (2015).

Hoeffding–Azuma inequality

Theorem 2.18 is due to Hoeffding (1963) if M_n is the partial sum of centered independent random variables and to Azuma (1967) when M_n is a martingale.

Theorem 2.18 *Let $(M_n, \mathcal{F}_n)_{n\geq 0}$ be a martingale with $M_0 = 0$, and $M_n^* = \max(0, M_1, \cdots, M_n)$. Assume that there exist a sequence $(B_n)_{n\geq 0}$ of real valued random variables adapted to $(\mathcal{F}_n)$ and a sequence of positive constants $(c_n)_{n\geq 1}$ such that for each $n \geq 1$*

$$B_{n-1} \leq M_n \leq B_{n-1} + c_n, \text{ a.s.} \tag{2.9}$$

Then, for any positive t,

$$\log\left(\mathbb{E}(\exp(tM_n))\right) \leq \frac{t^2 \sum_{i=1}^n c_i^2}{8}, \tag{2.10}$$

and consequently, for any $x \geq 0$,

$$\mathbb{P}(M_n^* \geq x) \leq \exp\left(-\frac{2x^2}{\sum_{i=1}^n c_i^2}\right). \tag{2.11}$$

Remark 2.19 *If $(d_k)_{k\in\mathbb{Z}}$ is a sequence of martingale differences adapted to a filtration $(\mathcal{F}_k)_{k\in\mathbb{Z}}$ and if there exist two sequences of positive constants $(a_n)_{n\geq 1}$ and $(b_n)_{n\geq 1}$ such that, for each $n \geq 1$,*

$$-a_n \leq d_n \leq b_n,$$

then applying Theorem 2.18 with $M_n = \sum_{i=1}^n d_i$, $B_{n-1} = M_{n-1} - a_n$ and $c_n = a_n + b_n$, we get that, for any positive t,

$$\log\left(\mathbb{E}(\exp(tM_n))\right) \leq \frac{t^2 \sum_{i=1}^n (a_i + b_i)^2}{8}, \tag{2.12}$$

and also that, for any $x \geq 0$,

$$\mathbb{P}\left(\max_{1\leq k\leq n} |M_k| \geq x\right) \leq 2\exp\left(-\frac{2x^2}{\sum_{i=1}^n (a_i + b_i)^2}\right). \tag{2.13}$$

Proof of Theorem 2.18 By Fact 2.2, for each $t > 0$, $(e^{tM_k}, \mathcal{F}_k)_{k\geq 0}$ is a submartingale. Therefore Doob's maximal inequality (see Theorem 2.5) gives that for any $x \geq 0$ and any $t > 0$,

$$\log\left(\mathbb{P}(M_n^* \geq x)\right) \leq \log\left(\exp(-xt)\mathbb{E}(\exp(tM_n))\right) = \log\left(\mathbb{E}(\exp(tM_n))\right) - xt.$$

Therefore

$$\log\left(\mathbb{P}(M_n^* \geq x)\right) \leq \inf_{t>0}\left(\log\left(\mathbb{E}(\exp(tM_n))\right) - xt\right). \tag{2.14}$$

By standard computations, (2.11) will then follow provided (2.10) holds.

Let us prove (2.10) by recurrence on n. First for $n = 0$, $M_0 = 0$, whence (2.10) holds true. Assume now that (2.10) holds for $n-1$. Let

$$Z_n = M_n - B_{n-1} - c_n/2. \tag{2.15}$$

Since $M_n - Z_n$ is $\mathcal{F}_{n-1}$–measurable, we get that

$$\begin{aligned}\log\left(\mathbb{E}(\exp(tM_n)|\mathcal{F}_{n-1})\right) &= \log\left(\mathbb{E}(\exp(tM_n - tZ_n)\exp(tZ_n)|\mathcal{F}_{n-1})\right) \\ &= t(B_{n-1} + c_n/2) + \log\left(\mathbb{E}(\exp(tZ_n)|\mathcal{F}_{n-1})\right) \ \text{a.s.}\end{aligned} \tag{2.16}$$

We shall now apply the following lemma whose proof is given later:

Lemma 2.20 *Let Z be a real random variable taking values in $[-b, b]$. For any σ-field $\mathcal{F}$ and any positive t,*

$$\log\left(\mathbb{E}(\exp(tZ)|\mathcal{F})\right) \leq t\mathbb{E}(Z|\mathcal{F}) + t^2b^2/2.$$

Since by (2.9), Z_n takes its values in $[-c_n/2, c_n/2]$ a.s., starting from (2.16) and applying the above lemma we derive that

$$\begin{aligned}\log\left(\mathbb{E}(\exp(tM_n)|\mathcal{F}_{n-1})\right) &\leq t(B_{n-1} + c_n/2) + t\left(\mathbb{E}(M_n|\mathcal{F}_{n-1}) - B_{n-1} - c_n/2\right) + t^2c_n^2/8 \\ &= t\mathbb{E}(M_n|\mathcal{F}_{n-1}) + t^2c_n^2/8 = tM_{n-1} + t^2c_n^2/8.\end{aligned}$$

Therefore

$$\mathbb{E}(\exp(tM_n)) = \mathbb{E}\left(\mathbb{E}(\exp(tM_n)|\mathcal{F}_{n-1})\right) \leq \exp(t^2c_n^2/8)\mathbb{E}(\exp(tM_{n-1})),$$

proving that if (2.10) holds at the rank $n-1$ then it does at the rank n. To end the proof it remains to prove Lemma 2.20. Without loss of generality, we can assume that $b = 1$ (if not divide Z by b and multiply t by b). Hence we write, $tZ = -t(1-Z)/2 + t(1+Z)/2$. Then, by convexity of the exponential function,

$$2\exp(tZ) \leq (1-Z)\exp(-t) + (1+Z)\exp(t).$$

Taking the conditional expectation with respect to $\mathcal{F}$, it follows that

$$\mathbb{E}(\exp(tZ)|\mathcal{F}) \leq \cosh(t) + \mathbb{E}(Z|\mathcal{F})\sinh(t).$$

We end the proof of the lemma by noticing that an application of the Taylor formula gives

$$\log(\cosh(t) + \mathbb{E}(Z|\mathcal{F})\sinh(t)) \leq t\mathbb{E}(Z|\mathcal{F}) + t^2/2,$$

(to see this, it suffices to notice that $f(0) = 1, f'(0) = \mathbb{E}(Z|\mathcal{F})$ and $(\log f(t))'' \leq 1$, where $f(t) = \cosh(t) + \mathbb{E}(Z|\mathcal{F})\sinh(t)$). ■

Bennett inequality

Theorem 2.21 is due to Bennett (1962) if M_n is the partial sum associated with centered independent random variables and to Freedman (1975) when M_n is a martingale. Notice that it has been extended by Pinelis (1994) when M_n is a martingale with values in 2-smooth Banach spaces.

Theorem 2.21 *Let $(M_n, \mathcal{F}_n)_{n\geq 0}$ be a martingale with $M_0 = 0$, and $M_n^* = \max(0, M_1, \cdots, M_n)$. Let $d_n = M_n - M_{n-1}$. Assume that there exists a positive constant c such that for each $n \geq 1$*

$$M_n \leq M_{n-1} + c, \text{ a.s.} \tag{2.17}$$

Assume that $\| \sum_{k=1}^n \mathbb{E}(d_k^2|\mathcal{F}_{k-1})\|_\infty \leq y_n$ a.s.. Then for each $x \geq 0$,

$$\mathbb{P}(M_n^* \geq x) \leq \exp\left(-\frac{y_n}{c^2}\, h\left(\frac{xc}{y_n}\right)\right), \tag{2.18}$$

where $h(u) = (1+u)\ln(1+u) - u$.

Remark 2.22 *Since for any $u \geq 0$, $h(u) \geq u\log(1+u)/2$ and $h(u) \geq u^2/(2+2u/3)$, under the assumptions of Theorem 2.21, the following inequalities are also valid:*

$$\mathbb{P}\left(\sup_{1\leq j\leq n} |M_j| \geq x\right) \leq 2\exp\left(-\frac{x}{2c}\ln\left(1 + \frac{xc}{y_n}\right)\right), \text{ and}$$

$$\mathbb{P}\left(\sup_{1\leq j\leq n} |M_j| \geq x\right) \leq 2\exp\left(\frac{-x^2}{2y_n + 2cx/3}\right)$$
$$\leq 2\max\left(\exp\left(\frac{-x^2}{4y_n}\right), \exp\left(\frac{-3x}{4c}\right)\right).$$

Proof of Theorem 2.21 Without loss of generality, we can assume that $c = 1$ (otherwise we divide the random variables by c). For any positive t, let

$$G_0 = 1 \text{ and } G_n = \exp\left(tM_n - (e^t - t - 1)\sum_{j=1}^n \mathbb{E}(d_j^2|\mathcal{F}_{j-1})\right).$$

We first show that $(G_n)_{n\geq 0}$ is a supermartingale adapted to $(\mathcal{F}_n)$. With this aim, we first notice that

$$\mathbb{E}(G_n|\mathcal{F}_{n-1}) = G_{n-1}\exp\left(\log(\mathbb{E}(e^{td_n}|\mathcal{F}_{n-1})) - \mathbb{E}(d_n^2|\mathcal{F}_{n-1})(e^t - t - 1)\right). \tag{2.19}$$

Now since $\log x \leq x - 1$ and $\mathbb{E}(d_n|\mathcal{F}_{n-1}) = 0$, $\log(\mathbb{E}(e^{td_n}|\mathcal{F}_{n-1})) \leq \mathbb{E}(e^{td_n} - td_n - 1|\mathcal{F}_{n-1})$. In addition, the function $g(u) := u^{-2}(e^u - 1 - u)$ for $u \neq 0$, $g(0) := \frac{1}{2}$, is increasing in $u \in \mathbb{R}$. Therefore, since $d_n \leq 1$,

$$e^{td_n} - td_n - 1 \leq d_n^2(e^t - t - 1),$$

implying that

$$\log(\mathbb{E}(e^{td_n}|\mathcal{F}_{n-1})) \leq (e^t - t - 1)\mathbb{E}(d_n^2|\mathcal{F}_{n-1}). \tag{2.20}$$

Starting from (2.19) and using (2.20), we then derive that $\mathbb{E}(G_n|\mathcal{F}_{n-1}) \leq G_{n-1}$ proving that $(G_n)_{n\geq 0}$ is a supermartingale. This implies that $\mathbb{E}(G_n) \leq \mathbb{E}(G_0) = 1$. Now since $\exp(tM_n) \leq G_n \exp\left(y_n(e^t - t - 1)\right)$, we obtain that for any positive t,

$$\mathbb{E}(\exp(tM_n)) \leq \exp\left((e^t - t - 1)y_n\right). \tag{2.21}$$

The theorem follows by using (2.21) in (2.14) and the minimization in t. ■

As a consequence of Theorem 2.21, we can deduce the following useful result:

Corollary 2.23 *With the notations and under the assumptions of Theorem 2.21, the following inequality is valid: For all $x, y > 0$,*

$$\mathbb{P}\left(\sup_{1\leq j\leq n}|M_j| \geq x, \sum_{j=1}^{n}\mathbb{E}(d_j^2|\mathcal{F}_{j-1}) \leq y\right) \leq 2\exp\left(-\frac{y}{c^2}h\left(\frac{xc}{y}\right)\right). \tag{2.22}$$

Proof Let $A_i = \{\sum_{j=1}^{i}\mathbb{E}(d_j^2|\mathcal{F}_{j-1}) \leq y\}$, and let $\widetilde{M}_j$ be the martingale $\widetilde{M}_j = \sum_{i=1}^{j} d_i I_{A_i}$. Clearly

$$\begin{aligned}\mathbb{P}\left(\sup_{1\leq j\leq n}|M_j| \geq x, \sum_{j=1}^{n}\mathbb{E}(d_j^2|\mathcal{F}_{j-1}) \leq y\right) &= \mathbb{P}\left(\sup_{1\leq j\leq n}|\widetilde{M}_j| \geq x, \sum_{j=1}^{n}\mathbb{E}(d_j^2|\mathcal{F}_{j-1}) \leq y\right)\\ &\leq \mathbb{P}\left(\sup_{1\leq j\leq n}|\widetilde{M}_j| \geq x\right).\end{aligned}$$

To conclude, it suffices to apply Theorem 2.21 to the martingale $\widetilde{M}_j$. ■

2.3 Central Limit Theorem for Triangular Arrays of Martingales

The following theorem is due to Brown and Eagleson (1971).

Theorem 2.24 *Assume that $(d_{n,i})_{1\leq i\leq n}$ is an array of martingale differences adapted to an array $(\mathcal{F}_{n,i})_{0\leq i\leq n}$ of filtrations. Suppose*

$$\sum_{j=1}^{k_n} \mathbb{E}(d_{n,j}^2|\mathcal{F}_{n,j-1}) \to^{\mathbb{P}} 1 \text{ as } n \to \infty \tag{2.23}$$

and for each $\varepsilon > 0$,

$$\sum_{j=1}^{k_n} \mathbb{E}\left(d_{n,j}^2 I(|d_{n,j}| > \varepsilon)|\mathcal{F}_{n,j-1}\right) \to^{\mathbb{P}} 0 \text{ as } n \to \infty. \tag{2.24}$$

Then $S_n = \sum_{j=1}^{k_n} d_{n,j}$ converges in distribution to a standard normal variable.

Proof of Theorem 2.24 The proof is based on the principle of conditioning going back to Jakubowski (1986) (see also de la Peña and Giné (1999, Ch. 7.1)). Let us first give the idea of this principle. To simplify the notations we will omit the index n and then write $d_j := d_{n,j}$ and $\mathcal{F}_j := \mathcal{F}_{n,j}$. Moreover, let

$$f_n(t) = \mathbb{E}[e^{itS_n}],\ f_n(t|cond) = \prod_{j=1}^{n} \mathbb{E}(e^{itd_j}|\mathcal{F}_{j-1}).$$

We establish the convergence of characteristic functions $f_n(t) \to e^{-t^2/2}$. To tackle the dependence of martingale differences, we first show (Part 1) that under certain technical conditions, for each t, $f_n(t)$ are asymptotically equivalent to $f_n(t|cond)$ (product of conditional characteristic functions). In part 2, we prove the convergence $f_n(t|cond) \to e^{-t^2/2}$. Finally, in part 3, we apply the truncation trick to justify the approach.

Part 1. We state this step in a slightly more general form as an auxiliary lemma. Let $X_j = e^{itd_j}$. Notice that

$$f_n(t) = \mathbb{E}[e^{itS_n}] = \mathbb{E}\left(\prod_{j=1}^{n} X_j\right), f_n(t|cond) = \prod_{j=1}^{n} \mathbb{E}(e^{itd_j}|\mathcal{F}_{j-1}) = \prod_{j=1}^{n} \mathbb{E}(X_j|\mathcal{F}_{j-1}).$$

Lemma 2.25 *Let X_j be $\mathcal{F}_j$ measurable variables such that $|X_j| \leq 1$. Assume that there is $\beta > 0$ such that for all $n \geq 1$,*

$$\left|\prod_{j=1}^{n} \mathbb{E}(X_j|\mathcal{F}_{j-1})\right| \geq \beta. \tag{2.25}$$

In addition assume that

$$\prod_{j=1}^{n} \mathbb{E}(X_j|\mathcal{F}_{j-1}) \to^{\mathbb{P}} L \text{ as } n \to \infty. \tag{2.26}$$

Then,

$$\left|\mathbb{E}\left(\prod_{j=1}^{n} X_j - L\right)\right| \to 0 \text{ as } n \to \infty. \tag{2.27}$$

Proof Notice that under (2.25), for all $n \geq 1$ and all $j \in \{1, ..., n\}$, $|\mathbb{E}(X_j|\mathcal{F}_{j-1})| \geq \beta$. Indeed, if not $\left|\prod_{j=1}^{n} \mathbb{E}(X_j|\mathcal{F}_{j-1})\right| < \beta$. Therefore by noticing that the r.v.'s X_j are adapted to $\mathcal{F}_j$ and by using (2.25) to avoid to divide by 0, we get that

$$\mathbb{E}\left(\frac{\prod_{j=1}^{n} X_j}{\prod_{j=1}^{n} \mathbb{E}(X_j|\mathcal{F}_{j-1})}\right) = 1. \tag{2.28}$$

Hence

$$\begin{aligned}
\left|\mathbb{E}\left(\prod_{j=1}^{n} X_j - L\right)\right| &= \left|\mathbb{E}\left(\prod_{j=1}^{n} X_j - L\frac{\prod_{j=1}^{n} X_j}{\prod_{j=1}^{n} \mathbb{E}(X_j|\mathcal{F}_{j-1})}\right)\right| \\
&\leq \mathbb{E}\left|1 - \frac{L}{\prod_{j=1}^{n} \mathbb{E}(X_j|\mathcal{F}_{j-1})}\right| \\
&\to 0,
\end{aligned}$$

since the sequence of variables $L/\prod_{j=1}^{n} \mathbb{E}(X_j|\mathcal{F}_{j-1})$ is bounded and tends to 1 in probability. ■

Part 2. Straightforward Taylor expansions. The goal in this part is to show convergence of conditional characteristic functions $f_n(t|cond) \to e^{-t^2/2}$.

Lemma 2.26 *Under the condition of Theorem 2.24,*

$$f_n(t|cond) = \prod_{j=1}^{n} \mathbb{E}(e^{itd_j}|\mathcal{F}_{j-1}) \to^{\mathbb{P}} e^{-t^2/2} \text{ as } n \to \infty. \tag{2.29}$$

Proof Take $\varepsilon \in]0, 2/t[$ and consider the set $B(n,\varepsilon) = \{\max_{1\le j\le n} \mathbb{E}(d_j^2|\mathcal{F}_{j-1}) \le \varepsilon^2\}$, that is w.l.o.g. each d_j is conditionally truncated, transformed to $d_j I(\mathbb{E}[d_j^2|\mathcal{F}_{j-1}] \le \varepsilon^2)$. Notice that condition (2.24) implies that

$$\mathbb{P}(\overline{B(n,\varepsilon)}) \to 0 \text{ as } n \to \infty. \tag{2.30}$$

Since both parts in (2.29) are bounded, w.l.o.g. we may work on $B(n,\varepsilon)$.

Throughout we apply the following elementary inequalities: For all $z_1, \cdots, z_m$ and $w_1, \cdots, w_m$ complex numbers of modulus at most 1, for every complex number u with $Re(u) \le 0$, and for every complex number z,

$$\left|\prod_{j=1}^m z_j - \prod_{j=1}^m w_j\right| \le \sum_{j=1}^m |z_j - w_j|,$$
$$\left|e^u - (1 + u + u^2/2)\right| \le \min\left\{|u|^2, |u|^3\right\},$$
$$|e^z - 1 - z| \le |z|^2 e^{|z|}.$$

Now, since $\mathbb{E}(d_j|\mathcal{F}_{j-1}) = 0$, we derive that on the set $B(n,\varepsilon)$,

$$\left|\prod_{j=1}^n \mathbb{E}(e^{itd_j}|\mathcal{F}_{j-1}) - \prod_{j=1}^n \left(1 - \frac{t^2\mathbb{E}(d_j^2|\mathcal{F}_{j-1})}{2}\right)\right|$$
$$\le \sum_{j=1}^n \left[t^2\mathbb{E}\left(d_j^2 I(|d_j| > \varepsilon)|\mathcal{F}_{j-1}\right) + \varepsilon|t|^3\mathbb{E}(d_j^2|\mathcal{F}_{j-1})\right]$$

and

$$\left|\prod_{j=1}^n e^{-t^2\mathbb{E}(d_j^2|\mathcal{F}_{j-1})/2} - \prod_{j=1}^n \left(1 - \frac{t^2\mathbb{E}(d_j^2|\mathcal{F}_{j-1})}{2}\right)\right| \le 3\varepsilon^2 \sum_{j=1}^n t^4\mathbb{E}(d_j^2|\mathcal{F}_{j-1}).$$

Taking into account (2.23) and (2.24) and letting n tend to infinity and then ε tend to zero, we obtain

$$\left|\prod_{j=1}^n \mathbb{E}(e^{itd_j}|\mathcal{F}_{j-1}) - \prod_{j=1}^n \left(1 - \frac{t^2\mathbb{E}(d_j^2|\mathcal{F}_{j-1})}{2}\right)\right| \to^{\mathbb{P}} 0$$

and

$$\left|\prod_{j=1}^n e^{-t^2\mathbb{E}(d_j^2|\mathcal{F}_{j-1})/2} - \prod_{j=1}^n \left(1 - \frac{t^2\mathbb{E}(d_j^2|\mathcal{F}_{j-1})}{2}\right)\right| \to^{\mathbb{P}} 0.$$

This together with (2.23) implies (2.29). ■

Part 3. End of the proof: Avoiding the assumption (2.25) *via the conditional truncation trick.* For all $j \in \{1, ..., n\}$, set

$$A_j = \left\{ \left| \prod_{k=1}^{j} \mathbb{E}(e^{itd_k} | \mathcal{F}_{k-1}) \right| > \frac{e^{-t^2/2}}{2} \right\}.$$

Notice first that by (2.29)

$$\mathbb{P}(A_n^c) = \mathbb{P}\left(\left| \prod_{k=1}^{n} \mathbb{E}(e^{itd_k} | \mathcal{F}_{k-1}) \right| \le \frac{e^{-t^2/2}}{2} \right) \to 0 \text{ as } n \to \infty. \tag{2.31}$$

Now, we apply the conditional truncation. Firstly, since $\mathbb{E}(e^{itS_n I_{A_n}}) = \mathbb{E}(e^{itS_n} I_{A_n} + I_{A_n^c})$,

$$\mathbb{E}(e^{itS_n}) = \mathbb{E}\left(e^{itS_n I_{A_n}}\right) + \mathbb{E}\left((e^{itS_n} - 1) I_{A_n^c}\right).$$

Secondly, since the A_j's are non-increasing in j, we derive that

$$\mathbb{E}\left(e^{itS_n I_{A_n}}\right) = \mathbb{E}\left(e^{it\sum_{j=1}^n d_j I_{A_j}} I_{A_n}\right) + \mathbb{E}(I_{A_n^c}).$$

Therefore

$$\mathbb{E}(e^{itS_n}) = \mathbb{E}\left(e^{it\sum_{j=1}^n d_j I_{A_j}}\right) + \mathbb{E}\left(e^{itS_n} I_{A_n^c}\right) - \mathbb{E}\left(e^{it\sum_{j=1}^n d_j I_{A_j}} I_{A_n^c}\right),$$

and hence

$$|\mathbb{E}(e^{itS_n}) - e^{-t^2/2}| \le 2\,\mathbb{P}(A_n^c) + \left|\mathbb{E}\left(e^{it\sum_{j=1}^n d_j I_{A_j}}\right) - e^{-t^2/2}\right|. \tag{2.32}$$

Using (2.31), the theorem will then follow if we can prove that

$$\lim_{n\to\infty} |\mathbb{E}\left(e^{it\sum_{j=1}^n d_j I_{A_j}}\right) - e^{-t^2/2}| = 0. \tag{2.33}$$

This can be achieved by applying Lemma 2.25 with $X_j = e^{itd_j I_{A_j}}$. To apply the lemma, we need to check its conditions (2.25) and (2.26).

Checking condition (2.25). We need to show that, for all t, there exists a $\beta > 0$ such that

$$\prod_{j=1}^{n} |\mathbb{E}(e^{itd_j I_{A_j}} | \mathcal{F}_{j-1})| > \beta. \tag{2.34}$$

With this aim, we first note that, for any $i \leq j$, $I_{A_i^c} I_{A_j^c} = I_{A_i^c}$ and $I_{A_i^c} I_{A_j} = 0$, which implies

$$\prod_{j=1}^{n} |\mathbb{E}(e^{itd_j I_{A_j}} | \mathcal{F}_{j-1})| = \prod_{j=1}^{n} \left(I_{A_j} |\mathbb{E}(e^{itd_j} | \mathcal{F}_{j-1})| + I_{A_j^c} \right)$$
$$= \sum_{k=0}^{n} \left(\prod_{j=1}^{k} I_{A_j} |\mathbb{E}(e^{itd_j} | \mathcal{F}_{j-1})| \right) \left(\prod_{j=k+1}^{n} I_{A_j^c} \right),$$

with the convention that $\prod_{i=p}^{q} = 1$ if $p > q$. Since the A_j's are non-increasing in j, using the convention that $I_{A_0} = I_{A_{n+1}^c} = 1$, we then get that

$$\prod_{j=1}^{n} |\mathbb{E}(e^{itd_j I_{A_j}} | \mathcal{F}_{j-1})| = \sum_{k=0}^{n} I_{A_k} I_{A_{k+1}^c} \prod_{j=1}^{k} |\mathbb{E}(e^{itd_j} | \mathcal{F}_{j-1})|$$
$$> \frac{e^{-t^2/2}}{2} \sum_{k=0}^{n} I_{A_k} I_{A_{k+1}^c}.$$

Since $\sum_{k=0}^{n} I_{A_k} I_{A_{k+1}^c} = 1$, it follows that

$$\prod_{j=1}^{n} |\mathbb{E}(e^{itd_j I_{A_j}} | \mathcal{F}_{j-1})| > \frac{e^{-t^2/2}}{2}, \tag{2.35}$$

proving (2.34).

Checking condition (2.26). The condition (2.26) applied to variables $X_j = e^{itd_j I_{A_j}}$ states

$$\prod_{j=1}^{n} \mathbb{E}(e^{itd_j I_{A_j}} | \mathcal{F}_{j-1}) \to^{\mathbb{P}} e^{-t^2/2} \text{ as } n \to \infty. \tag{2.36}$$

We note that $(d_j I_{A_j})_{j \geq 1}$ forms a sequence of martingale differences with respect to the σ-fields $\mathcal{F}_j$. In addition, since the A_j^c's are $\mathcal{F}_{j-1}$-measurable and non-decreasing in j,

$$\sum_{j=1}^{n} \mathbb{E}(d_j^2 I_{A_j^c} | \mathcal{F}_{j-1}) \leq I_{A_n^c} \sum_{j=1}^{n} \mathbb{E}(d_j^2 | \mathcal{F}_{j-1}).$$

By using (2.29) together with (2.23), this last inequality implies that

$$\sum_{j=1}^{n} \mathbb{E}(d_j^2 I_{A_j} | \mathcal{F}_{j-1}) \to^{\mathbb{P}} 1 \text{ as } n \to \infty.$$

Moreover, for each $\varepsilon > 0$, by using (2.24), we have that

$$\sum_{j=1}^{n} \mathbb{E}\left(d_j^2 I_{A_j} I(|d_j| I_{A_j} > \varepsilon) | \mathcal{F}_{j-1}\right) \leq \sum_{j=1}^{n} \mathbb{E}\left(d_j^2 I(|d_j| > \varepsilon) | \mathcal{F}_{j-1}\right) \to^{\mathbb{P}} 0 \text{ as } n \to \infty.$$

By Lemma 2.26 of part 2 applied to the sequence of martingale differences $(d_j I_{A_j})_{j\geq 1}$, (2.36) follows. This ends the proof of the theorem. ∎

As a corollary to Theorem 2.24, we obtain the following result:

Corollary 2.27 *Assume that $(X_{n,i})_{1\leq i\leq n}$ is an array of random variables adapted to an array of filtrations $(\mathcal{F}_{n,i})_{0\leq i\leq n}$. Suppose*

$$\sum_{j=1}^{n} \mathbb{E}(X_{n,j} | \mathcal{F}_{n,j-1}) \to^{\mathbb{P}} 0 \textit{ as } n \to \infty, \tag{2.37}$$

$$\sum_{j=1}^{n} \mathrm{Var}(X_{n,j} | \mathcal{F}_{n,j-1}) \to^{\mathbb{P}} 1 \textit{ as } n \to \infty \tag{2.38}$$

and for each $\varepsilon > 0$,

$$\sum_{j=1}^{n} \mathbb{E}\left(X_{n,j}^2 I(|X_{n,j}| > \varepsilon) | \mathcal{F}_{n,j-1}\right) \to^{\mathbb{P}} 0 \textit{ as } n \to \infty. \tag{2.39}$$

Then $S_n = \sum_{j=1}^{n} X_{n,j}$ converges in distribution to a standard normal variable.

Proof For any $1 \leq j \leq n$, let $d_{n,j} = X_{n,j} - \mathbb{E}(X_{n,j}|\mathcal{F}_{n,j-1})$. By (2.37), the corollary will follow if we can prove that $\sum_{j=1}^{n} d_{n,j}$ converges in distribution to a standard normal variable. With this aim, since $(d_{n,j})_{j\geq 1}$ is a sequence of martingale differences with respect to the filtrations $\mathcal{F}_{n,j}$, it suffices to verify the conditions of Theorem 2.24. The condition (2.23) is clearly satisfied by (2.38). The condition (2.24) is also satisfied by using (2.39) together with the following lemma that we apply with $X = X_{n,j}$, $\mathcal{F} = \mathcal{F}_{n,j-1}$.

Lemma 2.28 *Let X be a real random variable and $\mathcal{F}$ a sigma field. Let $Y = X - \mathbb{E}(X|\mathcal{F})$. For any $\varepsilon > 0$,*

$$\mathbb{E}\left(Y^2 I(|Y| > 4\varepsilon) | \mathcal{F}\right) \leq 12\, \mathbb{E}\left(X^2 I(|X| > \varepsilon) | \mathcal{F}\right). \tag{2.40}$$

It remains to prove the above lemma. With this aim, we first notice that for any positive reals a, b and ε, $(a+b)^2 I(a+b > 4\varepsilon) \leq 4a^2 I(a > 2\varepsilon) + 4b^2 I(b > 2\varepsilon)$. Therefore using additionally the fact that $(\mathbb{E}(X|\mathcal{F}))^2 \leq \mathbb{E}(X^2|\mathcal{F})$, we obtain that

$$Y^2 I(|Y| > 4\varepsilon) \leq 4X^2 I(|X| > 2\varepsilon) + 4\,\mathbb{E}(X^2 I(|\mathbb{E}(X|\mathcal{F})| > 2\varepsilon)|\mathcal{F}).$$

Hence,

$$\mathbb{E}\left(Y^2 I(|Y| > 4\varepsilon)|\mathcal{F}\right) \leq 8\,\mathbb{E}\left(X^2 I(|X| > \varepsilon)|\mathcal{F}\right) + 4\varepsilon^2 I(|\mathbb{E}(X|\mathcal{F})| > 2\varepsilon). \tag{2.41}$$

Notice now that $I(|\mathbb{E}(X|\mathcal{F})| > 2\varepsilon) \leq I(|\mathbb{E}(XI(|X| > \varepsilon)|\mathcal{F})| > \varepsilon)$, implying that

$$\varepsilon^2 I(|\mathbb{E}(X|\mathcal{F})| > 2\varepsilon) \leq |\mathbb{E}(XI(|X| > \varepsilon)|\mathcal{F})|^2 \leq \mathbb{E}\left(X^2 I(|X| > \varepsilon)|\mathcal{F}\right),$$

which together with (2.41) implies Lemma 2.28. ■

In the 1970s and 1980s many other sufficient conditions were derived for the conclusion of Theorem 2.24 (some of them are equivalent). For instance, as a consequence of Theorem 2.24, we obtain the following useful result, which in the form stated below is due to Gänssler and Haüsler (1979) and Rootzén (1980).

Theorem 2.29 *(Gänssler–Haüsler–Rootzén). Assume $(d_{n,i})_{1\leq i\leq n}$ is an array of martingale differences adapted to an array $(\mathcal{F}_{n,i})_{1\leq i\leq n}$ of filtrations. Suppose*

$$\mathbb{E}(\max_{1\leq j\leq n} |d_{n,j}|) \to 0 \text{ as } n \to \infty \tag{2.42}$$

and

$$\sum_{j=1}^{n} d_{n,j}^2 \to^{\mathbb{P}} 1 \text{ as } n \to \infty. \tag{2.43}$$

Then $S_n = \sum_{j=1}^n d_{n,j}$ converges in distribution to a standard normal variable.

Other sets of conditions implying the central limit theorem can be found in Helland (1982, Theorem 2.5). They are derived by using Lemma 2.28 and the two following general lemmas.

Lemma 2.30 *Let $(X_{n,i})_{i\geq 1,n\geq 1}$ be an array of real random variables adapted to an array $(\mathcal{F}_{n,i})_{i\geq 1,n\geq 1}$ of filtrations. Assume that $\max_{1\leq i\leq n} |X_{n,i}| \to^{\mathbb{P}} 0$. Then $\sum_{i=1}^n \mathbb{E}(X_{n,i}^2|\mathcal{F}_{n,i-1}) \to^{\mathbb{P}} 1$ implies $\sum_{i=1}^{k_n} X_{n,i}^2 \to^{\mathbb{P}} 1$. If we assume in addition that there is $c > 0$ such that $|X_{n,i}| \leq c$ for all n and all i, then $\sum_{i=1}^n X_{n,i}^2 \to^{\mathbb{P}} 1$ if and only if $\sum_{i=1}^n \mathbb{E}(X_{n,i}^2|\mathcal{F}_{n,i-1}) \to^{\mathbb{P}} 1$.*

Lemma 2.31 *Let $(X_{n,i})_{i\geq 1,n\geq 1}$ be an array of real random variables adapted to an array $(\mathcal{F}_{n,i})_{i\geq 1,n\geq 1}$ of filtrations. Assume that $\mathbb{E}(\max_{1\leq j\leq n} |X_{n,j}|) \to 0$. Then for any $\delta > 0$, $\sum_{i=1}^n \mathbb{E}(|X_{n,i}|I(|X_{n,i}| > \delta)|\mathcal{F}_{n,i-1}) \to^{\mathbb{P}} 0$.*

Note that Theorem 2.29 can be extended to infinite or double infinite sums as shown by the following corollary.

Corollary 2.32 *Assume $(d_{n,i})_{i\in\mathbb{Z}}$ is an array of martingale differences adapted to an array $(\mathcal{F}_{n,i})_{i\in\mathbb{Z}}$ of filtrations. Suppose that for each n*

$$\sum_{j\in\mathbb{Z}} \mathbb{E}(d_{n,j}^2) < \infty.$$

Assume in addition that

$$\mathbb{E}(\sup_{j\in\mathbb{Z}} |d_{n,j}|) \to 0 \text{ as } n \to \infty \tag{2.44}$$

and

$$\sum_{j\in\mathbb{Z}} d_{n,j}^2 \to^{\mathbb{P}} 1 \text{ as } n \to \infty. \tag{2.45}$$

Then $S_n = \sum_{j\in\mathbb{Z}} d_{n,j}$ converges in distribution to a standard normal variable.

Proof We have just to select N_n large enough such that

$$\lim_{n\to\infty} \sum_{|j|>N_n} \mathbb{E}(d_{n,j}^2) = 0.$$

The conditions of Theorem 2.29 are verified for $\sum_{|j|\le N_n} d_{n,j}$ under the assumptions imposed. ■

Comment 2.33 *A careful analysis of the proof of Theorem 2.24 reveals that if convergence (2.23) is replaced by the following one: There exists an a.s. finite random variable η^2 such that*

$$\sum_{j=1}^{n} \mathbb{E}(d_{n,j}^2|\mathcal{F}_{n,j-1}) \to^{\mathbb{P}} \eta^2 \text{ as } n \to \infty,$$

then $S_n = \sum_{j=1}^{n} d_{n,j}$ converges in distribution to Z where the r.v. Z has characteristic function $e^{-\eta^2 t^2/2}$.

A similar remark holds for Theorem 2.29 where condition (2.43) can be replaced by

$$\sum_{j=1}^{n} d_{n,j}^2 \to^{\mathbb{P}} \eta^2 \text{ as } n \to \infty.$$

2.4 Functional Central Limit Theorem for Triangular Arrays of Martingales

The next theorem is the functional form of the central limit theorem (CLT) for triangular arrays of martingale differences. Under the form stated below, it is due to Aldous (1978) with the condition

$$\sup_n \mathbb{E}(\max_{1\le i\le n} d_{n,i}^2) < \infty, \tag{2.46}$$

replacing the condition (2.47).

Theorem 2.34 *Assume $(d_{n,i})_{1\le i\le n}$ is an array of martingale differences adapted to an array $(\mathcal{F}_{n,i})_{1\le i\le n}$ of filtrations. Suppose*

$$\max_{1\le j\le n} |d_{n,j}| \text{ is uniformly integrable,} \tag{2.47}$$

and

$$\sum_{j=1}^{[nt]} d_{n,j}^2 \to^{\mathbb{P}} t \text{ as } n \to \infty \text{ for any } t \in [0, 1]. \tag{2.48}$$

Then $\{S_{[nt]} = \sum_{j=1}^{[nt]} d_{n,j}, t \in [0, 1]\}$ converges in distribution in $D([0, 1])$ to a standard Brownian motion.

Remark 2.35 *Another couple of conditions implying the functional central limit theorem can be found in Helland (1982, Theorem 3.2). For instance, using Lemma 2.30, the conclusion of Theorem 2.34 also holds if (2.47) is satisfied,*

$$\max_{1\le j\le n} |d_{n,j}| \to^{\mathbb{P}} 0 \text{ as } n \to \infty \tag{2.49}$$

and

$$\sum_{j=1}^{[nt]} \mathbb{E}(d_{n,j}^2|\mathcal{F}_{n,j-1}) \to^{\mathbb{P}} t \text{ as } n \to \infty \text{ for any } t \in [0, 1]. \tag{2.50}$$

Note also that (2.47) together with (2.49) are equivalent to (2.42).

Corollary 2.36 *Let $(D_i)_{i\in\mathbb{Z}}$ be a strictly stationary sequence of martingale differences adapted to the stationary non-decreasing filtration $(\mathcal{F}_i)_{i\in\mathbb{Z}}$. Assume $\mathbb{E}(D_0^2) < \infty$ and*

$$\left\| \frac{1}{n}\sum_{i=1}^{n} \mathbb{E}(D_i^2|\mathcal{F}_0) - \mathbb{E}(D_0^2) \right\|_1 \to 0 \text{ as } n \to \infty. \tag{2.51}$$

Then $\{n^{-1/2}\sum_{i=1}^{[nt]} D_i, t \in [0, 1]\}$ converges in distribution in $D([0, 1])$ to a Brownian motion with variance parameter $\mathbb{E}(D_0^2)$.

For ergodic martingale sequences the convergence (2.51) is straightforward, and we derive the following theorem.

Theorem 2.37 *Let $(D_i)_{i\in\mathbb{Z}}$ be a stationary and ergodic sequence of real-valued martingale differences in $\mathbb{L}^2$. Then $\{n^{-1/2}S_{[nt]} = n^{-1/2}\sum_{j=1}^{[nt]} D_i, t \in [0,1]\}$ converges in distribution in $D([0,1])$ to a Brownian motion with variance parameter $\mathbb{E}(D_0^2)$.*

Proof of Theorem 2.34 We first prove that (2.47) together with (2.48) imply (2.42). Indeed, due to (2.47), to prove (2.42) it suffices to prove that (2.49) holds. Now, it was quoted by Aldous (1978) on page 337, that the condition (2.48) implies (2.49). To prove it, let $B_{n,q} = \{\omega : |\sum_{i=1}^{[nj/q]} d_{n,i}^2 - j/q| \leq 1/q \text{ for each } j = 1, \cdots, q\}$. Then (2.48) entails that $\lim_{n\to\infty} \mathbb{P}(B_{n,q}) = 1$ for each q. In addition for $\omega \in B_{n,q}$,

$$\sum_{i=[n(j-1)/q]+1}^{[nj/q]} d_{n,i}^2(\omega) \leq 3/q, \quad \text{for each } j = 1, \cdots, q. \tag{2.52}$$

Therefore, $\max_{1\leq i\leq n} d_{n,i}^2 \leq 3/q$ on $B_{n,q}$ proving that (2.49) holds. Hence, the conditions of Theorem 2.34 imply those of Theorem 2.29.

Cramér–Wold device Let us prove now the convergence of the finite-dimensional distributions of $W_n := \{S_{[nt]}, t \in [0,1]\}$ to those of a standard Brownian motion. Applying the usual Cramér–Wold device, it suffices to show that for any positive integer N, any reals $a_1, \ldots, a_N$ and any $t_0 = 0 < t_1 < t_2 < \ldots < t_N \leq 1$,

$$\sum_{\ell=1}^{N} a_\ell (S_{[nt_\ell]} - (S_{[nt_{\ell-1}]}) \to^d \mathcal{N}(0, \sum_{\ell=1}^{N} a_\ell^2 (t_\ell - t_{\ell-1})) \text{ as } n \to \infty. \tag{2.53}$$

With this aim, we write that

$$\sum_{\ell=1}^{N} a_\ell (S_{[nt_\ell]} - (S_{[nt_{\ell-1}]}) = \sum_{i=1}^{n} b_{n,i} d_{n,i},$$

where $b_{n,i} = a_\ell I([nt_{\ell-1}] + 1 \leq i \leq [nt_\ell])$ for any $\ell = \{1, \ldots, N\}$ and $b_{n,i} = 0$ if $[nt_N] + 1 \leq i \leq n$. Notice that $(b_{n,i}d_{n,i})_{i\geq 1}$ is a sequence of martingale differences with respect to $(\mathcal{F}_{n,i})$. In addition, since we have proved that the condition (2.42) is satisfied for $d_{n,j}$ it is also obviously satisfied for $b_{n,j}d_{n,j}$ since the $b_{n,i}$'s are bounded by $\max_{1\leq \ell\leq N} |a_\ell|$.

To prove that $(b_{n,j}d_{n,j})_{j\geq 1}$ satisfies (2.43) we start by writing

$$\sum_{j=1}^{n} b_{n,j}^2 d_{n,j}^2 = \sum_{\ell=1}^{N} a_\ell^2 \sum_{j=[nt_{\ell-1}]+1}^{[nt_\ell]} d_{n,j}^2 \to^{\mathbb{P}} \sum_{\ell=1}^{N} a_\ell^2 (t_\ell - t_{\ell-1}) \text{ as } n \to \infty.$$

Therefore the sequence $(b_{n,i}d_{n,i})_{i\geq 1}$ satisfies the conditions of Theorem 2.29, and then (2.53) is satisfied.

Tightness Now, it remains to prove the tightness of $W_n := \{S_{[nt]}, t \in [0,1]\}$. Notice that by general probability arguments, we need to prove that, for any $\varepsilon > 0$,

$$\lim_{m\to\infty} \limsup_{n\to\infty} \sum_{i=1}^{m} \mathbb{P}\left(\sup_{(i-1)m^{-1}<t\leq im^{-1}} \left| \sum_{j=[n(i-1)m^{-1}]+1}^{[nt]} d_{n,j} \right| > \varepsilon \right) = 0. \tag{2.54}$$

With this aim, we start by defining, for any $i \geq 1$,

$$d'_{n,i} = d_{n,i} I(|d_{n,i}| \leq 1),\ X_{n,i} = d'_{n,i} - \mathbb{E}(d'_{n,i}|\mathcal{F}_{n,i-1}) \text{ and } Y_{n,i} = X_{n,i} I\left(\sum_{j=1}^{i-1} X_{n,j}^2 \leq 2 \right).$$

Taking into account Lemmas 2.30 and 2.31, one can prove that

$$\mathbb{P}(X_{n,i} \neq Y_{n,i} \text{ for some } i \leq n) \to 0 \text{ as } n \to \infty. \tag{2.55}$$

In addition, for any $\varepsilon > 0$,

$$\begin{aligned}
&\mathbb{P}\left(\max_{1\leq k\leq n} \left| \sum_{j=1}^{k} (d_{n,j} - X_{n,j}) \right| \geq 2\varepsilon \right) \\
&\leq \mathbb{P}\left(\sum_{i=1}^{n} |d_{n,i}| I(|d_{n,i}| > 1) \geq \varepsilon \right) + \mathbb{P}\left(\sum_{i=1}^{n} \mathbb{E}(|d_{n,i}| I(|d_{n,i}| > 1)|\mathcal{F}_{n,i-1}) \geq \varepsilon \right) \\
&\leq \mathbb{P}\left(\max_{1\leq j\leq n} |d_{n,j}| \geq \varepsilon \right) + \mathbb{P}\left(\sum_{i=1}^{n} \mathbb{E}(|d_{n,i}| I(|d_{n,i}| > 1)|\mathcal{F}_{n,i-1}) \geq \varepsilon \right).
\end{aligned}$$

The first term in the right-hand side converges to zero as n tends to infinity since we have previously shown that (2.48) implies (2.49). By using Lemma 2.31, the second term also vanishes as n tends to infinity (recall that we have proved that (2.47) together with (2.48) imply (2.42)). From these considerations together with (2.55), we deduce that for proving that (2.54) holds for the sequence $(d_{n,j})_{j\geq 1}$ it suffices to show that, for any $\varepsilon > 0$,

$$\lim_{m\to\infty} \limsup_{n\to\infty} \sum_{i=1}^{m} \mathbb{P}\left(\sup_{(i-1)m^{-1}<t\leq im^{-1}} \left| \sum_{j=[n(i-1)m^{-1}]+1}^{[nt]} Y_{n,j} \right| > \varepsilon \right) = 0. \tag{2.56}$$

With this aim, we first notice that $(Y_{n,i})_{i\geq 1}$ is a sequence of martingale differences with respect to $(\mathcal{F}_{n,i})$ satisfying, by using (2.48) and the previous considerations,

$$\sum_{j=1}^{[nt]} Y_{n,j}^2 \to^{\mathbb{P}} t \text{ as } n \to \infty, \text{ for any } t \in [0,1].$$

In addition,

$$\sum_{i=1}^{n} Y_{n,i}^2 \leq 2 + \max_{1\leq k\leq n} |X_{n,j}|^2 \leq 6.$$

Hence, for any $0 < s < t < 1$ and any positive real λ,

$$\mathbb{E}\left(\left|\sum_{j=[ns]+1}^{[nt]} Y_{n,j}\right| I\left(\left|\sum_{j=[ns]+1}^{[nt]} Y_{n,j}\right| \geq \lambda\right)\right) \leq \lambda^{-1}\mathbb{E}\left(\sum_{j=[ns]+1}^{[nt]} Y_{n,j}\right)^2$$
$$\leq \lambda^{-1}\sum_{i=1}^{n} \mathbb{E}(Y_{n,i})^2 \leq 6\lambda^{-1},$$

proving that for any $0 < s < t < 1$, $\left|\sum_{j=[ns]+1}^{[nt]} Y_{n,j}\right|$ is uniformly integrable. To finish the proof it remains to prove the following lemma.

Lemma 2.38 *Assume that $(Y_{n,i})_{1\leq i\leq n}$ is an array of martingale differences adapted to an array $(\mathcal{F}_{n,i})_{1\leq i\leq n}$ of filtrations. Suppose that $\max_{1\leq k\leq n} |Y_{n,k}|$ is uniformly integrable, that for any $t \in [0, 1]$, $\sum_{j=1}^{[nt]} Y_{n,j}^2 \to^{\mathbb{P}} t$ as $n \to \infty$, and that for any $0 < s < t < 1$, $\left|\sum_{j=[ns]+1}^{[nt]} Y_{n,j}\right|$ is uniformly integrable. Then (2.56) is satisfied.*

To prove the lemma, we first apply Proposition 2.9. Hence for any $\varepsilon > 0$ and any positive integer i,

$$\mathbb{P}\left(\sup_{(i-1)m^{-1}<t\leq im^{-1}} \left|\sum_{j=[n(i-1)m^{-1}]+1}^{[nt]} Y_{n,j}\right| > 4\varepsilon\right)$$
$$\leq \frac{1}{2\varepsilon}\mathbb{E}\left(\left|\sum_{j=[n(i-1)m^{-1}]+1}^{[nim^{-1}]} Y_{n,j}\right| I\left(\left|\sum_{j=[n(i-1)m^{-1}]+1}^{[nim^{-1}]} Y_{n,j}\right| > 2\varepsilon\right)\right).$$

Setting $h_\varepsilon(x) = (|x| - \varepsilon)_+$ and noticing that $|x|\mathbf{1}_{|x|>2\varepsilon} \leq 2h_\varepsilon(x)$, we then derive

$$\mathbb{P}\left(\sup_{(i-1)m^{-1}<t\leq im^{-1}} \left|\sum_{j=[n(i-1)m^{-1}]+1}^{[nt]} Y_{n,j}\right| > 4\varepsilon\right) \leq \frac{1}{\varepsilon}\mathbb{E}\left(h_\varepsilon\left(\sum_{j=[n(i-1)m^{-1}]+1}^{[nim^{-1}]} Y_{n,j}\right)\right).$$

Since $\sum_{j=[ns]+1}^{[nt]} Y_{n,j}^2 \to^{\mathbb{P}} (t - s)$ as $n \to \infty$, the conditions of the lemma imply that $\sum_{j=[n(i-1)m^{-1}]+1}^{[nim^{-1}]} Y_{n,j}$ converges in distribution to $m^{-1/2}N$ where N is a standard normal law. Hence by the continuous mapping theorem, $h_\varepsilon\left(\sum_{j=[n(i-1)m^{-1}]+1}^{[nim^{-1}]} Y_{n,j}\right) \to^d h_\varepsilon\left(m^{-1/2}N\right)$. In addition, by assumption, $\left(h_\varepsilon\left(\sum_{j=[n(i-1)m^{-1}]+1}^{[nim^{-1}]} Y_{n,j}\right)\right)_{n\geq 1}$ is an uniformly integrable family. Therefore, the convergence of moments theorem (see Theorem 1.12) entails that

$$\limsup_{n\to\infty} \mathbb{E}\left(h_\varepsilon\left(\sum_{j=[n(i-1)m^{-1}]+1}^{[nim^{-1}]} Y_{n,j}\right)\right) = \mathbb{E}\left(h_\varepsilon\left(m^{-1/2}N\right)\right).$$

Hence

$$\limsup_{n\to\infty} \mathbb{E}\left(h_\varepsilon\left(\sum_{j=[n(i-1)m^{-1}]+1}^{[nim^{-1}]} Y_{n,j}\right)\right) \leq \frac{1}{m\varepsilon}\mathbb{E}\left(N^2 I(|N| > m^{1/2}\varepsilon)\right).$$

It follows that for any $\varepsilon > 0$,

$$\begin{aligned}
&\lim_{m\to\infty} \limsup_{n\to\infty} \sum_{i=1}^{m} \mathbb{P}\left(\sup_{(i-1)m^{-1}<t\leq im^{-1}} \left|\sum_{j=[n(i-1)m^{-1}]+1}^{[nt]} Y_{n,j}\right| > 4\varepsilon\right) \\
&\leq \lim_{m\to\infty} \frac{1}{\varepsilon^2} \mathbb{E}\left(N^2 I(|N| > m^{1/2}\varepsilon)\right) = 0,
\end{aligned}$$

which ends the proof of the lemma and then of Theorem 2.24. ■

By inspecting the proof of Theorem 2.34, we can formulate a functional central limit theorem for the process

$$W_n(t) = \sum_{i=1}^{k_n(t)} d_{n,i},\ t \in [0, 1],$$

where $k_n(\cdot)$ is a sequence of integer-valued, non-decreasing and right-continuous functions defined on $[0, 1]$ and such that $k_n(0) = 0$ for all n. To soothe the notation, let $k_n = k_n(1)$.

Theorem 2.39 *Assume $(d_{n,i})_{1\leq i\leq k_n}$ is an array of martingale differences adapted to an array $(\mathcal{F}_{n,i})_{i\geq 0}$ of nested sigma fields. Let $\sigma(\cdot)$ be a non-negative function on $[0, 1]$ such that $\sigma^2(\cdot)$ is Lebesgue integrable. Suppose that the following conditions hold:*

$$\max_{1\leq j\leq k_n} |d_{n,j}| \text{ is uniformly integrable,} \tag{2.57}$$

and, for all $t \in [0, 1]$,

$$\sum_{j=1}^{k_n(t)} d_{n,j}^2 \to^{\mathbb{P}} \int_0^t \sigma^2(u)du \text{ as } n \to \infty. \tag{2.58}$$

Then $W_n = \{\sum_{j=1}^{k_n(t)} d_{n,j}, t \in [0, 1]\}$ converges in distribution in $D([0, 1])$ to $\left\{\int_0^t \sigma(u)dW(u), t \in [0, 1]\right\}$ where W is a standard Brownian motion.

Note that the deterministic sequence $k_n(t)$ can be replaced by a sequence $r_n(t)$ of a stopping times with respect to $\mathcal{F}_{n,i}$, $i = 0, 1, \ldots$, provided that the sample paths of $r_n(\cdot)$ are integer-valued, non-decreasing and right-continuous with $r_n(0) = 0$. We refer to Theorem 3.2 in Helland (1982) for such an extension.

2.5 Moderate Deviations for Martingales

Theorems 2.40 and 2.41 of this section are versions of the moderate deviations principle (MDP) of Theorem 3.1 in Puhalskii (1994) when we consider the partial sums or partial sums process associated with a triangular array of martingale difference sequences. We refer the reader to the Worms PhD thesis (2000) for more details concerning Puhalskii's proofs.

Theorem 2.40 *Assume $(d_{n,i})_{1 \le i \le n}$ is an array of martingale differences adapted to an array $(\mathcal{F}_{n,i})_{0 \le i \le n}$ of filtrations. Let a_n be a positive numerical sequence such that $a_n \to 0$. Suppose*

$$\max_{1 \le j \le n} \|d_{n,j}\|_\infty = o(\sqrt{a_n}) \text{ as } n \to \infty \tag{2.59}$$

and there exists $\sigma^2 \ge 0$ such that for any positive δ,

$$a_n \log \mathbb{P}\left(\left| \sum_{j=1}^{n} \mathbb{E}(d_{n,j}^2 | \mathcal{F}_{n,j-1}) - \sigma^2 \right| > \delta \right) \to -\infty \text{ as } n \to \infty. \tag{2.60}$$

Then $S_n = \sum_{j=1}^{n} d_{n,j}$ satisfies the MDP, namely the property (1.30), with speed a_n and rate function $I(t) = t^2/(2\sigma^2)$.

Concerning the functional form of the MDP we have:

Theorem 2.41 *Assume $(d_{n,i})_{1 \le i \le n}$ is an array of martingale differences adapted to an array $(\mathcal{F}_{n,i})_{0 \le i \le n}$ of filtrations. Let a_n be a positive numerical sequence such that $a_n \to 0$. Assume that (2.59) holds and there exists $\sigma^2 \ge 0$ such that for any $t \in [0, 1]$ and any positive δ,*

$$a_n \log \mathbb{P}\left(\left| \sum_{j=1}^{[nt]} \mathbb{E}(d_{n,j}^2 | \mathcal{F}_{n,j-1}) - t\sigma^2 \right| > \delta \right) \to -\infty \text{ as } n \to \infty. \tag{2.61}$$

For any $t \in [0, 1]$, let $W_n(t) = \sum_{i=1}^{[nt]} d_{n,i}$ and define $W_n = \{W_n(t), t \in [0, 1]\}$. Then $W_n(.)$ satisfies the moderate deviation principle in $D([0, 1])$, namely the property (1.31), with speed a_n and the good rate function $I_\sigma(\cdot)$ defined by

$$I_\sigma(h) = \frac{1}{2\sigma^2} \int_0^1 (h'(u))^2 du \tag{2.62}$$

if simultaneously $\sigma > 0$, $h(0) = 0$ and h is absolutely continuous, and $I_\sigma(h) = \infty$ otherwise.

Note that the rate function in the theorem above is inherited from the Wiener process with parameter σ^2.

For a strictly stationary sequence of bounded and non-degenerate martingale differences, we have

Corollary 2.42 *Let $(D_i)_{i\in\mathbb{Z}}$ be a strictly stationary sequence of nondegenerate martingale differences adapted to the stationary, non-decreasing filtration $(\mathcal{F}_i)_{i\in\mathbb{Z}}$. Assume $\|D_0\|_\infty < \infty$ and*

$$\left\| \frac{1}{n}\sum_{i=1}^{n} \mathbb{E}(D_i^2|\mathcal{F}_0) - \mathbb{E}(D_0^2) \right\|_\infty \to 0 \text{ as } n \to \infty. \tag{2.63}$$

Suppose in addition that $a_n \to 0$ and $na_n \to \infty$. Then $\{n^{-1/2}\sum_{i=1}^{[nt]} D_i, t \in [0,1]\}$ satisfies the moderate deviation principle in $D([0,1])$ with speed a_n and the good rate function $I_\sigma(\cdot)$ defined by (2.62) with $\sigma^2 = \mathbb{E}(D_0^2)$.

It should be noticed that the MDP principle is not true in general for a strictly stationary sequence of bounded martingale differences without a condition of the type (2.63) even if the sequence is ergodic. To illustrate this fact let us consider the following example. Let Y_k be the stationary discrete Markov chain with the state space $\mathbb{N}$ and the transition kernel given by $\mathbb{P}(Y_1 = j-1|Y_0 = j) = 1$ for all $j \geq 1$ and $\mathbb{P}(Y_1 = j|Y_0 = 0) = \mathbb{P}(\tau = j)$ for $j \in \mathbb{N}$ with $\mathbb{E}(\tau) < \infty$ and $\mathbb{P}(\tau = 1) > 0$, which implies that (Y_k) is ergodic. Let $X_k = \xi_k I_{(Y_k \neq 0)}$ where (ξ_k) is an i.i.d. sequence independent of (Y_k) and such that $\mathbb{P}(\xi_k = \pm 1) = 1/2$. Then (X_k) is a stationary ergodic martingale difference which is also a bounded function of an ergodic Markov chain. Straightforward computations show that if τ does not have a finite exponential moment then there exists a positive sequence $a_n \to 0$ with $na_n \to \infty$ for which $n^{-1/2}\sum_{i=1}^{n} X_i$ does not satisfy the property (1.30).

Proof of Theorem 2.40 We construct a sequence of stopping times $\tau(n)$, such that

$$a_n \log \mathbb{P}(S_n \neq S_{\tau(n)}) \to -\infty \quad \text{and} \quad \lim_{n\to\infty} a_n \log \mathbb{E}(\exp(tS_{\tau(n)}/\sqrt{a_n}) = \sigma^2 t^2/2 \text{ for all } t, \tag{2.64}$$

which proves the theorem by the exponential continuity and Gärtner–Ellis Lemma 1.24.

More exactly, our construction will in addition depend on two other parameters $1 \geq \delta > 0$ and $p > 1$ such that when $\delta \to 0$ and $p \to 1$, we derive (2.64).

Construction Fix n and omit it to simplify notation. For any positive integer m less than n and non zero real u, define

$$D_m = \sum_{k=1}^{m} \mathbb{E}(d_k^2|\mathcal{F}_{k-1}),\ G_m(u) = a_n \log \prod_{k=1}^{m} \mathbb{E}(e^{ud_k/\sqrt{a_n}}|\mathcal{F}_{k-1}),\ \varepsilon_n = a_n^{-1/2} \max_{1\leq j\leq n} \|d_{n,j}\|_\infty.$$

Similarly to the proof of the central limit theorem, it is easier to work with G_m rather than with the original exponential moment $\mathbb{E}\exp(uS_n/\sqrt{a_n})$. Indeed Taylor expansion together with the fact that $\mathbb{E}(d_i|\mathcal{F}_{i-1}) = 0$ and (2.59) imply that, for any $u \in \mathbb{R}$ and any positive integer m, there exist real-valued random variables $Z_k(u)$ such that $|Z_k(u)| \leq u^2(e^{|u|\varepsilon_n} - 1)\mathbb{E}(d_k^2|\mathcal{F}_{k-1})$ and for n large enough depending on u

$$G_m(u) = \frac{u^2}{2}\sum_{k=1}^{m}\mathbb{E}(d_k^2|\mathcal{F}_{k-1}) + \sum_{k=1}^{m} Z_k(u) \text{ a.s.}$$

To control its behavior we first introduce a "nuisance" event

$$R_n = R_{\delta,n} = \{|D_n - \sigma^2| > \delta\},$$

and the stopping time to avoid another class of similar "nuisance" events

$$\tau(n) = \min\left(n, \inf\{m \geq 1 : D_m \geq 1 + \sigma^2\}\right).$$

Observe that

$$|D_{m+1} - D_m| \leq a_n\varepsilon_n^2,\ |G_{m+1}(u) - G_m(u)| \leq |u|^2 a_n\varepsilon_n^2 \to 0, \quad (n \to \infty) \tag{2.65}$$

implying that for sufficiently large n, depending on u, δ, and thus w.l.o.g. for all n,

$$G_{\tau(n)}(u) \leq 3u^2(1+\sigma^2)/2. \tag{2.66}$$

Notice first that

$$\mathbb{P}\left(S_n \neq S_{\tau(n)}\right) = \mathbb{P}\left(\tau(n) < n\right).$$

For $0 < \delta \leq 1$, it follows by the construction that for n large enough

$$\max\left(\mathbb{P}(R_n), \mathbb{P}(\tau(n) < n)\right) \leq \mathbb{P}\left(\left|\sum_{j=1}^{n}\mathbb{E}(d_j^2|\mathcal{F}_{j-1}) - \sigma^2\right| > \delta/2\right). \tag{2.67}$$

Therefore by (2.60), we get that

$$a_n \log\mathbb{P}\left(S_n \neq S_{\tau(n)}\right) \to -\infty \text{ and } a_n \log\mathbb{P}\left(R_n\right) \to -\infty. \tag{2.68}$$

Now, it remains to show that the stopped random sums satisfy the Gärtner–Ellis Lemma 1.24 namely, for all $t \in \mathbb{R}$,

$$\lim_{n\to\infty} a_n \log\mathbb{E}\left(\exp(ta_n^{-1/2}S_{\tau(n)})\right) = \sigma^2 t^2/2. \tag{2.69}$$

For convenience, let

$$S_{\tau(n)}(t) = t\sqrt{a_n}S_{\tau(n)}.$$

By Hölder's inequality and (2.66)

$$\mathbb{E}\left(\exp(a_n^{-1}S_{\tau(n)}(t))I(R_n)\right) \leq \left(\mathbb{E}\left(e^{a_n^{-1}(S_{\tau(n)}(tp)-G_{\tau(n)}(tp))}e^{a_n^{-1}G_{\tau(n)}(tp)}\right)\right)^{1/p}\left(\mathbb{P}(R_n)\right)^{1-1/p}$$

$$\leq e^{\frac{1}{a_n p}\frac{3(1+\sigma^2)p^2t^2}{2}}\left(\mathbb{E}\left(e^{a_n^{-1}(S_{\tau(n)}(tp)-G_{\tau(n)}(tp))}\right)\right)^{1/p}\left(\mathbb{P}(R_n)\right)^{1-1/p}. \tag{2.70}$$

By Fact 2.4 applied to the martingale $\exp(a_n^{-1}(S_K(u) - G_K(u)))$,

$$\mathbb{E}\left(e^{a_n^{-1}(S_{\tau(n)}(u)-G_{\tau(n)}(u))}\right) = 1. \tag{2.71}$$

Therefore

$$a_n \log \mathbb{E}\left(\exp(a_n^{-1}S_{\tau(n)}(t))I(R_n)\right) \leq \frac{3(1+\sigma^2)pt^2}{2} + a_n(1-1/p)\log \mathbb{P}(R_n), \tag{2.72}$$

showing, by using the second part of (2.68), that

$$\lim_{n\to\infty} a_n \log \mathbb{E}\left(\exp(a_n^{-1}S_{\tau(n)}(t))I(R_n)\right) = -\infty. \tag{2.73}$$

Notice now that, by the Hölder's inequality as done in (2.70) together with (2.71), we derive

$$a_n \log \mathbb{E}\left(\exp(a_n^{-1}S_{\tau(n)}(t))I(\tau(n)<n)I(R_n^c)\right) \leq \frac{3(1+\sigma^2)pt^2}{2} + a_n(1-1/p)\log \mathbb{P}(\tau(n)<n).$$

Using (2.67) together with (2.60), we then obtain

$$\lim_{n\to\infty} a_n \log \mathbb{E}\left(\exp(a_n^{-1}S_{\tau(n)}(t))I(\tau(n)<n)I(R_n^c)\right) = -\infty. \tag{2.74}$$

Next introduce a "good" set

$$A_n = R_n^c \cap \{\tau(n) = n\}.$$

Note that

$$\begin{aligned}
\mathbb{E}\left(\exp(a_n^{-1}S_{\tau(n)}(t))I(\tau(n)=n)I(R_n^c)\right) &= \mathbb{E}\left(e^{a_n^{-1}(t\sqrt{a_n}S_n - G_n(t))}e^{a_n^{-1}G_n(t)}I(A_n)\right) \\
&\leq e^{a_n^{-1}((\delta+\sigma^2)t^2/2)}e^{a_n^{-1}(\delta+\sigma^2)(e^{|t|\varepsilon_n}-1)t^2}\,\mathbb{E}\left(e^{a_n^{-1}(t\sqrt{a_n}S_n - G_n(t))}\right) \\
&= e^{a_n^{-1}((\delta+\sigma^2)t^2/2)}e^{a_n^{-1}(\delta+\sigma^2)(e^{|t|\varepsilon_n}-1)t^2},
\end{aligned} \tag{2.75}$$

where we use again the martingale property (2.71).

Taking into account (2.73), (2.74) and (2.75), and letting $\delta \to 0$, we derive that

$$\limsup_{n\to\infty} a_n \log \mathbb{E}\left(\exp(a_n^{-1}S_{\tau(n)}(t))\right) \leq \sigma^2t^2/2.$$

To end the proof of (2.69), it remains to show that, for any $t \in \mathbb{R}$,

$$\liminf_{n\to\infty} a_n \log \mathbb{E}\left(\exp(a_n^{-1}S_{\tau(n)}(t))\right) \geq \sigma^2t^2/2. \tag{2.76}$$

With this aim, we first note that since $G_{\tau(n)}(t) \geq 0$ for all t and n large enough not depending on t,

$$\mathbb{E}\left(e^{a_n^{-1}(S_{\tau(n)}(t)-G_{\tau(n)}(t))}I(A_n^c)\right) \leq \mathbb{E}\left(\exp(a_n^{-1}S_{\tau(n)}(t))I(A_n^c)\right)$$
$$\leq \mathbb{E}\left(\exp(a_n^{-1}S_{\tau(n)}(t))I(R_n)\right) + \mathbb{E}\left(\exp(a_n^{-1}S_{\tau(n)}(t))I(\tau(n) < n)\right).$$

Applying Hölder's inequality as before, we get

$$a_n \log \mathbb{E}\left(\exp(a_n^{-1}S_{\tau(n)}(t))I(\tau(n) < n)\right) \leq \frac{3(1+\sigma^2)pt^2}{2} + a_n(1-1/p)\log \mathbb{P}\left(I(\tau(n) < n)\right),$$

This upper bound, together with (2.72), the second part of (2.68) and the fact that $a_n \log \mathbb{P}\left(I(\tau(n) < n)\right) \to -\infty$ by (2.67) and (2.60), entail that

$$\lim_{n\to\infty} a_n \log \mathbb{E}\left(e^{a_n^{-1}(S_{\tau(n)}(t)-G_{\tau(n)}(t))}I(A_n^c)\right) = -\infty. \tag{2.77}$$

We shall now use the following lemma whose proof is direct by using the fact that for any sequences of non-negative reals (b_n) and (c_n),

$$\liminf_{n\to\infty} a_n \log(b_n + c_n) \leq \max\{\liminf_{n\to\infty} a_n \log(b_n), \limsup_{n\to\infty} a_n \log(c_n)\}.$$

Lemma 2.43 *Let $(Y_n)_{n\geq 1}$ be a sequence of non-negative real-valued random variables defined on $(\Omega, \mathcal{A}, \mathbb{P})$ and let $(A_n)_{n\geq 1}$ be a sequence of events in $\mathcal{A}$. If for a real α,*

$$\liminf_{n\to\infty} a_n \log \mathbb{E}(Y_n) \geq \alpha \text{ and } \lim_{n\to\infty} a_n \log \mathbb{E}(Y_n I(A_n^c)) = -\infty,$$

then $\liminf_{n\to\infty} a_n \log \mathbb{E}(Y_n I(A_n)) \geq \alpha$.

Setting $Y_n = e^{a_n^{-1}(S_{\tau(n)}(t)-G_{\tau(n)}(t))}$, the martingale property (2.71) implies that

$$a_n \log \mathbb{E}(Y_n) = a_n \log \mathbb{E}\left(e^{a_n^{-1}(S_{\tau(n)}(t)-G_{\tau(n)}(t))}\right) = 0.$$

Therefore, by taking into account (2.77), the lemma applied with $\alpha = 0$ gives

$$\liminf_{n\to\infty} a_n \log \mathbb{E}\left(e^{a_n^{-1}(S_{\tau(n)}(t)-G_{\tau(n)}(t))}I(A_n)\right) \geq 0. \tag{2.78}$$

To show now the lower bound (2.76), we then observe that, for n large enough,

$$\mathbb{E}\left(\exp(a_n^{-1}S_{\tau(n)}(t)\right) \geq \mathbb{E}\left(\exp(a_n^{-1}S_{\tau(n)}(t)I(A_n)\right)$$
$$= \mathbb{E}\left(e^{a_n^{-1}(S_{\tau(n)}(t)-G_{\tau(n)}(t))}e^{a_n^{-1}G_n(t)}I(A_n)\right)$$
$$\geq e^{a_n^{-1}((-\delta+\sigma^2)t^2/2)}e^{-a_n^{-1}(\delta+\sigma^2)(e^{|t|\varepsilon_n}-1)t^2}\mathbb{E}\left(e^{a_n^{-1}(S_{\tau(n)}(t)-G_{\tau(n)}(t))}I(A_n)\right)$$

which implies, by taking into account (2.78),

$$\liminf_{n\to\infty} a_n \log \mathbb{E}\left(\exp(a_n^{-1} S_{\tau(n)}(t)\right) \geq -\delta + \sigma^2 t^2/2.$$

This shows (2.76) by letting δ going to zero. This ends the proof of Theorem 2.40. ■

Proof of Theorem 2.41 According to Section 1.3.4, the proof is divided in two steps. First we prove that the finite-dimensional laws satisfy the MDP with an appropriate good rate function, and after that we prove tightness in the sense of the moderate deviation principle. To soothe the notations we will omit the index n and then write $d_j := d_{n,j}$ and $\mathcal{F}_j := \mathcal{F}_{n,j}$.

Step 1. MDP for the finite-dimensional laws. We show first that for m a fixed integer and for any $0 = t_0 < t_1 < t_2 < \cdots < t_m \leq 1$, the m -tuple

$$(W_n(t_1), W_n(t_2), \dots, W_n(t_m))$$

satisfies the MDP in $\mathbb{R}^m$ with speed a_n and rate function given by

$$I_{t_1,\cdots,t_m}(u_1,\cdots,u_m) = \sup_{\lambda_1,\cdots,\lambda_m \in \mathbb{R}} \left(\sum_{j=1}^{m} \lambda_j u_j - \frac{\sigma^2}{2} \sum_{j,k=1}^{m} \lambda_j \lambda_k R(t_j, t_k) \right), \tag{2.79}$$

where $R(s,t) = t \wedge s$.

With this aim, we shall apply the Cramér–Wold device. For all integer $1 \leq \ell \leq m$, and for $\lambda_1, \dots, \lambda_m \in \mathbb{R}$, let us consider the decomposition

$$\sum_{\ell=1}^{m} \lambda_\ell W_n(t_\ell) = \sum_{j=1}^{n} c_j d_j, \tag{2.80}$$

where

$$c_j = \sum_{i=1}^{m} \sum_{\ell=i}^{m} \lambda_\ell 1_{[nt_{i-1}]+1 \leq j \leq [nt_i]}. \tag{2.81}$$

We apply Theorem 2.40 to the d_j defined as $c_j d_j$. We clearly have

$$\max_{1\leq j\leq n} ||c_j d_j||_\infty \leq \sum_{k=1}^{m} |\lambda_\ell| \max_{1\leq j\leq n} ||d_j||_\infty = o(a_n), \text{ as } n \to \infty.$$

Now

$$\sum_{j=1}^{n} c_j^2 \mathbb{E}(d_j^2 | \mathcal{F}_{j-1}) = \sum_{\ell=1}^{m} \left(\sum_{i=\ell}^{m} \lambda_i \right)^2 \sum_{j=[nt_{\ell-1}]+1}^{[nt_\ell]} \mathbb{E}(d_j^2 | \mathcal{F}_{j-1}).$$

Hence if (2.61) holds then we have for any $\delta > 0$,

$$a_n \log \mathbb{P}\left(\left| \sum_{j=1}^{n} c_j^2 \mathbb{E}(d_j^2 | \mathcal{F}_{j-1}) - \sigma^2 \sum_{\ell=1}^{m} \left(\sum_{i=\ell}^{m} \lambda_i \right)^2 (t_{\ell-1} - t_\ell) \right| > \delta \right) \to -\infty \text{ as } n \to \infty.$$

Hence, applying Theorem 2.40 together with the fact that

$$\sum_{\ell=1}^{m} \left(\sum_{i=\ell}^{m} \lambda_i \right)^2 (t_{\ell-1} - t_\ell) = \sum_{j,k=1}^{m} \lambda_j \lambda_k (t_j \wedge t_k),$$

the convergence of finite-dimensional distributions is proved.

Step 2. Exponential tightness. We turn now to prove the tightness in the sense of moderate deviations. Therefore we need to prove that (1.32) holds, i.e. for any $\epsilon > 0$,

$$\lim_{\eta \to 0} \limsup_{n \to \infty} a_n \log \mathbb{P}\left(\sup_{|s-t| \le \eta, 0 \le s,t \le 1} \sqrt{a_n} |W_n(t) - W_n(s)| \ge \epsilon \right) = -\infty.$$

But, by Inequality (1.23) of Lemma 1.15, for each $\epsilon > 0$,

$$\mathbb{P}\left(\sup_{|s-t| \le \eta, 0 \le s,t \le 1} \sqrt{a_n} |W_n(t) - W_n(s)| \ge 3\epsilon \right)$$
$$\le \sum_{i=1}^{m} \mathbb{P}\left(\sup_{(i-1)/m \le t < i/m} \sqrt{a_n} \left| W_n(t) - W_n\left(\frac{i-1}{m} \right) \right| \ge \epsilon \right),$$

where $m = [\eta^{-1}]$. Since $a_n \to 0$ as $n \to \infty$, to prove (1.32) it suffices to show that, for any $\epsilon > 0$,

$$\lim_{m \to \infty} \limsup_{n \to \infty} a_n \log \left(\sum_{i=1}^{m} \mathbb{P}\left(\sup_{(i-1)/m \le t < i/m} \sqrt{a_n} \left| \sum_{k=[n(i-1)/m]+1}^{[nt]} d_k \right| \ge \epsilon \right) \right) = -\infty. \tag{2.82}$$

With this aim, for any integer i in $[1, m]$, we write

$$\mathbb{P}\left(\sup_{(i-1)/m \le t < i/m} \sqrt{a_n} \left| \sum_{k=[n(i-1)/m]+1}^{[nt]} d_k \right| \ge \epsilon \right) \le I_m(i) + II_m(i)$$

where, with $y_m = \dfrac{\sigma^2}{m}(1 + \epsilon)$,

$$I_m(i) := \mathbb{P}\left(\sup_{(i-1)/m \le t < i/m} \sqrt{a_n} \left| \sum_{k=[n(i-1)/m]+1}^{[nt]} d_k \right| \ge \epsilon, \sum_{k=[n(i-1)/m]+1}^{[ni/m]} \mathbb{E}(d_k^2 | \mathcal{F}_{n,k-1}) \le y_m \right)$$

and

$$II_m(i) := \mathbb{P}\left(\sum_{k=[n(i-1)/m]+1}^{[ni/m]} \mathbb{E}(d_k^2|\mathcal{F}_{k-1}) > y_m\right).$$

Applying Inequality (2.22) and taking into account Remark 2.22, we get

$$I_m(i) \leq 2\exp\left(\frac{-\epsilon^2}{2a_n\left(y_m + \epsilon\max_{1\leq i\leq n}\|d_i\|_\infty/(3\sqrt{a_n})\right)}\right).$$

Therefore

$$a_n \log\sum_{i=1}^m I_m(i) \leq a_n\log(2m) - \frac{\epsilon^2}{2\left(y_m + \epsilon\max_{1\leq i\leq n}\|d_i\|_\infty/(3\sqrt{a_n})\right)}.$$

By taking into account that $a_n \to 0$ and condition (2.59), it follows that

$$\limsup_{n\to\infty} a_n \log\sum_{i=1}^m I_m(i) \leq -\frac{m\epsilon^2}{2\sigma^2(1+\epsilon)},$$

which tends to $-\infty$ by letting $m \to \infty$. On the other hand,

$$II_m(i) \leq \mathbb{P}\left(\left|\sum_{k=[n(i-1)/m]+1}^{[ni/m]} \mathbb{E}(d_k^2|\mathcal{F}_{k-1}) - \sigma^2/m\right| > \epsilon\sigma^2/m\right),$$

therefore,

$$a_n \log\sum_{i=1}^m II_m(i) \leq a_n\log(2m) + a_n \max_{1\leq i\leq m}\log\mathbb{P}\left(\left|\sum_{k=1}^{[ni/m]} \mathbb{E}(d_k^2|\mathcal{F}_{k-1}) - i\sigma^2/m\right| > \epsilon\sigma^2/(2m)\right).$$

Taking into account that $a_n \to 0$ and condition (2.61), it follows that

$$\limsup_{n\to\infty} a_n \log\sum_{i=1}^m II_m(i) \leq -\infty.$$

All these considerations show that (2.82) holds and ends the proof of the tightness of W_n.

End of the proof According to Theorem 3.2 in Arcones (2003-a), it follows from steps 1 and 2 that the process W_n satisfies the MDP in $D([0,1])$ with speed a_n and the good rate function given by

$$I_\sigma(z) = \sup\{I_{t_1,\ldots,t_m}(z(t_1),\ldots,z(t_m)) : t_1,\ldots,t_m \in [0,1], m \geq 1\}.$$

As it is shown in Arcones (2003-b), $I_\sigma(z)$ satisfies the equation (2.62). ∎

Proof of Corollary 2.42 To prove the result it suffices to verify that under the assumptions of Corollary 2.42, the sequence $(d_{n,i})_{1\le i\le n}$ of martingale differences defined by $d_{n,j} = n^{-1/2}D_j$ satisfies the conditions of Theorem 2.41. The condition (2.59) is clearly satisfied since $na_n \to \infty$. On another hand, by stationarity condition (2.61) is reduced, for any positive δ, to:

$$a_n \log \mathbb{P}\left(n^{-1}\left|\sum_{j=1}^{n}(\mathbb{E}(D_j^2|\mathcal{F}_{j-1}) - \mathbb{E}(D_j^2))\right| > \delta\right) \to -\infty \text{ as } n \to \infty. \tag{2.83}$$

To end the proof, it suffices to show that (2.83) holds under the assumptions of the corollary. With this aim we consider p a fixed positive integer and, for any k in $\mathbb{Z}$, we define

$$Z_k = \sum_{i=p(k-1)+1}^{pk} (\mathbb{E}(D_i^2|\mathcal{F}_{i-1}) - \mathbb{E}(D_i^2)).$$

Then,

$$\sum_{j=1}^{n}(\mathbb{E}(D_j^2|\mathcal{F}_{j-1}) - \mathbb{E}(D_j^2)) - \sum_{j=1}^{[n/p]} Z_j = \sum_{j=[n/p]p+1}^{n} (\mathbb{E}(D_j^2|\mathcal{F}_{j-1}) - \mathbb{E}(D_j^2)),$$

implying that

$$\left\|\sum_{j=1}^{n}(\mathbb{E}(D_j^2|\mathcal{F}_{j-1}) - \mathbb{E}(D_j^2)) - \sum_{j=1}^{[n/p]} Z_j\right\|_\infty \le 2p\|D_0\|_\infty^2.$$

Since $\|D_0\|_\infty < \infty$, the convergence (2.83) is reduced to prove that for any $\delta > 0$

$$a_n \log \mathbb{P}\left(n^{-1}\left|\sum_{k=1}^{[n/p]} Z_k\right| > \delta\right) \to -\infty \text{ as } n \to \infty. \tag{2.84}$$

Now, setting $G_k = \mathcal{F}_{pk-1}$, we have the following martingale decomposition:

$$\mathbb{P}\left(n^{-1}\left|\sum_{k=1}^{[n/p]} Z_k\right| > \delta\right) \le \mathbb{P}\left(n^{-1}\left|\sum_{k=1}^{[n/p]} (Z_k - \mathbb{E}(Z_k|G_{k-1}))\right| > \delta/2\right)$$
$$+ \mathbb{P}\left(n^{-1}\left|\sum_{k=1}^{[n/p]} \mathbb{E}(Z_k|G_{k-1})\right| > \delta/2\right). \tag{2.85}$$

By stationarity, if (2.63) holds then, for a certain $p = p_\delta$ sufficiently large and fixed, we have

$$n^{-1}\left\|\sum_{k=1}^{[n/p]}\mathbb{E}(Z_k|G_{k-1})\right\|_\infty \leq \frac{1}{p}\left\|\sum_{i=1}^{p}(\mathbb{E}(D_i^2|\mathcal{F}_0)-\mathbb{E}(D_i^2))\right\|_\infty \leq \frac{\delta}{4}$$

and so, the second term in the right-hand side of (2.85) vanishes. Applying the Hoeffding–Azuma Inequality (2.13) to the martingale part, we obtain that

$$\mathbb{P}\left(n^{-1}\left|\sum_{k=1}^{[n/p]}(Z_k-\mathbb{E}(Z_k|G_{k-1}))\right|>\delta/2\right)\leq 2\exp\left(\frac{-\delta^2 n^2}{32\sum_{k=1}^{[n/p]}\|Z_k\|_\infty^2}\right). \tag{2.86}$$

But, by stationarity,

$$\sum_{k=1}^{[n/p]}\|Z_k\|_\infty^2 \leq 4np\|D_0\|_\infty^4.$$

Therefore, by taking into account (2.86), and that $\|D_0\|_\infty<\infty$, $a_n\to 0$ and $na_n\to\infty$, we get

$$a_n\log\mathbb{P}\left(n^{-1}\left|\sum_{k=1}^{[n/p]}(Z_k-\mathbb{E}(Z_k|G_{k-1}))\right|>\delta/2\right)\to-\infty \text{ as } n\to\infty.$$

This ends the proof of the fact that (2.83) holds under the assumptions of Corollary 2.42. ■

3

Moment Inequalities via Martingale Methods

The aim of this chapter is to establish different kinds of moment inequalities for the partial sums (or the maximum of partial sums) associated with a large class of sequences of real-valued random variables that includes martingale sequences, mixingales and other dependent structures. Indeed all the bounds involve the moments of the conditional expectations of either the partial sums or of the individual random variables.

A common point in most of the proofs is essentially that martingale approximations are used allowing us to apply the well-known inequalities for the martingale part stated in Chapter 2.

Throughout the chapter, the original sequence $(X_k)_{k\in\mathbb{Z}}$ is mostly assumed to be adapted to a non-decreasing filtration $(\mathcal{F}_i)_i$. In addition, the following notations will be often used:

$$\mathbb{E}_k(X) := \mathbb{E}(X|\mathcal{F}_k),\ \mathbf{P}_k(X) = \mathbb{E}(X|\mathcal{F}_k) - \mathbb{E}(X|\mathcal{F}_{k-1}),$$

$$S_n = \sum_{j=1}^{n} X_j,\ S_n^* = \max_{1\le k\le n} |S_k|,$$

and, for $p > 1$, $\|\cdot\|_p$ will design the norm in $\mathbb{L}^p$.

3.1 Analysis of the Variance of Partial Sums in the Stationary Setting: The Dyadic Induction

Throughout this section $(X_k)_{k\in\mathbb{Z}}$ is a stationary sequence of real-valued random variables in $\mathbb{L}^2$ that are adapted to a non-decreasing filtration $(\mathcal{F}_k)_{k\in\mathbb{Z}}$. Moreover $\mathcal{G}$ will denote a stationary σ-field for $(X_k)_{k\in\mathbb{Z}}$ (see Chapter 1 for the definition). We will assume throughout the section that $\mathbb{E}(X|\mathcal{G}) = 0$ almost surely and we set

$$S_n = \sum_{i=1}^{n} X_i \text{ and } \bar{S}_n = S_{2n} - S_n.$$

Functional Gaussian Approximation for Dependent Structures. Florence Merlevède, Magda Peligrad and Sergey Utev. Oxford University Press (2019).
DOI: 10.1093/oso/9780198826941.001.0001

For any real-valued random variables X and Y, the following notations will be useful:

$$\begin{aligned}\mathbb{E}(X|\mathcal{G}) &= \mathbb{E}_{\mathcal{G}}(X),\ \|X\|_{\mathcal{G}} = \sqrt{\mathbb{E}_{\mathcal{G}}(X^2)},\\ \text{cov}_{\mathcal{G}}(X,Y) &= \mathbb{E}_{\mathcal{G}}(XY) - \mathbb{E}_{\mathcal{G}}(X)\mathbb{E}_{\mathcal{G}}(Y),\\ \text{Var}_{\mathcal{G}}(X) &= \mathbb{E}_{\mathcal{G}}(X^2) - (\mathbb{E}_{\mathcal{G}}(X))^2 (= \text{cov}_{\mathcal{G}}(X,X)),\\ \text{corr}_{\mathcal{G}}(X,Y) &= [\mathbb{E}_{\mathcal{G}}(XY) - \mathbb{E}_{\mathcal{G}}(X)\mathbb{E}_{\mathcal{G}}(Y)]/\|X\|_{\mathcal{G}}\|Y\|_{\mathcal{G}},\end{aligned}$$

provided that all the above conditional expectations are well defined.

The aim of this section is to give sufficient conditions for $n^{-1}\mathbb{E}_{\mathcal{G}}(S_n^2)$ to admit almost surely a finite limit, say η. With this aim, we start with the following lemma giving equivalent representations of this limit.

Lemma 3.1 *The following four statements are equivalent (all being treated almost surely):*

$$\text{(i)}\ \sum_{j=0}^{\infty} 2^{-j}\mathbb{E}_{\mathcal{G}}(S_{2^j}\bar{S}_{2^j}) < \infty,$$

$$\text{(ii)}\ \lim_{r\to\infty} \frac{\mathbb{E}_{\mathcal{G}}(S_{2^r}^2)}{2^r} = \mathbb{E}_{\mathcal{G}}(X_1^2) + \sum_{j=1}^{\infty} 2^{-j}\mathbb{E}_{\mathcal{G}}(S_{2^j}\bar{S}_{2^j}),$$

$$\text{(iii)}\ \prod_{j=1}^{\infty} \text{corr}_{\mathcal{G}}(S_{2^j}, \bar{S}_{2^j}) < \infty,$$

$$\text{(iv)}\ \lim_{r\to\infty} \frac{\mathbb{E}_{\mathcal{G}}(S_{2^r}^2)}{2^r} = \mathbb{E}_{\mathcal{G}}(X_1^2)\prod_{i=0}^{\infty}(1 + \text{corr}_{\mathcal{G}}(S_{2^i}, \bar{S}_{2^i})).$$

Proof of Lemma 3.1 Equivalence of (i) and (ii) follows from the additive type representation

$$\mathbb{E}_{\mathcal{G}}(S_{2^r}^2) = 2^r\mathbb{E}_{\mathcal{G}}(X_1^2) + \sum_{j=1}^{r-1} 2^{r-j}\mathbb{E}_{\mathcal{G}}(S_{2^j}\bar{S}_{2^j}).$$

Equivalence of (iii) and (iv) follows from the product type representation

$$\mathbb{E}_{\mathcal{G}}(S_{2^r}^2) = 2^r\mathbb{E}_{\mathcal{G}}(X_1^2)\prod_{i=0}^{r-1}(1 + \text{corr}_{\mathcal{G}}(S_{2^i}, \bar{S}_{2^i})).$$

Since the limit is unique, (ii) and (iv) are equivalent. ■

Lemma 3.2 *Assume in addition that $\mathcal{G}$ is $\mathbb{L}_2$-compatible with the stationary non-decreasing filtration $\{\mathcal{F}_k\}$, which means that for any integers i and j such that $j \geq i$,*

$$\mathbb{E}_{\mathcal{G}}(X_iX_j) = \mathbb{E}_{\mathcal{G}}(X_i\mathbb{E}(X_j|\mathcal{F}_i)). \tag{3.1}$$

Then, for every positive integer r, we have

$$\mathbb{E}_{\mathcal{G}}(S_{2^r}^2) \leq 2^r[\|X_1\|_{\mathcal{G}} + \Delta_r]^2, \text{ where } \Delta_r = \sum_{j=0}^{r-1} 2^{-j/2}\|\mathbb{E}(S_{2^j}|\mathcal{F}_0)\|_{\mathcal{G}}. \tag{3.2}$$

Proof We notice that, by stationarity,

$$\begin{aligned}\mathbb{E}_{\mathcal{G}}(S_{2n}^2) &= \mathbb{E}_{\mathcal{G}}(S_n^2) + \mathbb{E}_{\mathcal{G}}(\bar{S}_n^2) + 2\mathbb{E}_{\mathcal{G}}(S_n\bar{S}_n) \\ &= 2\mathbb{E}_{\mathcal{G}}(S_n^2) + 2\mathbb{E}_{\mathcal{G}}(S_n\bar{S}_n).\end{aligned}$$

Since $\mathcal{G}$ is $\mathbb{L}_2$-compatible

$$\mathbb{E}_{\mathcal{G}}(S_n\bar{S}_n) = \mathbb{E}_{\mathcal{G}}(S_n\mathbb{E}_n(\bar{S}_n)) \leq \|S_n\|_{\mathcal{G}}\left(\mathbb{E}_{\mathcal{G}}(\mathbb{E}_n^2(\bar{S}_n))\right)^{1/2}.$$

Hence, $\mathcal{G}$ being a stationary σ-field,

$$\mathbb{E}_{\mathcal{G}}(S_n\bar{S}_n) \leq \|S_n\|_{\mathcal{G}}\|\mathbb{E}_0(S_n)\|_{\mathcal{G}}.$$

So, overall,

$$\mathbb{E}_{\mathcal{G}}(S_{2n}^2) \leq 2\mathbb{E}_{\mathcal{G}}(S_n^2) + 2\|S_n\|_{\mathcal{G}}\|\mathbb{E}_0(S_n)\|_{\mathcal{G}}.$$

By recurrence on the first term, we then infer that for any positive integer r,

$$\mathbb{E}_{\mathcal{G}}(S_{2^r}^2) \leq 2^r\mathbb{E}_{\mathcal{G}}(X_1^2) + \sum_{i=1}^{r-1} 2^i\|S_{2^{r-i}}\|_{\mathcal{G}} \cdot \|\mathbb{E}(S_{2^{r-i}}|\mathcal{F}_0)\|_{\mathcal{G}}.$$

Notice that inequality (3.2) holds for $r = 0$. Let $r \geq 1$ and assume that inequality (3.2) holds as for any $k \leq r - 1$. Let us then show that it also holds for $k = r$. By the recurrence hypothesis, we then get

$$\begin{aligned}\mathbb{E}_{\mathcal{G}}(S_{2^r}^2) &\leq 2^r\|X_1\|_{\mathcal{G}}^2 + \sum_{i=1}^{r-1} 2^i 2^{(r-i)/2}\left(\|X_1\|_{\mathcal{G}} + \Delta_{r-i}\right)\|\mathbb{E}(S_{2^{r-i}}\|\mathcal{F}_0)\|_{\mathcal{G}} \\ &\leq 2^r\left[\|X_1\|_{\mathcal{G}}^2 + \sum_{i=1}^{r-1}\left(\|X_1\|_{\mathcal{G}} + \Delta_{r-i}\right)\frac{\|\mathbb{E}(S_{2^{r-i}}|\mathcal{F}_0)\|_{\mathcal{G}}}{2^{(r-i)/2}}\right].\end{aligned}$$

Therefore, since Δ_i is a non decreasing sequence,

$$\begin{aligned}\mathbb{E}_{\mathcal{G}}(S_{2^r}^2) &\leq 2^r(\|X_1\|_{\mathcal{G}} + \Delta_r)\left[\|X_1\|_{\mathcal{G}} + \sum_{i=1}^{r-1}\frac{\|\mathbb{E}(S_{2^{r-i}}|\mathcal{F}_0)\|_{\mathcal{G}}}{2^{(r-i)/2}}\right] \\ &= 2^r(\|X_1\|_{\mathcal{G}} + \Delta_r)^2,\end{aligned}$$

proving that (3.2) also holds at step r. This ends the proof of the lemma. ∎

The sufficient condition for the convergence To check one of the conditions stated in Lemma 3.1 with the help of Lemma 3.2, we assume that

$$\sum_{j=0}^{\infty} 2^{-j/2} \|\mathbb{E}(S_{2^j} | \mathcal{F}_0)\|_{\mathcal{G}} < \infty \text{ a.s.} \tag{3.3}$$

The following corollary is then valid.

Corollary 3.3 *If $\mathcal{G}$ is $\mathbb{L}_2$-compatible with the stationary non-decreasing filtration $\{\mathcal{F}_k\}$ and if condition (3.3) holds then $\Sigma_{j=0}^{\infty} 2^{-j} |\mathbb{E}_{\mathcal{G}}(S_{2^j} \bar{S}_{2^j})| < \infty$ almost surely, and thus (i)–(iv) of Lemma 3.1 hold.*

Now, we show that actually the limit holds for the whole sequence.

Lemma 3.4 *If $\mathcal{G}$ is $\mathbb{L}_2$-compatible with the stationary non-decreasing filtration $\{\mathcal{F}_k\}$ and if condition (3.3) holds then the following limit exists and is finite almost surely:*

$$\begin{aligned}
\eta := \lim \frac{\mathbb{E}_{\mathcal{G}}(S_n^2)}{n} &= \mathbb{E}_{\mathcal{G}}(X_1^2) + \sum_{j=1}^{\infty} 2^{-j} \mathbb{E}_{\mathcal{G}}(S_{2^j} \bar{S}_{2^j}) \\
&= \mathbb{E}(X_1^2) \prod_{i=0}^{\infty} (1 + \operatorname{corr}_{\mathcal{G}}(S_{2^i}, \bar{S}_{2^i})).
\end{aligned}$$

Proof Let n be a positive integer and r be an integer such that $2^{r-1} \leq n < 2^r$. We write its binary expansion

$$n = \sum_{k=0}^{r-1} 2^k b_k \text{ where } b_{r-1} = 1 \text{ and } b_k \in \{0, 1\} \text{ for } k = 0, \dots, r-2. \tag{3.4}$$

We then have the following representation:

$$S_n = \sum_{k=0}^{r-1} b_k T_{2^k} \text{ where } T_{2^k} = \sum_{i=n_{k-1}+1}^{n_k} X_i, \, n_k = \sum_{j=0}^{k} b_j 2^j \text{ and } n_{-1} = 0, \tag{3.5}$$

implying that

$$\mathbb{E}_{\mathcal{G}}[S_n^2] = \left(\sum_{i=0}^{r-1} b_i \mathbb{E}_{\mathcal{G}}[S_{2^i}^2] \right) + \left(\sum_{i \neq j = 0}^{r-1} b_i b_j \mathbb{E}_{\mathcal{G}}[T_{2^i} T_{2^j}] \right) := I_n + J_n.$$

By Corollary 3.3, as $n \to \infty$, $\mathbb{E}_{\mathcal{G}}[S_{2^j}^2]/2^j \to \eta$ a.s. which then implies that $I_n/n \to \eta$ a.s. It remains for us to prove that $J_n/n \to 0$ a.s. With this aim we first note that since $\mathcal{G}$ is $\mathbb{L}_2$-compatible with the stationary non-decreasing filtration $\{\mathcal{F}_k\}$, for $a < a+k \leq b < b+n$,

$$\mathbb{E}_{\mathcal{G}}[(S_{a+k} - S_a)(S_{b+n} - S_b)] = \mathbb{E}_{\mathcal{G}}[(S_{a+k} - S_a)\mathbb{E}_b(S_{b+n} - S_b)] \leq \|S_k\|_{\mathcal{G}} \|\mathbb{E}_0(S_n)\|_{\mathcal{G}}.$$

Hence, if $i < j < r$,

$$|\mathbb{E}_{\mathcal{G}}[T_{2^i} T_{2^j}]| \le \|S_{2^i}\|_{\mathcal{G}} \|\mathbb{E}_0(S_{2^j})\|_{\mathcal{G}} \le (C+\eta) 2^{i/2} \sqrt{n} \frac{\|\mathbb{E}_0(S_{2^j})\|_{\mathcal{G}}}{2^{j/2}}$$

and thus

$$|J_n| \le 2 \sum_{0 \le i < j \le r-1} |\mathbb{E}_{\mathcal{G}}[T_{2^i} T_{2^j}]| \le 2(C+\eta)\sqrt{n} \sum_{i=0}^{r-2} 2^{i/2} \sum_{j=i+1}^{r-1} \frac{\|\mathbb{E}_0(S_{2^j})\|_{\mathcal{G}}}{2^{j/2}}$$
$$\le 2(C+\eta)(\sqrt{2}+1)\sqrt{n} \sum_{j=1}^{r-1} \|\mathbb{E}_0(S_{2^j})\|_{\mathcal{G}}.$$

Therefore $J_n/n \to 0$ a.s. because

$$\lim_{K \to \infty} \sum_{j=K}^{\infty} 2^{-j/2} \|\mathbb{E}(S_{2^j}|\mathcal{F}_0)\|_{\mathcal{G}} = 0 \text{ a.s.}$$

This completes the proof of the lemma. ■

Now, we suggest a simple, sufficient and workable condition for (3.3).

Theorem 3.5 *If $\mathcal{G}$ is $\mathbb{L}_2$-compatible with the stationary non-decreasing filtration $\{\mathcal{F}_k\}$ and*

$$\sum_{n=0}^{\infty} n^{-3/2} \|\mathbb{E}(S_n|\mathcal{F}_0)\|_2 < \infty.$$

Then, the following limit exists and is finite almost surely: $\eta := \lim_{n\to\infty} \frac{\mathbb{E}_{\mathcal{G}}(S_n^2)}{n}$.

Proof Firstly, we observe that from the forthcoming Lemma 3.24

$$\sum_{n=0}^{\infty} n^{-3/2} \|\mathbb{E}(S_n|\mathcal{F}_0)\|_2 < \infty \iff \sum_{j=0}^{\infty} \frac{\|\mathbb{E}(S_{2^j}|\mathcal{F}_0)\|_2}{2^{j/2}} < \infty.$$

Moreover, since

$$\mathbb{E}\{\|\mathbb{E}(S_n|\mathcal{F}_0)\|_{\mathcal{G}}^2\} = \mathbb{E}\mathbb{E}\{[\mathbb{E}(S_n|\mathcal{F}_0)]^2|\mathcal{G}]\} = \mathbb{E}\{[\mathbb{E}(S_n|\mathcal{F}_0)]^2\}$$
$$= \|\mathbb{E}(S_n|\mathcal{F}_0)\|_2^2,$$

we have

$$\mathbb{E}\{\|\mathbb{E}(S_n|\mathcal{F}_0)\|_{\mathcal{G}}\} \le \left[\mathbb{E}\{\|\mathbb{E}(S_n|\mathcal{F}_0)\|_{\mathcal{G}}^2\}\right]^{1/2} \le \|\mathbb{E}(S_n|\mathcal{F}_0)\|_2.$$

This implies that

$$\mathbb{E}\left(\sum_{j=0}^{\infty} 2^{-j/2}\|\mathbb{E}(S_{2^j}|\mathcal{F}_0)\|_{\mathcal{G}}\right) < \infty \quad \text{and hence} \quad \sum_{j=0}^{\infty} 2^{-j/2}\|\mathbb{E}(S_{2^j}|\mathcal{F}_0)\|_{\mathcal{G}} < \infty \text{ a.s.,}$$

proving condition (3.3). The proof then follows by Lemma 3.4. ∎

3.2 Burkholder-type Inequalities

In this section we obtain Burkholder-type inequalities for sequences $(X_k)_{k\in\mathbb{Z}}$ that are not necessarily martingale difference sequences. For $p > 1$, let $C_p \geq 2$ be the minimal constant such that

$$\mathbb{E}\left[\max_{1\leq k\leq n} |Z_1 + \ldots + Z_k|^p\right] \leq C_p n^{p/2}\|Z_1\|_p^p \tag{3.6}$$

holds for all n and stationary martingale differences $\{Z_k\}$. By Corollary 2.14, $C_p < \infty$.

Further, by c_p we denote finite positive constants depending only on p.

3.2.1 *Burkholder-type Inequalities via Maxwell-Woodroofe type Characteristics*

We shall start by generalizing inequality (3.6) to stationary sequences that are adapted to a stationary non-decreasing filtration $(\mathcal{F}_k)_{k\in\mathbb{Z}}$. To this end, we introduce the following two quantities:

$$\delta_{n,p} = \sum_{j=1}^{n} j^{-3/2}\|\mathbb{E}(S_j|\mathcal{F}_0)\|_p \text{ and } \delta_{\infty,p} = \sum_{j=1}^{\infty} j^{-3/2}\|\mathbb{E}(S_j|\mathcal{F}_0)\|_p.$$

The next theorem was obtained by Peligrad, Utev and Wu (2007).

Theorem 3.6 *(A maximal $\mathbb{L}^p$ inequality) Let $(X_k)_{k\in\mathbb{Z}}$ be a stationary sequence adapted to a stationary non-decreasing filtration $(\mathcal{F}_k)_{k\in\mathbb{Z}}$. Assume that $\mathbb{E}(|X_1|^p) < \infty, p \geq 2$. Then*

$$\|S_n^*\|_p \leq C_p^{1/p} n^{1/2}(\|X_1\|_p + 12\delta_{n,p}). \tag{3.7}$$

When $(X_i)_{i\in\mathbb{Z}}$ is a stationary sequence of martingale differences with respect to the filtration $(\mathcal{F}_i)_{i\in\mathbb{Z}}$, then $\delta_{n,p} = 0$ and the inequality (3.7) reduces to the classical one (3.6).

From inequality (3.7), it is immediate that $\|S_n^*\|_p = O(\sqrt{n})$ if $\delta_{\infty,p} < \infty$. In order to comment on the minimality of the condition $\delta_{\infty,p} < \infty$, we construct, later on, a (renewal type) Markov chain example, showing that if this condition is barely altered the result fails to hold (see Section 17.1.2).

Since our method of proof for obtaining the maximal inequalities is based on martingale approximations, the dominant constant in all the inequalities are going to be the best known constants for the martingales.

From now on, the term mixingales will refer to sequences of random variables satisfying conditions imposed on conditional expectations of individual variables X_j. Using the triangular inequality $||\mathbb{E}(S_k|\mathcal{F}_0)||_p \le \sum_{i=1}^k ||\mathbb{E}(X_i|\mathcal{F}_0)||_p$ and the elementary bound $\sum_{k=j}^{\infty} k^{-3/2} \le 3j^{-1/2}$, valid for $j = 1, 2, \ldots$, we have the following inequality for stationary mixingales:

Corollary 3.7 *Let $(X_k)_{k\in\mathbb{Z}}$ be a stationary sequence adapted to a stationary non-decreasing filtration $(\mathcal{F}_k)_{k\in\mathbb{Z}}$. Assume that $\mathbb{E}(|X_1|^p) < \infty, p \ge 2$. Then*

$$||S_n^*||_p \le C_p^{1/p} n^{1/2} \left[||X_1||_p + 36 \sum_{k=1}^{n} k^{-1/2} ||\mathbb{E}(X_k|\mathcal{F}_0)||_p \right]. \tag{3.8}$$

The proof of Theorem 3.6 reveals that we can drop the stationarity assumption and get the following result.

Theorem 3.8 *Let $p \ge 2$ and $(X_k)_{k\in\mathbb{Z}}$ be a sequence of real-valued random variables in $\mathbb{L}^p$ and adapted to a filtration $(\mathcal{F}_k)_{k\in\mathbb{Z}}$. Let r be the unique positive integer such that $2^{r-1} \le n < 2^r$. Then,*

$$\begin{aligned} ||S_n^*||_p &\le (2c_p + 1) \left(\sum_{j=1}^{n} ||X_j||_p^2 \right)^{1/2} \\ &+ \sqrt{2}(2c_p + 1) \sum_{j=0}^{r-1} \left(\sum_{k=1}^{2^{r-j}} ||\mathbb{E}(S_{k2^j} - S_{(k-1)2^j} | \mathcal{F}_{(k-2)2^j+1})||_p^2 \right)^{1/2} . \end{aligned} \tag{3.9}$$

From the proof, we note that, when $p = 2$, the inequality holds with $c_p = 1$, namely:

$$||S_n^*||_2 \le 3 \left(\sum_{j=1}^{n} ||X_j||_2^2 \right)^{1/2} + 3\sqrt{2} \sum_{j=0}^{r-1} \left(\sum_{k=1}^{2^{r-j}} ||\mathbb{E}(S_{k2^j} - S_{(k-1)2^j} | \mathcal{F}_{(k-2)2^j+1})||_2^2 \right)^{1/2} . \tag{3.10}$$

When $p \in]1, 2[$, we can also obtain results in the spirit of the above theorems. Here we will only give the counterpart of Theorem 3.8 when $p \in]1, 2[$.

Theorem 3.9 *Let $p \in]1, 2[$ and $(X_k)_{k\in\mathbb{Z}}$ be a sequence of real-valued random variables in $\mathbb{L}^p$ and adapted to a filtration $(\mathcal{F}_k)_{k\in\mathbb{Z}}$. Let r be the unique positive integer such that $2^{r-1} \le n < 2^r$. Then,*

$$\|S_n^*\|_p \le (2c_p+1)\left(\sum_{j=1}^{n}\|X_j\|_p^p\right)^{1/p}$$
$$+\,2^{(p-1)/p}(2c_p+1)\sum_{j=0}^{r-1}\left(\sum_{k=1}^{2^{r-j}}\|\mathbb{E}(S_{k2^j}-S_{(k-1)2^j}|\mathcal{F}_{(k-2)2^j+1})\|_p^p\right)^{1/p}.$$

The proof is left to the reader since it follows the lines of the proof of Theorem 3.8 with obvious modifications, by using when necessary the case $p \in (1,2]$ of Theorem 2.13 instead of the case $p \ge 2$.

In the stationary setting, Theorem 3.9 is Theorem 3 in Wu and Zhao (2008).

3.2.2 *Burkholder-type Inequalities for Non-Stationary Sequences via Projective Criteria*

The next theorem gives moment inequalities in $\mathbb{L}^p$ under conditions involving the projective operators $\mathbf{P}_j$ defined, for a filtration $(\mathcal{F}_j)_{j\in\mathbb{Z}}$, by

$$\mathbf{P}_j(X) = \mathbb{E}(X|\mathcal{F}_j) - \mathbb{E}(X|\mathcal{F}_{j-1}). \tag{3.11}$$

It is a non-stationary and non-adapted version of Part 4 of Theorem 1 in Wu (2007). We refer also to Proposition 6 in Dedecker and Merlevède (2002) when $p = 2$.

Theorem 3.10 *Let $(\mathcal{F}_k)_{k\in\mathbb{Z}}$ be a filtration and $\{Y_k\}_{k\in\mathbb{Z}}$ be a sequence of random variables such that for all j, $\mathbb{E}(Y_j|\mathcal{F}_{-\infty}) = 0$ almost surely and Y_j is $\mathcal{F}_\infty$-measurable. Let $p \in]1,\infty[$. Assume that*

$$\|\mathbf{P}_{k-j}(Y_k)\|_p \le p_j \quad \textit{and} \quad \sum_{j=-\infty}^{\infty} p_j < \infty.$$

Let $\{g_k,\ k\in\mathbb{N}\}$ be a sequence of numbers and define,

$$S_k = \sum_{i=1}^{k} g_i Y_i. \tag{3.12}$$

Then, for $p > 1$,

$$\|\max_{1\le k\le n}|S_k|\|_p \le c_p\left(\sum_{i=1}^{n}|g_i|^{p\wedge 2}\right)^{1/(p\wedge 2)}\sum_{j=-\infty}^{\infty} p_j.$$

The next result is due to Dedecker and Rio (2000).

Theorem 3.11 *Let $(X_i)_{i\in\mathbb{Z}}$ be a sequence of square-integrable and centered random variables, adapted to a non-decreasing filtration $(\mathcal{F}_i)_{i\in\mathbb{Z}}$. Let $\tilde{S}_n = \max\{0, S_1, \dots, S_n\}$, λ be any non-negative real number and $\Gamma_k = (\tilde{S}_k > \lambda)$. We have*

$$\mathbb{E}((\tilde{S}_n - \lambda)_+^2) \le 4\sum_{k=1}^{n} \mathbb{E}(X_k^2 1_{\Gamma_k}) + 8\sum_{k=1}^{n-1} \|X_k 1_{\Gamma_k}\mathbb{E}(S_n - S_k|\mathcal{F}_k)\|_1.$$

As a corollary, we easily get the following result for a strictly stationary sequence of martingale differences.

Corollary 3.12 *Let $(d_i)_{i\in\mathbb{Z}}$ be a strictly stationary sequence of martingale differences in $\mathbb{L}^2$, adapted to a non-decreasing filtration $(\mathcal{F}_i)_{i\in\mathbb{Z}}$. Let $M_n = \sum_{i=1}^{n} d_i$ and $M_n^* = \max_{1\le k\le n} |M_k|$. Then*

$$\lim_{\lambda\to\infty} \sup_{n\ge 1} \mathbb{E}\left(\left(n^{-1/2}M_n^* - \lambda\right)_+^2\right) = 0,$$

i.e. $\left\{n^{-1}\max_{1\le k\le n} M_k^2\right\}_{n\ge 1}$ is an uniformly integrable family.

The following moment inequality is due to Dedecker and Doukhan (2003). It is a non-stationary version of Theorem 2.5 of Rio (2000).

Theorem 3.13 *Let p be a real number in $[2, \infty[$. Let $(X_i)_{i\in\mathbb{N}}$ be a sequence of centered real-valued random variables that are in $\mathbb{L}^p$. Let $(\mathcal{F}_i)_{i\in\mathbb{N}}$ be a non-decreasing sequence of σ-algebras such that X_i is $\mathcal{F}_i$-measurable. Define $S_n = X_1 + \cdots + X_n$ and*

$$b_{i,n} = \max_{i\le l\le n} \left\| X_i \sum_{k=i}^{l} \mathbb{E}(X_k|\mathcal{F}_i) \right\|_{p/2}.$$

For any integer $n \ge 0$, the following inequality holds

$$\|S_n\|_p \le \left(2(p-1)\sum_{i=1}^{n} b_{i,n}\right)^{1/2}. \tag{3.13}$$

3.2.3 *A Maximal Inequality à la Doob for Adapted Sequences*

Let $(X_i)_{i\in\mathbb{Z}}$ be a sequence of real-valued random variables that are adapted to a non-decreasing filtration $(\mathcal{F}_i)_{i\in\mathbb{Z}}$. If the sequence is in $\mathbb{L}^p$ for a $p > 1$ and is a sequence of martingale differences, that is: $\mathbb{E}(X_i|\mathcal{F}_{i-1}) = 0$ a.s. for any integer i, then the Doob maximal inequality (see Corollary 2.7) states that the study of the norm p of the maximum of its partial sums can be reduced to the study of the $\mathbb{L}^p$-norm of its partial sums. The next proposition, established in Merlevède and Peligrad (2013), is a generalization of the Doob maximal inequality for martingales to adapted sequences by adding a correction term which is zero for martingale differences sequences. The proof is based on chaining arguments.

Proposition 3.14 *Let X_i, $1 \leq i \leq 2^r$ be real-valued random variables in $\mathbb{L}^p$, $p > 1$ and $q = p/(p-1)$. Assume that the random variables are adapted to a non-decreasing filtration $(\mathcal{F}_i)_i$. Let $S_r = X_1 + \cdots + X_r$. We have that*

$$\| \max_{1 \leq m \leq 2^r} |S_m| \|_p \leq q \|S_{2^r}\|_p + q \sum_{l=0}^{r-1} \left(\sum_{k=1}^{2^{r-l}-1} \|\mathbb{E}_{k2^l}(S_{(k+1)2^l} - S_{k2^l})\|_p^p \right)^{1/p} . \tag{3.14}$$

Corollary 3.15 *In the stationary case, for any integer $r \geq 1$ and for any integer $n \in [2^{r-1}, 2^r[$,*

$$\| \max_{1 \leq i \leq n} |S_i| \|_p \leq 2q \max_{1 \leq m \leq n} \|S_m\|_p + (2^{1/p} q) n^{1/p} \sum_{l=0}^{r-1} 2^{-l/p} \|\mathbb{E}(S_{2^l} | \mathcal{F}_0)\|_p . \tag{3.15}$$

Moreover, due to the subadditivity of the sequence $\left(\|\mathbb{E}(S_n | \mathcal{F}_0)\|_p \right)_{n \geq 1}$, according to Lemma 3.24, we also have that for any positive integer n,

$$\| \max_{1 \leq i \leq n} |S_i| \|_p \leq 2q \max_{1 \leq m \leq n} \|S_m\|_p + \left(q \frac{2^{2+2/p}}{2^{1+1/p} - 1} \right) n^{1/p} \sum_{j=1}^{n} j^{-1-1/p} \|\mathbb{E}(S_j | \mathcal{F}_0)\|_p . \tag{3.16}$$

Furthermore inequalities (3.15) and (3.16) hold even if the variables are not centered.

Additional comments Inequality (3.14) is stated for real-valued random variables, but it holds also for variables taking values in a separable Banach $(\mathbb{B}, |\cdot|_{\mathbb{B}})$. In addition, as quoted in Dedecker, Merlevède and Pène (2013), in the case where the random variables are non-adapted to the underlying filtration $(\mathcal{F}_i)_i$, inequality (3.14) still holds if we add the following term in its right-hand side:

$$(q+1) \sum_{l=0}^{r} \left(\sum_{k=1}^{2^{r-l}} \|S_{k2^l} - S_{(k-1)2^l} - \mathbb{E}_{k2^l}(S_{k2^l} - S_{(k-1)2^l})\|_p^p \right)^{1/p} .$$

(See Proposition 3.2 and Remark 3.3 in Dedecker, Merlevède and Pène (2013).)

Note that Theorem 3.13 combined with Proposition 3.14 gives a maximal $\mathbb{L}^p$ inequality for the maximum of the partial sums. For instance in the stationary and bounded case, we get:

Theorem 3.16 *Assume that $(X_i)_{i \in \mathbb{Z}}$ is a strictly stationary sequence of centered real-valued random variables adapted to a stationary non-decreasing filtration $(\mathcal{F}_i)_{i \in \mathbb{Z}}$, and such that $\mathbb{P}(|X_0| \leq T) = 1$. Define $S_n = X_1 + \cdots + X_n$ and $S_n^* = \max_{1 \leq i \leq n} |S_i|$. Then, for any $p > 2$,*

$$\|S_n^*\|_p \leq C_p n^{1/2} T^{1/2} \left(\|X_0\|_{p/2}^{1/2} + \max_{1 \leq \ell \leq n} \|\mathbb{E}(S_\ell \,|\mathcal{F}_0)\|_{p/2}^{1/2} \right).$$

3.3 A Rosenthal-type Inequality for Stationary Sequences

The following Rosenthal-type inequality was obtained by Merlevède and Peligrad (2013). The case $2 < p \leq 3$ for the partial sum S_n is originated in Rio (2009).

Theorem 3.17 *Let $p > 2$ be a real number and let $(X_i)_{i\in\mathbb{Z}}$ be a stationary sequence adapted to a non-decreasing filtration $(\mathcal{F}_i)_i$. Then, for any positive integer n, the following inequality holds:*

$$\|S_n^*\|_p^p \leq c_p n \mathbb{E}(|X_1|)^p + cn \left(\sum_{k=1}^n \frac{1}{k^{1+1/p}} \|\mathbb{E}_0(S_k)\|_p \right)^p$$
$$+ n \left(\sum_{k=1}^n \frac{1}{k^{1+2\delta/p}} \|\mathbb{E}_0(S_k^2)\|_{p/2}^\delta \right)^{p/(2\delta)} \tag{3.17}$$

where $\delta = \min(1, 1/(p-2))$ and $c = 1$. When $p \geq 4$ we can take $c = 0$ by enlarging the constant involved.

In the traditional Rosenthal inequality for martingales (see Theorem 2.17), the lower bound and the upper one have the same order of magnitude. In general the upper bound obtained in Theorem 3.17 cannot be a lower bound for $\mathbb{E}(|S_n|^p)$ (up to some constants) without any additional assumptions. Indeed, if $X_k = Z_0 \circ T^{k-1} - Z_0 \circ T^k$ where Z_0 is $\mathcal{F}_0$-measurable and in $\mathbb{L}^p$, then $\mathbb{E}(|S_n|^p) \leq 2^p \|Z_0\|_p^p$, whereas the upper bound given in Theorem 3.17 is of order n.

It is interesting to indicate the monotonicity of the right-hand side of the inequality in δ. To be more precise, for any $0 < \delta \leq \gamma \leq 1$, the following inequality holds:

$$\left(\sum_{k=1}^n k^{-1-2\gamma/p} \|\mathbb{E}_0(S_k^2)\|_{p/2}^\gamma \right)^{1/\gamma} \leq 2^{(1+\gamma)(\gamma-\delta)/(\delta\gamma)} \left(\sum_{k=1}^n k^{-1-2\delta/p} \|\mathbb{E}_0(S_k^2)\|_{p/2}^\delta \right)^{1/\delta}.$$

To see this, we notice the subadditivity property

$$\|\mathbb{E}_0(S_{i+j}^2)\|_{p/2} \leq 2\|\mathbb{E}_0(S_i^2)\|_{p/2} + 2\|\mathbb{E}_0(S_j^2)\|_{p/2}$$

and apply then Item 3 of Lemma 3.24 with $C = 2$.

On the other hand, for any $0 < \delta < 1$ and any $\gamma > 1/\delta - 1$, by Hölder's inequality, there exists a positive constant C depending on p, γ and δ, such that

$$\left(\sum_{k=1}^n k^{-1-2\delta/p} \|\mathbb{E}_0(S_k^2)\|_{p/2}^\delta \right)^{p/(2\delta)} \leq C \left(\sum_{k=1}^n k^{-1-2/p} (\log k)^\gamma \|\mathbb{E}_0(S_k^2)\|_{p/2} \right)^{p/2}. \tag{3.18}$$

For the sake of applications, we indicate how to estimate the terms that appear in Theorem 3.17 in terms of individual summands. By subtracting $\mathbb{E}(S_k^2)$ and applying the triangle inequality we can reformulate all the inequalities in terms of the quantities $\mathbb{E}(S_k^2)$, $\|\mathbb{E}_0(S_k)\|_p$ and $\|\mathbb{E}_0(S_k^2) - \mathbb{E}(S_k^2)\|_{p/2}$. The next lemma proposes, then, a simple way to estimate these quantities in terms of coefficients, in the spirit of Gordin (1969).

Lemma 3.18 *Let $(X_i)_{i\in\mathbb{Z}}$ be a stationary sequence of real-valued random variables in $\mathbb{L}^p$ adapted to a non-decreasing filtration $(\mathcal{F}_i)$. Then the following estimates are valid:*

$$\mathbb{E}(S_k^2) \le 2k\sum_{j=0}^{k-1} |\mathbb{E}(X_0X_j)|, \tag{3.19}$$

$$\|\mathbb{E}_0(S_k)\|_p \le \sum_{\ell=1}^{k} \|\mathbb{E}_0(X_\ell)\|_p, \tag{3.20}$$

and for any real $p \ge 2$,

$$\begin{aligned}\|\mathbb{E}_0(S_k^2) - \mathbb{E}(S_k^2)\|_{p/2} &\le 2\sum_{i=1}^{k}\sum_{j=0}^{k-i} \|\mathbb{E}_0(X_iX_{i+j}) - \mathbb{E}(X_iX_{i+j})\|_{p/2} \\ &\le 2\sum_{i=1}^{k}\sum_{j=0}^{k-i} \sup_{\ell\ge 0} \|\mathbb{E}_0(X_iX_{i+\ell}) - \mathbb{E}(X_iX_{i+\ell})\|_{p/2} \wedge (2\|X_0\mathbb{E}_0(X_j)\|_{p/2}) \\ &\le 4\sum_{j=1}^{k} j\|X_0\mathbb{E}_0(X_j)\|_{p/2} + 2\sum_{i=1}^{k} i \sup_{j\ge i} \|\mathbb{E}_0(X_iX_j) - \mathbb{E}(X_iX_j)\|_{p/2}.\end{aligned} \tag{3.21}$$

Mixing coefficients are useful to continue the estimates from Lemma 3.18. We refer to the books by Bradley (2007, Theorem 4.13 via Remark 4.7, VI), Rio (2000, Theorem 2.5 and Appendix, Section C) and Dedecker *et al.* (2007, Remark 2.5 and Ch 3) for various estimates of the coefficients involved in Lemma 3.18 and examples. We also refer to Section 5 in Merlevède and Peligrad (2013) where Theorem 3.17 together with Lemma 3.18 has been used to study the $\mathbb{L}^p$-integrated risk of kernel estimators of the unknown marginal density of a stationary sequence that is assumed to be β-mixing in the weak sense, and to their Section 4.4 where the estimates of Lemma 3.18 have been computed when X_k is a functional of a linear process.

3.4 Maximal Exponential Inequalities

For the bounded case the optimal constants in the $\mathbb{L}^p$-martingale inequalities imply the following extension of Azuma's exponential inequality.

Theorem 3.19 *Let $(X_k)_{k\in\mathbb{Z}}$ be a stationary sequence adapted to a stationary non-decreasing filtration $(\mathcal{F}_k)_{k\in\mathbb{Z}}$ and such that $\|X_0\|_\infty < \infty$. For any $t \ge 0$, we have*

$$\mathbb{P}(S_n^* \geq t) \leq 4\sqrt{e}\exp\left(-t^2/2n(\|X_1\|_\infty + 12\delta_{n,\infty})^2\right), \tag{3.22}$$

where $\delta_{n,\infty} = \sum_{j=1}^{n} j^{-3/2}\|\mathbb{E}(S_j|\mathcal{F}_0)\|_\infty$.

In the stationary case this bound improves upon several corresponding inequalities established by Deddens, Peligrad and Yang (1987), Rio (2000) and also Dedecker and Prieur (2005).

In the next result, we bound the maximal exponential moment of the stationary sequence by using the projective operators (recall the definition (3.11)).

Theorem 3.20 *Let $(\mathcal{F}_k)_{k\in\mathbb{Z}}$ be a filtration and $\{Y_k\}_{k\in\mathbb{Z}}$ be a sequence of random variables such that for all j, $\mathbb{E}(Y_j|\mathcal{F}_{-\infty}) = 0$ almost surely and Y_j is $\mathcal{F}_\infty$-measurable. Assume that*

$$\|\mathbf{P}_{k-j}(Y_k)\|_\infty \leq p_j \quad \textit{and} \quad D := \sum_{j=-\infty}^{\infty} p_j < \infty.$$

Let $\{g_k,\ k \in \mathbb{N}\}$ be a sequence of numbers and define S_k by (3.12). Let $G_n^2 = \sum_{i=1}^{n} g_i^2$. Then, for any $t \geq 0$,

$$\mathbb{E}\exp\left(t \max_{1\leq k\leq n} S_k\right) \leq 4\exp\left(\frac{1}{2}G_n^2 D^2 t^2\right).$$

In particular,

$$\mathbb{P}\left(\max_{1\leq k\leq n}|S_k| \geq x\right) \leq 8\exp\left(-\frac{x^2}{2G_n^2 D^2}\right).$$

3.5 Proofs

3.5.1 Proofs of the Results of Section 3.2

Proof of Theorem 3.6 Let n, r be integers such that $n \geq 1, 2^{r-1} \leq n < 2^r$; let $K = 5/\sqrt{2}$. As a matter of fact, we shall prove a slightly stronger inequality

$$\|S_n^*\|_p \leq C_p^{1/p} n^{1/2}[\|X_1 - \mathbb{E}(X_1|\mathcal{F}_0)\|_p + K\Delta_{r,p}],$$

$$\text{where } \Delta_{r,p} = \Delta_{r,p}(X,\mathcal{F}) = \sum_{j=0}^{r-1} 2^{-j/2}\|\mathbb{E}(S_{2^j}|\mathcal{F}_0)\|_p. \tag{3.23}$$

First, we notice that $\Delta_{r,p} \leq \frac{2^{2+1/2}}{2^{3/2}-1}\delta_{n,p}$, which follows from Item 1 of Lemma 3.24 applied to the subadditive sequence $V_k = \|\mathbb{E}(S_k|\mathcal{F}_0)\|_p$ and $q = 1/2$. Then, Theorem 3.6 follows from (3.23) together with the inequality

$$\|X_1 - \mathbb{E}(X_1|\mathcal{F}_0)\|_p + K\Delta_{r,p} \le \|X_1\|_p + \left(K\frac{2^{2+1/2}}{2^{3/2}-1} + 1\right)\delta_{n,p}.$$

This explains the constant 12 involved in inequality (3.7) since $K \times \frac{2^{2+1/2}}{2^{3/2}-1} + 1 \le 12$.

We now prove (3.23) by induction on n. For $n = 1$

$$\|X_1\|_p \le \|X_1 - \mathbb{E}(X_1|\mathcal{F}_0)\|_p + \|\mathbb{E}(X_1|\mathcal{F}_0)\|_p = \|X_1 - \mathbb{E}(X_1|\mathcal{F}_0)\|_p + \Delta_{1,p}.$$

Assume that the inequality holds up to $n - 1$. By the triangle inequality

$$S_n^* \le \max_{1\le k\le n}\left|\sum_{i=1}^{k}[X_i - \mathbb{E}(X_i|\mathcal{F}_{i-1})]\right| + \max_{1\le k\le n}\left|\sum_{i=1}^{k}\mathbb{E}(X_i|\mathcal{F}_{i-1})\right|. \tag{3.24}$$

By the Burkholder maximal inequality (3.6),

$$\left\|\max_{1\le k\le n}\left|\sum_{i=1}^{k}(X_i - \mathbb{E}(X_i|\mathcal{F}_{i-1}))\right|\right\|_p \le C_p^{1/p}\sqrt{n}\|X_1 - \mathbb{E}(X_1|\mathcal{F}_0)\|_p. \tag{3.25}$$

To estimate the impact of the second term in (3.24) we start by writing $n = 2m$ or $n = 2m + 1$ according to a value odd or even of n. Notice that

$$\begin{aligned}\left\|\max_{1\le k\le n}\left|\sum_{i=1}^{k}\mathbb{E}(X_i|\mathcal{F}_{i-1})\right|\right\|_p &\le \left\|\max_{1\le k\le m}\left|\sum_{i=1}^{2k}\mathbb{E}(X_i|\mathcal{F}_{i-1})\right|\right\|_p \\ &+ \left\{\left\|\max_{0\le k\le m}|\mathbb{E}(X_{2k+1}|\mathcal{F}_{2k})|\right\|_p 1_{n=2m+1} + \left\|\max_{1\le k\le m}|\mathbb{E}(X_{2k-1}|\mathcal{F}_{2(k-1)})|\right\|_p 1_{n=2m}\right\}.\end{aligned} \tag{3.26}$$

The second term in the right-hand side of (3.26) is estimated in a trivial way:

$$\left\|\max_{0\le k\le m}|\mathbb{E}(X_{2k+1}|\mathcal{F}_{2k})|\right\|_p \le (m+1)^{1/p}\|\mathbb{E}(X_1|\mathcal{F}_0)\|_p. \tag{3.27}$$

For the first term in the right-hand side of (3.26) we apply the induction hypothesis to the stationary sequence $Y_0 = \mathbb{E}(X_{-1}|\mathcal{F}_{-2}) + \mathbb{E}(X_0|\mathcal{F}_{-1})$, the sigma algebra $\mathcal{G}_0 = \mathcal{F}_{-1}$ and the operator $\mathbb{T}^2$. Notice that the new filtration becomes $\{\mathcal{G}_i : i \in \mathbb{Z}\}$ where $\mathcal{G}_i = \mathcal{F}_{2i-1}$. By the induction hypothesis we obtain

$$\left\|\max_{1\le k\le m}\left|\sum_{i=1}^{2k}\mathbb{E}(X_i|\mathcal{F}_{i-1})\right|\right\|_p \le C_p^{1/p}m^{1/2}[\mathbb{E}|Y_1 - \mathbb{E}(Y_1|\mathcal{G}_0)|^p + K\Delta_{r-1,p}(Y,\mathcal{G})].$$

Clearly, $\|Y_1 - \mathbb{E}(Y_1|\mathcal{G}_0)\|_p \le 2\|\mathbb{E}(X_1|\mathcal{F}_{-1})\|_p \le 2\|\mathbb{E}(X_1|\mathcal{F}_0)\|_p$ and also

$$\begin{aligned}\Delta_{r-1,p}(Y,\mathcal{G}) &= \sum_{j=0}^{r-2} 2^{-j/2}\left\|\mathbb{E}\left(\sum_{k=1}^{2^j} Y_k|\mathcal{G}_0\right)\right\|_p \\ &= \sum_{j=0}^{r-2} 2^{-j/2}\|\mathbb{E}(S_{2^{j+1}}|\mathcal{F}_{-1})\|_p \\ &\le \sqrt{2}[\Delta_{r,p}(X,\mathcal{F}) - \|\mathbb{E}(X_1|\mathcal{F}_0)\|_p].\end{aligned}$$

And so

$$\begin{aligned}\left\|\max_{1\le k\le m}\left|\sum_{i=1}^{2k}\mathbb{E}(X_i|\mathcal{F}_{i-1})\right|\right\|_p \le\ & C_p^{1/p}\{4\sqrt{m}\|\mathbb{E}(X_1|\mathcal{F}_0)\|_p \\ & + \sqrt{2m}K\,[\Delta_{r,p}(X,\mathcal{F}) - \|\mathbb{E}(X_1|\mathcal{F}_0)\|_p]\}.\end{aligned}$$

This last relation, combined with (3.24), (3.25), (3.26), (3.27) and also with the fact that for all $m \ge 1$, we have $(m+1)^{1/p} \le \sqrt{m}C_p^{1/p}$ (since $C_p \ge 2$ for all $p \ge 2$), leads to

$$\begin{aligned}\|S_n^*\|_p &\le C_p^{1/p}\Big\{\sqrt{n}\|X_1 - \mathbb{E}(X_1|\mathcal{F}_0)\|_p + 5\sqrt{m}\|\mathbb{E}(X_1|\mathcal{F}_0)\|_p \\ &\quad + \sqrt{2m}K[\Delta_{r,p} - \|\mathbb{E}(X_1|\mathcal{F}_0)\|_p]\Big\} \\ &\le C_p^{1/p}\sqrt{n}[\|X_1 - \mathbb{E}(X_1|\mathcal{F}_0)\|_p + K\Delta_{r,p}]\end{aligned}$$

and so we take $K = 5/\sqrt{2}$ to guarantee the inequality. This completes the proof of the theorem. ■

Example 3.21 *By adapting the proof of Theorem 3.6 we derive the following version of inequality (3.23), which incorporates the Burkholder quadratic variation term:*

$$\mathbb{E}\left[\max_{1\le i\le n}|S_i|^p\right] \le p^p\left(\left\|\left(\sum_{i=1}^n (X_i - \mathbb{E}(X_i|\mathcal{F}_{i-1}))^2\right)^{1/2}\right\|_p + \sqrt{n}\left(\frac{5}{\sqrt{2}}\Delta_{r,p}\right)\right)^p,$$

where $p \ge 2$. To prove it, we replace the right-hand side of inequality (3.25) by the quadratic variation term and then carry on this term to the end of the proof.

Proof of Theorem 3.8 As for the proof of Theorem 3.6, the proof is done by induction. For $n = 1$, the inequality is clearly true. Assume that it holds up to $n-1$. Set $a_p = 2c_p + 1$. We start from the decomposition (3.24). Instead of (3.25), by using Theorem 2.13 together with Doob's maximal inequality, we write

$$\left\| \max_{1\le k\le n} \left| \sum_{i=1}^{k} (X_i - \mathbb{E}(X_i|\mathcal{F}_{i-1})) \right| \right\|_p \le c_p \left(\sum_{i=1}^{n} \|X_i - \mathbb{E}(X_i|\mathcal{F}_{i-1})\|_p^2 \right)^{1/2}$$
$$\le 2c_p \left(\sum_{i=1}^{n} \|X_i\|_p^2 \right)^{1/2}. \tag{3.28}$$

To estimate the second term in the right-hand side of (3.24), we still use the decomposition (3.26). Note that

$$\left\| \max_{0\le k\le m} |\mathbb{E}(X_{2k+1}|\mathcal{F}_{2k})| \right\|_p \mathbf{1}_{n=2m+1} + \left\| \max_{1\le k\le m} |\mathbb{E}(X_{2k-1}|\mathcal{F}_{2(k-2)})| \right\|_p \mathbf{1}_{n=2m}$$
$$\le \mathbf{1}_{n=2m+1} \left(\sum_{k=0}^{m} \|\mathbb{E}(X_{2k+1}|\mathcal{F}_{2k})\|_p^p \right)^{1/p} + \mathbf{1}_{n=2m} \left(\sum_{k=1}^{m} \|\mathbb{E}(X_{2k-1}|\mathcal{F}_{2(k-2)})\|_p^p \right)^{1/p}$$
$$\le \left(\sum_{i=1}^{n} \|X_i\|_p^p \right)^{1/p}.$$

Using the fact that $p \ge 2$, we get

$$\left\| \max_{0\le k\le m} |\mathbb{E}(X_{2k+1}|\mathcal{F}_{2k})| \right\|_p \mathbf{1}_{n=2m+1} + \left\| \max_{1\le k\le m} |\mathbb{E}(X_{2k-1}|\mathcal{F}_{2(k-2)})| \right\|_p \mathbf{1}_{n=2m}$$
$$\le \left(\sum_{i=1}^{n} \|X_i\|_p^2 \right)^{1/2}. \tag{3.29}$$

For the first term in the right-hand side of (3.26), we set

$$Y_i = \mathbb{E}(X_{2i-1}|\mathcal{F}_{2i-2}) + \mathbb{E}(X_{2i}|\mathcal{F}_{2i-1}),\ W_i = \sum_{j=1}^{i} Y_j \text{ and } \mathcal{G}_i = \mathcal{F}_{2i-1},$$

and we note that

$$\left\| \max_{1\le k\le m} \left| \sum_{i=1}^{2k} \mathbb{E}(X_i|\mathcal{F}_{i-1}) \right| \right\|_p = \left\| \max_{1\le k\le m} \left| \sum_{i=1}^{k} Y_i \right| \right\|_p.$$

By the induction hypothesis, noticing that $m < 2^{r-1}$,

$$\left\| \max_{1\le k\le m} \left| \sum_{i=1}^{k} Y_i \right| \right\|_p \le a_p \left(\sum_{j=1}^{m} \|Y_j\|_p^2 \right)^{1/2}$$
$$+ \sqrt{2}a_p \sum_{j=0}^{r-2} \left(\sum_{k=1}^{2^{r-1-j}} \|\mathbb{E}(W_{k2^j} - W_{(k-1)2^j}|\mathcal{G}_{(k-2)2^j+1})\|_p^2 \right)^{1/2}.$$

But

$$\|\mathbb{E}(W_{k2^j} - W_{(k-1)2^j}|\mathcal{G}_{(k-2)2^j+1})\|_p = \|\mathbb{E}(S_{k2^{j+1}} - S_{(k-1)2^{j+1}}|\mathcal{F}_{(k-2)2^{j+1}+1})\|_p.$$

On the other hand,

$$\sum_{j=1}^{m} \|Y_j\|_p^2 \leq 2\sum_{i=1}^{m} \left(\|\mathbb{E}(X_{2i-1}|\mathcal{F}_{2i-2})\|_p^2 + \|\mathbb{E}(X_{2i}|\mathcal{F}_{2i-1})\|_p^2\right) \leq 2\sum_{i=1}^{n} \|\mathbb{E}(X_i|\mathcal{F}_{i-1})\|_p^2.$$

Therefore,

$$\left\|\max_{1\leq k\leq m}\left|\sum_{i=1}^{2k} \mathbb{E}(X_i|\mathcal{F}_{i-1})\right|\right\|_p \leq \sqrt{2}a_p\left(\sum_{i=1}^{n} \|\mathbb{E}(X_i|\mathcal{F}_{i-1})\|_p^2\right)^{1/2}$$
$$+ \sqrt{2}a_p \sum_{j=1}^{r-1}\left(\sum_{k=1}^{2^{r-j}} \|\mathbb{E}(S_{k2^j} - S_{(k-1)2^j}|\mathcal{F}_{(k-2)2^j+1})\|_p^2\right)^{1/2},$$

which, since $n < 2^r$, gives

$$\left\|\max_{1\leq k\leq m}\left|\sum_{i=1}^{2k} \mathbb{E}(X_i|\mathcal{F}_{i-1})\right|\right\|_p$$
$$\leq \sqrt{2}a_p \sum_{j=0}^{r-1}\left(\sum_{k=1}^{2^{r-j}} \|\mathbb{E}(S_{k2^j} - S_{(k-1)2^j}|\mathcal{F}_{(k-2)2^j+1})\|_p^2\right)^{1/2}. \tag{3.30}$$

So, overall, starting from (3.24) and taking into account the upper bounds (3.28), (3.26), (3.29) and (3.30), inequality (3.9) follows, proving the induction hypothesis at step n. ■

Proof of Theorem 3.10 We start with the decomposition

$$Y_k = \sum_{j=-\infty}^{\infty} \mathbf{P}_{k-j}(Y_k). \tag{3.31}$$

Then

$$S_m = \sum_{j=-\infty}^{\infty}\sum_{k=1}^{m} g_k\mathbf{P}_{k-j}(Y_k) \text{ and } \max_{1\leq m\leq n}|S_m| \leq \sum_{j=-\infty}^{\infty} \max_{1\leq m\leq n}\left|\sum_{k=1}^{m} g_k\mathbf{P}_{k-j}(Y_k)\right|.$$

Thus,

$$\left\|\max_{1\leq k\leq n}|S_k|\right\|_p \leq \sum_{j=-\infty}^{\infty}\left\|\max_{1\leq m\leq n}\left|\sum_{k=1}^{m} g_k\mathbf{P}_{k-j}(Y_k)\right|\right\|_p.$$

Consider first the case $p \geq 2$. Since $\left(g_k \mathbf{P}_{k-j}(Y_k)\right)_{k\geq 1}$ is a sequence of martingale differences in $\mathbb{L}^p$, applying first Doob's maximal inequality (see Corollary 2.7) and after Theorem 2.13, we get

$$\| \max_{1\leq k\leq n} |S_k| \|_p \leq \frac{p}{p-1} c_p \sum_{j=-\infty}^{\infty} \left(\sum_{k=1}^{n} g_k^2 \|\mathbf{P}_{k-j}(Y_k)\|_p^2 \right)^{1/2},$$

where $c_p = (p-1)^{1/2}$. Then using that $\|\mathbf{P}_{k-j}(Y_k)\|_p \leq p_j$, we get the desired result.

For the case $p \in]1, 2[$, we use an appropriate part of the inequality given in Theorem 2.13.

This leads to

$$\begin{aligned} \| \max_{1\leq k\leq n} |S_k| \|_p &\leq 2^{2/p-1} \frac{p}{p-1} \sum_{j=-\infty}^{\infty} \left(\sum_{k=1}^{n} |g_k|^p \|\mathbf{P}_{k-j}(Y_k)\|_p^p \right)^{1/p} \\ &\leq 2^{2/p-1} \frac{p}{p-1} \sum_{j=-\infty}^{\infty} p_j \left(\sum_{k=1}^{n} |g_k|^p \right)^{1/p}. \end{aligned}$$

■

Proof of Theorem 3.11 The proof follows exactly the lines of Dedecker and Rio (2000). We give here the details for completeness. As in Garsia (1965):

$$(\tilde{S}_n - \lambda)_+^2 = \sum_{k=1}^{n} ((\tilde{S}_k - \lambda)_+^2 - (\tilde{S}_{k-1} - \lambda)_+^2). \tag{3.32}$$

Since the sequence $(\tilde{S}_k)_{k\geq 0}$ is non-decreasing, the summands in (3.32) are non-negative. Now

$$((\tilde{S}_k - \lambda)_+ - (\tilde{S}_{k-1} - \lambda)_+)((\tilde{S}_k - \lambda)_+ + (\tilde{S}_{k-1} - \lambda)_+) > 0$$

if and only if $S_k > \lambda$ and $S_k > \tilde{S}_{k-1}$. In that case $S_k = \tilde{S}_k$, whence

$$(\tilde{S}_k - \lambda)_+^2 - (\tilde{S}_{k-1} - \lambda)_+^2 \leq 2(S_k - \lambda)((\tilde{S}_k - \lambda)_+ - (\tilde{S}_{k-1} - \lambda)_+).$$

It follows that

$$\begin{aligned} (\tilde{S}_n - \lambda)_+^2 &\leq 2\sum_{k=1}^{n} (S_k - \lambda)(\tilde{S}_k - \lambda)_+ - 2\sum_{k=1}^{n} ((S_k - \lambda)(\tilde{S}_{k-1} - \lambda)_+) \\ &\leq 2(S_n - \lambda)_+(\tilde{S}_n - \lambda)_+ - 2\sum_{k=1}^{n} (\tilde{S}_{k-1} - \lambda)_+ X_k. \end{aligned}$$

Noting that

$$2(S_n - \lambda)_+(\tilde{S}_n - \lambda)_+ \leq \frac{1}{2}(\tilde{S}_n - \lambda)_+^2 + 2(S_n - \lambda)_+^2,$$

we infer that

$$(\tilde{S}_n - \lambda)_+^2 \le 4(S_n - \lambda)_+^2 - 4\sum_{k=1}^{n}(\tilde{S}_{k-1} - \lambda)_+ X_k. \tag{3.33}$$

In order to bound $(S_n - \lambda)_+^2$, we adapt the decomposition (3.32) and next we apply Taylor's formula:

$$\begin{aligned}(S_n - \lambda)_+^2 &= \sum_{k=1}^{n}((S_k - \lambda)_+^2 - (S_{k-1} - \lambda)_+^2)\\ &= 2\sum_{k=1}^{n}(S_{k-1} - \lambda)_+ X_k + 2\sum_{k=1}^{n} X_k^2 \int_0^1 (1-t)1_{S_{k-1}+tX_k>\lambda}\,dt.\end{aligned}$$

Since $1_{S_{k-1}+tX_k>\lambda} \le 1_{S_k^*>\lambda}$, it follows that

$$(S_n - \lambda)_+^2 \le 2\sum_{k=1}^{n}(S_{k-1} - \lambda)_+ X_k + \sum_{k=1}^{n} X_k^2 1_{S_k^*>\lambda}. \tag{3.34}$$

Hence, by (3.33) and (3.34),

$$(S_n^* - \lambda)_+^2 \le 4\sum_{k=1}^{n}\Big(2(S_{k-1} - \lambda)_+ - \big(S_{k-1}^* - \lambda\big)_+\Big)X_k + 4\sum_{k=1}^{n} X_k^2 1_{S_k^*>\lambda}.$$

Let $D_0 = 0$ and $D_k = 2(S_k - \lambda)_+ - (\tilde{S}_k - \lambda)_+$ for $k > 0$. Clearly

$$D_{k-1}X_k = \sum_{i=1}^{k-1}(D_i - D_{i-1})X_k.$$

Hence

$$(\tilde{S}_n - \lambda)_+^2 \le 4\sum_{i=1}^{n-1}(D_i - D_{i-1})(S_n - S_i) + 4\sum_{k=1}^{n} X_k^2 1_{\tilde{S}_k>\lambda}. \tag{3.35}$$

Since the random variables $D_i - D_{i-1}$ are $\mathcal{F}_i$-measurable, we have:

$$\begin{aligned}\mathbb{E}((D_i - D_{i-1})(S_n - S_i)) &= \mathbb{E}((D_i - D_{i-1})\mathbb{E}(S_n - S_i \mid \mathcal{F}_i))\\ &\le \mathbb{E}|(D_i - D_{i-1})\mathbb{E}(S_n - S_i \mid \mathcal{F}_i)|.\end{aligned} \tag{3.36}$$

It remains to bound $|D_i - D_{i-1}|$. If $(\tilde{S}_i - \lambda)_+ = (\tilde{S}_{i-1} - \lambda)_+$, then

$$|D_i - D_{i-1}| = 2|(S_i - \lambda)_+ - (S_{i-1} - \lambda)_+| \le 2|X_i|1_{\tilde{S}_i>\lambda},$$

because $D_i - D_{i-1} = 0$ whenever $S_i \leq \lambda$ and $S_{i-1} \leq \lambda$. Otherwise $S_i = \tilde{S}_i > \lambda$ and $S_{i-1} \leq \tilde{S}_{i-1} < S_i$, which implies that

$$D_i - D_{i-1} = (S_i - \lambda)_+ + (\tilde{S}_{i-1} - \lambda)_+ - 2(S_{i-1} - \lambda)_+.$$

Hence $D_i - D_{i-1}$ belongs to $[0, 2((S_i - \lambda)_+ - (S_{i-1} - \lambda)_+)\,]$. In any case

$$|D_i - D_{i-1}| \leq 2|X_i| 1_{\tilde{S}_i > \lambda},$$

which together with (3.35) and (3.36) implies Theorem 3.11. ∎

Proof of Corollary 3.12 Theorem 3.11 applied to the sequences $(d_i)_{i\in\mathbb{Z}}$ and $(-d_i)_{i\in\mathbb{Z}}$ gives, for any non-negative real λ,

$$\mathbb{E}\left(\left(n^{-1/2}M_n^* - \lambda\right)_+^2\right) \leq 8n^{-1}\sum_{k=1}^{n}\mathbb{E}\left(d_k^2 \mathbf{1}_{\{M_n^* > \lambda\sqrt{n}\}}\right). \tag{3.37}$$

Hence, for any $A > 0$,

$$\mathbb{E}\left(\left(n^{-1/2}M_n^* - \lambda\right)_+^2\right) \leq 8n^{-1}\sum_{k=1}^{n}\mathbb{E}\left(d_k^2 \mathbf{1}_{\{|d_k| > A\}}\right) + 8A^2\mathbb{P}\left(M_n^* > \lambda\sqrt{n}\right).$$

Hence, using stationarity and inequality (3.37) with $\lambda = 0$, we get

$$\sup_{n\geq 1}\mathbb{E}\left(\left(n^{-1/2}M_n^* - \lambda\right)_+^2\right) \leq 8\mathbb{E}\left(d_1^2 \mathbf{1}_{\{|d_1| > A\}}\right) + 8^2\frac{A^2}{\lambda^2},$$

which proves the desired result by letting λ tend to infinity and then A tend to infinity. ∎

Proof of Theorem 3.13 The proof follows the lines of Theorem 2.5 in Rio (2000, pages 46-47) and is done by induction. Our induction hypothesis at step $n-1$ is the following: For any $k \leq n-1$,

$$h_k(t) \leq K^p\left(\sum_{i=1}^{k-1} b_{i,k} + t b_{k,k}\right)^{p/2} \quad \text{where } K = \sqrt{2(p-1)}. \tag{3.38}$$

Since $K \geq 1$, the above inequality is clearly true for $k = 1$. Assume now that the induction hypothesis holds at range $n-1$ and that the inequality below is true

$$h_n(t) \leq p(p-1)\left(\sum_{k=1}^{n-1} b_{k,n}\int_0^1 (h_k(s))^{1-2/p}\,ds + b_{n,n}\int_0^t (h_n(s))^{1-2/p}\,ds\right); \tag{3.39}$$

then, using our induction hypothesis, it follows that

$$h_n(t) \le p(p-1)\left(\sum_{k=1}^{n-1} b_{k,n} \int_0^1 K^{p-2}\left(\sum_{i=1}^{k-1} b_{i,k} + s b_{k,k}\right)^{(p-2)/2} ds \right.$$
$$\left. + b_{n,n}\int_0^t (h_n(s))^{1-2/p}\, ds\right)$$
$$\le p(p-1)\left(K^{p-2}\sum_{k=1}^{n-1} b_{k,n} \int_0^1 \left(\sum_{i=1}^{k-1} b_{i,n} + s b_{k,n}\right)^{(p-2)/2} ds \right.$$
$$\left. + b_{n,n}\int_0^t (h_n(s))^{1-2/p}\, ds\right).$$

Integrating with respect to s, we get

$$b_{k,n}\int_0^1 \left(\sum_{i=1}^{k-1} b_{i,n} + s b_{k,n}\right)^{(p-2)/2} ds = \frac{2}{p}\left(\sum_{i=1}^{k} b_{i,n}\right)^{p/2} - \frac{2}{p}\left(\sum_{i=1}^{k-1} b_{i,n}\right)^{p/2},$$

implying that

$$\sum_{k=1}^{n-1} b_{k,n}\int_0^1 \left(\sum_{i=1}^{k-1} b_{i,n} + s b_{k,n}\right)^{(p-2)/2} ds = 2p^{-1}\left(\sum_{i=1}^{n-1} b_{i,n}\right)^{p/2}.$$

Therefore, since $K^2 = 2(p-1)$,

$$h_n(t) \le K^p\left(\sum_{i=1}^{n-1} b_{i,n}\right)^{p/2} + p(p-1)b_{n,n}\int_0^t (h_n(s))^{1-2/p}\, ds. \tag{3.40}$$

Let $H_n(t) = \int_0^t (h_n(s))^{1-2/p}\, ds$. The differential integral inequation (3.40) becomes

$$H_n'(s)\left(K^p\left(\sum_{i=1}^{n-1} b_{i,n}\right)^{p/2} + p(p-1)b_{n,n}H(s)\right)^{-1+2/p} \le 1.$$

Setting

$$R_n(s) = \left(K^p\left(\sum_{i=1}^{n-1} b_{i,n}\right)^{p/2} + p(p-1)b_{n,n}H(s)\right)^{2/p},$$

the previous inequality can be rewritten as

$$R_n'(s) \leq 2p^{-1}p(p-1)b_{n,n}.$$

Integrating between 0 and t, we derive

$$(h_n(t))^{2/p} - K^2 \sum_{i=1}^{n-1} b_{i,n} \leq R_n(t) - R_n(0) \leq 2tp^{-1}p(p-1)b_{n,n}.$$

Taking into account that $K^2 = 2(p-1)$, it follows that

$$(h_n(t))^{2/p} \leq K^2 \left(\sum_{i=1}^{n-1} b_{i,n} + tb_{n,n} \right),$$

showing that our induction hypothesis holds true at step n. To end the proof it suffices to prove (3.39). Setting $G(i,n,t) = X_i(t\mathbb{E}(X_n|\mathcal{F}_i) + \sum_{k=i}^{n-1} \mathbb{E}(X_k|\mathcal{F}_i))$ and applying Theorem 2.3 in Rio (2000) with $\psi(x) = |x|^p$, we get

$$\frac{h_n(t)}{p(p-1)} \leq \sum_{i=1}^{n-1} \int_0^1 \mathbb{E}\left(|S_{i-1} + sX_i|^{p-2} G(i,n,t)\right) ds + \int_0^t \mathbb{E}(|S_{n-1} + sX_n|^{p-2} X_n^2) ds. \tag{3.41}$$

Note that the function $t \to \mathbb{E}(|G(i,n,t)|^{p/2})$ is convex, so that for any t in $[0, 1]$, $\mathbb{E}(|G(i,n,t)|^{p/2}) \leq \mathbb{E}(|G(i,n,0)|^{p/2}) \vee \mathbb{E}(|G(i,n,1)|^{p/2}) \leq b_{i,n}^{p/2}$. Applying Hölder's inequality, we obtain

$$\mathbb{E}\left(|S_{i-1} + sX_i|^{p-2} G(i,n,t)\right) \leq (h_i(s))^{(p-2)/p} \, \|G(i,n,t)\|_{p/2} \leq (h_i(s))^{(p-2)/p} \, b_{i,n},$$

which ends the proof of (3.39) and then of the theorem. ■

Proof of Proposition 3.14 Denote $S_{2^r}^* = \max_{1 \leq i \leq 2^r} |S_i|$. For any $m \in [0, 2^r - 1]$, we have that

$$S_{2^r - m} = \mathbb{E}(S_{2^r}|\mathcal{F}_{2^r - m}) - \mathbb{E}(S_{2^r} - S_{2^r - m}|\mathcal{F}_{2^r - m}).$$

So,

$$S_{2^r}^* \leq \max_{0 \leq m \leq 2^r - 1} |\mathbb{E}(S_{2^r}|\mathcal{F}_{2^r - m})| + \max_{0 \leq m \leq 2^r - 1} |\mathbb{E}(S_{2^r} - S_{2^r - m}|\mathcal{F}_{2^r - m})|. \tag{3.42}$$

Since $(\mathbb{E}(S_{2^r}|\mathcal{F}_k))_{k \geq 1}$ is a martingale, we shall use Doob's maximal inequality (see Corollary 2.7) to deal with the first term in the right-hand side of (3.42). Hence

$$\| \max_{0 \leq m \leq 2^r - 1} |\mathbb{E}(S_{2^r}|\mathcal{F}_{2^r - m})| \|_p \leq q \|S_{2^r}\|_p. \tag{3.43}$$

Write now m in basis 2 as follows:

$$m = \sum_{i=0}^{r-1} b_i(m)2^i, \quad \text{where } b_i(m) = 0 \text{ or } b_i(m) = 1.$$

Set $m_l = \sum_{i=l}^{r-1} b_i(m)2^i$ and notice that

$$\begin{aligned}|\mathbb{E}(S_{2^r} - S_{2^r-m}|\mathcal{F}_{2^r-m})| &\le \sum_{l=0}^{r-1} |\mathbb{E}(S_{2^r-m_{l+1}} - S_{2^r-m_l}|\mathcal{F}_{2^r-m})| \\ &\le \sum_{l=0}^{r-1} \mathbb{E}\left(|\mathbb{E}(S_{2^r-m_{l+1}} - S_{2^r-m_l}|\mathcal{F}_{2^r-m_l})||\mathcal{F}_{2^r-m}\right).\end{aligned}$$

Now $m_l \neq m_{l+1}$ only if $b_l(m) = 1$, and in that case $m_l = k_{m,l}2^l$ with $k_{m,l}$ odd. It follows that

$$\begin{aligned}|\mathbb{E}(S_{2^r-m_{l+1}} - S_{2^r-m_l}|\mathcal{F}_{2^r-m_l})| &\le \max_{1\le k\le 2^{r-l}, k \text{ odd}} |\mathbb{E}(S_{2^r-(k-1)2^l} - S_{2^r-k2^l}|\mathcal{F}_{2^r-k2^l})| \\ &:= A_{r,l}.\end{aligned}$$

Therefore, using the fact that if $|X| \le |Y|$ then $\mathbb{E}(|X|\,|\mathcal{F}) \le \mathbb{E}(|Y|\,|\mathcal{F})$, we get

$$\|\max_{0\le m\le 2^r-1} |\mathbb{E}(S_{2^r} - S_{2^r-m}|\mathcal{F}_{2^r-m})|\|_p \le \sum_{l=0}^{r-1} \|\max_{0\le m\le 2^r-1} \mathbb{E}(A_{r,l}|\mathcal{F}_{2^r-m})\|_p$$

Notice that $(\mathbb{E}(A_{r,l}|\mathcal{F}_k))_{k\ge1}$ is a martingale, so by Doob's maximal inequality again, we obtain

$$\|\max_{0\le m\le 2^r-1} \mathbb{E}(A_{r,l}|\mathcal{F}_{2^r-m})\|_p \le q\|\mathbb{E}(A_{r,l}|\mathcal{F}_{2^r})\|_p \le q\|A_{r,l}\|_p.$$

Since

$$A_{r,l} \le \left(\sum_{k=1}^{2^{r-l}-1} |\mathbb{E}(S_{(k+1)2^l} - S_{k2^l}|\mathcal{F}_{k2^l})|^p\right)^{1/p},$$

we get overall

$$\begin{aligned}&\|\max_{0\le m\le 2^r-1} |\mathbb{E}(S_{2^r} - S_{2^r-m}|\mathcal{F}_{2^r-m})|\|_p \\ &\le q\sum_{l=0}^{r-1}\left(\sum_{k=1}^{2^{r-l}-1} \|\mathbb{E}(S_{(k+1)2^l} - S_{k2^l}|\mathcal{F}_{k2^l})\|_p^p\right)^{1/p}.\end{aligned} \tag{3.44}$$

To end the proof of Proposition 3.14, we start from (3.42) and consider the bounds (3.43) and (3.44). ■

Proof of Theorem 3.16 By using Theorem 3.13 combined with stationarity and the fact that the sequence is bounded, we get for all $m \geq 1$

$$\|S_m\|_p \leq (2(p-1))^{1/2} m^{1/2} T^{1/2} \left(\|X_0\|_{p/2}^{1/2} + \max_{1 \leq \ell \leq m} \|\mathbb{E}(S_\ell \,|\mathcal{F}_0)\|_{p/2}^{1/2} \right). \tag{3.45}$$

On the other hand

$$\sum_{j=1}^{n} j^{-1-1/p} \|\mathbb{E}(S_j|\mathcal{F}_0)\|_p \leq T^{1/2} \sum_{j=1}^{n} j^{-1/2-1/p} \|\mathbb{E}(S_j|\mathcal{F}_0)\|_{p/2}^{1/2}.$$

Therefore, since $p > 2$,

$$\sum_{j=1}^{n} j^{-1-1/p} \|\mathbb{E}(S_j|\mathcal{F}_0)\|_p \leq T^{1/2} n^{1/2-1/p} \sup_{1 \leq \ell \leq n} \|\mathbb{E}(S_\ell \,|\mathcal{F}_0)\|_{p/2}^{1/2} \left(1 + \frac{2p}{p-2}\right). \tag{3.46}$$

Combining (3.45) with inequality (3.16) and taking into account (3.46), we get the desired result. ■

3.5.2 *Proof of Theorem 3.17*

We shall prove inequality (3.17) in a slightly different form which, by an application of Lemma 3.24, will give the desired inequality: For any positive integer r and any integer n such that $2^{r-1} \leq n < 2^r$, (δ and c as above),

$$\mathbb{E}\left(\max_{1 \leq j \leq n} |S_j|^p\right) \ll n\mathbb{E}(|X_1|)^p + cn \left(\sum_{k=0}^{r-1} 2^{-k/p} \|\mathbb{E}_0(S_{2^k})\|_p \right)^p$$
$$+ n \left(\sum_{k=0}^{r-1} 2^{-2k\delta/p} \|\mathbb{E}_0(S_{2^k}^2)\|_{p/2}^{\delta} \right)^{p/(2\delta)}. \tag{3.47}$$

Indeed, by stationarity, for any positive integers i and j,

$$\|\mathbb{E}_0(S_{i+j})\|_p \leq \|\mathbb{E}_0(S_i)\|_p + \|\mathbb{E}_0(S_j)\|_p,$$

and also, for any $0 < \delta \leq 1$,

$$\|\mathbb{E}_0(S_{i+j}^2)\|_{p/2}^{\delta} \leq 2^{\delta} \|\mathbb{E}_0(S_i^2)\|_{p/2}^{\delta} + 2^{\delta} \|\mathbb{E}_0(S_j^2)\|_{p/2}^{\delta}.$$

Using Item 1 of Lemma 3.24, it follows that

$$\sum_{k=0}^{r-1} 2^{-k/p} \|\mathbb{E}_0(S_{2^k})\|_p \ll \sum_{k=1}^{n} k^{-1-1/p} \|\mathbb{E}_0(S_k)\|_p,$$

and

$$\sum_{k=0}^{r-1} 2^{-2k\delta/p} \|\mathbb{E}_0(S_{2^k}^2)\|_{p/2}^{\delta} \ll \sum_{k=1}^{n} k^{-1-2\delta/p} \|\mathbb{E}_0(S_k^2)\|_{p/2}^{\delta}. \tag{3.48}$$

We turn now to the proof of inequality (3.47). It is based on dyadic induction and involves several steps. With the notation $a_n = \|S_n\|_p$, the first step is to establish the following recurrence formula: For any real $p > 2$ and any positive integer r,

$$a_{2n}^p \le 2a_n^p + 2c_1 a_n^{p-1} \|\mathbb{E}_0(S_n)\|_p + 2c_2 a_n^{p-2\delta} \|\mathbb{E}_0(S_n^2)\|_{p/2}^{\delta}, \tag{3.49}$$

where c_1 and c_2 are positive constants depending only on p. Assume for the moment that the recurrence formula (3.49) is true. Let us then show how it implies (3.47).

Lemma 3.22 *Assume that for some $0 < \delta \le 1$ the recurrence formula (3.49) holds. Then the inequality (3.47) (and therefore also (3.17)) holds with the same δ.*

To prove the lemma we start from inequality (3.49) and apply a recurrence on the first term. We then obtain that, for any positive integer r,

$$a_{2^r}^p \le 2^r \left(a_{2^0}^p + c_1 \sum_{k=0}^{r-1} 2^{-k} a_{2^k}^{p-1} \|\mathbb{E}_0(S_{2^k})\|_p + c_2 \sum_{k=0}^{r-1} 2^{-k} a_{2^k}^{p-2\delta} \|\mathbb{E}_0(S_{2^k}^2)\|_{p/2}^{\delta} \right). \tag{3.50}$$

To establish the inequality (3.47) we first note that, due to the maximal inequality (3.15), it suffices to prove that the inequality is satisfied for $\max_{1\le j\le n} \mathbb{E}(|S_j|^p)$ instead of $\mathbb{E}\left(\max_{1\le j\le n} |S_j|^p\right)$. Therefore the goal is to establish that for any positive integer r and any integer n such that $2^{r-1} \le n < 2^r$,

$$\max_{1\le j\le n} \mathbb{E}\left(|S_j|^p\right) \ll n\mathbb{E}(|X_1|)^p + cn \left(\sum_{k=0}^{r-1} 2^{-k/p} \|\mathbb{E}_0(S_{2^k})\|_p \right)^p$$
$$+ n \left(\sum_{k=0}^{r-1} 2^{-2k\delta/p} \|\mathbb{E}_0(S_{2^k}^2)\|_{p/2}^{\delta} \right)^{p/(2\delta)}. \tag{3.51}$$

Starting from (3.50) and setting

$$B_r = \max_{0\le k\le r} (a_{2^k}^p / 2^k), \tag{3.52}$$

we get

$$B_r \le a_{2^0}^p + c_1 B_r^{1-1/p} \sum_{k=0}^{r-1} 2^{-k/p} \|\mathbb{E}_0(S_{2^k})\|_p + c_2 B_r^{1-2\delta/p} \sum_{k=0}^{r-1} 2^{-2k\delta/p} \|\mathbb{E}_0(S_{2^k}^2)\|_{p/2}^{\delta}.$$

Therefore, taking into account that either $B_r \le 3a_{2^0}^p$ or $B_r^{1/p} \le 3c_1 \sum_{k=0}^{r-1} 2^{-k/p} \|\mathbb{E}_0(S_{2^k})\|_p$ or $B_r^{2\delta/p} \le 3c_2 \sum_{k=0}^{r-1} 2^{-2k\delta/p} \|\mathbb{E}_0(S_{2^k}^2)\|_{p/2}^{\delta}$, we derive that

$$a_{2^r}^p \le 2^r \left(3a_{2^0}^p + \left(3c_1 \sum_{k=0}^{r-1} 2^{-k/p} \|\mathbb{E}_0(S_{2^k})\|_p \right)^p + \left(3c_2 \sum_{k=0}^{r-1} 2^{-2k\delta/p} \|\mathbb{E}_0(S_{2^k}^2)\|_{p/2}^{\delta} \right)^{p/2\delta} \right). \tag{3.53}$$

Let now $2^{r-1} \le n < 2^r$ and write its binary expansion as in (3.4). Then, using (3.5) and stationarity, it follows that

$$\|S_n\|_p \le \sum_{k=0}^{r-1} \|T_{2^k}\|_p \le \sum_{k=0}^{r-1} \|S_{2^k}\|_p,$$

where r is such that $2^{r-1} \le n < 2^r$. Then, by using (3.53) and the fact that $\sum_{k=0}^{r-1} 2^{k/p} \le 2^{r/p}/(2^{1/p} - 1)$, we derive the inequality (3.51) for $\mathbb{E}(|S_n|^p)$ and also for $\max_{1 \le j \le n} \mathbb{E}\left(|S_j|^p\right)$. As we indicate before, the inequality (3.47) follows now by the maximal inequality (3.15). This ends the proof of the lemma. ■

End of the proof of Theorem 3.17. The proof will be complete if we prove the recurrence formula (3.49). To establish it, we divide the proof in three cases according to the values of p. Denote $\bar{S}_n = X_{n+1} + \cdots + X_{2n}$. Consider first the case $p \ge 4$. With this aim we shall apply the following inequality: For any positive reals a and b and any real $q \ge 1$,

$$(a+b)^q \le a^q + b^q + 2^{q+1} a b^{q-1} + 2^{q+1} a^{q-1} b. \tag{3.54}$$

To prove the inequality we first assume $b/a \le 1$. Applying the Newton binomial formula we get

$$(a+b)^q \le a^q (1 + b/a)^{[q]+1} \le a^q \left(1 + \sum_{k=1}^{[q]+1} C_{[q]+1}^k \frac{b}{a} \right).$$

Hence, if $b/a \le 1$, $(a+b)^q \le a^q + 2^{q+1} a^{q-1} b$. Obviously, permuting the roles of a and b, we get, when $a/b \le 1$, that $(a+b)^q \le b^q + 2^{q+1} b^{q-1} a$. Adding the two last inequalities, (3.54) follows. Then, using inequality (3.54) with $q = p$, $a = |S_n|$, $b = |\bar{S}_n|$, taking the expectation and using stationarity, we obtain

$$a_{2n}^p \le 2a_n^p + 2^{p+1} \mathbb{E}\left(|S_n|^{p-1} |\bar{S}_n| + |\bar{S}_n|^{p-1} |S_n| \right).$$

Now set $x = p/2 - 1$. Note that since $p > 3$, $2x > 1$ and $p - 1 - 1/x > 0$. Therefore Hölder's inequality gives

$$\begin{aligned}\mathbb{E}\left(|S_n||\bar{S}_n|^{p-1}\right) &\le \| S_n|\bar{S}_n|^{1/x}\|_{2x}\| |\bar{S}_n|^{p-1-1/x}\|_{2x/(2x-1)} \\ &= \| S_n|\bar{S}_n|^{2/(p-2)}\|_{p-2} \left(\mathbb{E}(|\bar{S}_n|^p)\right)^{1-1/(p-2)} .\end{aligned}$$

Next, writing

$$\| S_n|\bar{S}_n|^{2/(p-2)}\|_{p-2} = \left(\mathbb{E}\left(|S_n|^{p-2}\mathbb{E}_n(\bar{S}_n^2)\right)\right)^{1/(p-2)},$$

we get by Hölder's inequality that

$$\| S_n|\bar{S}_n|^{2/(p-2)}\|_{p-2} \le \|S_n\|_p \|\mathbb{E}_n(\bar{S}_n^2)\|_{p/2}^{1/(p-2)} .$$

So, overall, taking into account the stationarity assumption, it follows that, for any $p > 3$,

$$\mathbb{E}\left(|S_n||\bar{S}_n|^{p-1}\right) \le a_n^{p-2/(p-2)} \|\mathbb{E}_0(S_n^2)\|_{p/2}^{1/(p-2)} . \tag{3.55}$$

On the other hand, using Hölder's inequality combined with stationarity, we infer that

$$\mathbb{E}(|S_n|^{p-1}|\bar{S}_n|) = \mathbb{E}(|S_n|^{p-1}\mathbb{E}_n^{1/2}(\bar{S}_n^2)) \le a_n^{p-1} \|\mathbb{E}_0(S_n^2)\|_{p/2}^{1/2}, \tag{3.56}$$

leading, since $p \ge 4$, to

$$\mathbb{E}(|S_n|^{p-1}|\bar{S}_n|) \le a_n^{p-2/(p-2)} \|\mathbb{E}_0(S_n^2)\|_{p/2}^{1/(p-2)} .$$

Therefore all these considerations show that when $p \ge 4$,

$$a_{2n}^p \le 2a_n^p + 2^{p+2} a_n^{p-2/(p-2)} \|\mathbb{E}_0(S_n^2)\|_{p/2}^{1/(p-2)},$$

establishing the recurrence formula (3.49) with $\delta = 1/(p-2)$, $c_1 = 0$, and $c_2 = 2^{p+1}$.

Assume now that $p \in]3, 4[$. Because of (3.56), an application of (3.54) will lead to the recurrence formula but with $\delta = 1/2$ rather than $1/(p-2)$. Therefore we shall instead use the following inequality: Let x and y be two real numbers and $3 < p \le 4$. Then

$$|x+y|^p \le |x|^p + |y|^p + p|x|^{p-1}\mathrm{sign}(x)y + \frac{p(p-1)}{2}|x|^{p-2}y^2 + \frac{2p}{(p-2)}|x||y|^{p-1} . \tag{3.57}$$

The proof of (3.57) is based on the Taylor integral formula of order 2 that gives

$$|x+y|^p - |x|^p = p|x|^{p-1}\mathrm{sign}(x)y + C_p^2|x|^{p-2}y^2 + 2C_p^2 y^2 \int_0^1 (1-t)(|x+ty|^{p-2} - |x|^{p-2})dt, \tag{3.58}$$

where $C_p^2 = p(p-1)/2$. Notice now that, for $3 < p \le 4$,

$$|x+ty|^{p-2} \le \frac{x^2 + 2|x||ty| + y^2}{(|x| + |ty|)^{4-p}} \le |x|^{p-2} + 2|x||ty|^{p-3} + |ty|^{p-2}.$$

Hence

$$\begin{aligned} &2C_p^2 y^2 \int_0^1 (1-t)(|x+ty|^{p-2} - |x|^{p-2})dt \\ &\le 2C_p^2 |y|^p \int_0^1 (1-t)t^{p-2}dt + 4C_p^2|x||y|^{p-1}\int_0^1 (1-t)t^{p-3}dt \\ &= 2|y|^p C_p^2 \frac{\Gamma(p-1)}{\Gamma(p+1)} + 4|x||y|^{p-1} C_p^2 \frac{\Gamma(p-2)}{\Gamma(p)} = |y|^p + \frac{2p}{(p-2)}|x||y|^{p-1}. \end{aligned} \tag{3.59}$$

Starting from (3.58) and using (3.59), the inequality (3.57) follows. Hence using it with $x = S_n$ and $y = \bar{S}_n$, taking the expectation and using Hölder's inequality and (3.55), we get by stationarity that for any positive integer r,

$$\begin{aligned} a_{2n}^p \le &2a_n^p + pa_n^{p-1}\|\mathbb{E}_0(S_n)\|_p + C_p^2 a_n^{p(1-2/p)}\|\mathbb{E}_0(S_n^2)\|_{p/2} \\ &+ 2p(p-2)^{-1}a_n^{p-2/(p-2)}\|\mathbb{E}_0(S_n^2)\|_{p/2}^{1/(p-2)}. \end{aligned}$$

Since $p > 3$, $\|\mathbb{E}_0(S_n^2)\|_{p/2} \le a_n^{2(p-3)/(p-2)}\|\mathbb{E}_0(S_n^2)\|_{p/2}^{1/(p-2)}$. It follows that

$$a_{2n}^p \le 2a_n^p + pa_n^{p-1}\|\mathbb{E}_0(S_n)\|_p + 4pa_n^{p-2/(p-2)}\|\mathbb{E}_0(S_n^2)\|_{p/2}^{1/(p-2)}.$$

Therefore (3.49) holds with $\delta = 1/(p-2)$, $c_1 = 2^{-1}p$, and $c_2 = 2p$.

When $2 < p \le 3$, (3.55) is not valid anymore, and we shall use the following inequality: Let x and y be two real numbers and $2 \le p \le 3$. Then

$$|x+y|^p \le |x|^p + |y|^p + p|x|^{p-1}\text{sign}(x)y + \frac{p(p-1)}{2}|x|^{p-2}y^2. \tag{3.60}$$

The above inequality comes from a Taylor expansion with integral rest (see Relation (3.3) in Rio (2009) for a complete proof). Applying it with $x = S_n$ and $y = \bar{S}_n$, taking the expectation and using stationarity and properties of the conditional expectation, we obtain

$$\mathbb{E}(|S_{2n}|^p) \le 2\mathbb{E}(|S_n|^p) + p\mathbb{E}(|S_n|^{p-1}\text{sign}(S_n)\mathbb{E}_n(\bar{S}_n)) + C_p^2\mathbb{E}(|S_n|^{p-2}\mathbb{E}_n(\bar{S}_n^2)),$$

where $C_p^2 = p(p-1)/2$. This inequality combined with Hölder's inequality gives

$$a_{2n}^p \le 2a_n^p + pa_n^{p-1}\|\mathbb{E}_0(S_n)\|_p + C_p^2 a_n^{p(1-2/p)}\|\mathbb{E}_0(S_n^2)\|_{p/2}$$

and therefore the recurrence formula (3.49) holds with $\delta = 1$, $c_1 = 2^{-1}p$, and $c_2 = 2^{-1}C_p^2$. This ends the proof of the recurrence formula for any $p > 2$ and then of the theorem. ■

3.5.3 Proofs of the Results of Section 3.4

Proof of Theorem 3.19 We are going to follow the proof of the $\mathbb{L}_p$ inequality from Theorem 3.6 for the variables in $\mathbb{L}_\infty$ and we shall point only the differences. Instead of Burkholder-type bound (3.6) for $\mathbb{L}_p$ moments of martingales with differences in $\mathbb{L}_p$ we shall use the following nearly optimal bound for martingales with differences in $\mathbb{L}_\infty$

$$\mathbb{E}\left[\max_{1\le k\le n}|Z_1+\ldots+Z_k|^p\right]\le p^p(p-1)^{-p}\mathbb{E}[|Z_1+\ldots+Z_n|^p]$$
$$\le p^p(p-1)^{-p}(\mathbb{E}|\eta|^p)n^{p/2}||Z_1||_\infty^p, \tag{3.61}$$

valid for all $p\ge 2$, integers n and stationary martingale differences $\{Z_k\}$, where η is a standard normal variable. The first part is the Doob maximal inequality (see Corollary 2.7). The second part follows from the Eaton–Hoeffding argument (Eaton, 1974, p 612-613) applied together with the Bentkus martingale generalization (2004, Lemma 4.3) of the Hunt comparison (1955) which is summarized in the following:

Example 3.23 *Let $\{Z_i\}$ be a martingale difference sequence. Then,*

$$\mathbb{E}\left|\sum_{j=1}^n Z_j\right|^p\le(\mathbb{E}|\eta|^p)\left(\sum_{j=1}^n||Z_j||_\infty^2\right)^{p/2}.$$

From (3.61), by using the approach of Theorem 3.6, since $p^p(p-1)^{-p}\le 4$, for $p\ge 2$, we derive the inequality

$$||S_n^*||_p\le 4^{1/p}||\eta||_p\sqrt{n}(||X_1||_\infty+12\delta_{n,\infty}).$$

To obtain the exponential inequality it remains to combine this bound with the last part of Theorem 2.4 in Rio (2000). ∎

Proof of Theorem 3.20 Starting with the decomposition (3.31), we write

$$Y_k=\sum_{j=-\infty}^{\infty}b_j\mathbf{P}_{k-j}(Y_k)/b_j,$$

where $b_j=p_j/D\ge||\mathbf{P}_{k-j}(Y_k)||_\infty/D$, for any $j\in\mathbb{Z}$. This gives

$$S_m=\sum_{j=-\infty}^{\infty}b_j\sum_{k=1}^m g_k\mathbf{P}_{k-j}(Y_k)/b_j.$$

Thus,

$$\max_{1\le m\le n}S_m\le\sum_{j=-\infty}^{\infty}b_j\max_{1\le m\le n}\sum_{k=1}^m\mathbf{P}_{k-j}(g_kY_k)/b_j=\sum_{j=-\infty}^{\infty}b_jM_n^{(j)}$$

where $M_n^{(j)}$ denotes $\max_{1\le m\le n}\sum_{k=1}^m g_k\mathbf{P}_{k-j}(Y_k)/b_j$.

Since $\exp(x)$ is convex and non–decreasing and $b_j \geq 0$ with $\sum_{j\in\mathbf{Z}} b_j = 1$,

$$\mathbb{E}\exp\left(t \max_{1\leq m\leq n} S_m\right) \leq \mathbb{E}\exp\left(\sum_{j=-\infty}^{\infty} b_j t M_n^{(j)}\right) \leq \sum_{j=-\infty}^{\infty} b_j \mathbb{E}\exp(tM_n^{(j)}).$$

Consider the martingale difference $U_k = g_k \mathbf{P}_{k-j}(Y_k)/b_j, j = 1, ..., n$. Since the variables $Z_k = \exp(t(U_1 + \cdots + U_k)/2)$ form a submartingale, Doob's inequality yields

$$\mathbb{E}\exp(tM_n^{(j)}) = \mathbb{E}\left(\max_{1\leq k\leq n} Z_k^2\right) \leq 4\mathbb{E}Z_n^2 = 4\mathbb{E}\exp(t(U_1 + \cdots + U_n)).$$

Applying Azuma's inequality (2.12) to the right-hand side, and noting that

$$||U_k||_\infty = |g_k|||\mathbf{P}_{k-j}(Y_k)||_\infty / b_j \leq |g_k| D,$$

we infer that

$$\mathbb{E}\exp(tM_n^{(j)}) \leq 4\exp(\frac{1}{2}G_n^2 D^2 t^2).$$

Since $\sum_{j\in\mathbf{Z}} b_j = 1$, we obtain that

$$\mathbb{E}\exp\left(t \max_{1\leq m\leq n} S_m\right) \leq 4\sum_{j\in\mathbf{Z}} b_j \exp\left(\frac{1}{2}G_n^2 D^2 t^2\right) = 4\exp\left(\frac{1}{2}G_n^2 D^2 t^2\right).$$

Next, to derive the one-sided probability inequality, we use the exponential bound with $t = x/(G_n^2 D^2)$, so

$$\mathbb{P}(\max_{1\leq m\leq n} S_m \geq x) \leq \mathbb{E}\exp(tM_n)\exp(-tx) = 4\exp\left(-\frac{x^2}{2G_n^2 D^2}\right).$$

Finally, to derive the two-sided inequality we observe that the stationary sequence $\{-Y_j\}$ also satisfies the conditions of the theorem. The proof is complete. ■

3.6 Facts about Subadditive Sequences

The next lemma deals with properties of subadditive sequences.

Lemma 3.24 *Let $(V_i)_{i\geq 0}$ be a sequence of nonnegative numbers such that $V_0 = 0$ and for all $i, j \geq 0$,*

$$V_{i+j} \leq C(V_i + V_j), \tag{3.62}$$

where $C \geq 1$ is a constant not depending on i and j. Then

1. *For any integer $r \geq 1$, any integer n satisfying $2^{r-1} \leq n < 2^r$ and any real $q \geq 0$*

$$\sum_{i=0}^{r-1} \frac{1}{2^{iq}} V_{2^i} \leq C 2^{q+2}(2^{q+1}-1)^{-1} \sum_{k=1}^{n} \frac{V_k}{k^{1+q}}.$$

2. *For any positive integers k and m and any real $q > 0$,*

$$\sum_{j=1}^{k} \frac{1}{j^q} V_{jm} \leq 2^{q+1} C q^{-1} m^{q-1} \sum_{\ell=1}^{m} \frac{1}{(\ell+m)^q} V_\ell + 2Cq^{-1} m^{q-1} \sum_{\ell=m+1}^{km} \frac{1}{\ell^q} V_\ell.$$

3. *Let $0 < \delta \leq \gamma \leq 1$. Then for any real $q \geq 0$,*

$$\left(\sum_{k=1}^{n} \frac{1}{k^{1+q\gamma}} V_k^\gamma \right)^{1/\gamma} \leq 2^{1/\delta - 1/\gamma} C^{(\gamma-\delta)/\delta} \left(\sum_{k=1}^{n} \frac{1}{k^{1+q\delta}} V_k^\delta \right)^{1/\delta}.$$

4. *When $C = 1$, for any real $q > 1$,*

$$\sum_{n \geq 1} \frac{1}{n^p} \max_{1 \leq i \leq n} V_i \leq \frac{1}{1 - 2^{1-p}} \sum_{j \geq 0} \frac{1}{2^{j(p-1)}} V_{2^j}.$$

5. *For any integer $r \geq 0$, any integer u satisfying $0 \leq u \leq r$ and any real $q > -1$,*

$$\sum_{l=u}^{r} 2^{-ql} V_{2^l} \leq C 2^{q+2}(2^{q+1}-1)^{-1} \left\{ 2^{-qu} \max_{1 \leq j \leq 2^u} V_j + \sum_{j=2^u+1}^{2^r} j^{-(q+1)} V_j \right\}.$$

Remark 3.25 *If $(V_i)_{i \geq 0}$ satisfies (3.62) with $C = 1$, then the sequence is said to be subadditive.*

Proof of Lemma 3.24 The condition (3.62) implies that for any integer k and any integer $0 \leq j \leq k$,

$$V_k \leq C(V_j + V_{k-j}) \text{ and then that } (k+1)V_k \leq 2C \sum_{j=1}^{k} V_j. \tag{3.63}$$

Therefore for $2^{r-1} \leq n < 2^r$,

$$\sum_{i=0}^{r-1} \frac{1}{2^{iq}} V_{2^i} \leq 2C \sum_{j=1}^{2^{r-1}} V_j \sum_{i:2^i \geq j} \frac{1}{2^{i(q+1)}},$$

proving Item 1. To prove Item 2, using again (3.63), it suffices to notice that

$$\sum_{j=1}^{k}\frac{1}{j^q}V_{jm} \leq 2C\sum_{j=1}^{k}\frac{1}{(1+jm)j^q}\sum_{\ell=1}^{jm}V_\ell$$
$$\leq 2Cm^{-1}(2m)^q\sum_{j=1}^{k}j^{-q-1}\sum_{\ell=1}^{m}\frac{1}{(\ell+m)^q}V_\ell + 2Cq^{-1}m^{q-1}\sum_{\ell=m+1}^{km}\frac{1}{\ell^q}V_\ell.$$

To prove Item 3, we first notice that (3.62) entails that

$$V_{i+j}^\gamma \leq C^\gamma(V_i^\gamma + V_j^\gamma) \text{ and then that } (k+1)V_k^\gamma \leq 2C^\gamma\sum_{j=1}^{k}V_j^\gamma.$$

Then for any real $q \geq 0$,

$$k^{-q(\gamma-\delta)}V_k^{\gamma-\delta} \leq 2^{1-\delta/\gamma}C^{\gamma-\delta}\left(\sum_{j=1}^{k}j^{-(1+q\gamma)}V_j^\gamma\right)^{1-\delta/\gamma}. \tag{3.64}$$

Writing that $k^{-(1+q\gamma)}V_k^\gamma = (k^{-(1+q\delta)}V_k^\delta)(k^{-q(\gamma-\delta)}V_k^{\gamma-\delta})$ and using (3.64), the following inequality holds

$$\sum_{k=1}^{n}\frac{1}{k^{1+q\gamma}}V_k^\gamma \leq 2^{1-\delta/\gamma}C^{\gamma-\delta}\left(\sum_{k=1}^{n}k^{-(1+q\delta)}V_k^\delta\right)\left(\sum_{j=1}^{n}j^{-(1+q\gamma)}V_j^\gamma\right)^{1-\delta/\gamma},$$

proving Item 3. To prove Item 4, we first write that for any positive integer i such that $i < 2^{r+1}$,

$$i = \sum_{k=0}^{r}a_k 2^k \text{ where } a_k \in \{0,1\}.$$

Therefore, taking into account (3.62) with $C = 1$, we get for $i \leq n < 2^{r+1}$,

$$V_i \leq \sum_{k=0}^{r}a_k V_{2^k} \leq \sum_{k=0}^{r}V_{2^k} \text{ and then } \max_{1\leq i\leq n}V_i \leq \sum_{k=0}^{r}V_{2^k}.$$

It follows that

$$\sum_{n\geq1}\frac{1}{n^p}\max_{1\leq i\leq n}V_i = \sum_{r=0}^{\infty}\sum_{n=2^r}^{2^{r+1}-1}\frac{1}{n^p}\max_{1\leq i\leq n}V_i \leq \sum_{r=0}^{\infty}2^{r(1-p)}\sum_{k=0}^{r}V_{2^k} = \sum_{k\geq0}V_{2^k}\sum_{r\geq k}2^{r(1-p)},$$

proving Item 4. To prove Item 5, we use (3.63) which gives

$$\sum_{l=u}^{r} 2^{-ql} V_{2^l} \leq 2C \sum_{l=u}^{r} 2^{-(q+1)l} \sum_{j=1}^{2^l} V_j = 2 \sum_{j=1}^{2^r} V_j \sum_{l \,:\, 2^l \geq \max(j, 2^u)} 2^{-(q+1)l}.$$

Hence

$$\sum_{l=u}^{r} 2^{-ql} V_{2^l} \leq 2C \sum_{j=1}^{2^u} V_j \sum_{l \,:\, 2^l \geq 2^u} 2^{-(q+1)l} + 2 \sum_{j=2^u+1}^{2^r} V_j \sum_{l \,:\, 2^l \geq j} 2^{-(q+1)l}$$

$$\leq 2^{q+2}(2^{q+1} - 1)^{-1} C \left\{ 2^{-(q+1)u} \sum_{j=1}^{2^u} V_j + \sum_{j=2^u+1}^{2^r} j^{-(q+1)} V_j \right\},$$

which leads to Item 5 and ends the proof of the lemma. ■

4

Gaussian Approximation via Martingale Methods

Gordin's seminal paper (1969) initiated a line of research in which limit theorems for stationary sequences are proved via appropriate approximations by stationary martingale difference sequences followed by an application of the corresponding limit theorem for such sequences. In this chapter, we first review different ways to get suitable martingale approximations and then derive the central limit theorem and its functional form for strictly stationary sequences in $\mathbb{L}^2$ under various type of projective criteria. More general normalizations than $\sqrt{n}$ will be investigated as well as the functional moderate deviation principle. We shall also address the question of the functional form of the central limit theorem for non-necessarily stationary sequences (see Section 4.4).

4.1 Martingale Approximations

4.1.1 General Martingale Approximations

Theorem 4.1 *Let $1<p<\infty$. Let $(X_n)_{n\in\mathbb{Z}}$ be a sequence of random variables in $\mathbb{L}^p(\mathbb{P})$, centered and adapted to a non-decreasing filtration $(\mathcal{F}_n)_{n\in\mathbb{Z}}$. Let $S_n = \sum_{k=1}^n X_k$. Assume that*

$$\sup_{n\geq 1}\sup_{m>n} \|\mathbb{E}(S_m - S_n|\mathcal{F}_n)\|_p < \infty. \tag{4.1}$$

Then $S_n = M_n - Z_n$ where $(M_n, \mathcal{F}_n)_{n\in\mathbb{N}}$ is a martingale in $\mathbb{L}^p(\mathbb{P})$ and $\sup_{n\geq 1} \|Z_n\|_p < \infty$.

Proof The first step is to construct a subsequence $(\ell_j)_{j\geq 1}$ such that

$$\left(\frac{1}{m}\sum_{j=1}^{m} \mathbb{E}\left(S_{\ell_j} - S_n|\mathcal{F}_n\right)\right)_{m\geq 1} \quad \text{is convergent to } Z_n \text{ in } \mathbb{L}^p \text{ for all integers } n. \tag{4.2}$$

Functional Gaussian Approximation for Dependent Structures. Florence Merlevède, Magda Peligrad and Sergey Utev. Oxford University Press (2019). © Florence Merlevède, Magda Peligrad and Sergey Utev 2019.
DOI: 10.1093/oso/9780198826941.001.0001

The proof follows the traditional diagonalization procedure and is rooted in Banach–Saks–Kakutani theorem. For $n = 1$, since $\sup_{m>1} \|\mathbb{E}(S_m - S_1|\mathcal{F}_1)\|_p < \infty$ and $\mathbb{L}^p$ is uniformly convex for $p \in]1, \infty[$, by the Banach–Saks–Kakutani theorem, there exists a subsequence $(n_{1,j})$ of $\mathbb{N}$ such that for any subsequence $\left(n'_{1,j}\right) \subset \left(n_{1,j}\right)$, we have

$$\frac{1}{m}\sum_{j=1}^{m} \mathbb{E}(S_{n'_{1,j}} - S_1|\mathcal{F}_1) \to Z_1 \text{ in } \mathbb{L}^p, \text{ as } m \to \infty.$$

For $n = 2$, we select by the same mechanism a subsequence $(n_{2,j}) \subset (n_{1,j})$ such that for any subsequence $(n'_{2,j}) \subset (n_{2,j})$, we have

$$\frac{1}{m}\sum_{j=1}^{m} \mathbb{E}\left(S_{n'_{2,j}} - S_2|\mathcal{F}_2\right) \to Z_2 \text{ in } \mathbb{L}^p, \text{ as } m \to \infty.$$

By an iterative procedure, we construct for all k a fixed integer, $(n_{k,j}) \subset (n_{k-1,j}) \subset (n_{k-2,j}) \subset \cdots \subset (n_{1,j})$ such that for any subsequence $(n'_{k,j}) \subset (n_{k,j})$,

$$\frac{1}{m}\sum_{j=1}^{m} \mathbb{E}(S_{n'_{k,j}} - S_k|\mathcal{F}_k) \to Z_k \text{ in } \mathbb{L}^p, \text{ as } m \to \infty.$$

Therefore, choosing $(\ell_j)_{j\geq 1} = (n_{j,j})_{j\geq 1}$, (4.2) follows.

Set $M_n = S_n + Z_n$. We show now that $(M_n, \mathcal{F}_n)_{n\in\mathbb{N}}$ is a martingale. First, M_n is clearly $\mathcal{F}_n$-measurable. Now, by conditioning by $\mathcal{F}_{n-1}$ in (4.2) we get on one hand that

$$\frac{1}{m}\sum_{j=1}^{m} \mathbb{E}(S_{\ell_j}|\mathcal{F}_{n-1}) \to \mathbb{E}(M_n|\mathcal{F}_{n-1}) \text{ in } \mathbb{L}^p, \text{ as } m \to \infty.$$

But on the other hand, we also have that

$$\frac{1}{m}\sum_{j=1}^{m} \mathbb{E}(S_{\ell_j}|\mathcal{F}_{n-1}) \to M_{n-1} \text{ in } \mathbb{L}^p, \text{ as } m \to \infty.$$

So overall, $\mathbb{E}(M_n|\mathcal{F}_{n-1}) = M_{n-1}$ a.s., proving that $(M_n, \mathcal{F}_n)_{n\in\mathbb{N}}$ is a martingale.

To end the proof of the theorem, we notice that

$$\|Z_n\|_p = \lim_{m\to\infty} \|\frac{1}{m}\sum_{j=1}^{m} \mathbb{E}(S_{\ell_j} - S_n|\mathcal{F}_n)\|_p \leq \sup_{n\geq 1}\sup_{k>n} \|\mathbb{E}(S_k - S_n|\mathcal{F}_n)\|_p < \infty. \quad \blacksquare$$

To prove Theorem 4.1 when $p = 1$, we cannot use the Banach–Saks–Kakutani theorem since $\mathbb{L}^1(\mathbb{P})$ is even not reflexive. However the following result holds.

Theorem 4.2 *Let $(X_n)_{n\in\mathbb{Z}}$ be a sequence of random variables in $\mathbb{L}^1(\mathbb{P})$, centered and adapted to a non-decreasing filtration $(\mathcal{F}_n)_{n\in\mathbb{Z}}$. Let $S_n = \sum_{k=1}^n X_k$. Assume that*

$$\left(|\mathbb{E}(S_m - S_n|\mathcal{F}_n)|\right)_{m>n\geq 1} \text{is uniformly integrable.} \tag{4.3}$$

Then $S_n = M_n - Z_n$ where $(M_n, \mathcal{F}_n)_{n\in\mathbb{N}}$ is a martingale in $\mathbb{L}^1(\mathbb{P})$ and $\sup_{n\geq 1} \|Z_n\|_1 < \infty$.

Proof By Komlós' theorem (1967) and the diagonalization procedure as done in the proof of Theorem 4.1, we can construct a subsequence $(\ell_j)_{j\geq 1}$ such that

$$\left(\frac{1}{m}\sum_{j=1}^m \mathbb{E}(S_{\ell_j} - S_n|\mathcal{F}_n)\right)_{m\geq 1} \quad \text{is convergent to } Z_n \text{ almost surely for all integers } n, \tag{4.4}$$

and Z_n is in $\mathbb{L}^1(\mathbb{P})$. Recall now that

$$X_n \to X \text{ in } \mathbb{L}^1 \iff X_n \to^{\mathbb{P}} X \text{ and } X_n \text{ is uniformly integrable.}$$

Since (4.3) implies that $\left(|m^{-1}\sum_{j=1}^m \mathbb{E}(S_{\ell_j} - S_n|\mathcal{F}_n)|\right)_{m\geq 1}$ is uniformly integrable, the convergence in (4.4) also holds in $\mathbb{L}^1$. We end the proof by following the arguments of the end of the proof of Theorem 4.1. ■

Remark 4.3 *If we assume stationarity then condition (4.3) can be weakened. To be more precise, the following statement holds: Let $(X_n)_{n\in\mathbb{Z}}$ be a stationary sequence of random variables in $\mathbb{L}^1(\mathbb{P})$, centered and adapted to a stationary non-decreasing filtration $(\mathcal{F}_n)_{n\in\mathbb{Z}}$. Let $S_n = \sum_{k=1}^n X_k$. Assume that*

$$\sup_{k>1} \|\mathbb{E}(S_k|\mathcal{F}_0)\|_1 < \infty. \tag{4.5}$$

Then $S_n = M_n - Z_n$ where $(M_n, \mathcal{F}_n)_{n\in\mathbb{N}}$ is a stationary martingale in $\mathbb{L}^1(\mathbb{P})$ and $\sup_{n\geq 1} \|Z_n\|_1 < \infty$. We refer the reader to the proof of Theorem 7 in Lin and Sine (1983).

We formulate now the following alternative to Theorem 4.1 that can be useful in the non-stationary setting. The proof is immediate and therefore left as an exercise.

Proposition 4.4 *Let $1 \leq p \leq \infty$. Let $(X_n)_{n\in\mathbb{Z}}$ be a sequence of random variables in $\mathbb{L}^p(\mathbb{P})$, centered and adapted to a non-decreasing filtration $(\mathcal{F}_n)_{n\in\mathbb{Z}}$. Let $S_n = \sum_{k=1}^n X_k$. Assume that for any $k > n$, there exists a sequence $(a_j)_{j\geq 0}$ of non negative reals such that*

$$\sum_{\ell\geq 0} a_\ell = A < \infty \text{ and } \|\mathbb{E}(X_k|\mathcal{F}_n)\|_p \leq a_{k-n}. \tag{4.6}$$

Then $S_n = M_n + R_n$ *where* $M_n = \sum_{k=1}^{n} D_k$ *with* $D_k = \sum_{j\geq k}(\mathbb{E}_k(X_j) - \mathbb{E}_{k-1}(X_j))$ *and* $R_n = Z_0 - Z_n$ *with* $Z_k = \sum_{j\geq k+1} \mathbb{E}_k(X_j)$. *In addition, for any* $k \in \mathbb{Z}$, D_k *and* Z_k *are in* $\mathbb{L}^p(\mathbb{P})$.

4.1.2 Stationary Martingale Approximation in $\mathbb{L}^2$

In this section we consider a stationary sequence of real-valued random variables $(X_n)_{n\in\mathbb{Z}}$ in $\mathbb{L}^2(\mathbb{P})$, centered and adapted to a non-decreasing and stationary filtration $(\mathcal{F}_n)_{n\in\mathbb{Z}}$. Let $S_n = \sum_{k=1}^{n} X_k$. It will be convenient to use the notation $\mathbb{E}(X|\mathcal{F}_k) = \mathbb{E}_k(X)$. Recalling the notation

$$\mathbf{P}_j(Y) = \mathbb{E}(Y|\mathcal{F}_j) - \mathbb{E}(Y|\mathcal{F}_{j-1})$$

we note that the the following representation is valid:

$$S_n = \mathbb{E}(S_n|\mathcal{F}_0) + \sum_{j=1}^{n} \mathbf{P}_j(S_n) = \mathbb{E}(S_n|\mathcal{F}_0) + \sum_{j=1}^{n} \mathbf{P}_j(S_n - S_{j-1}).$$

Let us assume we are able to show that

$$\mathbf{P}_1(S_n) \text{ is convergent in } \mathbb{L}^2 \text{ to a random variable } D_1. \tag{4.7}$$

Then clearly D_1 is a square integrable martingale difference. Furthermore, by stationarity, the convergence in (4.7) implies that for every $k \geq 1$

$$\mathbf{P}_k(S_{n+k} - S_{k-1}) \text{ is convergent in } \mathbb{L}^2 \text{ to } D_k.$$

We constructed in this way a sequence of martingale differences $(D_k)_{k\geq 1}$. Let

$$M_n = \sum_{j=1}^{n} D_j. \tag{4.8}$$

We would like to approximate the sequence of partial sums by this martingale in square mean. By using the fact that the martingale has orthogonal differences and using stationarity, we evaluate

$$\begin{aligned} ||S_n - M_n||_2^2 &= \left\| \sum_{j=1}^{n} (\mathbf{P}_j(S_n - S_{j-1}) - D_j) \right\|_2^2 + ||\mathbb{E}(S_n|\mathcal{F}_0)||_2^2 \\ &= \sum_{j=1}^{n} \left\| \mathbf{P}_1(S_{n-j+1}) - D_1 \right\|_2^2 + ||\mathbb{E}(S_n|\mathcal{F}_0)||_2^2 . \end{aligned}$$

Therefore

$$||S_n - M_n||_2^2 = \sum_{j=1}^{n} ||\mathbf{P}_1(S_j) - D_1||_2^2 + ||\mathbb{E}(S_n|\mathcal{F}_0)||_2^2. \tag{4.9}$$

Similarly

$$||S_n - M_n - \mathbb{E}(S_n|\mathcal{F}_0)||_2^2 = \sum_{j=1}^{n} ||\mathbf{P}_1(S_j) - D_1||_2^2$$

and

$$||S_n||_2^2 = ||\sum_{j=1}^{n} \mathbf{P}_j(S_n - S_{j-1})||_2^2 + ||\mathbb{E}(S_n|\mathcal{F}_0)||_2^2 = \sum_{j=1}^{n} ||\mathbf{P}_1(S_{n-j+1})||_2^2 + ||\mathbb{E}(S_n|\mathcal{F}_0)||_2^2$$

$$= \sum_{j=1}^{n} ||\mathbf{P}_1(S_j)||_2^2 + ||\mathbb{E}(S_n|\mathcal{F}_0)||_2^2. \tag{4.10}$$

These considerations prove the following theorem that is due to Heyde (1974). Further we shall give a more general result in the same spirit.

Theorem 4.5 *Assume that* (4.7) *holds and let* M_n *be defined by* (4.8). *Then, the following statements are equivalent*

$$\frac{1}{n}||\mathbb{E}(S_n|\mathcal{F}_0)||_2^2 \to 0, \tag{4.11}$$

$$\frac{1}{n}||S_n||_2^2 \to ||D_1||_2^2, \tag{4.12}$$

$$\frac{1}{n}||S_n - M_n||_2^2 \to 0. \tag{4.13}$$

Proof The fact that (4.13) implies (4.11) and (4.12) is obvious. Next, if (4.7) and (4.11) hold then, by (4.9), (4.13) holds and then (4.12) also. Now, if (4.7) and (4.12) hold then, by (4.10), (4.11) is satisfied and then (4.13) also. The proof is complete. ■

The condition

$$\mathbf{P}_1(S_n) \text{ is convergent in } \mathbb{L}^2 \text{ and } \frac{||S_n||_2^2}{n} \to ||D_1||_2^2 \tag{4.14}$$

is usually referred in the literature to as the Heyde's condition. Note that the first part of condition (4.14) alone is not sufficient for the central limit theorem as shown by Theorem 4 in Volný (1993).

The martingale approximation in (4.13) is a method used in many applications. For instance, we shall see, in Chapter 14, that this method can be successfully applied to obtain the central limit behavior of reversible Markov chains.

As a matter of fact if the convergence in (4.7) is weakened to a Cesàro convergence we can still obtain a martingale approximation, as we shall see below.

We introduce now the condition

$$\frac{1}{n}\sum_{j=1}^{n} \mathbf{P}_1(S_j) \text{ is convergent in } \mathbb{L}^2 \text{ to a random variable } D_1. \tag{4.15}$$

It is obvious that D_1 is a square integrable martingale difference. The convergence (4.15) implies that for every $k \geq 1$

$$\frac{1}{n}\sum_{j=1}^{n} \mathbf{P}_k(S_{j+k-1} - S_{k-1}) \text{ is convergent in } \mathbb{L}^2 \text{ to } D_k.$$

Notice that $(D_k)_{k\geq 1}$ is a stationary sequence of martingale differences in $\mathbb{L}^2$ and let

$$M_n = \sum_{j=1}^{n} D_j. \tag{4.16}$$

We are going now to prove the following theorem in the spirit of Dedecker, Merlevède and Volný (2007), Wu and Woodroofe (2004) and Peligrad (2010).

Theorem 4.6 *Assume (4.15), i.e. $n^{-1}\sum_{j=1}^{n} \mathbf{P}_1(S_j)$ is convergent in $\mathbb{L}^2$ to a random variable D_1. Then, statements (4.11), (4.12) and (4.13) are equivalent where, in these statements, the martingale M_n is constructed in (4.16).*

Moreover, if $(d_i)_{i\in\mathbb{Z}}$ is a strictly stationary sequence of martingale differences in $\mathbb{L}^2$, if $M'_n = \sum_{i=1}^{n} d_i$, the $\mathbb{L}^2$ martingale approximation (4.13): $n^{-1}\|S_n - M'_n\|_2^2 \to 0$, implies (4.15).

Proof As previously mentioned, the fact that (4.13) implies (4.11) and (4.12) is obvious. To prove that, under (4.15), the condition (4.11) or (4.12) implies (4.13) requires several steps.

Step 1. Triangular martingale representation for X_k and S_k.

Let m be a fixed positive integer. Let us then define

$$\theta_\ell^m = \frac{1}{m}\sum_{i=0}^{m-1} \mathbb{E}_\ell(X_\ell + \dots + X_{\ell+i})$$

and

$$D_\ell^m = \theta_\ell^m - \mathbb{E}_{\ell-1}(\theta_\ell^m) = \frac{1}{m}\sum_{i=0}^{m-1} \mathbf{P}_\ell(S_{\ell+i}) = \frac{1}{m}\sum_{i=0}^{m-1} \mathbf{P}_\ell(S_{\ell+i} - S_{\ell-1}). \tag{4.17}$$

Then, $(D_\ell^m)_{\ell\in\mathbb{Z}}$ is a triangular array of martingale differences adapted to the filtration $(\mathcal{F}_\ell)_{\ell\in\mathbb{Z}}$. By simple algebra,

$$\theta_\ell^m = X_\ell + \frac{1}{m}\sum_{i=1}^{m-1} \mathbb{E}_\ell(S_{\ell+i} - S_\ell).$$

Now, since $(\mathcal{F}_\ell)_{\ell\in\mathbb{Z}}$ is a non-decreasing filtration,

$$\theta_\ell^m = X_\ell + \frac{1}{m}\sum_{i=1}^{m-1} \mathbb{E}_\ell\left(\mathbb{E}_{\ell+1}(X_{\ell+1} + \dots + X_{\ell+i})\right) \tag{4.18}$$

suggesting relating the second term to $\theta^m_{\ell+1}$ which is

$$\theta^m_{\ell+1} = \frac{1}{m}\sum_{i=0}^{m-1} \mathbb{E}_{\ell+1}(X_{\ell+1} + \dots + X_{\ell+i+1}).$$

So

$$\begin{aligned}\mathbb{E}_\ell(\theta^m_{\ell+1}) &= \frac{1}{m}\sum_{i=0}^{m-1} \mathbb{E}_\ell(X_{\ell+1} + \dots + X_{\ell+i+1})\\ &= \frac{1}{m}\sum_{i=1}^{m-1} \mathbb{E}_\ell(X_{\ell+1} + \dots + X_{\ell+i}) + \frac{1}{m}\sum_{i=0}^{m-1} \mathbb{E}_\ell(X_{\ell+i+1}).\end{aligned}$$

Therefore we can express the above relation as

$$\frac{1}{m}\sum_{i=1}^{m-1} \mathbb{E}_\ell(X_{\ell+1} + \dots + X_{\ell+i}) = \mathbb{E}_\ell(\theta^m_{\ell+1}) - \frac{1}{m}\sum_{i=0}^{m-1} \mathbb{E}_\ell(X_{\ell+i+1}).$$

Using this equality in (4.18), we obtain

$$\theta^m_\ell = X_\ell + \mathbb{E}_\ell(\theta^m_{\ell+1}) - \frac{1}{m}\mathbb{E}_\ell(S_{\ell+m} - S_\ell)$$

and by (4.17), after we add and subtract $\theta^m_{\ell+1}$, we get

$$X_\ell = D^m_{\ell+1} + \theta^m_\ell - \theta^m_{\ell+1} + \frac{1}{m}\mathbb{E}_\ell(S_{\ell+m} - S_\ell). \tag{4.19}$$

This represents a triangular martingale decomposition for X_ℓ. In this martingale decomposition, the martingale difference $D^m_{\ell+1}$ is adapted to $\mathcal{F}_{\ell+1}$. It is convenient to point out a related martingale approximation, which helps us to approximate the partial sums with a martingale adapted to the same filtration. Notice that

$$\theta^m_\ell = X_\ell + \frac{1}{m}\sum_{i=1}^{m-1} \mathbb{E}_\ell(S_{\ell+i} - S_\ell) = X_\ell + \bar{\theta}^m_\ell,$$

$$\text{where } \bar{\theta}^m_\ell = \frac{1}{m}\sum_{i=1}^{m-1} \mathbb{E}_\ell(S_{\ell+i} - S_\ell).$$

Starting from (4.19) and using this notation we obtain

$$X_{\ell+1} = D^m_{\ell+1} + \bar{\theta}^m_\ell - \bar{\theta}^m_{\ell+1} + \frac{1}{m}\mathbb{E}_\ell(S_{\ell+m} - S_\ell). \tag{4.20}$$

Summing these relations and denoting

$$M_k^m = \sum_{i=1}^{k} D_i^m,$$

we obtain for every stationary sequence and for any positive integer m and k,

$$S_k = M_k^m + R_k^m, \tag{4.21}$$

$$\text{where } R_k^m = \bar{\theta}_0^m - \bar{\theta}_k^m + \frac{1}{m}\sum_{\ell=0}^{k-1} \mathbb{E}_\ell(S_{\ell+m} - S_\ell).$$

Notice that, by stationarity, for any $k \le m$

$$||R_k^m||_2 \le 3 \max_{1\le j\le m} ||\mathbb{E}_0(S_j)||_2. \tag{4.22}$$

Step 2. We show here that under (4.15) **and** (4.11) **the approximation in** (4.13) **holds.**

Let us consider the martingale defined in (4.16). By using (4.21) with $k = m = n$, we estimate

$$||S_n - M_n||_2^2 = ||M_n^n + R_n^n - M_n||_2^2.$$

By (4.22), the martingale properties and stationarity,

$$\begin{aligned}||S_n - M_n||_2^2 &\le 2\sum_{j=1}^{n} ||D_j^n - D_j||_2^2 + 2\times 3^2 \max_{1\le j\le n} ||\mathbb{E}_0(S_j)||_2^2 \\ &= 2n||D_1^n - D_1||_2^2 + 2\times 3^2 \max_{1\le j\le n} ||\mathbb{E}_0(S_j)||_2^2.\end{aligned}$$

Hence the approximation (4.13) follows under (4.15) and (4.11).

Step 3. We show here that under (4.15) **and** (4.12) **the approximation in**(4.13) **holds.**

Now we assume that condition (4.15) holds and construct the martingale as in (4.21). It remains to estimate

$$\frac{1}{n}\mathbb{E}\left(S_n - \sum_{i=1}^{n} D_i\right)^2 = \frac{1}{n}\left[\mathbb{E}(S_n^2) + \mathbb{E}(M_n^2) - 2\mathbb{E}\left(S_n \sum_{i=1}^{n} D_i\right)\right]. \tag{4.23}$$

By stationarity and orthogonality of the martingale differences,

$$\begin{aligned}\mathbb{E}\left(S_n \sum_{i=1}^{n} D_i\right) &= \sum_{i=1}^{n} \mathbb{E}\left(X_i \sum_{j=1}^{i} D_j\right) = \sum_{i=1}^{n}\sum_{j=1}^{i} \mathbb{E}(X_{i-j+1}D_1) \\ &= \mathbb{E}\left(D_1 \sum_{i=1}^{n}\sum_{j=1}^{i}\left(\mathbb{E}_1(X_j) - \mathbb{E}_0(X_j)\right)\right) = \mathbb{E}\left(D_1 \sum_{i=1}^{n} \mathbf{P}_1(S_i)\right).\end{aligned}$$

So, by definition (4.17) and stationarity,

$$\lim_{n\to\infty} \frac{1}{n}\mathbb{E}\left(S_n \sum_{i=1}^{n} D_i\right) = \lim_{n\to\infty} \mathbb{E}(D_1 D_1^n).$$

It is clear now that condition (4.15) implies $\mathbb{E}(D_1 D_1^n) \to \mathbb{E}(D_1^2)$, whence, by (4.23) and (4.12),

$$\frac{1}{n}||S_n - M_n||_2^2 \to 0,$$

which ends the proof of Step 3 and then of the first statement of the theorem.

To end the proof of the theorem, it remains to prove its second statement. Let $(D_\ell^m)_{\ell\in\mathbb{Z}}$ be the triangular array of martingale differences constructed in (4.17). Note that

$$\left\|D_1^n - d_1\right\|_2^2 = \frac{1}{n}\sum_{k=1}^{n} ||D_k^n - d_k||_2^2 = \frac{1}{n}||M_n^n - M_n'||_2^2 \le \frac{2}{n}||S_n - M_n'||_2^2 + \frac{2}{n}||S_n - M_n^n||_2^2.$$

Hence, by (4.22),

$$||D_1^n - d_1||_2^2 \le \frac{2}{n}||S_n - M_n'||_2^2 + \frac{18}{n}\max_{1\le j\le n}||\mathbb{E}_0(S_j)||_2^2.$$

Note now that if $n^{-1}||S_n - M_n'||_2^2 \to 0$ then (4.11) holds. Hence, if the $\mathbb{L}^2$ martingale approximation (4.13) holds with M_n', then $||D_1^n - d_1||_2^2 \to 0$, proving that condition (4.15) holds. ■

Let us mention that martingale approximations of the type obtained in this section are unique. Indeed if there are two approximating martingales, let say $M_n = \sum_{k=1}^{n} D_k$ and $M_n' = \sum_{k=1}^{n} D_k'$, satisfying

$$\lim_{n\to\infty}\frac{||S_n - M_n||_2}{\sqrt{n}} = \lim_{n\to\infty}\frac{||S_n - M_n'||_2}{\sqrt{n}} = 0,$$

then

$$\lim_{n\to\infty}\frac{||M_n - M_n'||_2}{\sqrt{n}} = ||D_1 - D_1'||_2 = 0.$$

So, $D_1 = D_1'$ a.s.

Remark 4.7 *Note that if we define for a fixed positive integer m,*

$$D_k^m = \mathbf{P}_k(S_{k+m-1} - S_{k-1}), \tag{4.24}$$

then the following alternative triangular martingale representations to (4.20) and (4.21) hold for X_{k+1} and S_k:

$$X_{k+1} = D_{k+1}^m + \mathbb{E}_k(S_{k+m} - S_k) - \mathbb{E}_{k+1}(S_{k+m+1} - S_{k+1}) + \mathbb{E}_{k+1}(X_{k+m+1}) \tag{4.25}$$

and

$$S_k = \sum_{\ell=1}^{k} D_\ell^m + \theta_0^m - \theta_k^m + \sum_{\ell=1}^{k} \mathbb{E}_\ell(X_{\ell+m}) \text{ where } \theta_\ell^m = \mathbb{E}_\ell(S_{\ell+m} - S_\ell). \tag{4.26}$$

4.2 The Central Limit Theorem for Stationary Sequences in $\mathbb{L}^1$ or in $\mathbb{L}^2$

We first state the following theorem which is essentially due to Gordin (1973) and gives sufficient conditions for a stationary sequence of random variables in $\mathbb{L}^1(\mathbb{P})$ to satisfy the central limit theorem (CLT).

Theorem 4.8 *Let $(X_n)_{n\in\mathbb{Z}}$ be a stationary and ergodic sequence of random variables in $\mathbb{L}^1(\mathbb{P})$, centered and adapted to a stationary non-decreasing filtration $(\mathcal{F}_n)_{n\in\mathbb{Z}}$. Let $S_n = \sum_{k=1}^n X_k$. Assume that (4.5) holds and that*

$$\liminf_{n\to\infty} \frac{\mathbb{E}|S_n|}{\sqrt{n}} < \infty. \tag{4.27}$$

Then $n^{-1/2}S_n$ converges in distribution to a centered normal variable (that may be degenerate).

Proof By Remark 4.3, (4.5) entails that there exists a stationary difference of martingale $(d_k)_{k\in\mathbb{Z}}$ in $\mathbb{L}^1(\mathbb{P})$ with respect to $(\mathcal{F}_n)_{n\in\mathbb{Z}}$ and such that $S_n = \sum_{k=1}^n d_k - Z_n$ where $\sup_{n\geq 1} \|Z_n\|_1 < \infty$. The theorem will follow by the usual CLT for a stationary and ergodic sequence of martingale differences (see Theorem 2.37) if we can prove that the d_k's are in $\mathbb{L}^2(\mathbb{P})$. This has been proved by Esseen and Janson (1985) and the arguments make use of the deviation probability of square functions of Burkholder (1979) (see Theorem 2.15). Indeed, assume for simplicity that $n^{-1/2}\mathbb{E}|S_n| \leq A < \infty$. Since $\sup_{n\geq 1} \|Z_n\|_1 < \infty$, we have that $n^{-1/2}\mathbb{E}|M_n| \leq B < \infty$. By Theorem 2.15, there exists a universal constant C, such that, for λ large enough,

$$\mathbb{P}\left(\left[\frac{1}{n}\sum_{k=1}^{n} d_k^2\right]^{1/2} > \lambda\right) \leq \frac{C}{\lambda}\frac{\mathbb{E}|M_n|}{\sqrt{n}} < 1. \tag{4.28}$$

Now by the ergodic theorem (still valid if $\mathbb{E}(d_0^2) = \infty$),

$$\frac{1}{n}\sum_{k=1}^{n} d_k^2 \to \mathbb{E}(d_0^2) \text{ almost surely as } n \to \infty.$$

If $\mathbb{E}(d_0^2) = \infty$, then $n^{-1}\sum_{k=1}^n d_k^2 \to \infty$ almost surely and then the left-hand side of (4.28) converges to 1 as $n \to \infty$, which is a contradiction. ■

When the stationary sequence of random variables is in $\mathbb{L}^2(\mathbb{P})$, the martingale approximation given in Theorem 4.6 together with the usual central limit theorem for a stationary and ergodic sequence of martingale differences (see Theorem 2.37) gives the following:

Theorem 4.9 *Let $(X_n)_{n\in\mathbb{Z}}$ be a stationary and ergodic sequence of random variables in $\mathbb{L}^2(\mathbb{P})$, centered and adapted to a stationary non-decreasing filtration $(\mathcal{F}_n)_{n\in\mathbb{Z}}$. Let $S_n = \sum_{k=1}^n X_k$. Assume that (4.15) holds and either condition (4.11) or condition (4.12) is satisfied. Then $n^{-1/2}S_n$ converges in distribution to a centered normal variable with variance $\sigma^2 = \mathbb{E}(D_1^2) = \lim_{n\to\infty} n^{-1}\mathbb{E}(S_n^2)$.*

As we shall show below, the conditions of Theorem 4.9 are satisfied if either

$$\mathbb{E}(X_0|\mathcal{F}_{-\infty}) = 0 \text{ a.s. and } \sum_{i\geq 0} \|\mathbf{P}_0(X_i)\|_2 < \infty, \tag{4.29}$$

or

$$\sum_{k\geq 1} \frac{\|\mathbb{E}_0(S_k)\|_2}{k^{3/2}} < \infty. \tag{4.30}$$

Condition (4.29) is usually referred in the literature to as Hannan's condition (1979), whereas condition (4.30) is referred to as the Maxwell–Woodroofe condition (2000). More precisely, the two following theorems hold:

Theorem 4.10 *(Hannan) Let $(X_n)_{n\in\mathbb{Z}}$ be a stationary and ergodic sequence of random variables in $\mathbb{L}^2(\mathbb{P})$, centered and adapted to a stationary non-decreasing filtration $(\mathcal{F}_n)_{n\in\mathbb{Z}}$. Let $S_n = \sum_{k=1}^n X_k$. If (4.29) holds then conditions (4.15) and (4.11) do. Consequently, if we assume in addition ergodicity of $(X_n)_{n\in\mathbb{Z}}$, the conclusion of Theorem 4.9 holds for $n^{-1/2}S_n$. In addition $\sigma^2 = \sum_{k\in\mathbb{Z}} \mathbb{E}(X_0X_k)$ where the series is absolutely convergent.*

Theorem 4.11 *(Maxwell–Woodroofe) Let $(X_n)_{n\in\mathbb{Z}}$ be a stationary sequence of random variables in $\mathbb{L}^2(\mathbb{P})$, centered and adapted to a stationary non-decreasing filtration $(\mathcal{F}_n)_{n\in\mathbb{Z}}$. Assume that (4.30) holds; then conditions (4.15) and (4.11) do too. Consequently, if we assume in addition ergodicity of $(X_n)_{n\in\mathbb{Z}}$, the conclusion of Theorem 4.9 holds for $n^{-1/2}S_n$.*

Proof of Theorem 4.10 Condition (4.29) entails clearly (4.15). Let us prove that (4.29) entails (4.11). Since $\mathbb{E}(X_0|\mathcal{F}_{-\infty}) = 0$ almost surely, we have $\mathbb{E}(S_n|\mathcal{F}_0) = \sum_{k\geq 0} \mathbf{P}_{-k}(S_n)$ almost surely. Hence, using that $\mathbb{E}(\mathbf{P}_{-k}(S_n)\mathbf{P}_{-\ell}(S_n)) = 0$ if $k \neq \ell$ together with stationarity, we get

$$\|\mathbb{E}(S_n|\mathcal{F}_0))\|_2^2 = \sum_{k\geq 0} \|\mathbf{P}_{-k}(S_n)\|_2^2 = \sum_{k\geq 0}\Big(\sum_{\ell=1}^{n} \|\mathbf{P}_{-k}(X_\ell)\|_2\Big)^2$$

$$\leq \sum_{k=0}^{n-1}\Big(\sum_{\ell\geq k+1} \|\mathbf{P}_{-\ell}(X_0)\|_2\Big)^2 + \sum_{k\geq n}\Big(\sum_{\ell=k+1}^{k+n} \|\mathbf{P}_{-\ell}(X_0)\|_2\Big)^2$$

$$\leq \sum_{k=0}^{n-1}\Big(\sum_{\ell\geq k+1}\|P_{-\ell}(X_0)\|_2\Big)^2 + \Big(\sum_{i\geq n+1}\|\mathbf{P}_{-i}(X_0)\|_2\Big)\sum_{k\geq n}\sum_{\ell=k+1}^{k+n}\|\mathbf{P}_{-\ell}(X_0)\|_2$$

$$\leq \sum_{k=0}^{n-1}\Big(\sum_{\ell\geq k+1}\|\mathbf{P}_{-\ell}(X_0)\|_2\Big)^2 + n\Big(\sum_{\ell\geq n+1}\|\mathbf{P}_{-\ell}(X_0)\|_2\Big)^2.$$

Therefore

$$\|\mathbb{E}(S_n|\mathcal{F}_0)\|_2^2 \leq 2\sum_{k=1}^{n}\Big(\sum_{\ell\geq k}\|\mathbf{P}_{-\ell}(X_0)\|_2\Big)^2. \tag{4.31}$$

The estimate (4.31) entails (4.11) provided that (4.29) holds. To end the proof, it remains to show that

$$\mathbb{E}(D_1^2) = \sum_{k\in\mathbb{Z}}\mathbb{E}(X_0X_k). \tag{4.32}$$

Notice first that since $\mathbb{E}(X_0|\mathcal{F}_{-\infty}) = 0$ a.s., $X_0 = \sum_{i=-\infty}^{0}\mathbf{P}_i(X_0)$. Therefore, by orthogonality,

$$\mathbb{E}(X_0X_k) = \mathbb{E}\Big(\sum_{i=-\infty}^{0}\mathbf{P}_i(X_0)\sum_{j=-\infty}^{k}\mathbf{P}_j(X_k)\Big) = \sum_{i=-\infty}^{0\wedge k}\mathbb{E}\left(\mathbf{P}_i(X_0)\mathbf{P}_i(X_k)\right).$$

Notice that this implies

$$|\mathbb{E}(X_0X_k)| \leq \sum_{i=-\infty}^{0\wedge k}\|\mathbf{P}_i(X_0)\|_2\|\mathbf{P}_{i-k}(X_0)\|_2,$$

and therefore

$$\sum_{k\in\mathbb{Z}}|\mathbb{E}(X_0X_k)| \leq \Big(\sum_{\ell\geq 0}\|\mathbf{P}_0(X_\ell)\|_2\Big)^2,$$

which is finite under (4.29). Moreover, by stationarity,

$$\sum_{k\in\mathbb{Z}}\mathbb{E}(X_0X_k) = \sum_{k\in\mathbb{Z}}\sum_{i=-\infty}^{0\wedge k}\mathbb{E}\left(\mathbf{P}_i(X_0)\mathbf{P}_i(X_k)\right) = \sum_{k\in\mathbb{Z}}\sum_{i=-(0\wedge k)}^{\infty}\mathbb{E}\left(\mathbf{P}_0(X_i)\mathbf{P}_0(X_{k+i})\right),$$

implying that

$$\sum_{k\in\mathbb{Z}}\mathbb{E}(X_0X_k) = \sum_{i\geq 0}\sum_{j\geq 0}\mathbb{E}\left(\mathbf{P}_0(X_i)\mathbf{P}_0(X_j)\right).$$

On the other hand,

$$\mathbb{E}(D_1^2) = \mathbb{E}\left(\sum_{i\geq 0} \mathbf{P}_0(X_i)\right)^2 = \sum_{i\geq 0}\sum_{j\geq 0} \mathbb{E}\left(\mathbf{P}_0(X_i)\mathbf{P}_0(X_j)\right).$$

The two last relations prove (4.32) and end the proof of the theorem. ■

Proof of Theorem 4.11 The proof follows from the following proposition (see Gordin and Peligrad (2011)).

Proposition 4.12 *Let $(X_n)_{n\in\mathbb{Z}}$ be a stationary and ergodic sequence of random variables in $\mathbb{L}^2(\mathbb{P})$, centered and adapted to a stationary non-decreasing filtration $(\mathcal{F}_n)_{n\in\mathbb{Z}}$. Let $S_n = \sum_{k=1}^n X_k$ and*

$$Y_i^m = \frac{1}{m}\mathbb{E}_i(S_{i+m} - S_i).$$

1. *If the following condition holds*

$$\lim_{m\to\infty} \limsup_{n\to\infty} \frac{1}{\sqrt{n}} \left\| \sum_{i=0}^{n-1} Y_i^m \right\|_2 = 0, \tag{4.33}$$

then (4.15) does.

2. *Under (4.30), condition (4.11) is satisfied and the following reinforcement of (4.33) holds:*

$$\lim_{m\to\infty} \limsup_{n\to\infty} \frac{1}{\sqrt{n}} \left\| \max_{1\leq k\leq n} \left| \sum_{i=0}^{k-1} Y_i^m \right| \right\|_2 = 0. \tag{4.34}$$

In the rest of the proof, we show Proposition 4.12 above. For any $k \geq 1$, let $D_k^m = m^{-1}\sum_{i=1}^m \mathbf{P}_k(S_{i+k-1} - S_{k-1})$ and $M_n^m = \sum_{k=1}^n D_k^m$. Let m' and m'' be positive integers. Notice that by stationarity

$$\|M_n^{m'} - M_n^{m''}\|_2^2 = n\|D_1^{m'} - D_1^{m''}\|_2^2.$$

On the other hand, starting from (4.21) we get

$$\|M_n^{m'} - M_n^{m''}\|_2^2 = \|R_n^{m'} - R_n^{m''}\|_2^2,$$

where $R_n^m = \frac{1}{m}\sum_{i=1}^{m-1}(\mathbb{E}_0(S_i) - \mathbb{E}_n(S_{n+i} - S_n)) + \sum_{k=0}^{n-1} Y_k^m$. Therefore

$$\|D_1^{m'} - D_1^{m''}\|_2^2 = \frac{1}{n}\|R_n^{m'} - R_n^{m''}\|_2^2.$$

Hence, by stationarity,

$$\|D_1^{m'} - D_1^{m''}\|_2 \le \frac{2m'}{n^{1/2}}\|X_0\|_2 + \frac{2m''}{n^{1/2}}\|X_0\|_2 + \frac{1}{\sqrt{n}}\left(\left\|\sum_{i=0}^{n-1} Y_i^{m'}\right\|_2 + \left\|\sum_{i=0}^{n-1} Y_i^{m''}\right\|_2\right),$$

giving

$$\|D_1^{m'} - D_1^{m''}\|_2 \le \limsup_{n\to\infty} \frac{1}{\sqrt{n}}\left(\left\|\sum_{i=0}^{n-1} Y_i^{m''}\right\|_2 + \left\|\sum_{i=0}^{n-1} Y_i^{m''}\right\|_2\right).$$

By (4.33), the limit when m' and m'' both tend to infinity is then zero, giving that D_1^m is a Cauchy sequence in $\mathbb{L}^2$ and is therefore convergent. This proves Item 1 of the proposition.

We prove now Item 2. By stationarity, for any nonnegative integers i and j,

$$\|\mathbb{E}_0(S_{i+j})\|_2 \le \|\mathbb{E}_0(S_i)\|_2 + \|\mathbb{E}_0(S_{j+i} - S_i)\|_2 \le \|\mathbb{E}_0(S_i)\|_2 + \|\mathbb{E}_0(S_j)\|_2,$$

proving that $(\|\mathbb{E}_0(S_i)\|_2)_{i\ge 0}$ is subadditive. Therefore, by Item 2 of Lemma 3.24, we have

$$\frac{1}{\sqrt{n}}\|\mathbb{E}_0(S_n)\|_2 \le 2^{5/2}\sum_{\ell=1}^{n} \frac{\|\mathbb{E}_0(S_\ell)\|_2}{(\ell+n)^{3/2}}.$$

This upper bound shows that condition (4.30) implies (4.11). It remains to show that (4.30) entails (4.33). With this aim, we apply Theorem 3.6 with $p=2$ to the stationary sequence $(Y_k^m)_{k\in\mathbb{Z}}$. This gives

$$\left\|\max_{1\le k\le n}\left|\sum_{i=0}^{k-1} Y_i^m\right|\right\|_2 \ll n^{1/2}(\|Y_0^m\|_2 + \Delta(Y_0^m)),$$

where

$$\Delta(Y_0^m) := \sum_{k=1}^{\infty} \frac{1}{k^{3/2}}\|\mathbb{E}_0(Y_0^m + \dots + Y_{k-1}^m)\|_2.$$

We first notice that $\|Y_0^m\|_2 \le m^{-1}\|\mathbb{E}_0(S_m)\|_2$. We estimate now $\Delta(Y_0^m)$. With this aim, it is convenient to use the decomposition

$$\Delta(Y_0^m) \le \sum_{k=1}^{m} \frac{1}{k^{3/2}}\|\mathbb{E}_0(Y_0^m + \dots + Y_{k-1}^m)\|_2 + \sum_{k=m+1}^{\infty} \frac{1}{k^{3/2}}\|\mathbb{E}_0(Y_0^m + \dots + Y_{k-1}^m)\|_2.$$

To estimate the first sum, notice that, by the properties of the conditional expectation, we have

$$\|\mathbb{E}_0(Y_0^m + \dots + Y_{k-1}^m)\|_2 \le k\|\mathbb{E}_0(Y_0^m)\|_2,$$

and then, since $\|\mathbb{E}_0(Y_0^m)\|_2 \leq \|\mathbb{E}_0(S_m)\|_2/m$, we have

$$\sum_{k=1}^{m} \frac{1}{k^{3/2}} \|\mathbb{E}_0(Y_0^m + \ldots + Y_{k-1}^m)\|_2 \leq \frac{1}{m} \sum_{k=1}^{m} \frac{\|\mathbb{E}_0(S_m)\|_2}{k^{1/2}} \ll \frac{1}{m^{1/2}} \|\mathbb{E}_0(S_m)\|_2 .$$

To estimate the second sum we also apply the properties of the conditional expectation and write this time that, for any $k \geq m+1$,

$$\|\mathbb{E}_0(Y_0^m + \cdots + Y_{k-1}^m)\|_2 \leq \frac{1}{m} \left\| \sum_{i=0}^{m-1} \mathbb{E}_0(S_{i+k} - S_i) \right\|_2 \leq \|\mathbb{E}_0(S_k)\|_2 .$$

Then,

$$\sum_{k=m+1}^{\infty} \frac{1}{k^{3/2}} \|\mathbb{E}_0(Y_0^m + \ldots + Y_{k-1}^m)\|_2 \leq \sum_{k=m+1}^{\infty} \frac{\|\mathbb{E}_0(S_k)\|_2}{k^{3/2}} ,$$

and overall

$$\Delta(Y_0^m) \ll \frac{1}{m^{1/2}} \|\mathbb{E}_0(S_m)\|_2 + \sum_{k=m+1}^{\infty} \frac{\|\mathbb{E}_0(S_k)\|_2}{k^{3/2}} .$$

We conclude that, for any strictly positive integers n and m,

$$\frac{1}{\sqrt{n}} \left\| \max_{1 \leq k \leq n} \left| \sum_{i=0}^{k-1} Y_i^m \right| \right\|_2 \ll \frac{\|\mathbb{E}_0(S_m)\|_2}{m^{1/2}} + \sum_{k=m+1}^{\infty} \frac{\|\mathbb{E}_0(S_k)\|_2}{k^{3/2}} . \tag{4.35}$$

Letting m tend to infinity, the last term converges to zero by (4.30), as well as the first one in the right-hand side since we have already proved that (4.11) holds under (4.30). This ends the proof of Proposition 4.12. ■

Remark 4.13 *Even if the conditions (4.29) and (4.30) are of different nature they both are satisfied if*

$$\sum_{k \geq 1} \frac{\|\mathbb{E}_0(X_k)\|_2}{k^{1/2}} < \infty . \tag{4.36}$$

The fact that (4.36) implies (4.30) follows directly by contracting the norm and interverting the sums. To show that (4.36) implies (4.29), we first note that (4.36) implies that $\lim_{k\to\infty} \|\mathbb{E}_{-k}(X_0)\|_2 = 0$ *and then the first part of (4.29). In addition,*

$$\sum_{n \geq 1} \|\mathbf{P}_0(X_n)\|_2 = \sum_{k \geq 1} \sum_{n \geq k} \frac{1}{n} \|\mathbf{P}_0(X_n)\|_2 \leq 2 \sum_{k \geq 1} \left(\sum_{n \geq k} \frac{1}{(n+1)^2} \right)^{1/2} \left(\sum_{n \geq k} \|\mathbf{P}_0(X_n)\|_2^2 \right)^{1/2}$$

$$\leq 2 \sum_{k \geq 1} \left(\frac{1}{k} \sum_{n \geq k} \|\mathbf{P}_0(X_n)\|_2^2 \right)^{1/2} = 2 \sum_{k \geq 1} \left(\frac{1}{k} \sum_{n \geq k} \|\mathbf{P}_{-n}(X_0)\|_2^2 \right)^{1/2} .$$

We notice now that

$$\sum_{n\geq k} \|\mathbf{P}_{-n}(X_0)\|_2^2 = \left\| \sum_{n\geq k} \mathbf{P}_{-n}(X_0) \right\|_2^2 = \|\mathbb{E}_{-k}(X_0) - \mathbb{E}_{-\infty}(X_0)\|_2^2 \leq \|\mathbb{E}_{-k}(X_0)\|_2^2 .$$

Hence

$$\sum_{n\geq 1} \|\mathbf{P}_0(X_n)\|_2 \leq 2 \sum_{k\geq 1} \frac{\|\mathbb{E}_0(X_k)\|_2}{k^{1/2}} .$$

4.3 On the Functional Central Limit Theorem for Stationary Sequences in $\mathbb{L}^2$

A natural question is the following: "Are the conditions of Theorems 4.8 or 4.9 sufficient to get the functional form of the central limit theorem?" The answer to this question is negative (see Section 17.2.2).

However, as we shall see, under the conditions of Theorems 4.10 or 4.11, the functional form of the central limit theorem also holds. This follows from the following very general theorem (see Gordin and Peligrad (2011)).

Theorem 4.14 *Let $(X_n)_{n\in\mathbb{Z}}$ be a stationary sequence of random variables in $\mathbb{L}^2(\mathbb{P})$, centered and adapted to a stationary non-decreasing filtration $(\mathcal{F}_n)_{n\in\mathbb{Z}}$, and let $S_n = \sum_{k=1}^n X_k$. Assume that (4.34) holds. Then*

$$\left\| \max_{1\leq k\leq n} |S_k - M_k| \right\|_2 = o(n^{1/2}) \text{ as } n \to \infty, \tag{4.37}$$

where $M_n = \sum_{i=1}^n D_i$ and $(D_i)_{i\in\mathbb{Z}}$ is the stationary sequence of martingale differences in $\mathbb{L}^2$ defined by (4.15). Moreover $\{\max_{1\leq k\leq n} S_k^2/n\}_{n\geq 1}$ is an uniformly integrable family.

Proof of Theorem 4.14 By using (4.21), we have, for any positive integer k and any fixed integer m,

$$|S_k - M_k| \leq |M_k^m - M_k| + |\bar{\theta}_0^m - \bar{\theta}_k^m| + \left| \sum_{i=0}^{k-1} Y_i^m \right| ,$$

where we recall that

$$Y_i^m = \frac{1}{m}\mathbb{E}_i(S_{i+m} - S_i) \text{ and } \bar{\theta}_k^m = \frac{1}{m} \sum_{i=1}^{m-1} \mathbb{E}_k(S_{k+i} - S_k). \tag{4.38}$$

Then,

$$\frac{1}{\sqrt{n}} \| \max_{1\leq k\leq n} |S_k - M_k| \|_2 \leq \frac{1}{\sqrt{n}} \| \max_{1\leq k\leq n} |M_k^m - M_k| \|_2$$
$$+ \frac{1}{\sqrt{n}} \|\bar{\theta}_0^m\|_2 + \frac{1}{\sqrt{n}} \| \max_{1\leq k\leq n} |\bar{\theta}_k^m| \|_2 + \frac{1}{\sqrt{n}} \left\| \max_{1\leq k\leq n} \left| \sum_{i=0}^{k-1} Y_i^m \right| \right\|_2 .$$

By Doob's maximal inequality (see Corollary 2.7) and stationarity, we have

$$\frac{1}{\sqrt{n}}\|\max_{1\le k\le n}|M_k^m - M_k|\|_2 \le 2\|D_0^m - D_0\|_2.$$

For fixed m, since $(\bar{\theta}_k^m)_{k\in\mathbb{Z}}$ is a stationary sequence of square integrable random variables, for any $A > 0$, we have

$$\begin{aligned}\frac{1}{n}\mathbb{E}\left(\max_{1\le k\le n}|\bar{\theta}_k^m|^2\right) &\le \frac{A^2}{n} + \frac{1}{n}\sum_{k=1}^n \mathbb{E}\left(|\bar{\theta}_k^m|^2 I(|\theta_k^m| > A)\right)\\ &= \frac{A^2}{n} + \mathbb{E}\left(|\bar{\theta}_0^m|^2 I(|\bar{\theta}_0^m| > A)\right)\end{aligned}$$

and then clearly

$$\lim_{n\to\infty}\frac{1}{\sqrt{n}}\|\max_{1\le k\le n}|\bar{\theta}_k^m|\|_2 = 0.$$

Therefore,

$$\limsup_{n\to\infty}\frac{1}{\sqrt{n}}\|\max_{1\le k\le n}|S_k - M_k|\|_2 \le 2\|D_0^m - D_0\|_2 + \limsup_{n\to\infty}\frac{1}{\sqrt{n}}\left\|\max_{1\le k\le n}\left|\sum_{i=0}^{k-1} Y_i^m\right|\right\|_2.$$

The convergence (4.37) follows by letting $m \to \infty$, from the fact that $D_0^m \to D_0$ in $\mathbb{L}^2$ and that (4.34) is assumed.

Now the uniform integrability of $\{\max_{1\le k\le n} S_k^2/n\}_{n\ge 1}$ follows from (4.37) and the uniform integrability of $\{\max_{1\le k\le n} M_k^2/n\}_{n\ge 1}$ (see for instance Corollary 3.12). ∎

Combining Theorem 4.14 with Theorems 2.37 and 1.18, we get sufficient conditions for the functional form of the central limit theorems for stationary sequences.

Theorem 4.15 *Let $(X_n)_{n\in\mathbb{Z}}$ be a stationary and ergodic sequence of random variables in $\mathbb{L}^2(\mathbb{P})$, centered and adapted to a stationary non-decreasing filtration $(\mathcal{F}_n)_{n\in\mathbb{Z}}$, and let $S_n = \sum_{k=1}^n X_k$. Assume that (4.34) holds. Then $\{n^{-1/2}S_{[nt]},\ t \in [0, 1]\}$ converges in distribution in $D([0, 1])$ to a Brownian motion with parameter $\sigma^2 = \lim_{n\to\infty} n^{-1}\mathbb{E}(S_n^2)$.*

Moreover $\{\max_{1\le k\le n} S_k^2/n\}_{n\ge 1}$ is an uniformly integrable family.

Part 2 of Proposition 4.12 shows that the Maxwell–Woodroofe condition (4.30) implies condition (4.34), which implies that the functional form of Theorem 4.11 also holds (the fact that the functional form of Theorem 4.11 holds has been proved by Peligrad and Utev (2005) with another proof).

Theorem 4.16 *(Peligrad and Utev) Let $(X_n)_{n\in\mathbb{Z}}$ be a stationary and ergodic sequence of random variables in $\mathbb{L}^2(\mathbb{P})$, centered and adapted to a stationary non-decreasing filtration $(\mathcal{F}_n)_{n\in\mathbb{Z}}$. Let $S_n = \sum_{k=1}^n X_k$. Assume (4.30) holds, i.e.*

$$\sum_{k\ge 1}\frac{\|\mathbb{E}_0(S_k)\|_2}{k^{3/2}} < \infty.$$

Then $\{n^{-1/2}S_{[nt]}, t \in [0,1]\}$ converges in distribution in $D([0,1])$ to a Brownian motion with parameter $\sigma^2 = \lim_{n\to\infty} n^{-1}\mathbb{E}(S_n^2)$. Moreover $\{\max_{1\le k\le n} S_k^2/n\}_{n\ge1}$ is an uniformly integrable family.

As we will show, Hannan's projection condition (4.29) implies (4.34). Hence, we can also deduce from Theorem 4.15 the functional form of Theorem 4.10, which is due to Hannan (1979).

Theorem 4.17 *(Hannan) Let $(X_n)_{n\in\mathbb{Z}}$ be a stationary ergodic sequence of random variables in $\mathbb{L}^2(\mathbb{P})$, centered and adapted to a stationary non-decreasing filtration $(\mathcal{F}_n)_{n\in\mathbb{Z}}$. Assume the projection condition (4.29) holds, i.e.*

$$\mathbb{E}(X_0|\mathcal{F}_{-\infty}) = 0 \text{ a.s. and } \sum_{i\ge0} \|\mathbf{P}_0(X_i)\|_2 < \infty.$$

Then $\{n^{-1/2}S_{[nt]}, t \in [0,1]\}$ converges in distribution in $D([0,1])$ to a Brownian motion with parameter $\sigma^2 = \lim_{n\to\infty} n^{-1}\mathbb{E}(S_n^2)$. Furthermore $\sigma^2 = \sum_{k\in\mathbb{Z}} \mathbb{E}(X_0X_k)$ where the series is absolutely convergent. Moreover $\{\max_{1\le k\le n} S_k^2/n\}_{n\ge1}$ is an uniformly integrable family.

Proof of Theorem 4.17 To prove that (4.29) implies (4.34), we apply Theorem 3.10 with $p=2$ and $g_k = 1$ to $(Y_k^m)_{k\ge0}$ defined by (4.38), namely:

$$Y_i^m = \frac{1}{m}\mathbb{E}_i(S_{i+m} - S_i).$$

By construction, stationarity and the triangle inequality, we get

$$\frac{1}{\sqrt{n}}\left\|\max_{1\le j\le n}\left|\sum_{k=0}^{j-1} Y_k^m\right|\right\|_2 \le 2\sum_{i=1}^{\infty} \|\mathbb{E}_{-i}(Y_0^m) - \mathbb{E}_{-i-1}(Y_0^m)\|_2$$
$$\le \frac{2}{m}\sum_{i=1}^{\infty}\sum_{k=1}^{m} \|\mathbb{E}_{-i}(X_k) - \mathbb{E}_{-i-1}(X_k)\|_2.$$

Now, by stationarity and the change of order of summation,

$$\frac{1}{\sqrt{n}}\left\|\max_{1\le j\le n}\left|\sum_{k=0}^{j-1} Y_k^m\right|\right\|_2 \le \frac{2}{m}\sum_{k=1}^{m}\sum_{j=k+1}^{\infty} \|\mathbb{E}_{-j}(X_0) - \mathbb{E}_{-j-1}(X_0)\|_2.$$

To verify condition (4.34) we let $n \to \infty$ followed by $m \to \infty$. Notice that the term in the right-hand side of the previous inequality tends to 0 as $m \to \infty$ by (4.29). ■

Finally, we state another invariance principle result, established by Dedecker and Rio (2000), who applied the Lindeberg method. Our proof is based on the maximal inequality stated in Theorem 3.11 and the forthcoming Theorem 4.32 (see Section 4.5.1).

Theorem 4.18 *Let $(X_n)_{n\in\mathbb{Z}}$ be a stationary ergodic sequence of random variables in $\mathbb{L}^2(\mathbb{P})$, centered and adapted to a stationary non-decreasing filtration $(\mathcal{F}_n)_{n\in\mathbb{Z}}$. Assume that*

$$\mathbb{E}(X_0 S_n|\mathcal{F}_0) \text{ is convergent in } \mathbb{L}^1 \text{ to a random variable } \mu \text{ as } n \to \infty. \tag{4.39}$$

Then, $\lim_{n\to\infty} \sigma_n^2/n = c := \mathbb{E}(X_0^2) + 2\sum_{i\geq 1}\mathbb{E}(X_0X_i)$. Moreover, if $c > 0$, then $\{n^{-1/2}S_{[nt]}$, $t\in[0,1]\}$ converges in distribution in $D([0,1])$ to a Brownian motion with parameter c. In addition $\{\max_{1\leq k\leq n} S_k^2/n\}_{n\geq 1}$ is an uniformly integrable family.

Let us mention that, in the statement of Theorem 4.18, the ergodicity assumption is not required in its full generality and can be weakened. Indeed, ergodicity can be replaced by 2-ergodicity as defined in Definition 1.2 in Dedecker-Saulière (2017), namely: A stationary sequence $(X_n)_{n\in\mathbb{Z}}$ will be said to be 2-ergodic if, for any bounded mesurable function f from $\mathbb{R}^2$ to $\mathbb{R}$, and any non-negative integer k,

$$\lim_{n\to\infty}\frac{1}{n}\sum_{i=1}^{n} f(X_i, X_{i+k}) = \mathbb{E}\big(f(X_0, X_k)\big) \text{ almost surely.}$$

Note also that, if we assume that $\lim_{n\to\infty}\alpha_{2,X}(n) = 0$, where the coefficients $\alpha_{2,X}(n)$ are defined in (5.13), then the sequence $(X_n)_{n\in\mathbb{Z}}$ is 2-ergodic (see e.g. the arguments developed on page 131 in Dedecker-Prieur (2007)).

4.4 On the Functional Central Limit Theorem for Non-Stationary Sequences in $\mathbb{L}^2$

4.4.1 *Statements*

Let us first assume the following general setting:

(C1) Assume $(X_{n,j})_{0\leq j\leq n}$ is an array of centered random variables that are square integrable and adapted to an array of sigma fields $(\mathcal{F}_{n,j})_{1\leq j\leq n}$, with $\mathcal{F}_{n,j}\subset\mathcal{F}_{n,j+1}$ for all $0\leq j\leq n-1$.

Introduce the notations: $S_{n,j} = \sum_{k=1}^{j} X_{n,k}$, $S_n = S_{n,n}$ and for $0\leq j\leq n$ denote

$$A_{n,j} = \mathbb{E}(S_n - S_{n,j}|\mathcal{F}_{n,j}), \tag{4.40}$$

and the projection operators

$$\mathbf{P}_{n,j}X = \mathbb{E}(X|\mathcal{F}_{n,j}) - \mathbb{E}(X|\mathcal{F}_{n,j-1}).$$

As a consequence of Theorem 2.34, we formulate:

Corollary 4.19 *Assume (C1) and* $\sup_{n\geq 1} \mathbb{E}S_n^2 \leq C < \infty$*. Also assume that*

$$\max_{0\leq j\leq n} |A_{n,j}| \to^{\mathbb{P}} 0 \text{ as } n \to \infty \tag{4.41}$$

and there exists a positive constant c^2 *such that for all* $t \in [0, 1]$,

$$\sum_{j=1}^{[nt]} (\mathbf{P}_{n,j} S_n)^2 \to^{\mathbb{P}} tc^2 \text{ as } n \to \infty. \tag{4.42}$$

Then $\left\{\sum_{i=1}^{[nt]} X_{n,i}, t \in [0, 1]\right\}$ *converges in distribution in* $D([0, 1])$ *to* cW *where* W *is a standard Brownian motion.*

Concerning condition (4.42), we have the following:

Proposition 4.20 *Under* $\sup_{n\geq 1} \mathbb{E}S_n^2 \leq C < \infty$ *and (4.41), condition (4.42) is equivalent to: There exists a constant* c^2 *such that for any* $t \in [0, 1]$,

$$\sum_{j=1}^{[nt]} (X_{n,j}^2 + 2X_{n,j} A_{n,j}) \to^{\mathbb{P}} tc^2 \text{ as } n \to \infty. \tag{4.43}$$

We also have the following proposition:

Proposition 4.21 *Let* $(X_i)_{i\geq 1}$ *be a sequence of real-valued random variables in* $\mathbb{L}^2$ *and set* $S_n = \sum_{i=1}^n X_i$ *for* $n \geq 1$ *and* $S_0 = 0$*. Let* $(\mathcal{F}_i)_{i\geq 0}$ *be a non-decreasing sequence of* σ*-algebras such that* X_i *is* $\mathcal{F}_i$*-measurable for any* $i \geq 1$*. Assume that, for any* $\varepsilon > 0$,

$$\lim_{n\to\infty} \frac{1}{\sqrt{n}} \sum_{i=1}^n \mathbb{E}\left(|X_i| 1_{|X_i|>\varepsilon\sqrt{n}}\right) = 0, \tag{4.44}$$

and that

$$\sum_{k\geq 1} k^{-3/2} a(k) < \infty \text{ where } a(k) = \max_{i\geq 0} \|\mathbb{E}(S_{k+i} - S_i|\mathcal{F}_i)\|_2. \tag{4.45}$$

Then

$$\frac{\max_{0\leq j\leq n} |\mathbb{E}(S_n - S_j|\mathcal{F}_j)|}{\sqrt{n}} \to^{\mathbb{P}} 0 \text{ as } n \to \infty. \tag{4.46}$$

In addition if (4.45) holds and $\sup_{n\geq 1} n^{-1} \sum_{j=1}^n \mathbb{E}(X_j^2) \leq C < \infty$*, then*

$$\sup_{n\geq 1} n^{-1} \|\max_{1\leq i\leq n} |S_i|\|_2^2 \leq C' < \infty. \tag{4.47}$$

Corollary 4.19 gives the following non-stationary version of Theorem 4.16:

Theorem 4.22 *Let $(X_n)_{n\geq 1}$ be a sequence of random variables in $\mathbb{L}^2(\mathbb{P})$, centered and adapted to a non-decreasing filtration $(\mathcal{F}_n)_{n\geq 1}$. Let $S_n = \sum_{k=1}^n X_k$. Suppose that $\sup_{n\geq 1} n^{-1}\sum_{j=1}^n \mathbb{E}(X_j^2) \leq C < \infty$, (4.44) and (4.45) hold. Assume in addition that there exists a constant c^2 such that for any $t \in [0, 1]$,*

$$\frac{1}{n}\sum_{i=1}^{[nt]} (\mathbf{P}_i(S_n))^2 \to^{\mathbb{P}} tc^2 \text{ as } n \to \infty, \tag{4.48}$$

where $\mathbf{P}_i(S_n) = \mathbb{E}_i(S_n) - \mathbb{E}_{i-1}(S_n)$. Then $\left\{n^{-1/2}\sum_{k=1}^{[nt]} X_k, t \in [0, 1]\right\}$ converges in distribution in $D([0, 1])$ to cW where W is a standard Brownian motion.

Remark 4.23 *According to Proposition 4.20, the conclusion of Theorem 4.22 still holds if we replace condition (4.48) by: For any $t \in [0, 1]$,*

$$\frac{1}{n}\sum_{k=1}^{[nt]}\left(X_k^2 + 2X_k \sum_{\ell=k+1}^{n} \mathbb{E}_k(X_\ell)\right) \to^{\mathbb{P}} tc^2 \text{ as } n \to \infty. \tag{4.49}$$

For use in applications, the condition (4.50) below can be easier to verify than (4.49). Our next result is in this direction.

Theorem 4.24 *Let $(X_n)_{n\geq 1}$ be a sequence of random variables in $\mathbb{L}^2(\mathbb{P})$, centered, and adapted to a non-decreasing filtration $(\mathcal{F}_n)_{n\geq 1}$. Let $S_n = \sum_{k=1}^n X_k$. Suppose that $\sup_{n\geq 1} n^{-1}\sum_{j=1}^n \mathbb{E}(X_j^2) \leq C < \infty$, (4.44) and (4.45) hold. Assume in addition that there exists a constant c^2 such that, for any $t \in [0, 1]$ and any $\varepsilon > 0$,*

$$\lim_{m\to\infty} \limsup_{n\to\infty} \mathbb{P}\left(\left|\frac{1}{n}\sum_{k=1}^{[nt]}\left(X_k^2 + \frac{2}{m}X_k \sum_{i=1}^{m-1} \mathbb{E}_k(S_{k+i} - S_k)\right) - tc^2\right| > \varepsilon\right) = 0. \tag{4.50}$$

Then $\left\{n^{-1/2}\sum_{k=1}^{[nt]} X_k, t \in [0, 1]\right\}$ converges in distribution in $D([0, 1])$ to cW where W is a standard Brownian motion.

We refer to Theorem 3.1 in Merlevède et al. (2019) for an extension of this result to general triangular arrays.

Remark 4.25 *Note that condition (4.45) holds if*

$$\sum_{k\geq 1} k^{-1/2} \sup_{i\geq 0} \|\mathbb{E}(X_{k+i}|\mathcal{F}_i)\|_2 < \infty. \tag{4.51}$$

We mention also the following proposition which allows to verify condition (4.50) in many situations:

Proposition 4.26 *Assume* $\sup_{n\geq 1} n^{-1}\sum_{j=1}^{n}\mathbb{E}(X_j^2)\leq C<\infty$,

$$\lim_{n\to\infty}\sup_{j\geq k\geq 0}\|\mathbb{E}_k(X_{k+n}X_{j+n})-\mathbb{E}_0(X_{k+n}X_{j+n})\|_1=0, \tag{4.52}$$

and that, for any $\varepsilon>0$,

$$\lim_{n\to\infty}\frac{1}{n}\sum_{k=1}^{n}\mathbb{E}\{X_k^2 I(|X_k|>\varepsilon\sqrt{n})\}=0. \tag{4.53}$$

Suppose in addition that there exists a positive constant c^2 *such that for any* $t\in[0,1]$ *and any* $\varepsilon>0$,

$$\lim_{m\to\infty}\limsup_{n\to\infty}\mathbb{P}\left(\left|\frac{1}{n}\sum_{k=1}^{[nt]}\left\{\mathbb{E}_0(X_k^2)+\frac{2}{m}\sum_{i=1}^{m-1}\mathbb{E}_0(X_k(S_{k+i}-S_k))\right\}-tc^2\right|>\varepsilon\right)=0. \tag{4.54}$$

Then condition (4.50) *holds with the same* c^2.

4.4.2 *Proofs*

Proof of Corollary 4.19 Note that

$$\mathbb{E}(\max_{1\leq k\leq n}(\mathbf{P}_{n,k}S_n)^2)\leq\mathbb{E}\left(\sum_{j=1}^{n}(\mathbf{P}_{n,j}S_n)^2\right)=\mathbb{E}(S_n-\mathbb{E}(S_n|\mathcal{F}_{n,0}))^2\leq\mathbb{E}(S_n^2)\leq C<\infty.$$

Under condition (4.42), we can then apply Theorem 2.29 to the triangular array of martingale differences $\{d_{n,i}\}_{1\leq i\leq n}$ where $d_{n,i}=\mathbf{P}_{n,i}S_n$. This entails that $\left\{\sum_{i=1}^{[nt]}\mathbf{P}_{n,i}S_n, t\in[0,1]\right\}$ converges in distribution in $D([0,1])$ to cW where W is a standard Brownian motion. But since

$$\sum_{i=1}^{k}X_{n,i}=S_{n,k}=\mathbb{E}(S_{n,k}|\mathcal{F}_{n,k})\text{ and }\sum_{i=1}^{k}\mathbf{P}_{n,i}S_n=\mathbb{E}(S_n|\mathcal{F}_{n,k})-\mathbb{E}(S_n|\mathcal{F}_{n,0}),$$

we get

$$\begin{aligned}\sup_{t\in[0,1]}\left|\sum_{i=1}^{[nt]}X_{n,i}-\sum_{i=1}^{[nt]}\mathbf{P}_{n,i}S_n\right|&=\max_{1\leq k\leq n}|\mathbb{E}(S_n-S_{n,k}|\mathcal{F}_{n,k})-\mathbb{E}(S_n|\mathcal{F}_{n,0})|\\&\leq\max_{1\leq k\leq n}|\mathbb{E}(S_n-S_{n,k}|\mathcal{F}_{n,k})|+|\mathbb{E}(S_n|\mathcal{F}_{n,0})|\leq 2\max_{0\leq k\leq n}|\mathbb{E}(S_n-S_{n,k}|\mathcal{F}_{n,k})|\\&=2\max_{0\leq j\leq n}|A_{n,j}|.\end{aligned}$$

Therefore under (4.41), the corollary follows. ∎

Proof of Proposition 4.20 For simplicity we drop the index n in the notation, so $X_j = X_{n,j}$, $A_j = A_{n,j}$, $d_j = d_{n,j} = \mathbf{P}_{n,j} S_j$. Note first that, for $1 \le j \le n$,

$$d_j := \mathbb{E}(S_n|\mathcal{F}_{n,j}) - \mathbb{E}(S_n|\mathcal{F}_{n,j-1}) = \mathbf{P}_{n,j}(S_n - S_{j-1}) := X_j + A_j - A_{j-1}. \tag{4.55}$$

Therefore,

$$d_j^2 = (X_j^2 + 2X_jA_j) + A_j^2 - A_{j-1}^2 + 2(A_{j-1} - X_j - A_j)A_{j-1},$$

and, again using (4.55), we obtain that, for any $t \in [0, 1]$,

$$\sum_{j=1}^{[nt]} d_j^2 = \sum_{j=1}^{[nt]} (X_j^2 + 2X_jA_j) - 2\sum_{j=1}^{[nt]} d_jA_{j-1} + A_{[nt]}^2 - A_0^2. \tag{4.56}$$

Note that the two last terms are convergent in probability to zero by taking into account condition (4.41). Therefore conditions (4.42) and (4.43) will be equivalent if we can prove that, for any $t \in [0, 1]$,

$$\sum_{j=1}^{[nt]} d_jA_{j-1} \to^{\mathbb{P}} 0, \ \text{as } n \to \infty. \tag{4.57}$$

With this aim, we use a truncation argument. Let $\varepsilon > 0$ and denote $A_j^\varepsilon = A_j I(|A_j| \le \varepsilon)$. For any $t \in [0, 1]$ and $a > 0$,

$$\begin{aligned}\mathbb{P}\left(\left|\sum_{j=1}^{[nt]} d_jA_{j-1}\right| > a\right) &\le \mathbb{P}\left(\max_{1\le j\le n} |A_j| > \varepsilon\right) + \mathbb{P}\left(\left|\sum_{j=1}^{[nt]} d_jA_{j-1}^\varepsilon\right| > a\right)\\ &\le \mathbb{P}\left(\max_{1\le j\le n} |A_j| > \varepsilon\right) + \frac{1}{a^2}\mathbb{E}\left(\sum_{j=1}^{[nt]} d_jA_{j-1}^\varepsilon\right)^2.\end{aligned}$$

But $(d_jA_{j-1}^\varepsilon)_{j\ge1}$ is a sequence of martingale differences. Therefore

$$\mathbb{E}\left(\sum_{j=1}^{[nt]} d_jA_{j-1}^\varepsilon\right)^2 = \sum_{j=1}^{[nt]} \mathbb{E}(d_jA_{j-1}^\varepsilon)^2 \le \varepsilon^2 \sum_{j=1}^{n} \mathbb{E}(d_j)^2 = \varepsilon^2\mathbb{E}\left(\sum_{j=1}^{n} d_j\right)^2 \le \varepsilon^2\mathbb{E}(S_n^2) \le C\varepsilon^2.$$

So, overall, for any $a > 0$,

$$\mathbb{P}\left(\left|\sum_{j=1}^{n} d_jA_{j-1}\right| > a\right) \le \mathbb{P}\left(\max_{1\le j\le n} |A_j| > \varepsilon\right) + \frac{C\varepsilon^2}{a^2}.$$

Then, we take into account that $\max_{1\le j\le n}|A_j|$ is negligible in probability by (4.41) and we conclude the convergence to 0 by letting $n\to\infty$ followed by $\varepsilon\to 0$. This ends the proof of (4.57) and then of the proposition. ■

Proof of Proposition 4.21 Let r be the only integer such that $2^{r-1}+1\le n\le 2^r$. Note that

$$\max_{0\le j\le n}\left|\mathbb{E}\left(S_n-S_j|\mathcal{F}_j\right)\right|\le\max_{0\le j\le n}\left|\mathbb{E}\left(S_{2^r}-S_n|\mathcal{F}_j\right)\right|+\max_{0\le j\le 2^r}\left|\mathbb{E}\left(S_{2^r}-S_j|\mathcal{F}_j\right)\right|.$$

Therefore, for any $\epsilon>0$,

$$\mathbb{P}\left(\max_{0\le j\le n}|\mathbb{E}(S_n-S_j|\mathcal{F}_j)|\ge 4\epsilon\sqrt{n}\right)\le\mathbb{P}\left(\max_{0\le j\le n}|\mathbb{E}(S_{2^r}-S_n|\mathcal{F}_j)|\ge\epsilon\sqrt{n}\right)$$
$$+\,\mathbb{P}\left(\max_{0\le j\le 2^r}|\mathbb{E}(S_{2^r}-S_j|\mathcal{F}_j)|\ge 3\epsilon\sqrt{n}\right).\tag{4.58}$$

Since $(\mathbb{E}(S_{2^r}-S_n|\mathcal{F}_j))_{j\ge0}$ is a martingale, we shall use Doob's maximal inequality to deal with the first term on the right-hand side. This gives

$$\mathbb{P}\left(\max_{0\le j\le n}|\mathbb{E}(S_{2^r}-S_n|\mathcal{F}_j)|\ge\epsilon\sqrt{n}\right)\le\frac{1}{\epsilon^2 n}\|\mathbb{E}(S_{2^r}-S_n|\mathcal{F}_n)\|_2^2.\tag{4.59}$$

To deal with the second term on the right-hand side of (4.58), we shall follow the lines of the proof of Proposition 16 in Merlevède *et al.* (2012) by writing first

$$\max_{0\le j\le 2^r}\left|\mathbb{E}(S_{2^r}-S_j|\mathcal{F}_j)\right|=\max_{0\le m\le 2^r}\left|\mathbb{E}(S_{2^r}-S_{2^r-m}|\mathcal{F}_{2^r-m})\right|.$$

Write now m in base 2 as follows:

$$m=\sum_{i=0}^{r}b_i(m)2^i,\quad\text{where } b_i(m)=0\text{ or } b_i(m)=1.$$

Set $m_l=\sum_{i=l}^{r}b_i(m)2^i$. With this notation $m_0=m$. Let u be an integer such that $0\le u\le r$ and write that

$$|\mathbb{E}(S_{2^r}-S_{2^r-m}|\mathcal{F}_{2^r-m})|\le|\mathbb{E}(S_{2^r-m_u}-S_{2^r-m}|\mathcal{F}_{2^r-m})|+|\mathbb{E}(S_{2^r}-S_{2^r-m_u}|\mathcal{F}_{2^r-m})|.$$

Notice first that

$$\mathbb{P}\left(\max_{0\le m\le 2^r}|\mathbb{E}(S_{2^r-m_u}-S_{2^r-m}|\mathcal{F}_{2^r-m})|\ge 2\epsilon\sqrt{n}\right)$$
$$\le\mathbb{P}\left(\max_{0\le m\le 2^r}\sum_{j=2^r-m+1}^{2^r-m_u}|\mathbb{E}(X_j|\mathcal{F}_{2^r-m})|\ge 2\epsilon\sqrt{n}\right).$$

But, by using the fact that $m - m_u \le 2^u$,

$$\sum_{j=2^r-m+1}^{2^r-m_u} |X_j| \le \epsilon\sqrt{n} + \sum_{j=2^r-m+1}^{2^r-m_u} |X_j| 1_{\{|X_j| \ge \epsilon 2^{-u}\sqrt{n}\}}.$$

Hence,

$$\begin{aligned}
&\mathbb{P}\Big(\max_{0\le m\le 2^r} |\mathbb{E}(S_{2^r-m_u} - S_{2^r-m}|\mathcal{F}_{2^r-m})| \ge 2\epsilon\sqrt{n}\Big)\\
&\le \mathbb{P}\Big(\max_{0\le m\le 2^r} \sum_{j=2^r-m+1}^{2^r-m_u} \mathbb{E}(|X_j| 1_{\{|X_j|\ge \epsilon 2^{-u}\sqrt{n}\}}|\mathcal{F}_{2^r-m}) \ge \epsilon\sqrt{n}\Big)\\
&\le \mathbb{P}\Big(\max_{0\le m\le 2^r} \sum_{j=1}^{2^r} \mathbb{E}(|X_j| 1_{\{|X_j|\ge \epsilon 2^{-u}\sqrt{n}\}}|\mathcal{F}_{2^r-m}) \ge \epsilon\sqrt{n}\Big).
\end{aligned}$$

Since $\left(\sum_{j=1}^{2^r} \mathbb{E}(|X_j| 1_{\{|X_j|\ge \epsilon 2^{-u}\sqrt{n}\}}|\mathcal{F}_k)\right)_{k\ge 0}$ is a martingale, Doob's maximal inequality implies that

$$\mathbb{P}\Big(\max_{0\le m\le 2^r} |\mathbb{E}(S_{2^r-m_u} - S_{2^r-m}|\mathcal{F}_{2^r-m})| \ge 2\epsilon\sqrt{n}\Big) \le \frac{1}{\epsilon\sqrt{n}} \sum_{i=1}^{2^r} \mathbb{E}(|X_i| 1_{\{|X_i|\ge \epsilon 2^{-u}\sqrt{n}\}}). \tag{4.60}$$

On the other hand, following the proof of Proposition 3.14, for any $m \in \{0, ..., 2^r\}$ and any $p \ge 1$, we get

$$\left|\mathbb{E}(S_{2^r} - S_{2^r-m_u}|\mathcal{F}_{2^r-m})\right| \le \sum_{l=u}^{r} \mathbb{E}(A_{r,l}|\mathcal{F}_{2^r-m}),$$

where

$$A_{r,l} = \max_{1\le k\le 2^{r-l}, k \text{ odd}} |\mathbb{E}(S_{2^r-(k-1)2^l} - S_{2^r-k2^l}|\mathcal{F}_{2^r-k2^l})|.$$

Notice now that $\left(\sum_{l=u}^{r} \mathbb{E}(A_{r,l}|\mathcal{F}_k)\right)_{k\ge 0}$ is a martingale, so by Doob's maximal inequality again,

$$\mathbb{P}\Big(\max_{0\le m\le 2^r} |\mathbb{E}(S_{2^r} - S_{2^r-m_u}|\mathcal{F}_{2^r-m})| \ge \epsilon\sqrt{n}\Big) \le \frac{1}{\epsilon^2 n}\Big\|\sum_{l=u}^{r} \mathbb{E}(A_{r,l}|\mathcal{F}_{2^r})\Big\|_2^2 \le \frac{1}{\epsilon^2 n}\Big(\sum_{l=u}^{r} \|A_{r,l}\|_2\Big)^2.$$

Since

$$A_{r,l} \le \left(\sum_{k=1}^{2^{r-l}-1} \left|\mathbb{E}(S_{2^r-(k-1)2^l} - S_{2^r-k2^l}|\mathcal{F}_{2^r-k2^l})\right|^2\right)^{1/2},$$

it follows that

$$\mathbb{P}\left(\max_{0\le m\le 2^r-1}\left|\mathbb{E}(S_{2^r}-S_{2^r-m_u}|\mathcal{F}_{2^r-m})\right|\ge\epsilon\sqrt{n}\right)$$
$$\le\frac{1}{\epsilon^2 n}\left(\sum_{l=u}^{r}\left(\sum_{k=1}^{2^{r-l}-1}\|\mathbb{E}(S_{(k+1)2^l}-S_{k2^l}|\mathcal{F}_{k2^l})\|_2^2\right)^{1/2}\right)^2. \tag{4.61}$$

Considering the bounds (4.60) and (4.61), the following inequality holds: For any integer $u\in[0,r]$,

$$\mathbb{P}\left(\max_{0\le j\le 2^r}|\mathbb{E}(S_{2^r}-S_j|\mathcal{F}_j)|\ge 3\epsilon\sqrt{n}\right)\le\frac{1}{\epsilon\sqrt{n}}\sum_{i=1}^{2^r}\mathbb{E}\left(|X_i|1_{\{|X_i|\ge\epsilon\sqrt{n}/2^u\}}\right)$$
$$+\frac{1}{\epsilon^2 n}\left(\sum_{l=u}^{r}\left(\sum_{k=1}^{2^{r-l}-1}\|\mathbb{E}(S_{(k+1)2^l}-S_{k2^l}|\mathcal{F}_{k2^l})\|_2^2\right)^{1/2}\right)^2. \tag{4.62}$$

Setting

$$a(k)=\max_{i\ge 0}\|\mathbb{E}(S_{k+i}-S_i|\mathcal{F}_i)\|_2,$$

it follows that

$$\left(\sum_{l=u}^{r}\left(\sum_{k=1}^{2^{r-l}-1}\|\mathbb{E}(S_{(k+1)2^l}-S_{k2^l}|\mathcal{F}_{k2^l})\|_2^2\right)^{1/2}\right)^2\le 2^r\left(\sum_{l=u}^{r}2^{-l/2}a(2^l)\right)^2. \tag{4.63}$$

Therefore starting from (4.58) and considering the upper bounds (4.59), (4.62) and (4.63), we derive that, for any integer $u\in[0,r]$,

$$\mathbb{P}\left(\max_{0\le j\le n}|\mathbb{E}(S_n-S_j|\mathcal{F}_j)|\ge 4\epsilon\sqrt{n}\right)\le\frac{1}{\epsilon\sqrt{n}}\sum_{i=1}^{2^r}\mathbb{E}\left(|X_i|1_{\{|X_i|\ge\epsilon\sqrt{n}/2^u\}}\right)$$
$$+\frac{a^2(2^r-n)}{\epsilon^2 n}+\frac{2}{\epsilon^2}\left(\sum_{l=u}^{r}2^{-l/2}a(2^l)\right)^2. \tag{4.64}$$

Note now that $(a(k))_{k\ge 0}$ satisfies the following property: For any $k,\ell\ge 0$,

$$a(k+\ell)\le a(k)+a(\ell). \tag{4.65}$$

Indeed, by the triangle inequality, for any $k,\ell\ge 0$,

$$a(k+\ell)\le\max_{i\ge 0}\|\mathbb{E}(S_{k+\ell+i}-S_{k+i}|\mathcal{F}_i)\|_2+\max_{i\ge 0}\|\mathbb{E}(S_{k+i}-S_i|\mathcal{F}_i)\|_2.$$

But, by the properties of the conditional expectation,

$$\|\mathbb{E}(S_{k+\ell+i} - S_{k+i}|\mathcal{F}_i)\|_2 \le \|\mathbb{E}(S_{k+\ell+i} - S_{k+i}|\mathcal{F}_{k+i})\|_2 \le a(\ell),$$

which ends the proof of (4.65). From this subadditivity property and Item 5 of Lemma 3.24, we get

$$\sum_{l=u}^{r} 2^{-l/2} a(2^l) \le 2^{5/2}(2^{3/2}-1)^{-1}\left\{2^{-u/2}\max_{1\le j\le 2^u} a(j) + \sum_{j=2^u+1}^{2^r} j^{-3/2} a(j)\right\}. \tag{4.66}$$

This upper bound together with (4.64) entail

$$\mathbb{P}\left(\max_{0\le j\le n} |\mathbb{E}(S_n - S_j|\mathcal{F}_j)| \ge 4\epsilon\sqrt{n}\right) \le \frac{1}{\epsilon\sqrt{n}}\sum_{i=1}^{2^r}\mathbb{E}\left(|X_i|1_{\{|X_i|\ge\epsilon\sqrt{n}/2^u\}}\right)$$
$$+\frac{a^2(2^r-n)}{\epsilon^2 n} + \frac{C}{\epsilon^2}\left(\sum_{j=2^u+1}^{2^r} j^{-3/2}a(j) + 2^{-u/2}\max_{1\le j\le 2^u} a(j)\right)^2. \tag{4.67}$$

We use now the following lemma whose proof is given later.

Lemma 4.27 *For every $\gamma > 0$ and $n \ge 1$,*

$$\frac{1}{n^\gamma}\max_{1\le k\le n} a(k) \le c_\gamma \sum_{k=n+1}^{6n}\frac{1}{k^{\gamma+1}}a(k),$$

where $c_\gamma = 2^{2\gamma+3}$.

This lemma combined with (4.67) gives that, for any integer $u \in [0, r]$ and any $\epsilon > 0$,

$$\mathbb{P}\left(\max_{0\le j\le n}\left|\mathbb{E}\left(S_n - S_j|\mathcal{F}_j\right)\right| \ge 4\epsilon\sqrt{n}\right) \ll \frac{1}{\epsilon\sqrt{n}}\sum_{i=1}^{2n}\mathbb{E}\left(|X_i|1_{\{|X_i|\ge\epsilon\sqrt{n}/2^u\}}\right) + \frac{1}{\epsilon^2}\left(\sum_{j\ge 2^u} j^{-3/2}a(j)\right)^2. \tag{4.68}$$

Taking into account conditions (4.44) and (4.45), and letting first n tend to infinity followed by u, the condition (4.46) follows.

We continue the proof of the proposition by proving its second part, namely: If $\sup_{n\ge1} n^{-1}\sum_{j=1}^n \mathbb{E}(X_j^2) \le C < \infty$ and (4.45) is satisfied, then (4.47) holds. With this aim, note that inequality (3.10) implies

$$n^{-1/2}\|\max_{1\le i\le n}|S_i|\|_2 \le 3n^{-1/2}\left(\sum_{j=1}^n \|X_j\|_2^2\right)^{1/2} + 6\sum_{j\ge0} 2^{-j/2}a(2^j). \tag{4.69}$$

Inequality (4.47) then follows by taking into account that $\sup_{n\geq 1} n^{-1}\sum_{i=1}^{n}\mathbb{E}(X_i^2) \leq C < \infty$, and by noticing that by (4.66) (used with $u = 0$), condition (4.45) implies that $\sum_{l\geq 0} 2^{-l/2}a(2^l) < \infty$.

To end the proof of the proposition, it remains to prove Lemma 4.27.

Proof of Lemma 4.27 We start by proving that in addition to satisfying (4.65), $(a(j))_{j\geq 0}$ satisfies the following property: For any $n, k \geq 0$,

$$a(n) \leq a(k+n) + a(k). \tag{4.70}$$

To prove it, we use again the triangle inequality which gives

$$a(n) \leq \max_{i\geq 0} \|\mathbb{E}(S_{n+k+i} - S_i|\mathcal{F}_i)\|_2 + \max_{i\geq 0} \|\mathbb{E}(S_{n+k+i} - S_{n+i}|\mathcal{F}_i)\|_2.$$

Inequality (4.70) then follows by taking into account the properties of the conditional expectation that entail

$$\|\mathbb{E}(S_{n+k+i} - S_{n+i}|\mathcal{F}_i)\|_2 \leq \|\mathbb{E}(S_{n+k+i} - S_{n+i}|\mathcal{F}_{n+i})\|_2 \leq a(k).$$

We follow now the lines of the proof of Lemma 19 in Merlevède *et al.* (2012). Note first that, for any $n \geq 1$,

$$\frac{1}{n^\gamma} = \frac{1}{n^{\gamma+1}}\left(\sum_{k=n+1}^{2n} 1\right) \leq 2^{\gamma+1}\sum_{k=n+1}^{2n}\frac{1}{k^{\gamma+1}}.$$

By taking into account (4.70), it follows that

$$\begin{aligned}\frac{1}{n^\gamma}a(n) &\leq 2^{\gamma+1}\sum_{k=n+1}^{2n}\frac{1}{k^{\gamma+1}}a(k+n) + 2^{\gamma+1}\sum_{k=n+1}^{2n}\frac{1}{k^{\gamma+1}}a(k)\\ &\leq 2^{2\gamma+2}\sum_{k=n+1}^{2n}\frac{1}{(k+n)^{\gamma+1}}a(k+n) + 2^{\gamma+1}\sum_{k=n+1}^{2n}\frac{1}{k^{\gamma+1}}a(k).\end{aligned}$$

Therefore

$$\frac{1}{n^\gamma}a(n) \leq 2^{2\gamma+2}\sum_{l=n+1}^{3n}\frac{1}{l^{\gamma+1}}a(l). \tag{4.71}$$

By (4.70), we have that for any integer $k \geq 0$

$$a(k) \leq a(k+n) + a(n),$$

so that

$$\max_{1\le k\le n} a(k) \le \max_{n\le k\le 2n} a(k) + a(n) \le 2 \max_{n\le k\le 2n} a(k).$$

The result then follows by the inequality (4.71) applied for each k, $n \le k \le 2n$. ■

Proof of Theorem 4.24 The proof will follow from Theorem 4.22 together with the proposition below.

Proposition 4.28 *Let $(X_i)_{i\in\mathbb{N}}$ be a sequence of real-valued random variables in $\mathbb{L}^2$ and set $S_n = \sum_{i=1}^n X_i$ for $n \ge 1$ and $S_0 = 0$. Let $(\mathcal{F}_i)_{i\ge 0}$ be a non-decreasing sequence of σ-algebras such that X_i is $\mathcal{F}_i$-measurable for any $i \ge 1$. Suppose that $\sup_{n\ge 1} n^{-1}\sum_{j=1}^n \mathbb{E}(X_j^2) \le C < \infty$. Assume in addition that (4.44), (4.45) and (4.50) are satisfied. Then condition (4.48) holds.*

Proof of Proposition 4.28 Let m be a fixed positive integer. Let us define

$$\theta_\ell^m = \frac{1}{m}\sum_{i=1}^{m-1} \mathbb{E}_\ell(X_{\ell+1} + \dots + X_{\ell+i}),\ D_\ell^m = \frac{1}{m}\sum_{i=0}^{m-1} \mathbf{P}_\ell(S_{\ell+i}),$$

and

$$Y_\ell^m = \frac{1}{m}\mathbb{E}_\ell(S_{\ell+m} - S_\ell),\ R_k^m = \sum_{\ell=0}^{k-1} Y_\ell^m.$$

Then, $(D_\ell^m)_{\ell\ge 1}$ is an array of martingale differences adapted to the filtration $(\mathcal{F}_\ell)_{\ell\in\mathbb{Z}}$ and, according to (4.20), we have the decomposition

$$X_\ell = D_\ell^m + \theta_{\ell-1}^m - \theta_\ell^m + Y_{\ell-1}^m. \tag{4.72}$$

Also, for any positive integer m and k, we have

$$S_k = M_k^m + \theta_0^m - \theta_k^m + R_k^m. \tag{4.73}$$

To prove the proposition, we write, for any $t \in [0, 1]$ and any positive integer m,

$$\begin{aligned}\sum_{j=1}^{[nt]}(\mathbf{P}_j(S_n))^2 &= \sum_{j=1}^{[nt]}(\mathbf{P}_j(S_n) - D_j^m)^2 + 2\sum_{j=1}^{[nt]}(\mathbf{P}_j(S_n) - D_j^m)D_j^m + \sum_{j=1}^{[nt]}(D_j^m)^2 \\ &= 2\sum_{j=1}^{[nt]}(\mathbf{P}_j(S_n) - D_j^m)(\mathbf{P}_j S_n) - \sum_{j=1}^{[nt]}(\mathbf{P}_j(S_n) - D_j^m)^2 + \sum_{j=1}^{[nt]}(D_j^m)^2.\end{aligned}$$

Note that, by Cauchy–Schwarz's inequality, for any $t \in [0, 1]$,

$$\begin{aligned}
&n^{-1}\mathbb{E}\left|\sum_{j=1}^{[nt]}\left(\mathbf{P}_j(S_n) - D_j^m\right)\left(\mathbf{P}_j(S_n)\right)\right| \\
&\leq n^{-1}\mathbb{E}\left\{\left(\sum_{j=1}^{n}(\mathbf{P}_j(S_n) - D_j^m)^2\right)^{1/2}\left(\sum_{j=1}^{n}(\mathbf{P}_j(S_n))^2\right)^{1/2}\right\} \\
&\leq \left(n^{-1}\sum_{j=1}^{n}\mathbb{E}(\mathbf{P}_j(S_n) - D_j^m)^2\right)^{1/2}\left(n^{-1}\sum_{j=1}^{n}\mathbb{E}(\mathbf{P}_j(S_n))^2\right)^{1/2}.
\end{aligned}$$

But $n^{-1}\sum_{j=1}^{n}\mathbb{E}(\mathbf{P}_j(S_n))^2 = n^{-1}\mathbb{E}(S_n - \mathbb{E}_0(S_n))^2 \leq n^{-1}\mathbb{E}(S_n^2) \leq C$ by taking into account the second part of Proposition 4.21. So, to prove that condition (4.48) is satisfied, it suffices to show that

$$\lim_{m\to\infty}\limsup_{n\to\infty} n^{-1}\sum_{j=1}^{n}\mathbb{E}(\mathbf{P}_j(S_n) - D_j^m)^2 = 0 \tag{4.74}$$

and that, for any $t \in [0, 1]$ and any $\varepsilon > 0$,

$$\lim_{m\to\infty}\limsup_{n\to\infty}\mathbb{P}\left(\left|\frac{1}{n}\sum_{j=1}^{[nt]}(D_j^m)^2 - c^2t\right| \geq \varepsilon\right) = 0. \tag{4.75}$$

To prove the first convergence we write

$$\mathbf{P}_j(S_n) - D_j^m = \mathbf{P}_j(S_n) - \frac{1}{m}\sum_{u=0}^{m-1}\mathbf{P}_j(S_{u+j}) = \frac{1}{m}\sum_{u=0}^{m-1}\mathbf{P}_j(S_n - S_{u+j}).$$

By orthogonality,

$$\sum_{j=1}^{n}\mathbb{E}\left(\frac{1}{m}\sum_{u=0}^{m-1}\mathbf{P}_j(S_n - S_{u+j})\right)^2 = \frac{1}{m^2}\mathbb{E}\left(\sum_{j=1}^{n}\sum_{u=0}^{m-1}\mathbf{P}_j(S_n - S_{u+j})\right)^2.$$

For all u fixed, $0 \leq u \leq m-1$, we have

$$\begin{aligned}
\sum_{j=1}^{n}\mathbf{P}_j(S_n - S_{u+j}) &= \sum_{j=1}^{n}\left(\mathbb{E}_j(S_n - S_{j+u}) - \mathbb{E}_{j-1}(S_n - S_{j-1+u}) + \mathbb{E}_{j-1}(X_{j+u})\right) \\
&= \mathbb{E}_n(S_n - S_{n+u}) - \mathbb{E}_0(S_n - S_u) + \sum_{j=1}^{n}\mathbb{E}_{j-1}(X_{j+u}).
\end{aligned}$$

So,

$$\sum_{j=1}^{n}\sum_{u=0}^{m-1}\mathbf{P}_j(S_n - S_{u+j}) = \sum_{u=0}^{m-1}\left(\mathbb{E}_n(S_n - S_{n+u}) - \mathbb{E}_0(S_n - S_u) + \sum_{j=1}^{n}\mathbb{E}_{j-1}(X_{j+u})\right).$$

Therefore,

$$\frac{1}{n}\sum_{j=1}^{n}\mathbb{E}(\mathbf{P}_j(S_n) - D_j^m)^2 = \frac{1}{nm^2}\left\|\sum_{u=0}^{m-1}\left(\mathbb{E}_n(S_n - S_{n+u}) - \mathbb{E}_0(S_n - S_u) + \sum_{j=1}^{n}\mathbb{E}_{j-1}(X_{j+u})\right)\right\|_2^2$$
$$\leq \frac{3}{n}\left(\frac{1}{m}\sum_{u=0}^{m-1}a(u)\right)^2 + \frac{3}{n}\left(\frac{1}{m}\sum_{u=0}^{m-1}a(n-u)\right)^2 + \frac{3}{n}\left\|\sum_{j=0}^{n-1}Y_j^m\right\|_2^2,$$

where we recall that for any $k \geq 0$, $a(k) = \sup_{i\geq 0}\|\mathbb{E}(S_{k+i} - S_i|\mathcal{F}_i)\|_2$. The first term on the right-hand side obviously converges to zero as $n \to \infty$. For the second term, we note that

$$\frac{1}{m\sqrt{n}}\sum_{u=0}^{m-1}a(n-u) \leq \frac{1}{\sqrt{n}}\max_{1\leq k\leq n}a(k),$$

which converges to zero as $n \to \infty$ by using Lemma 4.27 and taking into account condition (4.45). Hence, the convergence (4.74) is reduced to prove that

$$\lim_{m\to\infty}\limsup_{n\to\infty} n^{-1}\left\|\sum_{j=0}^{n-1}Y_j^m\right\|_2^2 = 0,$$

which follows from the next lemma together with condition (4.45).

Lemma 4.29 *There exists a positive constant C such that, for all positive integers $n \geq 1$ and $m \leq n$,*

$$\frac{1}{n^{1/2}}\left\|\max_{1\leq j\leq n}\left|\sum_{k=0}^{j-1}Y_k^m\right|\right\|_2 \leq C\sum_{k=m+1}^{6n}\frac{a(k)}{k^{3/2}}. \tag{4.76}$$

Note that the above lemma is a generalization of inequality (4.35) to the non-stationary setting. Its proof is given later in this section.

To end the proposition, it remains to prove (4.75). Starting from the decomposition (4.72), we have

$$\begin{aligned}(D_\ell^m)^2 &= (X_\ell - \theta_{\ell-1}^m + \theta_\ell^m)^2 - 2(X_\ell - \theta_{\ell-1}^m + \theta_\ell^m)Y_{\ell-1}^m + (Y_{\ell-1}^m)^2\\ &= (X_\ell - \theta_{\ell-1}^m + \theta_\ell^m)^2 - 2D_\ell^m Y_{\ell-1}^m - (Y_{\ell-1}^m)^2\\ &= (X_\ell^2 + 2X_\ell\theta_\ell^m) + (\theta_\ell^m)^2 - (\theta_{\ell-1}^m)^2 - 2D_\ell^m(Y_{\ell-1}^m + \theta_{\ell-1}^m) - 2\theta_{\ell-1}^m Y_{\ell-1}^m - (Y_{\ell-1}^m)^2.\end{aligned}$$

It follows that, for any $t \in [0, 1]$,

$$\sum_{\ell=1}^{[nt]} (D_\ell^m)^2 = \sum_{\ell=1}^{[nt]} (X_\ell^2 + 2X_\ell \theta_\ell^m) + (\theta_{[nt]}^m)^2 - (\theta_0^m)^2$$
$$- 2\sum_{\ell=1}^{[nt]} D_\ell^m (Y_{\ell-1}^m + \theta_{\ell-1}^m) - 2\sum_{\ell=1}^{[nt]} \theta_{\ell-1}^m Y_{\ell-1}^m - \sum_{\ell=1}^{[nt]} (Y_{\ell-1}^m)^2. \tag{4.77}$$

Note that

$$\frac{1}{\sqrt{n}} \left(\|\theta_{[nt]}^m\|_2 + \|\theta_0^m\|_2 \right) \le \frac{2}{\sqrt{n}} \max_{1 \le i \le m-1} a(i) \to 0, \text{ as } n \to \infty.$$

Note now that $n^{-1} \sum_{\ell=1}^{n} \mathbb{E}(Y_{\ell-1}^m)^2 \le m^{-2} a^2(m)$, which converges to zero as $m \to \infty$ by taking into account Lemma 4.27 and condition (4.45). This shows also that, for any $\varepsilon > 0$,

$$\lim_{m \to \infty} \limsup_{n \to \infty} \mathbb{P}(\max_{0 \le \ell \le n-1} |Y_\ell^m| > \varepsilon \sqrt{n}) = 0.$$

Proceeding, as in the proof of Proposition 4.20, it follows that the martingale transform $n^{-1} \sum_{\ell=1}^{[nt]} D_\ell^m Y_{\ell-1}^m$ has a negligible contribution since, by (4.73),

$$\frac{1}{n} \sum_{\ell=1}^{n} \mathbb{E}(D_\ell^m)^2 = \frac{1}{n} \mathbb{E} \left(\sum_{\ell=1}^{n} D_\ell^m \right)^2 = \frac{1}{n} \mathbb{E} \left(S_n - \theta_0^m + \theta_n^m - R_n^m \right)^2$$
$$\le \frac{4}{n} \mathbb{E}(S_n^2) + \frac{8}{nm} \max_{1 \le i \le m} a^2(i) + \frac{4}{n} \mathbb{E}(R_n^m)^2,$$

which is less than a constant non-depending on (n, m) by taking into account the second part of Proposition 4.21, Lemma 4.27 and Lemma 4.29. Similarly, the martingale transform $n^{-1} \sum_{\ell=1}^{[nt]} D_\ell^m \theta_{\ell-1}^m$ has a negligible contribution since for any m fixed,

$$\frac{1}{\sqrt{n}} \max_{0 \le \ell \le n} |\theta_\ell^m| \to^{\mathbb{P}} 0, \text{ as } n \to \infty.$$

Indeed

$$\frac{1}{\sqrt{n}} \max_{0 \le \ell \le n} |\theta_\ell^m| \le \frac{1}{\sqrt{n}} \max_{0 \le \ell \le 2^r} |\mathbb{E}_\ell (S_{2^r} - S_\ell)| + \frac{1}{m\sqrt{n}} \max_{0 \le \ell \le 2^r} \sum_{i=1}^{m-1} |\mathbb{E}_\ell (S_{2^r+i} - S_{\ell+i})|$$
$$+ \frac{1}{m\sqrt{n}} \max_{0 \le \ell \le 2^r} \sum_{i=1}^{m-1} |\mathbb{E}_\ell (S_{2^r+i} - S_{2^r})| , \tag{4.78}$$

where r is the unique positive integer such that $2^{r-1} < n \leq 2^r$. Note that the sequence $\big(\sum_{i=1}^{m-1} |\mathbb{E}_\ell(S_{2^r+i} - S_{2^r})|\big)_{\ell\geq 0}$ is a submartingale with respect to the filtration $(\mathcal{F}_\ell)_{\ell\geq 0}$. By Doob's maximal inequality (see Theorem 2.5), it follows that, for any $\varepsilon > 0$,

$$\mathbb{P}\left(\max_{0\leq\ell\leq 2^r} \sum_{i=1}^{m-1} |\mathbb{E}_\ell(S_{2^r+i} - S_{2^r})| \geq \varepsilon\sqrt{nm}\right)$$
$$\leq \frac{1}{m^2 n\varepsilon^2}\mathbb{E}\left(\sum_{i=1}^{m-1} |\mathbb{E}_{2^r}(S_{2^r+i} - S_{2^r})|\right)^2 \leq \frac{1}{n\varepsilon^2}\left(\frac{1}{m}\sum_{i=1}^{m-1} a(i)\right)^2 \to 0, \text{ as } n \to \infty.$$

To prove now that the two first terms on the right-hand side of (4.78) converge to zero in probability as $n \to \infty$, it suffices to use the first part of Proposition 4.21 since (4.44) and (4.45) are assumed. So, overall, starting from (4.77) and taking into account the above considerations and condition (4.50), the proof of (4.75) is reduced to show that, for any $t \in [0, 1]$ and any $\varepsilon > 0$,

$$\lim_{m\to\infty} \limsup_{n\to\infty} \mathbb{P}\left(\left|\frac{1}{n}\sum_{\ell=1}^{[nt]} \theta^m_{\ell-1} Y^m_{\ell-1}\right| \geq \varepsilon\right) = 0.$$

But

$$\mathbb{E}\left|\frac{1}{n}\sum_{\ell=1}^{[nt]} \theta^m_{\ell-1} Y^m_{\ell-1}\right| \leq \frac{1}{n}\sum_{\ell=1}^{n} \|\theta^m_{\ell-1}\|_2 \|Y^m_{\ell-1}\|_2 \leq \frac{1}{m^2} a(m)\left(\sum_{i=1}^{m-1} a(i)\right) \leq \frac{1}{m}\max_{1\leq i\leq m} a^2(i),$$

which converges to zero as $m \to \infty$ taking into account Lemma 4.27 and condition (4.45). This ends the proof of (4.75).

To finish the proof of Proposition 4.28, it remains to prove Lemma 4.29.

Proof of Lemma 4.29 In order to prove (4.76), we apply inequality (3.10) to the sequence $(Y^m_k)_{k\in\mathbb{Z}}$. Recalling the notation $R^m_k = \sum_{\ell=0}^{k-1} Y^m_\ell$, we get

$$\left\|\max_{1\leq j\leq n}\left|\sum_{k=0}^{j-1} Y^m_k\right|\right\|_2 \leq 3\left(\sum_{k=0}^{n-1} \|Y^m_k\|_2^2\right)^{1/2} + 3\sqrt{2}\sum_{r=0}^{d}\left(\sum_{k=1}^{2^{d-r}} \|\mathbb{E}(R^m_{k2^r} - R^m_{(k-1)2^r} | \mathcal{F}_{(k-2)2^r})\|_2^2\right)^{1/2}, \tag{4.79}$$

where d is the unique positive integer such that $2^{d-1} \leq n < 2^d$ (so $d \leq [\log_2(2n)]$). We first notice that

$$\sum_{k=0}^{n-1} \|Y^m_k\|_2^2 \leq \frac{n}{m^2} a^2(m). \tag{4.80}$$

We estimate now the last term of (4.79). With this aim, it is convenient to use the decomposition

$$\sum_{r=0}^{d}\left(\sum_{k=1}^{2^{d-r}}\left\|\mathbb{E}\left(R^m_{k2^r}-R^m_{(k-1)2^r}|\mathcal{F}_{(k-2)2^r}\right)\right\|_2^2\right)^{1/2}=\sum_{r=0}^{b}\left(\sum_{k=1}^{2^{d-r}}\left\|\mathbb{E}\left(R^m_{k2^r}-R^m_{(k-1)2^r}|\mathcal{F}_{(k-2)2^r}\right)\right\|_2^2\right)^{1/2}$$

$$+\sum_{r=b+1}^{d}\left(\sum_{k=1}^{2^{d-r}}\left\|\mathbb{E}\left(R^m_{k2^r}-R^m_{(k-1)2^r}|\mathcal{F}_{(k-2)2^r}\right)\right\|_2^2\right)^{1/2},$$

where b is the unique positive integer such that $2^{b-1}\leq m<2^b$. To estimate the first sum on the right-hand side, notice that, by the properties of the conditional expectation, we have

$$\begin{aligned}\left\|\mathbb{E}\left(R^m_{k2^r}-R^m_{(k-1)2^r}|\mathcal{F}_{(k-2)2^r}\right)\right\|_2 &\leq \sum_{\ell=0}^{2^r-1}\left\|\mathbb{E}\left(Y^m_{\ell+(k-1)2^r}|\mathcal{F}_{(k-2)2^r}\right)\right\|_2\\ &\leq \frac{1}{m}\sum_{\ell=0}^{2^r-1}\left\|\mathbb{E}\left(S_{\ell+(k-1)2^r+m}-S_{\ell+(k-1)2^r}|\mathcal{F}_{(k-2)2^r}\right)\right\|_2\\ &\leq \frac{1}{m}\sum_{\ell=0}^{2^r-1}\left\|\mathbb{E}\left(S_{\ell+(k-1)2^r+m}-S_{\ell+(k-1)2^r}|\mathcal{F}_{(k-1)2^r+\ell}\right)\right\|_2\leq\frac{2^r}{m}a(m).\end{aligned}$$

Therefore,

$$\sum_{r=0}^{b}\left(\sum_{k=1}^{2^{d-r}}\left\|\mathbb{E}(R^m_{k2^r}-R^m_{(k-1)2^r}|\mathcal{F}_{(k-2)2^r})\right\|_2^2\right)^{1/2}\leq 2^{d/2}\frac{a(m)}{m}\sum_{r=0}^{b}2^{r/2}\leq\frac{2\sqrt{2}}{\sqrt{2}-1}\sqrt{n}\frac{a(m)}{\sqrt{m}}.$$

To estimate the second sum we also apply the properties of the conditional expectation and write this time

$$\left\|\mathbb{E}(R^m_{k2^r}-R^m_{(k-1)2^r}|\mathcal{F}_{(k-2)2^r})\right\|_2\leq\frac{1}{m}\sum_{u=0}^{m-1}\|\mathbb{E}(S_{k2^r+u}-S_{(k-1)2^r+u}|\mathcal{F}_{(k-1)2^r+u})\|_2\leq a(2^r).$$

Hence

$$\sum_{r=b+1}^{d}\left(\sum_{k=1}^{2^{d-r}}\left\|\mathbb{E}(R^m_{k2^r}-R^m_{(k-1)2^r}|\mathcal{F}_{(k-2)2^r})\right\|_2^2\right)^{1/2}\leq 2^{d/2}\sum_{r=b+1}^{d}2^{-r/2}a(2^r).$$

So, overall,

$$\sum_{r=0}^{d}\left(\sum_{k=1}^{2^{d-r}}\left\|\mathbb{E}(R^m_{k2^r}-R^m_{(k-1)2^r}|\mathcal{F}_{(k-2)2^r})\right\|_2^2\right)^{1/2} \le \frac{2\sqrt{2}}{\sqrt{2}-1}\sqrt{n}\frac{a(m)}{\sqrt{m}}+\sqrt{2n}\sum_{r=b+1}^{d}2^{-r/2}a(2^r). \tag{4.81}$$

Starting from (4.79) and taking into account the upper bounds (4.80) and (4.81), we derive

$$n^{-1/2}\left\|\max_{1\le j\le n}\left|\sum_{k=0}^{j-1}Y_k^m\right|\right\|_2 \le 3\left(1+\frac{4}{\sqrt{2}-1}\right)\frac{a(m)}{\sqrt{m}}+6\sum_{r=b+1}^{d}2^{-r/2}a(2^r),$$

which proves the lemma by taking into account (4.66) and Lemma 4.27. ∎

Proof of Proposition 4.26 Clearly condition (4.54) will imply (4.50) with the same c^2, if we can prove that, for any $t\in[0,1]$ and any fixed non-negative integer ℓ,

$$\lim_{n\to\infty} n^{-1}\left\|\sum_{k=1}^{[nt]}(\mathbb{E}_k(X_kX_{k+\ell})-\mathbb{E}_0(X_kX_{k+\ell}))\right\|_1 = 0. \tag{4.82}$$

With this aim, we first write the following decomposition: For any positive fixed integer b (less than $[nt]$),

$$\begin{aligned}\sum_{k=1}^{[nt]}(\mathbb{E}_k(X_kX_{k+\ell})-\mathbb{E}_0(X_kX_{k+\ell})) &= \sum_{k=1}^{b}(\mathbb{E}_k(X_kX_{k+\ell})-\mathbb{E}_0(X_kX_{k+\ell}))\\ &\quad+\sum_{k=b+1}^{[nt]}(\mathbb{E}_k(X_kX_{k+\ell})-\mathbb{E}_0(X_kX_{k+\ell})).\end{aligned}$$

But,

$$n^{-1}\sum_{k=1}^{b}\|\mathbb{E}_k(X_kX_{k+\ell})-\mathbb{E}_0(X_kX_{k+\ell})\|_1 \le 2n^{-1}\sum_{k=1}^{b}\|X_k\|_2\|X_{k+\ell}\|_2 \le 2n^{-1}\sum_{k=1}^{b+\ell}\|X_k\|_2^2.$$

Hence, for any $\varepsilon>0$,

$$n^{-1}\sum_{k=1}^{b}\|\mathbb{E}_k(X_kX_{k+\ell})-\mathbb{E}_0(X_kX_{k+\ell})\|_1 \le 2\left\{\varepsilon^2(b+\ell)+n^{-1}\sum_{k=1}^{b+\ell}\mathbb{E}\left(X_k^2 1_{|X_k|>\varepsilon\sqrt{n}}\right)\right\},$$

which converges to zero as $n \to \infty$ followed by $\varepsilon \to 0$, by taking into account condition (4.53). Now

$$\sum_{k=b+1}^{[nt]} (\mathbb{E}_k(X_k X_{k+\ell}) - \mathbb{E}_0(X_k X_{k+\ell})) = \sum_{k=b+1}^{[nt]} (\mathbb{E}_k(X_k X_{k+\ell}) - \mathbb{E}_{k-b}(X_k X_{k+\ell})) + \sum_{k=b+1}^{[nt]} (\mathbb{E}_{k-b}(X_k X_{k+\ell}) - \mathbb{E}_0(X_k X_{k+\ell})).$$

Taking into account condition (4.52), we have

$$\lim_{b\to\infty} \limsup_{n\to\infty} n^{-1} \left\| \sum_{k=b+1}^{[nt]} (\mathbb{E}_{k-b}(X_k X_{k+\ell}) - \mathbb{E}_0(X_k X_{k+\ell})) \right\|_1 = 0.$$

We show now that

$$\lim_{b\to\infty} \limsup_{n\to\infty} n^{-1} \left\| \sum_{k=b+1}^{[nt]} (\mathbb{E}_k(X_k X_{k+\ell}) - \mathbb{E}_{k-b}(X_k X_{k+\ell})) \right\|_1 = 0. \tag{4.83}$$

Together with the convergences proved above, this will show that (4.82) is satisfied.

To prove (4.83), we fix a positive real ε and write

$$\left\| \sum_{k=b+1}^{[nt]} (\mathbb{E}_k(X_k X_{k+\ell}) - \mathbb{E}_{k-b}(X_k X_{k+\ell})) \right\|_1 \le \left\| \sum_{k=b+1}^{[nt]} \left(\mathbb{E}_k(Y'_{k,\ell}) - \mathbb{E}_{k-b}(Y'_{k,\ell})\right) \right\|_1 + \left\| \sum_{k=b+1}^{[nt]} \left(\mathbb{E}_k(Y''_{k,\ell}) - \mathbb{E}_{k-b}(Y''_{k,\ell})\right) \right\|_1$$

where

$$Y'_{k,\ell} = X_k X_{k+\ell} \mathbf{1}_{|X_k X_{k+\ell}| \le \varepsilon^2 n} \text{ and } Y''_{k,\ell} = X_k X_{k+\ell} \mathbf{1}_{|X_k X_{k+\ell}| > \varepsilon^2 n}.$$

Note now that the following inequalities are valid: For any reals a and b and any positive real M,

$$|ab| \mathbf{1}_{\{|ab|>M\}} \le 2^{-1} \left(|a^2 + b^2| 1_{\{|a^2+b^2|>2M\}}\right) \le a^2 \mathbf{1}_{\{a^2>M\}} + b^2 \mathbf{1}_{\{b^2>M\}}. \tag{4.84}$$

Hence,

$$n^{-1} \left\| \sum_{k=b+1}^{[nt]} \left(\mathbb{E}(Y''_{k,\ell}|\mathcal{F}_k) - \mathbb{E}(Y''_{k,\ell}|\mathcal{F}_{k-b})\right) \right\|_1 \le 4n^{-1} \sum_{k=1}^{n+\ell} \mathbb{E}(X_k^2 1_{|X_k|>\varepsilon\sqrt{n}}),$$

which together with condition (4.53) implies that

$$\lim_{n\to\infty} n^{-1}\left\|\sum_{k=b+1}^{[nt]}\left(\mathbb{E}_k(Y''_{k,\ell})-\mathbb{E}_{k-b}(Y''_{k,\ell})\right)\right\|_1 = 0. \tag{4.85}$$

On the other hand,

$$\sum_{k=b+1}^{[nt]}\left(\mathbb{E}_k(Y'_{k,\ell})-\mathbb{E}_{k-b}(Y'_{k,\ell})\right)=\sum_{k=b+1}^{[nt]}\sum_{j=0}^{b-1}\mathbf{P}_{k-j}(Y'_{k,\ell})=\sum_{j=0}^{b-1}\sum_{k=b+1}^{[nt]}\mathbf{P}_{k-j}(Y'_{k,\ell}),$$

where we recall $\mathbf{P}_j = \mathbb{E}(\cdot|\mathcal{F}_j)-\mathbb{E}(\cdot|\mathcal{F}_{j-1})$. Since $(\mathbf{P}_{k-j}(Y'_{k,\ell}))_{k\geq 1}$ is a sequence of martingale differences,

$$\left\|\sum_{k=b+1}^{[nt]}\left(\mathbb{E}_k(Y'_{k,\ell})-\mathbb{E}_{k-b}(Y'_{k,\ell})\right)\right\|_1\leq\sum_{j=0}^{b-1}\left\|\sum_{k=b+1}^{[nt]}\mathbf{P}_{k-j}(Y'_{k,\ell})\right\|_2\leq\sum_{j=0}^{b-1}\left(\sum_{k=b+1}^{[nt]}\|\mathbf{P}_{k-j}(Y'_{k,\ell})\|_2^2\right)^{1/2}.$$

But

$$\|\mathbf{P}_{k-j}(Y'_{k,\ell})\|_2^2\leq\|\mathbb{E}_{k-j}(Y'_{k,\ell})\|_2^2\leq\varepsilon^2 n\|X_kX_{k+\ell}\|_1\leq 2^{-1}\varepsilon^2 n(\|X_k\|_2^2+\|X_{k+\ell}\|_2^2).$$

Therefore,

$$n^{-1}\left\|\sum_{k=b+1}^{[nt]}\left(\mathbb{E}_k(Y'_{k,\ell})-\mathbb{E}_{k-b}(Y'_{k,\ell})\right)\right\|_1\leq b\,\varepsilon\sup_{n\geq 1}\left(\frac{1}{n}\sum_{k=b+1}^{n+\ell}\|X_k\|_2^2\right)^{1/2},$$

which converges to zero by letting ε going to 0. This last convergence together with (4.85) entail (4.83) and then (4.82). This ends the proof of the proposition. ■

4.5 Toward a More General Normalization

In previous sections, we have surveyed several aspects of the central limit theorem and its weak invariance principle under conditions ensuring that the variance of partial sums is linear in n. More precisely, we treated the case when

$$\lim_{n\to\infty}\frac{\mathrm{Var}(S_n)}{n}=\sigma^2<\infty.$$

The central limit theorem and its invariance principle might hold also under more general conditions than treated in Sections 4.2 and 4.3. The price to pay is a change of normalization to either

$$b_n=\sigma_n=\mathrm{Stdev}(S_n)\quad\text{or}\quad b_n=\sqrt{\frac{\pi}{2}}\,\mathbb{E}|S_n|\,.$$

These normalizations were used in many papers involving uniformly mixing structures starting with a paper of Ibragimov (1975) and followed by works of Dehling, Denker and Philipp (1986) or Peligrad (1990), among others.

Concerning the normalization $b_n = \sqrt{\frac{\pi}{2}}\,\mathbb{E}|S_n|$, Merlevède and Peligrad (2000) and Merlevède (2003) studied the invariance principle for strongly mixing sequences under the condition:

$$\liminf_{n\to\infty} \frac{\mathrm{Var}(S_n)}{n} > 0.$$

The results in both these papers are extremely sharp. The sufficient conditions based on a combination of mixing rates and the quantile function of a variable in the sequence are minimal for the CLT as shown by very sharp counterexamples in Bradley (1997–a).

The central limit theorem and its invariance principle for martingale-like sequences under a normalization different from $\sqrt{n}$ came more recently in the focus of research of several mathematicians. In this direction we would like to mention the papers by Dedecker and Merlevède (2002), Wu and Woodroofe (2004) and Merlevède and Peligrad (2006).

When the normalization is $\sigma_n = \mathrm{Stdev}(S_n)$ and $W_n = S_{[nt]}/\sigma_n$ converges weakly to the Brownian motion, then, necessarily σ_n^2 has the representation $\sigma_n^2 = nh(n)$, with $h(n)$ a slowly varying function in the strong sense at infinity, meaning that there exists a continuous function $f : (0, \infty) \to (0, \infty)$ such that $f(n) = h(n)$ for all $n \in \mathbb{N}$, and $f(x)$ is slowly varying as $x \to \infty$. It is natural then to give sufficient conditions for this representation of σ_n^2.

Proposition 4.30 *Assume that $(X_i)_{i\in\mathbb{Z}}$ is a stationary sequence of centered and square integrable random variables, adapted to a stationary non-decreasing filtration $(\mathcal{F}_i)_{i\in\mathbb{Z}}$. Assume that $\sigma_n^2 \to \infty$, as $n \to \infty$, and that one of the conditions below is satisfied:*

$$(i) \qquad \|\mathbb{E}(S_n|\mathcal{F}_0)\|_2 = o(\sigma_n) \text{ as } n \to \infty \tag{4.86}$$

or

$$(ii) \qquad \liminf_{n\to\infty} \frac{\mathrm{Var}(S_n)}{n} > 0 \quad \text{and} \quad n\mathbb{E}(X_0X_n) \to 0 \text{ as } n \to \infty.$$

Then

$$\sigma_n^2 = nh(n) \text{ where } h(n) \text{ is slowly varying in the strong sense at infinity.} \tag{4.87}$$

We refer to Theorem 8.13 in Bradley (2007) or also to Lemma 1 in Wu and Woodroofe (2004) for proofs of these facts.

The first result we would like to mention is in the spirit of Theorem 1 in Dedecker and Merlevède (2002) and is a version of Theorem 1 in Merlevède and Peligrad (2006). The result will be applied later on to linear processes and strong mixing sequences.

Theorem 4.31 *Let $(X_i)_{i\in\mathbb{Z}}$ be a stationary sequence of square integrable and centered random variables, adapted to a stationary non-decreasing filtration $(\mathcal{F}_i)_{i\in\mathbb{Z}}$. Set $S_n := \sum_{k=1}^n X_k$ and $\sigma_n^2 = \mathrm{Var}(S_n)$. Assume that $\sigma_n^2 \to \infty$,*

$$\frac{S_n^2}{\sigma_n^2} \text{ is an uniformly integrable family} \tag{4.88}$$

$$\sigma_n^{-1}\mathbb{E}(S_n|\mathcal{F}_0) \to^{\mathbb{P}} 0 \text{ as } n \to \infty \tag{4.89}$$

and

$$\lim_{n\to\infty} \mathbb{E}\left(\frac{S_n^2}{\sigma_n^2}|\mathcal{F}_0\right) = 1 \text{ in } \mathbb{L}_1. \tag{4.90}$$

Then $\sigma_n^{-1}S_n$ converges in distribution to N, where N is a standard Gaussian r.v.

Note that by standard arguments, (4.88) and (4.89) imply (4.86) (and then also (4.87)).

Proof of Theorem 4.31 We apply Corollary 2.27 to super big blocks of random variables. More exactly, for a fixed positive integer m, let

$$S_{n,j} := S_{n,j}^{(m)} = \sum_{i=[n/m](j-1)+1}^{[n/m]j} X_i,$$

$$\mathcal{F}_{n,j} := \mathcal{F}_{n,j}^{(m)} = \mathcal{F}_{j[n/m]} \text{ and } \mathbb{E}_j(X) = \mathbb{E}(X|(\mathcal{F}_{[n/m]j}).$$

By the conditions of the theorem, we can choose sufficiently slow growing to infinity $m = m_n$ such that as $n \to \infty$,

$$\left|\frac{m\sigma_{[n/m]}^2}{\sigma_n^2} - 1\right| \to 0, \ \frac{m}{\sigma_n} \to 0, \ \frac{\sigma_n}{\sigma_{[n/m]}} \to \infty, \tag{4.91}$$

$$m\left\|\mathbb{E}_0\left(\frac{S_{n,1}}{\sigma_{[n/m]}}\right)\right\|_2^2 \to 0, \mathbb{E}\left|\mathbb{E}_0\left(\frac{S_{n,1}^2}{\sigma_{[n/m]}^2} - 1\right)\right| \to 0. \tag{4.92}$$

For the chosen sequence of m_n, we shall apply Corollary 2.27 to the sum $\sigma_n^{-1}S_{m[n/m]}$ with

$$\sigma_n^{-1}S_{m[n/m]} = \sum_{j=1}^m X_{n,j} \quad \text{with stationary connected summands}$$

$$X_{n,j} = S_{n,j}^{(m)}/\sigma_n, \ j = 1 \ldots, m.$$

First, note that the second convergence in (4.91) together with stationarity entail

$$\sigma_n^{-1}\mathbb{E}|S_n - S_{m[n/m]}| \le \mathbb{E}|X_1|\frac{m}{\sigma_n} \to 0 \text{ as } n \to \infty.$$

Now, we need to check the following three conditions,

$$I_n = \sum_{j=1}^{m} \mathbb{E}_{j-1}(X_{n,j}) \to^{\mathbb{P}} 0 \text{ as } n \to \infty, \tag{4.93}$$

$$J_n = \sum_{j=1}^{m} \mathrm{Var}_{j-1}(X_{n,j}) \to^{\mathbb{P}} 1 \text{ as } n \to \infty \tag{4.94}$$

and for each $\varepsilon > 0$,

$$Q_n = \sum_{j=1}^{m} \mathbb{E}_{j-1}\left(X_{n,j}^2 I(|X_{n,j}| > \varepsilon)\right) \to^{\mathbb{P}} 0 \text{ as } n \to \infty. \tag{4.95}$$

It is easier to prove the $\mathbb{L}^1$ convergence rather than the convergence in probability.

By stationarity, construction and the first convergence in (4.92) and (4.91), we get

$$\mathbb{E}|I_n| \le m\mathbb{E}|\mathbb{E}_0(X_{n,1})| = \frac{m\sigma_{[n/m]}}{\sigma_n}\left\|\mathbb{E}_0\left(\frac{S_{n,1}}{\sigma_{[n/m]}}\right)\right\|_1 \le \sqrt{m}\frac{\sigma_{[n/m]}}{\sigma_n}\sqrt{m}\left\|\mathbb{E}_0\left(\frac{S_{n,1}}{\sigma_{[n/m]}}\right)\right\|_2 \to 0$$

proving first convergence (4.93).

The convergence (4.94) follows from

$$\begin{aligned}
\mathbb{E}|J_n - 1| &= \frac{1}{\sigma_n^2}\mathbb{E}\left|\left\{\sum_{j=1}^{m} \mathbb{E}_{j-1}(S_{n,j}^2) - \mathbb{E}_{j-1}^2(S_{n,j})\right\} - \sigma_n^2\right| \\
&\le \frac{m\sigma_{[n/m]}^2}{\sigma_n^2}\mathbb{E}\left|\mathbb{E}_0\left(\frac{S_{n,1}^2}{\sigma_{[n/m]}^2} - 1\right)\right| + \frac{m\sigma_{[n/m]}^2}{\sigma_n^2}\mathbb{E}\left|\mathbb{E}_0\left(\frac{S_{n,1}}{\sigma_{[n/m]}}\right)\right|^2 + \left|\frac{m\sigma_{[n/m]}^2}{\sigma_n^2} - 1\right|,
\end{aligned}$$

which converges to zero as $n \to \infty$ by (4.92) and (4.91). Finally, the convergence (4.95) follows from

$$\begin{aligned}
\mathbb{E}(Q_n) &= m\mathbb{E}[X_{n,1}^2 I(|X_{n,1}| > \varepsilon)] \\
&= \frac{m\sigma_{[n/m]}^2}{\sigma_n^2}\mathbb{E}\left(\frac{S_{[n/m]}^2}{\sigma_{[n/m]}^2} I\left(\frac{|S_{[n/m]}|}{\sigma_{[n/m]}} > \varepsilon\frac{\sigma_n}{\sigma_{[n/m]}}\right)\right) \\
&\to 0 \text{ as } n \to \infty,
\end{aligned}$$

since $(S_k^2/\sigma_k^2)_{k\geq 1}$ is an uniformly integrable family and in addition $\dfrac{\sigma_n}{\sigma_{[n/m]}} \to \infty$ and $\dfrac{m\sigma_{[n/m]}^2}{\sigma_n^2} \to 1$ as $n \to \infty$ by the construction (4.91). ■

Adding tightness, we extend Theorem 4.31 to the invariance principle.

Theorem 4.32 *Assume that the conditions of Theorem 4.31 are satisfied and in addition, suppose the tightness condition is satisfied in the form*

$$\lim_{\lambda\to\infty} \limsup_{n\to\infty} \lambda^2 \mathbb{P}\left(\max_{1\leq i\leq n} \frac{|S_i|}{\sigma_n} \geq \lambda\right) = 0. \tag{4.96}$$

Then $\sigma_n^{-1}W_n$ converges in distribution to W in $D([0,1])$, where W is a standard Brownian motion.

Clearly conditions (4.88) and (4.96) are satisfied if

$$\frac{\max_{1\leq i\leq n} S_i^2}{\sigma_n^2} \text{ is uniformly integrable.} \tag{4.97}$$

The following corollary is useful in many applications.

Corollary 4.33 *Suppose that $\sigma_n^2 \to \infty$. Assume that the conditions (4.89) and (4.90) are satisfied and that for some $t > 2$,*

$$\mathbb{E}|S_n|^t \leq K(\mathrm{Var}(S_n))^{t/2}, \textit{ for all } n \geq 1. \tag{4.98}$$

Then $\sigma_n^{-1}W_n$ converges in distribution to W in $D([0,1])$, where W is a standard Brownian motion.

To prove it, notice first that by Proposition 4.30, $\sigma_n^2 = nh(n)$, with $h(n)$ a slowly varying function. Indeed, it is clear that (4.98) implies (4.88), which together with (4.89) entails (4.86). Moreover by applying the maximal inequality given in Lemma 1.20, it follows that (4.97) holds under (4.98). The corollary follows now from Theorem 4.32. ■

Finally, we end this section with the following lemma that can be useful to verify condition (4.89).

Lemma 4.34 *Assume that (4.87) holds and that*

$$\|\mathbb{E}(S_n|\mathcal{F}_{-n})\|_1 = o(\sigma_n) \textit{ as } n \to \infty, \tag{4.99}$$

then (4.89) holds.

To prove it, we start with the binary expansion

$$n = \sum_{k=0}^{r-1} a_k 2^k \text{ where } a_{r-1} = 1 \text{ and } a_k \in \{0,1\}.$$

Then, we have the following representation

$$S_n = \sum_{j=0}^{r-1} a_j T_{2^j} \text{ where } T_{2^j} = S_{n_j} - S_{n_{j-1}},\ n_j = \sum_{k=0}^{j} a_k 2^k,\ n_{-1} = 0.$$

Applying this representation, by the properties of the norm, we easily derive that

$$\| \mathbb{E}(S_n|\mathcal{F}_0)\|_1 \le \sum_{j=0}^{r-1} \| \mathbb{E}(S_{2^j}|\mathcal{F}_0)\|_1. \tag{4.100}$$

Now, by stationarity,

$$\| \mathbb{E}(S_{2^j}|\mathcal{F}_0)\|_1 \le \| \mathbb{E}(S_{2^{j-1}}|\mathcal{F}_0)\|_1 + \| \mathbb{E}(S_{2^{j-1}}|\mathcal{F}_{-2^{j-1}})\|_1.$$

It follows by recurrence that

$$\| \mathbb{E}(S_{2^j}|\mathcal{F}_0)\|_1 \le \sum_{k=1}^{j-1} \| \mathbb{E}(S_{2^k}|\mathcal{F}_{-2^k})\|_1.$$

Starting from this inequality and using (4.99), we derive that

$$\| \mathbb{E}(S_{2^j}|\mathcal{F}_0)\|_1 \le \sum_{k=1}^{N} \sigma_{2^k} + \epsilon_N \sigma_{2^j} \sum_{k=N+1}^{j-1} \frac{\sigma_{2^k}}{\sigma_{2^j}},$$

where ϵ_N is such that $\lim_{N\to\infty} \epsilon_N = 0$.

Now since $\sigma_n^2 = nh(n)$, by the properties of the slowly varying functions, we infer that $\sum_{k=N+1}^{j-1} \frac{\sigma_{2^k}}{\sigma_{2^j}} < C$, where C is a constant not depending on N and j. Then, by first letting j tend to infinity and after N, it follows easily that

$$\| \mathbb{E}(S_{2^j}|\mathcal{F}_0)\|_1 = o(\sigma_{2^j}). \tag{4.101}$$

Now starting from (4.100) and using (4.101) together with the preceding arguments it follows that $\| \mathbb{E}(S_n|\mathcal{F}_0)\|_1 = o(\sigma_{2^{r-1}})$. Since $2^{r-1} \le n < 2^r$, it follows from the fact that $\sigma_n^2 = nh(n)$ and the properties of the slowly varying functions that $\sigma_{2^{r-1}} = O(\sigma_n)$. This completes the proof of the lemma. ■

4.5.1 *Proof of Theorem 4.18*

The proof of this theorem is divided in several steps and follows from an application of Theorem 4.32.

Step 1. For any integer $k \geq 1$, let $c_k = \sigma_k^2 - \sigma_{k-1}^2$. By stationarity

$$c_k = \mathbb{E}(X_0^2) + 2\sum_{i=1}^{k-1} \mathbb{E}(X_0 X_i).$$

Therefore, under (4.39), c_k converges to c. On another hand $\frac{\sigma_n^2}{n} = \frac{1}{n}\sum_{k=1}^{n} c_k$. Hence via Cesàro's lemma, $\lim_{n\to\infty} \sigma_n^2/n = c$. We assume now in all the rest of the proof that $c > 0$.

Step 2. We prove here that (4.97) is satisfied. With this aim and since $c > 0$, it suffices to show that

$$\lim_{\lambda\to\infty} \limsup_{n\to\infty} n^{-1}\, \mathbb{E}((S_n^* - \lambda\sqrt{n})_+^2) = 0, \tag{4.102}$$

where $S_n^* = \max_{1\leq i\leq n} |S_i|$. From Theorem 3.11 applied to the sequences $(X_i)_{i\in\mathbb{Z}}$ and $(-X_i)_{i\in\mathbb{Z}}$ and stationarity we derive that

$$\mathbb{E}\left(\left(S_n^* - \lambda\sqrt{n}\right)_+^2\right) \leq 8\, n\, \mathbb{E}\left(X_0^2\, \mathbf{I}_{S_n^*>\lambda\sqrt{n}}\right) + 16\sum_{k=1}^{n-1} \mathbb{E}\left|X_0\, \mathbb{E}\left(S_k|\mathcal{F}_0\right)\, \mathbf{I}_{S_n^*>\lambda\sqrt{n}}\right|. \tag{4.103}$$

Now the condition (4.39) entails that for each $k \geq 1$, $\mathbb{E}\,|X_0\, \mathbb{E}(S_k|\mathcal{F}_0)| \leq K$ where K is a positive constant. Hence, from (4.103) with $\lambda = 0$ we get that

$$n^{-1}\, \mathbb{E}(S_n^*)^2 \leq 8\, \mathbb{E}(X_0^2) + 16K := C,$$

implying that

$$\mathbb{P}(S_n^* > \lambda\sqrt{n}) \leq C\lambda^{-2}, \tag{4.104}$$

which converges to zero as λ tends to infinity.

Hence from (4.104), (4.103) and (4.39), we easily get that

$$n^{-1}\, \mathbb{E}((S_n^* - \lambda\sqrt{n})_+^2) \leq \delta(\lambda)$$

for some non-increasing function δ satisfying $\lim_{\lambda\to\infty} \delta(\lambda) = 0$. This proves (4.102) and then (4.97).

Step 3. We prove here condition (4.89). Since $\lim_{n\to\infty} \sigma_n^2/n = c > 0$, by Lemma 4.34, (4.89) will hold if we can prove that (4.99) is satisfied, namely:

$$\|\, \mathbb{E}(S_n|\mathcal{F}_{-n})\|_1 = o\left(\sqrt{n}\right) \text{ as } n \to \infty.$$

To prove the convergence above, we first notice that $\mathbb{E}(|X_0|) > 0$. Indeed if not then $X_k = 0$ a.s. for all k and we will have $c = 0$. Now, for any N, a fixed positive integer less than n, we set

$$A_N = \frac{1}{N} \sum_{i=1}^{N} |X_{-i}| - \mathbb{E}(|X_0|).$$

We then write

$$\begin{aligned}
\frac{1}{\sqrt{n}} \|\mathbb{E}(|X_0|)\mathbb{E}(S_n|\mathcal{F}_{-n})\|_1 &\le \frac{1}{\sqrt{n}} \|\mathbb{E}_{-n}(S_n A_N)\|_1 + \frac{1}{N\sqrt{n}} \sum_{i=1}^{N} \|\mathbb{E}_{-n}(|X_{-i}|S_n)\|_1 \\
&\le \frac{1}{\sqrt{n}} \left\| \mathbb{E}_{-n}(S_n A_N) \right\|_1 + \frac{1}{N\sqrt{n}} \sum_{i=1}^{N} \left\| X_{-i} \mathbb{E}_{-i}(S_n) \right\|_1 .
\end{aligned} \tag{4.105}$$

From stationarity and (4.39), we infer that the second term on the right-hand side tends to zero as n tends to infinity. Applying first Cauchy–Schwarz's inequality and next the $\mathbb{L}^2$-ergodic theorem, we easily deduce that the first term on the right-hand side is as small as we wish by choosing N large enough. This ends the proof of Step 3.

Step 4. We finish the proof by proving that (4.90) holds, namely:

$$\lim_{n\to\infty} \|\sigma_n^{-2}\, \mathbb{E}(S_n^2|\mathcal{F}_0) - 1\,\|_1 = 0.$$

Notice first that

$$\sigma_n^{-2}\, \mathbb{E}(S_n^2|\mathcal{F}_0) = \sigma_n^{-2} \sum_{i=1}^{n} \mathbb{E}(X_i^2|\mathcal{F}_0) + 2\sigma_n^{-2} \sum_{i=1}^{n-1} \mathbb{E}(X_i(S_n - S_i)|\mathcal{F}_0).$$

By the fact that $\lim_{n\to\infty} \frac{\sigma_n^2}{n} = c > 0$ and since the sequence is ergodic and the $\mathbb{L}^1$ convergence is preserved under conditioning, Theorem 1.3 leads to

$$\sigma_n^{-2} \sum_{i=1}^{n} \mathbb{E}(X_i^2|\mathcal{F}_0) \to c^{-1}\, \mathbb{E}(X_0^2) \text{ in } \mathbb{L}^1, \text{ as } n \to \infty. \tag{4.106}$$

Hence to show that (4.90) holds, it remains to prove that

$$\left\| \sigma_n^{-2} \sum_{i=1}^{n-1} \mathbb{E}(X_i(S_n - S_i)|\mathcal{F}_0) - \eta \right\|_1 \to 0, \text{ as } n \to \infty, \tag{4.107}$$

where $\eta = 2^{-1}(1 - c^{-1}\, \mathbb{E}(X_0^2))$. Observe that, by stationarity and the properties of the conditional expectation,

$$\left\| \sigma_n^{-2} \sum_{i=1}^{n-1} \mathbb{E}(X_i(S_n - S_i)|\mathcal{F}_0) - \eta \right\|_1 \le \sigma_n^{-2} \sum_{i=1}^{n-1} \|X_0 \mathbb{E}_0(S_i) - \mu\|_1 + \mathbb{E}\left| \sigma_n^{-2} \sum_{i=1}^{n-1} \mathbb{E}_0(\mu \circ T^i) - \eta \right|, \tag{4.108}$$

where T is the bi-measurable transformation such that $X_k = X_0 \circ T^k$. According to (4.39), we get

$$\sigma_n^{-2} \sum_{i=1}^{n-1} \|X_0 \mathbb{E}_0(S_i) - \mu\|_1 \le \frac{K(N)}{\sigma_n^2} + \varepsilon_N \frac{n}{\sigma_n^2}, \tag{4.109}$$

where $K(N)$ is a constant only depending on N and ε_N is such that $\lim_{N\to\infty} \varepsilon_N = 0$. This shows that the first term on the right-hand side of (4.108) converges to zero as n tends to infinity. Now, using once again that the sequence is ergodic and the fact that the $\mathbb{L}^1$ convergence is preserved under conditioning, Theorem 1.3 leads to

$$\frac{1}{\sigma_n^2} \sum_{i=1}^{n-1} \mathbb{E}_0(\mu \circ T^i) \to c^{-1}\mathbb{E}(\mu) \text{ in } \mathbb{L}^1, \text{ as } n \to \infty.$$

This shows that the last term on the right-hand side of (4.108) also converges to zero as n tends to infinity since

$$\mathbb{E}(X_0)^2 + 2\mathbb{E}(\mu) = \lim_{n\to\infty} \frac{\mathbb{E}(S_n^2)}{n} = c,$$

so $\eta = c^{-1}\mathbb{E}(\mu)$. This ends the proof of Step 4 and completes the proof of the theorem. ■

4.6 Moderate Deviations

In what follows, by $\|X\|_\infty$ we denote the $\mathbb{L}^\infty$-norm, that is the smallest u such that $\mathbb{P}(|X| > u) = 0$. For a sequence $(X_i)_{i\in\mathbb{Z}}$ of random variables, let

$$S_n = \sum_{j=1}^{n} X_j \text{ and } W_n(t) = n^{-1/2} \sum_{i=1}^{[nt]} X_i.$$

Theorem 4.35 *Let $(X_i)_{i\in\mathbb{Z}}$ be a strictly stationary sequence of real-valued random variables adapted to a non-decreasing and stationary filtration $(\mathcal{F}_i)_{i\in\mathbb{Z}}$. Assume that $\mathbb{E}(X_0|\mathcal{F}_{-\infty}) = 0$ a.s. Assume in addition that*

$$\sum_{j\ge 0} \|\mathbf{P}_0(X_j)\|_\infty < \infty, \tag{4.110}$$

and

$$\lim_{n\to\infty} n^{-1}\|\mathbb{E}(S_n^2|\mathcal{F}_0) - \mathbb{E}(S_n^2)\|_\infty = 0. \tag{4.111}$$

Then $n^{-1}\mathbb{E}(S_n^2)$ converges to $\sigma^2 = \sum_{j\in\mathbb{Z}} \mathbb{E}(X_0X_j)$ and, for all positive sequences a_n with $a_n \to 0$ and $na_n \to \infty$, the normalized partial sums processes $W_n(.)$ satisfy the moderate deviation principle in $D([0,1])$, namely the property (1.31), with speed a_n and the good rate function $I_\sigma(\cdot)$ defined by

$$I_\sigma(h) = \frac{1}{2\sigma^2}\int_0^1 (h'(u))^2du \tag{4.112}$$

if simultaneously $\sigma > 0$, $h(0) = 0$ and h is absolutely continuous, and $I_\sigma(h) = \infty$ otherwise.

Remark 4.36 *In view of applications, it should be noted that if $\mathbb{E}(X_0|\mathcal{F}_{-\infty}) = 0$ a.s. and (4.110) holds then the condition (4.111) is satisfied as soon as for all $j \geq 0$,*

$$\lim_{n\to\infty} \|\frac{1}{n}\sum_{i=1}^{n}[\mathbb{E}(X_iX_{i+j}|\mathcal{F}_0) - \mathbb{E}(X_0X_j)]\|_\infty = 0. \tag{4.113}$$

From the above remark, it follows that condition (4.111) is weaker than the corresponding condition (6) required in Theorem 3 of Dedecker et al. (2009).

Alternatively to the theorem above, the following result holds.

Theorem 4.37 *Let $(X_i)_{i\in\mathbb{Z}}$ be a strictly stationary sequence of real-valued random variables adapted to a non-decreasing stationary filtration $(\mathcal{F}_i)_{i\in\mathbb{Z}}$. Assume that $\|X_0\|_\infty < \infty$. In addition, assume that*

$$\sum_{n=1}^{\infty} n^{-3/2}\|\mathbb{E}(S_n|\mathcal{F}_0)\|_\infty < \infty, \tag{4.114}$$

and that (4.111) holds. Then $\lim_{n\to\infty} n^{-1}\mathbb{E}(S_n^2) = \sigma^2$ and, for all positive sequences a_n with $a_n \to 0$ and $na_n \to \infty$, the normalized partial sums processes $W_n(.)$ satisfy the moderate deviation principle in $D([0,1])$ with speed a_n and the good rate function $I_\sigma(\cdot)$ defined by (4.112).

The following corollary gives simplified conditions for the MDP principle, which are easy to verify in several examples.

Corollary 4.38 *Let $(X_i)_{i\in\mathbb{Z}}$ be a strictly stationary sequence of real-valued random variables adapted to a non-decreasing stationary filtration $(\mathcal{F}_i)_{i\in\mathbb{Z}}$. Assume that $\|X_0\|_\infty < \infty$. In addition, assume that*

$$\sum_{n=1}^{\infty} n^{-1/2}\|\mathbb{E}(X_n|\mathcal{F}_0)\|_\infty < \infty, \tag{4.115}$$

and that for all i, j ≥ 1,

$$\lim_{n\to\infty} ||\mathbb{E}(X_iX_j|\mathcal{F}_{-n}) - \mathbb{E}(X_iX_j)||_\infty = 0. \tag{4.116}$$

Then the conclusion of Theorem 4.35 *holds with* $\sigma^2 = \sum_{k\in\mathbf{Z}} \mathbb{E}(X_0X_k)$.

In what follows we only give the proof of Theorem 4.35. For a proof of Theorem 4.37 and its corollary we refer to the paper of Dedecker et al. (2009).

Proof of Theorem 4.35 Under the conditions of the theorem, condition (4.29) obviously holds. Therefore, combining Theorems 4.6 and 4.10 we get that $n^{-1}\mathbb{E}(S_n^2)$ converges to $\sigma^2 = \sum_{j\in\mathbb{Z}} \mathbb{E}(X_0X_j)$.

To prove the result we shall proceed in two steps. First we set

$$D_j = \sum_{\ell\geq j} \mathbf{P}_j(X_\ell) \text{ for any } j \in \mathbb{Z} \text{ and } M_n = \sum_{j=1}^{n} D_j \text{ for any } n \geq 1. \tag{4.117}$$

Clearly $(D_j)_{j\in\mathbb{Z}}$ is a stationary sequence of martingale differences with respect to the filtration $(\mathcal{F}_j)_{j\in\mathbb{Z}}$. In addition, under (4.110), $||D_0||_\infty \leq \sum_{j\geq 0} ||\mathbf{P}_0(X_j)||_\infty < \infty$. The first step of the proof consists of showing that $\{n^{-1/2}\sum_{i=1}^{[nt]} D_i, t \in [0,1]\}$ and $W_n(.)$ have the same behavior in term of the moderate deviations principle and then, according to the exponential equivalence in Lemma 1.25, that, for any $\delta > 0$,

$$\limsup_{n\to\infty} a_n \log \mathbb{P}\Big(\sqrt{a_n} \max_{1\leq k\leq n} \Big|\sum_{i=1}^{k} X_i - \sum_{i=1}^{k} D_i\Big| \geq \delta\sqrt{n}\Big) = -\infty. \tag{4.118}$$

The second step of the proof consists then of showing that, for all positive sequences a_n with $a_n \to 0$ and $na_n \to \infty$, the normalized partial sum processes $\{n^{-1/2}\sum_{i=1}^{[nt]} D_i, t \in [0,1]\}$ satisfy the moderate deviation principle in $D([0, 1])$, namely the property (1.31) with the rate function $I_\sigma(\cdot)$ defined in Theorem 4.35.

Let us start by showing (4.118). Fix a positive integer m and recall the martingale approximation (4.26) given in Remark 4.7

$$S_k = \sum_{\ell=1}^{k} D_\ell^m + \theta_0^m - \theta_k^m + \sum_{\ell=1}^{k} \mathbb{E}_\ell(X_{\ell+m}), \tag{4.119}$$

where

$$D_\ell^m = \mathbf{P}_\ell(S_{\ell+m-1} - S_{\ell-1}) \text{ and } \theta_\ell^m = \mathbb{E}_\ell(S_{\ell+m} - S_\ell). \tag{4.120}$$

Since $M_k = \sum_{j=1}^k D_j$, from (4.119), it follows that

$$S_k - M_k = \sum_{\ell=1}^{k} (D_\ell^m - D_\ell) + \theta_0^m - \theta_k^m + \sum_{\ell=1}^{k} \mathbb{E}_\ell(X_{\ell+m}). \tag{4.121}$$

Recall that since $\mathbb{E}(X_0|\mathcal{F}_{-\infty}) = 0$ a.s., $X_0 = \sum_{i\geq 0} \mathbf{P}_{-i}(X_0)$ a.s. Hence, by stationarity,

$$\|\theta_0^m\|_\infty \leq m\|X_0\|_\infty \leq m \sum_{i\geq 0} \|\mathbf{P}_{-i}(X_0)\|_\infty. \tag{4.122}$$

Similarly, by stationarity, we obtain

$$\max_{1\leq k\leq n} \|\theta_k^m\|_\infty \leq m \max_{2\leq k\leq n+m} \|X_k\|_\infty \leq m \sum_{i\geq 0} \|\mathbf{P}_{-i}(X_0)\|_\infty. \tag{4.123}$$

Therefore, starting from (4.121) and taking into account (4.122), (4.123), and $a_n \to 0$, we conclude that to prove (4.118) it suffices to show that, for any $\delta > 0$,

$$\limsup_{m\to\infty} \limsup_{n\to\infty} a_n \log \mathbb{P}\left(\sqrt{a_n} \max_{1\leq k\leq n} \left|\sum_{\ell=1}^{k} (D_\ell^m - D_\ell + \mathbb{E}_\ell(X_{\ell+m}))\right| \geq \delta\sqrt{n}\right) = -\infty. \tag{4.124}$$

Next we set $Y_\ell = D_\ell^m - D_\ell + \mathbb{E}_\ell(X_{\ell+m})$ and note that by stationarity, for any $j \geq 1$,

$$\|\mathbf{P}_{\ell-j}(Y_\ell)\|_\infty = \|\mathbf{P}_{\ell-j}(X_{\ell+m})\|_\infty = \|\mathbf{P}_0(X_{j+m})\|_\infty$$

and

$$\|\mathbf{P}_\ell(Y_\ell)\|_\infty = \|D_\ell^m - D_\ell + \mathbf{P}_\ell(X_{\ell+m})\|_\infty \leq \sum_{i\geq m} \|\mathbf{P}_0(X_i)\|_\infty + \|\mathbf{P}_0(X_m)\|_\infty.$$

Also we get $\|\mathbf{P}_{\ell+j}(Y_\ell)\|_\infty = 0$ for any $j \geq 1$. Therefore, using Theorem 3.20 with $D = 2\sum_{i\geq m} \|\mathbf{P}_0(X_i)\|_\infty$, we get, for any positive real x,

$$\mathbb{P}\left(\max_{1\leq k\leq n} \left|\sum_{\ell=1}^{k} (D_\ell^m - D_\ell + \mathbb{E}_\ell(X_{\ell+m}))\right| \geq x\right) \leq 8 \exp\left(-\frac{x^2}{2n\left(2\sum_{j\geq m} \|\mathbf{P}_0(X_j)\|_\infty\right)^2}\right),$$

implying that for any $\delta > 0$,

$$a_n \log \mathbb{P}\left(\sqrt{a_n} \max_{1\le k\le n}\left|\sum_{\ell=1}^{k}(D_\ell^m - D_\ell + \mathbb{E}_\ell(X_{\ell+m}))\right| \ge \delta\sqrt{n}\right)$$
$$\le a_n \log(8) - \frac{\delta^2}{8\left(\sum_{j\ge m}\|\mathbf{P}_0(X_j)\|_\infty\right)^2}.$$

This shows (4.124) (and then (4.118)) by letting first n tend to infinity and after m and by considering condition (4.110).

We turn now to the second step of the proof consisting of showing that, for all positive sequences a_n with $a_n \to 0$ and $na_n \to \infty$, the normalized partial sum processes $\{n^{-1/2}\sum_{i=1}^{[nt]} D_i, t\in[0,1]\}$ satisfy the moderate deviation principle in $D([0,1])$ with the rate function $I_\sigma(\cdot)$. With this aim we shall apply Corollary 2.42. Indeed, under (4.110), $\|D_0\|_\infty \le \sum_{j\ge 0}\|\mathbf{P}_0(X_j)\|_\infty < \infty$. In addition, as we have already mentioned, $\lim_{n\to\infty} n^{-1}\mathbb{E}(S_n^2) = \sigma^2 = \mathbb{E}(D_0^2)$. Therefore it suffices to show that condition (2.63) is satisfied under the conditions of the theorem. With this aim, we consider a fixed positive integer m and we note that by stationarity, for any $j \ge 1$,

$$\|\mathbb{E}_0((D_j - D_j^m)(D_j + D_j^m))\|_\infty \le \|D_0 - D_0^m\|_\infty \|D_0 + D_0^m\|_\infty \le 2\sum_{k\ge 0}\|\mathbf{P}_0(X_k)\|_\infty \sum_{\ell\ge m}\|\mathbf{P}_0(X_\ell)\|_\infty,$$

where D_j^m is defined in (4.120). Therefore by condition (4.110), to prove (2.63), it suffices to show that

$$\lim_{m\to\infty}\limsup_{n\to\infty}\frac{1}{n}\left\|\sum_{j=1}^{n}(\mathbb{E}((D_j^m)^2|\mathcal{F}_0) - \mathbb{E}((D_j^m)^2))\right\|_\infty = 0. \tag{4.125}$$

With this aim, we first notice

$$\sum_{j=1}^{n}(\mathbb{E}((D_j^m)^2|\mathcal{F}_0) - \mathbb{E}((D_j^m)^2)) = \mathbb{E}((M_n^m)^2|\mathcal{F}_0) - \mathbb{E}((M_n^m)^2),$$

where $M_n^m = \sum_{j=1}^{n} D_j^m$. Using the martingale approximation (4.119), we get

$$n^{-1}\|\sum_{j=1}^{n}(\mathbb{E}((D_j^m)^2|\mathcal{F}_0) - \mathbb{E}((D_j^m)^2))\|_\infty \le n^{-1}\|\mathbb{E}(S_n^2|\mathcal{F}_0) - \mathbb{E}(S_n^2)\|_\infty$$
$$+ 4n^{-1}\|\mathbb{E}(S_nR_n^m|\mathcal{F}_0)\|_\infty + 2n^{-1}\|\mathbb{E}((R_n^m)^2|\mathcal{F}_0)\|_\infty, \tag{4.126}$$

where

$$R_n^m = \theta_0^m - \theta_n^m + \sum_{\ell=1}^{n} \mathbb{E}_\ell(X_{\ell+m}). \tag{4.127}$$

By the Cauchy–Schwarz inequality and Jensen's inequality,

$$\begin{aligned} n^{-1}\|\mathbb{E}(S_n R_n^m|\mathcal{F}_0)\|_\infty &\le n^{-1}\|\mathbb{E}_0(S_n^2)\mathbb{E}_0((R_n^m)^2)\|_\infty^{1/2} \\ &\le n^{-1}\|\mathbb{E}_0(S_n^2) - \mathbb{E}(S_n^2)\|_\infty^{1/2}\|\mathbb{E}_0((R_n^m)^2)\|_\infty^{1/2} + n^{-1}\sqrt{\mathbb{E}(S_n^2)}\|\mathbb{E}_0((R_n^m)^2)\|_\infty^{1/2}. \end{aligned} \tag{4.128}$$

Recall that under (4.110), $\lim_{n\to\infty} n^{-1}\mathbb{E}(S_n^2) = \sigma^2$. So, starting from (4.126) and considering (4.128), it follows that, under (4.111), proving (4.125) is reduced to showing that

$$\lim_{m\to\infty} \limsup_{n\to\infty} \frac{1}{n}\|\mathbb{E}((R_n^m)^2|\mathcal{F}_0)\|_\infty = 0. \tag{4.129}$$

Since $\mathbb{E}(X_0|\mathcal{F}_{-\infty}) = 0$ a.s., $\mathbb{E}_\ell(X_{\ell+m}) = \sum_{i=-\infty}^{\ell} \mathbf{P}_i(X_{\ell+m})$ a.s. So, by orthogonality,

$$\mathbb{E}_0\left(\left(\sum_{\ell=1}^{n} \mathbb{E}_\ell(X_{\ell+m})\right)^2\right) = \sum_{\ell=1}^{n}\sum_{k=1}^{n}\sum_{i=-\infty}^{\ell\wedge k} \mathbb{E}_0\left(\mathbf{P}_i(X_{\ell+m})\mathbf{P}_i(X_{k+m})\right).$$

Hence

$$\left\|\mathbb{E}_0\left(\left(\sum_{\ell=1}^{n} \mathbb{E}_\ell(X_{\ell+m})\right)^2\right)\right\|_\infty \le 2\sum_{k=1}^{n}\sum_{\ell=k}^{n}\sum_{i=-\infty}^{k} \|\mathbf{P}_i(X_{\ell+m})\|_\infty\|\mathbf{P}_i(X_{k+m})\|_\infty,$$

implying by stationarity that

$$\begin{aligned} n^{-1}\left\|\mathbb{E}_0\left(\left(\sum_{\ell=1}^{n} \mathbb{E}_\ell(X_{\ell+m})\right)^2\right)\right\|_\infty &\le 2n^{-1}\sum_{k=1}^{n}\sum_{\ell=k}^{n}\sum_{i=-\infty}^{k} \|\mathbf{P}_0(X_{\ell-i+m})\|_\infty\|\mathbf{P}_0(X_{k-i+m})\|_\infty \\ &\le 2\left(\sum_{\ell\ge m} \|\mathbf{P}_0(X_\ell)\|_\infty\right)^2, \end{aligned}$$

which converges to zero as m tends to infinity since (4.110) is assumed. This convergence, together with (4.122) and (4.123), shows that (4.129) holds. This ends the proof of the theorem. ■

5

Dependence Coefficients for Sequences

In this chapter we survey the traditional mixing coefficients and much weaker dependence characteristics, illustrated by numerous examples.

5.1 Traditional Mixing Coefficients

5.1.1 *Definitions*

Let $(\Omega, \mathcal{A}, \mathbb{P})$ be a probability space and let $\mathcal{U}$ and $\mathcal{V}$ be two σ-algebras of $\mathcal{A}$. In order to estimate the correlation between $\mathcal{U}$ and $\mathcal{V}$, various coefficients are used in the literature, such as:

$$\alpha(\mathcal{U}, \mathcal{V}) = \sup\{| \mathbb{P}(U \cap V) - \mathbb{P}(U)\,\mathbb{P}(V)| \,:\, U \in \mathcal{U},\ V \in \mathcal{V}\},$$

$$\beta(\mathcal{U}, \mathcal{V}) = \mathbb{E}\big(\sup\{|\mathbb{P}(V|\mathcal{U}) - \mathbb{P}(V)| \,:\, V \in \mathcal{V}\}\big),$$

$$\phi(\mathcal{U}, \mathcal{V}) = \sup\{|\mathbb{P}(V|U) - \mathbb{P}(V)| \,:\, U \in \mathcal{U}, \mathbb{P}(U) > 0, V \in \mathcal{V}\},$$

$$\rho(\mathcal{U}, \mathcal{V}) = \sup\{|\mathrm{corr}(X, Y)| \,:\, X \in \mathbb{L}^2(\mathcal{U}), Y \in \mathbb{L}^2(\mathcal{V})\}.$$

In the definition of the coefficient $\beta(\mathcal{U}, \mathcal{V})$, we assume implicitly that $\mathcal{V}$ is separable and also the existence of a regular conditional probability $(\mathbb{P}(V|\mathcal{U}), V \in \mathcal{V})$ of events in $\mathcal{V}$, given the σ-algebra $\mathcal{U}$ (see Proposition 3.22 in Bradley (2007) for more explanations). Between the coefficients defined above, the following comparison inequalities hold:

$$2\alpha(\mathcal{U}, \mathcal{V}) \le \beta(\mathcal{U}, \mathcal{V}) \le \phi(\mathcal{U}, \mathcal{V}) \text{ and } 4\alpha(\mathcal{U}, \mathcal{V}) \le \rho(\mathcal{U}, \mathcal{V}) \le 2\phi^{1/2}(\mathcal{U}, \mathcal{V})$$

(see Bradley (2007, vol.1) and the references therein). Note that $\alpha(\mathcal{U}, \mathcal{V})$ is usually called the strong mixing coefficient of Rosenblatt (1956) between two σ-algebras $\mathcal{U}$ and $\mathcal{V}$. An equivalent definition of this coefficient, based on the conditional expectation, is given in Bradley (2007), Theorem 4.4, Item (a2):

$$\alpha(\mathcal{U}, \mathcal{V}) = \frac{1}{4} \sup\{\|\mathbb{E}(Y|\mathcal{U})\|_1 / \|Y\|_\infty \,:\, \text{where } Y \text{ is } \mathcal{V}\text{-measurable with } \mathbb{E}(Y) = 0\}. \quad (5.1)$$

Functional Gaussian Approximation for Dependent Structures. Florence Merlevède, Magda Peligrad and Sergey Utev. Oxford University Press (2019).
DOI: 10.1093/oso/9780198826941.001.0001

Let us indicate that $\beta(\mathcal{U}, \mathcal{V})$ is the β-mixing coefficient (or the absolute regularity coefficient) introduced by Volkonskii and Rozanov (1959), whereas $\phi(\mathcal{U}, \mathcal{V})$ is the ϕ-mixing coefficient of Ibragimov (1962). The ρ-mixing coefficient was introduced by Kolmogorov and Rozanov (1960). It is also known under the terminology of maximum coefficient of correlation.

From these coefficients, we can quantify at which rate the events in the distant future of a discrete time process $\mathbf{X} = (X_i)_{i\in\mathbb{Z}}$ of real-valued random variables (or with values in a Polish space S) become independent of the past by considering the sequences of its strong mixing (or α-mixing), β-mixing, ϕ-mixing or ρ-mixing coefficients. These are defined respectively as follows: Let $\mathcal{F}_i^j = \sigma(X_i, \dots, X_j)$,

$$\alpha(n) = \sup_{k\in\mathbb{Z}} \alpha(\mathcal{F}_{-\infty}^k, \mathcal{F}_{n+k}^{\infty}), \; \beta(n) = \sup_{k\in\mathbb{Z}} \beta(\mathcal{F}_{-\infty}^k, \mathcal{F}_{n+k}^{\infty}) \tag{5.2}$$

$$\phi(n) = \sup_{k\in\mathbb{Z}} \phi(\mathcal{F}_{-\infty}^k, \mathcal{F}_{n+k}^{\infty}), \; \rho(n) = \sup_{k\in\mathbb{Z}} \rho(\mathcal{F}_{-\infty}^k, \mathcal{F}_{n+k}^{\infty}). \tag{5.3}$$

The sequence $(X_k)_{k\in\mathbb{Z}}$ is said to be strong mixing if $\lim_{n\to\infty} \alpha(n) = 0$, β-mixing (or absolutely regular) if $\lim_{n\to\infty} \beta(n) = 0$, ϕ-mixing if $\lim_{n\to\infty} \phi(n) = 0$ and ρ-mixing if $\lim_{n\to\infty} \rho(n) = 0$.

An important stronger condition than ρ-mixing is the interlaced ρ^*-mixing, defined by $\lim_{n\to\infty} \rho^*(n) = 0$ with

$$\rho^*(n) = \sup_{Q,Q^*\subset\mathbb{N}} \rho(\mathcal{F}_Q, \mathcal{F}_{Q^*}); \; d(Q, Q^*) \geq n,$$

where for every $Q \subset \mathbb{N}$ we denote $\mathcal{F}_Q = \sigma(X_i, i \in Q)$ and d is the distance between the sets Q, Q^*,

$$d(S, T) = \min_{x\in Q, y\in Q^*} |x - y|.$$

Under strict stationarity, the ρ^*-mixing condition goes back at least to Stein (1972) and to Rosenblatt (1972).

As we mentioned first for the strong mixing coefficients, it is interesting to give equivalent definitions of the mixing coefficients associated with the discrete time process $\mathbf{X} = (X_i)_{i\in\mathbb{Z}}$ of real-valued random variables in terms of supremum over the class of functions which are measurable and bounded by one. Let us denote by $\mathbb{X}_n = (X_k)_{k\geq n}$ and by $\mathcal{F}_j = \sigma(X_i, i \leq j)$. Finally set B_1 as the class of functions from $\mathbb{R}^\infty$ to $\mathbb{R}$, measurable and bounded by one, and denote by $\mathbb{L}_2$ the class of measurable functions from $\mathbb{R}^\infty$ to $\mathbb{R}$, such that $f(\mathbb{X}_{n+k})$ is square integrable. We then have

$$\alpha(n) = \frac{1}{4} \sup_{k\in\mathbb{Z}} \sup_{f\in B_1} \|\mathbb{E}(f(\mathbb{X}_{n+k})|\mathcal{F}_k) - \mathbb{E}f(\mathbb{X}_{n+k})\|_1,$$

$$\beta(n) = \frac{1}{2} \sup_{k \in \mathbb{Z}} \| \sup_{f \in B_1} |\mathbb{E}\left(f(\mathbb{X}_{n+k}) \,|\mathcal{F}_k\right) - \mathbb{E}f(\mathbb{X}_{n+k})\,| \|_1,$$

$$\rho(n) = \sup_{k \in \mathbb{Z}} \sup_{f \in \mathbb{L}_2} \|\mathbb{E}(f(\mathbb{X}_{n+k})|\mathcal{F}_k) - \mathbb{E}f(\mathbb{X}_{n+k})\|_2,$$

$$\phi(n) = \frac{1}{2} \sup_{k \in \mathbb{Z}} \sup_{f \in B_1} \|\mathbb{E}(f(\mathbb{X}_{n+k})|\mathcal{F}_k) - \mathbb{E}f(\mathbb{X}_{n+k})\|_\infty,$$

where the conditional expectations are defined via the regular conditional distribution of $\mathbb{X}_{n+k}$ given $\mathcal{F}_k$.

The case of Markov chains In the case when $\mathbf{X} = (X_i)_{i \geq 0}$ forms a Markov chain, the definitions of mixing conditions are significantly simplified in the sense that we can replace in their definition $\sigma(X_i, i \leq j)$ by $\sigma(X_j)$ and $\sigma(X_i, i \geq j + n)$ by $\sigma(X_{j+n})$ (see Chapter 7 in Bradley 2007, vol 1). In addition, note that if $\mathbf{X} = (X_i)_{i \in \mathbb{Z}}$ is a stationary Markov process with Kernel operator K and invariant measure ν, then the coefficients above simplify as follows:

$$\alpha(n) = \frac{1}{4} \sup_{\|f\|_\infty \leq 1} \nu\left(|K^n(f) - \nu(f)|\right),\ \beta(n) = \frac{1}{2} \nu\left(\sup_{\|f\|_\infty \leq 1} |K^n(f) - \nu(f)| \right)$$

and

$$\rho(n) = \sup_{\|f\|_{\nu,2} \leq 1} \|K^n(f) - \nu(f)\|_{\nu,2},\ \phi(n) = \frac{1}{2} \sup_{\|f\|_\infty \leq 1} \|K^n(f) - \nu(f)\|_{\nu,\infty}.$$

Furthermore, we have $\rho(n) \leq \rho^n(1)$ and $2\phi(n) \leq (2\phi(1))^n$. If $\phi(n) < 1$ for some $n \geq 1$, then $\phi(n) \to 0$ at least exponentially fast, as $n \to \infty$. The original use of these inequalities is hard to trace. Their proofs are given in Theorem 7.5 in Bradley (2007, vol 1).

5.1.2 *Examples of Mixing Processes*

We give here some examples of mixing processes. Other examples are given for instance in the monograph by Doukhan (1994). We also refer the reader to the survey of Bradley (2005).

1) Let us consider the following causal linear process

$$X_k = \sum_{j \geq 0} a_j \varepsilon_{k-j}, k \in \mathbb{Z},$$

where $|a_j| \leq C\rho^j$ with $\rho \in [0, 1]$ and the ε_k's are i.i.d. random variables with mean zero, finite second moment and density with respect to the Lebesgue measure. Then the series above converges in $\mathbb{L}^2$, and $(X_k)_{k \in \mathbb{Z}}$ is geometrically β-mixing meaning that its sequence of β-mixing coefficients decreases to zero exponentially fast (see Mokkadem (1988)).

2) The following example comes from Doukhan, Massart and Rio (1994). Let $a > 0$ and let λ denote the Lebesgue measure on $[0, 1]$. Define the probability laws μ and ν by

$$\mu = (1 + a)x^a \lambda \text{ and } \nu = ax^{a-1}\lambda.$$

We define now a strictly stationary Markov chain by defining its transition probabilities $K(x, A)$ as follows:

$$K(x, A) = (1 - x)\delta_x(A) + x\mu(A),$$

where δ_x denotes the Dirac measure at point x. Then ν is the unique invariant probability measure of the chain with transition probabilities $K(x, \cdot)$. Let $(X_i)_{i\in\mathbb{Z}}$ be the stationary Markov chain on $[0, 1]$ with transition probabilities $K(x, \cdot)$ and law ν. According to Lemma 2 in Doukhan, Massart and Rio (1994),

$$\beta_n := \frac{1}{2}\nu\left(\sup_{\|f\|_\infty \le 1} |K^n(f) - \nu(f)|\right) \le 3\,\mathbb{E}_\nu(T^{[n/2]}),$$

where $T(x) = 1 - x$ and $\mathbb{E}_\nu$ denotes the expectation with respect to ν. Noting that for any $b > -1$,

$$\int_0^1 (1 - x)^k x^b \, dx = k^{-(b+1)} \int_0^k (1 - x/k)^k x^b \, dx$$

and using the fact that for any $x \in [0, 1]$, $\log(1 - x) \le -x$, we get that

$$\int_0^1 (1 - x)^k x^b \, dx \le k^{-(b+1)} \int_0^k e^{-x} x^b \, dx \le k^{-(b+1)}\Gamma(b + 1),$$

implying that

$$\mathbb{E}_\nu(T^k) \le a\Gamma(a)k^{-a}.$$

Therefore, for any $n \ge 1$, $\beta(n) \le Cn^{-a}$ where C is a constant depending only on a.

3) More generally, let $\mathbf{X} = (X_i)_{i\in\mathbb{Z}}$ be a strictly stationary (real) Markov chain and let ν denote the distribution of X_0 on $\mathbb{R}$. We shall say that the Markov chain is Harris recurrent if there exists a Borelian set A with $\nu(A) = 1$ such that for every $x \in A$ and every Borelian set B such that $\nu(B) > 0$, one has

$$\mathbb{P}(X_n \in B \text{ for infinitely many } n \ge 1 | X_0 = x) = 1.$$

Then strictly stationary, Harris recurrent and aperiodic Markov chains are β-mixing (see Orey (1959) and also Corollary 21.7 in Bradley (2007, vol. 2)). Note also that for strictly stationary Markov chains, the notion of geometric ergodicity is equivalent to the fact that $\beta(n)$ converges to zero exponentially fast. Finally, as stated in Theorem 21.5 in Bradley (2007, vol. 2), for a strictly stationary (real) Markov chain $\mathbf{X} = (X_i)_{i\in\mathbb{Z}}$ with invariant measure ν, Harris recurrence is equivalent to irreducibility in the following sense: The Markov chain is said to be irreducible if there exists a Borelian set A with $\nu(A) = 1$ such that for every $x \in A$ and every Borelian set B such that $\nu(B) > 0$, there exists $n \in \mathbb{N}$ such that $\mathbb{P}(X_n \in B|X_0 = x) > 0$. Moreover, these notions are equivalent to the fact that the chain is ergodic and its β-mixing coefficients are such that $\beta(n) < 1$ for some $n \geq 1$.

4) For irreducible, positive recurrent and aperiodic discrete Markov chains having an atom, it is also possible to link the rate of convergence of $\beta(n)$ to zero with the tail of the distribution of the return times to the atom. This has been made clear by Davydov (1973). We also refer to Chapter 30 of Bradley (2007).

To give an example, let $p > 1$ and consider a simple renewal type Markov chain on $\mathbb{N}$, jumping from 0 to $n > 0$ with probability $p_{0,n} := 1/(\zeta(p+1)n^{p+1})$ and from $n > 0$ to $n-1$ with probability 1. This Markov chain has an invariant probability measure ν given by $\nu\{n\} = \sum_{i \geq n} d/i^{p+1}$ for $n > 0$ and $\nu\{0\} = \nu\{1\}$, where $d > 0$ is chosen so that ν is of mass 1. This chain is irreducible, aperiodic and Harris positively recurrent. In addition, if we denote by K its associated Markov transition operator,

$$\beta(n) = \frac{1}{2}\nu\Big(\sup_{\|f\|_\infty \leq 1} |K^n(f) - \nu(f)| \Big) \leq C_1 \sum_{j \geq n} \sum_{k \geq j+1} p_{0,k} \leq C_2 n^{1-p}.$$

5) *The computation of Ibragimov coefficient $\phi(n)$ in the Markovian case.* If $(X_k)_{k\geq 1}$ is a Markov sequence the definition simplifies as indicated at the end of Section 5.1.1:

$$\phi(1) = \max_{k \geq 1} \phi(\sigma(X_k), \sigma(X_{k+1})), \; 2\phi(n) \leq (2\phi(1))^n.$$

This shows that the burden is to estimate $\phi(1)$ and especially to establish that $\phi(1) < 1$. We mention that

$$\delta(\sigma(X_k), \sigma(X_{k+1})) = 1 - \sup_{B \in \sigma(X_{k+1})} [\text{ess sup}\, \mathbb{P}(B|\sigma(X_k)) - \text{ess inf}\, \mathbb{P}(B|\sigma(X_k))]$$

is the famous Dobrushin coefficient of ergodicity and, by relation (1.1.2) in Iosifescu and Theodorescu (1969), the following inequalities hold:

$$[1 - \delta(\sigma(X_k), \sigma(X_{k+1}))]/2 \leq \phi(\sigma(X_k), \sigma(X_{k+1})) \leq 1 - \delta(\sigma(X_k), \sigma(X_{k+1})).$$

If $(X_k)_{1 \leq k \leq n}$ are discrete random variables, we denote by

$$q_{ij}^{(k)} = \mathbb{P}(X_{k+1} = j | X_k = i).$$

Then by Proposition 1.2.3 in Iosifescu and Theodorescu (1969),

$$\phi(\sigma(X_k), \sigma(X_{k+1})) \leq \frac{1}{2} \sup_{i,j} \sum_{\ell} |q_{i\ell}^{(k)} - q_{j\ell}^{(k)}|.$$

Now, assume we have a strictly stationary Markov chain $(X_k)_{k\in\mathbb{Z}}$ satisfying the following form of the Doeblin condition: There is a Borelian set A with $\mathbb{P}(X_0 \in A) = 1$ and there is an $\varepsilon \in (0, 1)$ such that, for every $x \in A$,

$$\mathbb{P}(X_1 \in B|X_0 = x) \leq 1 - \varepsilon,$$

as soon as $\mathbb{P}(X_0 \in B) \leq \varepsilon$. In this case we have $\phi(1) \leq 1 - \varepsilon$ (see for instance Section 21.23 in Bradley (2007), vol. 2). Here, because of stationarity, ϕ is defined as $\phi(\sigma(X_0), \sigma(X_1))$.

In many situations, the following two quantities are relevant to the computation of the coefficient ϕ:

$$\psi^*(1) = \sup_{A,B} \frac{\mathbb{P}(X_0 \in A, X_1 \in B)}{\mathbb{P}(X_0 \in A)(X_1 \in B)} \quad \text{and} \quad \psi'(1) = \inf_{A,B} \frac{\mathbb{P}(X_0 \in A, X_1 \in B)}{\mathbb{P}(X_0 \in A)(X_1 \in B)},$$

where in the definitions above we interpret 0/0 as 0. Then, by Proposition 5.2 in Bradley (2007), vol. 1, we have $1 \leq \psi^*(1) \leq \infty$ and $0 \leq \psi'(1) \leq 1$. Moreover,

$$\phi(1) \leq 1 - \frac{1}{\psi^*(1)} \quad \text{and} \quad \phi(1) \leq 1 - \psi'(1).$$

These inequalities are practical in many situations. For instance, it is convenient to specify a stationary Markov chain $(X_k)_{k\in\mathbb{Z}}$ with marginal distribution function F by a copula $C(x_0, x_1)$:

$$\mathbb{P}(X_0 \leq x_0, X_1 \leq x_1) = C(F(x_0), F(x_1)).$$

If for some $0 < \delta \leq 1$ the absolutely continuous part of the copula $C(x_0, x_1)$ has a density $c(x_0, x_1) \geq \delta$, then we have $\psi' \geq \delta$ and $\phi \leq 1 - \delta$. For instance, for the Marshall–Olkin copula

$$C_\alpha(x_0, x_1) = \min(x_0 x_1^{1-\alpha}, x_1 x_0^{1-\alpha}),\ 0 \leq \alpha < 1,$$

we have $c(x_0, x_1) \geq 1 - \alpha$ and therefore $\phi \leq \alpha$. For detailed formulation of the ϕ-mixing coefficients in terms of copulas and further examples see for instance Longla and Peligrad (2012).

Let us now give an estimate of the coefficient ϕ for a specific example related to number theory. For every irrational number x in $(0, 1)$ there is a unique sequence of positive integers $x_1, x_2, x_3, \ldots$ such that the following continued fraction expansion holds:

$$x = \cfrac{1}{x_1 + \cfrac{1}{x_2 + \cfrac{1}{x_3 + \cdots}}}.$$

If we introduce on $[0, 1]$ the Gauss probability measure with the density $f(x) = (\ln 2)^{-1}(1+x)^{-1}$, then the sequence $(x_1, x_2, x_3, \ldots)$ is strictly stationary. We know from Lemma 2.1 in Philipp (1988) that for this sequence $\psi^*(1) < 1.8$ and then $\phi(1) \leq 1 - (\psi^*(1))^{-1} \leq 0.45$.

6) For Gaussian processes, there are special implications between the various kinds of mixing. For instance if $(X_k)_{k\in\mathbb{Z}}$ is a Gaussian stationary process which is ϕ-mixing then it is m-dependent (see Ibragimov (1961)). On the other hand, for Gaussian processes, α-mixing and ρ-mixing are equivalent since in addition to $4\alpha(k) \leq \rho(k)$ we also have $\rho(k) \leq 2\pi\alpha(k)$ for all $k \geq 1$. These properties can be found in Ibragimov and Rozanov (1970) where some equivalences between the behavior of the spectral density of a non-degenerate stationary Gaussian process $(X_k)_{k\in\mathbb{Z}}$ and the fact that it is either α-mixing, β-mixing or ϕ-mixing (and then m-dependent) are proved. For instance a stationary Gaussian process is α-mixing if and only if it has a spectral density function h on the unit circle T (recall (1.18)) of the form

$$h(t) = |p(t)|^2 \exp[u(t) + \tilde{v}(t)],$$

where p is a polynomial, u and v are continuous real functions, and $\tilde{v}$ is the conjugate function of v. It is absolutely regular if and only if its spectral density function h on T is of the form

$$h(t) = |p(t)|^2 \exp\left(\sum_{j=-\infty}^{\infty} a_j t^j\right),$$

where p is a polynomial whose roots lie on the unit circle and $\sum_{j=-\infty}^{\infty} |j||a_j|^2 < \infty$. The ϕ-mixing condition is equivalent to the following fact: "$(X_k)_{k\in\mathbb{Z}}$ has a spectral density function h on T of the form $h(t) = |p(t)|^2$ where p is a polynomial". We also mention that Rosenblatt (1972 and 1985, pp. 73–74, Theorem 7 and Lemma 2) showed that if a stationary Gaussian sequence has a continuous positive spectral density, then it is ρ^*-mixing.

Furthermore in the computation of the ρ-mixing coefficients it is enough to use only linear functions instead of all the functions. More precisely,

$$\rho(\mathcal{F}_Q, \mathcal{F}_{Q^*}) = \sup \operatorname{corr}\left(\sum_{j\in Q} a_j X_j, \sum_{j\in Q^*} a_j X_j\right),$$

where the supremum is taken over all choices of constants a_j and we recall that, for every $Q \subset \mathbb{N}$, $\mathcal{F}_Q := \sigma(X_i, i \in Q)$.

In addition, the spectral density f of a stationary Gaussian sequence provides an useful tool to bound the ρ^*-mixing coefficient. For instance, according to Bradley (1992) (see also Remark 3 in Bryc and Smolenski (1993)), if $0 < m < f < M$, one has the following inequality: $\rho^*(1) \leq 1 - m/M < 1$.

5.1.3 *Coupling Properties of the Mixing Coefficients*

One of the most useful techniques in obtaining limit theorems for partial sums associated with a sequence of random variables which is mixing in some sense is the coupling of the initial sequence with an independent one. Generally speaking a coupling theorem enables us to replace the initial dependent sequence by an independent one, having the same marginals.

Moreover, a variable in the newly constructed sequence is independent of the past of the initial sequence and it is closed, in some sense, to the variable having the same rank. This approximation has to be carried out in a manner that permits limit theorems for the independent random variables to carry over directly to the dependent ones.

The idea of coupling goes back at least to Doeblin (1937) for the case of the Markov chains with a countable state space and was further extended for more general Markov processes by Vasershtein (1969) and Griffeath (1975). The problem of coupling was also considered in ergodic theory, in the context of Bernoulli shifts, by Ornstein (1974) and Shields (1973).

This method has gained popularity for obtaining limit theorems for dependent random variables since the publication of two articles by Berkes and Philipp (1977, 1979). Their approach is based on estimates of the Prohorov-distance and the Strassen–Dudley Theorem (Strassen (1965), Dudley (1968)). This method was successfully exploited by many authors including Berkes and Philipp (1977–78), Philipp (1979), Kuelbs and Philipp (1980), Dehling and Philipp (1982), Berger (1982), Dabrowski (1982), Eberlein (1983), Dudley and Philipp (1983), and Borovkova, Burton and Dehling (2001), among others.

Let us start with a well-known coupling result which characterizes the coefficient β of absolutely regularity.

Theorem 5.1 *Let X and Y be random variables defined on $(\Omega, \mathcal{A}, \mathbb{P})$ with values in a Polish space S. Let $\sigma(X)$ be a σ-algebra generated by X and U be a random variable uniformly distributed on $[0, 1]$ independent of (X, Y). Then there exists a random variable Y^* measurable with respect of $\sigma(X) \vee \sigma(Y) \vee \sigma(U)$, independent of X and distributed as Y, and such that*

$$\mathbb{P}(Y \neq Y^*) = \beta(\sigma(X), \sigma(Y)).$$

This theorem is frequently named Berbee's Lemma (1979). Related results can be found in works by Goldstein (1979), Berkes and Philipp (1979), Schwarz (1980), Bryc (1982), Dehling and Philipp (1982) and, in the context of ergodic theory, in Ornstein (1974) and Shields (1973). In all these papers, the constructions of the variable Y^* are equivalent.

We shall sketch here the construction by Bryc (1982) which is versatile and leads to further interesting estimates for the distance between Y and Y^*.

Because the variables are S-valued, by completeness, according to the result in Bryc (1982), we can reduce the problem to the case of discrete S-valued random variables. Leaving the technical aspect apart, the main step of the construction is to define the following probabilistic structure.

With the notation $e_{x,y} := \mathbb{P}(X = x, Y = y) - \mathbb{P}(X = x)\mathbb{P}(Y = y)$, we have

$$\mathbb{P}(X = x, Y = y, Y^* = z) = \begin{cases} \dfrac{e^+_{x,y} e^-_{x,z}}{\sum_t e^+_{x,t}} & \text{if } z \neq y \\ \mathbb{P}(X = x)\mathbb{P}(Y = y) & \text{if } z = y \text{ and } e_{x,y} \geq 0 \\ \mathbb{P}(X = x, Y = y) & \text{if } z = y \text{ and } e_{x,y} < 0. \end{cases}$$

It is easy to verify that with this construction, we have

1) Y^* distributed as Y,

2) Y^* independent of X,
and for all A, B and C Borel sets of S,

3) $\mathbb{P}(X \in A, Y \in B, Y \neq Y^*) = \sum_{x \in A} \sum_{y \in B} \left[\mathbb{P}(X = x, Y = y) - \mathbb{P}(X = x)\mathbb{P}(Y = y)\right]^+$,

4) $\mathbb{P}(X \in A, Y^* \in C, Y \neq Y^*) = \sum_{x \in A} \sum_{y \in C} \left[\mathbb{P}(X = x, Y = y) - \mathbb{P}(X = x)\mathbb{P}(Y = y)\right]^-$.

By this construction, one can easily obtain the following proposition.

Proposition 5.2 *Let Y be a random variable defined on $(\Omega, \mathcal{T}, \mathbb{P})$ with values in a Polish space (S, d). Let p be a positive number and s in S. Let Y^* be constructed as in Theorem 5.1. Then, for all $x \geq 0$*

(i) $\mathbb{P}\left(Y \neq Y^* | \sigma(X)\right) \leq \phi(\sigma(X), \sigma(Y))$ *a.s.,*

(ii) $\mathbb{P}\left(d(Y, Y^*) \geq x\right) \leq 2\phi(\sigma(X), \sigma(Y))\mathbb{P}\left(d(Y, s) \geq x/2\right)$,
and if we assume that $d^p(Y, s)$ is integrable, then

(iii) $\mathbb{E}\left(d^p(Y, Y^*)\right) \leq 2^p\phi(\sigma(X), \sigma(Y))\mathbb{E}\left(d^p(Y, s)\right)$.

Point (*ii*) comes from Bryc (1982). Point (*i*) is probably well-known (a proof can be found in Peligrad (2001)). Point (*iii*) is obtained by integrating with respect to x in (*ii*).

The following proposition aims at estimating the expectation of the distance between Y and Y^* in terms of the absolutely regular coefficient. Before stating it, let us recall the following definition: For any non-negative random variable W, define the "upper tail" quantile function via

$$Q_W(u) = \inf\{t \geq 0 : \mathbb{P}(W > t) \leq u\}.$$

With this notation, we have:

Proposition 5.3 *Let X and Y be two random variables defined on $(\Omega, \mathcal{T}, \mathbb{P})$ with values in a Polish space S. Let p be a positive number and s in S and assume that $d^p(Y, s)$ is integrable. Let Y^* be constructed as in Theorem 5.1. Then for every positive p*

$$\mathbb{E}\left(d^p(Y, Y^*)\right) \leq 2^{p+2} \int_0^{\beta(X,Y)} Q_{d^p(Y,s)}(u)du.$$

(See Merlevède and Peligrad (2002) for a proof.) In the case when $p \leq 2$, with another approach the constant is sharper, namely 2^{2p}.

Comment 5.4

1) *If* $\mathbb{E}\left(d^{p+\delta}(Y,s)\right) < +\infty$ *for a* $\delta > 0$*, then*

$$\mathbb{E}\left(d^p(Y,Y^*)\right) \leq 2^{p+2}\beta^{\frac{\delta}{p+\delta}}(X,Y)\left(\mathbb{E}\left(d^{p+\delta}(Y,s)\right)\right)^{\frac{p}{p+\delta}}.$$

2) *Since*

$$\mathbb{E}\left(d^p(Y,s)\mathbf{1}\left(d(Y,s) > Q_{d(Y,s)}(\beta(X,Y))\right)\right) \leq \int_0^{\beta(X,Y)} Q_{d^p(Y,s)}(u)du$$
$$\leq \mathbb{E}\left(d^p(Y,s)\mathbf{1}\left(d(Y,s) \geq Q_{d(Y,s)}(\beta(X,Y))\right)\right),$$

we get, in particular, that

$$\mathbb{E}\left(d^p(Y,Y^*)\right) \leq 2^{p+2}\mathbb{E}\left(d^p(Y,s)\mathbf{1}\left(d(Y,s) \geq Q_{d(Y,s)}(\beta(X,Y))\right)\right).$$

Let us mention now the following result of Bradley (1983): "For two σ-fields $\mathcal{A}$ and $\mathcal{B}$ such that one of them (say $\mathcal{A}$) is completely atomic with exactly N atoms, one has $\beta(\mathcal{A},\mathcal{B}) \leq (8N)^{1/2}\alpha(\mathcal{A},\mathcal{B})$." By using this result, we can easily see that some coupling results involving the strong mixing coefficient can be obtained for Y taking its values in S and having a finite numbers of values, and the result will strongly depend on N. For deriving more general strong approximation theorems for random variables taking their values in a Polish space, under the strong mixing condition, we refer to Bradley (1983).

Iterating the results above, we can formulate the next theorem which gathers a variety of coupling results for S-valued sequences. The upper bounds for the discrepancy of the variables with the same rank are given in terms of mixing coefficients.

Theorem 5.5 *Let* $(X_n, n \geq 1)$ *be a sequence of random variables with values in a Polish space* (S,d)*. Denote by* $\mathcal{F}_1^n = \sigma(X_1, \ldots, X_n)$*. Then, we can redefine* $\{X_n, n \geq 1\}$ *onto a richer probability space (which supports a sequence* $(\delta_n, n \geq 1)$ *of independent uniformly* $[0,1]$*-distributed random variables and independent of* $\{X_n, n \geq 1\}$*) together with a sequence* $\{X_n^*, n \geq 1\}$ *of independent random variables such that for each* $1 \leq m \leq n$*, we have:*

a) X_m^* *is independent of* $\mathcal{F}_1^{m-1}$*,*

b) X_m^* *has the same distribution as* X_m*,*

c) $\mathbb{P}\left(X_m \neq X_m^* | \mathcal{F}_1^{m-1}\right) \leq \phi(\mathcal{F}_1^{m-1}, \sigma(X_m))$*, a.s.*

d) $\mathbb{P}(X_m \neq X_m^*) = \beta\left(\mathcal{F}_1^{m-1}, \sigma(X_m)\right)$*,*

e) $\mathbb{P}(d(X_m, X_m^*) \geq x) \leq 2\phi(\mathcal{F}_1^{m-1}, \sigma(X_m))\mathbb{P}(d(X_m, s) \geq x/2)$*, for all* $x \geq 0$

and if $\mathbb{E}(d^p(X_m, s)) < \infty$ *for a positive* p *and* $s \in S$*,*

f) $\mathbb{E}(d^p(X_m, X_m^*)) \leq 2^{p+2}\int_0^{\beta(\mathcal{F}_1^{m-1}, \sigma(X_m))} Q_{d^p(X_m,s)}(u)du.$

Remark 5.6 *Of course, now the following question arises: "Does the point (f) of the above-mentioned approximation theorem also hold in the context of strongly mixing sequences with values in an infinite dimensional space?"*

Following Dehling (1983), it turns out that the answer is negative without any additional assumptions. In his paper he has constructed an example of strongly mixing sequences of ℓ^2-valued random variables X_n with values in an infinite-dimensional space, which cannot be approximated by independent random variables X_n^* in such a way that $X_n - X_n^*$ converges to zero in probability.

By integrating (c) of Theorem 5.5 on the set $\bigcap_{i=1}^{m}(X_i \neq X_i^*)$, we easily deduce by an iterative procedure the following useful corollary which estimates the joint distributions of discrepancies between $(X_i)_{i\geq 1}$ and $(X_i^*)_{i\geq 1}$. Before stating it we need to introduce the following coefficient of dependence: Let $(X_n, n \geq 1)$ be a sequence of random variables with values in a Polish space S. For $m \geq n$ we denote by $\mathcal{F}_n^m$ the sigma algebra generated by the variables $(X_n, X_{n+1}, \cdots, X_m)$ and set

$$\tilde{\phi}_k := \sup_{m\geq 1} \phi(\mathcal{F}_1^m, \mathcal{F}_{m+k}^{m+k}) = \sup_{m\geq 1} \sup\left\{|\mathbb{P}(B|A) - \mathbb{P}(B)|; \mathbb{P}(A) \neq 0, A \in \mathcal{F}_1^m, B \in \mathcal{F}_{m+k}^{m+k}\right\}. \tag{5.4}$$

The definition of $\tilde{\phi}_k$ is using the sigma algebra generated by only one random variable in the future after k steps. This coefficient has already been used in many papers including Deddens, Peligrad and Yang (1987) and Rio (2000).

The next result is proven in Peligrad (2001).

Corollary 5.7 *Under the setting of Theorem 5.5, we obtain that: For all $1 \leq i_1 < i_2 < \cdots < i_m$,*

$$\mathbb{P}\left(\bigcap_{j=1}^{m} X_{i_j} \neq X_{i_j}^*\right) \leq \tilde{\phi}_1^m. \tag{5.5}$$

Moreover for $1 \leq \ell \leq n$

$$\mathbb{P}\left(X_i \neq X_i^* \text{ at least } \ell \text{ times for } 1 \leq i \leq n\right) \leq \frac{4}{\sqrt{2\pi}} \frac{2^{n+1}}{\sqrt{n}} \frac{\tilde{\phi}_1^\ell}{1 - \tilde{\phi}_1}. \tag{5.6}$$

Due to Remark 5.6, the case of strong mixing processes has to be analyzed in the real case and with a different approach that is called conditional quantile transform.

In order to construct Y^* we shall use, as in Rio (1995), the method based on the conditional quantile transform, introduced and studied by Major (1978).

Let us denote: $F(y) = \mathbb{P}(Y \leq y), F_{\mathcal{A}}(y) = \mathbb{P}(Y \leq y|\mathcal{A})$, and

$$V = F_{\mathcal{A}}(Y - 0) + U(F_{\mathcal{A}}(Y) - F_{\mathcal{A}}(Y - 0)),$$

where U is a random variable uniformly distributed on $[0, 1]$ and independent of Y and $\mathcal{A}$.

By Lemma F.1 in Rio (2000), page 161, we can see that $P(V \leq x|\mathcal{A}) = x$ *a.s.*, and, as a consequence, by integrating this expression, we deduce that V is independent of $\mathcal{A}$ and uniformly distributed on $[0,1]$. In addition:

$$F_{\mathcal{A}}^{-1}(V) = Y \text{a.s.},$$

where we use the notation of generalized inverse: $h^{-1}(y) = \inf\{t \in R : h(t) \geq y\}$.

We now define:

$$Y^* = F^{-1}(V).$$

According to Major (1978), this choice of Y^* minimizes the distance in $\mathbb{L}^1$ between Y and any other variable distributed as Y and independent of $\mathcal{A}$.

By the above notations and considerations, since V is uniformly distributed on $[0, 1]$, clearly:

$$\mathbb{E}|Y^* - Y| = \mathbb{E}|F^{-1}(V) - F_{\mathcal{A}}^{-1}(V)| = \mathbb{E}\left(\int_0^1 |F^{-1}(u) - F_{\mathcal{A}}^{-1}(u)|du\right). \tag{5.7}$$

In order to obtain an upper bound for (5.7), we shall introduce the following weaker version of the coefficients of dependence: Let $(\Omega, \mathcal{A}, \mathbb{P})$ be a probability space and let X be a real random variable defined on Ω. Let $\mathcal{B}$ be a σ-algebra of $\mathcal{A}$ and let $\mathcal{T}(X)$ be the family of events of the form $\{X < x\}$ and $\{X > x\}$ for $x \in \mathbb{R}$. Then define the "weak" α-mixing coefficient (see Section 1.3 in Rio (2000)) by:

$$\begin{aligned}\overline{\alpha}(\mathcal{B}, X) &= \sup_{B \in \mathcal{B}, x \in \mathbb{R}} |\mathbb{P}((X > x) \cap B) - \mathbb{P}(X > x)\mathbb{P}(B)| \\ &= \frac{1}{2} \sup_{x \in \mathbb{R}} \mathbb{E}\left(|\mathbb{P}(X \leq x|\mathcal{B}) - \mathbb{P}(X \leq x)|\right),\end{aligned} \tag{5.8}$$

and the "weak" reverse-ϕ-mixing coefficient by:

$$\overline{\phi^r}(\mathcal{B}, X) = \sup_{B \in \mathcal{B}, C \in \mathcal{T}(X)} |\mathbb{P}(C \cap B) - \mathbb{P}(C)\mathbb{P}(B)|/\mathbb{P}(C)$$

The next result is due to Peligrad (2002).

Theorem 5.8 *Let Y be a real-valued integrable random variable that is defined on a probability space $(\Omega, \mathcal{A}, \mathbb{P})$. Suppose $\mathcal{B}$ is a sub σ-algebra of $\mathcal{A}$. Suppose U is a uniform random variable on $[0, 1]$ that is independent on the σ-algebra generated by $\mathcal{B}$ and Y. Then, there is a random variable Y^*, which is measurable with respect to the σ-algebra $\mathcal{B} \vee \sigma(Y) \vee \sigma(U)$ with the following properties:*

(a) *Y^* is independent of the σ-algebra $\mathcal{B}$*

(b) *Y^* has the same distribution as Y*

$$(c)\ \mathbb{E}|Y - Y^*| \leq 4 \int_0^{\overline{\alpha}(\mathcal{B},Y)} Q_{|Y|}(u)du$$

$$(d)\ \mathbb{E}|Y - Y^*| \leq 2\overline{\phi^r}(\mathcal{B}, Y)\mathbb{E}|Y|.$$

Comments

1) Point (d) improves on the constant used in the proof of Corollary 2.4 of Bryc (1982).
2) Point (c) extends the coupling result of Rio (1995) who considered the case where Y is bounded. Indeed, suppose that Y is bounded, say $a \leq Y \leq b$ a.s., where $a < b$ are real numbers. Let $C := (b - a)/2$. Subtracting the constant $(a + b)/2$ from Y (and from Y^*), we may assume without loss of generality that $-C \leq Y \leq C$ a.s., that is we can assume $a = -C$ and $b = C$. Now applying point (c), since $|Y| \leq C$ a.s. and hence $Q_{|Y|}(u) \leq C$ for all $u \in (0, 1)$, we get

$$\mathbb{E}|Y - Y^*| \leq 4C\bar{\alpha}(\mathcal{A}, Y) = 2(b - a)\bar{\alpha}(\mathcal{A}, Y),$$

which is exactly the result of Rio (Theorem 4, 1995).
3) The constants in the formulations (c) and (d) of Theorem 5.8 are as sharp as possible in the sense that it is possible to exhibit an example which attains exactly the bounds. To be more precise, Bradley, in a personal communication, provided us with the following example: Let W and Y be random variables that take only the values 1 and -1 and such that

$$\mathbb{P}(W = Y = 1) = \mathbb{P}(W = Y = -1) = 3/10 \text{ and}$$
$$\mathbb{P}(W = 1 \text{ and } Y = -1) = \mathbb{P}(W = -1 \text{ and } Y = 1) = 1/5\,.$$

Note that the distribution of (W, Y) is the same as that of (Y, W). Take $\mathcal{B} = \sigma(W)$, and suppose that Y^* is a random variable which is independent of $\mathcal{B}$ and has the same distribution as Y. Some calculations yield that $\mathbb{E}|Y - Y^*| \geq 1/5$, and that

$$\overline{\alpha}(\mathcal{B}, Y) = 0.05 \text{ and } \overline{\phi^r}(\mathcal{B}, Y) = 0.1.$$

From easy computations, it follows that the bounds in Theorem 5.8 are exactly attained in this particular case.

Concerning the case of a sequence of real random variables, by iteration, Theorem 5.8 gives the following useful result.

Theorem 5.9 *Let $(X_n, n \geq 1)$ be a sequence of real-valued random variables. Denote by $\mathcal{F}_1^n = \sigma(X_1, ..., X_n)$. Then, we can redefine $\{X_n, n \geq 1\}$ onto a richer probability space (which supports a sequence $(\delta_n, n \geq 1)$ of independent uniformly [0,1]-distributed random variables and independent of $\{X_n, n \geq 1\}$) together with a sequence $\{X_n^*, n \geq 1\}$ of independent random variables such that for each $1 \leq m \leq n$, we have*

a) X_m^* *is independent of* $\mathcal{F}_1^{m-1}$,

b) X_m^* *has the same distribution as* X_m,

c) $\mathbb{E}|X_m - X_m^*| \leq 4 \int_0^{\overline{\alpha}(\mathcal{F}_1^{m-1}, X_m)} Q_{|X_m|}(u)du$,

d) $\mathbb{E}|X_m - X_m^*| \leq 2\overline{\phi}^r(\mathcal{F}_1^{m-1}, X_m)\mathbb{E}|X_m|$.

Applications of coupling results are various and important. For instance Rio (2000) used point (c) in Theorem 5.9 to prove a Fuk–Nagaev type inequality for strongly mixing sequences (see the paper by Fuk and Nagaev (1971) in the independent case). Uniform laws of averages of dependent sequences which extend the classical laws of large number from a single function to a collection of such functions can be obtained by using coupling and known results for independent random variables. We refer to Chapter 16 where uniform laws of large numbers are stated under various dependence conditions.

5.1.4 *Examples of Non-Mixing Processes*

Unfortunately, a lot of Markov chains, even very simple, are known not to be strong mixing. Take for instance

$$X_k = \sum_{i=0}^{\infty} \frac{\varepsilon_{k-i}}{2^{i+1}}, \; k \in \mathbb{Z}, \tag{5.9}$$

where (ε_i) is an i.i.d. sequence of r.v.'s such that $\mathbb{P}(\varepsilon_0 = 0) = \mathbb{P}(\varepsilon_0 = 1) = 1/2$. This is a Markov chain with invariant measure λ (the Lebesgue measure on) $[0, 1]$ and transition Markov operator given by:

$$K(f)(x) = \frac{1}{2}\left(f\left(\frac{x}{2}\right) + f\left(\frac{x+1}{2}\right)\right).$$

This Markov chain is not strong mixing! Indeed, for any $k \in \mathbb{Z}$, we have $2X_{k+1} = X_k + \varepsilon_{k+1}$, hence X_k is the fractional part of $2X_{k+1}$ implying that $\sigma(X_k) \subset \sigma(X_{k+1})$ and then, by iteration, $\sigma(X_k) \subset \sigma(X_j, j \geq k+n)$ for any $n \geq 0$. Therefore

$$\frac{1}{4} \geq \alpha(n) \geq \sup_k \alpha(\sigma(X_k), \sigma(X_k)) = \frac{1}{4}.$$

This well-known example shows that the assumption that the ε_k's have a common density w.r.t. the Lebesgue measure in the first example of Section 5.1.2 is crucial. It is noteworthy to indicate that the fact that $\alpha(n) = 1/4$ for all $n \geq 0$ has been observed by Andrews (1984). However, the paper of Rosenblatt (1980) implicitly showed that that particular $AR(1)$ process is not strongly mixing. Indeed it is proved on page 267 of Rosenblatt's paper that this particular $AR(1)$ process is (strictly stationary, nondegenerate and) purely deterministic in one direction of time, and it had been well known for a couple of decades before that, that a strictly stationary, nondegenerate strongly mixing process is purely non-deterministic in both directions of time. Let us indicate that, with a longer, more roundabout argument, Ibragimov and Linnik (1971), pages 360–2, had shown even earlier that that particular $AR(1)$ process is not strongly mixing.

As we shall see in Section 5.2, this process is, however, α-dependent in the weak sense and for any $k \geq 1$, $(\alpha_{k,\mathbf{X}}(n))_{n\geq 1}$ decreases exponentially fast. In addition, it is related with dynamical systems as follows.

Define the following transformation of the interval $[0, 1]$: $\theta(x) = 2x - [2x] = 2x(\text{mod})\,1$. For this map, there is only one absolutely continuous invariant probability ν which is the uniform law of the interval $[0, 1]$. Let us consider its associated Perron–Frobenius operator K which is defined as follows: For any h in $\mathbb{L}^1([0, 1], \nu)$ and any f in $\mathbb{L}^\infty([0, 1], \nu)$

$$\nu(K(h) \cdot f) = \nu(h \cdot f \circ \theta). \tag{5.10}$$

This is a Markov operator and if we consider the Markov chain $\mathbf{Y} = (Y_i)_{i\in\mathbb{Z}}$ with invariant probability measure ν and Markov operator K defined by (5.10), it can be proved that the process $(Y_k)_k$ is equal in distribution to the autoregressive process defined by (5.9). In addition, the following equality in law holds $(\theta^n, \dots, \theta^2, \theta) \stackrel{d}{=} (Y_1, \dots, Y_{n-1}, Y_n)$ (where in the left-hand side the law is meant under ν and in the right-hand side the law is meant under $\mathbb{P}$). Therefore, to study the limiting of the partial sums $\sum_{k=1}^n f \circ \theta^k$ it is enough to study the one of $\sum_{k=1}^n f(Y_k)$.

We consider now intermittent maps and in particular the LSV map (Liverani, Saussol and Vaienti, 1999), defined as follows:

$$\text{For } 0 < \gamma < 1, \quad \theta(x) = \begin{cases} x(1 + 2^\gamma x^\gamma) & \text{if } x \in [0, 1/2] \\ 2x - 1 & \text{if } x \in [1/2, 1]. \end{cases} \tag{5.11}$$

If $\gamma \geq 1$, there is no absolutely continuous invariant probability whereas if $\gamma \in]0, 1[$, there is only one absolutely continuous invariant probability ν. Its density h satisfies $0 < c \leq h(x)/x^{-\gamma} \leq C < \infty$. As previously, we can associate a Markov chain $\mathbf{Y} = (Y_i)_{i\in\mathbb{Z}}$ with invariant probability measure ν and Markov operator K defined by (5.10). The following equality in law still holds: $(\theta^n, \dots, \theta^2, \theta) \stackrel{d}{=} (Y_1, \dots, Y_{n-1}, Y_n)$ and since

$$\max_{1\leq k\leq n} \left| \sum_{\ell=k}^{n} (f \circ \theta^\ell - \nu(f)) \right| \stackrel{d}{=} \max_{1\leq k\leq n} \left| \sum_{\ell=1}^{k} (f(Y_\ell) - \nu(f)) \right|,$$

many limit theorems for the partial sums associated with the observables of the iterates of the map can be deduced from those obtained for the partial sums associated with the Markov chain. For this LSV map, once again the associated Markov chain is not strong mixing. It is however α-dependent (see Section 5.2) and its coefficients of α-dependence decrease arithmetically with a rate depending on γ.

5.2 Weak Dependence Coefficients

As we have seen previously, the mixing coefficients have many properties such as coupling, but many simple Markov chains are neither β- nor α-mixing. The aim of this section is to introduce less stringent coefficients which are easier to compute.

5.2.1 *Definitions*

In all this section $\mathbf{X} := (X_k)_{k\in\mathbb{Z}}$ will be a sequence of real-valued random variables.

The α-dependent coefficients

In the previous section, we introduced a "weak" α-mixing coefficient (see definition (5.8)). From now on, this "weak" α-mixing coefficient will be called the α-dependent coefficient and we have, with the definition below, $\alpha_{1,\mathbf{X}}(n) = 2\sup_{j\in\mathbb{Z}} \overline{\alpha}(\mathcal{F}_j, X_{n+j})$ where $\mathcal{F}_j = \sigma(X_i, i \le j)$.

Definition 5.10 *For a sequence $\mathbf{X} = (X_i)_{i\in\mathbb{Z}}$ of real-valued random variables, for any $n \in \mathbb{N}$, let*

$$\alpha_{1,\mathbf{X}}(n) = \sup_{j\in\mathbb{Z}} \sup_{x\in\mathbb{R}} \left\| \mathbb{E}\left(\mathbf{1}_{X_{n+j}\le x} | \mathcal{F}_j\right) - \mathbb{E}(\mathbf{1}_{X_{n+j}\le x}) \right\|_1 . \tag{5.12}$$

As previously mentioned, these weak dependence coefficients have been defined in Rio (2000) (see his definitions (1.10c) and (2.30)) and can be extended if necessary with a finite number of variables in the future as follows: For any integrable random variable Z, let $Z^{(0)} = Z - \mathbb{E}(Z)$ and define for any positive integer k,

$$\alpha_{k,\mathbf{X}}(n) = \sup_{j\in\mathbb{Z}} \max_{1\le l\le k} \sup_{n+j\le i_1\le\ldots\le i_l} \alpha(\mathcal{F}_j, (X_{i_1}, \ldots, X_{i_l})), n \ge 0, \tag{5.13}$$

where for any random variable $V = (V_1, \cdots, V_k)$ with values in $\mathbb{R}^k$ and any σ-algebra $\mathcal{F}$,

$$\alpha(\mathcal{F}, V) = \sup_{(x_1,\ldots,x_k)\in\mathbb{R}^k} \left\| \mathbb{E}\left(\prod_{j=1}^{k} (\mathbf{1}_{V_j\le x_j})^{(0)} | \mathcal{F} \right) - \mathbb{E}\left(\prod_{j=1}^{k} (\mathbf{1}_{V_j\le x_j})^{(0)} \right) \right\|_1 .$$

For any positive integer k, the weak dependence coefficients $\alpha_{k,\mathbf{X}}(n)$ are weaker than the usual α-mixing coefficients defined in (5.2).

Between these coefficients, the following relation holds: For any positive integers k and n, $\alpha_{k,\mathbf{Y}}(n) \le 2\alpha(n)$. In addition, the order of magnitude of the α-dependent coefficient as defined in (5.13) may be computed for instance for many Markov chains associated with dynamical systems that fail to be strongly mixing in the sense of Rosenblatt (we refer to the paper by Dedecker, Gouëzel and Merlevède (2010)).

Remark 5.11 *Contrary to the strong mixing case where any functional of a strong mixing sequence is still strong mixing with the same strong mixing coefficients, any function of a α-dependent sequence $\mathbf{X} = (X_i)_{i\in\mathbb{Z}}$ is not necessarily α-dependent (meaning that its dependency coefficients do not necessarily tend to zero). However, the first thing to notice is that, if f is monotonic on some open interval and 0 elsewhere, and if $\mathbf{Y} = (f(X_i))_{i\in\mathbb{Z}}$, then for any positive integer k, $\alpha_{k,\mathbf{Y}}(n) \le 2^k\alpha_{k,\mathbf{X}}(n)$.*

The β-dependent coefficients

To simplify the exposition, we consider a stationary sequence of real-valued random variables $\mathbf{X} = (X_i)_{i\in\mathbb{Z}}$ and we denote by $\mathcal{F}_\ell = \sigma(X_i, i \le \ell)$. Straightforward extensions of the definition below can be obtained in the non-stationary setting.

Definition 5.12 *Let P be the law of X_0 and $P_{(X_i,X_j)}$ be the law of (X_i, X_j). Let $P_{X_k|X_0}$ be the conditional distribution of X_k given X_0, $P_{X_k|\mathcal{F}_\ell}$ be the conditional distribution of X_k given $\mathcal{F}_\ell$, and $P_{(X_i,X_j)|\mathcal{F}_\ell}$ be the conditional distribution of (X_i, X_j) given $\mathcal{F}_\ell$. Define the functions $f_t = \mathbf{1}_{[-\infty,t]}$, and $f_t^{(0)} = f_t - P(f_t)$. Define the random variables*

$$b(X_0, k) = \sup_{t\in\mathbb{R}} |P_{X_k|X_0}(f_t) - P(f_t)|,$$
$$b_1(\mathcal{F}_\ell, k) = \sup_{t\in\mathbb{R}} |P_{X_k|\mathcal{F}_\ell}(f_t) - P(f_t)|,$$
$$b_2(\mathcal{F}_\ell, i, j) = \sup_{(s,t)\in\mathbb{R}^2} |P_{(X_i,X_j)|\mathcal{F}_\ell}(f_t^{(0)} \otimes f_s^{(0)}) - P_{(X_i,X_j)}(f_t^{(0)} \otimes f_s^{(0)})|.$$

Define now the coefficients

$$\beta(\sigma(X_0), X_k) = \mathbb{E}(b(X_0, k)), \quad \beta_{1,\mathbf{X}}(k) = \mathbb{E}(b_1(\mathcal{F}_0, k)), \tag{5.14}$$

$$\textit{and} \quad \beta_{2,\mathbf{X}}(k) = \max\{\beta_{1,\mathbf{X}}(k), \sup_{i>j\ge k} \mathbb{E}(b_2(\mathcal{F}_0, i, j))\}. \tag{5.15}$$

Note that $\alpha_{1,\mathbf{X}}(k) \le \beta_{1,\mathbf{X}}(k) \le \beta_{2,\mathbf{X}}(k)$. For any $\ell \ge 3$, the coefficients $\beta_{\ell,\mathbf{X}}(k)$ can be defined in a similar way.

The ϕ-dependent coefficients

Replacing in definition (5.13) the $\mathbb{L}^1$-norm with the $\mathbb{L}^\infty$-norm, we define the ϕ-dependent coefficients. For any positive integer k, let then

$$\phi_{k,\mathbf{X}}(n) = \sup_{j\in\mathbb{Z}} \max_{1\le l\le k} \sup_{n+j\le i_1\le\ldots\le i_l} \phi(\mathcal{F}_j, (X_{i_1}, \ldots, X_{i_l})), n \ge 0, \tag{5.16}$$

where for any random variable $V = (V_1, \cdots, V_k)$ with values in $\mathbb{R}^k$ and any σ-algebra $\mathcal{F}$,

$$\phi(\mathcal{F}, V) = \sup_{(x_1,\ldots,x_k)\in\mathbb{R}^k} \left\| \mathbb{E}\left(\prod_{j=1}^k (\mathbf{1}_{V_j\le x_j})^{(0)} | \mathcal{F}\right) - \mathbb{E}\left(\prod_{j=1}^k (\mathbf{1}_{V_j\le x_j})^{(0)}\right)\right\|_\infty,$$

where we recall that for any integrable random variable Z, $Z^{(0)} = Z - \mathbb{E}(Z)$.

5.2.2 *The Weak Dependence Coefficients in the Markov Chains Setting*

If $\mathbf{X} = (X_i)_{i\in\mathbb{Z}}$ is a stationary Markov process with Kernel operator K and invariant measure ν, then its $\alpha_{1,\mathbf{X}}(n)$, $\beta_{1,\mathbf{X}}(n)$ and $\phi_{1,\mathbf{X}}(n)$ coefficients can be rewritten as follows: Let BV_1 be the class of bounded variation functions h such that $|h|_v \leq 1$ (where $|h|_v = \|dh\|_v$ is the variation norm of the measure dh). Then

$$\alpha_{1,\mathbf{X}}(n) = \frac{1}{2}\sup_{f\in BV_1} \nu\left(|K^n(f) - \nu(f)|\right),\ \beta_{1,\mathbf{X}}(n) = \frac{1}{2}\nu\left(\sup_{f\in BV_1} |K^n(f) - \nu(f)|\right)$$

and

$$\phi_{1,\mathbf{X}}(n) = \frac{1}{2}\sup_{f\in BV_1} \|K^n(f) - \nu(f)\|_{\nu,\infty}.$$

Now, for any $k \geq 2$, $\alpha_{k,\mathbf{X}}(n)$ (resp. $\beta_{k,\mathbf{X}}(n)$, resp. $\phi_{k,\mathbf{X}}(n)$) will have the same order of magnitude as $\alpha_{1,\mathbf{X}}(n)$ (resp. $\beta_{1,\mathbf{X}}(n)$, resp. $\phi_{1,\mathbf{X}}(n)$) if the space BV_1 is invariant under K, i.e., there exists a positive constant C such that, for any function f in BV_1,

$$|K^n(f)|_v \leq C|f|_v, \quad \text{for all } n \geq 1. \tag{5.17}$$

5.2.3 *Examples*

Functions of an i.i.d. sequence We follow here the approach of Doukhan and Louhichi (1999). Let $(\xi_i)_{i\in\mathbb{Z}}$ be a sequence of i.i.d. random variables with values in a measurable space $\mathcal{X}$. Let H be a function defined on a subset of $\mathcal{X}^{\mathbb{N}}$ and with values in $\mathbb{R}$ and such that $H(\xi_0, \xi_{-1}, \xi_{-2}, \ldots,)$ is defined almost surely. Let us define the sequence $(X_n)_{n\in\mathbb{Z}}$ by setting

$$X_n = H(\xi_n, \xi_{n-1}, \xi_{n-2}, \ldots).$$

Let $(\xi_i')_{i\in\mathbb{Z}}$ be an independent copy of $(\xi_i)_{i\in\mathbb{Z}}$. Let X_n^* be the random variable defined by

$$X_n^* = H(\xi_n, \ldots, \xi_1, \xi_0', \xi_{-1}', \ldots).$$

It follows easily that X_n^* is distributed as X and is independent of $\mathcal{F}_0 = \sigma(X_i, i \leq 0)$. For all $p > 0$ (p can be infinite), we consider the sequence $(a_{i,p})_{i>0}$ defined as follows:

$$a_{i,p} = \left(\mathbb{E}(|X_i - X_i^*|^p)\right)^{1/p} \quad \text{and} \quad a_{i,\infty} = \|X_i - X_i^*\|_\infty. \tag{5.18}$$

In Dedecker and Prieur (2005) the following bounds have been proved: If X_0 is assumed to have a continuous cumulative distribution function, with modulus of continuity w, and if we set g_p as the function defined by $g_p(y) = y(w(y))^{1/p}$, then for all $0 < p < \infty$,

$$\alpha_{1,\mathbf{X}}(n) \leq \beta_{1,\mathbf{X}}(n) \leq 2\left(\frac{a_{n,p}}{g_p^{-1}(a_{n,p})}\right)^p$$

and

$$\alpha_{1,\mathbf{X}}(n) \leq \beta_{1,\mathbf{X}}(n) \leq \phi_{1,\mathbf{X}}(n) \leq w(a_{n,\infty}).$$

In particular if X_0 has a density w.r.t. the Lebesgue measure bounded by K, then

$$\beta_{1,\mathbf{X}}(n) \leq 2(Ka_{n,p})^{\frac{p}{p+1}} \text{ and } \phi_{1,\mathbf{X}}(n) \leq Ka_{n,\infty}.$$

For causal linear processes defined by $X_n = \sum_{j\geq 0} b_j \xi_{n-j}$, for any $n \in \mathbb{Z}$, we have for any $p \geq 1$,

$$a_{n,p} \leq 2\|\xi_0\|_p \sum_{j\geq n} |b_j| \quad \text{and} \quad a_{n,2} = \sqrt{2\mathrm{Var}(\xi_0)\sum_{j\geq n} b_j^2}.$$

Note that the upper bounds above are obtained without any assumption on the distribution of ξ_0 (except moments conditions) and are valid for many processes which are not α-mixing. For instance if $b_j = 2^{-j-1}$ and ξ_0 is a sequence of r.v.'s with Bernoulli distribution of parameter $1/2$, then X_0 is uniformly distributed over $[0, 1]$ and then $\phi_{1,\mathbf{X}}(n) \leq 2^{-n}$.

We also refer to the paper by Dedecker and Prieur (2007, Section 6) where the coefficients $\beta_{2,\mathbf{X}}(n)$ and $\phi_{2,\mathbf{X}}(n)$ have been computed for such models.

Random iterative models Let $(X_n)_{n\geq 0}$ be a stationary Markov chain such that

$$X_n = F(X_{n-1}, \xi_n), \tag{5.19}$$

where F is a measurable function and $(\xi_i)_{i>0}$ is a sequence of i.i.d. r.v.'s which is in addition independent of X_0. Let X_0^* be a random variable distributed as X_0 and independent of $(X_0, (\xi_i)_{i>0})$. We define then X_n^* from X_0^* and $(\xi_i)_{i>0}$ as in (5.19). The variable X_n^* is distributed as X_n and is independent of $\mathcal{F}_0 = \sigma(X_0)$. If we define $a_{i,p}$ as in (5.18), then we obtain for $\alpha_{1,\mathbf{X}}(n)$, $\beta_{1,\mathbf{X}}(n)$ and $\phi_{1,\mathbf{X}}(n)$ the same bounds as in the previous example.

More precisely, let μ be the distribution of X_0 and $(X_n^x)_{n\geq 0}$ be the chain starting from x. With these notations, for all $0 < p < \infty$,

$$a_{i,p} = \left(\iint \mathbb{E}(|X_i^x - X_i^y|^p)\mu(dx)\mu(dy)\right)^{1/p}. \tag{5.20}$$

For instance, if there exists a sequence $(d_{i,p})_{i\geq 0}$ such that

$$\left(\mathbb{E}(|X_i^x - X_i^y|^p)\right)^{1/p} \leq d_{i,p}|x-y|,$$

then $a_{i,p} \leq d_{i,p}(\mathbb{E}(|X_0 - X_0^*|^p))^{1/p}$. In particular, if

$$\left(\mathbb{E}(|F(x,\xi_0) - F(y,\xi_0)|^p)\right)^{1/p} \leq \kappa|x-y| \tag{5.21}$$

for some $\kappa < 1$, we can take $d_{i,p} = \kappa^i$ (see again Dedecker and Prieur (2005)). Once again, we refer to the paper by Dedecker and Prieur (2007, Section 6) for the upper bounds concerning the coefficients $\beta_{2,\mathbf{X}}(n)$ and $\phi_{2,\mathbf{X}}(n)$.

An important example is the autoregressive process defined by $X_n = f(X_{n-1}) + \xi_n$, for some κ-Lipschitz function f. If $|X_0|$ has a moment of order p, then the following upper bound holds: $a_{i,p} \leq \kappa^i(\mathbb{E}(|X_0 - X_0^*|^p))^{1/p}$. Note that we do not require assumptions on the distribution of ξ_0. We refer also to the papers by Diaconis and Freedman (1999) and Wu and Shao (2004) for various examples of iterative random maps.

Expanding maps of $[0, 1]$ Let $\theta : [0, 1] \mapsto [0, 1]$ be a map which preserves a probability ν on $[0, 1]$, and denote by $X_i = \theta^i$. Note that $(X_i)_{i\geq 0}$ is a strictly stationary sequence of random variables from $([0, 1], \nu)$ in $[0, 1]$ but not mixing in the sense of Rosenblatt (except if ν is trivial).

In many cases we can show that for any h, a bounded variation function, and any function f in $\mathbb{L}^1([0, 1], \nu)$,

$$|\mathrm{cov}(h(X_0), f(X_n))| \leq C\rho^n \|f(X_n)\|_1 |h|_v, \tag{5.22}$$

for some constant $C > 0$ and $\rho \in]0, 1[$. If (5.22) holds we can then deduce that

$$\phi(\sigma(X_n), X_0) \leq C\rho^n.$$

For transformations which are called uniformly expanding, the inequality (5.22) comes from the contraction properties of the transition operator K (defined by (5.10)) of a Markov chain $\mathbf{Y} = (Y_i)_{i\in\mathbb{Z}}$ with invariant probability measure ν. Indeed, for some uniformly expanding maps, it can be shown (see Lasota and Yorke (1973)) that, for any h, a bounded variation function, there exist some constants $C > 0$ and $\rho \in]0, 1[$ such that

$$\|K^n(h) - \nu(h)\|_\infty \leq C\rho^n |h|_v,$$

and inequality (5.22) follows.

For the intermittent maps (non uniformly expanding) such as the LSV map defined by (5.11), it can be shown that that there exist positive constants A and B such that, for any $n \geq 1$,

$$\frac{A}{n^{(1-\gamma)/\gamma}} \leq \alpha(\sigma(X_n), X_0) = \alpha(\sigma(Y_0), Y_n) \leq \frac{B}{n^{(1-\gamma)/\gamma}},$$

and that (5.17) is satisfied (see Dedecker, Gouëzel and Merlevède (2010)).

Note also that it has been proved in Dedecker, Dehling and Taqqu (2015) that a similar upper bound holds for the β-dependent coefficient. More precisely, they proved (see their appendix) that there exists a positive constant C such that for any $n > 0$,

$$\beta_{1,\mathbf{Y}}(n) = \frac{1}{2}\nu\left(\sup_{f\in BV_1} |K^n(f) - \nu(f)|\right) \leq \frac{C}{n^{(1-\gamma)/\gamma}}.$$

6

Moment Inequalities and Gaussian Approximation for Mixing Sequences

Although the "mixing assumption" is nowadays considered as a rather restrictive condition, it is a powerful tool, which possesses strong properties such as preservation under functional transform, meaning that any measurable function of a mixing sequence is still mixing.

Note that the mixing assumption is still often assumed in econometric literature, often to check the mixingale properties (see Wooldridge and White (1988) for instance).

Also, it is apparent that approximation by Markov chains, or m-dependent variables, has become a powerful tool in the analysis of dynamical systems.

Moment inequalities whose upper bounds are expressed in terms of norms of conditional expectations lead to good moment inequalities in case of α-dependent sequences or of strong mixing sequences. However, when we consider ρ-mixing and ϕ-mixing sequences, this method does not lead to the optimal moment inequalities and other techniques have to be implemented.

6.1 Probability and Moment Inequalities for ρ-mixing Sequences

In this section we consider the sequence of maximal correlation coefficients $(\rho(n))_{n\geq 0}$ associated with a sequence $(X_k)_{k\in\mathbb{Z}}$ of real-valued random variables and defined by (5.3) in Section 5.1.1. Associated with this sequence, we set, for any positive integer n,

$$C(\rho, n) = \exp\left(2\sum_{j=0}^{[\log_2 n]} \rho(2^{j/3})\right), \tag{6.1}$$

where we use the convention that $\rho(x) = \rho([x])$ for any $x \geq 0$. By c, κ, possibly with subscripts, we denote absolute constants.

Functional Gaussian Approximation for Dependent Structures. Florence Merlevède, Magda Peligrad and Sergey Utev. Oxford University Press (2019).
DOI: 10.1093/oso/9780198826941.001.0001

6.1.1 *Analysis of the Variance: A Twist of the Dyadic Induction*

The main result in this part is as follows.

Theorem 6.1 *Let $(X_k)_{k\in\mathbb{Z}}$ be a sequence of centered real-valued random variables in $\mathbb{L}^2$. Let $(\rho(n))_{n\geq 0}$ be its associated maximal correlation coefficients sequence. Let $S_n = \sum_{i=1}^n X_i$. Then*

$$\|S_n\|_2^2 \leq \kappa_1 C(\rho, n) \sum_{j=1}^{n} \|X_j\|_2^2. \tag{6.2}$$

One can take $\kappa_1 = 8\,\mathrm{e}^{2^{20}}$. As an immediate corollary, we obviously have:

Corollary 6.2 $\|S_n\|_2^2 \leq n\,\kappa_1 C(\rho, n) \max_{1\leq j\leq n} \|X_j\|_2^2$.

Before proving Theorem 6.1 in its generality, it is instructive to first show how it can be proven *under the additional assumption of stationarity*. So let us assume for the moment that $(X_k)_{k\in\mathbb{Z}}$ is a strictly stationary sequence of centered real-valued random variables in $\mathbb{L}^2$, and still denote by $(\rho(n))_{n\geq 0}$ its associated maximal correlation coefficients sequence. We shall see that in this case proving that there exists a positive universal constant κ_1 such that

$$\|S_n\|_2^2 \leq n\,\kappa_1 C(\rho, n)\|X_0\|_2^2$$

can be done by using the dyadic technique as already used in Sections 3.1 and 3.3. With this aim, for any positive integers k and n, we set

$$S(k,n) = \sum_{i=k+1}^{n} X_i \text{ and } \sigma_n^2 = \|S_n\|_2^2. \tag{6.3}$$

By stationarity, we have

$$\sigma_{2n}^2 = 2\sigma_n^2 + 2\mathbb{E}(S_n S(n, 2n)).$$

Next we consider a nonnegative integer m less than n and, writing $S(n, 2n) = S(n, n+m) + S(n+m, 2n)$, by using stationarity and the definition of the maximal correlation coefficients, we get that

$$|\mathbb{E}(S_n S(n, 2n))| \leq \sigma_n\sigma_m + \rho(m+1)\sigma_n\sigma_{n-m}\,.$$

But,

$$\sigma_{n-m} \leq \sigma_n + \sigma_m \text{ and } \sigma_m \leq m\|X_0\|_2. \tag{6.4}$$

So, overall, the following upper bound holds:

$$\sigma_{2n}^2 \leq 2\sigma_n^2(1+\rho(m+1)) + 2m\sigma_n\|X_0\|_2(1+\rho(m+1)). \tag{6.5}$$

Let now r be a positive integer. The inequality above with $n = 2^{r-1}$ and $m = [2^{r/3}]$ gives

$$\begin{aligned}\sigma_{2^r}^2 &\leq 2\sigma_{2^{r-1}}^2(1+\rho(2^{r/3})) + 2\sigma_{2^{r-1}}2^{r/3}\|X_0\|_2(1+\rho(2^{r/3}))\\ &\leq 2\left(\sigma_{2^{r-1}} + 2^{r/3-1}\|X_0\|_2\right)^2(1+\rho(2^{r/3})).\end{aligned}$$

So, setting

$$B_{2^r} = \frac{\sigma_{2^r}}{2^{r/2}}, \tag{6.6}$$

we get

$$B_{2^r} \leq B_{2^{r-1}}\Big(1+\frac{1}{2}\rho(2^{r/3})\Big) + 2^{-r/6}\frac{\|X_0\|_2}{\sqrt{2}}\Big(1+\frac{1}{2}\rho(2^{r/3})\Big). \tag{6.7}$$

By a recurrence on the term $B_{2^{r-1}}$ and setting

$$\varepsilon_{2^r} = 1+\frac{1}{2}\rho(2^{r/3}) \text{ and } \delta_{2^r} = 2^{-r/6}\frac{\|X_0\|_2}{\sqrt{2}}\Big(1+\frac{1}{2}\rho(2^{r/3})\Big),$$

we infer that for any positive integer r,

$$B_{2^r} \leq B_{2^0}\prod_{i=1}^{r}\varepsilon_{2^i} + \sum_{i=1}^{r}\delta_{2^i}\prod_{j=i+1}^{r}\varepsilon_{2^j},$$

with the usual convention that $\prod_{j=u}^{v} = 1$ if $v < u$. Therefore

$$B_{2^r} \leq \|X_0\|_2\left(1+\frac{1}{\sqrt{2}(2^{1/6}-1)}\right)\exp\left(2^{-1}\sum_{j=1}^{r}\rho(2^{j/3})\right). \tag{6.8}$$

To end the proof we consider the binary expansion of n as in (3.4). Then, using (3.5) and stationarity, it follows that

$$\|S_n\|_2 \leq \sum_{k=0}^{r-1}\|T_{2^k}\|_2 \leq \sum_{k=0}^{r-1}\|S_{2^k}\|_2,$$

where r is such that $2^{r-1} \leq n < 2^r$. Then, by using (6.8) and the fact that the following inequality is valid: $\sum_{k=0}^{r-1}2^{k/2} \leq 2^{r/2}/(2^{1/2}-1)$, we derive that

$$\|S_n\|_2^2 \leq C\,n\,\|X_0\|_2^2\exp\left(\sum_{j=0}^{[\log_2 n]}\rho(2^{j/3})\right), \tag{6.9}$$

where we can take $C = 2\left(\sqrt{2}+1\right)^2\left(1+\frac{1}{\sqrt{2}(2^{1/6}-1)}\right)^2$.

This proof reveals that, to use the usual dyadic method, the stationarity assumption is crucial at different places. Obviously, if no stationarity is assumed then an inequality such as (6.5) fails to hold since we can neither link $\mathbb{E}(S(n,2n))^2$ with σ_n^2 nor use the subadditivity of σ_n as is done in (6.4). In the "non-stationary" case, the division by 2 is usually not appropriate so instead of trying to get an inequality such as (6.8), we shall try to establish for a certain extremal characteristic, say Z_n, that there exist two non-increasing sequences of nonnegative reals $(\varepsilon_n)_{n\geq 1}$ and $(\delta_n)_{n\geq 1}$ and a sequence of positive integers $(T_n)_{n\geq 1}$ such that for any positive integer n

$$Z_n \leq \max_{1\leq \ell\leq k} \{Z_{T_\ell}(1+\varepsilon_\ell)+\delta_\ell\}.$$

Obviously, in the stationary case T_k grows like $k/2$. As we shall see in the proof Theorem 6.1 the adapted extremal characteristic is defined as follows: Define $Z_0 = 1$ and, for $n \geq 1$, let

$$Z_n = \sup_{0\leq a<\infty} \max_{1\leq i\leq n} \left\{ \|S(a,a+i)\|_2 \Bigg/ \left(\sum_{j=a+1}^{a+i} \|X_j\|_2^2 \right)^{1/2} \right\},$$

where $S(a, a+i)$ is defined in (6.3). It follows immediately from the construction that

$$\|S_n\|_2^2 \leq Z_n^2 \sum_{j=1}^{n} \|X_j\|_2^2. \tag{6.10}$$

Clearly, the theorem will follow if we can show that

$$Z_n^2 \leq \kappa_1 C(\rho, n). \tag{6.11}$$

With this aim, we shall analyze the properties of (Z_n). Obviously $(Z_n)_{n\geq 1}$ is a non-decreasing sequence of nonnegative numbers but we can even say more, as stated in the next lemma.

Lemma 6.3 *(Properties of Z_n) For any $k \geq 1$, the following assertions hold:*

$$1 = Z_0 \leq Z_{k-1} \leq Z_k \leq k^{1/2} \tag{6.12}$$

and there exists a sequence of positive integers $(T_k)_{k\geq 1}$ satisfying $T_k \leq 2^{-1}k(1+k^{-1/3})$ for any $k \geq 1$ and such that

$$Z_k \leq \max_{1\leq \ell\leq k} \{Z_{T_\ell}\left(1+2^{-1}\rho(\ell^{1/3})+2^{14}\ell^{-1/9}\right)\}. \tag{6.13}$$

As we shall see, the proof of this lemma is based on the following combinatoric trick which is due to Utev (1991):

Lemma 6.4 *Let $(x_n)_{n\geq 1}$ be a sequence of nonnegative real numbers. For any positive integer k, define $m = [k^{1/3}]$,*

$$A_k = \left\{ u \in \mathbf{N} \, : \, 0 \leq u \leq k \text{ and } \sum_{i=um+1}^{(u+1)m} x_i \leq (m/k)^{1/3} \sum_{j=1}^{k} x_j \right\}$$

and

$$B_k = \{ u \in \mathbf{N} : 0 \leq u \leq k \text{ and } 2^{-1}(k - m^2) \leq um \leq 2^{-1}k \}.$$

Then for any $k \geq k_0 = 2^{14}$, $A_k \cap B_k \neq \emptyset$.

Moreover, the property (6.13) combined with the following lemma (proved in Utev (1991)) will allow us to prove (6.11) and then Theorem 6.1.

Lemma 6.5 *Let $(a_n)_{n\geq 1}$ be a non-decreasing sequence of non-negative reals. Assume that there exist two non-increasing sequences of non-negative real numbers $(\varepsilon_n)_{n\geq 1}$ and $(\delta_n)_{n\geq 1}$, and a sequence of positive integers $(T_n)_{n\geq 1}$ such that for any $k \geq 1$*

$$T_k \leq 2^{-1}k(1 + k^{-1/3}) \text{ and } a_k \leq \max_{1\leq \ell \leq k} \{ a_{T_\ell}(1 + \varepsilon_\ell) + \delta_\ell \}.$$

Then

$$a_n \leq \left(a_{n_0} + 2 \sum_{j=0}^{[\log_2 n]-1} \delta_{2^j} \right) \exp\left(2 \sum_{j=0}^{[\log_2 n]-1} \varepsilon_{2^j} \right), \tag{6.14}$$

for all $n \geq 1$ where one can take $n_0 = 2^3$. The following convention is used: $\sum_{\ell=0}^{i} = 0$ for any $i < 0$.

Indeed, property (6.13) shows that the conditions of the lemma are satisfied with $a_n = Z_n$, $\varepsilon_n = 2^{-1}\rho(n^{1/3}) + 2^{14}n^{-1/9}$ and $\delta_n = 0$. Therefore, for any $n \geq 1$,

$$Z_n \leq Z_8 \exp\left(\sum_{j=0}^{[\log_2 n]} (\rho(2^{j/3}) + 2^{15}2^{-j/9}) \right).$$

By (6.12), $Z_8 \leq 2\sqrt{2}$. Moreover $\sum_{j\geq 0} 2^{-j/9} \leq 2^4$. Therefore

$$Z_n \leq 2\sqrt{2}\, e^{2^{19}} \exp\left(\sum_{j=0}^{[\log_2 n]} \rho(2^{j/3}) \right),$$

which gives (6.11) with $\kappa_1 = 8\, e^{2^{20}}$.

To end the proof of Theorem 6.1, it remains to prove Lemmas 6.3, 6.4 and 6.5.

Proof of Lemma 6.3 By construction, the sequence $(Z_n)_{n\geq 0}$ is non-decreasing. Moreover, we clearly have $\|S(a, a+i)\|_2^2 \leq i \sum_{j=a+1}^{a+i} \|X_j\|_2^2$, ending the proof of Property (6.12).

Let us prove the assertion (6.13). With this aim, we proceed by recurrence. Let $k_0 = 2^{14}$ and $T_k = 1$ for $k = 1,...,k_0$. Then for any positive integer $k \leq k_0$ one obtains

$$Z_k \leq k_0 \leq \max_{1\leq \ell \leq k} \{Z_{T_\ell}(1 + 2^{14}\ell^{-1/9})\}, \tag{6.15}$$

since for any $\ell \in \{1,...,k_0\}$, $Z_{T_\ell} = Z_1 = 1$. This proves the assertion for any positive integer $k \leq k_0$. Now assume that $k > k_0$ and that for any $i \leq k-1$, the assertion (6.13) holds. Let us then prove it at the step k. Obviously the sequence T_k should be constructed so that, for any $a \geq 0$ and any $i = 1, ..., k$,

$$\|S(a, a+i)\|_2 \leq \left(\sum_{j=a+1}^{a+i} \|X_j\|_2^2\right)^{1/2} \max_{1\leq \ell \leq k} \{Z_{T_\ell}\left(1 + 2^{-1}\rho(\ell^{1/3}) + 2^{14}\ell^{-1/9}\right)\}.$$

Without loss of generality we may consider $a = 0$. Taking into account the recurrence assumption, it suffices then to construct T_k so that

$$\|S_k\|_2 \leq \left(\sum_{j=1}^{k} \|X_j\|_2^2\right)^{1/2} \left(Z_{T_k}\left(1 + 2^{-1}\rho(k^{1/3}) + 2^{14}k^{-1/9}\right)\right). \tag{6.16}$$

With this aim, we use Lemma 6.4 with $x_i = \|X_i\|_2^2$. Since $k > k_0 = 2^{14}$, we can fix $u \in A_k \cap B_k$ and write

$$\|S_k\|_2 \leq \|S_{um} + S(um+m, k)\|_2 + \|S(um, um+m)\|_2. \tag{6.17}$$

According to the definition of the maximal correlation coefficients $(\rho(n))_{n\geq 0}$, we get

$$\begin{aligned}\|S_{um} + S(um+m, k)\|_2^2 &\leq \|S_{um}\|_2^2 + \|S(um+m, k)\|_2^2 \\ &+ 2\,\rho(m)\|S_{um}\|_2\|S(um+m, k)\|_2.\end{aligned} \tag{6.18}$$

Hence, starting from (6.17) and taking into account (6.18) and the definition of Z_n,

$$\begin{aligned}\|S_k\|_2 &\leq Z_m \left(\sum_{i=um+1}^{(u+1)m} \|X_i\|^2\right)^{1/2} \\ &+ \left(Z_{um}^2 x^2 + Z_{k-um-m}^2 y^2 + 2xy\rho(m) Z_{um} Z_{k-um-m}\right)^{1/2},\end{aligned} \tag{6.19}$$

where

$$x = \left(\sum_{i=1}^{um} \|X_i\|_2^2\right)^{1/2} \quad \text{and } y = \left(\sum_{i=um+m+1}^{k} \|X_i\|_2^2\right)^{1/2}.$$

Set now $T_k = \max(um + m, k - um)$. Note that since $u \in B_k$ and $k \geq 8$, it follows that T_k is a positive integer satisfying $T_k \leq 2^{-1}k(1 + k^{-1/3})$. Using this selection of T_k in (6.19), the fact that (Z_n) is a non-decreasing sequence and that, since $u \in A_k \cap B_k$,

$$\sum_{i=um+1}^{(u+1)m} \|X_i\|_2^2 \leq (m/k)^{1/3} \sum_{i=1}^{k} \|X_i\|_2^2,$$

we get

$$\|S_k\|_2 \leq Z_{T_k} \left\{ \left(\sum_{j=1}^{k} \|X_j\|_2^2\right)^{1/2} (m/k)^{1/6} + \left(x^2 + y^2 + 2xy\rho(m)\right)^{1/2} \right\}.$$

Then, using in addition the elementary inequality $2xy \leq x^2 + y^2$ and the fact that $x^2 + y^2 \leq \sum_{j=1}^{k} \|X_j\|_2^2$, we derive

$$\begin{aligned} \|S_k\|_2 &\leq Z_{T_k} \left(\sum_{j=1}^{k} \|X_j\|_2^2\right)^{1/2} \left((m/k)^{1/6} + (1 + \rho(m))^{1/2}\right) \\ &\leq Z_{T_k} \left(\sum_{j=1}^{k} \|X_j\|_2^2\right)^{1/2} \left(1 + 2^{-1}\rho([k^{1/3}]) + k^{-1/9}\right). \end{aligned}$$

Hence, inequality (6.16) and thus Lemma 6.3 are proved. ■

Proof of Lemma 6.4 We first note that

$$\text{Card}(B_k) \geq \text{Card}\left\{u \in \mathbf{N} : 0 \leq u \leq k \text{ and } \left[\frac{k}{2m}\right] + 1 - \left[\frac{m}{2}\right] \leq u \leq \left[\frac{k}{2m}\right]\right\},$$

implying that

$$\text{Card}(B_k) \geq \left[\frac{m}{2}\right]. \tag{6.20}$$

Let now

$$\bar{A}_k = \left\{u \in \mathbf{N} : 0 \leq u \leq k \text{ and } \sum_{i=um+1}^{(u+1)m} x_i > (m/k)^{1/3} \sum_{j=1}^{k} x_j\right\}.$$

Note that, since $k \geq k_0$,

$$\mathrm{Card}(\bar{A}_k \cap B_k) \leq$$
$$\mathrm{Card}\left\{u \in \mathbf{N} : 0 \leq u \leq (k-m)/m \text{ and } \sum_{i=um+1}^{(u+1)m} x_i > (m/k)^{1/3} \sum_{j=1}^{k} x_j\right\}.$$

Therefore

$$\mathrm{Card}(\bar{A}_k \cap B_k) \leq (k/m)^{1/3}, \tag{6.21}$$

otherwise it would be possible to write $S = \sum_{j=1}^{k} x_j$ as a sum of more than $(k/m)^{1/3}$ terms of the type $\sum_{i=jm+1}^{(j+1)m} x_i$ (with j such that $(j+1)m \leq k$) each satisfying $\sum_{i=jm+1}^{(j+1)m} x_i > (m/k)^{1/3} S$, which is clearly wrong. To end the proof of the lemma, we write

$$\mathrm{Card}(A_k \cap B_k) = \mathrm{Card}(B_k \backslash (\bar{A}_k \cap B_k)) = \mathrm{Card}(B_k) - \mathrm{Card}(\bar{A}_k \cap B_k).$$

Hence, taking into account (6.20) and (6.21),

$$\mathrm{Card}(A_k \cap B_k) \geq \left[\frac{m}{2}\right] - (k/m)^{1/3}.$$

Using the fact that $[m/2] \geq 2^{-1}k^{1/3} - 3/2$ and that $m \geq 2^{-1}k^{1/3}$ (since $k^{1/3} \geq 2$), it follows that

$$\mathrm{Card}(A_k \cap B_k) \geq \frac{k^{1/3}}{2}\left(1 - \frac{2^{4/3}}{k^{1/9}}\right) - \frac{3}{2}.$$

Hence $\mathrm{Card}(A_k \cap B_k) > 0$ provided that $k \geq k_0 = 2^{14}$. This ends the proof of the lemma. ■

Proof of Lemma 6.5 We shall prove that for any integer $r \geq 0$,

$$a_{2^r} \leq \left(a_{n_0} + 2\sum_{j=0}^{r-2} \delta_{2^j}\right) \exp\left(2\sum_{j=0}^{r-2} \varepsilon_{2^j}\right) := E_r, \tag{6.22}$$

with the convention that $E_0 = E_1 = a_{n_0}$. Together with the fact that $(a_n)_{n\geq 1}$ is a non-decreasing sequence, inequality (6.22) will imply the lemma. To prove (6.22), we shall proceed by recurrence. Obviously, since $(a_n)_{n\geq 1}$ is non-decreasing and $(\varepsilon_n)_{n\geq 1}$ and $(\delta_n)_{n\geq 1}$ are sequences of non-negative numbers, inequality (6.22) is true for any r such that $2^r \leq n_0 = 2^3$. Let now k be an integer such that $k \geq 4$. Assuming that the assertion (6.22) holds for any $r \leq k$, let us prove that it also holds at step $k+1$, namely that $a_{2^{k+1}} \leq E_{k+1}$. We have

$$a_{2^{k+1}} \le \max_{1\le \ell \le 2^{k+1}} \{a_{T_\ell}(1+\varepsilon_\ell)+\delta_\ell\} \le \max_{0\le j\le k+1} \max_{[2^{j-1}]+1\le \ell \le 2^j} \{a_{T_\ell}(1+\varepsilon_\ell)+\delta_\ell\}.$$

Since $(\varepsilon_n)_{n\ge 1}$ and $(\delta_n)_{n\ge 1}$ are non-increasing, it follows that

$$a_{2^{k+1}} \le \max_{0\le j\le k+1} \Big\{(1+\varepsilon_{[2^{j-1}]+1}) \max_{[2^{j-1}]+1\le \ell\le 2^j} a_{T_\ell} + \delta_{[2^{j-1}]+1}\Big\}. \tag{6.23}$$

Note now that since $(a_n)_{n\ge 1}$ is non-decreasing and since for $\ell \le 2^j$, $T_\ell \le 2^j$, we have

$$\max_{0\le j\le k} \Big\{(1+\varepsilon_{[2^{j-1}]+1}) \max_{[2^{j-1}]+1\le \ell\le 2^j} a_{T_\ell} + \delta_{[2^{j-1}]+1}\Big\} \le \max_{0\le j\le k} \{a_{2^j}\left(1+\varepsilon_{[2^{j-1}]+1}\right) + \delta_{[2^{j-1}]+1}\}.$$

So by the recurrence assumption,

$$\begin{aligned}\max_{0\le j\le k} &\Big\{(1+\varepsilon_{[2^{j-1}]+1}) \max_{[2^{j-1}]+1\le \ell\le 2^j} a_{T_\ell} + \delta_{[2^{j-1}]+1}\Big\} \\ &\le \max_{0\le j\le k} \left\{\left(a_{n_0} + 2\sum_{\ell=0}^{j-2} \delta_{2^\ell}\right) \exp\left(2\sum_{\ell=0}^{j-2} \varepsilon_{2^\ell} + \varepsilon_{[2^{j-1}]+1}\right) + \delta_{[2^{j-1}]+1}\right\} \\ &\le \max_{0\le j\le k} \left(a_{n_0} + 2\sum_{\ell=0}^{j-2} \delta_{2^\ell} + \delta_{[2^{j-1}]+1}\right) \exp\left(2\sum_{\ell=0}^{j-2} \varepsilon_{2^\ell} + \varepsilon_{[2^{j-1}]+1}\right),\end{aligned} \tag{6.24}$$

where we have used the fact that the ε_k's are non-negative. Therefore, since $(\varepsilon_n)_{n\ge 1}$ and $(\delta_n)_{n\ge 1}$ are non-increasing,

$$\max_{0\le j\le k} \Big\{(1+\varepsilon_{[2^{j-1}]+1}) \max_{[2^{j-1}]+1\le \ell\le 2^j} a_{T_\ell} + \delta_{[2^{j-1}]+1}\Big\} \le E_{k+1}. \tag{6.25}$$

Starting from (6.23) and taking into account (6.25), we can conclude that the lemma will be proved if we can show that

$$(1+\varepsilon_{2^k+1}) \max_{2^k+1\le \ell\le 2^{k+1}} a_{T_\ell} + \delta_{2^k+1} \le E_{k+1}. \tag{6.26}$$

With this aim, we start by noting that for any integer i such that $i \ge k_0 = 21$,

$$T_i \le 2^{-1} i (1 + i^{-1/3}) \le [2^{-1/2} i]. \tag{6.27}$$

Therefore, since $k \ge 4$, for any $\ell \le 2^{k+1}$,

$$T_\ell \le 2^{-1} 2^{k+1} (1 + 2^{-(k+1)/3}) \le [2^{k+1/2}]. \tag{6.28}$$

Since $(a_n)_{n\geq 1}$ is non-decreasing, it follows that

$$(1+\varepsilon_{2^k+1}) \max_{2^k+1\leq \ell\leq 2^{k+1}} a_{T_\ell} \leq a_{[2^{k+1/2}]}(1+\varepsilon_{2^k+1}). \tag{6.29}$$

We write now

$$\begin{aligned} a_{[2^{k+1/2}]} \leq \max\Bigg\{ & \max_{2^k+1\leq \ell\leq [2^{k+1/2}]} \{a_{T_\ell}(1+\varepsilon_\ell)+\delta_\ell\}, \\ & \max_{0\leq i\leq k} \max_{[2^{i-1}]+1\leq \ell\leq 2^i} \{a_{T_\ell}(1+\varepsilon_\ell)+\delta_\ell\}\Bigg\}. \end{aligned} \tag{6.30}$$

Using the fact that $(\varepsilon_n)_{n\geq 1}$ and $(\delta_n)_{n\geq 1}$ are non-increasing and (6.24), we get

$$\begin{aligned} &(1+\varepsilon_{2^k+1}) \max_{0\leq i\leq k} \max_{[2^{i-1}]+1\leq \ell\leq 2^i} \{a_{T_\ell}(1+\varepsilon_\ell)+\delta_\ell\} \\ &\leq (1+\varepsilon_{2^k+1}) \max_{0\leq i\leq k} \Big\{(1+\varepsilon_{[2^{i-1}]+1}) \max_{[2^{i-1}]+1\leq \ell\leq 2^i} a_{T_\ell} + \delta_{[2^{i-1}]+1}\Big\} \\ &\leq (1+\varepsilon_{2^k+1}) \max_{0\leq i\leq k} \Big(a_{n_0} + 2\sum_{\ell=0}^{i-2} \delta_{2^\ell} + \delta_{[2^{i-1}]+1}\Big) \exp\Big(2\sum_{\ell=0}^{i-2} \varepsilon_{2^\ell} + \varepsilon_{[2^{i-1}]+1}\Big). \end{aligned}$$

Therefore, using again the fact that $(\varepsilon_n)_{n\geq 1}$ and $(\delta_n)_{n\geq 1}$ are non-increasing,

$$\begin{aligned} &(1+\varepsilon_{2^k+1}) \max_{0\leq i\leq k} \max_{[2^{i-1}]+1\leq \ell\leq 2^i} \{a_{T_\ell}(1+\varepsilon_\ell)+\delta_\ell\} + \delta_{2^k+1} \\ &\leq \max_{0\leq i\leq k} \Big(a_{n_0} + 2\sum_{\ell=0}^{i-2} \delta_{2^\ell} + 2\delta_{2^{i-1}}\Big) \exp\Big(2\sum_{\ell=0}^{i-2} \varepsilon_{2^\ell} + 2\varepsilon_{2^{i-1}}\Big) \leq E_{k+1}. \end{aligned} \tag{6.31}$$

Starting from (6.29) and considering (6.30) and (6.31), it follows that (6.26) will be proved if one can show that

$$(1+\varepsilon_{2^k+1}) \max_{2^k+1\leq \ell\leq [2^{k+1/2}]} \{a_{T_\ell}(1+\varepsilon_\ell)+\delta_\ell\} + \delta_{2^k+1} \leq E_{k+1}. \tag{6.32}$$

Using again the monotonicity properties of $(a_n)_{n\geq 1}$, $(\varepsilon_n)_{n\geq 1}$ and $(\delta_n)_{n\geq 1}$ together with (6.27) with $[2^{k+1/2}]$ in place of i (notice that $[2^{k+1/2}] \geq 21$ since $k\geq 4$), we get

$$\begin{aligned} &(1+\varepsilon_{2^k+1}) \max_{2^k+1\leq \ell\leq [2^{k+1/2}]} \{a_{T_\ell}(1+\varepsilon_\ell)+\delta_\ell\} \\ &\leq a_{2^k}(1+\varepsilon_{2^k+1})^2 + (1+\varepsilon_{2^k+1})\delta_{2^k+1} \leq a_{2^k}(1+\varepsilon_{2^{k-1}})^2 + (1+\varepsilon_{2^{k-1}})\delta_{2^{k-1}}. \end{aligned}$$

So, by the recurrence assumption and the fact that, for any $x\geq 0$, $(1+x)^2 \leq \exp(2x)$, we get

$$
\begin{aligned}
&(1+\varepsilon_{2^k+1}) \max_{2^k+1\le \ell \le [2^{k+1/2}]} \{a_{T_\ell}(1+\varepsilon_\ell)+\delta_\ell\} + \delta_{2^k+1} \\
&\le \left(a_{n_0} + 2\sum_{j=0}^{k-2} \delta_{2^j}\right) \exp\left(2\sum_{j=0}^{k-1} \varepsilon_{2^j}\right) + (1+\varepsilon_{2^{k-1}})\delta_{2^{k-1}} + \delta_{2^{k-1}} \\
&\le \left(a_{n_0} + 2\sum_{j=0}^{k-2} \delta_{2^j}\right) \exp\left(2\sum_{j=0}^{k-1} \varepsilon_{2^j}\right) + 2\delta_{2^{k-1}} \exp(\varepsilon_{2^{k-1}}),
\end{aligned}
$$

which proves (6.32) since the ε_k's as well as the δ_k's are non-negative. This ends the proof of the lemma. ∎

6.1.2 An Extension of the Variance Inequality

Let $\{X_i\}_{i\in\mathbb{Z}}$ and $\{Y_i\}_{i\in\mathbb{Z}}$ be two sequences of square integrable zero mean real-valued random variables. Set

$$
x(a,b) = \sum_{a<i\le b} X_i \text{ and } y(a,b) = \sum_{a<i\le b} Y_i,
$$

where by convention, sums over empty sets are equal to zero. Let us define ρ-type coefficients between various combinations of partial sums of X_i, and Y_i by: For any $k \ge 0$,

$$
\begin{aligned}
u_1(k) &= \sup\left\{|\operatorname{corr}\left(x(j,j+t), y(j+t+s, j+t+s+u)\right)| : s\ge k, j\ge 0, t\ge 1, u\ge 0\right\},\\
u_2(k) &= \sup\left\{|\operatorname{corr}\left(x(j,j+t), x(j+t+s, j+t+s+u)\right)| : s\ge k, j\ge 0, t\ge 1, u\ge 0\right\},\\
u_3(k) &= \sup\left\{|\operatorname{corr}\left(y(j,j+t), y(j+t+s, j+t+s+u)\right)| : s\ge k, j\ge 0, t\ge 1, u\ge 0\right\},
\end{aligned}
$$

and finally let us define the max coefficient

$$
u(k) = \max\{u_1(k), u_2(k), u_3(k)\}. \tag{6.33}
$$

In addition, introduce the following constants

$$
C(u,n) = \exp\left(2\sum_{j=0}^{[\log_2 n]} u(2^{j/3})\right) \text{ and } r(u,n) = C(u,n)\sum_{j=0}^{[\log_2 n]} u(2^j).
$$

The main result of this part is the following:

Theorem 6.6 *There exists an absolute constant c_1 such that for all nonnegative integers a, n we have*

$$
\left|\sum_{i=a+1}^{a+n} \mathbb{E}\left(X_i y(i, a+n)\right)\right| \le c_1 r(u,n) \left(\sum_{j=a+1}^{a+n} \|X_j\|_2^2 \sum_{j=a+1}^{a+n} \|Y_j\|_2^2\right)^{1/2}.
$$

Choice of the constant. It follows from the proof that one can take $c_1 = 8 + 2(2^4 + 1)\kappa_1$ with $\kappa_1 = 8\,\mathrm{e}^{2^{20}}$.

Before proving the theorem, we notice that by inspecting the proof of Theorem 6.1, it reveals that the result holds under a weaker ρ-type coefficient defined for partial sums. More exactly, the following result holds:

Lemma 6.7 *For any nonnegative integers a, n,*

$$\Big\| \sum_{i=a+1}^{a+n} X_i \Big\|_2^2 \le \kappa_1 C(u,n) \sum_{i=a+1}^{a+n} \|X_i\|_2^2 \text{ and } \Big\| \sum_{i=a+1}^{a+n} Y_i \Big\|_2^2 \le \kappa_1 C(u,n) \sum_{i=a+1}^{a+n} \|Y_i\|_2^2,$$

where $\kappa_1 = 8\,\mathrm{e}^{2^{20}}$.

Now, we are ready to proceed to the proof of the theorem.

Proof of Theorem 6.6 The proof is very similar to the proof of Theorem 6.1. It is then based on the extremal approach applied to the analysis of

$$a_k := \sup_{0 \le a < \infty} \max_{1 \le j \le k} \frac{|\Delta(a, a+j)|}{D(a, a+j)}, \ k \in \mathbb{N}^*, \quad \text{where } 0/0 = 0,$$

$$\Delta(a,b) = \sum_{i=a+1}^{b} \mathbb{E}[X_i y(i,b)] \text{ and } D(a,b) = \Big(\sum_{i=a+1}^{b} \|X_i\|_2^2 \sum_{i=a+1}^{b} \|Y_i\|_2^2 \Big)^{1/2}.$$

Let v be a positive integer such that $v \le n$ and set $m = [v^{1/3}]$. Let r be a non-negative integer and assume first that $v \ge 2^{14}$. According to Lemma 6.4, there exists a non-negative integer $u \le v$ such that

$$\sum_{i=r+um+1}^{r+(u+1)m} \|Y_i\|_2^2 \le (m/v)^{1/3} \sum_{j=r+1}^{r+v} \|Y_j\|_2^2 \text{ and } 2^{-1}(v - m^2) \le um \le 2^{-1} v. \tag{6.34}$$

For such an integer u, let us write the following simple identity which is the starting point of the proof:

$$\begin{aligned} \Delta(r, r+v) &= \Delta(r, r+um) + \Delta(r+um, r+v) \\ &+ \mathbb{E}[x(r, r+um) y(r+um, r+um+m)] + \mathbb{E}[x(r, r+um) y(r+um+m, r+v)]. \end{aligned}$$

Let $T(v) = \max(um, v - um)$. Notice that from the above identity, and since (a_k) is not decreasing,

$$\begin{aligned}|\Delta(r, r+v)| &\le |\Delta(r, r+um)| + |\Delta(r+um, r+v)| \\ &\quad + |\mathbb{E}(x(r, r+um)y(r+um, r+um+m))| \\ &\quad + |\mathbb{E}(x(r, r+um)y(r+um+m, r+v))| \\ &\le a_{T(v)}D(r, r+um) + a_{T(v)}D(r+um, r+v) \\ &\quad + u(m)\left\|\sum_{i=r+1}^{r+um} X_i\right\|_2 \left\|\sum_{i=r+um+m+1}^{r+v} Y_i\right\|_2 + u(0)\left\|\sum_{i=r+1}^{r+um} X_i\right\|_2 \left\|\sum_{i=r+um+1}^{r+um+m} Y_i\right\|_2 .\end{aligned}$$

Applying the variance inequalities of Lemma 6.7 to each partial sums in the third and fourth terms and leaving the first two terms unchanged for the moment, we derive

$$\begin{aligned}|\Delta(r, r+v)| &\le a_{T(v)}D(r, r+um) + a_{T(v)}D(r+um, r+v) \\ &\quad + u(m)\kappa_1 C(u,n)\left(\sum_{i=r+1}^{r+um} \|X_i\|_2^2\right)^{1/2}\left(\sum_{i=r+(u+1)m+1}^{r+v} \|Y_i\|_2^2\right)^{1/2} \\ &\quad + u(0)\kappa_1 C(u,n)\left(\sum_{i=r+1}^{r+um} \|X_i\|_2^2\right)^{1/2}\left(\sum_{i=r+um+1}^{r+um+m} \|Y_i\|_2^2\right)^{1/2} \\ &= I_1 + I_2 + I_3 + I_4, \text{ say.}\end{aligned}$$

From the elementary inequality: $\sqrt{(u_1+u_2)(v_1+v_2)} \ge (u_1v_1)^{1/2} + (u_2v_2)^{1/2}$, valid for any non-negative reals u_1, u_2, v_1, v_2, we get that $D(\cdot, \cdot)$ satisfies the following superadditivity property: for any non-negative integers a, b, c with $a \le b \le c$,

$$D(a,b) + D(b,c) \le D(a,c),$$

implying that the sum of the two first terms is bounded above by

$$I_1 + I_2 \le a_{T(v)}D(r, r+v).$$

Obviously

$$I_3 \le u(m)K_n D(r, r+v) \text{ where } K_n = \kappa_1 C(u,n).$$

On the other hand, by the first part of (6.34), we get (since $m \le v^{1/3}$)

$$I_4 \le u(0)K_n\left(\sum_{i=r+1}^{r+um} \|X_i\|_2^2\right)^{1/2}\left(v^{-2/9}\sum_{j=r+1}^{r+v} \|Y_j\|_2^2\right)^{1/2} \le u(0)K_n v^{-1/9}D(r, r+v).$$

Taking into account the above estimates, we derive

$$|\Delta(r, r+v)| \leq D(r, r+v)[a_{T(v)} + K_n u(m) + K_n u(0) v^{-1/9}].$$

Since $m = [v^{1/3}]$, we get overall that for any integer $v \geq 2^{14}$,

$$a_v \leq a_{T(v)} + u([v^{1/3}])K_n + u(0)K_n v^{-1/9}.$$

We consider now the case when $1 \leq v < 2^{14}$. Notice first that, for any non-negative integers r, b,

$$|\Delta(r, r+b)| \leq u(0) b D(r, r+b).$$

Hence, for any non-negative integer k,

$$a_k \leq u(0)k.$$

Take

$$\delta_v = u([v^{1/3}])K_n + u(0)K_n v^{-1/9} = \kappa_1 C(u, n)\left\{u([v^{1/3}]) + u(0)v^{-1/9}\right\}. \tag{6.35}$$

Then, for any n fixed, $(\delta_v)_{v\geq 1}$ is a non-increasing sequence of non-negative real numbers. In addition, for $1 \leq v < 2^{14}$, since $C(u, n) \geq 1$ and $\kappa_1 \geq 2^{14} \times 4$, we have

$$\delta_v \geq 2^{14} u(0) \times 4 \times 2^{-14/9} \geq u(0) 2^{14} \geq u(0)v \geq a_v.$$

Overall, for any $v \geq 1$, taking $T(v) = 1$ when $1 \leq v < 2^{14}$ and recalling that $T(v) = \max(um, v - um)$ when $v \geq 2^{14}$, we get that, for any positive integer $v \leq n$,

$$a_v \leq a_{T(v)} + \delta_v,$$

where δ_v is defined by (6.35).

To complete the proof of the theorem, we note that by the second part of (6.34), the selected sequence $(T(k))_{k\geq 1}$ satisfies: $T(k) \leq 2^{-1}k(k + k^{-1/3})$. Therefore, applying Lemma 6.5, we get that for any $n \geq 1$,

$$a_n \leq a_{n_0} + 2 \sum_{j=0}^{[\log_2 n]-1} \delta_{2^j} \leq a_{n_0} + 2\kappa_1 C(u, n)\left\{\sum_{j=0}^{[\log_2 n]-1} u([2^{j/3}]) + u(0)2^4\right\},$$

where $n_0 = 8$. Since $a_8 \leq 8u(0)$ and $C(u, n) \geq 1$, this gives

$$\begin{aligned} a_n &\leq 8u(0) + 2(2^4 + 1)\kappa_1 C(u, n) \sum_{j=0}^{[\log_2 n]} u([2^{j/3}]) \\ &\leq \{8 + 2(2^4 + 1)\kappa_1\} r(u, n). \end{aligned}$$

■

Remark 6.8 *If the X_j's are orthogonal, then, by inspecting the proof above, we get that there exists an absolute constant c_2 such that for all non-negative integers a, n we have*

$$\left|\sum_{i=a+1}^{a+n} \mathbb{E}(X_i y(i, a+n))\right| \leq c_2 \tilde{r}(u,n) \left(\sum_{j=a+1}^{a+n} \|X_j\|_2^2 \sum_{j=a+1}^{a+n} \|Y_j\|_2^2 \right)^{1/2}$$

$$\text{with } \tilde{r}(u,n) = C^{1/2}(u,n) \sum_{j=0}^{[\log_2 n]} u(2^j).$$

One can take $c_2 = 8 + 2(2^4+1)\kappa_1^{1/2}$ with $\kappa_1 = 8\,e^{2^{20}}$.

Comment 6.9 *Inspecting the proof, it is not difficult to see that Theorems 6.1 and 6.6 can be easily extended to variables taking values in separable Hilbert spaces, the absolute values being replaced by the norm of the Hilbert space and the product between variables by the inner product.*

6.1.3 *The Maximal Version of the Variance Inequality under ρ-mixing*

As we shall see, starting from inequality (3.10), the following maximal version of inequality (6.2) holds.

Theorem 6.10 *Let $(X_k)_{k\in\mathbb{Z}}$ be a sequence of centered real-valued random variables in $\mathbb{L}^2$. Let $(\rho(n))_{n\geq 0}$ be its associated maximal correlation coefficients sequence. Let $S_n = \sum_{i=1}^n X_i$. Then*

$$\| \max_{1\leq k\leq n} |S_k| \|_2 \leq 3C^{1/2}(\rho,n)(1+\sqrt{2\kappa_1})\left(1+\sum_{j=0}^{[\log_2 n]} \rho(2^j)\right)\left(\sum_{j=1}^{n} \|X_j\|_2^2\right)^{1/2}, \tag{6.36}$$

where $C(\rho, n)$ is defined by (6.1) and $\kappa_1 = 8\,e^{2^{20}}$.

Note that the inequality above is more accurate than the one obtained in Shao in 1995. Indeed, his Lemma 2.4 states a similar inequality but with $n \max_{1\leq i\leq n} \|X_j\|_2^2$ instead of $\sum_{j=1}^n \|X_j\|_2^2$.

Proof of Theorem 6.10 Let us apply inequality (3.10) with $X_{i,n} = a_{i,n}X_i$ where $(X_k)_{k\in\mathbb{Z}}$ is a sequence of centered real-valued random variables in $\mathbb{L}^2$ with sequence of maximal correlation coefficients $(\rho(n))_{n\geq 0}$, $\mathcal{F}_k = \sigma(X_j, j \leq k)$ and $a_{i,n} = 1$ if $i \leq n$ and 0 otherwise. Let $S_k = \sum_{i=1}^k X_{i,n}$. Setting r as the unique positive integer such that $2^{r-1} \leq n < 2^r$, we then get

$$\| \max_{1\leq k\leq n} |S_k| \|_2 \leq 3\left(\sum_{j=1}^{n} \|X_j\|_2^2\right)^{1/2} + 3\sqrt{2}\sum_{j=0}^{r-1}\left(\sum_{k=1}^{[n2^{-j}]} \|\mathbb{E}(S_{k2^j} - S_{(k-1)2^j} | \mathcal{F}_{(k-2)2^j+1})\|_2^2 \right.$$
$$\left. + \|\mathbb{E}(S_n - S_{[n2^{-j}]2^j} | \mathcal{F}_{([n2^{-j}]-1)2^j+1})\|_2^2\right)^{1/2}.$$

We denote now $\mathbb{E}_j(\cdot) = \mathbb{E}(\cdot|\mathcal{F}_j)$. By the definition of the maximal correlation coefficients $(\rho(n))_{n\geq 0}$, we have

$$\begin{aligned}\|\mathbb{E}_{(k-2)2^j+1}(S_{k2^j} - S_{(k-1)2^j})\|_2^2 &\leq \rho(2^j)\|\mathbb{E}_{(k-2)2^j+1}(S_{k2^j} - S_{(k-1)2^j})\|_2 \\ &\times \|S_{k2^j} - S_{(k-1)2^j}\|_2,\end{aligned}$$

implying that

$$\|\mathbb{E}_{(k-2)2^j+1}(S_{k2^j} - S_{(k-1)2^j})\|_2 \leq \rho(2^j)\|S_{k2^j} - S_{(k-1)2^j}\|_2.$$

Using Theorem 6.1, it follows that for any non-negative integer $j \leq r-1$,

$$\|\mathbb{E}_{(k-2)2^j+1}(S_{k2^j} - S_{(k-1)2^j})\|_2 \leq \sqrt{\kappa_1 C(\rho, n)}\rho(2^j)\left(\sum_{\ell=(k-1)2^j+1}^{k2^j} \|X_\ell\|_2^2\right)^{1/2}.$$

Similarly, we get that for any non-negative integer $j \leq r-1$,

$$\|\mathbb{E}_{([n2^{-j}]-1)2^j+1}(S_n - S_{[n2^{-j}]2^j})\|_2 \leq \sqrt{\kappa_1 C(\rho, n)}\rho(2^j)\left(\sum_{\ell=[n2^{-j}]2^j+1}^{n} \|X_\ell\|_2^2\right)^{1/2}.$$

Therefore, for any non-negative integer $j \leq r-1$,

$$\begin{aligned}&\sum_{k=1}^{[n2^{-j}]} \|\mathbb{E}(S_{k2^j} - S_{(k-1)2^j}|\mathcal{F}_{(k-2)2^j+1})\|_2^2 + \|\mathbb{E}(S_n - S_{[n2^{-j}]2^j}|\mathcal{F}_{([n2^{-j}]-1)2^j+1})\|_2^2 \\ &\quad\leq \kappa_1 C(\rho, n)\rho^2(2^j)\sum_{\ell=1}^{n} \|X_\ell\|_2^2.\end{aligned}$$

So, overall,

$$\|\max_{1\leq k\leq n} |S_k|\|_2 \leq 3\left\{1 + \sqrt{2\kappa_1 C(\rho, n)}\left(\sum_{j=0}^{r-1} \rho(2^j)\right)\right\}\left(\sum_{j=1}^{n} \|X_j\|_2^2\right)^{1/2}.$$

Since $C(\rho, n) \geq 1$, this gives the result. ■

As a direct application of the maximal inequality (6.36), we obtain the following convergence of series result.

Theorem 6.11 *Let $(X_k)_{k\in\mathbb{Z}}$ be a sequence of centered real-valued random variables in $\mathbb{L}^2$. Let $(\rho(n))_{n\geq 0}$ be its associated maximal correlation coefficients sequence. Assume that $\sum_{k\geq 1} \rho(2^k) < \infty$. Then*

(1) *If* $\sum_{n\geq 1} \mathbb{E}(X_n^2) < \infty$ *then the series* $\sum_{n\geq 1} X_n$ *converges almost surely.*

(2) *Let* $(a_n)_{n\geq 1}$ *be a non-decreasing sequence of positive reals such that* $a_n \to \infty$*, as* $n \to \infty$*. If* $\sum_{n\geq 1} a_n^{-2}\mathbb{E}(X_n^2) < \infty$ *then* $a_n^{-1}\sum_{k=1}^{n} X_k \to 0$ *almost surely, as* $n \to \infty$*.*

Proof The second statement of the theorem follows from the first one and the Kronecker lemma. Let us prove the first statement. By (6.36) and the fact that $\sum_{k\geq 1}\rho(2^k) < \infty$, we derive that there exists a positive constant K such that for any $\varepsilon > 0$,

$$\sup_{n>0}\left\{\mathbb{P}\left(\max_{m\leq k\leq m+n}\left|\sum_{i=m}^{k} X_i\right| > \varepsilon\right)\right\} \leq \varepsilon^{-2}\sup_{n>0}\left\{\mathbb{E}\left(\max_{m\leq k\leq m+n}\left|\sum_{i=m}^{k} X_i\right|^2\right)\right\}$$
$$\leq K\varepsilon^{-2}\sum_{j=m}^{\infty}\mathbb{E}(X_j^2) \to 0,$$

as $m \to \infty$. This ends the proof of the theorem. ■

Comment 6.12 *Theorem 6.11 has been obtained by Utev (1991) via the probability version of (6.36). More exactly, combining Theorem 6.1 with Theorem 6.6 and its Remark 6.8, he obtained the following version of the Kolmogorov maximal inequality: There exists an universal positive constant* c_3 *such that for any* $x > 0$*,*

$$\mathbb{P}\left(\max_{1\leq k\leq n}|S_k| \geq x\right) \leq x^{-2}c_3C(\rho, n)\left(1 + \sum_{j=0}^{[\log_2 n]}\rho(2^j)\right)^2\sum_{j=1}^{n}\mathbb{E}(X_j^2), \tag{6.37}$$

where $C(\rho, n)$ *is defined in* (6.1) *and we can take* $c_3 = 2\kappa_1 + 4(8 + 2(2^4 + 1)\sqrt{\kappa_1})^2$ *with* $\kappa_1 = 8\,\mathrm{e}^{2^{20}}$*. Obviously, inequality* (6.37) *is implied by inequality* (6.36) *but is enough to prove Theorem* 6.11.

A direct proof of inequality (6.37) Set

$$E_k = \{|S_k| \geq x, |S_i| < x, i = 1, \ldots, k-1\}.$$

As for the proof of the Kolmogorov maximal inequality in the case of independent r.v.'s, we first write that for any $x > 0$,

$$\mathbb{P}\left(\max_{1\leq k\leq n}|S_k| \geq x\right) \leq x^{-2}\sum_{k=1}^{n}\mathbb{E}\left(S_k^2 1_{E_k}\right). \tag{6.38}$$

We are going to apply Theorem 6.6 and its Remark 6.8 with

$$x(a, b) = \sum_{k=a+1}^{b} S_k\mathbf{1}_{E_k} \quad\text{and}\quad y(a, b) = \sum_{k=a+1}^{b} X_k.$$

Notice that the r.v.'s $S_k 1_{E_k}$ are orthogonal, and so $u_1(k) = 0, k \geq 1$. Next, $u_2(k) \leq \rho(k)$ from the condition of the theorem. Moreover, since $S_k 1_{E_k}$ is $\mathcal{F}_k$-measurable,

$$\begin{aligned}|\mathbb{E}[x(a,b)y(b+k,c)]| &\leq \rho(k)\{\mathrm{Var}[x(a,b)]\mathrm{Var}[y(b+k,c)]\}^{1/2}\\ &\leq \rho(k)\|x(a,b)\|_2\|y(b+k,c)\|_2,\end{aligned}$$

implying that $u_3(k) \leq \rho(k)$ and overall

$$u(k) = \max(u_1(k), u_2(k), u_3(k)) \leq \rho(k).$$

Thus from Remark 6.8 of Theorem 6.6 and the elementary inequality $2xy \leq 2x^2 + y^2/2$, we derive

$$2\left|\sum_{k=1}^{n} \mathbb{E}\left(S_k 1_{E_k} \sum_{j=k+1}^{n} X_j\right)\right| \leq \frac{1}{2}\sum_{k=1}^{n} \mathbb{E}\left(S_k^2 1_{E_k}\right) + 2(c_2 \tilde{r}(\rho, n))^2 \sum_{j=1}^{n} \|X_j\|_2^2. \tag{6.39}$$

Then, noticing that

$$\mathbb{E}(S_n^2) \geq \sum_{k=1}^{n} \mathbb{E}\left(S_n^2 1_{E_k}\right) \geq \sum_{k=1}^{n} \mathbb{E}\left(S_k^2 1_{E_k}\right) + 2\sum_{k=1}^{n} \mathbb{E}\left(S_k 1_{E_k} \sum_{j=k+1}^{n} X_j\right),$$

we get

$$\sum_{k=1}^{n} \mathbb{E}\left(S_k^2 1_{E_k}\right) \leq 2\mathbb{E}(S_n^2) + (2c_2 \tilde{r}(\rho, n))^2 \sum_{j=1}^{n} \|X_j\|_2^2.$$

Starting from (6.38) and taking into account the inequality above, we obtain

$$\mathbb{P}\left(\max_{1\leq k\leq n} |S_k| \geq x\right) \leq x^{-2}\left(2\mathbb{E}(S_n^2) + (2c_2 \tilde{r}(\rho, n))^2 \sum_{j=1}^{n} \|X_j\|_2^2\right),$$

and then the desired result by using inequality (6.2) to give an upper bound of $\mathbb{E}(S_n^2)$. ∎

6.1.4 *Rosenthal-Type Inequality under ρ-mixing*

Next theorem gives a Rosenthal-type inequality under ρ-mixing and stationarity.

Theorem 6.13 *Let $(X_k)_{k\in\mathbb{Z}}$ be a strictly stationary sequence of centered real-valued random variables in $\mathbb{L}^p$ for some $p \geq 2$. Let $(\rho(n))_{n\geq 0}$ be its associated maximal correlation coefficients sequence. Assume that $\rho(n) \to 0$ as $n \to \infty$. Let $S_n = \sum_{i=1}^{n} X_i$. Then there exists a positive constant $C = C(p, \rho(\cdot))$ depending only on p and $\rho(\cdot)$ such that, for any positive integer n,*

$$\left\| \max_{1\le k\le n} |S_k| \right\|_p \le Cn^{1/p}\,\|X_1\|_p \exp\left(C \sum_{i=0}^{[\log_2 n]} \rho^{2/p}(2^i) \right)$$
$$+ Cn^{1/2}\,\|X_1\|_2 \exp\left(C \sum_{i=0}^{[\log_2 n]} \rho(2^i) \right).$$

Theorem 6.13 can be extended to non-stationary sequences by replacing $\|X_1\|_p$ (resp. $\|X_1\|_2$) by $\max_{1\le k\le n} \|X_k\|_p$ (resp. $\max_{1\le k\le n} \|X_k\|_2$). In this situation, the proof is more delicate. We refer to Shao (1995) for a proof of Theorem 6.13 in this case.

Proof of Theorem 6.13 We start by proving that, for any $p \ge 2$ and any integer $r \ge 0$, there exists a positive constant $K = K(p, \rho(\cdot))$ depending only on p and $\rho(\cdot)$ such that

$$\|S_{2^r}\|_p \le K2^{r/p}\|X_1\|_p \exp\left(K \sum_{i=0}^{r-1} \left(\rho(2^{i/(p+1)})\right)^{2/p} \right)$$
$$+ K2^{r/2}\|X_1\|_2 \exp\left(2^{-1} \sum_{\ell=1}^{r} \rho(2^{\ell/3}) \right). \tag{6.40}$$

By(6.8), the inequality above holds for $p = 2$. Let us then prove that it also holds for $p > 2$. Let m be a positive integer and

$$\bar{S}_n = \sum_{i=n+m+1}^{2n+m} X_i.$$

From the equation

$$S_{2n} = S_n + \bar{S}_n + \sum_{i=n+1}^{n+m} X_i - \sum_{i=2n+1}^{2n+m} X_i,$$

we get, by stationarity,

$$\|S_{2n}\|_p \le \|S_n + \bar{S}_n\|_p + 2m\|X_1\|_p. \tag{6.41}$$

Recall now the following elementary inequality (3.54) that has been used (and proved) in the proof of Theorem 3.17: For any positive reals a and b and any real $q \ge 1$,

$$(a+b)^q \le a^q + b^q + 2^{q+1}ab^{q-1} + 2^{q+1}a^{q-1}b.$$

Applying it with $a = |S_n|$ and $b = |\bar{S}_n|$ and taking the expectation, we get

$$\|S_n + \bar{S}_n\|_p^p \le \|S_n\|_p^p + \|\bar{S}_n\|_p^p + 2^{p+1}\mathbb{E}\left(|S_n| \times |\bar{S}_n|^{p-1}\right) + 2^{p+1}\mathbb{E}\left(|S_n|^{p-1} \times |\bar{S}_n|\right).$$

Hence, taking into account the stationarity assumption, it follows that

$$\|S_n + \bar{S}_n\|_p^p \le 2\|S_n\|_p^p + 2^{p+1}\mathbb{E}\left(|S_n| \times |\bar{S}_n|^{p-1}\right) + 2^{p+1}\mathbb{E}\left(|S_n|^{p-1} \times |\bar{S}_n|\right).$$

To handle the two last terms in the above inequality, we recall the following covariance inequality which is due to Bradley and Bryc (1985).

Lemma 6.14 *Let $r, s \geq 1$ with $1/r + 1/s = 1$. Let k be a positive integer. Suppose $X \in \mathbb{L}^r(\mathcal{F}_1^k)$ and $Y \in \mathbb{L}^s(\mathcal{F}_{k+n}^{\infty})$ where $\mathcal{F}_i^j = \sigma(X_i, \dots, X_j)$. Then*

$$|\mathrm{cov}(X, Y)| \leq 10\,(\rho(n))^{2\left(\frac{1}{r}\wedge\frac{1}{s}\right)}\,\|X\|_r\|Y\|_s.$$

Applying the lemma above, we derive

$$\mathbb{E}\left(|S_n| \times |\bar{S}_n|^{p-1}\right) \leq \|S_n\|_1\|\bar{S}_n\|_{p-1}^{p-1} + 10\,(\rho(m))^{2/p}\,\|S_n\|_p\|\bar{S}_n\|_p^{p-1}$$

and

$$\mathbb{E}\left(|S_n|^{p-1} \times |\bar{S}_n|\right) \leq \|S_n\|_{p-1}^{p-1}\|\bar{S}_n\|_1 + 10\,(\rho(m))^{2/p}\,\|S_n\|_p^{p-1}\|\bar{S}_n\|_p.$$

So, overall, by stationarity,

$$\|S_n + \bar{S}_n\|_p^p \leq 2\|S_n\|_p^p\left(1 + 10 \times 2^{p+1}\,(\rho(m))^{2/p}\right) + 2^{p+2}\|S_n\|_2\|S_n\|_{p-1}^{p-1}. \tag{6.42}$$

Starting from (6.41) and considering (6.42), we get

$$\|S_{2n}\|_p \leq 2^{1/p}\|S_n\|_p\left(1 + 10 \times 2^{p+1}\,(\rho(m))^{2/p}\right)^{1/p} + 2m\|X_1\|_p$$
$$+ 2^{1+2/p}\|S_n\|_2^{1/p}\|S_n\|_{p-1}^{1-1/p}.$$

Writing this inequality for $n = 2^{r-1}$ and $m = [n^{1/(p+1)}]$ and setting

$$a_i = \left(1 + 10 \times 2^{p+1}\left(\rho(2^{i/(p+1)})\right)^{2/p}\right)^{1/p},$$

we then have

$$\|S_{2^r}\|_p \leq 2^{1/p}a_{r-1}\|S_{2^{r-1}}\|_p + 2 \times 2^{(r-1)/(p+1)}\|X_1\|_p$$
$$+ 2^{1+2/p}\|S_{2^{r-1}}\|_2^{1/p}\|S_{2^{r-1}}\|_{p-1}^{1-1/p}. \tag{6.43}$$

By recurrence on the term $\|S_{2^{r-1}}\|_p$, we get that for any integer k such that $0 \leq k \leq r-1$,

$$\|S_{2^r}\|_p \leq 2^{(r-k)/p}\|S_{2^k}\|_p\prod_{i=k}^{r-1}a_i + 2\|X_1\|_p\sum_{j=1}^{r-k}2^{(r-j)/(p+1)}2^{(j-1)/p}\prod_{i=1}^{j-1}a_{r-i}$$
$$+ 2^{1+2/p}\sum_{j=1}^{r-k}\|S_{2^{r-j}}\|_2^{1/p}\|S_{2^{r-j}}\|_{p-1}^{1-1/p}2^{(j-1)/p}\prod_{i=1}^{j-1}a_{r-i},$$

with the convention that $\prod_{i=1}^{0}a_{r-i} = 1$. Therefore, the following inequality is valid: for any integer k such that $0 \leq k \leq r-1$,

$$\|S_{2^r}\|_p \le 2^{r/p}\left(2^{-k/p}\|S_{2^k}\|_p + b_1 2^{-k/(p(p+1))}\|X_1\|_p\right)\exp\left(b_2\sum_{i=0}^{r-1}\left(\rho(2^{i/(p+1)})\right)^{2/p}\right)$$
$$+ 2^{1+2/p}\sum_{j=1}^{r-k}\|S_{2^{r-j}}\|_2^{1/p}\|S_{2^{r-j}}\|_{p-1}^{1-1/p}2^{(j-1)/p}\prod_{i=1}^{j-1}a_{r-i}, \tag{6.44}$$

where $b_1 = 2^{1-1/p}\dfrac{2^{1/(p(p+1))}}{2^{1/(p(p+1))}-1}$ and $b_2 = 10p^{-1}2^{p+1}$.

Assume now that $p\in]2,3]$. Starting from (6.44), noticing that $\|S_{2^{r-j}}\|_{p-1}\le\|S_{2^{r-j}}\|_2$, using (6.8) and taking into account the fact that $(a_i)_{i\ge 0}$ is non-increasing, we get that, for any non-negative integer $k\le r-1$,

$$\|S_{2^r}\|_p \le 2^{r/p}\left(2^{-k/p}\|S_{2^k}\|_p + b_1 2^{-k/(p(p+1))}\|X_1\|_p\right)\exp\left(b_2\sum_{i=0}^{r-1}\left(\rho(2^{i/(p+1)})\right)^{2/p}\right)$$
$$+ 2^{1+1/p}\|X_1\|_2\left(1+\frac{1}{\sqrt{2}(2^{1/6}-1)}\right)\exp\left(2^{-1}\sum_{\ell=1}^{r}\rho(2^{\ell/3})\right)\sum_{j=1}^{r-k}2^{(r-j)/2}2^{j/p}(a_k)^{j-1}.$$

Choose now k_0 such that

$$1+10\times 2^{p+1}\left(\rho(2^{k_0/(p+1)})\right)^{2/p}\le\left(\frac{3}{2}\right)^{(p-2)/2}.$$

This is always possible since $\rho(n)\to 0$. Using the fact that $\|S_{2^{k_0}}\|_p \le 2^{k_0}\|X_1\|_p$, we derive

$$\|S_{2^r}\|_p \le 2^{r/p}\|X_1\|_p\left(2^{k_0}+b_1\right)\exp\left(b_2\sum_{i=0}^{r-1}\left(\rho(2^{i/(p+1)})\right)^{2/p}\right) + 2^{r/2}\times 2^{1+1/p}\|X_1\|_2$$
$$\times\frac{2^{(p-2)/(2p)}}{4^{(p-2)/(2p)}-3^{(p-2)/(2p)}}\left(1+\frac{1}{\sqrt{2}(2^{1/6}-1)}\right)\exp\left(2^{-1}\sum_{\ell=1}^{r}\rho(2^{\ell/3})\right).$$

This proves that inequality (6.40) holds for any $p\in]2,3]$.

Assume now that $p\in]3,4]$. Starting from (6.44) and taking into account that $\|S_{2^{r-j}}\|_{p-1}$ (resp. $\|S_{2^{r-j}}\|_2$) satisfies inequality (6.40) (resp. (6.8)) and that $(a_i)_{i\ge 0}$ is non-increasing, we get that for any integer k such that $0\le k\le r-1$,

$$\|S_{2^r}\|_p \le 2^{r/p}\left(2^{-k/p}\|S_{2^k}\|_p + b_1 2^{-k/(p(p+1))}\|X_1\|_p\right)\exp\left(b_2\sum_{i=0}^{r-1}\left(\rho(2^{i/(p+1)})\right)^{2/p}\right)$$
$$+ 2^{r/2}c^{1/p}C_p^{1-1/p}2^{1+1/p}\|X_1\|_2\exp\left(2^{-1}\sum_{\ell=1}^{r}\rho(2^{\ell/3})\right)\sum_{j=1}^{r-k}\frac{2^{j/p}}{2^{j/2}}(a_k)^{j-1}$$
$$+ 2^{3r/(2p)}c^{1/p}C_p^{1-1/p}2^{1+1/p}\|X_1\|_2^{1/p}\|X_1\|_{p-1}^{1-1/p}$$
$$\times\exp\left((2p)^{-1}\sum_{\ell=1}^{r}\rho(2^{\ell/3})\right)\left((p-1)p^{-1}C_p\sum_{i=0}^{r-1}\left(\rho(2^{i/p})\right)^{2/(p-1)}\right)\sum_{j=1}^{r-k}2^{-j/(2p)}(a_k)^{j-1}, \tag{6.45}$$

where $c = 1 + \frac{1}{\sqrt{2}(2^{1/6}-1)}$ and $C_p = K(p-1, \rho(\cdot))$. Set now

$$\alpha = \frac{p-3}{p-2}.$$

Since $p > 3$, $\alpha \in]0, 1[$. In addition, $p - 1 = 2(1-\alpha) + p\alpha$. Therefore, Hölder's inequality implies that

$$\|X_1\|_{p-1}^{p-1} \le \|X_1\|_2^{2(1-\alpha)} \times \|X_1\|_p^{\alpha p},$$

giving

$$\|X_1\|_2^{1/p} \|X_1\|_{p-1}^{1-1/p} \le \|X_1\|_2^{1-\alpha} \times \|X_1\|_p^{\alpha}.$$

In addition, $\frac{3}{2p} = \frac{\alpha}{p} + \frac{1-\alpha}{2}$. Hence $2^{3r/(2p)} = 2^{\alpha r/p} \times 2^{r(1-\alpha)/2}$. Using now the following inequality: $a^{\alpha} b^{1-\alpha} \le \alpha a + (1-\alpha) b$ for non-negative reals a, b and any $\alpha \in [0, 1]$, we then infer that

$$\begin{aligned}
&2^{3r/(2p)} \|X_1\|_2^{1/p} \|X_1\|_{p-1}^{1-1/p} \times \exp\left((2p)^{-1} \sum_{\ell=1}^{r} \rho(2^{\ell/3}) \right) \\
&\times \exp\left((p-1)p^{-1} C_p \sum_{i=0}^{r-1} \left(\rho(2^{i/p})\right)^{2/(p-1)} \right) \\
&\le (1-\alpha) 2^{r/2} \|X_1\|_2 \exp\left((2p)^{-1} \sum_{\ell=1}^{r} \rho(2^{\ell/3}) \right) \\
&+ \alpha 2^{r/p} \|X_1\|_p \exp\left(\alpha^{-1} (p-1) p^{-1} C_p \sum_{i=0}^{r-1} \left(\rho(2^{i/p})\right)^{2/(p-1)} \right) \\
&\times \exp\left((2p)^{-1} \sum_{\ell=1}^{r} \rho(2^{\ell/3}) \right).
\end{aligned} \tag{6.46}$$

Inserting (6.46) into (6.45) and choosing $k = k_0$, where k_0 is the smallest integer such that

$$a_{k_0} = \left(1 + 10 \times 2^{p+1} \left(\rho(2^{k_0/(p+1)}) \right)^{2/p} \right)^{1/p} \le \left(\frac{3}{2} \right)^{1/(2p)},$$

we then get that (6.40) also holds for any $p \in]3, 4]$. Proceeding by induction, we infer that inequality (6.40) holds for any $p > 2$.

Consider now the binary expansion of n as in (3.4). Then, for r such that $2^{r-1} \le n < 2^r$, using (3.5), stationarity and (6.40), it follows that for any $p > 2$,

$$\|S_n\|_p \le \sum_{k=0}^{r-1} \|T_{2^k}\|_p \le \sum_{k=0}^{r-1} \|S_{2^k}\|_p$$
$$\le K\|X_1\|_p \exp\Big(K\sum_{i=0}^{r-1}\big(\rho(2^{i/(p+1)})\big)^{2/p}\Big)\sum_{k=0}^{r-1} 2^{k/p}$$
$$+ K\|X_1\|_2 \exp\Big(2^{-1}\sum_{\ell=1}^{r}\rho(2^{\ell/3})\Big)\sum_{k=0}^{r-1} 2^{k/2}.$$

Since $\sum_{k=0}^{r-1} 2^{k/2} \le 2^{r/2}/(2^{1/2}-1) \le \sqrt{2n}/(2^{1/2}-1)$ and $\sum_{k=0}^{r-1} 2^{k/p} \le 2^{1/p}n^{1/p}/(2^{1/p}-1)$, we derive

$$\|S_n\|_p \le n^{1/p}K2^{1/p}(2^{1/p}-1)^{-1}\|X_1\|_p \exp\Big(K\sum_{i=0}^{r-1}\big(\rho(2^{i/(p+1)})\big)^{2/p}\Big)$$
$$+ n^{1/2}K\sqrt{2}(\sqrt{2}+1)\|X_1\|_2 \exp\Big(2^{-1}\sum_{\ell=1}^{r}\rho(2^{\ell/3})\Big). \tag{6.47}$$

To end the proof (in the case $p > 2$ since for $p = 2$ Theorem 6.13 is derived from inequality (6.36)), we shall use inequality (3.15) of Corollary 3.15. The term $\|S_m\|_p$ is bounded with the help of (6.47). To handle the term $\sum_{l=0}^{r-1} 2^{-l/p}\|\mathbb{E}(S_{2^l}|\mathcal{F}_0)\|_p$ where r is the positive integer satisfying $n \in [2^{r-1}, 2^r[$ and $\mathcal{F}_0 = \sigma(X_j, j \le 0)$, we proceed as follows. We first write that

$$\sum_{l=0}^{r-1} 2^{-l/p}\|\mathbb{E}(S_{2^l}|\mathcal{F}_0)\|_p \le \sum_{l=0}^{r-1} 2^{-l/p}\|S_{2^{l/(p+1)}}\|_p + \sum_{l=0}^{r-1} 2^{-l/p}\|\mathbb{E}(S_{2^l} - S_{2^{l/(p+1)}}|\mathcal{F}_0)\|_p. \tag{6.48}$$

Let

$$Z_0 = |\mathbb{E}_0(S_{2^l} - S_{2^{l/(p+1)}})|^{p-1}\mathrm{sign}(\mathbb{E}_0(S_{2^l} - S_{2^{l/(p+1)}})).$$

Lemma 6.14 gives

$$\|\mathbb{E}(S_{2^l} - S_{2^{l/(p+1)}}|\mathcal{F}_0)\|_p^p = \mathrm{cov}(Z_0, S_{2^l} - S_{2^{l/(p+1)}})$$
$$\le 10\big(\rho(2^{l/(p+1)})\big)^{2/p}\|Z_0\|_{p/(p-1)}\|S_{2^l} - S_{2^{l/(p+1)}}\|_p.$$

But $\|Z_0\|_{p/(p-1)} = \|\mathbb{E}(S_{2^l} - S_{2^{l/(p+1)}}|\mathcal{F}_0)\|_p^{p-1}$. Therefore

$$\|\mathbb{E}(S_{2^l} - S_{2^{l/(p+1)}}|\mathcal{F}_0)\|_p \le 10\big(\rho(2^{l/(p+1)})\big)^{2/p}\|S_{2^l} - S_{2^{l/(p+1)}}\|_p. \tag{6.49}$$

Starting from (6.48), it follows that

$$\sum_{l=0}^{r-1} 2^{-l/p} \|\mathbb{E}(S_{2^l}|\mathcal{F}_0)\|_p \le \sum_{l=0}^{r-1} 2^{-l/p} \|S_{2^{l/(p+1)}}\|_p$$
$$+ 10 \sum_{l=0}^{r-1} 2^{-l/p} \left(\rho(2^{l/(p+1)})\right)^{2/p} \|S_{2^l} - S_{2^{l/(p+1)}}\|_p. \tag{6.50}$$

We end the proof by inserting (6.40) in (6.50) and by taking into account that for any $p > 2$,

$$\sum_{l=0}^{r-1} 2^{-l/p} 2^{l/2} \left(\rho(2^{l/(p+1)})\right)^{2/p} \le \sum_{l=0}^{r-1} 2^{-l/p} 2^{l/2} \le 2^{r/2} 2^{-r/p} \times \frac{1}{2^{(p-2)/(2p)} - 1}.$$ ■

We mention next a lemma which will be useful later in the text (in Section 6.2).

Lemma 6.15 *Let $(X_k)_{k\in\mathbb{Z}}$ be a sequence of centered real-valued random variables in $\mathbb{L}^2$. Let $(\rho(n))_{n\ge 0}$ be its associated maximal correlation coefficients sequence. Assume that there exists a positive integer p such that $\rho(p) < 1$. Then*

$$\max_{1\le k\le n} \mathbb{E}(S_k^2) \le 2(1-\rho(p))^{-1}[\mathbb{E}(S_n^2) + p^2 \max_{1\le k\le n} \mathbb{E}(X_k^2)].$$

Proof Note that for $n - p < k \le n$ we have

$$\|S_k\|_2 \le \|S_n\|_2 + p \max_{1\le k\le n} \|X_k\|_2. \tag{6.51}$$

On the other hand, for $k \le n - p$, by the definition of the ρ-mixing coefficient we obtain

$$\mathbb{E}[(S_k + (S_n - S_{k+p}))^2] \ge \mathbb{E}(S_k^2) + \mathbb{E}[(S_n - S_{k+p})^2] - 2\rho(p)\|S_k\|_2\|S_n - S_{k+p}\|_2$$
$$\ge (1-\rho(p))(\mathbb{E}(S_k^2) + \mathbb{E}[(S_n - S_{k+p})^2]) \ge (1-\rho(p))\mathbb{E}(S_k^2).$$

Obviously,

$$\|S_k + S_n - S_{k+p}\|_2 \le \|S_n\|_2 + p \max_{1\le k\le n} \|X_k\|_2.$$

By combining the latter two inequalities, for $k \le n - p$ we obtain

$$\mathbb{E}(S_k^2) \le 2(1-\rho(p))^{-1}[\mathbb{E}(S_n^2) + p^2 \max_{1\le k\le n} \mathbb{E}(X_k^2)]. \tag{6.52}$$

Now, the desired result follows by (6.51) and (6.52). ■

Concerning now strictly stationary ρ-mixing sequence, the following result is valid:

Theorem 6.16 *Let $(X_n)_{n\in\mathbb{Z}}$ be a strictly stationary ρ-mixing sequence of centered random vari--ables and assume that, for some $\delta > 0$, we have $\mathbb{E}(|X_0|^{2+\delta}) < \infty$. Suppose in addition that $\sigma_n^2 = \mathbb{E}(S_n^2) \to \infty$. Then there is a positive constant C such that for all $n \geq 1$*

$$\left\| \max_{1\leq k\leq n} |S_k| \right\|_{2+\delta} \leq C\sigma_n \,.$$

Proof By Theorem 11.1 in Bradley (2007) we know that there is a constant C such that for all $n \geq 1$

$$||S_n||_{2+\delta} \leq C\sigma_n \,. \tag{6.53}$$

Recalling that σ_n^2 has the representation $\sigma_n^2 = nh(n)$ with $h(t)$ a slowly varying function on the real line (see Theorem 8.13 in Bradley (2007)), the theorem follows by applying Lemma 1.20. ∎

6.2 Probability and Moment Maximal Inequalities under ϕ-mixing

In this section, we investigate probability and moment maximal inequalities when the sequence of ϕ-mixing coefficients $(\phi(n))_{n\geq 0}$ associated with a sequence $(X_k)_{k\in\mathbb{Z}}$ of real-valued random variables, and defined by (5.3) in Section 5.1.1, satisfies some conditions.

We start by recalling the following relation between the $\rho(n)$ and the $\phi(n)$-ceofficients as given in Lemma 1.17 in Iosifescu and Theodorescu (1969):

$$\rho(n) \leq 2\phi^{1/2}(n). \tag{6.54}$$

Theorem 6.17 *Let $(X_k)_{k\in\mathbb{Z}}$ be a sequence of real-valued random variables and $(\phi(n))_{n\geq 0}$ its associated ϕ-mixing coefficients. Let p be a positive integer such that $\phi(p) < 1/2$. Then there exists a constant $c_{\phi(p)}$, depending on $\phi(p)$ only, such that for any $r \geq 1$ and all $1 \leq q \leq r$, the following inequality holds:*

$$\mathbb{E}(\max_{1\leq k\leq n} |S_k|^r) \leq (c_{\phi(p)} r)^r \left(p^r \mathbb{E} \max_{1\leq k\leq n} |X_k|^r + \max_{1\leq k\leq n} (\mathbb{E}|S_k|^q)^{r/q} \right),$$

where $S_k = \sum_{i=1}^k X_i$.

Remark 6.18 *The result still holds if the random variables take values in a separable Banach space $(\mathbb{B}, ||\cdot||_{\mathbb{B}})$. In this case the absolute values have to be replaced by the norm $||\cdot||_{\mathbb{B}}$.*

The proof of Theorem 6.17 is based on the following lemma. To state it we need the following notation: For fixed positive integers n, p and a positive real number x_0, let

$$\eta_{p,x_0,n} = \phi(p) + \max_{0\le i\le n} \mathbb{P}(|S_n - S_i| \ge x_0),$$

where $S_0 = 0$.

Lemma 6.19 *Assume that there are positive integers n, p and a positive real number x_0 such that $\eta_{p,x_0,n} := \eta < 1$. Then for any positive real numbers x and a, we have:*

$$\mathbb{P}(\max_{1\le k\le n} |S_k| \ge x + 2a + 2x_0) \le \eta(1-\eta)^{-1}\mathbb{P}(\max_{1\le k\le n} |S_k| \ge x)$$
$$+ 2(1-\eta)^{-1}\mathbb{P}(\max_{1\le k\le n} |X_k| > a/(p+1)).$$

Proof Let x_1 be a positive real number. Let

$$R(t) = \mathbb{P}\left(\max_{1\le k\le n} |S_k| \ge t\right) \text{ and } E_k(t) = \left\{\max_{i\le k-1} |S_i| < t \le |S_k|\right\}.$$

Notice that

$$\left\{\max_{1\le k\le n} |S_k| \ge t\right\} = \bigcup_{k=1}^{n} E_k(t),\ E_k(t) \cap E_s(t) = \emptyset,\ k \neq s,$$

and that, for any $1 \le k \le n-p$,

$$\{|S_k| \ge x_1 + a + x_0\} \cap \{\max_{1\le k\le n} |X_k| \le a/p\} \cap \{|S_n| \le x_1\} \subseteq \{|S_n - S_{k+p}| \ge x_0\}.$$

Taking $x_2 = x_1 + a + x_0$, we then derive

$$\begin{aligned}
I &:= \sum_{k=1}^{n} \mathbb{P}\left(E_k(x_2) \cap \{\max_{1\le i\le n} |X_i| \le a/p\} \cap \{|S_n| \le x_1\}\right)\\
&= \sum_{k=1}^{n-p} \mathbb{P}\left(E_k(x_2) \cap \{\max_{1\le i\le n} |X_i| \le a/p\} \cap \{|S_n| \le x_1\}\right)\\
&\le \sum_{k=1}^{n-p} \mathbb{P}\left(E_k(x_2) \cap \{|S_n - S_{k+p}| \ge x_0\}\right)\\
&\le \left(\sum_{k=1}^{n} \mathbb{P}(E_k(x_2))\right) \max_{1\le k\le n-p} \mathbb{P}\left(|S_n - S_{k+p}| \ge x_0 | E_k(x_2)\right)\\
&= R(x_2) \max_{1\le k\le n-p} \mathbb{P}\left(|S_n - S_{k+p}| \ge x_0 | E_k(x_2)\right).
\end{aligned}$$

Since, for any $1 \le k \le n-p$,

$$\mathbb{P}\left(|S_n - S_{k+p}| \ge x_0 | E_k(x_2)\right) \le \phi(p) + \mathbb{P}(|S_n - S_{k+p}| \ge x_0) \le \eta,$$

we get, overall,

$$I \leq \eta R(x_2)$$

from which we derive

$$\begin{aligned} R(x_2) &\leq \mathbb{P}(|S_n| > x_1) + \mathbb{P}\left(\max_{1\leq k\leq n} |X_k| > a/p\right) + I \\ &\leq \mathbb{P}\left(|S_n| > x_1\right) + \mathbb{P}(\max_{1\leq k\leq n} |X_k| > a/p) + \eta R(x_2). \end{aligned}$$

Therefore, the following maximal inequality holds for $\eta < 1$:

$$\mathbb{P}\left(\max_{1\leq k\leq n} |S_k| \geq x_1 + a + x_0\right) \leq (1-\eta)^{-1}\left(\mathbb{P}(|S_n| > x_1) + \mathbb{P}(\max_{1\leq k\leq n} |X_k| > a/p)\right). \tag{6.55}$$

On the other hand, for any positive real number x and any positive integers k and p such that $k \leq n-p$, we have

$$\{|S_{k-1}| < x\} \cap \{\max_{1\leq k\leq n} |X_k| \leq a/(p+1)\} \cap \{|S_n| > x + a + x_0\} \subseteq \{|S_n - S_{k+p}| > x_0\},$$

and if $n-p+1 \leq k \leq n$,

$$\{|S_{k-1}| < x\} \cap \{\max_{1\leq k\leq n} |X_k| \leq a/(p+1)\} \cap \{|S_n| > x + a + x_0\} = \emptyset.$$

And similarly to the previous arguments,

$$\begin{aligned} &\mathbb{P}(|S_n| > x + a + x_0) \leq \mathbb{P}(\max_{1\leq k\leq n} |X_k| > a/(p+1)) \\ &\quad + \sum_{k=1}^{n} \mathbb{P}(E_k(x) \cap \{\max_{1\leq i\leq n} |X_i| \leq a/(p+1)\} \cap \{|S_n| > x + a + x_0\}) \\ &\leq \mathbb{P}(\max_{1\leq k\leq n} |X_k| > a/(p+1)) + \sum_{k=1}^{n-p} \mathbb{P}(E_k(x) \cap \{|S_n - S_{k+p}| > x_0\}) \\ &\leq \mathbb{P}(\max_{1\leq k\leq n} |X_k| > a/(p+1)) + \sum_{i=1}^{n} \mathbb{P}(E_i(x)) \max_{1\leq k\leq n-p} \mathbb{P}(|S_n - S_{k+p}| > x_0 | E_k(x)). \end{aligned}$$

Hence

$$\mathbb{P}(|S_n| > x + a + x_0) \leq \mathbb{P}(\max_{1\leq k\leq n} |X_k| > a/(p+1)) + \eta \mathbb{P}(\max_{1\leq k\leq n} |S_k| \geq x).$$

Taking $x_1 = x + a + x_0$ and substituting the preceding inequality into relation (6.55), we finish the proof of the lemma. ■

Proof of Theorem 6.17 We are now in position to prove the theorem. Let x_0 be a positive real such that $\eta_{p,x_0,n} := \eta < 1$. Let $V > 2$ be a positive real number that will be chosen later. Setting $R(u) = \mathbb{P}(\max_{1\le k\le n} |S_k| \ge u)$, we have

$$\mathbb{E}(\max_{1\le k\le n} |S_k|^r) = r\int_0^\infty u^{r-1}R(u)du \le (Vx_0)^r + r\int_{Vx_0}^\infty u^{r-1}R(u)du$$
$$\equiv (Vx_0)^r + J.$$

Let α be a positive real and set

$$\Delta = \exp(2r(\alpha + (V-2)^{-1})).$$

Make the change of variable

$$u = x(1+2\alpha) + 2x_0.$$

We are going to apply Lemma 6.19. Taking into account that $u \ge V\,x_0$ implies $x \ge x_0(V-2)/(1+2\alpha)$, we derive

$$J \le (1+2\alpha)^r(1+2(V-2)^{-1})^{r-1}r\int_0^\infty x^{r-1}R(x(1+2\alpha)+2x_0)dx$$
$$\le \Delta\left(\eta(1-\eta)^{-1}r\int_0^\infty x^{r-1}R(x)dx + 2(1-\eta)^{-1}r\right.$$
$$\left.\times\int_0^\infty x^{r-1}\mathbb{P}(\max_{1\le k\le n}|X_k| > \alpha x/(p+1))dx\right)$$
$$\le \Delta\left(\eta(1-\eta)^{-1}\mathbb{E}(\max_{1\le k\le n}|S_k|^r) + 2(1-\eta)^{-1}\mathbb{E}(\max_{1\le k\le n}|X_k|^r)((p+1)/\alpha)^r\right).$$

Overall,

$$\mathbb{E}\max_{1\le k\le n}|S_k|^r(1-\Delta\eta(1-\eta)^{-1}) \le (Vx_0)^r + 2\Delta(1-\eta)^{-1}\mathbb{E}(\max_{1\le k\le n}|X_k|^r)((p+1)/\alpha)^r$$

and it remains to take

$$x_0 = \varepsilon^{-1/q}\max_{1\le k\le n}(\mathbb{E}|S_k|^q)^{1/q},\ \alpha = \varepsilon/r,\ V = r/\varepsilon,\ \varepsilon = (1-2\phi(p))/N,$$

where N is a sufficiently large number. The theorem is proved. ■

Actually, the estimate derived in the theorem, is optimal up to a constant. More exactly there exists a positive constant $c = c_r$ such that for all $1 \le q \le r$,

$$c\left(\mathbb{E}(\max_{1\le k\le n}|X_k|^r) + \max_{1\le k\le n}(\mathbb{E}|S_k|^q)^{r/q}\right) \le \mathbb{E}(\max_{1\le k\le n}|S_k|^r).$$

Using Theorem 6.17 together with the relation (6.54) and Theorem 6.1, the following maximal moment inequality follows:

Corollary 6.20 *Let $(X_k)_{k\in\mathbb{Z}}$ be a sequence of centered real-valued random variables in $\mathbb{L}^2$. Let $(\phi(n))_{n\geq 0}$ be its associated ϕ-mixing coefficients sequence. Assume that*

$$\phi = \sum_{k=1}^{\infty} \phi^{1/2}(2^k) < \infty .$$

Then, there exists a positive constant k_ϕ depending only on ϕ such that, for any $r \geq 2$,

$$\mathbb{E}(\max_{1\leq k\leq n} |S_k|^r) \leq (rk_\phi)^r \left(\sum_{i=1}^{n} \mathbb{E}(|X_i|^r) + \left(\sum_{i=1}^{n} \mathbb{E}(X_i^2) \right)^{r/2} \right).$$

By taking into account again relation (6.54) and then by combining Theorem 6.17 with Lemma 6.15 we also obtain the following corollary:

Corollary 6.21 *Assume that there exists a positive integer p such that $\phi(p) < 1/4$. Assume that $\mathbb{E}(X_i) = 0, i = 1, \cdots, n$. Then, for $t \geq 2$, the following inequality holds:*

$$\mathbb{E}(\max_{1\leq k\leq n} |S_k|^t) \leq (c'_{\phi(p)}t)^t(p^t\mathbb{E}(\max_{1\leq k\leq n} |X_k|^t) + (\mathbb{E}(S_n^2))^{t/2}),$$

where $c'_{\phi(p)}$ is a constant depending on $\phi(p)$ only.

Obviously we also have:

Corollary 6.22 *Let $X_1, \cdots, X_n$ be m-dependent, $\mathbb{E}(X_i) = 0, i = 1, \dots, n$. There exists an absolute constant c, such that, for $t \geq 2$, the following inequality holds:*

$$\mathbb{E}(\max_{1\leq k\leq n} |S_k|^t) \leq (ct)^t \left[m^t\mathbb{E}(\max_{1\leq k\leq n} |X_k|^t) + (\mathbb{E}(S_n^2))^{t/2} \right].$$

In addition, a direct application of Theorem 6.17 with $q = 1$ gives:

Corollary 6.23 *Assume that there exists a positive integer p such that $\phi(p) < 1/2$. Then $\mathbb{E}\left(\sum_{i=1}^n |X_i|\right)^2 \leq c(\phi, p)\left(\sum_{i=1}^n \mathbb{E}|X_i|^2 + \left[\sum_{i=1}^n \mathbb{E}|X_i|\right]^2\right)$, where $c(\phi, p)$ depends only on $\phi(p)$ and p.*

As a matter of fact we can also give an uniform integrability result which is obtained from Lemma 6.19. Recall that uniform integrability is essential for proving functional central limit theorem.

Proposition 6.24 *Assume that there exists a positive integer p such that $\phi(p) < 1/26$. Suppose in addition that $\mathbb{E}X_i = 0, i \geq 1$. Then*

$$\frac{\max_{1\leq k\leq n} |S_k|^2}{\sigma_n^2} \text{ is uniformly integrable if and only if } \frac{\max_{1\leq k\leq n} |X_k|^2}{\sigma_n^2} \text{ is uniformly integrable}$$

where $\sigma_n^2 = \mathbb{E}(S_n^2)$.

Proof The fact that $\sigma_n^{-2}\max_{1\le k\le n}|S_k|^2$ is uniformly integrable implies that $\sigma_n^{-2}\max_{1\le k\le n}|X_k|^2$ is uniformly integrable is obvious. Let us prove the converse implication, so let us assume that $\sigma_n^{-2}\max_{1\le k\le n}|X_k|^2$ is uniformly integrable. Note that by Markov inequality, Lemma 6.15 and relation (6.54), we have

$$\eta = \eta_{p,A\sigma_n,n} = \phi(p) + \max_{0\le i\le n}\mathbb{P}(|S_n - S_i| \ge A\sigma_n)$$
$$\le \phi(p) + \frac{4}{A^2\sigma_n^2}2(1-2\phi^{1/2}(p))^{-1}[\mathbb{E}(S_n^2) + p^2\max_{1\le k\le n}\mathbb{E}(X_k^2)].$$

Therefore, since we have assumed that $\sigma_n^{-2}\max_{1\le k\le n}|X_k|^2$ is uniformly integrable,

$$\lim_{A\to\infty}\limsup_{n\to\infty}\eta_{p,A\sigma_n,n} \le \phi(p) < 1/26.$$

Denote $M_n = \max_{1\le k\le n}|S_k|$ and $N_n = \max_{1\le k\le n}|X_k|$. For A and n large enough, we obtain, from Lemma 6.19, that for all $x > A$,

$$\mathbb{P}(M_n \ge 5x\sigma_n) \le \eta(1-\eta)^{-1}\mathbb{P}(M_n \ge x\sigma_n) + 2(1-\eta)^{-1}\mathbb{P}(N_n > x\sigma_n/(p+1)).$$

Therefore, by using the formula

$$\mathbb{E}\{X^2 I(|X| > A)\} = A^2\mathbb{P}(|X| > A) + 2\int_A^\infty x\mathbb{P}(|X| > x)dx\,,$$

we obtain

$$\frac{\mathbb{E}(M_n^2 I(M_n > 5A\sigma_n))}{25\sigma_n^2} \le \frac{\eta}{1-\eta}\frac{\mathbb{E}(M_n^2 I(M_n > A\sigma_n))}{\sigma_n^2}$$
$$+\frac{2(p+1)^2}{1-\eta}\frac{\mathbb{E}(N_n^2 I(N_n > A\sigma_n/(p+1)))}{\sigma_n^2}.$$

Denoting $L = \lim_{A\to\infty}\limsup_{n\to\infty}\mathbb{E}(M_n^2 I(M_n > A\sigma_n)/\sigma_n^2$ and passing to the limit above, we obtain that the uniform integrability of N_n^2/σ_n^2 implies

$$\frac{1}{25}L \le \frac{\phi(p)}{1-\phi(p)}L.$$

This gives $L = 0$ since $\phi(p) < 1/26$. This ends the proof of the proposition. ∎

We shall prove next a proposition where we compare the tails of maximum of a sequence with the corresponding ones of independent identically distributed random variables with the same marginals in terms of the ϕ-mixing coefficient. It is a non-stationary version of Proposition 3.1 in Peligrad (1990).

Lemma 6.25 *Consider now a sequence (X_k^*) of independent random variables where each X_k^* is distributed as X_k. Then for all real numbers x,*

$$(1-\phi(1))\mathbb{P}(\max_{1\le k\le n}X_k^* > x) \le \mathbb{P}(\max_{1\le k\le n}X_k > x) \le (1+\phi(1))\mathbb{P}(\max_{1\le k\le n}X_k^* > x), \quad (6.56)$$

and therefore, for any $r \geq 0$,

$$(1-\phi(1))\mathbb{E}(\max_{1\leq k\leq n}|X_k^*|^r) \leq \mathbb{E}(\max_{1\leq k\leq n}|X_k|^r) \leq (1+\phi(1))\mathbb{E}(\max_{1\leq k\leq n}|X_k^*|^r).$$

Proof In what follows, we shall use the following notations:

$$N_{n-j} = \max_{1+j\leq k\leq n} X_k,\ j = 0, \cdots, n-1,$$

$$B_j = \mathbb{P}(X_j \leq x),\ A_j = (1+\phi(1))\mathbb{P}(X_j > x),\ A_j^* = (1-\phi(1))\mathbb{P}(X_j > x),\ j = 1, \cdots, n.$$

We first notice that, by the definition of the ϕ-mixing coefficient, we have

$$\begin{aligned}\mathbb{P}(\max_{1\leq k\leq n} X_k > x) &= \mathbb{P}(\max_{2\leq k\leq n} X_k > x) + \mathbb{P}(\max_{2\leq k\leq n} X_k \leq x, X_1 > x)\\ &\leq \mathbb{P}(N_{n-1} > x) + \{\phi(1) + \mathbb{P}(N_{n-1} \leq x)\}\,\mathbb{P}(X_1 > x)\\ &\leq (1+\phi(1))\mathbb{P}(X_1 > x) + \mathbb{P}(N_{n-1} > x)\mathbb{P}(X_1 \leq x) = A_1 + B_1\mathbb{P}(N_{n-1} > x).\end{aligned}$$

By iteration we get

$$\mathbb{P}(\max_{1\leq k\leq n} X_k > x) \leq \sum_{i=1}^{n-1} A_i \prod_{k=1}^{i-1} B_k + \prod_{k=1}^{n-1} B_k\mathbb{P}(X_n > x) \leq \sum_{i=1}^{n} A_i \prod_{k=1}^{i-1} B_k.$$

But $\prod_{k=1}^{i-1} B_k = \mathbb{P}(\max_{1\leq k\leq i-1} X_k^* \leq x)$ implying that

$$\begin{aligned}\sum_{i=1}^{n} A_i \prod_{k=1}^{i-1} B_k &= (1+\phi(1))\sum_{i=1}^{n}\mathbb{P}(\max_{1\leq k\leq i-1} X_k^* \leq x, X_i^* > x)\\ &= (1+\phi(1))\mathbb{P}(\max_{1\leq k\leq i} X_k^* > x).\end{aligned}$$

This proves the right-hand side of inequality (6.56). To prove the left-hand side, we use similar arguments. We then notice that, again by the definition of the ϕ-mixing coefficient, we have

$$\begin{aligned}\mathbb{P}(\max_{1\leq k\leq n} X_k > x) &= \mathbb{P}(\max_{2\leq k\leq n} X_k > x) + \mathbb{P}(\max_{2\leq k\leq n} X_k \leq x, X_1 > x)\\ &\geq \mathbb{P}(N_{n-1} > x) + \{\mathbb{P}(N_{n-1} \leq x) - \phi(1)\}\,\mathbb{P}(X_1 > x)\\ &\geq (1-\phi(1))\mathbb{P}(X_1 > x) + \mathbb{P}(N_{n-1} > x)\mathbb{P}(X_1 \leq x) = A_1^* + B_1\mathbb{P}(N_{n-1} > x).\end{aligned}$$

So, by iteration and taking into account the previous considerations,

$$\mathbb{P}(\max_{1\le k\le n} X_k > x) \ge \sum_{i=1}^{n-1} A_i^* \prod_{k=1}^{i-1} B_k + \prod_{k=1}^{n-1} B_k \mathbb{P}(X_n > x)$$

$$\ge \sum_{i=1}^{n} A_i^* \prod_{k=1}^{i-1} B_k = (1-\phi(1))\mathbb{P}(\max_{1\le k\le n} X_k^* > x).$$

This concludes the proof of the lemma. ■

The left-hand side of inequality (6.56) has a meaning only if $\phi(1) < 1$. In order to use (6.56) when we only know that $\lim_{n\to\infty}\phi(n) = 0$, we give in what follows a kindred relation valid in the strictly stationary case.

Lemma 6.26 *Suppose (X_k) is a strictly stationary sequence of real-valued random variables and let (X_k^*) be a i.i.d. sequence with the same marginal distribution. Then for all real numbers x and very positive integer $p \le n$,*

$$(1-\phi(p))\mathbb{P}\left(\max_{1\le k\le [n/p]} X_k^* > x\right) \le \mathbb{P}\left(\max_{1\le k\le n} X_k > x\right) \le p(1+\phi(p))\mathbb{P}\left(\max_{1\le k\le [n/p]+1} X_k^* > x\right).$$

and therefore for any $r \ge 1$

$$(1-\phi(p))\mathbb{E}\left(\max_{1\le k\le [n/p]} |X_k^*|^r\right) \le \mathbb{E}\left(\max_{1\le k\le n} |X_k|^r\right) \le p(1+\phi(p))\mathbb{E}\left(\max_{1\le k\le [n/p]+1} |X_k^*|^r\right).$$

Proof This result follows by applying Lemma 6.25 via the following inequality

$$\mathbb{P}\left(\max_{1\le j\le [n/p]} X_{pj} > x\right) \le \mathbb{P}\left(\max_{1\le k\le n} X_k > x\right) \le p\mathbb{P}\left(\max_{0\le j\le [n/p]} X_{pj+1} > x\right),$$

thanks to stationarity. ■

6.3 Moment Inequalities for α-dependent Sequences

Theorems 3.11, 3.13 and 3.17 give moment inequalities whose upper bounds are expressed in terms of conditional expectations. Estimates of these upper bounds can be computed for various classes of weakly dependent sequences. In this section, we consider the class of α-dependent sequences associated with a sequence $\mathbf{X} = (X_i)_{i\in\mathbb{Z}}$ of real-valued random variables as defined in Section 5.2. So, for any $n \in \mathbb{N}$, $\alpha_{1,\mathbf{X}}(n)$ is defined in (5.12) and more generally $\alpha_{k,\mathbf{X}}(n)$ is defined in (5.13), for any positive integer k.

To formulate the results of this section, let us introduce some notations: We write $(X_i) \prec X$ if there exists a positive random variable X such that $Q_X \ge \sup_{k\ge1} Q_{|X_k|}$ where $Q_{|Y|}$ is the quantile function of $|Y|$, that is, the càdlàg inverse function of the tail function $\mathbb{P}(|Y| > t)$. For any positive integer k, let also

$$\alpha_{k,\mathbf{X}}^{-1}(u) = \inf\{i \in \mathbb{N} : \alpha_{k,\mathbf{X}}(i) \le u\} = \sum_{i\ge0} \mathbf{1}_{u<\alpha_{k,\mathbf{X}}(i)}.$$

In addition, throughout this section, $\mathcal{F}_j = \sigma(X_i, i \le j)$ and we will often use the notation $\mathbb{E}_j(\cdot) = \mathbb{E}(\cdot|\mathcal{F}_j)$.

6.3.1 Burkholder-type Inequalities

The next result is a Burkholder-type inequality for the partial sums associated with a α-dependent sequence of real-valued random variables. It is slightly more general than Theorem 2.5 (b) of Rio (2000) since strict stationarity of the sequence is not required. The proof comes from an application of Theorem 3.13.

Theorem 6.27 *Let $(X_i)_{i\in\mathbb{Z}}$ be a sequence of centered real-valued random variables and let $(\alpha_{1,\mathbf{X}}(n))_{n\ge 0}$ be its weak dependence coefficients defined by (5.12). Assume that $(X_i) \prec X$ and let $Q = Q_X$. Then, for any $p \ge 2$,*

$$\|S_n\|_p \le (4n(p-1))^{1/2}\left(\int_0^1 (\alpha_{1,\mathbf{X}}^{-1}(u) \wedge n)^{p/2} Q^p(u)du\right)^{1/p}. \tag{6.57}$$

Moreover, if $p > 2$,

$$\|\max_{1\le i\le n} |S_i|\|_p \le c_p n^{1/2}\left(\int_0^1 (\alpha_{1,\mathbf{X}}^{-1}(u) \wedge (2n))^{p/2} Q^p(u)du\right)^{1/p}, \tag{6.58}$$

where $c_p = 2p\sqrt{2}(p-1)^{-1/2} + 2p(p-1)^{-1}(2^{(p-2)/(2p)} - 1)^{-1}$.

Note that, according to Annex C in Rio (2000), for any positive real number q and any non-negative real r,

$$\int_0^1 (\alpha_{1,\mathbf{X}}^{-1}(u) \wedge n)^q Q^r(u)du \le q\max(1, 2^{1-q})\sum_{k=0}^{n-1}(k+1)^{q-1}\int_0^{\alpha_{1,\mathbf{X}}(k)} Q^r(u)du. \tag{6.59}$$

Hence, for any $p \ge 2$,

$$\int_0^1 (\alpha_{1,\mathbf{X}}^{-1}(u) \wedge n)^{p/2} Q^p(u)du \le 2^{-1}p\sum_{k=0}^{n-1}(k+1)^{p/2-1}\int_0^{\alpha_{1,\mathbf{X}}(k)} Q^p(u)du.$$

Proof of Theorem 6.27 The proof of (6.57) is based on an application of Theorem 3.13. We shall then prove that

$$b_{i,n} = \max_{i\le l\le n}\left\|X_i \sum\nolimits_{k=i}^{l} \mathbb{E}(X_k|\mathcal{F}_i)\right\|_{p/2} \le 2\left(\int_0^1 (\alpha_{1,\mathbf{X}}^{-1}(u) \wedge n)^{p/2} Q^p(u)du\right)^{2/p}. \tag{6.60}$$

To prove the inequality above, we first need to recall some important tools. The first one is Rio's covariance inequality (see Rio (1993)):

Lemma 6.28 *Let X and Y be two real-valued random variables in* $\mathbb{L}^2$. *Then*

$$|\mathrm{cov}(X,Y)| \leq 2\int_0^{2\bar{\alpha}(X,Y)} Q_{|X|}(u)Q_{|Y|}(u)du \leq 4\int_0^{\bar{\alpha}(X,Y)} Q_{|X|}(u)Q_{|Y|}(u)du,$$

where $\bar{\alpha}(X,Y) = \sup_{(s,t)\in\mathbb{R}^2} |\mathrm{cov}(1_{X\leq s}, 1_{Y\leq t})|$.

Remark 6.29 *It should be noticed that since* $\bar{\alpha}(X,Y) \leq \alpha(\sigma(X),\sigma(Y))$, *Rio's covariance inequality contains Ibragimov's covariance inequality (1962) for bounded real-valued random variables as well as Davydov's inequality (1968). We refer to Chapter 1 of Rio (2000) for a proof of Lemma 6.28.*

The second tool needed to be recalled concerns some properties involving quantile functions. It is Lemma 2.1 in Rio (2000).

Lemma 6.30 *Let* $Z_1, \ldots, Z_p$ *be positive real-valued random variables. Then*

(a) $\int_0^1 Q_{Z_1Z_2}(u)Q_{Z_3}(u)\ldots Q_{Z_p}(u)du \leq \int_0^1 Q_{Z_1}(u)Q_{Z_2}(u)Q_{Z_3}(u)\ldots Q_{Z_p}(u)du$

and

(b) $\int_0^1 Q_{Z_1+Z_2}(u)Q_{Z_3}(u)\ldots Q_{Z_p}(u)du \leq \int_0^1 (Q_{Z_1}(u)+Q_{Z_2}(u))Q_{Z_3}(u)\ldots Q_{Z_p}(u)du.$

Let us now prove inequality (6.60). With this aim, we set

$$c_{i,n} = \|X_i \sum_{k=i}^{n} |\mathbb{E}(X_k|\mathcal{F}_i)|\,\|_{p/2}$$

and first note that

$$b_{i,n}^{p/2} \leq c_{i,n}^{p/2} = \sum_{k=i}^{n} \mathbb{E}(Z_{i,k,n}X_i\mathbb{E}(X_k|\mathcal{F}_i)) = \sum_{k=i}^{n} \mathbb{E}(Z_{i,k,n}X_iX_k),$$

where $Z_{i,k,n} = Y_{i,n}\mathrm{sign}\,(X_i\mathbb{E}(X_k|\mathcal{F}_i))$ with $Y_{i,n} = |X_i \sum_{\ell=i}^{n} |\mathbb{E}(X_\ell|\mathcal{F}_i)|\,|^{p/2-1}$. Since X_k is centered, it follows that

$$c_{i,n}^{p/2} \leq \sum_{k=i}^{n} \mathrm{cov}(Z_{i,k,n}X_i, X_k).$$

Hence, applying Lemma 6.28, we get

$$c_{i,n}^{p/2} \leq 2\sum_{k=i}^{n}\int_0^{2\bar{\alpha}(Z_{i,k,n}X_i,X_k)} Q_{Y_{i,n}|X_i|}(u)Q_{|X_k|}(u)du.$$

But

$$2\bar{\alpha}(Z_{i,k,n}X_i, X_k) \leq 2\sup_{A\in\mathcal{F}_i} \bar{\alpha}(1_A, X_k) \leq \alpha_{1,\mathbf{X}}(k-i).$$

This consideration together with Item (a) of Lemma 6.30 give

$$c_{i,n}^{p/2} \le 2\sum_{k=i}^{n} \int_0^{\alpha(k-i)} Q_{Y_{i,n}}(u)Q^2(u)du. \tag{6.61}$$

If $p = 2$ then $Q_{Y_{i,n}}(u) = 1$ and in this case inequality (6.60) follows by taking into account that, for any $i \ge 0$,

$$\sum_{k=i}^{n} \mathbf{1}_{u<\alpha_{1,\mathbf{X}}(k-i)} = \sum_{k=0}^{n-i} \mathbf{1}_{u<\alpha_{1,\mathbf{X}}(k)} \le \alpha_{1,\mathbf{X}}^{-1}(u) \wedge n. \tag{6.62}$$

Now for $p > 2$, starting from (6.61), taking into account (6.62) and using Hölder's inequality, we get

$$c_{i,n}^{p/2} \le 2\left(\int_0^1 Q_{Y_{i,n}}^{p/(p-2)}(u)du\right)^{(p-2)/p} \left(\int_0^1 (\alpha_{1,\mathbf{X}}^{-1}(u) \wedge n)^{p/2} Q^p(u)du\right)^{2/p}.$$

To see that this inequality entails (6.60), it suffices to note that

$$\left(\int_0^1 Q_{Y_{i,n}}^{p/(p-2)}(u)du\right)^{(p-2)/p} = \left(\mathbb{E}\left(|Y_{i,n}|^{p/(p-2)}\right)\right)^{(p-2)/p} = c_{i,n}^{p/2-1}.$$

We turn now to the proof of the maximal inequality (6.58). This inequality will be derived from an application of Proposition 3.14 by taking into account inequality (6.57) together with the following upper bound: For any non-negative integers k and l,

$$\|\mathbb{E}_{k2^l}(S_{(k+1)2^l} - S_{k2^l})\|_p \le 2\left(\int_0^1 (\alpha_{1,\mathbf{X}}^{-1}(u) \wedge 2^l)^p Q^p(u)du\right)^{1/p}. \tag{6.63}$$

Indeed, if r is the positive integer such that $2^{r-1} < n \le 2^r$, Proposition 3.14 combined with inequality (6.57) gives

$$\begin{aligned} \|\max_{1\le m\le n} |S_m|\|_p &\le a_p n^{1/2}\left(\int_0^1 (\alpha_{1,\mathbf{X}}^{-1}(u) \wedge (2n))^{p/2} Q^p(u)du\right)^{1/p} \\ &+ \frac{p}{p-1}\sum_{l=0}^{r-1}\left(\sum_{k=1}^{2^{r-l}-1} \|\mathbb{E}_{k2^l}(S_{(k+1)2^l} - S_{k2^l})\|_p^p\right)^{1/p}, \end{aligned} \tag{6.64}$$

where $a_p = 2p\sqrt{2}(p-1)^{-1/2}$. Now, the upper bound (6.63) entails that

$$\sum_{l=0}^{r-1}\left(\sum_{k=1}^{2^{r-l}-1}\|\mathbb{E}_{k2^l}(S_{(k+1)2^l}-S_{k2^l})\|_p^p\right)^{1/p}\leq 2\times 2^{r/p}\sum_{l=0}^{r-1}2^{-l/p}\left(\int_0^1(\alpha^{-1}(u)\wedge 2^l)^p Q^p(u)du\right)^{1/p}$$

$$\leq 2\times 2^{r/p}\left(\int_0^1(\alpha^{-1}(u)\wedge 2^{r-1})^{p/2}Q^p(u)du\right)^{1/p}\sum_{l=0}^{r-1}2^{-l/p}2^{l/2}. \tag{6.65}$$

Therefore, if $p > 2$,

$$\sum_{l=0}^{r-1}\left(\sum_{k=1}^{2^{r-l}-1}\|\mathbb{E}_{k2^l}(S_{(k+1)2^l}-S_{k2^l})\|_p^p\right)^{1/p}\leq 2\times 2^{r/2}$$

$$\times\frac{1}{2^{(p-2)/(2p)}-1}\left(\int_0^1(\alpha^{-1}(u)\wedge n)^{p/2}Q^p(u)du\right)^{1/p}. \tag{6.66}$$

Inserting (6.66) in (6.64) gives the maximal inequality (6.58). To complete the proof it remains to show inequality (6.63). We note that

$$\|\mathbb{E}_{k2^l}(S_{(k+1)2^l}-S_{k2^l})\|_p^p=\mathbb{E}\left(U_{k,l}\text{sign}(\mathbb{E}_{k2^l}(S_{(k+1)2^l}-S_{k2^l}))(S_{(k+1)2^l}-S_{k2^l})\right),$$

where $U_{k,l} = |\mathbb{E}_{k2^l}(S_{(k+1)2^l}-S_{k2^l})|^{p-1}$. Since the X_i's are centered, an application of Lemma 6.28 together with the fact that

$$2\bar{\alpha}(U_{k,l}\text{sign}(\mathbb{E}_{k2^l}(S_{(k+1)2^l}-S_{k2^l})),X_i)\leq 2\sup_{A\in\mathcal{F}_{k2^l}}\bar{\alpha}(1_A,X_i)\leq\alpha_{1,\mathbf{X}}(i-k2^l),$$

gives

$$\|\mathbb{E}_{k2^l}(S_{(k+1)2^l}-S_{k2^l})\|_p^p\leq 2\sum_{i=k2^l+1}^{(k+1)2^l}\int_0^{\alpha_{1,\mathbf{X}}(i-k2^l)}Q_{U_{k,l}}(u)Q_{|X_i|}(u)du$$

$$\leq 2\int_0^1(\alpha_{1,\mathbf{X}}^{-1}(u)\wedge 2^l)Q_{U_{k,l}}(u)Q(u)du.$$

Hölder's inequality then entails that

$$\|\mathbb{E}_{k2^l}(S_{(k+1)2^l}-S_{k2^l})\|_p^p\leq 2\left(\int_0^1 Q_{U_{k,l}}^{p/(p-1)}(u)du\right)^{(p-1)/p}\left(\int_0^1(\alpha_{1,\mathbf{X}}^{-1}(u)\wedge 2^l)^p Q^p(u)du\right)^{1/p}.$$

Since $\left(\int_0^1 Q_{U_{k,l}}^{p/(p-1)}(u)du\right)^{(p-1)/p} = \|\mathbb{E}_{k2^l}(S_{(k+1)2^l}-S_{k2^l})\|_{p-1}^p$, the inequality above leads to (6.63). ■

The next result completes the maximal inequality (6.58) in the case where $p = 2$. In this case, the inequality comes from an application of Theorem 3.11.

Theorem 6.31 *Let $(X_i)_{i\in\mathbb{Z}}$ be a sequence of centered real-valued random variables and let $(\alpha_{1,\mathbf{X}}(n))_{n\geq 0}$ be its weak dependence coefficients defined by (5.12). Assume that $(X_i) \prec X$ and let $Q = Q_X$. Then*

$$\mathbb{E}\left(\max_{1\leq i\leq n} S_i^2\right) \leq 32\, n \int_0^1 (\alpha_{1,\mathbf{X}}^{-1}(u) \wedge n) Q^2(u) du.$$

Proof Using Theorem 3.11 with $X_i, -X_i$ and $\lambda = 0$, we get:

$$\mathbb{E}\left(\max_{1\leq i\leq n} S_i^2\right) \leq 16 \sum_{i=1}^{n} \sum_{k=i}^{n} \mathbb{E}\,|X_i \mathbb{E}(X_k|\mathcal{F}_i)\,|.$$

But,

$$\mathbb{E}\,|X_i \mathbb{E}(X_k|\mathcal{F}_i)\,| = \mathbb{E}\left(|X_i| \mathrm{sign}(\mathbb{E}_i(X_k)) X_k\right).$$

Therefore, as has already been used, since $\mathbb{E}(X_k) = 0$, an application of Lemma 6.28 gives: For any integers $k \geq i$,

$$\mathbb{E}\left|X_i \mathbb{E}(X_k|\mathcal{F}_i)\right| \leq 2 \int_0^{\alpha_{1,\mathbf{X}}(k-i)} Q^2(u) du\,.$$

So overall,

$$\mathbb{E}\left(\max_{1\leq i\leq n} S_i^2\right) \leq 32\, n \sum_{k=0}^{n-1} \int_0^{\alpha_{1,\mathbf{X}}(k)} Q^2(u) du,$$

which ends the proof of the theorem. ∎

6.3.2 *Rosenthal-type Inequality*

In the stationary case, starting now from Theorem 3.17, we have:

Theorem 6.32 *Let $p > 2$ be a real number and let $(X_i)_{i\in\mathbb{Z}}$ be a stationary sequence of centered real-valued random variables. Let $(\alpha_{1,\mathbf{X}}(n))_{n\geq 0}$ and $(\alpha_{2,\mathbf{X}}(n))_{n\geq 0}$ be its weak dependence coefficients of order 1 and 2 respectively defined by (5.12) and (5.13). Let $Q = Q_{|X_0|}$. Then, for any positive integer n, the following inequality holds:*

$$\begin{aligned}\|\max_{1\leq j\leq n} |S_j|\|_p &\ll n^{1/2}\left(\int_0^1 (\alpha_{1,\mathbf{X}}^{-1}(u) \wedge n) Q^2(u) du\right)^{1/2} \\ &+ n^{1/p} \sum_{k=1}^{n} \frac{1}{k^{1+1/p}} \left(\int_0^1 (\alpha_{1,\mathbf{X}}^{-1}(u) \wedge k)^p Q^p(u) du\right)^{1/p} \\ &+ n^{1/p}\left(\sum_{k=1}^{n} \frac{1}{k^{1+2\delta/p}} \left(\int_0^1 (\alpha_{2,\mathbf{X}}^{-1}(u) \wedge k)^p Q^p(u) du\right)^{2\delta/p}\right)^{1/(2\delta)}\end{aligned}$$

where $\delta = \min(1, 1/(p-2))$.

Remark 6.33 *Let us consider the stationary case and $p > 2$. If*

$$\sum_{k\geq 0}(k+1)^{p/2-1}\int_0^{\alpha_{1,\mathbf{X}}(k)} Q^p(u)du < \infty, \tag{6.67}$$

then by Theorem 6.27,

$$\mathbb{E}\left(\max_{1\leq j\leq n}|S_j|^p\right) \ll n^{p/2}. \tag{6.68}$$

Note that by an application of Theorem 6.32, the moment estimate (6.68) also holds under the weaker condition: There exists a positive constant C such that for any $k \geq 0$,

$$\int_0^{\alpha_{2,\mathbf{X}}(k)} Q^p(u)du \leq \frac{C}{(k+1)^{p/2}}. \tag{6.69}$$

To see this it suffices to notice that, under (6.69), by using (6.59),

$$\int_0^1 (\alpha_{2,\mathbf{X}}^{-1}(u)\wedge k)^p Q^p(u)du \ll Ck^{p/2}.$$

Proof of Theorem 6.32 The Rosenthal-type inequality follows easily from Theorem 3.17 by taking into account the estimates below: For any positive integer k,

$$\|\mathbb{E}_0(S_k)\|_p \leq 2\left(\int_0^1 (\alpha_{1,\mathbf{X}}^{-1}(u)\wedge k)^p Q^p(u)du\right)^{1/p} \tag{6.70}$$

and

$$\|\mathbb{E}_0(S_k^2)\|_{p/2} \leq 2^6\left(\int_0^1 (\alpha_{2,\mathbf{X}}^{-1}(u)\wedge k)^p Q^p(u)du\right)^{2/p} + 4k\int_0^1 (\alpha_{1,\mathbf{X}}^{-1}(u)\wedge k)Q^2(u)du. \tag{6.71}$$

To prove (6.70), we set $Y_{k,0} := \sum_{i=1}^k |\mathbb{E}_0(X_i)|$ and write

$$\|\mathbb{E}_0(S_k)\|_p \leq \|Y_{k,0}\|_p = \left(\sum_{i=1}^k \mathbb{E}\left(Y_{k,0}^{p-1}\operatorname{sign}(\mathbb{E}_0(X_i))X_i\right)\right)^{1/p}.$$

Since $\mathbb{E}(X_i) = 0$, we infer from Lemma 6.28 and Item (a) of Lemma 6.30 that

$$\|Y_{k,0}\|_p^p \leq 2\sum_{i=1}^{k}\int_0^{\alpha_{1,\mathbf{X}}(i)} Q_{Y_{k,0}}^{p-1}(u)Q(u)du \leq 2\int_0^1 (\alpha_{1,\mathbf{X}}^{-1}(u)\wedge k)Q_{Y_{k,0}}^{p-1}(u)Q(u)du.$$

An application of Hölder's inequality then gives

$$\|Y_{k,0}\|_p^p \leq 2\left(\int_0^1 Q_{Y_{k,0}}^p(u)du\right)^{(p-1)/p}\left(\int_0^1 (\alpha_{1,\mathbf{X}}^{-1}(u)\wedge k)^p Q^p(u)du\right)^{1/p}.$$

This ends the proof of (6.70) since $\int_0^1 Q_{Y_{k,0}}^p(u)Q(u)du = \|Y_{k,0}\|_p^p$. We prove now the upper bound (6.71).

With this aim, we first write

$$|\mathbb{E}_0(S_k^2)| \leq 2\sum_{i=1}^{k}\sum_{j=0}^{k-i}|\mathbb{E}_0(X_iX_{i+j})| \leq 2A_{k,0} + 2B_{k,0} + 2\sum_{i=1}^{k}\sum_{j=0}^{i-1}|\mathbb{E}(X_iX_{i+j})|, \tag{6.72}$$

where

$$A_{k,0} = \sum_{i=1}^{k}\sum_{j=i}^{k-i}|\mathbb{E}_0(X_iX_{i+j})|$$

and

$$B_{k,0} = \sum_{i=1}^{k}\sum_{j=0}^{i-1}|\mathbb{E}_0(X_iX_{i+j}) - \mathbb{E}(X_iX_{i+j})|.$$

As has already been proved, by using Lemma 6.28, we get

$$\sum_{i=1}^{k}\sum_{j=0}^{i-1}|\mathbb{E}(X_iX_{i+j})| \leq 2k\int_0^1 (\alpha_{1,\mathbf{X}}^{-1}(u)\wedge k)Q^2(u)du. \tag{6.73}$$

On the other hand,

$$\|A_{k,0}\|_{p/2}^{p/2} = \sum_{i=1}^{k}\sum_{j=i}^{k-i}\mathbb{E}\left(A_{k,0}^{p/2-1}\operatorname{sign}(\mathbb{E}_0(X_iX_{i+j}))X_iX_{i+j}\right).$$

Since $\mathbb{E}(X_{i+j}) = 0$, an application of Lemma 6.28 and of Item (a) of Lemma 6.30 gives

$$\|A_{k,0}\|_{p/2}^{p/2} \leq 2\sum_{i=1}^{k}\sum_{j=i}^{k-i}\int_0^{\alpha_{1,\mathbf{X}}(j)} Q_{A_{k,0}}^{p/2-1}(u)Q^2(u)du$$

$$\leq 2\sum_{j=1}^{k-1} j\int_0^{\alpha_{1,\mathbf{X}}(j)} Q_{A_{k,0}}^{p/2-1}(u)Q^2(u)du \leq 2\int_0^1 (\alpha_{1,\mathbf{X}}^{-1}(u)\wedge k)^2 Q_{A_{k,0}}^{p/2-1}(u)Q^2(u)du.$$

Hölder's inequality together with the fact that $\int_0^1 Q_{A_{k,0}}^{p/2}(u)Q(u)du = \|A_{k,0}\|_{p/2}^{p/2}$ implies that

$$\|A_{k,0}\|_{p/2} \le 2\left(\int_0^1 (\alpha_{1,\mathbf{X}}^{-1}(u) \wedge k)^p Q^p(u)du\right)^{2/p}. \tag{6.74}$$

We handle now $\|B_{k,0}\|_{p/2}$. With this aim, we set

$$Z_{k,0} = B_{k,0}^{p/2-1}\operatorname{sign}(\mathbb{E}_0(X_iX_{i+j}) - \mathbb{E}(X_iX_{i+j}))$$

and write

$$\|B_{k,0}\|_{p/2}^{p/2} = \sum_{i=1}^{k}\sum_{j=0}^{i-1}\mathbb{E}\left(Z_{k,0}(X_iX_{i+j} - \mathbb{E}(X_iX_{i+j}))\right).$$

To estimate the term in the right-hand side, we use Proposition A.1 and Lemma A.1 in Dedecker and Rio (2008) rather than Lemma 6.28. We then infer that

$$\begin{aligned}\mathbb{E}\left(Z_{k,0}(X_iX_{i+j} - \mathbb{E}(X_iX_{i+j}))\right) &= \mathbb{E}\left((Z_{k,0} - \mathbb{E}(Z_{k,0}))X_iX_{i+j}\right)\\ &\le 2^4\int_0^{2^{-1}\bar{\alpha}(Z_{k,0},X_i,X_{i+j})} Q_{|Z_{k,0}|}(u)Q^2(u)du,\end{aligned}$$

where

$$\begin{aligned}&\bar{\alpha}(Z_{k,0}, X_i, X_{i+j})\\ &\quad= \sup_{(s,t,u)\in\mathbb{R}^3}\left|\mathbb{E}\left((1_{Z_{k,0}\le s} - \mathbb{E}(1_{Z_{k,0}\le s}))(1_{X_i\le t} - \mathbb{E}(1_{X_i\le s}))\ (1_{X_{i+j}\le u} - \mathbb{E}(1_{X_{i+j}\le u}))\right)\right|.\end{aligned}$$

But for any non-negative integers i and j,

$$2\bar{\alpha}(Z_{k,0}, X_i, X_{i+j}) \le \alpha_{2,\mathbf{X}}(i).$$

So, overall,

$$\begin{aligned}\|B_{k,0}\|_{p/2}^{p/2} &\le 2^4\sum_{i=1}^{k} i\int_0^{\alpha_{2,\mathbf{X}}(i)} Q_{B_{k,0}}^{p/2-1}(u)Q^2(u)du\\ &\le 2^4\int_0^1 (\alpha_{2,\mathbf{X}}^{-1}(u) \wedge k)^2 Q_{B_{k,0}}^{p/2-1}(u)Q^2(u)du.\end{aligned}$$

Using Hölder's inequality, as before, we derive

$$\|B_{k,0}\|_{p/2} \le 2^4\left(\int_0^1 (\alpha_{2,\mathbf{X}}^{-1}(u) \wedge k)^p Q^p(u)du\right)^{2/p}. \tag{6.75}$$

Starting from (6.72) and taking into account (6.73), (6.74) and (6.75), together with the fact that $\alpha_{1,\mathbf{X}}^{-1}(u) \leq \alpha_{2,\mathbf{X}}^{-1}(u)$, the upper bound (6.71) follows. ■

6.3.3 Comments

Since any function of a strong mixing sequence is still strong mixing with the same strong mixing coefficients, Theorems 6.27, 6.31 and 6.32 can be used to get an upper bound for the moments of the partial sums associated with $Y_i = f(X_i) - \mathbb{E}(f(X_i))$ for any measurable function f provided that we replace the weak dependence coefficients $\alpha_{1,\mathbf{X}}(k)$ and $\alpha_{2,\mathbf{X}}(k)$ by the strong mixing ones. Let us, however, mention the Rosenthal inequality obtained by Rio (2000) (see his Theorem 6.3) which is more precise in the case of strong mixing sequences than the one obtained via Theorem 6.32. Rio's Rosenthal inequality is obtained via a Fuk–Nagaev-type inequality (see Theorem 6.2 in Rio (2000)) for α-mixing sequences. The proof of this Fuk–Nagaev-type inequality uses a blocking technique and the coupling property of α-mixing coefficients. Rio's Rosenthal-type inequality cannot then be fully extended to α-dependent sequences by using the same technique of proof. However, in a very recent paper, Dedecker and Merlevède (2017-a), by using blocking, martingale approximation and Theorem 3.17 when S_k is a martingale, have extended Rio's Rosenthal-type inequality to α-dependent sequences. Before stating their result, let us make some comments. Contrary to the strong mixing case, any function of a α-dependent sequence $\mathbf{X} = (X_i)_{i\in\mathbb{Z}}$ is not necessarily α-dependent (meaning that its dependency coefficients do no necessarily tend to zero). Hence, we need to impose some constraints on the observables. As in Remark 5.11, if f is monotonic on some open interval and 0 elsewhere, and if $\mathbf{Y} = (f(X_i))_{i\in\mathbb{Z}}$, then for any positive integer k, $\alpha_{k,\mathbf{Y}}(n) \leq 2^k \alpha_{k,\mathbf{X}}(n)$. As a consequence of this remark, Theorems 6.27, 6.31 and 6.32 also hold for $\sum_{k=1}^{n}(f(X_i) - \mathbb{E}(f(X_i))$, where f is monotonic on a single interval, and we can easily extend them by linearity to convex combinations of such functions. Such classes are given in Definition 1.9 in Dedecker, Gouëzel and Merlevède (2010) and can be described as follows:

Definition 6.34 *If μ is a probability measure on $\mathbb{R}$ and $Q = H^{-1}$ is an integrable quantile function, let $\widetilde{\mathrm{Mon}}(Q,\mu)$ be the set of functions g which are monotonic on some open interval of $\mathbb{R}$ and null elsewhere and such that $\mu(|g| > t) \leq H(t)$ for any $t \in \mathbb{R}^+$. Let $\tilde{\mathcal{F}}(Q,\mu)$ be the closure in $\mathbb{L}^1(\mu)$ of the set of functions which can be written as $\sum_{\ell=1}^{L} a_\ell f_\ell$, where $\sum_{\ell=1}^{L} |a_\ell| \leq 1$ and f_ℓ belongs to $\widetilde{\mathrm{Mon}}(Q,\mu)$.*

Recall that $Q = H^{-1}$ is a quantile function meaning that for $u \in [0, 1]$, $Q(u) = \inf\{t \geq 0 : H(t) \leq u\}$ (with the convention that $\inf\{\emptyset\} = \infty$), where H is a tail function, that is a function from $\mathbb{R}^+$ to $[0, 1]$ which is non-increasing, right-continuous and converges to zero at infinity.

For this class of functions, Dedecker and Merlevède (2017-a) have obtained the following Rosenthal-type inequality for the maximum of partial sums of α-dependent sequences for all powers $p \geq 2$.

Theorem 6.35 *Let $p \geq 2$ and let Q be a quantile function in $\mathbb{L}^p$. Let $\mathbf{Y} = (Y_i)_{i\in\mathbb{Z}}$ be a strictly stationary sequence of real-valued random variables. Denote by P_{Y_0} the distribution of Y_0.*

Let $X_i = f(Y_i) - \mathbb{E}(f(Y_i))$, where f belongs to $\tilde{\mathcal{F}}(Q, P_{Y_0})$ and let $S_n = \sum_{k=1}^n X_k$. Then, there exists a positive constant c only depending on p such that for any positive integer n,

$$\left\| \max_{1 \le k \le n} |S_k| \right\|_p^p \le cn^{p/2} \left(\sum_{k=0}^{n-1} \int_0^{\alpha_{1,\mathbf{Y}}(k)} Q^2(u)du \right)^{p/2} + cn \sum_{k=0}^{n-1} (k+1)^{p-2} \int_0^{\alpha_{2,\mathbf{Y}}(k)} Q^p(u)du.$$

Moreover, concerning the maximum of partial sums of β-dependent sequences for all powers $p > 2$, the following inequality is valid and has been obtained by Dedecker and Merlevède (2017-b).

Proposition 6.36 *Let $(Y_i)_{i \in \mathbb{Z}}$ be a strictly stationary sequence of real-valued random variables. For any $p > 2$ and any positive integer n, there exists a non-negative $\mathcal{F}_0$-measurable random variable $A_0(n, p)$ satisfying $\mathbb{E}(A_0(n, p)) \le \sum_{k=1}^n k^{p-2} \beta_{2,\mathbf{Y}}(k)$ and such that for any BV function h from $\mathbb{R}$ to $\mathbb{R}$, letting $X_i = h(Y_i) - \mathbb{E}(h(Y_i))$ and $S_n = \sum_{k=1}^n X_k$, we have*

$$\mathbb{E}\left(\max_{1 \le k \le n} |S_k|^p \right) \ll n^{p/2} \left(\sum_{i=0}^{n-1} |\mathrm{cov}(X_0, X_i)| \right)^{p/2} + n\|dh\|^{p-1} \mathbb{E}\left(|h(Y_0)| A_0(n, p) \right) + n\|dh\|^{p-1} \mathbb{E}\left(|h(Y_0)| \right) \sum_{k=0}^{n} (k+1)^{p-2} \beta_{2,\mathbf{Y}}(k), \tag{6.76}$$

where $\|dh\|$ is the total variation norm of the measure dh.

The proof of this proposition is also based on blocking technique, martingale approximation and Theorem 3.17 when S_k is a martingale.

Remark 6.37 *For $p = 2$, using Theorem 3.11, the inequality can be simplified as follows. There exists a non-negative $\mathcal{F}_0$-measurable random variable $A_0(n, 2)$ satisfying $\mathbb{E}(A_0(n, 2)) \le \sum_{k=1}^n \beta_{1,\mathbf{Y}}(k)$ and such that for any BV function h from $\mathbb{R}$ to $\mathbb{R}$, letting $X_i = h(Y_i) - \mathbb{E}(h(Y_i))$ and $S_n = \sum_{k=1}^n X_k$, we have*

$$\mathbb{E}\left(\max_{1 \le k \le n} S_k^2 \right) \ll n\|dh\| \mathbb{E}\left(|h(Y_0)| (1 + A_0(n, 2)) \right). \tag{6.77}$$

6.4 Invariance Principle for Strong Mixing Sequences

In this section, we investigate the functional central limit theorem question when we consider a strictly stationary sequence of centered random variables $(X_k)_{k \in \mathbb{Z}}$ in $\mathbb{L}^2$ which is additionally strong mixing in the sense that

$$\alpha_n = \alpha(\mathcal{F}_{-\infty}^0, \mathcal{F}_n^\infty) \to 0,$$

as $n \to \infty$, with $\mathcal{F}_i^j = \sigma(X_i, \dots, X_j)$ and where the coefficient α between two σ-algebras has been defined in Section 5.1.1. Note that in this case the process $(X_k)_{k\in\mathbb{Z}}$ is ergodic. Let $S_n = \sum_{k=1}^n X_k$ and $\sigma_n^2 = \mathrm{Var}(S_n)$.

As a corollary of our Theorem 4.31, we shall formulate the above result whose first part is originated by Denker (1986).

Theorem 6.38 *A strictly stationary, centered and strongly mixing sequence with $\sigma_n \to \infty$ satisfies the central limit theorem under the normalization σ_n (i.e. $S_n/\sigma_n \to^d N(0,1)$) if and only if $(\sigma_n^{-2}S_n^2)$ is an uniformly integrable family. Furthermore if $\max_{1\le k\le n} S_k^2/\sigma_n^2$ is uniformly integrable then $W_n = \{S_{[nt]}/\sigma_n, t \in [0,1]\} \to^d W$ in $D([0,1])$.*

Proof The fact that $(\sigma_n^{-2}S_n^2)_{n\ge1}$ needs to be a uniformly integrable family for the central limit theorem to hold comes directly from the moments convergence theorem (see Theorem 1.12). To prove the counterpart of Theorem 6.38, we shall apply Theorem 4.31 with $\mathcal{F}_k = \sigma(X_i, i \le k)$. Consider N a fixed integer less than n, A a positive real and define

$$U_{n,N} = Y'_{n,N} - \mathbb{E}(Y'_{n,N}) \text{ where } Y'_{n,N} = \frac{S_n - S_N}{\sigma_n} 1_{|S_n - S_N| \le A\sigma_n}$$

and

$$V_{n,N} = \sigma_n^{-1}(S_n - S_N) - U_{n,N}.$$

Using the uniform integrability of $(\sigma_n^{-2}S_n^2)$ and the fact that $\sigma_n \to \infty$, it is easy to see that

$$\lim_{A\to\infty} \limsup_{N\to\infty} \limsup_{n\to\infty} \|\sigma_n^{-1}S_N + V_{n,N}\|_1 = 0.$$

Therefore condition (4.89) will hold if one can prove that

$$\lim_{A\to\infty} \limsup_{N\to\infty} \limsup_{n\to\infty} \|\mathbb{E}(U_{n,N}|\mathcal{F}_0)\|_1 = 0. \tag{6.78}$$

Note that $U_{n,N}$ is centered, $\sigma(X_j, j \ge N+1)$-measurable and such that $\|U_{n,N}\|_\infty \le 2A$. Therefore by using (5.1),

$$\|\mathbb{E}(U_{n,N}|\mathcal{F}_0)\|_1 \le 8A\alpha_N,$$

which converges to zero as N tends to infinity. This ends the proof of (6.78) and then of the fact that condition (4.89) holds.

Let us show now that condition (4.90) is satisfied. Using once again the definition (5.1) of the strong mixing coefficients, we have

$$\|\mathbb{E}_0((Y'_{n,N})^2) - \mathbb{E}((Y'_{n,N})^2)\|_1 \le 8A^2\alpha_N,$$

which converges to zero as N tends to infinity. To show that condition (4.90) holds, it is then sufficient to prove that

$$\lim_{A\to\infty} \limsup_{N\to\infty} \limsup_{n\to\infty} \|\sigma_n^{-2}S_n^2 - (Y'_{n,N})^2\|_1 = 0.$$

This follows from the uniform integrability of $(\sigma_n^{-2}S_n^2)$ and the fact that $\sigma_n \to \infty$. This ends the proof of the fact that for a strictly stationary, centered and strongly mixing sequence with $\sigma_n \to \infty$ and such that $(\sigma_n^{-2}S_n^2)$ is an uniformly integrable family, the central limit theorem holds under the normalization σ_n.

To prove the second part of the theorem, we proceed as follows. If we assume that $\max_{1\le k\le n} S_k^2/\sigma_n^2$ is uniformly integrable then (1.25) holds, which implies that $W_n(t)$ is tight in $C([0,1])$. Since the sequence $W_n(t)$ is tight in $C([0,1])$, it is easy to show that it has asymptotically independent increments. This is so since, if $0 \le t_0 < t_1 < \ldots < t_k \le 1$,

$$\left|\mathbb{P}\left(\bigcap_{i=1}^{k}\{S_{[nt_i]} - S_{[nt_{i-1}]} \in A_i\}\right) - \prod_{i=1}^{k}\mathbb{P}(\{S_{[nt_i]} - S_{[nt_{i-1}]} \in A_i\})\right| \le k\alpha_{[nu_k]},$$

where $u_k = \min_{1\le j\le k}(t_j - t_{j-1})$. Now, the CLT implies that the finite-dimensional distributions are multivariate normal. ■

To complete Theorem 6.38, let us mention that a result of Ibragimov (1962) (see Theorem 18.1.1 in Ibragimov and Linnik (1971)) tells us that, if a strictly stationary strongly mixing sequence satisfies the central limit theorem with the normalization σ_n with $\sigma_n \to \infty$, then necessarily σ_n^2 has the representation $\sigma_n^2 = nh(n)$ where $h(n)$ is a slowly varying function in the strong sense, and $(\sigma_n^{-2}S_n^2)$ is a uniformly integrable family.

To formulate the next result we recall that for a real-valued random variable Z, the function $Q_{|Z|}$ defined on $[0,1]$ denotes the càdlàg inverse of the tail function $t \mapsto \mathbb{P}(|Z| > t)$. As an application of Theorem 4.18, we get the following weak invariance principle due to Doukhan, Massart and Rio (1994).

Theorem 6.39 *Let $(X_i)_{i\in\mathbb{Z}}$ be a strictly stationary and strongly mixing sequence of centered random variables. Assume that*

$$\sum_{k=0}^{\infty}\int_0^{\alpha_k} Q_{|X_0|}^2(u)du < \infty. \tag{6.79}$$

Then σ_n^2/n converges to $\sigma^2 = \sum_{k\in\mathbb{Z}} \mathbb{E}(X_0X_k)$ and if $\sigma^2 > 0$ then the invariance principle of Theorem 4.32 holds. Moreover $\{\max_{1\le k\le n} S_k^2/n\}_{n\ge1}$ is an uniformly integrable family.

Condition (6.79) is often referred to in the literature as the Doukhan–Massart–Rio condition. We refer to Theorem 2 in Doukhan, Massart and Rio (1994) and to Bradley (1997-a) for a discussion on the optimality of this condition. We only mention that if for some $0 < \delta \le \infty$, we have $\|X_0\|_{2+\delta} < \infty$ and $\sum_{k\ge1} k^{2/\delta}\alpha_k < \infty$, then condition (6.79) is satisfied.

Proof of Theorem 6.39 To prove the theorem, we shall apply Corollary 4.18 with $\mathcal{F}_k = \sigma(X_i, i \le k)$ (the notation $\mathbb{E}_0(\cdot) = \mathbb{E}(\cdot|\mathcal{F}_0)$ will be used throughout the proof).

Therefore it suffices to show that under (6.79) condition (4.39) holds. Note that obviously this latter condition is satisfied if

$$\sum_{k\geq 0} ||X_0\mathbb{E}_0(X_k)||_1 < \infty. \tag{6.80}$$

Let now $Y_0 = |X_0|\text{sign}(\mathbb{E}_0(X_k))$. We clearly have

$$||X_0\mathbb{E}_0(X_k)||_1 = \mathbb{E}(Y_0\mathbb{E}_0(X_k)) = \mathbb{E}(Y_0X_k) = \text{cov}(Y_0, X_k), \tag{6.81}$$

where for the last equality we use the fact that X_k is centered. Using Rio's covariance inequality (see Lemma 6.28) in (6.81) and taking into account that $Q_{Y_0}(u) \leq Q_{|X_0|}(u)$ for any $u \in [0, 1]$ and that $\alpha(\sigma(X_i, i \leq 0), \sigma(X_k)) \leq \alpha_k$, we get

$$||X_0\mathbb{E}_0(X_k)||_1 \leq 4\int_0^{\alpha_k} Q^2_{|X_0|}(u)du, \tag{6.82}$$

which shows that (6.80) holds if (6.79) does. ■

According to Theorem 6.38, if we do not impose the uniform integrability condition (4.88) then the central limit theorem cannot hold under the normalization σ_n. However if we relax condition (6.79) to

$$\lim_{n\to\infty} \frac{1}{\sigma_n^2}\sum_{i=1}^{n} i\int_0^{\alpha_i} Q^2_{|X_0|}(u)du = 0, \tag{6.83}$$

then the following result holds:

Theorem 6.40 *Let $(X_i)_{i\in\mathbb{Z}}$ be a strictly stationary and strongly mixing sequence of centered random variables in $\mathbb{L}^2$. Assume that $\sigma_n^2 \to \infty$ and that condition (6.83) holds. Then $B_n^{-1}W_n$ converges in distribution to W in $D([0, 1])$, where $B_n := \sqrt{(\pi/2)}\mathbb{E}|S_n|$ and W is a standard Brownian motion.*

To prove the theorem, we shall apply the following technical propositions, motivated by Theorem 18.4.1 in Ibragimov and Linnik (1971). The proofs are left to the reader but the details can be found in the proofs of Propositions 3 and 4 in Merlevède and Peligrad (2006). Below, we use the notation $\mathbb{E}_k(\cdot) = \mathbb{E}(\cdot|\mathcal{F}_k)$.

Proposition 6.41 *Let $(\mathcal{F}_i)_{i\in\mathbb{Z}}$ and $(X_i)_{i\in\mathbb{Z}}$ be as in Theorem 4.31. Assume that (4.87) holds. Assume in addition that there exists a sequence of positive integers $p_n \to \infty$ such that $p_n = o(n)$ and satisfying with $k_n = [n/p_n]$,*

(a) $\sqrt{k_n}\,\dfrac{||\mathbb{E}_{-p_n}(S_{p_n})||_1}{\sigma_{p_n}} \to 0$, *as* $n \to \infty$,

(b) *for any* $\varepsilon > 0$, $\sigma_{p_n}^{-2}\,\mathbb{E}\left(S_{p_n}^2\,\mathbf{1}\left(|S_{p_n}| \geq \varepsilon\sigma_{p_n}\sqrt{k_n}\right)\right) \to 0$, *as* $n \to \infty$,

(c) $||\,\mathbb{E}_0(\sigma_{p_n}^{-2}S_{p_n}^2) - 1||_1 \to 0$ *as* $n \to \infty$.

(d) $||\sigma_{p_n}^{-1}\mathbb{E}_0(S_{p_n})||_2 \to 0$ *as* $n \to \infty$.

Then $B_n^{-1}S_n$ converges in distribution to N where $B_n = \sqrt{k_n}\sigma_{p_n}$ or $\sqrt{(\pi/2)}\,\mathbb{E}|S_n|$ and N is a standard Gaussian random variable.

We can also derive the functional version of Proposition 6.41. With the notation of Proposition 6.41, the following result holds.

Proposition 6.42 *Assume that the conditions of Proposition 6.41 are satisfied and that in addition, for all positive ε,*

$$\lim_{n\to\infty} k_n\,\mathbb{P}\left(\max_{1\le i\le p_n} \frac{|\sum_{j=1}^{i} X_j|}{\sigma_{p_n}} \ge \varepsilon\sqrt{k_n}\right) = 0. \tag{6.84}$$

Then $B_n^{-1}W_n$ converges in distribution to W in $D([0, 1])$, where $B_n = \sqrt{k_n}\sigma_{p_n}$ or $\sqrt{(\pi/2)}\,\mathbb{E}|S_n|$ and W is a standard Brownian motion.

Proof of Theorem 6.40 In order to prove Theorem 6.40, we shall verify the conditions of Proposition 6.42. For the sake of clarity, we have divided the proof in several steps.

We first prove that condition (4.86) holds. Together with the assumption that $\sigma_n^2 \to \infty$ and Proposition 4.30, this will imply both (4.87) and Item (d) of Proposition 6.41. With this aim, notice that

$$\frac{\mathbb{E}\left(\mathbb{E}(S_n|\mathcal{F}_0)\right)^2}{\sigma_n^2} \le \frac{2}{\sigma_n^2}\sum_{i=1}^{n}\sum_{j=1}^{i} \mathbb{E}|X_j\,\mathbb{E}(X_i|\mathcal{F}_0)|.$$

According to the computations leading to the inequality (4.7) in Rio (2000), we infer that

$$\frac{\mathbb{E}\left(\mathbb{E}(S_n|\mathcal{F}_0)\right)^2}{\sigma_n^2} \le \frac{8}{\sigma_n^2}\sum_{i=1}^{n} i\int_0^{\alpha_i} Q_{|X_0|}^2(u)du, \tag{6.85}$$

which converges to zero provided that condition (6.83) holds.

We prove now that (4.90) holds (and then Item (c) of Proposition 6.41). By using stationarity, we first write

$$\begin{aligned}
\|\mathbb{E}_0(S_n^2) - \mathbb{E}(S_n^2)\|_1 &\le 2\sum_{i=1}^{n}\sum_{j=0}^{n-i} \|\mathbb{E}_0(X_iX_{i+j}) - \mathbb{E}(X_iX_{i+j})\|_1 \\
&\le 2\sum_{i=1}^{n}\sum_{j=0}^{n-i} \sup_{\ell\ge 0} \|\mathbb{E}_0(X_iX_{i+\ell}) - \mathbb{E}(X_iX_{i+\ell})\|_1 \wedge (2\|X_0\mathbb{E}_0(X_j)\|_1) \\
&\le 4\sum_{j=1}^{n} j\|X_0\mathbb{E}_0(X_j)\|_1 + 2\sum_{i=1}^{n} i \sup_{j\ge i} \|\mathbb{E}_0(X_iX_j) - \mathbb{E}(X_iX_j)\|_1.
\end{aligned} \tag{6.86}$$

Let now $Z_0 = \text{sign}(\mathbb{E}_0(X_iX_j) - \mathbb{E}(X_iX_j))$. We clearly have

$$\|\mathbb{E}_0(X_iX_j) - \mathbb{E}(X_iX_j)\|_1 = \text{cov}(Z_0, X_iX_j).$$

Using Lemma 6.28 and taking into account that $Q_{Z_0}(u) \le 1$ for any $u \in [0,1]$ and that for every $A \in (0,1)$: $\int_0^A Q_{|X_iX_j|}(u)du \le \int_0^A Q^2_{|X_0|}(u)du$, we derive that

$$\sup_{j\ge i} \|\mathbb{E}_0(X_iX_j) - \mathbb{E}(X_iX_j)\|_1 \le 4\int_0^{\alpha_i} Q^2_{|X_0|}(u)du. \tag{6.87}$$

Starting from (6.86) and taking into account (6.82) and (6.87), we get

$$\sigma_n^{-2}\|\mathbb{E}_0(S_n^2) - \mathbb{E}(S_n^2)\|_1 \le 24\,\sigma_n^{-2}\sum_{i=1}^n i\int_0^{\alpha_i} Q^2_{|X_0|}(u)du,$$

which shows that (4.90) holds if (6.83) does.

To prove now that Items (a) and (b) of Proposition 6.41 hold as well as (6.84), we need to select a sequence of integers p_n which converges to infinity such that $p_n = o(n)$ and satisfying the conditions mentioned above. With this aim, let us start with some useful remarks: Since $\sigma_n^2 \to \infty$, one has (with $Q = Q_{|X_0|}$) that

$$\lim_{u\to 0+} Q(u) > 0 \text{ (the limit may be } \infty). \tag{6.88}$$

Indeed, since $\sigma_n^2 \to \infty$, one has that $\mathbb{P}(|X_0| > 0) > 0$. (Otherwise, one would have $X_0 = 0$ a.s., and $S_n = 0$ a.s. for all $n \ge 1$, contradicting the fact that $\sigma_n^2 \to \infty$). For $u_1 = (1/2)\,\mathbb{P}(|X_0| > 0)$, one has that $u_1 < \mathbb{P}(|X_0| > 0)$ and hence $0 < Q(u_1)$. Now (6.88) follows from the fact that Q is a non-increasing function. Set now

$$T_n = Q(\alpha_n). \tag{6.89}$$

By combining (6.88) with the fact that $\alpha_n \to 0$, it follows that

$$\lim_{n\to\infty} T_n > 0 \text{ (the limit may be } \infty). \tag{6.90}$$

We are now in a position to give the selection of the sequence of integers p_n: Let p_n be the greatest non-negative integer such that

$$\frac{p_n^3\,T_{p_n}^2}{\sigma_{p_n}^2} \le n. \tag{6.91}$$

Using (6.90) and the fact that $\sigma_n^2 = nh(n)$, it is easy to see that $p_n \to \infty$, as $n \to \infty$ (and hence $p_n = 0$ for at most finitely many n), and also that $p_n = o(n)$. In addition the definition of p_n obviously implies

$$\frac{(p_n+1)^3\,T^2_{p_n+1}}{\sigma^2_{p_n+1}} > n. \tag{6.92}$$

With the above selection of p_n, let us prove that Item (a) of Proposition 6.41 holds. We have

$$\sqrt{k_n}\frac{\|\mathbb{E}_{-p_n}(S_{p_n})\|_1}{\sigma_{p_n}} \leq \sqrt{k_n}\frac{p_n}{\sigma_{p_n}}\|\mathbb{E}_{-p_n}(X_0)\|_1 \leq 4\sqrt{k_n}\frac{p_n}{\sigma_{p_n}}\int_0^{\alpha_{p_n}} Q_{|X_0|}(u)du,$$

where, for the last inequality, we have used similar arguments as the ones leading to (6.82). Using (6.92), it follows that

$$\sqrt{k_n}\frac{\|\mathbb{E}_{-p_n}(S_{p_n})\|_1}{\sigma_{p_n}} \ll \frac{p_n^2 Q(\alpha_{p_n})}{\sigma^2_{p_n}}\int_0^{\alpha_{p_n}} Q_{|X_0|}(u)du \ll \frac{p_n^2}{\sigma^2_{p_n}}\int_0^{\alpha_{p_n}} Q^2_{|X_0|}(u)du,$$

where, for the last inequality, we used that Q is non-increasing. Now since $(\alpha_n)_{n\geq 0}$ is a non-increasing sequence, we have that for any integer $i \in [1, p_n]$,

$$\int_0^{\alpha_{p_n}} Q^2_{|X_0|}(u)du \leq \int_0^{\alpha_i} Q^2_{|X_0|}(u)du. \tag{6.93}$$

So overall,

$$\sqrt{k_n}\frac{\|\mathbb{E}_{-p_n}(S_{p_n})\|_1}{\sigma_{p_n}} \ll \frac{p_n^2 Q(\alpha_{p_n})}{\sigma^2_{p_n}}\int_0^{\alpha_{p_n}} Q_{|X_0|}(u)du \ll \frac{1}{\sigma^2_{p_n}}\sum_{i=1}^{p_n} i\int_0^{\alpha_i} Q^2_{|X_0|}(u)du,$$

proving that Item (a) of Proposition 6.41 holds under (6.83).

To end the proof, we note that Item (b) of Proposition 6.41 and condition (6.84) are obviously both satisfied if the following condition holds: For any $\varepsilon > 0$,

$$\sigma_{p_n}^{-2}\mathbb{E}\left(\max_{1\leq i\leq p_n} S_i^2\,\mathbf{I}\left(\sigma_{p_n}^{-1}\max_{1\leq i\leq p_n}|S_i| \geq \varepsilon\sqrt{n/p_n}\right)\right) \to 0 \text{ as } n \to \infty. \tag{6.94}$$

We then end the proof by verifying condition (6.94). With this aim, we need to use a truncation argument. We truncate the variables X_i in the following way:

$$X_i' = \begin{cases} X_i 1\left(|X_i| \leq T_{p_n}\right) - \mathbb{E}X_i 1\left(|X_i| \leq T_{p_n}\right) & \text{if } X_i \text{ is an unbounded sequence} \\ X_i & \text{if ess.sup}|X_i| = T \text{ a.s.} \end{cases}$$

$$\text{and} \tag{6.95}$$

$$X_i'' = \begin{cases} X_i 1\left(|X_i| > T_{p_n}\right) - \mathbb{E}X_i 1\left(|X_i| > T_{p_n}\right) & \text{if } X_i \text{ is an unbounded sequence} \\ 0 & \text{if ess.sup}|X_i| = T \text{ a.s.} \end{cases}$$

Define $S'_n = \sum_{i=1}^n X'_i$ and $S''_n = \sum_{i=1}^n X''_i$. Now observe that for any $\varepsilon > 0$

$$
\begin{aligned}
&\sigma_{p_n}^{-2}\mathbb{E}\left(\max_{1\le i\le p_n} S_i^2 1\left(\sigma_{p_n}^{-1}\max_{1\le i\le p_n}|S_i| \ge \varepsilon(n/p_n)^{1/2}\right)\right) \\
&= \sigma_{p_n}^{-2}\mathbb{E}\left(\max_{1\le i\le p_n}(S'_i + S''_i)^2\,\mathbf{I}\left(\sigma_{p_n}^{-1}\max_{1\le i\le p_n}|S'_i + S''_i| \ge \varepsilon(n/p_n)^{1/2}\right)\right) \\
&\le 2\sigma_{p_n}^{-2}\mathbb{E}\left(\max_{1\le i\le p_n}(S'_i)^2 1\left(\sigma_{p_n}^{-1}\max_{1\le i\le p_n}|S'_i| \ge \frac{\varepsilon}{2}(n/p_n)^{1/2}\right)\right) \\
&\quad + 2\sigma_{p_n}^{-2}\,\mathbb{E}\left(\max_{1\le i\le p_n}(S''_i)^2\right).
\end{aligned}
\tag{6.96}
$$

We first treat the last term of the inequality. Using Theorem 3.11 with $X''_i, - X''_i$ and $\lambda = 0$, we get:

$$
\sigma_{p_n}^{-2}\mathbb{E}\left(\max_{1\le i\le p_n}(S''_i)^2\right) \le 8\sigma_{p_n}^{-2}\sum_{i=1}^{p_n}\mathbb{E}(X''_i)^2 + 16\sigma_{p_n}^{-2}\sum_{i=1}^{p_n}\mathbb{E}\left|X''_i\mathbb{E}(S''_{p_n} - S''_i|\mathcal{F}_i)\right|.
$$

Hence by stationarity

$$
\sigma_{p_n}^{-2}\mathbb{E}\left(\max_{1\le i\le p_n}(S''_i)^2\right) \le 16\,\sigma_{p_n}^{-2}p_n^2\mathbb{E}(X''_0)^2 \le 16\,\sigma_{p_n}^{-2}p_n^2\mathbb{E}\left(X_0^2 1(|X_0| > T_{p_n})\right).
$$

Observe that

$$
Q_{|X_0|1(|X_0|>T_{p_n})}(u) = Q_{|X_0|}(u)1(u \le \alpha_{p_n}).
\tag{6.97}
$$

It follows that

$$
\mathbb{E}\left(X_0^2 1(|X_0| > T_{p_n})\right) = \int_0^{\alpha_{p_n}} Q_{|X_0|}^2(u)du.
$$

So, overall, using (6.93), we get

$$
\sigma_{p_n}^{-2}\mathbb{E}\left(\max_{1\le i\le p_n}(S''_i)^2\right) \le 32\,\sigma_{p_n}^{-2}\sum_{i=1}^{p_n} i\int_0^{\alpha_i} Q_{|X_0|}^2(u)du.
$$

Hence if (6.83) holds

$$
\lim_{n\to\infty}\sigma_{p_n}^{-2}\mathbb{E}\left(\max_{1\le i\le p_n}(S''_i)^2\right) = 0.
\tag{6.98}
$$

Now we compute the first term in (6.96). We obviously have for all $\varepsilon > 0$,

$$\sigma_{p_n}^{-2}\mathbb{E}\left(\max_{1\le i\le p_n} S_i'^2 1\left(\max_{1\le i\le p_n}|S_i'| \ge \varepsilon\sigma_{p_n}(n/p_n)^{1/2}\right)\right) \le \frac{1}{\varepsilon^2}\frac{p_n\mathbb{E}(\max_{1\le i\le p_n} S_i'^4)}{n\sigma_{p_n}^4}.$$

Theorem 3.16 with $p = 4$ together with (6.91) ensure that

$$\frac{p_n}{n\sigma_{p_n}^4}\mathbb{E}\left(\max_{1\le i\le p_n} S_i'^4\right) \ll \frac{q_n^3 T_{p_n}^2}{n\sigma_{p_n}^4}\left\{\mathbb{E}(X_0^2) + \sup_{\ell\in[1,p_n]}\mathbb{E}\left(\mathbb{E}(S_\ell'|\mathcal{F}_0)\right)^2\right\}$$
$$\ll \frac{1}{\sigma_{p_n}^2}\left\{\mathbb{E}(X_0^2) + \sup_{\ell\in[1,p_n]}\mathbb{E}\left(\mathbb{E}(S_\ell'|\mathcal{F}_0)\right)^2\right\}. \qquad (6.99)$$

Now notice that

$$\sigma_{p_n}^{-2}\sup_{\ell\in[1,p_n]}\mathbb{E}\left(\mathbb{E}(S_\ell'|\mathcal{F}_0)\right)^2 \le 2\sigma_{p_n}^{-2}\sup_{\ell\in[1,p_n]}\mathbb{E}\left(\mathbb{E}(S_\ell|\mathcal{F}_0)\right)^2 + 2\sigma_{p_n}^{-2}\sup_{\ell\in[1,p_n]}\mathbb{E}(S_\ell'')^2. \qquad (6.100)$$

The last term of the above inequality has been shown to tend to zero (see (6.98)). The first term in the inequality also converges to zero with the same arguments as those leading to the fact that (6.83) implies (4.86). Then we have shown that for all $\varepsilon > 0$,

$$\lim_{n\to\infty}\sigma_{p_n}^{-2}\mathbb{E}\left(\max_{1\le i\le p_n} S_i'^2 1\left(\max_{1\le i\le p_n}|S_i'| \ge \varepsilon\sigma_{p_n}(n/p_n)^{1/2}\right)\right) = 0, \qquad (6.101)$$

which combined with (6.96) and (6.98) entails that (6.94) holds. This ends the proof of Theorem 6.40. ■

6.5 Gaussian Approximation under Lindeberg's Condition

Let $\{X_{k,n} : k = 1, ..., n\}$ be a triangular array of square integrable zero mean real-valued random variables. For a fixed n, we define

$$S_n = \sum_{i=1}^{n} X_{i,n},\ S_{a,b} = \sum_{i=a+1}^{b} X_{i,n} \text{ and } S_A = \sum_{i\in A} X_{i,n}.$$

The aim of this section is to give sufficient conditions on the triangular array $\{X_{k,n} : k = 1, ..., n\}$ for the CLT to hold, i.e. $S_n \to N(0,\sigma^2)$, as $n \to \infty$. We shall start with a very general lemma and then give an application to ρ-mixing triangular arrays.

6.5.1 A General Result

Following the dyadic coupling version of the extremal approach as suggested in Utev (1990), which essentially consists of the ingredients listed below, we shall first give a general result on the asymptotic behavior of S_n in terms of pseudo metrics on the class of real-valued random variables. The parts **I** and **II** below give the list of the main ingredients for this approach to apply.

I: Pseudo metric Let $d(X,Y)$ be a pseudo metric on the class of real-valued random variables meaning that the two properties below are satisfied:

(1) $d(X, Y) = d(X', Y')$ if $X =^d X'$ and $Y =^d Y'$ (which means that the pseudo metric is actually defined on distributions).

(2) The distance properties: $d(X, Y) \geq 0$, $d(X, X) = 0$ and $d(X, Y) \leq d(X, Z) + d(Z, Y)$. We assume also that there exists a positive constant c_1 such that

$$|\mathrm{Var}(X) - \mathrm{Var}(Y)| \leq c_1 d(X, Y), \tag{6.102}$$

and a positive constant c_2 such that if $\mathbb{E}[X^k] = \mathbb{E}[Y^k]$ for $k = 1, 2$, then

$$d(X, Y) \leq c_2\{\mathbb{E}|X|^3 + \mathbb{E}|Y|^3\}. \tag{6.103}$$

(3) (*Convolution bound*) For two sequences $(X_i)_{i\geq 1}$ and $(Y_i)_{i\geq 1}$ of independent real-valued random variables,

$$d\left(\sum_{i=1}^{n} X_i, \sum_{i=1}^{n} Y_i\right) \leq \sum_{i=1}^{n} d(X_i, Y_i). \tag{6.104}$$

(4) If $N_1 \sim \mathcal{N}(0, a^2)$ and $N_2 \sim \mathcal{N}(0, b^2)$, there exists a positive constant c_3 such that

$$d(N_1, N_2) \leq c_3|a^2 - b^2|. \tag{6.105}$$

Comment 6.43

(i) *The word "pseudo" in the definition allows for the possibility to have $d(X, Y) = 0$ when $X \neq Y$ in the distribution sense.*

(ii) *Note that $d(X, Y) = |Var(X) - Var(Y)|$ on the subclass of variables with zero means is a pseudo metric satisfying all the above properties.*

(iii) *Note that (6.104) implies in particular that*

$$d(X + Y,\ X + Y') \leq d(Y, Y'), \tag{6.106}$$

where X, Y and Y' are independent. Now this last property implies the convolution bound by assuming, without loss of generality, that all variables in question are *mutually independent and applying the Lindeberg trick. More exactly, let $T_k = Y_k + \dots + Y_n$, $S_k = X_1 + \dots + X_k$. Then*

$$d\left(\sum_{i=1}^{n} X_i, \sum_{i=1}^{n} Y_i\right) = d(S_n, T_1) \le \sum_{j=n}^{1} d(S_{j-1} + T_{j+1} + X_j, S_{j-1} + T_{j+1} + Y_j)$$

and hence

$$\le \sum_{i=n}^{1} d(X_i, Y_i) = \sum_{i=1}^{n} d(X_i, Y_i).$$

II: Coupling bounds for the pseudo metric $d(X, Y)$

Let n_0 be a positive integer. We assume that for any subinterval $\Delta = \{r+1, \dots, r+k\}$, where $n_0 \le k \le n$ and $0 \le r \le n-k$, there exists a splitting point $a(k)$ such that, setting $\Delta_1 = \{r+1, \dots, r+a(k)\}$ and $\Delta_2 = \{r+a(k)+1, \dots, r+k\}$,

$$d(S_\Delta, S_{\Delta_1} + S'_{\Delta_2}) \le \delta_k D_{r,r+k}, \tag{6.107}$$

where S'_{Δ_2} is independent of S_{Δ_1} and distributed as S_{Δ_2}, and δ_k, $(D_{\ell,\ell+k})$ and $a(k)$ have the following properties:

(i) $(\delta_j)_{j\ge 1}$ is a non-increasing, non-negative sequence with

$$\sum_{j=0}^{\infty} \delta_{2^j} < \infty. \tag{6.108}$$

(ii) $(D_{r,t})_{0<r<t}$ is a superadditive, non-decreasing, bounded and non-negative sequence, in the sense that, for any $1 \le r < t < u < \infty$,

$$0 \le D_{r,t} + D_{t,u} \le D_{r,u} \le D_\infty < \infty. \tag{6.109}$$

(iii) The splitting point $a := a(k)$ makes the intervals Δ_1 and Δ_2 with almost equal cardinal, in the sense that there exist reals $c > 0$ and $0 < q < 1$ such that

$$\frac{k}{2}\left(1 - \frac{c}{k^q}\right) \le a(k) \le \frac{k}{2}\left(1 + \frac{c}{k^q}\right). \tag{6.110}$$

The following dynamical bound lemma will be useful (the proof will be given at the end of this section):

Lemma 6.44 *Let $c > 0$ and $0 < q < 1$. Define two functions on positive reals:*

$$h(t) = \frac{t}{2}\left(1 - \frac{c}{t^q}\right) \quad \text{and} \quad H(t) = \frac{t}{2}\left(1 + \frac{c}{t^q}\right).$$

Let f^k be the k-th iteration of a function f. Then, for any positive integer k,

$$\frac{t}{2^k} \geq h^k(t) \geq \frac{t}{2^k}\left(1 - \sum_{\ell=0}^{k-1} \frac{c}{(t/2^\ell)^q}\right) \quad \text{when } t/2^k \geq t_0, \tag{6.111}$$

$$\frac{x}{2^k} \leq H^k(x) \leq \frac{x}{2^k} \prod_{\ell=0}^{k-1}\left(1 + \frac{c}{(x/2^\ell)^q}\right) \quad \text{when } x > 0, \tag{6.112}$$

where t_0 is defined as a turning point from which h becomes increasing, i.e. $0 = h'(t_0) = (1/2)(1 - c(1-q)t_0^{-q})$.

We are now in a position to give the following very general CLT result.

Lemma 6.45 *Under assumptions (6.102)–(6.110), and if*

$$\mathrm{Var}(S_n) = 1 \text{ for all } n \geq 1 \text{ and } \sum_{i=1}^{n} \mathbb{E}|X_{i,n}|^3 = L_n \to 0, \text{ as } n \to \infty, \tag{6.113}$$

the following convergence holds:

$$d(S_n, N(0,1)) \to 0, \text{ as } n \to \infty. \tag{6.114}$$

Note that the second part of condition (6.113) is usually referred to in the literature as the Lyapunov condition.

Proof of Lemma 6.45 The proof is based on induction. Let $\varkappa = (2^q - 1)/(2c)$ and n be large enough (at least greater than $4\max\left(2n_0, 2t_0, \varkappa^{-1/q}\right)$ where we recall that t_0 is the turning point from which h becomes increasing). Let r be the unique non-negative integer such that $2^r \leq n < 2^{r+1}$. Let

$$K_r = \min\left(\sup\{\ell \in \mathbb{N} \,:\, 2^{r-\ell} \geq 2\max(t_0, n_0)\}, \sup\{\ell \in \mathbb{N} \,:\, 2^{\ell q} \leq \varkappa 2^{rq}\}\right).$$

Note that $K_r \geq 1$ since n is large enough. Let us also make the following remark that will be useful in the following: For any positive integer $j \leq K_r$,

$$h^j(2^r) \geq \max(n_0, t_0). \tag{6.115}$$

Indeed, for any positive integer $j \leq K_r$, we have $2^{r-j} \geq t_0$ and $2^{jq} \leq \varkappa 2^{rq}$. Therefore, by (6.111),

$$\begin{aligned} h^j(2^r) &\geq \frac{2^r}{2^j}\left(1 - \sum_{\ell=0}^{j-1} \frac{c}{(2^r/2^\ell)^q}\right) = \frac{2^r}{2^j}\left(1 - \frac{c}{2^{rq}}\frac{2^{jq}-1}{2^q-1}\right) \geq \frac{2^r}{2^j}\left(1 - \frac{1}{2^{rq}}\frac{2^{jq}}{2\varkappa}\right) \\ &\geq \frac{1}{2} \times \frac{2^r}{2^j}. \end{aligned} \tag{6.116}$$

Since $2^j \leq 2^{K_r} \leq 2^{r-1}/\max(n_0, t_0)$, (6.116) entails (6.115).

We shall now prove by induction that we can construct a sequence of sets $\{P_k\}_{1\leq k\leq K_r}$ such that, for any $k \in \{1, ...K_r\}$, the $P_k := \{r_{k,0}, r_{k,1}, \dots, r_{k,2^k}\}$ are subsets of $\{0, ..., n\}$ with $2^k + 1$ elements such that $0 = r_{k,0} < r_{k,1} < \dots < r_{k,2^k} = n$, which are non-decreasing in the sense that $P_{k-1} \subseteq P_k$ and for which we have several properties. To state them, we set

$$\Delta_{0,1} = \{1, \dots, n\}$$

and, for any positive integer $k \leq K_r$ and any $i = 1, ..., 2^k$,

$$c_{k,i} = r_{k,i} - r_{k,i-1} \text{ and } \Delta_{k,i} = \{r_{k,i-1} + 1, \dots, r_{k,i}\}.$$

With the notation above, note that $\Delta_{0,1} = \bigcup_{i=1}^{2^k} \Delta_{k,i}$. Next, for any positive integer $k \leq K_r$, the P_k's must satisfy the following conditions:

$$h^k(2^r) \leq \min_{1\leq i\leq 2^k} c_{k,i} \leq \max_{1\leq i\leq 2^k} c_{k,i} \leq H^k(2^{r+1}) \tag{6.117}$$

and

$$d\left(\sum_{i=1}^{2^{k-1}} Y_{k-1,i}, \sum_{i=1}^{2^k} Y_{k,i}\right) \leq \delta_{2^{r-k}} D_\infty, \tag{6.118}$$

where for any fixed non-negative integer ℓ, $\{Y_{\ell,i}\}_{i=1,\dots,2^\ell}$ is a family of mutually independent real-valued random variables such that, for any $i = 1, ..., 2^\ell$, $Y_{\ell,i} =^d S_{\Delta_{\ell,i}}$.

Let us prove by induction that such a construction is possible. With this aim, we start by noticing that for $k = 1$, we can select $r_{1,1} = a(n)$ where $a(n)$ is the splitting point such that (6.107) and (6.110) are satisfied. Indeed, by (6.110), we have

$$h(n) \leq r_{1,1} = c_{1,1} \leq H(n) \text{ and } h(n) \leq n - r_{1,1} = r_{1,2} - r_{1,1} = c_{1,2} \leq H(n).$$

Since $n \geq 2t_0$ (which is the turning point from which h becomes increasing) and $2^r \leq n < 2^{r+1}$, we have $h(n) \geq h(2^r)$ and $H(n) \leq H(2^{r+1})$. This proves that, for $k = 1$, (6.117) is satisfied when $r_{1,1} = a(n)$. In addition, since $n \geq n_0$, by property (6.107) and by denoting by $S'_{\Delta_{1,2}}$ a real-valued random variable distributed as $S_{\Delta_{1,2}}$ and independent of $S_{\Delta_{1,1}}$, we have, for $r_{1,1} := a(n)$,

$$d\left(Y_{0,1}, Y_{1,1} + Y_{1,2}\right) = d\left(S_{\Delta_{0,1}}, S_{\Delta_{1,1}} + S'_{\Delta_{1,2}}\right) \leq \delta_n D_{0,n} \leq \delta_{2^{r-1}} D_\infty,$$

proving that (6.118) holds when $r_{1,1} := a(n)$.

Now we prove that, if we assume that we can construct, for any positive integer $k \leq j < K_r$, subsets $P_k = \{r_{k,0}, r_{k,1}, \dots, r_{k,2^k}\}$ of $\{0, ..., n\}$ such that $0 = r_{k,0} < r_{k,1} < \dots < r_{k,2^k} = n$, which are non-decreasing and satisfy the properties (6.117) and (6.118), then we can construct a subset $P_{j+1} = \{r_{j+1,0}, r_{j+1,1}, \dots, r_{j+1,2^{j+1}}\}$ of $\{0, ..., n\}$ such that

$0 = r_{j+1,0} < r_{j+1,1} < \ldots < r_{j+1,2^{j+1}} = n$, $P_j \subseteq P_{j+1}$ and which satisfies the properties (6.117) and (6.118) with $k = j + 1$. To do it, for each $i = 1, \ldots, 2^j$, we split each set $\Delta_{j,i}$ into two disjoint subsets $\Delta_{j+1,2i-1}$ and $\Delta_{j+1,2i}$ such that $\Delta_{j,i} = \Delta_{j+1,2i-1} \cup \Delta_{j+1,2i}$ according to (6.107). To be more precise,

$$\Delta_{j+1,2i-1} = \{r_{j,i-1} + 1, \ldots, r_{j,i-1} + a(c_{j,i})\}$$
$$\text{and } \Delta_{j+1,2i} = \{r_{j,i-1} + a(c_{j,i}) + 1, \ldots, r_{j,i}\},$$

where $a(c_{j,i})$ is such that

$$h(c_{j,i}) \le a(c_{j,i}) \le H(c_{j,i}),$$

and

$$d(S_{\Delta_{j,i}}, S_{\Delta_{j+1,2i-1}} + S'_{\Delta_{j+1,2i}}) \le \delta_{c_{j,i}} D_{r_{j,i-1}, r_{j,i}}, \tag{6.119}$$

where $S'_{\Delta_{j+1,2i}}$ is independent of $S_{\Delta_{j+1,2i-1}}$ and distributed as $S_{\Delta_{j+1,2i}}$. Next, for any $i = 1, \ldots, 2^j$,

$$r_{j+1,2i-1} = r_{j,i-1} + a(c_{j,i}) \text{ and } r_{j+1,2i} = r_{j,i},$$

and we notice that for any $\ell = 1, \ldots, 2^{j+1}$, $\Delta_{j+1,\ell} = \{r_{j+1,\ell-1} + 1, \ldots, r_{j+1,\ell}\}$ with $0 = r_{j+1,0} < r_{j+1,1} < \ldots < r_{j+1,2^{j+1}} = n$. In addition, we have the following inclusion: $P_{j+1} = \{r_{j+1,0}, r_{j+1,1}, \ldots, r_{j+1,2^{j+1}}\} \subseteq P_j$. On the other hand, for any $i = 1, \ldots, 2^j$, note that $|\Delta_{j+1,2i-1}| = a(c_{j,i})$ and $|\Delta_{j+1,2i}| = c_{j,i} - a(c_{j,i})$. Moreover, for any $i = 1, \ldots, 2^{j+1}$, $|\Delta_{j+1,i}| = c_{j+1,i}$. Therefore,

$$\min_{1\le i\le 2^{j+1}} c_{j+1,i} = \min_{1\le i\le 2^j} \min(a(c_{j,i}), c_{j,i} - a(c_{j,i}))$$
$$\text{and } \max_{1\le i\le 2^{j+1}} c_{j+1,i} = \max_{1\le i\le 2^j} \max(a(c_{j,i}), c_{j,i} - a(c_{j,i})).$$

But, according to (6.110),

$$h(c_{j,i}) \le \min(a(c_{j,i}), c_{j,i} - a(c_{j,i})) \le \max(a(c_{j,i}), c_{j,i} - a(c_{j,i})) \le H(c_{j,i}).$$

Hence,

$$\min_{1\le i\le 2^j} h(c_{j,i}) \le \min_{1\le i\le 2^{j+1}} c_{j+1,i} \le \max_{1\le i\le 2^{j+1}} c_{j+1,i} \le \max_{1\le i\le 2^j} H(c_{j,i}).$$

Provided that $\min_{1\le i\le 2^j} c_{j,i} \ge t_0$, it follows that

$$h(\min_{1\le i\le 2^j} c_{j,i}) \le \min_{1\le i\le 2^{j+1}} c_{j+1,i} \le \max_{1\le i\le 2^{j+1}} c_{j+1,i} \le H(\max_{1\le i\le 2^j} c_{j,i}).$$

Taking into account (6.117) and since $h^j(2^r) \geq t_0$ for $j \leq K_r$, we derive

$$h^{j+1}(2^r) \leq \min_{1\leq i\leq 2^{j+1}} c_{j+1,i} \leq \max_{1\leq i\leq 2^{j+1}} c_{j+1,i} \leq H^{j+1}(2^r),$$

showing that the subset P_{j+1} satisfies (6.117) with $k = j + 1$.

Let us prove now that (6.118) holds with $k = j + 1$ assuming that it holds for any positive integer $k \leq j < K_r$. Recall that for any positive integer ℓ, $\{Y_{\ell,i}\}_{i=1,\dots,2^\ell}$ denotes a family of mutually independent real-valued random variables such that, for any $i = 1, \dots, 2^\ell$, $Y_{\ell,i} =^d S_{\Delta_{\ell,i}}$. Therefore, (6.119) implies that

$$d(Y_{j,i}, Y_{j+1,2i-1} + Y_{j+1,2i}) \leq \delta_{c_{j,i}} D_{r_{j,i-1}, r_{j,i}}.$$

Applying the convolution bound (6.104), we then derive

$$\begin{aligned} d\left(\sum_{i=1}^{2^j} Y_{j,i}, \sum_{i=1}^{2^{j+1}} Y_{j+1,i}\right) &= d\left(\sum_{i=1}^{2^j} Y_{j,i}, \sum_{i=1}^{2^j}(Y_{j+1,2i-1} + Y_{j+1,2i})\right) \\ &\leq \sum_{i=1}^{2^j} d(Y_{j,i}, Y_{j+1,2i-1} + Y_{j+1,2i}) \\ &\leq \sum_{i=1}^{2^j} \delta_{c_{j,i}} D_{r_{j,i-1}+1, r_{j,i}} \leq \left(\max_{1\leq i\leq 2^j} \delta_{c_{j,i}}\right) \sum_{i=1}^{2^j} D_{r_{j,i-1}, r_{j,i}}. \end{aligned}$$

Therefore, by the superadditivity property (6.109),

$$d\left(\sum_{i=1}^{2^j} Y_{j,i}, \sum_{i=1}^{2^{j+1}} Y_{j+1,i}\right) \leq \left(\max_{1\leq i\leq 2^j} \delta_{c_{j,i}}\right) D_{r_{j,0}, r_{j,2^j}} \leq \left(\max_{1\leq i\leq 2^j} \delta_{c_{j,i}}\right) D_\infty.$$

But $(\delta_\ell)_{\ell\geq 1}$ is a non-increasing sequence and by the induction hypothesis (6.117), $h^j(2^r) \leq \min_{1\leq i\leq 2^j} c_{j,i}$. Hence,

$$d\left(\sum_{i=1}^{2^j} Y_{j,i}, \sum_{i=1}^{2^{j+1}} Y_{j+1,i}\right) \leq \delta_{h^j(2^r)} D_\infty.$$

Taking into account (6.116) and again the fact that $(\delta_\ell)_{\ell\geq 1}$ is a non-increasing sequence, we get

$$d\left(\sum_{i=1}^{2^j} Y_{j,i}, \sum_{i=1}^{2^{j+1}} Y_{j+1,i}\right) \leq \delta_{2^{r-j-1}} D_\infty,$$

which proves that (6.118) holds with $k = j + 1$. This ends the proof of the fact that, for any positive integer $k \leq K_r$, we can construct subsets $P_k = \{r_{k,0}, r_{k,1}, \dots, r_{k,2^k}\}$

of $\{0, ..., n\}$ such that $0 = r_{k,0} < r_{k,1} < ... < r_{k,2^k} = n$, which are non-decreasing and satisfy the properties (6.117) and (6.118).

Now fix an integer m such that $0 \le r - m \le K_r$. Observe first that, by the triangle inequality for the pseudo metric,

$$\begin{aligned} d(S_n, N(0,1)) &= d(S_{\Delta_{0,1}}, N(0,1)) \\ &\le d\left(S_{\Delta_{0,1}}, \sum_{i=1}^{2} Y_{1,i}\right) + d\left(\sum_{i=1}^{2} Y_{1,i}, N(0,1)\right) = d\left(Y_{0,1}, \sum_{i=1}^{2} Y_{1,i}\right) \\ &\quad + d\left(\sum_{i=1}^{2} Y_{1,i}, N(0,1)\right). \end{aligned}$$

Iterating this procedure, we get that

$$d(S_n, N(0,1)) \le \sum_{k=0}^{r-m-1} d\left(\sum_{i=1}^{2^k} Y_{k,i}, \sum_{j=1}^{2^{k+1}} Y_{k+1,j}\right) + d\left(\sum_{i=1}^{2^{r-m}} Y_{r-m,i}, N(0,1)\right).$$

Using (6.118), it follows that

$$\begin{aligned} d(S_n, N(0,1)) &\le D_\infty \sum_{k=0}^{r-m-1} \delta_{2^{r-k-1}} + d\left(\sum_{i=1}^{2^{r-m}} Y_{r-m,i}, N(0,1)\right) \\ &\le D_\infty \sum_{k \ge m} \delta_{2^k} + d\left(\sum_{i=1}^{2^{r-m}} Y_{r-m,i}, N(0,1)\right). \end{aligned} \tag{6.120}$$

Now, for any fixed m, by construction, the random variables $(Y_{r-m,i})_{i=1,...,2^{r-m}}$ are mutually independent and such that $Y_{r-m,i} =^d S_{\Delta_{r-m,i}}$. Note that $S_{\Delta_{r-m,i}} = \sum_{\ell=r_{r-m,i-1}+1}^{r_{r-m,i}} X_{\ell,n}$. In addition, by (6.117),

$$c_{r-m,i} = r_{r-m,i} - r_{r-m,i-1} \le H^{r-m}(2^{r+1}).$$

But, by (6.112) of Lemma 6.44,

$$\begin{aligned} H^{r-m}(2^{r+1}) &\le \frac{2^{r+1}}{2^{r-m}} \exp\left(c\,2^{-q(r+1)} \sum_{k=0}^{r-m-1} 2^{kq}\right) \\ &\le 2^{m+1} \exp\left(c2^{-q(m+1)}(2^q - 1)^{-1}\right) := t_m. \end{aligned}$$

Hence,

$$L_{n,m} := \sum_{i=1}^{2^{r-m}} \mathbb{E}|Y_{r-m,i}|^3 \le t_m^2 \sum_{i=1}^{2^{r-m}} \sum_{\ell=r_{r-m,i-1}+1}^{r_{r-m,i}} \mathbb{E}|X_{\ell,n}|^3 \le t_m^2 \sum_{i=1}^{n} \mathbb{E}|X_{i,n}|^3.$$

Hence, by the second part of condition (6.113),

$$L_{n,m} \to 0, \text{ as } n \to \infty. \tag{6.121}$$

Let now $\{Z_{r-m,i}\}_{i=1,\ldots,2^{r-m}}$ be a triangular array of independent centered Gaussian variables with the same variances as $Y_{r-m,i}$. Since

$$\mathbb{E}|N(0, \mathrm{Var}(Y))|^3 \le 3|\mathrm{Var}(Y)|^{3/2} \le 3\mathbb{E}|Y|^3,$$

for any random variable Y, taking into account the convolution bound (6.104) and the upper bound (6.103), we derive that

$$\begin{aligned} d\left(\sum_{i=1}^{2^{r-m}} Y_{r-m,i}, \sum_{i=1}^{2^{r-m}} Z_{r-m,i}\right) &\le c_2 \sum_{i=1}^{2^{r-m}} \left(\mathbb{E}|Y_{r-m,i}|^3 + \mathbb{E}|Z_{r-m,i}|^3\right) \\ &\le 4c_2 \sum_{i=1}^{2^{r-m}} \mathbb{E}|Y_{r-m,i}|^3 = 4c_2 L_{n,m}. \end{aligned}$$

Hence, taking into account (6.121), we get

$$d\left(\sum_{i=1}^{2^{r-m}} Y_{r-m,i}, \sum_{i=1}^{2^{r-m}} Z_{r-m,i}\right) \to 0, \text{ as } n \to \infty. \tag{6.122}$$

Starting from (6.120) and taking into account (6.122), we get

$$\limsup_{n\to\infty} d(S_n, N(0,1)) \le D_\infty \sum_{k\ge m} \delta_{2^k} + \limsup_{n\to\infty} d\left(\sum_{i=1}^{2^{r-m}} Z_{r-m,i}, N(0,1)\right). \tag{6.123}$$

Proceeding as we did to get (6.120), note now that

$$d\left(S_n, \sum_{i=1}^{2^{r-m}} Y_{r-m,i}\right) \le \sum_{k=0}^{r-m-1} d\left(\sum_{i=1}^{2^k} Y_{k,i}, \sum_{i=1}^{2^{k+1}} Y_{k+1,i}\right) \le D_\infty \sum_{k\ge m} \delta_{2^k}.$$

Therefore, since $\mathrm{Var}\left(\sum_{i=1}^{2^{r-m}} Y_{r-m,i}\right) = \mathrm{Var}\left(\sum_{i=1}^{2^{r-m}} Z_{r-m,i}\right)$, taking into account the fact that $\sum_{j\ge 0} \delta_{2^j} < \infty$ together with the lower bound (6.102), we get

$$\lim_{m\to\infty} \limsup_{n\to\infty} \left|\mathrm{Var}(S_n) - \mathrm{Var}\left(\sum_{i=1}^{2^{r-m}} Z_{r-m,i}\right)\right| \to 0. \tag{6.124}$$

But since $\mathrm{Var}(S_n) = 1$ and $\sum_{i=1}^{2^{r-m}} Z_{r-m,i} \sim \mathcal{N}\left(0, \mathrm{Var}\left(\sum_{i=1}^{2^{r-m}} Z_{r-m,i}\right)\right)$, property (6.105) implies that

$$d\left(\sum_{i=1}^{2^{r-m}} Z_{r-m,i}, N(0,1)\right) \leq c_3 \left|\mathrm{Var}(S_n) - \mathrm{Var}\left(\sum_{i=1}^{2^{r-m}} Z_{r-m,i}\right)\right|,$$

which combined with (6.124) entails that

$$\lim_{m\to\infty} \limsup_{n\to\infty} d\left(\sum_{i=1}^{2^{r-m}} Z_{r-m,i}, N(0,1)\right) = 0. \tag{6.125}$$

Starting from (6.123) and taking into account (6.125) together with the fact that $\sum_{j\geq 0} \delta_{2^j} < \infty$, the convergence (6.114) is proved. ■

To finish this section, it remains to prove Lemma 6.44.

Proof of Lemma 6.44 Obviously H is non-decreasing on $\mathbb{R}_+^*$. For any $t > 0$, note that

$$\frac{t}{2} \leq H(t) = \frac{t}{2}(1 + c/t^q),$$

which proves (6.112) for $k = 1$ and implies also that for any integer $j \geq 1$,

$$\frac{H^{j-1}(t)}{2} \leq H(H^{j-1}(t)) = \frac{H^{j-1}(t)}{2}\left(1 + c/(H^{j-1}(t))^q\right).$$

Let j be a positive integer and assume that (6.112) holds for any $1 \leq k \leq j-1$. Then

$$\frac{t}{2^j} = \frac{t/2^{j-1}}{2} \leq H(H^{j-1}(t)) = H^j(t).$$

To prove the upper bound of (6.112) when $k = j$, we note that, by the induction hypothesis,

$$\begin{aligned} H^j(t) = H(H^{j-1}(t)) &\leq \frac{H^{j-1}(t)}{2}\left(1 + c/(H^{j-1}(t))^q\right) \\ &\leq \frac{1}{2} \times \frac{t}{2^{j-1}} \prod_{\ell=0}^{j-2}\left(1 + \frac{c}{(t/2^\ell)^q}\right)\left(1 + c/(H^{j-1}(t))^q\right). \end{aligned}$$

Hence using the fact that by the induction hypothesis, $H^{j-1}(t) \geq t/2^{j-1}$, we get

$$H^j(t) \leq \frac{t}{2^j} \prod_{\ell=0}^{j-2}\left(1 + \frac{c}{(t/2^\ell)^q}\right)\left(1 + c/(t/2^{j-1})^q\right) = \frac{t}{2^j} \prod_{\ell=0}^{j-1}\left(1 + \frac{c}{(t/2^\ell)^q}\right),$$

which proves (6.112) when $k = j$. The bound (6.112) is then proved by induction.

Now, let us deal with the function h. The function is non-decreasing for $1 - c(1-q)x^{-q} \geq 0$. Here the upper bound $h^j(t) \leq t/2^j$ for any positive integer j is immediate. To prove the lower bound we write

$$\frac{t}{2} \geq h(t) = \frac{t}{2}(1 - c/t^q),$$

which proves (6.111) for $k = 1$ and implies also that for any integer $j \geq 1$,

$$h(h^{j-1}(t)) = \frac{h^{j-1}(t)}{2} - \frac{c}{2}(h^{j-1}(t))^{1-q}.$$

Let j be a positive integer and assume that (6.111) holds for any $1 \leq k \leq j-1$. Then, by monotonicity, we have, for $t/2^j \geq t_0$,

$$\begin{aligned} h^j(t) = h(h^{j-1}(t)) &\geq \frac{t}{2^j}\left(1 - \sum_{k=0}^{j-2} \frac{c}{(t/2^k)^q}\right) - \frac{c}{2}\frac{1}{(t/2^{j-1})^{q-1}} \\ &= \frac{t}{2^j}\left(1 - \sum_{k=0}^{j-1} \frac{c}{(t/2^k)^q}\right). \end{aligned}$$

The proof is complete. ■

6.5.2 *Application: CLT for ρ-mixing Triangular Arrays under Lindeberg's Condition*

In this section we consider a triangular array $\{X_{k,n} : k = 1, ..., n\}$ of square integrable zero mean real-valued random variables which is ρ-mixing in the sense that its sequence of uniform ρ-mixing coefficients tends to 0, more exactly:

$$\rho(k) = \sup_{n \geq 1} \max_{1 \leq j \leq n-k} \rho\left(\sigma(X_{t,n}, 1 \leq t \leq j), \sigma(X_{t,n}, j+k \leq t \leq n)\right) \to 0, \text{ as } k \to \infty,$$

where $\sigma(X_t, t \in A)$ is the σ-field generated by the r.v.'s X_t with indices in A, and we recall that the maximal correlation coefficient $\rho(\mathcal{U}, \mathcal{V})$ between two σ-algebras has been defined in Section 5.1.1.

For such a triangular array, the following lemma is valid:

Lemma 6.46 *Consider a pseudo metric d on the class of real-valued random variables satisfying the convolution property (6.104). Suppose in addition that for any pairs of real-valued square integrable random variables X and Y, the pseudo metric d satisfies the two additional properties: There exists a positive constant c_d depending only on the pseudo metric d such that*

$$d(X+Y, X) \leq c_d \|Y\|_2 (\|X\|_2 + \|Y\|_2), \tag{6.126}$$

and, for Y' a copy of Y independent of X,

$$d(X+Y, X+Y') \le c_d\,\rho(\sigma(X), \sigma(Y))\|X\|_2\|Y\|_2. \tag{6.127}$$

Then, assuming that

$$\sum_{k\ge 0} \rho(2^k) < \infty, \tag{6.128}$$

the coupling bounds (6.107) hold with D a non-negative superadditive function and $(\delta_k)_{k\ge 1}$ satisfying (6.108). In particular, if we assume, in addition, that the pseudo metric satisfies the conditions (6.102), (6.103) and (6.105) and if $\sup_{n\ge 1}\sum_{k=1}^n \mathbb{E}(X_{k,n}^2) < \infty$ and (6.113) is satisfied then the CLT convergence result (6.114) holds.

Proof of Lemma 6.46 For any integers a, b such that $0 \le a < b \le n$, we define

$$D_{a,b} := \kappa_1 C(\rho) \sum_{j=a+1}^{b} \|X_{j,n}\|_2^2, \tag{6.129}$$

where $\kappa_1 = 8\,e^{2^{20}}$ and $C(\rho) = \exp\left(2\sum_{j\ge 0}\rho(2^{j/3})\right)$. Note that D is a non-negative superadditive function.

Now, let $n_0 = 2^{14}$ and k, r, n be integers such that $n_0 \le k \le n$ and $0 \le r \le n-k$. Let $m = [k^{1/3}]$. According to Lemma 6.4, there is an integer u such that

$$0 \le u \le k \text{ and } 2^{-1}(k - m^2) \le um \le 2^{-1}k,$$

and

$$\sum_{i=um+1}^{(u+1)m} \|X_{i+r,n}\|_2^2 \le (m/k)^{1/3} \sum_{j=1}^{k} \|X_{j+r,n}\|_2^2. \tag{6.130}$$

For such a u, we start with the following decomposition:

$$S_{r,r+k} = S_{r,r+um} + S_{r+um,r+um+m} + S_{r+um+m,r+k} = X + s + Y, \text{ say.}$$

Let Y' be a copy of Y independent of X. By applying (6.127) and then Theorem 6.1 and condition (6.128), we derive

$$\begin{aligned} d(X+Y, X+Y') &\le c_d\rho(m)\|X\|_2\|Y\|_2 \le c_d\,\rho(m) D_{r,r+um}^{1/2} D_{r+um+m,r+k}^{1/2} \\ &\le c_d\,\rho(m) D_{r,r+k}, \end{aligned} \tag{6.131}$$

where at the last step we applied that $um \le k$ and D is non-decreasing.

Notice that the variable s is relatively small. Applying first property (6.126), and then Theorem 6.1 and condition (6.128) again, and finally observing that by the construction (6.130), $||s||_2^2 \leq k^{-2/9}D_{r,r+k}$ and $um + m \leq k$, we derive

$$\begin{aligned} d(X+s+Y, X+Y) &\leq c_d||s||_2(||s||_2 + ||X+Y||_2) \leq c_d||s||_2(||s||_2 + ||X||_2 + ||Y||_2) \\ &\leq c_d D^{1/2}_{r+um,r+um+m}(D^{1/2}_{r+um,r+um+m} + D^{1/2}_{r,r+um} + D^{1/2}_{r+um+m,r+k}) \\ &\leq 3c_d k^{-1/9} D_{r,r+k}. \end{aligned} \tag{6.132}$$

Now, let Z' be a copy of $Z = s + Y$ independent of X. Then, the convolution property (6.104) and a similar argument as in the proof of (6.132) imply that

$$\begin{aligned} d(X+Z', X+Y') \leq d(Z, Y) = d(s+Y, Y) &\leq c_d||s||_2(||s||_2 + ||Y||_2) \\ &\leq 2c_d k^{-1/9} D_{r,r+k}. \end{aligned} \tag{6.133}$$

Now, by using the triangular inequality and taking into account the bounds (6.131), (6.132) and (6.133), we derive

$$\begin{aligned} d(X+s+Y, X+(s+Y)') = d\left(S_{r,r+k}, S_{r,r+um} + S'_{r+um,r+k}\right) &\leq \delta_k D_{r,r+k} \\ \text{with } \delta_k = c_d\left(\rho([2^{k/3}] + 5k^{-1/9}\right). \end{aligned}$$

The sequence $(\delta_k)_{k\geq 1}$ is non-increasing and satisfies the summability condition (6.108) since (6.128) is assumed. This proves the coupling bound (6.107).

To end the proof of the lemma it suffices to apply Lemma 6.45 by noticing that if $\sup_{n\geq 1} \sum_{k=1}^{n} \mathbb{E}(X^2_{k,n}) < \infty$ then the function D defined in (6.129) is bounded. ■

Lemma 6.46 together with the technical lemma below give Theorem 6.48 concerning the CLT of a ρ-mixing triangular array under the Lindeberg condition (this theorem is due to Utev (1990)). The same approach also works in case of associated variables (see Peligrad and Utev (1997)).

Lemma 6.47 *For any two real-valued random variables in $\mathbb{L}^2$ and any real number t fixed, let*

$$d(X, Y) = d_1(X, Y) + d_2(X, Y), \tag{6.134}$$

where

$$d_1(X, Y) = |\mathbb{E}[e^{itX}] - \mathbb{E}[e^{itY}]| \text{ and } d_2(X, Y) = |\mathrm{Var}[X] - \mathrm{Var}[Y]|.$$

Then d is a pseudo metric on the class of real-valued random variables satisfying (6.102) with $c_1 = 1$, (6.103) with $c_2 = |t|^3/6$, (6.105) with $c_3 = 1 + t^2/2$ and also the convolution bound (6.104). In addition d satisfies the properties (6.126) and (6.127).

Theorem 6.48 *Let $\{X_{k,n} : k = 1, ..., n\}$ be a triangular array of square integrable zero mean real-valued variables such that (6.128) holds. Let $S_n = \sum_{k=1}^{n} X_{k,n}$ and $\sigma_n^2 = \mathrm{Var}(S_n)$. Assume in addition that*

$$\sigma_n^2 = 1 \quad \textit{and} \quad \sup_n \sum_{i=1}^{n} \mathbb{E}(X_{i,n}^2) < \infty,$$

and that for any $\varepsilon > 0$,

$$L_n(\varepsilon) = \sum_{i=1}^{n} \mathbb{E}\left(X_{i,n}^2 \mathbf{1}_{\{|X_{i,n}| \geq \varepsilon\}}\right) \to 0, \ \ \textit{as } n \to \infty. \tag{6.135}$$

Then S_n converges in distribution to a standard normal distribution.

For a functional version of this result, we refer to Theorem 4.1 in Merlevède et al. (2019).

Proof of Theorem 6.48 We first notice that we can always find a sequence $(\varepsilon_n)_{n\geq 1}$ of positive reals such that

$$\varepsilon_n \to 0 \text{ and } L_n(\varepsilon_n) \to 0, \text{ as } n \to \infty. \tag{6.136}$$

Consider now the truncated array $\{X'_{k,n} \ : \ k = 1, \dots, n\}$ with

$$X'_{k,n} = X_{k,n}\mathbf{1}_{\{|X_{k,n}|<\varepsilon_n\}} - \mathbb{E}\left(X_{k,n}\mathbf{1}_{\{|X_{k,n}|<\varepsilon_n\}}\right).$$

For any integer k, let $X''_{k,n} = X_{k,n} - X'_{k,n}$. Note that by Theorem 6.1 and condition (6.128), there exists a finite positive constant C such that

$$\mathbb{E}\left(\sum_{k=1}^{n} X''_{k,n}\right)^2 \leq C \sum_{k=1}^{n} \mathbb{E}\left(X_{k,n}^2 \mathbf{1}_{\{|X_{k,n}|\geq\varepsilon_n\}}\right) = CL_n(\varepsilon_n).$$

Hence, by (6.136), $\mathrm{Var}\left(\sum_{k=1}^{n} X''_{k,n}\right) \to 0$, as $n \to \infty$ which proves that $\sum_{k=1}^{n} X''_{k,n}$ has a negligible contribution for the convergence in distribution of S_n. In addition, since $\sigma_n^2 = 1, \mathrm{Var}\left(\sum_{k=1}^{n} X'_{k,n}\right) \to 1$, as $n \to \infty$. Therefore, denoting $\sigma'_n = \sqrt{\mathrm{Var}\left(\sum_{k=1}^{n} X'_{k,n}\right)}$, we note that $\{Y_{k,n} : k = 1, \dots, n\}$ with

$$Y_{k,n} = \frac{1}{\sigma'_n} X'_{k,n}$$

forms a triangular array of square integrable zero mean real-valued random variables whose associated sequence of uniform ρ-mixing coefficients satisfies condition (6.128). In addition, $\mathrm{Var}\left(\sum_{k=1}^{n} Y_{k,n}\right) = 1$ and

$$\sum_{k=1}^{n} \mathbb{E}(|Y_{k,n}|^3) \leq 2\frac{\varepsilon_n}{(\sigma'_n)^3} \sum_{i=1}^{n} \mathbb{E}(X_{i,n}^2) \to 0, \text{ as } n \to \infty,$$

since $\sup_n \sum_{i=1}^{n} \mathbb{E}(X_{i,n}^2) < \infty$ by assumption, $\varepsilon_n \to 0$ and $\sigma'_n \to 1$. Therefore, applying Lemma 6.46 to the triangular array $\{Y_{k,n} \ : \ k = 1, \dots, n\}$ and taking into account

Lemma 6.47, it follows that $\sum_{k=1}^{n} Y_{k,n}$ converges in distribution to a standard normal distribution. Together with the fact that $\sigma_n' \to 1$, as $n \to \infty$, and that $\sum_{k=1}^{n} X''_{k,n}$ has a negligible contribution for the convergence in distribution of S_n, the theorem follows. ■

To end this section, it remains to prove Lemma 6.47.

Proof of Lemma 6.47 Since d_1 and d_2 are both pseudo metrics, their sum is still a pseudo metric. Note that d satisfies (6.102) with $c_1 = 1$. Consider now two real-valued random variables X and Y such that $\mathbb{E}[X^k] = \mathbb{E}[Y^k]$ for $k = 1, 2$. Clearly $d_2(X, Y) = 0$ and, by the elementary inequality $|e^{iz} - (1 + iz - z^2/2)| \le |z|^3/6$,

$$\begin{aligned} d_1(X, Y) &= |\mathbb{E}[\, e^{itX} - (1 + itX - t^2X^2/2)] - \mathbb{E}[\, e^{itY} - (1 + itY - t^2Y^2/2)]| \\ &\le (|t|^3/6)(\mathbb{E}|X|^3 + \mathbb{E}|Y|^3), \end{aligned}$$

which proves that d satisfies (6.103) with $c_2 = |t|^3/6$. In addition, if $N_1 \sim \mathcal{N}(0, a^2)$ and $N_1 \sim \mathcal{N}(0, b^2)$, then $d_2(N_1, N_2) = |a^2 - b^2|$ and

$$d_1(N_1, N_2) = \left| e^{-t^2a^2/2} - e^{-t^2b^2/2} \right| \le \frac{t^2}{2}|a^2 - b^2|.$$

Therefore d satisfies (6.105) with $c_3 = 1 + t^2/2$.

Let us prove now that d satisfies the convolution bound (6.104). According to Item (iii) of Comment 6.43, it suffices to prove that for any independent real-valued random variables X, Y and Y', (6.106) holds. Obviously, by independence,

$$\begin{aligned} d(X + Y, X + Y') &= d_1(X + Y, X + Y') + d_2(X + Y, X + Y') \\ &= |\mathbb{E}[e^{itX}]||\mathbb{E}[e^{itY}] - \mathbb{E}[e^{itY'}]| + |\mathrm{Var}(Y) - \mathrm{Var}(Y')| \le d(Y, Y'), \end{aligned}$$

proving (6.106).

We prove now that d satisfies the properties (6.126) and (6.127). With this aim, we shall prove that each of the pseudo metrics d_1 and d_2 satisfies these properties. Note first that, by the definition of the maximal correlation coefficient,

$$\begin{aligned} |\mathbb{E}[e^{it(X+Y)}] - \mathbb{E}[e^{itX}]\mathbb{E}[e^{itY}]| &= |\mathbb{E}\{(e^{itX} - \mathbb{E}[e^{itX}])(e^{itY} - \mathbb{E}[e^{itY}])\}| \\ &\le 2\rho(\sigma(X), \sigma(Y))\sqrt{\mathbb{E}|e^{itX} - \mathbb{E}[e^{itX}]|^2}\sqrt{\mathbb{E}|e^{itY} - \mathbb{E}[e^{itY}]|^2} \\ &\le 8t^2\rho(\sigma(X), \sigma(Y))\|X\|_2\|Y\|_2, \end{aligned}$$

proving (6.127) for d_1. Assuming now that $\mathbb{E}[Y] = 0$ (we are considering centered variables),

$$\begin{aligned} |\mathbb{E}[\, e^{it(X+Y)}] - \mathbb{E}[\, e^{itX}]| &= |\mathbb{E}\{(\, e^{itX})(\, e^{itY} - 1)\}| \\ &= |\mathbb{E}\{(e^{itX} - 1)(\, e^{itY} - 1)\} + \mathbb{E}\{(\, e^{itY} - (1 + itY))\}| \end{aligned}$$

$$\leq |t|^2\|X\|_2\|Y\|_2 + (|t|^2/2)\|Y\|_2^2$$
$$\leq t^2\|Y\|_2(\|X\|_2 + \|Y\|_2), \tag{6.137}$$

proving (6.126) for d_1.

We turn now to the pseudo metric d_2. Establishing the bounds properties (6.126) and (6.127) for d_2 is very similar as for d_1 but somewhat easier. Indeed,

$$|\text{Var}[X+Y] - (\text{Var}[X] + \text{Var}[Y])| = 2|\text{cov}(X,Y)| \leq 2\rho(\sigma(X), \sigma(Y))\|X\|_2\|Y\|_2$$

and

$$|\text{Var}[X+Y] - \text{Var}[X]| = |2\text{cov}(X,Y) + \text{Var}[Y]| \leq 2\|Y\|_2(\|X\|_2 + \|Y\|_2),$$

which respectively proves (6.127) and (6.126). ■

6.6 Invariance Principle for Stationary ρ-mixing Sequences

Here we investigate the functional central limit theorem question when we consider a strictly stationary sequence of centered random variables $(X_k)_{k\in\mathbb{Z}}$ in $\mathbb{L}^2$ and which are additionally ρ-mixing in the sense that

$$\rho(n) = \rho(\mathcal{F}_{-\infty}^0, \mathcal{F}_n^\infty) \to 0,$$

as $n \to \infty$, with $\mathcal{F}_i^j = \sigma(X_i, \dots, X_j)$ and where the coefficient ρ between two σ-algebras has been defined in Section 5.1.1. Note that in this case the process $(X_k)_{k\in\mathbb{Z}}$ is ergodic.

Applying Theorem 4.16 together with the following lemma applied with $p = 2$ and $\alpha = 1/2$, we derive Theorem 6.50.

Lemma 6.49 *Let $p \geq 2$ and $(X_n)_{n\in\mathbb{Z}}$ be a strictly stationary sequence of centered real-valued random variables in $\mathbb{L}^p(\mathbb{P})$. Let $S_n = \sum_{k=1}^n X_k$ and $(\rho(n))_{n\geq 0}$ be its associated sequence of ρ-mixing coefficients. Assume that for some $\alpha \in]0, 1/2]$,*

$$\sum_{i\geq 0} 2^{i(1/2-\alpha)}\rho^{2/p}(2^i) < \infty. \tag{6.138}$$

Then

$$\sum_{n\geq 1} \frac{\|\mathbb{E}(S_n|\mathcal{F}_0)\|_p}{n^{1+\alpha}} < \infty. \tag{6.139}$$

Theorem 6.50 *Let $(X_n)_{n\in\mathbb{Z}}$ be a strictly stationary and ρ-mixing sequence of centered real-valued random variables in $\mathbb{L}^2(\mathbb{P})$. Let $S_n = \sum_{k=1}^n X_k$. Assume that*

$$\sum_{k\geq 1} \rho(2^k) < \infty. \tag{6.140}$$

Then $\{n^{-1/2}S_{[nt]}, t \in [0, 1]\}$ *converges in distribution in* $D([0, 1])$ *to a Brownian motion with parameter* $\sigma^2 = \lim_{n\to\infty} n^{-1}\mathbb{E}(S_n^2)$.

Note that the central limit theorem under (6.140) has been proved by Ibragimov in 1975, but the invariance principle remained an open question for a long time and led to intensive efforts on the part of many mathematicians to solve this question. The invariance principle, under (6.140), was finally proved by Shao in 1988. The fact that condition (6.140) is essentially optimal in the context of the central limit theorem and its functional form is due to Bradley (2007), vol. 1, page 367, and vol. 3, Theorem 34.13. Bradley's example shows that if (6.140) fails then $S_n/\|S_n\|_2$ might have non-degenerate non-normal distributions as weak limit points.

Proof of Lemma 6.49 We start by noticing that since $\left(\|\mathbb{E}_0(S_n)\|_p\right)_{n\geq 1}$ is a subadditive sequence, it follows from Lemma 3.24 that proving (6.139) is equivalent to proving that

$$\sum_{k\geq 0} 2^{-\alpha k}\|\mathbb{E}(S_{2^k}|\mathcal{F}_0)\|_p < \infty. \tag{6.141}$$

To prove the convergence above, we first note that, by stationarity, for any $k \geq 0$,

$$\|\mathbb{E}(S_{2^{k+1}}|\mathcal{F}_0)\|_p \leq \|\mathbb{E}(X_1|\mathcal{F}_0)\|_p + \sum_{j=0}^{k} \|\mathbb{E}(S_{2^{j+1}} - S_{2^j}|\mathcal{F}_0)\|_p.$$

Now to evaluate $\|\mathbb{E}_0(S_{2^{j+1}} - S_{2^j})\|_p$, we proceed as we did to get (6.49). Hence, we derive

$$\|\mathbb{E}(S_{2^{j+1}} - S_{2^j}|\mathcal{F}_0)\|_p \leq 10\left(\rho(2^j)\right)^{2/p}\|S_{2^{j+1}} - S_{2^j}\|_p = 10\left(\rho(2^j)\right)^{2/p}\|S_{2^j}\|_p.$$

Next we use Theorem 6.13 to bound $\|S_{2^j}\|_p$. Therefore, since by (6.138), $\sum_{i\geq 0}\rho^{2/p}(2^i) < \infty$, it follows that there exists a positive constant C such that for any $j \geq 0$,

$$\|\mathbb{E}(S_{2^{j+1}} - S_{2^j}|\mathcal{F}_0)\|_p \leq C2^{j/2}\left(\rho(2^j)\right)^{2/p}.$$

So, overall, for any $k \geq 0$,

$$\|\mathbb{E}(S_{2^{k+1}}|\mathcal{F}_0)\|_p \leq \|X_1\|_p + C\sum_{j=0}^{k} 2^{j/2}\left(\rho(2^j)\right)^{2/p},$$

implying that

$$\sum_{k\geq 0} 2^{-\alpha k}\|\mathbb{E}(S_{2^k}|\mathcal{F}_0)\|_p \leq \frac{2^\alpha}{2^\alpha - 1}\|X_1\|_p + \frac{C}{2^\alpha - 1}\sum_{i\geq 0} 2^{i(1/2-\alpha)}\rho^{2/p}(2^i).$$

This ends the proof of (6.141) and then of the lemma. ■

6.7 Central Limit Theorem for ϕ-mixing Sequences and Discussion on Ibragimov Conjecture

We consider a strictly stationary sequence of centered random variables $(X_k)_{k\in\mathbb{Z}}$ in $\mathbb{L}^2$ and which are additionally ϕ-mixing in the sense that

$$\phi(n) = \phi(\sigma(X_\ell; \ell \le 0), \sigma(X_j; j \ge n)) \to 0,$$

as $n \to \infty$, where the coefficient ϕ between two σ-algebras has been defined in Section 5.1.1.

We denote by $S_n = \sum_{k=1}^{n} X_k$ and start with a discussion on Ibragimov's conjecture (1962) which can be formulated as follows.

Conjecture 6.51 *Assume $(X_k)_{k\in\mathbb{Z}}$ is a stationary ϕ-mixing sequence such that $\mathbb{E}(X_0) = 0$ and $\mathbb{E}(X_0^2) < \infty$. Denote by $\sigma_n^2 = \mathbb{E}(S_n^2)$ and assume $\sigma_n^2 \to \infty$. Then $(S_n/\sigma_n)_{n\ge 1}$ converges in distribution to a standard normal variable.*

This conjecture was reformulated in Iosifescu (1980) to include the functional central limit theorem. For x real denote by $[x]$ the integer part of x and introduce

$$W_n(t) = S_{[nt]}/\sigma_n, 0 \le t \le 1,$$

a random element of $D([0, 1])$, the space of functions defined on $[0, 1]$, which are continuous from the right and have left-hand limits. We endow, as usual in this monograph, $D([0, 1])$ with uniform topology.

Conjecture 6.52 *Assume that the conditions of Conjecture 6.51 are satisfied. Then W_n is weakly convergent to W, where W denotes the standard Brownian motion on $[0, 1]$.*

We shall argue that the Ibragimov–Iosifescu conjecture holds under the additional condition assumed in Peligrad (1985):

$$\liminf_n \sigma_n^2/n > 0. \tag{6.142}$$

This follows from the following theorem:

Theorem 6.53 *Assume that the conditions of Conjecture 6.51 are satisfied. Suppose in addition that the following condition is satisfied:*

$$\frac{n}{\sigma_n^2}\mathbb{E}(X_1^2 I(|X_1| > \varepsilon\sigma_n)) \to 0. \tag{6.143}$$

Then W_n is weakly convergent to W, where W denotes the standard Brownian motion on $[0, 1]$.

Proof We start by mentioning that

$$\frac{1}{\sigma_n^2}\mathbb{E}(\max_{1\le k\le n} X_k^2) \le \varepsilon + \frac{n}{\sigma_n^2}\mathbb{E}(X_1^2 I(|X_1| > \varepsilon\sigma_n)).$$

Therefore

$$\frac{1}{\sigma_n^2}\mathbb{E}(\max_{1\le k\le n} X_k^2) \to 0 \text{ as } n \to \infty.$$

By Proposition 6.24 we obtain that the family $(\max_{1\le k\le n} S_k^2/\sigma_n^2)$ is uniformly integrable and by Theorem 6.38 the functional CLT holds. ∎

Other results related to the Ibragimov conjecture can be found, for instance, in Peligrad (1985, 1990, 1993), Samur (1984), Hahn, Kuelbs and Samur (1987), Berkes and Philipp (1998) and Utev (1990). They are all based on inequalities for maximum of partial sums and on Hoffman–Jorgensen type inequalities (see Peligrad (1985)), which are valid for ϕ_k sufficiently small for some $k \ge 1$. These inequalities also lead to Rosenthal-type inequalities.

Under ρ-mixing we have also the following theorem, which is due to Ibragimov (1975) (see Theorem 3.1 and its proof there):

Theorem 6.54 *Assume that $(X_n)_{n\in\mathbb{Z}}$ is a centered, stationary, ρ-mixing process such that $\sigma_n^2 \to \infty$. Assume in addition that $\mathbb{E}(|X_1|^{2+\delta}) < \infty$, for some $\delta > 0$. Then W_n is weakly convergent to W, where W denotes the standard Brownian motion on [0, 1].*

Proof First of all, by Theorem 6.16 we know that there is a constant C such that for all $n \ge 1$

$$\|\max_{1\le k\le n} |S_k|\|_{2+\delta} \le C\sigma_n .$$

This implies that the family $(\max_{1\le k\le n} S_k^2/\sigma_n^2)$ is uniformly integrable and it remains to apply Theorem 6.38 to obtain the result. ∎

It is worth noticing that the stationarity assumption can be replaced by weaker forms of stationarity in all functional CLT for strong mixing sequences. For instance, concerning the central limit theorem question, as we have seen in Section 6.5.2 and in particular in Theorem 6.48, the stationarity can be weakened to Lindeberg's condition.

7

Weakly Associated Random Variables

$\mathbb{L}^2$-Bounds and Approximation by Independent Structures

The purpose of this chapter is to introduce the notion of weakly associated (negatively or positively) random variables and to develop tools that will allow us, in the next chapters, to give estimation of moments of partial sums, maximal inequalities and asymptotic results with both Gaussian and non-gaussian limits. As we shall see these results shed light on the asymptotic behavior of numerous examples such as exchangeable variables, certain Gaussian processes, empirical process, classes of Markov chains and determinantal process. They are also useful to study stochastic processes that are functionals of two independent processes mentioned above.

7.1 Definition of the Weak Associated Coefficients between Two Random Variables

To restrict the dependence and to measure the distance from independence the authors have over the years used various conditions on the pairs and vectors of random variables (X, Y). For a pair of random variables the dependence is usually defined by measuring the $\text{cov}(f(X), g(Y))$ for f and g in some classes of functions. For various classes of functions we define various dependence measures. For instance, strong mixing coefficients are based on the families $f(X) = I(X \in A)$, $g(Y) = I(Y \in B)$, where A, B are Borelians, whereas associated variables are defined by $\text{cov}(f(X), g(Y)) \geq 0$ using f and g increasing such that the covariance makes sense.

In this chapter we develop tools for studying processes that feature a weak form of positive or negative dependence. To describe the concept we shall start by introducing the coefficients of dependence between two random variables.

$$r' = r'(X, Y) = \sup \text{corr}\,(f(X), g(Y)) \tag{7.1}$$

Functional Gaussian Approximation for Dependent Structures. Florence Merlevède, Magda Peligrad and Sergey Utev. Oxford University Press (2019). © Florence Merlevède, Magda Peligrad and Sergey Utev 2019.
DOI: 10.1093/oso/9780198826941.001.0001

where the supremum is taken over all non-decreasing functions f and g for which the correlation makes sense. We shall refer to r' as the coefficient of r'—association between X and Y. When $r' < 1$ we shall call X and Y weakly negatively dependent or simply r'-negatively dependent (r'-ND).

When $r' = 0$, X and Y are called in the literature negatively associated (or negatively dependent) variables. This definition goes back to Lehmann (1966), who introduced the notion of "negatively quadrant dependent" that further generated the important field of negatively associated processes as introduced by Alam and Saxena (1981) and Joag-Dev and Prochan (1983), and studied in many papers by Newman (1984), Shao (2000), Matula (2003) and Pemantle (2000), among others.

Notice that always $r' \leq 1$. Obviously $r' < 1$ is a restriction of dependence. All the variables that are negatively associated and a large class of positively associated ones will satisfy this condition. Under the additional restriction that all the correlations above are positive, r' becomes a measure of the strength of positive association between X and Y.

This coefficient is also related to the maximum coefficient of correlation defined as

$$\rho(X, Y) = \sup(\mathrm{corr}\,(f(X), g(Y))) \tag{7.2}$$

where the supremum is taken after all measurable functions f and g. Clearly $\rho(X, Y) = \sup(|\mathrm{corr}\,(f(X), g(Y))|)$. By comparing to the condition $\rho(X, Y) < 1$, the condition $r'(X, Y) < 1$ can be viewed as a restricted (to non-decreasing functions only) one-sided condition.

In a similar way, we introduce the coefficient of modified positive dependence, that allows also for negative dependence but not too strong:

$$a' = \sup -\mathrm{corr}(f(X), g(Y)) \tag{7.3}$$

where the supremum is taken over all non-decreasing functions f and g for which the correlation makes sense. Notice that always $a' \geq -1$. When $a' = 0$, X and Y are called positively associated, as introduced by Lehmann (1966), a notion that was further developed into the field of positive association studied by Esary, Proschan and Walkup (1967), and further extensively studied in Newman (1980, 1984), Newman and Wright (1981, 1982), Cox and Grimmett (1984), Burton, Dabrowski and Dehling (1986), Birkel (1987, 1993) and many others.

When $a' > -1$ we shall call X and Y weakly positively dependent or simply a'-positively dependent (a'-PD).

As we shall see in the next section, the coefficients r' and a' are well adapted to estimate the following quantity

$$\mathbf{H}(X, Y) = \int_{-\infty}^{\infty}\int_{-\infty}^{\infty} |H(x, y)|dxdy, \tag{7.4}$$

where

$$H(x,y) = H_{(X,Y)}(x,y) = \mathbb{P}(X > x, Y > y) - \mathbb{P}(X > x)P(Y > y).$$

Notice that

$$\operatorname{cov}(X,Y) = \int_{-\infty}^{\infty}\int_{-\infty}^{\infty} H(x,y)dxdy$$

(see Lehmann (1966)).

Note also that the quantity $\mathbf{H}(X,Y)$ can be easily estimated under various conditions. For instance if X and Y are positively quadrant dependent, i.e. $H(x,y) \geq 0$ for any x and y, then

$$\mathbf{H}(X,Y) = \int_{-\infty}^{\infty}\int_{-\infty}^{\infty} H(x,y)dxdy = \operatorname{cov}(X,Y).$$

If X and Y are negatively quadrant dependent, then

$$\mathbf{H}(X,Y) = \int_{-\infty}^{\infty}\int_{-\infty}^{\infty} -H(x,y)dxdy = -\operatorname{cov}(X,Y).$$

On another hand, if we define

$$\alpha_{p,q} = \sup_{x>0,y>0} \frac{|\mathbb{P}(X \geq x, Y \geq y) - \mathbb{P}(X \geq x)\mathbb{P}(Y \geq y)|}{\mathbb{P}^{1/p}(X \geq x)\mathbb{P}^{1/q}(Y \geq y)},$$

where the supremum extends for all positive x and y with $\mathbb{P}(X \geq x)\mathbb{P}(Y \geq y) > 0$, $1 < p < \infty$ and $1/p + 1/q = 1$, and if we assume that $X \geq 0$ and $Y \geq 0$ then we have the following sharp (up to a constant) estimate

$$\int_0^{\infty}\int_0^{\infty} |H(x,y)|dxdy \leq 13\max(p,q)\alpha_{p,q}(1 - \log\alpha_{p,q})\|X\|_p\|Y\|_q.$$

(See Lemma 4.14 and Theorem 4.15 in Bradley (2007)).

Our first goal is to evaluate $\mathbf{H}(X,Y) = \int_{-\infty}^{\infty}\int_{-\infty}^{\infty}|H(x,y)|dxdy$ for more general coefficients than $\alpha_{p,q}$.

Throughout the chapter we use the following notations:

$$\|X\|_p = (\mathbb{E}[|X|^p])^{1/p},\ \operatorname{cov}(X,Y) = \mathbb{E}(XY) - \mathbb{E}(X)\mathbb{E}(Y),$$
$$\operatorname{corr}(X,Y) = \operatorname{cov}(X,Y)/(\|X\|_2\|Y\|_2),\ \|f\|_\infty = \sup_x |f(x)|, C_f = \|f''\|_\infty,$$
$$H(x,y) = \mathbb{P}(X > x, Y > y) - \mathbb{P}(X > x)\mathbb{P}(Y > y),$$
$$m(X,Y) = \sup_{g\in\mathbb{L}^2(Y)} (\operatorname{corr}(X, g(Y)))$$

7.2 Approximation by Independent Pair

We shall start by introducing and analyzing several new coefficients of dependence which incorporate both the notion of mixing type dependence and the notion of negative or positive association.

7.2.1 *Weak Negative Dependence*

A related negative dependence coefficient is the coefficient of quadrant dependence

$$r^* = r^*(X, Y) = \sup_{x,y} \operatorname{corr}(I(X > x), I(Y > y)). \tag{7.5}$$

Now, notice that both coefficients r' and r^* are non-negative and $r^* \leq r'$. Also, for any non-decreasing functions f and g,

$$r^*(f(X), g(Y)) \leq r^*(X, Y) \quad \text{and} \quad r^*(-X, -Y) = r^*(X, Y). \tag{7.6}$$

In the following lemma we compare r' and r^* in the other direction, and show that r' is small provided r^* is. Moreover, if $r^*(X, Y)$ is small, then the variables X and Y are weakly dependent in some sense (point (c) in the next lemma).

Lemma 7.1

(a) *With the above notations,*

$$\mathbf{H}(X, Y) \leq -\operatorname{cov}(X, Y) + 104r^*(1 - \log r^*)\|X\|_2\|Y\|_2. \tag{7.7}$$

(b) *In particular,* $r' \leq 52r^*(1 - \log r^*)$.

(c) *Moreover, let* $f : \mathbb{R}^2 \to \mathbb{R}$ *be a twice differentiable function with* $|\partial^2 f(x, y)/\partial x \partial y| \leq C_f$ *and let* Y' *be a variable distributed as* Y *and independent of* X. *Then,*

$$\begin{aligned} |\mathbb{E}[f(X, Y)] - \mathbb{E}[f(X, Y')]| &\leq C_f \mathbf{H}(X, Y) \\ &\leq C_f\{-\operatorname{cov}(X, Y) + 104r^*(1 - \log r^*)\|X\|_2\|Y\|_2\}. \end{aligned} \tag{7.8}$$

Proof We begin by proving (a). Having in mind the elementary equalities:

$$\begin{aligned} &\mathbb{P}(X > x, Y > y) - \mathbb{P}(X > x)\mathbb{P}(Y > y) = \mathbb{P}(X \leq x, Y \leq y) - \mathbb{P}(X \leq x)\mathbb{P}(Y \leq y) \\ &= \mathbb{P}(X > x)\mathbb{P}(Y \leq y) - \mathbb{P}(X > x, Y \leq y) = \mathbb{P}(X \leq x)\mathbb{P}(Y > y) - \mathbb{P}(X \leq x, Y > y), \end{aligned} \tag{7.9}$$

we decompose the integral into four integrals on the quadrant domains defined by $D_1 = (0, \infty) \times (0, \infty)$, $D_2 = (-\infty, 0) \times (-\infty, 0)$, $D_3 = (0, \infty) \times (-\infty, 0)$, $D_4 = (-\infty, 0) \times (0, \infty)$.

Consider for example the integral on $D_1 = (0, \infty) \times (0, \infty)$. Since

$$H(x, y) \leq \min(\mathbb{P}(X > x), \mathbb{P}(Y > y), r^*[\mathbb{P}(X > x)\mathbb{P}(Y > y)]^{1/2}),$$

it is easy to see that

$$\int_0^\infty \int_0^\infty |H(x, y)| dxdy$$

$$\leq \int_0^\infty \int_0^\infty |H(x, y) - \min\{\mathbb{P}(X > x), \mathbb{P}(Y > y), r^* \mathbb{P}^{1/2}(X > x)\mathbb{P}^{1/2}(Y > y)| dxdy$$

$$+ \int_0^\infty \int_0^\infty \min\{\mathbb{P}(X > x), \mathbb{P}(Y > y), r^* \mathbb{P}^{1/2}(X > x)\mathbb{P}^{1/2}(Y > y)\} dxdy$$

$$= 2 \int_0^\infty \int_0^\infty \min\{\mathbb{P}(X > x), \mathbb{P}(Y > y), r^* \mathbb{P}^{1/2}(X > x)\mathbb{P}^{1/2}(Y > y)\} dxdy$$

$$- \int_0^\infty \int_0^\infty H(x, y) dxdy = 2I - Q, \text{ say.}$$

To estimate I, we apply Lemma 4.14 in Bradley (2007, vol.1) to derive

$$I \leq 13r^*(1 - \log r^*)||X||_2||Y||_2.$$

Other terms are bounded similarly by using (7.9) and (7.6). And so, by combining the estimates together, we prove (a), since

$$\text{cov}(X, Y) = \int_{-\infty}^\infty \int_{-\infty}^\infty H(x, y) dxdy = \int_{-\infty}^\infty \int_{-\infty}^\infty \text{cov}(I(X > x), I(Y > y)) dxdy. \tag{7.10}$$

Then, it follows directly from (a) that

$$\text{cov}(X, Y) \leq 52r^*(1 - \log r^*)||X||_2||Y||_2. \tag{7.11}$$

Now, statement (b) easily follows from (a) and (7.11) applied together with (7.6) and the fact that the function $x(1 - \log x)$ is increasing on $(0, 1]$.

Finally, statement (c) follows from (a) and the fact that, by integration by parts,

$$\mathbb{E}[f(X, Y)] - \mathbb{E}[f(X, Y')] = \int_{-\infty}^\infty \int_{-\infty}^\infty H(x, y) \frac{\partial^2 f(x, y)}{\partial x \partial y} dxdy. \qquad \blacksquare$$

Note that the following corollary follows easily from Item (c) of Lemma 7.1 or from Formula (7.10):

Corollary 7.2 *If $r^*(X, Y) = 0$ and $\text{cov}(X, Y) = 0$ then X and Y are independent.*

7.2.2 *Weak Positive Dependence*

A related coefficient of dependence is

$$a^* = \sup_{x,y} -\text{corr}(I(X > x), I(Y > y)). \tag{7.12}$$

Both coefficients a' and a^* are non-negative and $a^* \leq a'$. Similarly to the r'-negative dependence case, we derive Lemma 7.3 below whose proof repeats the proof of the previous lemma with minor changes. The main difference is that in the proof we use the following fact:

$$-H(x,y) \leq \min(\mathbb{P}(X > x), \mathbb{P}(Y > y), a^*[\mathbb{P}(X > x)\mathbb{P}(Y > y)]^{1/2}),$$

and then

$$\begin{aligned}
&\int_0^\infty \int_0^\infty |H(x,y)|dxdy \\
&\leq \int_0^\infty \int_0^\infty |H(x,y) + \min\{\mathbb{P}(X > x), \mathbb{P}(Y > y), a^*\mathbb{P}^{1/2}(X > x)\mathbb{P}^{1/2}(Y > y)|dxdy \\
&\quad + \int_0^\infty \int_0^\infty \min\{\mathbb{P}(X > x), \mathbb{P}(Y > y), a^*\mathbb{P}^{1/2}(X > x)\mathbb{P}^{1/2}(Y > y)\}dxdy \\
&= 2\int_0^\infty \int_0^\infty \min\{\mathbb{P}(X > x), \mathbb{P}(Y > y), a^*\mathbb{P}^{1/2}(X > x)\mathbb{P}^{1/2}(Y > y)\}dxdy \\
&\quad + \int_0^\infty \int_0^\infty H(x,y)dxdy.
\end{aligned}$$

We have then:

Lemma 7.3

(a) *With the above notation,*

$$\mathbf{H}(X,Y) \leq \text{cov}(X,Y) + 104a^*(1 - \log a^*)\|X\|_2\|Y\|_2. \tag{7.13}$$

(b) *In particular,* $a' \leq 52a^*(1 - \log a^*)$.

(c) *Moreover, let* $f: \mathbb{R}^2 \to \mathbb{R}$ *be a twice differentiable function with* $|\partial^2 f(x,y)/\partial x \partial y| \leq C_f$ *and let* Y' *be a variable distributed as* Y *and independent of* X. *Then,*

$$|\mathbb{E}[f(X,Y)] - \mathbb{E}[f(X,Y')]| \leq C_f\{\text{cov}(X,Y) + 104a^*(1 - \log a^*)\|X\|_2\|Y\|_2\}. \tag{7.14}$$

7.2.3 *The Weak Martingale Coefficient*

For various examples it can be useful to express $\mathbf{H}(X, Y)$ as a covariance with the help of the coefficient $m(X, Y)$ defined as

$$m(X,Y) = \sup_{g \in \mathbb{L}_2(Y)} \text{corr}(X, g(Y)). \tag{7.15}$$

Notice that for $\mathbb{E}(X) = 0$

$$m(X, Y) = ||\mathbb{E}(X|Y)||_2/||X||_2.$$

(See Lemma 4.3 in Bradley (2007)).

Lemma 7.4 *Assume X and Y are square integrable real-valued random variables and that X is centered. Then*

$$\mathbf{H}(X, Y) \leq 2m(X, Y)||X||_2||Y||_2 \leq 2||\mathbb{E}(X|Y)||_2||Y||_2.$$

Moreover, let $f: \mathbb{R}^2 \to \mathbb{R}$ be a twice differentiable function with $|\partial^2 f(x, y)/\partial x\partial y| \leq C_f$ and let Y' be a random variable distributed as Y and independent of X. Then,

$$\begin{aligned}|\mathbb{E}[f(X, Y)] - \mathbb{E}[f(X, Y')]| &\leq 2C_f m(X, Y)||X||_2||Y||_2 \\ &\leq 2C_f||\mathbb{E}(X|Y)||_2||Y||_2.\end{aligned}$$

Proof Let (X', Y') be an independent copy of (X, Y). Denote by

$$A_x = \{y \in \mathbb{R} \ : \ \mathbb{E}(I(X < x \leq X')I(Y < y \leq Y')) > 0\}$$

and observe that

$$\begin{aligned}&\int_{-\infty}^{\infty}\int_{-\infty}^{\infty} |H_{X,Y}(x, y)|dxdy \\ &= \frac{1}{2}\int_{-\infty}^{\infty}\int_{-\infty}^{\infty} |\mathbb{E}(I(X < x \leq X')I(Y < y \leq Y'))|dxdy \\ &= \frac{1}{2}\int_{-\infty}^{\infty}\int_{-\infty}^{\infty} \mathbb{E}\left(I(X < x \leq X')I(Y < y \leq Y')(I(A_x(y)) - I(A_x^c(y)))\right) dydx \\ &= \frac{1}{2}\mathbb{E}\left(\int_{-\infty}^{\infty} \mathbb{E}(I(X < x \leq X')|Y, Y')\left\{\int_{-\infty}^{\infty} I(Y < y \leq Y')(I(A_x(y)) - I(A_x^c(y)))dy\right\}dx\right) \\ &\leq \frac{1}{2}\mathbb{E}\left|\mathbb{E}\left(\int_{-\infty}^{\infty} I(X < x \leq X')dx|Y, Y'\right)\left\{\int_{-\infty}^{\infty} I(Y < y \leq Y')dy\right\}\right| \\ &= \frac{1}{2}\mathbb{E}\,|\mathbb{E}\,(X' - X|Y, Y')\,(Y' - Y)\,| = \mathbb{E}|(Y - Y')\mathbb{E}\,(X|Y)|.\end{aligned}$$

Since $\mathbb{E}(X) = 0$, it follows that

$$\begin{aligned}\int_{-\infty}^{\infty}\int_{-\infty}^{\infty} |H_{X,Y}(x, y)|dxdy &\leq \mathbb{E}|\{Y + \mathbb{E}|Y|\}\mathbb{E}(X|Y)| \\ &= \mathbb{E}\left([|Y + \mathbb{E}|Y||](1_{\{\mathbb{E}(X|Y)>0\}} - 1_{\{\mathbb{E}(X|Y)\leq 0\}})X\right) := \mathrm{cov}(h(Y), X).\end{aligned}$$

To end the proof notice that $|h(Y)| \leq |Y + \mathbb{E}|Y||$. ■

7.3 Preservation of the Weak Negative or Positive Dependence under Convolution

The following lemma will allow us to use the symmetrization method as a tool for weakly negatively or positively dependent variables. It is also important for studying processes whose evolution involves several independent components. Recall that the coefficients r', a' and m have been respectively defined in (7.1), (7.3) and (7.15).

Lemma 7.5 *Suppose that $u(x, y)$ and $v(x, y)$ are measurable and coordinatewise non-decreasing functions. Assume that the vectors $X = (X_1, X_2)$ and $Y = (Y_1, Y_2)$ are independent. Then*

$$r'(u(X_1, Y_1), v(X_2, Y_2)) \leq \max(r'(X_1, X_2), r'(Y_1, Y_2)),$$

$$a'(u(X_1, Y_1), v(X_2, Y_2)) \leq \max(a'(X_1, X_2), a'(Y_1, Y_2)),$$

$$r^*(X_1 + Y_1, X_2 + Y_2) \leq \max(r^*(X_1, X_2), r^*(Y_1, Y_2))$$

and

$$m((X_1, Y_1), (X_2, Y_2)) \leq m(X_1, X_2) + m(Y_1, Y_2).$$

Proof Let $f(x)$ and $g(y)$ be two non-decreasing functions such that

$$\mathbb{E}[f(u(X_1, Y_1))] = 0 = \mathbb{E}[g(v(X_2, Y_2))]$$

and $\mathbb{E}f^2(X_1) < \infty$, and so on. Let $\mathbb{E}_Y$ and Var_Y be the conditional expectation and variance, i.e. $\mathrm{Var}_Y(W) = \mathbb{E}[W^2|Y] - (\mathbb{E}[W|Y])^2$. Then, obviously

$$\mathrm{cov}[f(u(X_1, Y_1)), g(v(X_2, Y_2))] = \mathbb{E}(\mathbb{E}_Y[f(u(X_1, Y_1))g(v(X_2, Y_2))])$$

and, by the definition of r'

$$\begin{aligned} \mathbb{E}_Y[f(u(X_1, Y_1))g(v(X_2, Y_2))] &\leq \mathbb{E}_Y[f(u(X_1, Y_1))]\mathbb{E}_Y[g(v(X_2, Y_2))] \\ &+ r'(X_1, X_2)\sqrt{\mathrm{Var}_Y[f(u(X_1, Y_1))]}\sqrt{\mathrm{Var}_Y[g(v(X_2, Y_2))]} \quad \text{almost surely.} \end{aligned}$$

Since the functions $\mathbb{E}[f(u(X_1, y))]$ and $\mathbb{E}[g(v(X_2, y))]$ are both non-decreasing, it then follows that

$$\begin{aligned} \mathbb{E}(\mathbb{E}_Y[f(u(X_1, Y_1))]\mathbb{E}_Y[g(v(X_2, Y_2))]) &\leq \mathbb{E}(\mathbb{E}_Y[f(u(X_1, Y_1))])\mathbb{E}(\mathbb{E}_Y[g(v(X_2, Y_2))]) \\ &+ r'(Y_1, Y_2)\sqrt{\mathrm{Var}\ \mathbb{E}_Y[f(u(X_1, Y_1))]}\sqrt{\mathrm{Var}\ \mathbb{E}_Y[g(v(X_2, Y_2))]}. \end{aligned}$$

Note that, for $i=1,2$,

$$\mathbb{E}(\mathbb{E}_Y[f(u(X_i,Y_i))]) = 0 \text{ and } \mathrm{Var}\mathbb{E}_Y[f(u(X_i,Y_i))] = \mathbb{E}(\mathbb{E}_Y[f(u(X_i,Y_i))])^2.$$

Denote

$$a = \mathbb{E}_Y[f^2(u(X_1,Y_1))],\ b = (\mathbb{E}_Y[f(u(X_1,Y_1))])^2,$$
$$c = \mathbb{E}_Y[g^2(v(X_2,Y_2))],\ d = (\mathbb{E}_Y[g(v(X_2,Y_2))])^2,$$

where by Jensen's inequality $a \geq b$ and $c \geq d$ almost surely.

By combining these inequalities we derive

$$\begin{aligned}
\mathrm{cov}[f(u(X_1,Y_1)), g(v(X_2,Y_2))] &\leq r'(Y_1,Y_2)\sqrt{\mathrm{Var}\ \mathbb{E}_Y[f(u(X_1,Y_1))]}\sqrt{\mathrm{Var}\ \mathbb{E}_Y[g(v(X_2,Y_2))]}\\
&+ r'(X_1,X_2)\mathbb{E}\sqrt{\mathrm{Var}_Y[f(u(X_1,Y_1))]}\sqrt{\mathrm{Var}_Y[g(v(X_2,Y_2))]}\\
&\leq \max(r'_X, r'_Y)\left(\mathbb{E}[\sqrt{a-b}\sqrt{c-d}] + \sqrt{\mathbb{E}[b]}\sqrt{\mathbb{E}[d]}\right).
\end{aligned}$$

By the Cauchy inequality

$$\begin{aligned}
\mathbb{E}[\sqrt{a-b}\sqrt{c-d}] + \sqrt{\mathbb{E}[b]}\sqrt{\mathbb{E}[d]} &\leq \sqrt{\mathbb{E}[a]-\mathbb{E}[b]}\sqrt{\mathbb{E}[c]-\mathbb{E}[d]]} + \sqrt{\mathbb{E}[b]}\sqrt{\mathbb{E}[d]}\\
&\leq \sqrt{\mathbb{E}[a]}\sqrt{\mathbb{E}[c]},
\end{aligned}$$

where at the last stage we have applied an elementary inequality: For positive reals x, y, u, v: $2\sqrt{xyuv} \leq xy + uv$. This ends the proof of the first inequality of the lemma. The other inequalities can be obtained similarly. ∎

Remark 7.6 *Assume $u(x,y)$ and $v(x,y)$ are non-decreasing on x only. Then, the inequalities of Lemma 7.5 are obviously true in the following form*

$$r'(u(X_1,Y_1), v(X_2,Y_2)) \leq \max(r'(X_1,X_2), \rho(Y_1,Y_2))$$

and

$$a'(u(X_1,Y_1), v(X_2,Y_2)) \leq \max(a'(X_1,X_2), \rho(Y_1,Y_2)),$$

where $\rho(Y_1, Y_2)$ is defined in (7.2).

7.4 $\mathbb{L}^2$-bounds via Interlaced Quantities

In this section we shall give bounds for the variance of partial sums of random variables in terms of the strength of negative or positive dependence.

7.4.1 Definitions of the Linear Negative or Positive Dependence Coefficients

We introduce below *the coefficients of weak linear negative and positive dependence*. For a finite or infinite N, let $X = \{X_i\ ;\ i = 1, \dots, N\}$ be a set of real-valued random variables. For any subset A of $\{1, \dots, N\}$, set $S_A = \sum_{i \in A} X_i$ and define, for any positive integer j,

$$r'_j = \sup r'(S_A, S_B), \ \ a'_j = \sup a'(S_A, S_B), \tag{7.16}$$

and

$$r^*_j = \sup r^*(S_A, S_B), \ \ a^*_j = \sup a^*(S_A, S_B), \tag{7.17}$$

where the supremum is taken over all non-intersecting subsets $A, B \subset \mathbf{N}$ with the Euclidean distance larger than j, and the coefficients r', r^*, a' and a^* are as introduced above.

7.4.2 Estimates of the Variance under Near Association Type Conditions

The general bounds

To have more insight into the variances we shall first give two results which are due to Bradley (2007, vol.1, Lemmas 8.21 and 8.22).

Lemma 7.7 *Suppose $0 \leq r < 1$. Suppose that $\{Y_1, Y_2, \dots, Y_n\}$ is a vector of square integrable centered random variables such that for any non-empty subsets $Q \subset \{1, 2, \dots, n\}$ and $Q^* = \{1, 2, \dots, n\} \backslash Q$,*

$$\operatorname{corr}\left(\sum_{k \in Q} Y_k, \sum_{k \in Q^*} Y_k \right) \leq r.$$

Then,

$$\mathbb{E}\left(\sum_{k=1}^{n} Y_k \right)^2 \leq \frac{1+r}{1-r} \sum_{k=1}^{n} \mathbb{E}(Y_k^2).$$

We remark here that this lemma also works if r^*, defined by using indicator functions, is sufficiently small to assure $r < 1$ (since $r \leq 52\, r^*(1 - \log r^*)$).

Proof of Lemma 7.7 Denote by $S_Q = \sum_{k \in Q} Y_k$, $S_{Q^*} = \sum_{k \in Q^*} Y_k$, $\sigma_n^2 = \mathbb{E}\left(\sum_{k=1}^{n} Y_k\right)^2$, $\sigma_Q^2 = \mathbb{E}\left(S_Q^2\right)$, $\sigma_{Q^*}^2 = \mathbb{E}\left(S_{Q^*}^2\right)$. With this notation and the definition of r we easily obtain

$$\begin{aligned} \sigma_Q^2 + \sigma_{Q^*}^2 &= \sigma_n^2 - 2\operatorname{cov}(S_Q, S_{Q^*}) \geq \sigma_n^2 - 2r\|S_Q\|_2 \times \|S_{Q^*}\|_2 \\ &\geq \sigma_n^2 - r(\sigma_Q^2 + \sigma_{Q^*}^2). \end{aligned}$$

Whence

$$\sigma_Q^2 + \sigma_{Q^*}^2 \geq \frac{1}{1+r} \sigma_n^2.$$

On the other hand, by the same type of arguments and the above inequality,

$$\mathbb{E}(S_Q - S_{Q^*})^2 = \sigma_Q^2 + \sigma_{Q^*}^2 - 2\text{cov}(S_Q, S_{Q^*}) \geq (1-r)(\sigma_Q^2 + \sigma_{Q^*}^2) \geq \frac{1-r}{1+r}\sigma_n^2.$$

Let $\{\varepsilon_i\}_{1\leq i\leq n}$ be a vector independent of $\{Y_i\}_{1\leq i\leq n}$ such that the variables ε_i are independent with $\mathbb{P}(\varepsilon_i = 1) = \mathbb{P}(\varepsilon_i = -1) = 1/2, i = 1, ..., n$. By the independence between $\{Y_i\}_{1\leq i\leq n}$ and the Rademacker sequence we have

$$\sum_{k=1}^n \mathbb{E}(Y_k^2) = \mathbb{E}\left(\sum_{k=1}^n \varepsilon_k Y_k\right)^2 = \frac{1}{2^n} \sum_{Q\subset\{1,2,...n\}} \mathbb{E}\left(S_Q - S_{Q^*}\right)^2 \geq \frac{1-r}{1+r}\sigma_n^2,$$

and the result follows. ∎

Similarly we have

Lemma 7.8 *Suppose $0 \leq a < 1$. Suppose that $\{Y_1, Y_2, ..., Y_n\}$ is a vector of square integrable centered random variables such that for any non-empty subsets $Q \subset \{1, 2, \dots, n\}$ and $Q^* = \{1, 2, ..., n\}\backslash Q$,*

$$\text{corr}\left(\sum_{k\in Q} Y_k, \sum_{k\in Q^*} Y_k\right) \geq -a.$$

Then,

$$\mathbb{E}\left(\sum_{k=1}^n Y_k\right)^2 \geq \frac{1-a}{1+a}\sum_{k=1}^n \mathbb{E}(Y_k^2).$$

The following lemma summarizes the variance inequalities under association type dependence conditions.

Lemma 7.9 *Let $X = \{X_1, ..., X_n\}$ be a set of real-valued random variables with finite second moments and zero means. Then, for any positive integer m and any $A \subseteq \{1, \dots, n\}$,*

$$\frac{1-a_1'}{1+a_1'}\sum_{k\in A} \mathbb{E}(X_k^2) \leq \mathbb{E}\left(\sum_{k\in A} X_k\right)^2 \leq m\frac{1+r_m'}{1-r_m'}\sum_{k\in A} \mathbb{E}(X_k^2). \tag{7.18}$$

The upper bound is non-trivial if $r_m' < 1$. It also works if r_m^* is sufficiently small to assure $r_m' < 1$ (since $r' \leq 52r^*(1-\log r^*)$). The lower bound is non-trivial if $a_1' < 1$. A similar remark holds for the coefficients a' and a^*, by taking into account that $a' \leq 52a^*(1-\log a^*)$.

Proof of Lemma 7.9 Let $S_A = \sum_{k\in A} X_k$. The upper bound follows by applying the representation

$$S_A = S_{A(0)} + \dots + S_{A(j-1)}, \text{ where } A(t) = \{i \in A : i = k + mt \text{ for some integer } m\}.$$

and using the triangular inequality. ∎

Asymptotically linear growth of $\mathrm{Var}(S_n)$ under approximate association

We discuss the asymptotically linear growth of the variance of an $\mathbb{L}^2$-stationary sequence of random variables.

Proposition 7.10 *Let $X = (X_k)_{k\geq 1}$ be a $\mathbb{L}^2$-stationary sequence of centered real-valued random variables and assume that there exists a non-negative sequence $r_n \to 0$ such that for any two non-intersecting subsets $A, B \subset \mathbf{N}^*$ with the Euclidean distance larger than n*

$$\mathrm{corr}\left(\sum_{k\in A} X_k, \sum_{k\in B} X_k\right) \leq r_n.$$

Then

$$\lim_{n\to\infty} \frac{\mathbb{E}\left(\sum_{k=1}^{n} X_k\right)^2}{n} \text{ exists.}$$

The proof follows from a slightly stronger Property 7.12 (see below).

With a similar proof we have:

Proposition 7.11 *Let $X = (X_k)_{k\geq 1}$ be a $\mathbb{L}^2$-stationary sequence of centered real-valued random variables and assume that there exists a non-negative sequence $a_n \to 0$ such that for any two non-intersecting subsets $A, B \subset \mathbf{N}^*$ with the Euclidean distance larger than n*

$$\mathrm{corr}\left(\sum_{k\in A} X_k, \sum_{k\in B} X_k\right) \geq -a_n.$$

Assume in addition that there is a constant K such that for every finite subset A

$$\mathbb{E}\left(\sum_{k\in A} X_k\right)^2 \leq K \sum_{k\in A} \mathbb{E}(X_k)^2.$$

Then

$$\lim_{n\to\infty} \frac{\mathbb{E}\left(\sum_{k=1}^{n} X_k\right)^2}{n} \text{ exists.}$$

Proof of Propositions 7.10 and 7.11 They follow from the slightly stronger Property 7.12 by taking into account Lemma 7.9. ■

A Hilbert space reformulation

Consider a Hilbert space H with the inner (or dot) product $< x, y > = x \cdot y$ and the generated norm $|x| = \sqrt{< x, x >}$.

We call the sequence $\{x_i, i \in \mathbb{Z}\}$ stationary if for all m

$$< x_i, x_j > = < x_{i+m}, x_{j+m} > .$$

Now, for any finite set B of $\mathbb{Z}$ such that $B = \bigcup_{i=1}^m A_i$ where $A_i \cap A_j = \emptyset$ if $i \neq j$, define

$$\Delta_{B,(A_i)_{1\le i\le m}} = \min_{1\le i<j\le m} d(A_i, A_j)$$

if B contains at least two elements and $\Delta_{B,B} = \infty$ if B contains at most 1 element.

Property 7.12 *Let $\{x_i, i \in \mathbb{Z}\}$ be a stationary sequence in H and assume that there exists a positive constant K such that for any finite set B,*

$$\left|\sum_{i\in B} x_i\right|^2 \le K \sum_{i\in B} |x_i|^2.$$

In addition, assume that there exists a positive valued sequence $\varepsilon(k) \to 0$ as $k \to \infty$ and such that we have one of the following properties: For any finite set B of $\mathbb{Z}$ such that $B = \bigcup_{i=1}^m A_i$ where $A_i \cap A_j = \emptyset$ if $i \neq j$,
(i) either

$$\left|\sum_{i\in B} x_i\right|^2 \le [1 + \varepsilon(\Delta_{B,(A_i)_{1\le i\le m}})] \sum_{i=1}^m \left|\sum_{k\in A_i} x_k\right|^2, \tag{7.19}$$

(ii) or

$$\left|\sum_{i\in B} x_i\right|^2 \ge [1 - \varepsilon(\Delta_{B,(A_i)_{1\le i\le m}})] \sum_{i=1}^m \left|\sum_{k\in A_i} x_k\right|^2. \tag{7.20}$$

Then, there exists a finite positive number c^2 such that

$$\lim_{n\to\infty} \frac{|x_1 + \dots + x_n|^2}{n} = c^2.$$

Proof of Property 7.12 We only prove the first part (the other one is left to the reader). We are using the blocking technique. We start by fixing two integers p and q, and writing $n = (p+q)k_n + b$, with $0 \le b < p + q$, (at the end we choose $p, q \to \infty$, such that $q = o(p)$). Finally, let

$$\sigma_n^2 = |x_1 + \dots + x_n|^2.$$

We partition the n variables into blocks. We start with a large block of size p followed by a small block of size q and so on. We are going to have $2k + 1$ blocks where $k = k_n = [n/(p+q)]$; ([.] =integer part) and the last block contains at most $p + q$ variables. Denote the sums in the large blocks of size p by $\{y_i\}_{1\le i\le k_n}$ and the sums in the small blocks of size q by $\{z_i\}_{1\le i\le k}$. The last block is denoted by z_{k+1}.

The final decomposition will look like

$$S_n = y_1 + z_1 + \dots + y_k + z_k + z_{k+1}.$$

Notice that the variables $\{y_i\}_{1\le i\le k}$ have a gap of size q between two consecutive variables. And so, by assumption,

$$|y_1 + \dots + y_k|^2 \le (1 + \varepsilon(q))(|y_1|^2 + \dots + |y_k|^2) = (1 + \varepsilon(q))k\sigma_p^2$$

and

$$|z_1 + \dots + z_{k+1}|^2 \le K(kq + (p + q))|x_1|^2. \tag{7.21}$$

By the triangle inequality

$$\begin{aligned}\sigma_n &\le |y_1 + \dots + y_k| + |z_1 + \dots + z_{k+1}| \\ &\le \{k(1 + \varepsilon(q))\}^{1/2}\sigma_p \;+\; \{K(kq + (p + q))|x_1|^2\}^{1/2},\end{aligned}$$

which gives

$$\frac{\sigma_n}{n^{1/2}} \le \{(1 + \varepsilon(q))\}^{1/2}\sigma_p\left(\frac{k_n^{1/2}}{n^{1/2}}\right) + \frac{1}{\sqrt{n}}\{K(k_n q + (p + q))|x_1|^2\}^{1/2}.$$

By letting n go to infinity, by the definition of k_n we obtain, for all p and q

$$\limsup_{n\to\infty} \frac{\sigma_n}{n^{1/2}} \le \{(1 + \varepsilon(q))p/(p + q)\}^{1/2}\{\sigma_p/p^{1/2}\} + \{Kq/(p + q)\}^{1/2}|x_1|.$$

By selecting $q = o(p)$, as p tends to infinity and taking the inferior limit over p we obtain

$$\limsup_{n\to\infty} \frac{\sigma_n}{n^{1/2}} \le \liminf_{n\to\infty} \frac{\sigma_n}{n^{1/2}},$$

which implies the existence of $\lim_{n\to\infty} n^{-1/2}\sigma_n$. This limit is finite since, by assumption, there exists a finite constant K such that $\sigma_n^2 \le Kn|x_1|^2$. ■

7.5 Approximation Inequalities for *n* Variables

7.5.1 *A First Approximation Inequality*

Let N be a finite or infinite index. Provided that, for any subset $A \in \{1, \dots, N\}$, S_A is identically distributed as $-S_A$, Lemma 7.13 below shows that it is possible to approximate the partial sums S_n by the partial sum associated with conditionally independent symmetric variables.

Lemma 7.13 *For a finite or infinite index N, let $X = \{X_i\,, i = 1, ..., N\}$ be a set of real-valued random variables with finite second moments and such that, for any subset $A \in \{1, ..., N\}$, S_A is identically distributed as $-S_A$. Suppose that $f : \mathbb{R} \to \mathbb{R}$ is a twice differentiable function satisfying $|f''(x)| \le C_f$. In addition, let $\{\varepsilon_1, ..., \varepsilon_n\}$ denote i.i.d. variables that are independent of X and such that $\mathbb{P}(\varepsilon_1 = \pm 1) = 1/2$. For any positive integer n less than N, define $S_n = X_1 + ... + X_n$ and $S_{n,\varepsilon} = X_1\varepsilon_1 + ... + X_n\varepsilon_n$. Let r_1^* and a_1^* be the weak linear dependence coefficients associated with X and defined by (7.17). Then*

$$|\mathbb{E}[f(S_n)] - \mathbb{E}[f(S_{n,\varepsilon})]|$$
$$\le C_f\left[-\frac{1}{2}(\mathrm{Var}(S_n) - \mathrm{Var}(S_{n,\varepsilon})) + 104r_1^*(1 - \log r_1^*)(\mathrm{Var}(S_n) + \mathrm{Var}(S_{n,\varepsilon}))\right], \quad (7.22)$$

where $\mathrm{Var}(S_{n,\varepsilon}) = \sum_{i=1}^n \mathbb{E}(X_i^2)$. *Also,*

$$|\mathbb{E}[f(S_n)] - \mathbb{E}[f(S_{n,\varepsilon})]|$$
$$\le C_f\left[\frac{1}{2}(\mathrm{Var}(S_n) - \mathrm{Var}(S_{n,\varepsilon})) + 104a_1^*(1 - \log a_1^*)(\mathrm{Var}(S_n) + \mathrm{Var}(S_{n,\varepsilon}))\right]. \quad (7.23)$$

Proof We shall first prove (7.22). Fix a subset $Q \in \{1, ..., n\}$ and let Q^* be its complement, i.e. $Q^* = \{1, 2, ..., n\}\backslash Q$. Let S'_{Q^*} be a variable which is identically distributed as S_{Q^*} and independent of the sequence X. By applying twice part (c) of Lemma 7.1 with the variables $X = S_Q$, $Y = S_{Q^*}$ and functions $f(x, y) = f(x + y)$ and also $f(x, y) = f(x - y)$ we obtain

$$|\mathbb{E}[f(S_Q + S_{Q^*})] - \mathbb{E}[f(S_Q + S'_{Q^*})]| \le C_f[-\mathrm{cov}(S_Q, S_{Q^*}) + \delta_{Q,Q^*}] \quad (7.24)$$

and

$$|\mathbb{E}[f(S_Q - S_{Q^*})] - \mathbb{E}[f(S_Q - S'_{Q^*})]| \le C_f[-\mathrm{cov}(S_Q, S_{Q^*}) + \delta_{Q,Q^*}], \quad (7.25)$$

where

$$\delta_{Q,Q^*} = 104r_1^*(1 - \log r_1^*)||S_Q||_2||S'_{Q^*}||_2.$$

By combining (7.24) and (7.25), and using the fact that

$$S_Q + S'_{Q^*} =^d S_Q - S'_{Q^*} \text{ and } S_Q + S_{Q^*} = S_n,$$

we then derive

$$|\mathbb{E}[f(S_n)] - \mathbb{E}[f(S_Q - S_{Q^*})]| \le 2C_f[-\mathrm{cov}(S_Q, S_{Q^*}) + \delta_{Q,Q^*}]$$
$$= C_f\left[\frac{1}{2}(-\mathrm{Var}(S_n) + \mathrm{Var}(S_Q - S_{Q^*})) + 2\delta_{Q,Q^*}\right].$$

Hence, noticing that

$$\mathbb{E}(f(S_{n,\varepsilon})) = \frac{1}{2^n} \sum_{Q\subset\{1,\dots,n\}} \mathbb{E}(f(S_Q - S_{Q^*})),$$

it follows that

$$|\mathbb{E}[f(S_n)] - \mathbb{E}[f(S_{n,\varepsilon})]| \leq C_f\left\{\frac{1}{2}(-\mathrm{Var}(S_n) + \mathrm{Var}(S_{n,\varepsilon})) + \frac{2}{2^n}\sum_{Q\subset\{1,\dots,n\}} \delta_{Q,Q^*}\right\}.$$

Notice that since $2S_Q = S_n + (S_Q - S_{Q^*})$ and $2S_{Q^*} = S_n + (S_{Q^*} - S_Q)$,

$$\begin{aligned} ||S_Q||_2||S'_{Q^*}||_2 = ||S_Q||_2||S_{Q^*}||_2 &\leq \frac{1}{4}\left(||S_n||_2 + ||S_Q - S_{Q^*}||_2\right)^2 \\ &\leq \frac{1}{2}\left(||S_n||_2^2 + ||S_Q - S_{Q^*}||_2^2\right). \end{aligned}$$

Hence summing over $Q \subset \{1, \dots, n\}$ and dividing by 2^n, we have

$$\begin{aligned} \frac{1}{2^n}\sum\nolimits_Q ||S_Q||_2||S'_{Q^*}||_2 \leq \frac{1}{2^n}\sum\nolimits_Q \mathbb{E}(S_Q^2) &\leq \frac{1}{2}\mathrm{Var}(S_n) + \frac{1}{2^{n+1}}\sum\nolimits_Q \mathbb{E}((S_Q - S_{Q^*})^2) \\ &= \frac{1}{2}(\mathrm{Var}(S_n) + \mathrm{Var}(S_{n,\varepsilon})). \end{aligned}$$

This completes the proof of (7.22).

Inequality (7.23) involving a_1^* is shown in a similar way by using part (c) of Lemma 7.3 which gives

$$|\mathbb{E}[f(S_n)] - \mathbb{E}[f(S_Q - S_{Q^*})]| \leq 2C_f\{\mathrm{cov}(S_Q, S_{Q^*}) + \delta'_{Q,Q^*}\},$$

where $\delta'_{Q,Q^*} = 104a_1^*(1 - \log a_1^*)||S_Q||_2||S_{Q^*}||_2$. ■

7.5.2 *A Second Approximation Inequality*

Now, we are going to approximate our partial sums by sums of independent variables. The bound is not so flexible as in the first approximation but it can be useful when we consider the approximation by a process with independent increments (see for instance the end of Section 10.1).

Lemma 7.14 *For a finite or infinite N, let $X = \{X_i\ ;\ i = 1, \dots, N\}$ be a set of real-valued random variables with finite second moments and such that S_A is identically distributed as $-S_A$ for any subset $A \subset \{1, \dots, N\}$. Suppose that $f : \mathbb{R}^2 \to \mathbb{R}$ is a twice differentiable function and assume that $|f''(x)| \leq C_f$. In addition, by $\{X'_1, \dots, X'_n\}$ denote independent variables such that X'_i is distributed as X_i for each $i = 1, \dots, n$. Define $S_n = X_1 + \dots + X_n$ and $\hat{S}_n = X'_1 + \dots + X'_n$. Let $(r_j^*)_{j\geq 1}$ and $(a_j^*)_{j\geq 1}$ be the sequence of weak linear dependence coefficients* *associated with X and defined by* (7.17). *Assume that there is a constant K such that for every finite subset A*

$$\mathbb{E}\left(\sum_{k\in A} X_k\right)^2 \le K\sum_{k\in A}\mathbb{E}(X_k^2). \tag{7.26}$$

Then, for $2^k < n \le 2^{k+1}$, *we have*

$$|\mathbb{E}[f(S_n)] - \mathbb{E}[f(\hat{S}_n)]| \le C_f\Big[-\frac{1}{2}(\mathrm{Var}(S_n) - \mathrm{Var}(\hat{S}_n)) + 104K\mathrm{Var}(\hat{S}_n)\sum_{i=0}^{k}(r_{2^i}^*)(1-\log r_{2^i}^*)\Big]. \tag{7.27}$$

A similar bound holds by using the coefficient of positive dependence:

$$|\mathbb{E}[f(S_n)] - \mathbb{E}[f(\hat{S}_n)]| \le C_f\Big[\frac{1}{2}(\mathrm{Var}(S_n) - \mathrm{Var}(\hat{S}_n)) + 104K\mathrm{Var}(\hat{S}_n)\sum_{i=0}^{k}(a_{2^i}^*)(1-\log a_{2^i}^*)\Big]. \tag{7.28}$$

Proof Let us prove (7.27); the proof of the other statement is the same. We are going to use induction. The initial step (so for $k = 0$) follows from inequality (7.8) of Lemma 7.1 by using the identity

$$2\,\mathrm{cov}(X, Y) = \mathrm{Var}(X+Y) - \mathrm{Var}(X+Y'), \tag{7.29}$$

where X and Y' are independent and Y' is distributed as Y.

To justify the induction step, we assume that (7.27) holds for any $k' < k$. Let $2^k < n \le 2^{k+1}$. Without restricting the generality we assume $n = 2m$ (if n is odd we just add a last variable equal to 0). Denote

$$A = \{1, 3, .., 2m-1\},\ \ B = \{2, 4, 6, ..., 2m\}.$$

Let $\tilde{X} = \{\tilde{X}_1, \dots, \tilde{X}_n\}$ be an independent copy of X and set $\tilde{S}_B = \sum_{i\in B}\tilde{X}_i$. Then, from inequality (7.8) of Lemma 7.1 applied with $X = S_A$ and $Y = S_B$, $Y' = \tilde{S}_B$, by (7.29) and (7.26), we obtain

$$\begin{aligned}|\mathbb{E}[f(S_A+S_B)] - \mathbb{E}[f(S_A+\tilde{S}_B)]| &= |\mathbb{E}[f(S_n)] - \mathbb{E}[f(S_A+\tilde{S}_B)]|\\ &\le C_f\Big[-\frac{1}{2}(\mathrm{Var}(S_A+S_B) - \mathrm{Var}(S_A+\tilde{S}_B)) + 104r_1^*(1-\log r_1^*)\|S_A\|_2\|S_B\|_2\Big]\\ &\le C_f\Big[-\frac{1}{2}(\mathrm{Var}(S_n) - \mathrm{Var}(S_A+\tilde{S}_B)) + 104Kr_1^*(1-\log r_1^*)\|\hat{S}_n\|_2^2\Big].\end{aligned}$$

Now, we define a new set of variables

$$Z_1 = X_1 + \tilde{X}_2,\ Z_2 = X_3 + \tilde{X}_4, \dots,\ Z_m = X_{2m-1} + \tilde{X}_{2m}.$$

Then the new set of variables $Z = \{Z_1, \dots, Z_m\}$ satisfies the condition of Lemma 7.14 with $k' = k - 1$, $T_m = Z_1 + \dots + Z_m = S_A + \tilde{S}_B$ and $\hat{T}_m =^d \hat{S}_n$ where $\hat{T}_m = \sum_{i=1}^m Z_i'$ with $Z_1', \dots, Z_m'$ independent r.v.'s such that $Z_i' =^d Z_i$. Moreover, by Lemma 7.5 we have $r_{2^i}^*(Z) \leq r_{2^{i+1}}^*(X)$ for all $1 \leq i \leq k-1$. With the notation $c(x) = x(1 - \log x)$, for $x > 0$, by the induction hypothesis, it follows

$$\begin{aligned}
|\mathbb{E}[f(S_A + \tilde{S}_B)] - \mathbb{E}[f(\hat{S}_n)]| &= |\mathbb{E}[f(T_m)] - \mathbb{E}[f(\hat{T}_m)]| \\
&\leq C_f\left[-\frac{1}{2}(\mathrm{Var}(T_m) - \mathrm{Var}(\hat{T}_m)) + 104K\|\hat{T}_m\|_2^2 \sum_{i=0}^{k-1} c(r_{2^i}^*(Z))\right] \\
&\leq C_f\left[-\frac{1}{2}(\mathrm{Var}(S_A + \tilde{S}_B) - \mathrm{Var}(\hat{S}_n)) + 104K\|\hat{S}_n\|_2^2 \sum_{i=0}^{k-1} c(r_{2^{i+1}}^*)\right] \\
&= C_f\left[-\frac{1}{2}(\mathrm{Var}(S_A + \tilde{S}_B) - \mathrm{Var}(\hat{S}_n)) + 104K\|\hat{S}_n\|_2^2 \sum_{i=1}^{k} c(r_{2^i}^*)\right]
\end{aligned}$$

and combining the bounds together, we have proved that (7.27) holds at step k provided it holds at step $k - 1$.

The proof of the coupling inequality in the second part of the lemma is similar, with the difference that we use Lemma 7.3 instead of Lemma 7.1. ■

8

Maximal Moment Inequalities for Weakly Negatively Dependent Variables

The aim of this chapter is to give sufficient conditions in terms of weak dependence coefficients for the partial sums or the maximum of partial sums of real-valued random variables to satisfy moment inequalities such as Khintchine or Rosenthal inequalities. We start with general probability and moment inequalities for Lipschitz functions of weakly negatively dependent variables. Besides being interesting in themselves, they will be key tools to prove moment inequalities for the partial sums associated with weakly negatively dependent variables.

8.1 General Moment Inequalities for Weakly Negatively Dependent Variables

To treat partial sums and the maximum of partial sums, it is convenient to work with Lipschitz functions with respect to a certain seminorm $||x||$ on $\mathbb{R}^n$. More exactly, a function $f : \mathbb{R}^n \to \mathbb{R}$ is called Lipschitz or $||\cdot||$–Lipschitz if

$$|f(x) - f(y)| \leq K||x - y|| \tag{8.1}$$

for a finite positive $K = K_f$ depending on f. We refer to K as a Lipschitz coefficient of the function f.

Lemma 8.1 *Assume $X = \{X_1, ..., X_n\}$ is a vector of real-valued random variables and let $\{\varepsilon_i\}_{1\leq i\leq n}$ be a vector independent on X such that the variables ε_i are independent with $\mathbb{P}(\varepsilon_i = 1) = \mathbb{P}(\varepsilon_i = -1) = 1/2$, $i = 1, ..., n$. Put $\varepsilon X = (\varepsilon_1 X_1, ..., \varepsilon_n X_n)$. Then, for any seminorm $||\cdot||$ and non-decreasing positive Lipschitz function $f : \mathbb{R}^n \to \mathbb{R}^+$ with Lipschitz coefficient K, and for any positive numbers x and y,*

Functional Gaussian Approximation for Dependent Structures. Florence Merlevède, Magda Peligrad and Sergey Utev. Oxford University Press (2019).
DOI: 10.1093/oso/9780198826941.001.0001

$$\mathbb{P}(f(X) > x + 6y) \le \mathbb{P}(f(X) > x)[r^* + \mathbb{P}(f(X) > x)] + 6\,\mathbb{P}(K\|\varepsilon X\| > y), \tag{8.2}$$

where

$$r^* = r^*[f(2X_Q), f(2X_{Q^*})] = \sup_{x,y} \operatorname{corr}\left(1_{\{f(2X_Q)>x\}}, 1_{\{f(2X_{Q^*})>y\}}\right), \tag{8.3}$$

with $Q \subset \{1, 2, \dots, n\}$, $Q^* = \{1, 2, \dots, n\}\backslash Q$ *and* $X_Q = (Y_1, Y_2, \dots, Y_n)$ *where* $Y_i = X_i$ *for* $i \in Q$ *and* $Y_i = 0$ *for* $i \notin Q$.

Proof Since $f(X) = f([X_{Q^*} - X_Q] + 2X_Q) \le f(2X_Q) + K\|X_Q - X_{Q^*}\|$, we have

$$\mathbb{P}(f(X) > x + 6y) \le \mathbb{P}(K\|X_Q - X_{Q^*}\| > 3y) + \mathbb{P}(f(2X_Q) > x + 3y).$$

We continue our estimate by analyzing the last term.

$$\begin{aligned}\mathbb{P}(f(2X_Q) > x + 3y) = \mathbb{P}(&f(2X_Q) > x + 3y, f(2X_{Q^*}) > x + y) \\ &+ \mathbb{P}(f(2X_Q) > x + 3y, f(2X_{Q^*}) \le x + y) = I + II.\end{aligned}$$

The second term, II, satisfies

$$II \le \mathbb{P}(f(2X_Q) - f(2X_{Q^*}) > 2y) \le \mathbb{P}(K\|X_Q - X_{Q^*}\| > y).$$

To estimate I, using the definition of $r^*[f(2X_Q), f(2X_{Q^*})]$, we first notice that

$$\begin{aligned}I \le\ & \mathbb{P}(f(2X_Q) > x + 3y) \times \mathbb{P}(f(2X_{Q^*}) > x + y) \\ &+ r^*[\mathbb{P}(f(2X_Q) > x + 3y)]^{1/2}[\mathbb{P}(f(2X_{Q^*}) > x + y)]^{1/2}.\end{aligned}$$

Since $f(2X_Q) = f(X + X_Q - X_{Q^*}) \le f(X) + K\|X_Q - X_{Q^*}\|$, we derive

$$\mathbb{P}(f(2X_Q) > x + 3y) \le \mathbb{P}(f(X) > x) + \mathbb{P}(K\|X_Q - X_{Q^*}\| > 3y)$$

and, similarly,

$$\mathbb{P}(f(2X_{Q^*}) > x + y) \le \mathbb{P}(f(X) > x) + \mathbb{P}(K\|X_Q - X_{Q^*}\| > y).$$

By combining estimates together, we get that, for any positive numbers x and y,

$$\begin{aligned}\mathbb{P}(f(X) > x + 6y) \le\ & 2\mathbb{P}(K\|X_Q - X_{Q^*}\| > y) + \left(\mathbb{P}(f(X) > x) + \mathbb{P}(K\|X_Q - X_{Q^*}\| > y)\right)^2 \\ &+ r^* \times \left(\mathbb{P}(f(X) > x) + \mathbb{P}(K\|X_Q - X_{Q^*}\| > y)\right) \\ \le\ & (5 + r^*)\mathbb{P}(K\|X_Q - X_{Q^*}\| > y) + \mathbb{P}(f(X) > x)[r^* + \mathbb{P}(f(X) > x)].\end{aligned}$$

To finish the proof, we sum over all the subsets Q, then divide by 2^n, and notice that

$$(1/2^n)\sum_Q \mathbb{P}(K\|X_Q - X_{Q^*}\| > y) = \mathbb{P}(K\|\varepsilon X\| > x),$$

and also that $r^* \leq 1$. ■

Lemma 8.2 *Under the notations of Lemma 8.1, assume that $r^* < 1$. Let $A > 0$,*

$$\delta = (1 - r^*)/2 \quad \textit{and assume that} \quad \mathbb{P}(f(X) > A) \leq \delta.$$

Let $p > 0$ and define $\alpha = (1/6)[(1 + \delta)^{1/p} - 1]$ and $C = 6(K(1+6\alpha)/\alpha)^p$. Then the following inequalities hold:

$$\mathbb{E}[f^p(X)1_{\{f(X)>(1+6\alpha)A\}}] \leq (1 - \delta^2)\mathbb{E}[f^p(X)1_{\{f(X)>A\}}] + C\mathbb{E}[\|\varepsilon X\|^p 1_{\{\|\varepsilon X\|>A\alpha/K\}}], \quad (8.4)$$

and also

$$\mathbb{E}[f^p(X)] \leq \delta^{-2}(1 + \delta)A^p + C\delta^{-2}\mathbb{E}[\|\varepsilon X\|^p]. \quad (8.5)$$

Proof First, we notice that

$$r^* + \mathbb{P}(f(X) > A) \leq r^* + (1 - r^*)/2 \leq 1 - (1 - r^*)/2 = 1 - \delta.$$

We apply now inequality (8.2) with $y = \alpha x$ and $x > A$ and obtain

$$\begin{aligned}\mathbb{P}(f(X) > x[1 + 6\alpha]) &\leq \mathbb{P}(f(X) > x)[r^* + \mathbb{P}(f(X) > A)] + 6\,\mathbb{P}(K\|\varepsilon X\| > \alpha x)\\ &\leq (1 - \delta)\mathbb{P}(f(X) > x) + 6\,\mathbb{P}(K\|\varepsilon X\| > \alpha x). \quad (8.6)\end{aligned}$$

For proving (8.4) we shall use the well-known formula below obtained by integration by parts: For any positive integrable random variable U and positive number a,

$$\mathbb{E}(U^p 1_{\{U>a\}}) = a^p\mathbb{P}(U > a) + p\int_a^\infty x^{p-1}\mathbb{P}(U > x)dx.$$

We apply this equality with $U = f(X)/[1 + 6\alpha]$ and $a = A$.

Multiplying (8.6) by px^{p-1} and integrating from A to $+\infty$, we derive

$$\begin{aligned}p\int_A^\infty x^{p-1}\mathbb{P}(f(X) > x[1 + 6\alpha])\,dx &\leq (1 - \delta)p\int_A^\infty x^{p-1}\mathbb{P}(f(X) > x)\,dx\\ &+ 6p\int_A^\infty x^{p-1}\mathbb{P}(K\|\varepsilon X\| > \alpha x)\,dx. \quad (8.7)\end{aligned}$$

By (8.6) written with $x = A$, we have

$$A^p\mathbb{P}(f(X) > A[1+6\alpha]) \le (1-\delta)A^p\mathbb{P}(f(X) > A) + 6A^p\mathbb{P}(K\|\varepsilon X\| > \alpha A),$$

which, added to the inequality (8.7), leads to

$$[1+6\alpha]^{-p}G(A[1+6\alpha]) \le (1-\delta)G(A) + I \tag{8.8}$$

where

$$G(w) = \mathbb{E}[f^p(X)1_{\{f(X)>w\}}] \text{ and } I = 6(K/\alpha)^p\mathbb{E}[\|\varepsilon X\|^p 1_{\{\|\varepsilon X\|>A\alpha/K\}}].$$

This completes the proof of (8.4) since $[1+6\alpha]^p(1-\delta) = 1-\delta^2$.

To prove (8.5), we notice that by (8.4),

$$\begin{aligned}\mathbb{E}(f^p(X)) &\le (1+6\alpha)^pA^p + \mathbb{E}[f^p(X)1_{\{f(X)>(1+6\alpha)A\}}]\\ &\le (1+\delta)A^p + (1-\delta^2)\mathbb{E}(f^p(X)) + C\mathbb{E}(\|\varepsilon X\|^p),\end{aligned}$$

and the result follows. ■

8.2 A Khintchine–Marcinkiewicz–Zygmund type Inequality

This section deals with moment inequalities under weak linear *negative* dependence. From Lemma 8.2, we easily generalize the Khintchine inequality to random variables satisfying $r_1' < 1$.

Proposition 8.3 *Assume that $X = (X_1, ..., X_n)$ is a vector of centered random variables with $r_1' < 1$, where r_1' is defined by (7.16). Let $2 \le t < \infty$. Then there exists a constant $c_{t,\delta}$ which depends on t and $\delta = (1-r_1')/2$ such that*

$$\mathbb{E}|S_n|^t \le c_{t,\delta}\mathbb{E}\left|\sum_{i=1}^n X_i^2\right|^{t/2}. \tag{8.9}$$

Proof Let $x_+ = \max(x, 0)$. Since

$$\mathbb{E}|S_n|^t = \mathbb{E}(S_n)_+^t + \mathbb{E}(-S_n)_+^t$$

and since

$$-X = (-X_1, \dots, -X_n)$$

satisfies

$$r_1'(X) = r_1'(-X),$$

instead of the function $|x|^t$ we can consider the coordinatewise non-decreasing function $(x_+)^t$. Also since $r_1^* \leq r_1' < 1$, Lemma 8.2 holds also with r_1' instead of r_1^* and $\delta = (1 - r_1')/2$ instead of $\delta = (1 - r_1^*)/2$. We then apply Lemma 8.2 with $p = t$, $f((x_1, \dots, x_n)) = (\sum_{i=1}^n x_i)_+$, the seminorm $\|\varepsilon X\| = |\sum_{i=1}^n \varepsilon_i X_i|$ and $A^2 = \delta^{-1}[\mathbb{E}(S_n^2)]$, where $\delta = (1 - r_1')/2$ and obtain by (8.5) that

$$\mathbb{E}[(S_n)_+^t] \leq \delta^{-2}(1 + \delta)A^t + \delta^{-2}C\mathbb{E}[\|\varepsilon X\|^t].$$

Note that $\sum_{i=1}^n \varepsilon_i X_i$ is a martingale with respect to the sigma algebra generated by $\{(\varepsilon_k, X_k)\}_{1\leq k\leq n}$. Hence by the Burkholder–Khintchine inequality (see Theorem 2.12), the last term is bounded by

$$\mathbb{E}\left(\left|\sum_{i=1}^n \varepsilon_i X_i\right|^t\right) \leq C_t \mathbb{E}\left(\left|\sum_{i=1}^n X_i^2\right|^{t/2}\right), \tag{8.10}$$

where C_t is a positive constant depending only on t. By combining these inequalities we obtain:

$$\mathbb{E}[(S_n)_+^t] \leq \delta^{-2-t/2}(1 + \delta)\mathbb{E}^{t/2}(S_n^2) + c_{t,\delta}\mathbb{E}\left(\left|\sum_{i=1}^n X_i^2\right|^{t/2}\right). \tag{8.11}$$

The inequality (8.9) follows easily from (8.11) and the estimation of $\mathbb{E}(S_n^2)$ in Lemma 7.9, that holds when $r_1' < 1$. To complete the proof we shall use the inequality $\mathbb{E}(X_1^2 + \dots + X_n^2)^{t/2} \geq [\mathbb{E}(X_1^2) + \dots + \mathbb{E}(X_n^2)]^{t/2}$ valid for $t \geq 2$. ■

Proposition 8.4 *If in addition to the conditions of Proposition* 8.3 *we assume that* $(|X_i|^t)_{1\leq i\leq n}$ *is an uniformly integrable family, then* $\{(|S_n|/\sqrt{n})^t\}_{1\leq i\leq n}$ *is uniformly integrable.*

Proof Denote by

$$L(A) = \limsup_{n\to\infty} \mathbb{E}\left(\frac{|S_n|^t}{n^{t/2}} 1_{\{|S_n|>n^{1/2}A\}}\right).$$

We have to show that $\lim_{A\to\infty} L(A) = 0$. Note that $L(A)$ is finite since, by Proposition 8.3 and the fact that $(|X_i|^t)_{1\leq i\leq n}$ is an uniformly integrable family, $\{\mathbb{E}(|S_n|/\sqrt{n})^t\}_{1\leq i\leq n}$ is uniformly bounded. It is also non-increasing in A and therefore convergent. Denote by $L = \lim_{A\to\infty} L(A)$. Let $\{\varepsilon_i\}_{1\leq i\leq n}$ be a vector independent of $\{X_i\}_{1\leq i\leq n}$ such that the variables ε_i are independent with $\mathbb{P}(\varepsilon_i = 1) = \mathbb{P}(\varepsilon_i = -1) = 1/2$, $i = 1, \dots, n$. Let

$$S_n^\varepsilon = \sum_{i=1}^n \varepsilon_i X_i.$$

By applying (8.4) to $X = S_n/\sqrt{n}$ and $f(x) = x_+/\sqrt{n}$ and passing to the limit $A \to \infty$, we then deduce

$$L \leq (1 - \delta^2)L + C \lim_{A\to\infty} \limsup_{n\to\infty} n^{-t/2}\mathbb{E}[|S_n^\varepsilon|^t 1_{\{|S_n^\varepsilon|>A\alpha n^{-1/2}\}}].$$

Therefore

$$L \leq \frac{C}{\delta^2} \lim_{A\to\infty} \limsup_{n\to\infty} n^{-t/2}\mathbb{E}[|S_n^\varepsilon|^t \mathbf{1}_{\{|S_n^\varepsilon|>A\alpha n^{-1/2}\}}].$$

To end the proof it remains to show that $\{(|S_n^\varepsilon|/\sqrt{n})^t\}_{n\geq 1}$ is uniformly integrable. With this aim, we shall use a truncation argument. Let A and B be positive real numbers. Define

$$X_i' = X_i I(|X_i| \leq B) \text{ and } X_i'' = X_i - X_i'.$$

Let

$$S_n' = \sum_{i=1}^n \varepsilon_i X_i' \text{ and } S_n'' = \sum_{i=1}^n \varepsilon_i X_i''.$$

Observe that by the convexity of $|x|^t$ and the linearity of the expectation,

$$\mathbb{E}\left(\frac{|S_n^\varepsilon|^t}{n^{t/2}} 1_{\{|S_n^\varepsilon|>2An^{1/2}\}}\right) \leq 2^t A^{-t} n^{-t} \mathbb{E}\left(|S_n'|^{2t}\right) + 2^t n^{-t/2}\mathbb{E}\left(|S_n''|^t\right).$$

Applying again the Burkholder–Khintchine inequality, we get

$$\mathbb{E}\left(|S_n'|^{2t}\right) \leq c_t \mathbb{E}\left(\sum_{i=1}^n (X_i')^2\right)^t \leq c_t n^t B^{2t},$$

$$\mathbb{E}\left(|S_n''|^t\right) \leq c_t \mathbb{E}\left(\sum_{i=1}^n \left(X_i''\right)^2\right)^{t/2} \leq c_t n^{t/2} \max_{1\leq i\leq n} \mathbb{E}\left(|X_i|^t \mathbf{1}_{\{|X_i|>B\}}\right).$$

So, overall,

$$\mathbb{E}\left(\frac{|S_n^\varepsilon|^t}{n^{t/2}} 1_{\{|S_n^\varepsilon|>2An^{1/2}\}}\right) \leq 2^t c_t \left(A^{-t}B^{2t} + \max_{1\leq i\leq n} \mathbb{E}\left(|X_i|^t 1_{\{|X_i|>B\}}\right)\right).$$

The uniform integrability of the sequence $\{(|S_n^\varepsilon|/\sqrt{n})^t\}_{n\geq 1}$ then follows by letting $B = A^{1/4}$ and letting A tend to ∞. This ends the proof of the proposition. ■

8.3 Rosenthal Moment Inequalities

To derive more powerful inequalities of similar type to von Bahr and Esseen (1965) and Rosenthal (1970) moment inequalities, as well as maximal inequalities, we need a more restrictive coefficient of negative dependence than the coefficient of weak linear negative dependence used before.

8.3.1 Definition of r-negatively Dependent Vectors

Definition 8.5 *Let $X = (X_1, \ldots, X_n)$ be a $\mathbb{R}^n$-valued random vector and assume that there exists a non-negative $r < 1$ such that for any two non-intersecting subsets $Q, Q^* \subseteq \{1, \ldots, n\}$ and any two coordinatewise non-decreasing functions v and w*

$$\operatorname{corr}(v(X_Q), w(X_{Q^*})) \leq r < 1, \tag{8.12}$$

where $X_Q = (Y_1, Y_2, \ldots, Y_n)$ with $Y_i = X_i$ for $i \in Q$ and $Y_i = 0$ for $i \notin Q$. The vector X is then said to be rND (r-negatively dependent). An infinite collection of variables is said to be a rND sequence if every finite subcollection is rND.

Comments

1) If $r = 0$ in (8.12), this is exactly the definition of *negative association* as introduced by Alam and Saxena (1981) and Joag-Dev and Prochan (1983), and studied in many papers by Newman (1984), Shao (2000), Pemantle (2000) and Matula (2003) among others. To shorten the notation, we then call X an *ND* vector (resp. an *ND* sequence) an *0ND* vector (resp. a *0ND* sequence), and we have $ND \equiv NA$.
   As pointed out by Joag-Dev and Proschan (1983), a number of well-known multivariate distributions are negatively associated such as multinomial, multivariate geometric, Dirichlet, negatively correlated normal distribution, permutation distribution, distributions encountered in random sampling without replacement and joint distributions of ranks. The paper by Pemantle (2000) contains an impressive survey of negative dependence and a variety of examples and conjectures. Shao (2000) obtained various moment inequalities for this structure.
2) If in addition of assuming $r < 1$ in (8.12), we suppose also that all the correlations in (8.12) are positive, we get a subclass of (positively) associated random vectors. Associated processes were studied first by Newman (1980) and further extensively studied by many other mathematicians. Papers by Newman and Wright (1982), Newman (1984) and Birkel (1987, 1993) among many others contain limit theorems, maximal inequalities, estimation of higher moments basically obtained under conditions imposed on $u(m) = \sup_k \sum_{|k-i| \geq m} \operatorname{cov}(X_k, X_i)$. Our definition provides an alternative way of restricting the dependence for an associated vector.

The next remark gathers some useful facts about *rND*-vectors.

Remark 8.6 *Assume that the $\mathbb{R}^n$-valued random vector $X = (X_1, \ldots, X_n)$ satisfies (8.12). Then so does the vector $-X$ and the vector $X_g = (g_1(X_1), \ldots, g_n(X_n))$, where all the functions g_i, $i = 1, \ldots, n$ are non-decreasing. Moreover if X and Y satisfy (8.12) and are independent so does the vector $U_g = (g_1(X_1, Y_1), \ldots, g_n(X_n, Y_n))$, where all the functions g_i, $i = 1, \ldots, n$ are coordinatewise non-decreasing.*

8.3.2 *Rosenthal-type Moment Inequality for the Partial Sums*

Lemma 8.7 *Assume that the centered $\mathbb{R}^n$-valued random vector $X = (X_1, ..., X_n)$ satisfies* (8.12). *Let $1 \le t < \infty$. Then there exist positive constants $c_{t,\delta}$ and $c'_{t,\delta}$ which depend only on t and $\delta = (1 - r)/2$ such that*

$$\mathbb{E}|S_n|^t \le c_{t,\delta} L_t, \ 1 \le t \le 2 \ \textit{and} \ \mathbb{E}|S_n|^t \le c'_{t,\delta} \max(L_t, L_2^{t/2}), \ t \ge 2,$$

where

$$L_t = \sum_{i=1}^{n} \mathbb{E}|X_i|^t, t \ge 1. \tag{8.13}$$

Proof In the proof we apply Lemma 8.2 with the seminorm $\|x\| = |x_1 + ... + x_n|$ and functional $f(x) = (\sum_{i=1}^n x_i)_+ = \max\left(\sum_{i=1}^n x_i, 0\right)$.

To estimate the moments of order $t \ge 1$, we shall use condition (8.12) in its full power, by using a truncation argument.

We consider first the case $1 \le t \le 2$. Similarly to the proof of Proposition 8.3, it is enough to show that

$$\mathbb{E}(S_n)_+^t \le c_{t,\delta} L_t.$$

Fix $y > 0$ and let $f_y(x) = (x \wedge y) \vee (-y)$. Define $S_{n,y} = f_y(X_1) + ... + f_y(X_n)$. We notice that

$$\mathbb{P}((S_n)_+ > y) \le \left[\sum_{i=1}^{n} \mathbb{P}(|X_i| > y)\right] + y^{-2}[\mathbb{E}(S_{n,y})_+]^2 + y^{-2}\text{Var}(S_{n,y})_+.$$

By Markov's inequality, the first term is bounded by $L_t y^{-t}$. Next, since $\mathbb{E}(X_i) = 0$,

$$|\mathbb{E}[f_y(X_i)]| = |\mathbb{E}\left(X_i I_{(|X_i| \le y)} + y I_{(X_i > y)} - y I_{(X_i < -y)}\right| \le \mathbb{E}(X_i I_{(|X_i| > y)}) + y I_{(|X_i| > y)}.$$

Hence $|\mathbb{E}[f_y(X_i)]| \le 2y^{1-t}\mathbb{E}|X_i|^t$. Therefore $y^{-2}[\mathbb{E}(S_{n,y})]^2 \le 4(L_t/y^t)^2$. On the other hand, by Lemma 7.9, the third term is bounded by

$$y^{-2}\delta^{-1}\left[\sum_{i=1}^{n} \mathbb{E}(f_y^2(X_i))\right] = \delta^{-1}\left[\sum_{i=1}^{n} \mathbb{E}\left(\frac{X_i^2}{y^2} I_{|X_i| \le y)}\right) + \sum_{i=1}^{n} \mathbb{P}(|X_i| > y)\right] \le \delta^{-1} y^{-t} L_t,$$

where we have used that $t \le 2$. Taking $y = 4\delta^{-2/t} L_t^{1/t}$, we get

$$\mathbb{P}((S_n)_+ > y) \le L_t/y^t + 4(L_t/y^t)^2 + \delta^{-1} y^{-t} L_t \le \delta.$$

By using (8.5) with $A = y$ and the Burkholder–Khintchine inequality for martingales (see Theorem 2.12), we then derive that, for $1 \le t \le 2$,

$$\mathbb{E}[(S_n)_+^t] \le \delta^{-2}\left((1 + \delta)y^t + CC_t \mathbb{E}\left|\sum_{i=1}^{n} X_i^2\right|^{t/2}\right) \le c_{t,\delta} L_t$$

and the proof of the case $1 \le t \le 2$ is complete.

We turn now to the case $t > 2$. With this aim, we apply the induction argument on i such that $2^i < t \leq 2^{i+1}$. Let $i = 1$. As before, without loss of generality we work with $\mathbb{E}[(S_n)_+^t]$ and show that there exists a positive constant depending only on t and δ such that

$$\mathbb{E}[(S_n)_+^t] \leq c_{t,\delta} \max(L_t, L_2^{t/2}). \tag{8.14}$$

Take $A^2 = L_2/\delta^2$. Then, by Markov's inequality and Lemma 7.7

$$\mathbb{P}(|S_n| > A) \leq \mathbb{E}[S_n^2]/A^2 \leq L_2\delta^{-1}/(L_2\delta^{-2}) = \delta.$$

Further, we proceed in similar fashion as in the case $1 \leq t \leq 2$. By using (8.5) and the Burkholder–Khintchine inequality for martingales (see Theorem 2.12), we get

$$\mathbb{E}[(S_n)_+]^t \leq \delta^{-2-t}(1+\delta)L_2^{t/2} + \delta^{-2}CC_t\mathbb{E}\left|\sum_{i=1}^n X_i^2\right|^{t/2}.$$

For the second term J, say, we first bound it by

$$\begin{aligned}
J &\leq \delta^{-2}CC_t3^{t/2-1}\left(\sum_{i=1}^n \mathbb{E}[X_i^2]\right)^{t/2} \\
&+ \delta^{-2}CC_t3^{t/2-1}\mathbb{E}\left|\sum\nolimits_{i=1}^n \{(X_i)_+^2 - \mathbb{E}[(X_i)_+^2]\}\right|^{t/2} \\
&+ \delta^{-2}CC_t3^{t/2-1}\mathbb{E}\left|\sum\nolimits_{i=1}^n \{(-X_i)_+^2 - \mathbb{E}[(-X_i)_+^2]\}\right|^{t/2} \\
&= J_1 + J_2 + J_3.
\end{aligned}$$

It is then immediate that

$$J_1 \leq \delta^{-2}CC_t3^{t/2-1}L_2^{t/2}.$$

To estimate J_2 and J_3 we may apply the first part of Lemma 8.7 (because $1 < t/2 \leq 2$) and by the technical Remark 8.6, the random variables $\{(X_i)_+^2 - \mathbb{E}[(X_i)_+^2], i = 1, ..., n\}$ are r-negatively dependent and so are $\{(-X_i)_+^2 - \mathbb{E}[(-X_i)_+^2], i = 1, ..., n\}$. Therefore, there exists a positive constant c depending only t and δ such that

$$\begin{aligned}
J_2 + J_3 &\leq c\sum_{i=1}^n \left(\mathbb{E}|(X_i)_+^2 - \mathbb{E}[(X_i)_+^2]|^{t/2} + \mathbb{E}|(-X_i)_+^2 - \mathbb{E}[(-X_i)_+^2]|^t\right) \\
&\leq 2^{t+1}c\sum_{i=1}^n \mathbb{E}|X_i|^t = 2^{t+1}cL_t.
\end{aligned} \tag{8.15}$$

Combining together the above estimates, we prove (8.14) in the case $2 \leq t \leq 4$.

Assume now that $2^{i+1} < t \le 2^{i+2}$ where $i \ge 1$ so $t > 4$. The induction step from i to $i+1$ is proved in the similar fashion. The main difference is in the last step where we apply the induction assumption and therefore Lemma 8.7 to the r-negatively dependent variables $\{(X_i)_+^2 - \mathbb{E}[(X_i)_+^2], i = 1, ..., n\}$ and $\{(-X_i)_+^2 - \mathbb{E}[(-X_i)_+^2], i = 1, ..., n\}$. And using the bound (8.15) on the Lyapunov characteristics L_t' of the transformed variables via the Lyapunov characteristics L_t of original variables we then derive

$$J_2 + J_3 \le c \max[L_t, L_4^{t/4}],$$

where c is a positive constant depending only t and δ. To finish the proof of the induction step we remind that for any positive integers u, v and any $\alpha \in [0, 1]$: $L_{\alpha u+(1-\alpha)v} \le L_u^{\alpha} L_v^{1-\alpha}$. Hence, since $2\frac{t-4}{t-2} + t\frac{2}{t-2} = 4$,

$$\begin{aligned} L_4^{t/4} \le L_2^{t(t-4)/(4(t-2))} \cdot L_t^{2t/(4(t-2))} &\le [\max(L_2^{t/2}, L_t)]^{(t-4)/(2(t-2))+t/(2(t-2))} \\ &= [\max(L_2^{t/2}, L_t)]. \end{aligned}$$

■

8.4 Maximal Moment Inequalities for Partial Sums

Now, we are ready to state and prove the key maximal inequality of this chapter which is the extension of Lemma 8.7 to the maximum of partial sums.

Theorem 8.8 *Assume that the centered $\mathbb{R}^n$-valued random vector $X = (X_1, ..., X_n)$ satisfies (8.12). Let $1 \le t < \infty$. Then there exist positive constants $c_{t,\delta}$ and $c_{t,\delta}'$ which depend only on t and $\delta = (1-r)/2$ such that*

$$\mathbb{E}[\max_{1\le k\le n} |S_k|^t] \le c_{t,\delta} L_t,\ 1 \le t \le 2 \ \text{and}\ \mathbb{E}[\max_{1\le k\le n} |S_k|^t] \le c_{t,\delta}' \max(L_t, L_2^{t/2}),\ t > 2, \tag{8.16}$$

where L_t is defined in (8.13).

Proof Let $x_+ = \max(x, 0)$ and define

$$f(x_1, \dots, x_n) = \max_{1\le k\le n}(x_1 + \dots + x_k)_+ \quad \text{and} \quad \|(x_1, \dots, x_n)\| = \max_{1\le k\le n} |x_1 + \dots + x_k|.$$

Notice that the function f is coordinatewise non-decreasing and satisfies the Lipschitz condition with the Lipschitz coefficient 1 with respect to the norm $\|(x_1, ..., x_n)\|$.

Observe that

$$\mathbb{E}[\max_{1\le k\le n} |S_k|^t] \le \mathbb{E}(f(X))^t + \mathbb{E}(f(-X))^t.$$

By the inequality (8.5) of Lemma 8.2 applied separately to $f(X)$ and $f(-X)$ with $A = \mathbb{E}\max_{1\le k\le n} |S_k|/\delta$, and Burkholder-Khintchine inequality (see Theorem 2.12), we obtain that there exists a positive constant $c_{t,\delta}^{(1)}$ depending only on t and δ such that

$$\mathbb{E}[\max_{1\le k\le n} |S_k|^t] \le c^{(1)}_{t,\delta}\left(\mathbb{E}[\max_{1\le k\le n} |S_k|/\delta]^t + \mathbb{E}\,|X_1^2 + \ldots + X_n^2|^{t/2}\right). \tag{8.17}$$

To estimate the second term in the right-hand side of (8.17), we apply separately the moment inequality of Lemma 8.7 to the variables $\{(X_i)_+^2 - \mathbb{E}[(X_i)_+^2], i = 1, \ldots, n\}$ and $\{(-X_i)_+^2 - \mathbb{E}[(-X_i)_+^2], i = 1, \ldots, n\}$. By taking into account Remark 8.6 we then derive that there exists a positive constant $C_{t/2,\delta}$ depending only on t and δ such that

$$\mathbb{E}\,|X_1^2 + \ldots + X_n^2|^{t/2} \le C^t_{t/2,\delta} Q_t,$$

where $Q_t = L_t$, when $1 \le t \le 2$ and $Q_t = \max(L_t, L_2^{t/2})$, when $t > 2$.

Therefore, inequality (8.17) implies

$$\mathbb{E}[\max_{1\le k\le n} |S_k|^t] \le c^{(2)}_{t,\delta}\left[\left(\mathbb{E}\max_{1\le k\le n} |S_k|\right)^t + Q_t\right]. \tag{8.18}$$

From (8.18) it follows that in order to prove (8.16) it remains to show that there exists a positive constant $c^{(3)}_{t,\delta}$ depending only on δ and t such that

$$\mathbb{E}[\max_{1\le k\le n} |S_k|] \le c^{(3)}_{t,\delta}[\mathbb{E}(X_1^2) + \ldots + \mathbb{E}(X_n^2)]^{t/2},$$

which is proved in Lemma 8.9 below completing the proof. ∎

Lemma 8.9 *Suppose that $(X_1, \ldots, X_n)$ is a square integrable centered $\mathbb{R}^n$-valued random vector satisfying (8.12). Then there exists a positive constant $K = K(r)$ such that*

$$\mathbb{E}\max_{1\le i\le n}\left|\sum_{j=1}^{i} X_j\right| \le K\sqrt{\sum_{j=1}^{n} \mathbb{E}X_j^2}\,.$$

Proof Let $\sigma_n^2 = \sum_{i=1}^n \mathbb{E}(X_i^2)$ and define

$$a_n = \sup_X\left(\mathbb{E}\max_{1\le i\le n}\left|\sum_{j=1}^{i} X_j\right| / \sigma_n\right), \tag{8.19}$$

where the supremum is taken over all centered and square integrable random vectors $X = (X_1, \ldots, X_n)$ such that $\sigma_n^2 > 0$ and satisfying (8.12) with the same $r < 1$.

Let M be a fixed positive integer (less than n) that will be specified later in this section, $\varepsilon = 1/\sqrt{M}$ and

$$g_{\varepsilon\sigma_n}(x) = (x \wedge \varepsilon\sigma_n) \vee (-\varepsilon\sigma_n).$$

Start with the decomposition

$$\Sigma_{i\in A} X_i = \Sigma_{i\in A}\xi_i + \Sigma_{i\in A}\eta_i,$$

where, for $i = 1, \ldots, n$,

$$\xi_i = g_{\varepsilon\sigma_n}(X_i) - \mathbb{E}g_{\varepsilon\sigma_n}(X_i) \text{ and } \eta_i = X_i - \xi_i.$$

Note that

$$\mathbb{E}\max_{1\le j\le n}\left|\sum_{i=1}^{j}\eta_i\right| \le \sum_{i=1}^{n}\mathbb{E}|\eta_i| \le 2\sum_{i=1}^{n}\mathbb{E}(X_i - \varepsilon\sigma_n)_+ .$$

Hence

$$\mathbb{E}\max_{1\le j\le n}\left|\sum_{i=1}^{j}\eta_i\right| \le \frac{2\sqrt{M}}{\sigma_n}\sum_{i=1}^{n}\mathbb{E}(X_i^2) = 2\sqrt{M}\sigma_n .$$

It follows that

$$\mathbb{E}\max_{1\le i\le n}\left|\sum_{j=1}^{i}X_j\right| \le \mathbb{E}\max_{1\le i\le n}\left|\sum_{j=1}^{i}\xi_j\right| + 2\sqrt{M}\sigma_n .$$

To estimate $\mathbb{E}\max_{1\le i\le n}\left|\sum_{j=1}^{i}\xi_j\right|$ we shall use a blocking procedure. Take $m_0 = 0$ and define the integers m_i recursively by

$$m_i = \min\left\{m, m > m_{i-1}, \sum_{j=m_{i-1}+1}^{m}\mathbb{E}(\xi_j^2) > (1/M)\sigma_n^2\right\}.$$

Note that, if we denote by ℓ the number of integers produced by this procedure, i.e. m_0, $m_1, \ldots, m_{\ell-1}$, we have

$$1 \ge \left\{\sum_{i=1}^{\ell-1}\sum_{j=m_{i-1}+1}^{m_i}\mathbb{E}(\xi_j^2)\right\}/\sigma_n^2 > (\ell-1)/M \text{ so that } \ell \le M.$$

For convenience we put $m_\ell = n$. For $1 \le i \le \ell - 1$, define

$$Z_i = \sum_{j=m_{i-1}+1}^{m_i}\xi_j, \text{ and let } Z_\ell = \sum_{j=m_{\ell-1}+1}^{n}\xi_j.$$

Observe that

$$\mathbb{E}\max_{1\le i\le n}\left|\sum_{j=1}^{i}\xi_j\right| \le \mathbb{E}\max_{1\le k\le \ell}\left|\sum_{i=1}^{k}Z_i\right| + \mathbb{E}\max_{1\le k\le \ell}\left(\max_{m_{k-1}<j<m:k}\left|\sum_{i=m_{k-1}+1}^{j}\xi_i\right|\right).$$

Therefore,

$$\begin{aligned}\mathbb{E}\max_{1\le i\le n}\left|\sum_{j=1}^{i}\xi_j\right| &\le \sum_{k=1}^{\ell}\mathbb{E}\,|Z_k| + \left\{\sum_{k=1}^{\ell}\mathbb{E}\max_{m_{k-1}<j<m_k}\left|\sum_{i=m_{k-1}+1}^{j}\xi_i\right|^4\right\}^{1/4}\\ &= I + I\!I.\end{aligned} \tag{8.20}$$

By the Cauchy–Schwarz inequality and Lemma 8.7 with $t = 2$, we get

$$\mathbb{E}|Z_k| \le (\mathbb{E}Z_k^2)^{1/2} \le \left(c_r \sum_{j=m_{k-1}+1}^{m_k} \mathbb{E}\xi_j^2\right)^{1/2},$$

where c_r is a positive constant depending only on r. Hence, by the Cauchy inequality for sequences and the fact that $\ell \le M$, we derive

$$I \le c_r^{1/2}\sum_{k=1}^{\ell}\left(\sum_{i=m_{k-1}+1}^{m_k}\mathbb{E}\xi_i^2\right)^{1/2} \le [c_r M]^{1/2}\sigma_n.$$

To estimate the second term II in (8.20) we apply inequality (8.18). It follows that there exists a positive constant c_r' depending only on r such that

$$\begin{aligned}&\mathbb{E}\max\nolimits_{m_{k-1}<j<m_k}\left|\sum\nolimits_{i=m_{k-1}+1}^{j}\xi_i\right|^4\\ &\le c_r'\left[\mathbb{E}^4\max\nolimits_{m_{k-1}<j<m_k}\left|\sum\nolimits_{i=m_{k-1}+1}^{j}\xi_i\right|\right.\\ &\left.+\sum\nolimits_{j=m_{k-1}+1}^{m_k-1}\mathbb{E}(\xi_j^4) + \left(\sum\nolimits_{j=m_{k-1}+1}^{m_k-1}\mathbb{E}(\xi_j^2)\right)^2\right].\end{aligned} \tag{8.21}$$

We estimate each term in the right-hand side of (8.21) separately.

By the definition of m_k,

$$\sum_{j=m_{k-1}+1}^{m_k-1}\mathbb{E}(\xi_j^2) \le \frac{\sigma_n^2}{M},$$

so that by using our notation (8.19) for a_n and the definition of ξ_k, we obtain

$$\sum_{j=m_{k-1}+1}^{m_k-1}\mathbb{E}(\xi_j^4) \le 4\varepsilon^2\sigma_n^2\sum_{j=m_{k-1}+1}^{m_k-1}\mathbb{E}(\xi_j^2) \le \frac{4\sigma_n^4}{M^2}$$

and

$$\mathbb{E}^4 \max_{m_{k-1}<j<m_k} \left| \sum_{i=m_{k-1}+1}^{j} \xi_i \right| \leq a_n^4 \left(\sum_{j=m_{k-1}+1}^{m_k-1} \mathbb{E}(\xi_j^2) \right)^2 \leq \frac{a_n^4 \sigma_n^4}{M^2}.$$

Overall, by (8.21), the above considerations and the fact that $\ell \leq M$, we obtain

$$II^4 \leq c_r' \left[\frac{a_n^4 \sigma_n^4}{M} + \frac{5\sigma_n^4}{M} \right].$$

Now by (8.20) and our estimates for I and II, we get

$$\mathbb{E} \max_{1 \leq i \leq n} \left| \sum_{j=1}^{i} X_j \right| \leq \mathbb{E} \max_{1 \leq i \leq n} \left| \sum_{j=1}^{i} \xi_j \right| + 2\sqrt{M}\sigma_n \leq [c_r M]^{1/2} \sigma_n$$

$$+ (c_r')^{1/4} \left[\frac{a_n \sigma_n}{M^{1/4}} + \frac{5^{1/4} \sigma_n}{M^{1/4}} \right] + 2M^{1/2} \sigma_n.$$

Therefore, by the definition (8.19) and setting $D_r = \max(1, c_r, c_r')$, we get

$$a_n \leq (D_r/M)^{1/4} a_n + (D_r^{1/2} + 2)M^{1/2} + (5D_r/M)^{1/4}.$$

Now select $M = M(r)$ to be

$$M = [16D_r] + 1.$$

Since $M \geq 16D_r$ and $D_r \geq 1$ implying that $M \leq 17D_r$, we obtain

$$a_n \leq \frac{a_n}{2} + 17^{1/2}(D_r^{1/2} + 2)D_r^{1/2} + (5/16)^{1/4} := \frac{a_n}{2} + \frac{K}{2},$$

which implies the desired result. ■

8.5 The Weak Law of the Large Numbers Type Arguments

Convergence of quadratic characteristics such as quadratic variation $\sum_k (X_{t_{k+1}} - X_{t_k})^2$ is an important tool in the analysis of stochastic processes.

In this part, we develop tools in order to prove such a convergence, via the asymptotic weak dependence properties. In our first Lemma 8.10, we prove the weak law of large numbers using the coefficients of weak linear dependence r^* and a^* defined in (7.17). Then, in Lemma 8.11, we treat the convergence for the triangular arrays under the stronger condition (8.22). These results will be later used for proving the CLT and IP.

8.5.1 Convergence for Near-Stationary Sequences

Lemma 8.10 *Let $\{X_i\,; i = 1, 2, ...\}$ be a centered, uniformly square integrable sequence of random variables such that either*

(i) $r_j^* \to 0$ *as* $j \to \infty$ *or*

(ii) $a_j^* \to 0$ *as* $j \to \infty$ *and in addition* $Var(S_n)/n^2 \to 0$ *as* $n \to \infty$.

Then, for any positive integer m,

$$\frac{1}{n}\sum_{i=1}^{n}[(S_{i+m} - S_i)^2 - \mathbb{E}(S_{i+m} - S_i)^2] \to^{\mathbb{P}} 0 \text{ as } n \to \infty,$$

where $S_k = \sum_{i=1}^{k} X_i$ *for* $k \geq 1$, *and* $S_0 = 0$.

Proof Fix m and let $S_{i,m} = S_{i+m} - S_i$, $i = 0, 1, 2, \ldots$. To prove the lemma it is enough to prove that

$$\frac{1}{n}\sum_{i=1}^{n}[(S_{i,m})_+^2 - \mathbb{E}(S_{i,m})_+^2] \to^{\mathbb{P}} 0 \text{ and } \frac{1}{n}\sum_{i=1}^{n}[(S_{i,m})_-^2 - \mathbb{E}(S_{i,m})_-^2] \to^{\mathbb{P}} 0 \text{ as } n \to \infty.$$

Let us prove only the first part, the second part being proved similarly. For a positive integer M, define $f(x) = \min(M, \max(x, 0))$ and notice that, by the fact that $\{X_i\}$ is uniformly square integrable, it is enough to show that

$$\frac{1}{n}\sum_{i=1}^{n}[f^2(S_{i,m}) - \mathbb{E}(f^2(S_{i,m}))] \to^{\mathbb{P}} 0 \text{ as } n \to \infty.$$

Under the condition $r_j^* \to 0$ as $j \to \infty$, by Item (b) of Lemma 7.1 we get $r_j' \to 0$ as $j \to \infty$ and then we apply Lemma 7.9 to the truncated variables $f^2(S_{i,m})$.

In order to establish (ii), we first apply Lemma 7.3 Item (c), to derive, for $k \geq i + m$,

$$\text{cov}(f^2(S_{i,m}), f^2(S_{k,m})) \leq 4M^2[\text{cov}(S_{i,m}, S_{k,m}) + m^2 \max_{1 \leq i \leq n} \|X_i\|_2^2 b(k - i - m)],$$

where $b(k) = 104a_k^*(1 - \ln a_k^*)$.

Thus we obtain

$$\text{Var}\left(\sum_{i=1}^{n} f^2(S_{i,m})\right) = 2\sum_{i=1}^{n-m}\sum_{k=i+m}^{n} \text{cov}(f^2(S_{i,m}), f^2(S_{k,m}))$$
$$+ \sum_{i=1}^{n} \text{Var}(f^2(S_{i,m})) + 2\sum_{i=1}^{n}\sum_{k=i+1}^{(i+m-1)\wedge n} \text{cov}(f^2(S_{i,m}), f^2(S_{k,m}))$$

$$\leq 4M^2 \left\{ 2 \sum_{i=1}^{n-m} \sum_{k=i+m}^{n} \text{cov}(S_{i,m}, S_{k,m}) + 2m^2 \max_{1 \leq i \leq n} ||X_i||_2^2 \sum_{k=0}^{n-1} b(k) \right\} + nM^4(2m+1)$$

$$\leq 4M^2 \left\{ \text{Var}\left(\sum_{i=1}^{n} S_{i,m} \right) + 2 \sum_{i=1}^{n} \sum_{k=i+1}^{(i+m-1)\wedge n} |\text{cov}(S_{i,m}, S_{k,m})| + 2m^2 \max_{1 \leq i \leq n} ||X_i||_2^2 \sum_{k=0}^{n-1} b(k) \right\}$$
$$+ nM^4(2m+1)$$

$$\leq 4M^2 \left\{ \text{Var}\left(\sum_{i=1}^{n} S_{i,m} \right) + 2m^2 \max_{1 \leq i \leq n} ||X_i||_2^2 \sum_{k=0}^{n-1} b(k) \right\} + nM^4(2m+1) + 8nM^2m^3 \max_{1 \leq i \leq n} ||X_i||_2^2.$$

This implies the lemma since $b(n) \to 0$ and $\text{Var}(S_n)/n^2 \to 0$ as $n \to \infty$. ■

8.5.2 *Convergence under rND Condition*

The condition below generalizes to triangular arrays the *rND* condition (8.12).

***rND* condition** Given a triangular array $Y_{1n}, ..., Y_{nn}$, assume there exists $r < 1$ such that for any n, any two non-intersecting subsets $Q, Q^* \subseteq \{1, \dots, n\}$ and any two coordinatewise non-decreasing functions v and w

$$\text{corr}(v(X_Q), w(X_{Q^*})) \leq r < 1, \tag{8.22}$$

where for the given n, $X_Q = (Y_1, Y_2, ..., Y_n)$ with $Y_i = Y_{in}$ for $i \in Q$ and $Y_i = 0$ for $i \notin Q$. A triangular array $Y_{1n}, ..., Y_{nn}$ satisfying (8.22) will be said to be an *rND* triangular array.

For a triangular array $Y_{1n}, ..., Y_{nn}$ satisfying condition (8.22), the following weak law of large numbers is valid.

Lemma 8.11 *Assume that the triangular array $X_n = (Y_{1n}, ..., Y_{nn})$ of non-negative variables satisfies (8.22). In addition assume that, for any $a > 0$,*

$$Q_n(a) := \mathbb{E}\left(Y_{1n} 1_{\{Y_{1n} \geq a\}}\right) + \dots + \mathbb{E}\left(Y_{nn} 1_{\{Y_{nn} \geq a\}}\right) \to 0 \quad \textit{and}$$
$$A := \sup_n (\mathbb{E}(Y_{1n}) + \dots + \mathbb{E}(Y_{nn})) < \infty.$$

Then,

$$\mathbb{E}\left| \sum_{i=1}^{n} (Y_{in} - \mathbb{E}[Y_{in}]) \right| \to 0 \quad \textit{as } n \to \infty. \tag{8.23}$$

Proof First, we find $a_n \to 0$ such that $Q_n(a_n) \to 0$ as $n \to \infty$. Then, by truncating the variables Y_{in} at the level a_n, and by setting

$$Y'_{in} = (Y_{in} \wedge a_n) \text{ and } Y''_{in} = Y_{in} - Y'_{in},$$

we write

$$R_n := \sum_{i=1}^{n}(Y_{in} - \mathbb{E}[Y_{in}]) = \sum_{i=1}^{n}(Y'_{in} - \mathbb{E}[Y'_{in}]) + \sum_{i=1}^{n}(Y''_{in} - \mathbb{E}[Y''_{in}])$$
$$= I_n + J_n.$$

Clearly, $\mathbb{E}|J_n| \le 2Q_n(a_n) \to 0$ as $n \to \infty$. On the other hand, by Lemma 7.9 we derive

$$\mathbb{E}[I_n^2] \le c\sum_{i=1}^{n}\mathbb{E}[Y'^2_{in}] \le ca_nA \to 0$$

as $n \to \infty$, and so $\mathbb{E}|R_n| \le \mathbb{E}|J_n| + \sqrt{\mathbb{E}[I_n^2]} \le 2Q_n(a_n) + \sqrt{ca_nA} \to 0$, as $n \to \infty$ proving the lemma. ■

8.5.3 *rND-generalization Principle*

Let $Y = \{Y_i\ ;\ i \in \mathbb{Z}\}$ be a sequence of variables satisfying (8.12). Let $X = \{X_i\ ;\ i \in \mathbb{Z}\}$ be a sequence of independent variables with the same marginals, i.e. X_i is distributed as Y_i. Suppose that the sequence X satisfies property P, say, and in addition, its proof is based only on maximal moment inequality (8.16) or properties that follows from it. Then, Y also satisfies P.

Examples of these properties P are:

Property 8.12 *(rND-version of the Borel–Cantelli lemma) Assume that A_i are measurable sets in a probability space and the sequence of variables $Y = \{I_{A_j}; i \in \mathbb{Z}\}$ satisfies condition (8.12). Then $\sum I_{A_j} = \infty$ if and only if $\sum P(A_j) = \infty$.*

Property 8.13 *(rND-version of the Kolmogorov three series theorem) Let $Y = \{Y_i; i \in \mathbb{Z}\}$ be a sequence of variables satisfying (8.12). $\sum Y_i$ converges almost surely if there exists a constant $c > 0$ (finite or infinite) such that the following three conditions hold:*

$$(i)\ \sum \mathbb{E}\left(Y_i I_{(|Y_i| \le c)}\right) \text{ converges,}$$
$$(ii)\ \sum \mathrm{Var}((Y_i I_{(|Y_i| \le c)}) < \infty,$$
$$(iii)\ \sum \mathbb{P}(|Y_i| > c) < \infty.$$

Property 8.14 *(rND-version of the strong law of large numbers) Assume that $\{Y_i\}$ is uniformly integrable and satisfies (8.12). Then*

$$\frac{Y_1 + \ldots + Y_n}{n} - \frac{\mathbb{E}[Y_1] + \ldots + \mathbb{E}[Y_n]}{n} \to 0 \text{ a.s.}$$

We leave the proofs to the reader.

Ergodicity

Let us consider a stationary sequence $\{Y_i\}$ which satisfies the condition (8.12). We are going to show that the sequence is ergodic.

Lemma 8.15 *Let $\{Y_i\}$ be a stationary sequence which satisfies condition* (8.12). *Then, $\{Y_i\}$ is ergodic.*

Proof To prove ergodicity, we fix an integer k and introduce the subspace B_k of functions $f : \mathbb{R}^k \to \mathbb{R}$, such that $\mathbb{E}|f(Y_1, \dots, Y_k)| < \infty$, and need to show that

$$\frac{1}{n}\sum_{i=1}^{n} f(Y_{i+1}, \dots, Y_{i+k}) \to \mathbb{E}[f(Y_1, \dots, Y_k)] \text{ as } n \to \infty. \tag{8.24}$$

Let ν denote the joint measure of the vector $(Y_1, \dots, Y_k)$. Notice that B_k is a closed linear subspace in the Banach space $\mathbb{L}^1(\mathbb{R}^k, \nu)$. This shows that in order to prove (8.24) it is enough to consider f being of the product form $f(x_1, \dots, x_k) = f_1(x_1)\cdots f_k(x_k)$ where the functions $f_i : \mathbb{R} \to \mathbb{R}$ are bounded, non-decreasing and non-negative. Then, the sequence $W_i = f_1(Y_{i+1})\cdots f_k(Y_{i+k})$ satisfies condition (8.12), and (8.24) follows from the strong law of large numbers for the stationary *rND* sequences mentioned in the *rND*-generalization principle. ■

8.5.4 *Beurling–Malliavin Density of Weakly Negatively Dependent Random Sequences*

In this part, motivated by Lyons (2003) and following Seip and Ulanovskii (1997), we compute the Beurling–Malliavin density of an *rND* sequence and apply it to find the density for the strictly stationary case. In particular, we partly extend the corresponding result by Lyons (2003), who proved the negative association property (usually referred to as the negative dependence property throughout this monograph) for the class of determinantal probability measures and among others applied it to compute the Beurling–Malliavin density of a stationary sequence with a determinantal probability measure.

To define the Beurling–Malliavin density of a set $S = \{\lambda_n\}_{n\in\mathbb{Z}}$ of real numbers, we say that the ordered system of intervals $\{(a_n, b_n); n \in \mathbb{Z}^+\}$ is substantial if either $0 < a_1 < b_1 \le a_2 < b_2 \dots$ and $\sum a_n^{-2}(b_n - a_n)^2 = \infty$ or $0 > b_1 > a_1 \ge b_2 > a_2 \dots$ and $\sum |b_n|^{-2}(b_n - a_n)^2 = \infty$. If S has a bounded subsequence, then the Beurling–Malliavin density of the set S, denoted $BM(S)$ is defined by $BM(S) = \infty$. Otherwise, $BM(S)$ is the supremum over all $R \ge 0$ with the property that there exists a substantial sequence of intervals $\{(a_n, b_n)\ ;\ n \in \mathbb{Z}^+\}$ such that $S \cap (a_n, b_n)$ contains at least $R(b_n - a_n)$ elements.

Following Seip and Ulanovskii (1997), we now introduce the notion of the associated Beurling–Malliavin density of the non-negative measure μ denoted by $d(\mu)$. If μ is locally infinite on $\mathbb{R} - \{0\}$ (that is there exists an interval $(a, b) \subset \mathbb{R}^+ \cup \mathbb{R}^-$ with $\mu(a, b) = \infty$), we let $d(\mu) = \infty$. Otherwise $d(\mu)$ is the supremum over all $R \ge 0$ with the property that there exists a substantial sequence of intervals $\{(a_n, b_n)\ ;\ n \in \mathbb{Z}^+\}$ such that the following inequality holds: $\mu((a_n, b_n)) \ge R(b_n - a_n)$ for every $n \in \mathbb{Z}^+$. Seip and Ulanovskii (1997) pointed out that the two notations coincide for the counting measure $\mu_S(A) = \sum_{s \in S \cap A} \delta_s$, where δ_s is the unit measure concentrated at s.

Lemma 8.16 *Let $S = \{\xi_n \ ; \ n \in \mathbb{Z}\}$ be a sequence of real-valued random variables such that for any interval (a, b) of reals the sequence $\{1_{\{\xi_n \in (a,b)\}}\}_{n \in \mathbb{Z}}$ satisfies the rND condition (8.12). Define the measure by $\mu(A) = \sum_{n \in \mathbb{Z}} \mathbb{P}(\xi_n \in A)$. Then, $BM(S) = d(\mu)$ almost surely.*

Proof The proof of this lemma comes from the *rND*-generalization principle. Indeed, examining the proof of Seip and Ulanovskii (1997), we notice that the independence assumption was used to derive the moment inequality in their Lemma 1, which follows from Fatou's lemma and our moment inequality for *rND* sequences as stated in Lemma 8.7. In addition, they have used the Borel–Cantelli lemma. And so their proof is easily adaptable to the *rND* case which completes the proof of the lemma. ■

Applied to stationary measures, we derive the following corollary, which partly extends the corresponding result of Lyons (2003, Corollary 7.14) to strictly stationary *rND* sequences. In addition, Lyons (2003, Theorem 7.15) proved that the discrete determinantal point processes defined by positive contractions have trivial tails. This fact was independently discovered by Shirai and Takahashi (2003, Theorem 1.4) and it can be also derived from Matula (2003), who established the strong mixing property for negatively associated integer valued variables. By Proposition 10.15, the ergodicity is not true in general for arbitrary determinantal measures.

Corollary 8.17 *Let $\{Y_n \ ; \ n \in \mathbb{Z}\}$ be a strictly stationary sequence satisfying the rND condition (8.12). Assume that $\mathbb{P}(Y_n \in \{0, 1\}) = 1$ and $\mathbb{P}(Y_n = 1) = p$. Let S be the set of all n where $Y_n = 1$. Then, $BM(S) = p$ almost surely. Moreover, the sequence $\{Y_n \ ; \ n \in \mathbb{Z}\}$ is ergodic.*

Proof As was pointed in Lyons (2003), the first part of the corollary directly follows from Lemma 8.16 by taking $\xi_n = nY_n$ and considering only accumulation on the positive part of the real line. Actually, the proof by Lyons (2003, p.194) can be adapted as well. The only difference is that instead of using the exponential bound of his Corollary 6.7, we take a $\delta \in (0, 1/4)$ and we use the moment inequality as stated in Lemma 8.7 with $t > 4$ to get the inequality $|\text{Card}(S \cap [1, k]) - pk| \leq ck^{1-\delta}$ for all but finitely many k a.s., and then the other lines follow with minor changes.

Finally, the ergodicity has been already proved in Section 8.5.3. ■

8.6 A Useful Fact Concerning the *rND* Condition

The aim of this section is to give an alternative characterization of *rND*-vectors (hence satisfying condition (8.12)) and of *rND* triangular arrays (hence satisfying condition (8.22)). The next lemma shows that we can relax the family of functions for which we should verify (8.12) or (8.22) by considering only coordinatewise non-decreasing functions v and w which are in addition *bounded and continuous*.

Lemma 8.18 *Let X and Y be two vectors with values in $\mathbb{R}^k$ and $\mathbb{R}^d$ respectively. Assume that there exists a positive r such that*

$$\operatorname{corr}\left(f(X), g(Y)\right) \leq r, \tag{8.25}$$

for all bounded, continuous and coordinatewise non-decreasing functions f and g. Then (8.25) also holds for all coordinatewise non-decreasing functions f and g provided that $\|f(X)\|_2 < \infty$ and $\|g(Y)\|_2 < \infty$.

This lemma is in the spirit of Theorem 3.3 in Esary *et al.* (1967) (see also Theorem 1.32 in Oliveira (2012)). Indeed, this theorem states that a vector X of $\mathbb{R}^k$ is *associated* provided that $\operatorname{cov}(f(X), g(X)) \geq 0$ for all coordinatewise non-decreasing continuous and bounded functions $f, g : \mathbb{R}^k \to \mathbb{R}$.

The arguments developed in Section 3 of Esary *et al.* (1967) cannot, however, be fully adapted to prove our Lemma 8.18 and we propose below a proof mainly based on *non-decreasing sets.*

Proof of Lemma 8.18 Throughout the proof, we use the notation $\|\xi\| = \sqrt{\operatorname{Var}(\xi)}$ for any square integrable real-valued random variable ξ. For m a positive integer, we denote also by $\mathcal{ND}_m$ the class of coordinatewise non-decreasing functions from $\mathbb{R}^m$ to $\mathbb{R}$.

Assume first that we can restrict the proof to the following statement:

Lemma 8.19 *Let X and Y be two vectors with values in $[0, 1]^k$ and $[0, 1]^d$ respectively. Assume that there exists a positive r such that (8.25) holds for all bounded, continuous and coordinatewise non-decreasing functions f and g with real values. Then (8.25) also holds for all coordinatewise non-decreasing functions f and g such that $0 \leq f, g \leq 1$.*

Then the proof of Lemma 8.18 will consist of proving Lemma 8.19 (step 1) and then in showing that, indeed, we can assume without loss of generality that X and Y can be assumed to take their values in $[0, 1]^k$ and $[0, 1]^d$ respectively and that f and g can be assumed, in addition to being coordinatewise non-decreasing, to take their values in $[0, 1]$.

Step 1. We prove here Lemma 8.19. All along this step, X and Y are two vectors with values in $[0, 1]^k$ and $[0, 1]^d$ respectively. Let M be a positive integer and set $\delta = 1/M$. As a matter of fact, by standard arguments, since $0 \leq f, g \leq 1$, we only need to show that there exists a positive r (not depending on M) such that

$$\operatorname{corr}\left(\tilde{f}(X), \tilde{g}(Y)\right) \leq r, \tag{8.26}$$

for all coordinatewise non-decreasing functions f and g such that $0 \leq f, g \leq 1$, where

$$\tilde{f}(x) = \sum_{j=1}^{M} \delta 1\{f(x) \geq \delta j\} \text{ and } \tilde{g}(x) = \sum_{\ell=1}^{M} \delta 1\{g(x) \geq \delta \ell\}.$$

In what follows, j and ℓ are fixed. Let $C_j = \{x \in [0, 1]^k : f(x) \geq \delta j\}$ and note that it is a *non-decreasing set* in the sense that if $x \in C_j$ and $y \geq x$ then $y \in C_j$. Recall that $y \geq x$

means that if $x = (x_1, ..., x_k)$ and $y = (y_1, ..., y_k)$ then $y_i \geq x_i$ for all $i = 1, ..., k$. A similar observation can be done for the sets $D_\ell = \{x \in [0,1]^d : g(x) \geq \delta\ell\}$.

Fix $\varepsilon > 0$ (the same for any j and any ℓ). It is possible to choose a compact set $K^\varepsilon_{C_j}$ included in C_j such that $\mathbb{P}(X \in C_j \backslash K^\varepsilon_{C_j}) < \varepsilon$. Now, for any set $A \subseteq [0,1]^k$, let $Q(A)$ be the smallest non-decreasing set that includes A, that is $Q(A) := \{y \in [0,1]^k :$ there exists $x \in A$ such that $x \leq y\}$. Set $H^\varepsilon_{C_j} = Q(K^\varepsilon_{C_j})$. Hence $K^\varepsilon_{C_j} \subseteq H^\varepsilon_{C_j} \subseteq C_j$ and then $\mathbb{P}(X \in C_j \backslash H^\varepsilon_{C_j}) < \varepsilon$. Similarly, we can choose a compact set $K^\varepsilon_{D_\ell}$ included in D_ℓ such that $\mathbb{P}(Y \in D_\ell \backslash H^\varepsilon_{D_\ell}) < \varepsilon$ where $H^\varepsilon_{D_\ell} = Q(K^\varepsilon_{D_\ell})$. Therefore,

$$\operatorname{cov}\left(1\{X \in C_j\}, 1\{Y \in D_\ell\}\right) = \operatorname{cov}\left(1\{X \in H^\varepsilon_{C_j}\}, 1\{Y \in H^\varepsilon_{D_\ell}\}\right)$$
$$+ \operatorname{cov}\left(1\{X \in C_j \backslash H^\varepsilon_{C_j}\}, 1\{Y \in D_\ell\}\right) + \operatorname{cov}\left(1\{X \in H^\varepsilon_{C_j}\}, 1\{Y \in D_\ell \backslash H^\varepsilon_{D_\ell}\}\right).$$

But

$$\operatorname{cov}\left(1\{X \in C_j \backslash H^\varepsilon_{C_j}\}, 1\{Y \in D_\ell\}\right) + \operatorname{cov}\left(1\{X \in H^\varepsilon_{C_j}\}, 1\{Y \in D_\ell \backslash H^\varepsilon_{D_\ell}\}\right)$$
$$\leq \sqrt{\mathbb{P}(X \in C_j \backslash H^\varepsilon_{C_j})} + \sqrt{\mathbb{P}(Y \in D_\ell \backslash H^\varepsilon_{D_\ell})} \leq 2\sqrt{\varepsilon}.$$

Therefore, if we can prove that there exists a positive r (not depending on (M, ε)) such that

$$\operatorname{cov}\left(\sum_{j=1}^{M} \delta \mathbf{1}\left\{X \in H^\varepsilon_{C_j}\right\}, \sum_{\ell=1}^{M} \delta \mathbf{1}\left\{Y \in H^\varepsilon_{D_\ell}\right\}\right)$$
$$\leq r \left\| \sum_{j=1}^{M} \delta \mathbf{1}\left\{X \in C_j\right\} \right\| \left\| \sum_{\ell=1}^{M} \delta \mathbf{1}\left\{Y \in D_\ell\right\} \right\| + R(\varepsilon), \tag{8.27}$$

with $R(\varepsilon) \to 0$, as $\varepsilon \to 0$, then (8.26) will follow by letting $\varepsilon \to 0$ and the proof of Lemma 8.19 will be complete. To prove (8.27), we proceed as follows. Since $H^\varepsilon_{C_j} = Q(K^\varepsilon_{C_j})$ and $K^\varepsilon_{C_j}$ is compact, $H^\varepsilon_{C_j}$ is a closed set (see Nachbin (1965, page 44) where it is proved that that non-decreasing compact generated sets are closed). Note that indicator function of a closed set is upper semi-continuous. Let $\alpha \in]0,1[$, U be a random vector of $[0,1]^k$ with continuous density on a compact of $[0,1]^k$ and define

$$h^\varepsilon_{\alpha,C_j}(x) = \mathbb{E}(1\{x + \alpha U \in H^\varepsilon_{C_j}\}).$$

The function $h^\varepsilon_{\alpha,C_j}$ defined above is continuous (by Lebesgue theorem), bounded and coordinatewise non-decreasing. In addition, for any $x \in [0,1]^k$,

$$0 \leq h^\varepsilon_{\alpha,C_j}(x) - 1\{x \in H^\varepsilon_C\} \to 0, \quad \text{as } \alpha \to 0,$$

implying by Lebesgue's dominated convergence theorem that

$$||h^{\varepsilon}_{\alpha,C_j}(X) - \mathbf{1}\{X \in H^{\varepsilon}_{C_j}\}||_1 \to 0, \text{ as } \alpha \to 0, \tag{8.28}$$

Similarly, the function $h^{\varepsilon}_{\alpha,D_\ell}$ defined on $[0, 1]^d$ by

$$h^{\varepsilon}_{\alpha,D_\ell}(x) = \mathbb{E}(1\{x + \alpha V \in H^{\varepsilon}_{D_\ell}\}),$$

where V, a random vector of $[0, 1]^d$ with continuous density on a compact of $[0, 1]^d$, is continuous, bounded and coordinatewise non-decreasing, and satisfies

$$\lim_{\alpha\to 0} \left\| h^{\varepsilon}_{\alpha,D_\ell}(Y) - 1\{Y \in H^{\varepsilon}_{D_\ell}\} \right\|_1 = 0.$$

Next, setting

$$h^{\varepsilon}_{\alpha,C}(x) = \sum_{j=1}^{M} \delta h^{\varepsilon}_{\alpha,C_j}(x) \text{ and } h^{\varepsilon}_{\alpha,D}(x) = \sum_{\ell=1}^{M} \delta h^{\varepsilon}_{\alpha,D_\ell}(x),$$

we write

$$\begin{aligned}
\operatorname{cov}\left(\sum_{j=1}^{M} \delta 1\{X \in H^{\varepsilon}_{C_j}\}, \sum_{\ell=1}^{M} \delta 1\{Y \in H^{\varepsilon}_{D_\ell}\}\right) &= \operatorname{cov}\left(h^{\varepsilon}_{\alpha,C}(X), h^{\varepsilon}_{\alpha,D}(Y)\right) \\
&+ \sum_{j=1}^{M}\sum_{\ell=1}^{M} \delta^2 \Big\{\operatorname{cov}\left(h^{\varepsilon}_{\alpha,C_j}(X), 1\{Y \in H^{\varepsilon}_{D_\ell}\} - h^{\varepsilon}_{\alpha,D_\ell}(Y)\right) \\
&+ \operatorname{cov}\left(1\{X \in H^{\varepsilon}_{C_j}\} - h^{\varepsilon}_{\alpha,C_j}(X), 1\{Y \in H^{\varepsilon}_{D_\ell}\}\right)\Big\}.
\end{aligned}$$

Therefore, since (8.25) is assumed to hold for all bounded, continuous and coordinatewise non-decreasing functions, and that $h^{\varepsilon}_{\alpha,C}(x)$ and $h^{\varepsilon}_{\alpha,D}(x)$ are, there exists a positive r (not depending on (M, ε, α)) such that

$$\operatorname{cov}\left(\sum_{j=1}^{M} \delta 1\{X \in H^{\varepsilon}_{C_j}\}, \sum_{\ell=1}^{M} \delta 1\{Y \in H^{\varepsilon}_{D_\ell}\}\right) \leq r||h^{\varepsilon}_{\alpha,C}(X)||||h^{\varepsilon}_{\alpha,D}(Y)|| + R_1(\varepsilon, \alpha, M),$$

where

$$R_1(\varepsilon, \alpha, M) := \sum_{j=1}^{M}\sum_{\ell=1}^{M} \delta^2 \left\{||1\{X \in H^{\varepsilon}_{C_j}\} - h^{\varepsilon}_{\alpha,C_j}(X)|| + ||1\{Y \in H^{\varepsilon}_{D_\ell}\} - h^{\varepsilon}_{\alpha,D_\ell}(Y)||\right\}.$$

Note that since $||\xi|| \leq ||\xi||_2$ and $||1\{X \in H^{\varepsilon}_{C_j}\} - h^{\varepsilon}_{\alpha,C_j}(X)||_\infty \leq 1$,

$$\|1\{X \in H^{\varepsilon}_{C_j}\} - h^{\varepsilon}_{\alpha,C_j}(X)\| \leq \|1\{X \in H^{\varepsilon}_{C_j}\} - h^{\varepsilon}_{\alpha,C_j}(X)\|_1^{1/2} \to 0, \text{ as } \alpha \to 0,$$

by (8.28). Similarly $\|1\{Y \in H^{\varepsilon}_{D_\ell}\} - h^{\varepsilon}_{\alpha,D_\ell}(Y)\| \to 0$, as $\alpha \to 0$. Therefore $R_1(\varepsilon, \alpha, M) \to 0$, as $\alpha \to 0$. On the other hand,

$$\begin{aligned}
\left\|h^{\varepsilon}_{\alpha,C}(X)\right\|^2 &= \left\|\sum_{j=1}^{M} \delta 1\{X \in C_j\}\right\|^2 + \left\|\sum_{j=1}^{M} \delta 1\{X \in C_j\} - h^{\varepsilon}_{\alpha,C}(X)\right\|^2 \\
&\quad - 2\mathrm{cov}\left(\sum_{j=1}^{M} \delta 1\{X \in C_j\} - h^{\varepsilon}_{\alpha,C}(X), \sum_{j=1}^{M} \delta 1\{X \in C_j\}\right) \\
&\leq \left\|\sum_{j=1}^{M} \delta 1\{X \in C_j\}\right\|^2 + 3\left\|\sum_{j=1}^{M} \delta 1\{X \in C_j\} - h^{\varepsilon}_{\alpha,C}(X)\right\|_1^{1/2} \\
&\leq \left\|\sum_{j=1}^{M} \delta 1\{X \in C_j\}\right\|^2 + 3\sqrt{\varepsilon} + 3\left\|\sum_{j=1}^{M} \delta 1\{X \in H^{\varepsilon}_{C_j}\} - h^{\varepsilon}_{\alpha,C}(X)\right\|_1^{1/2},
\end{aligned}$$

where we have used that for any $j \in \{1, \ldots, M\}$, $\mathbb{P}(X \in C_j \backslash H^{\varepsilon}_{C_j}) < \varepsilon$ for the last inequality. Similarly,

$$\|h^{\varepsilon}_{\alpha,D}(Y)\|^2 \leq \left\|\sum_{\ell=1}^{M} \delta 1\{Y \in D_\ell\}\right\|^2 + 3\sqrt{\varepsilon} + 3\left\|\sum_{\ell=1}^{M} \delta 1\{Y \in H^{\varepsilon}_{D_\ell}\} - h^{\varepsilon}_{\alpha,D}(Y)\right\|_1^{1/2}.$$

Hence

$$\|h_{\varepsilon,C}(X)\|\|h_{\varepsilon,D}(Y)\| \leq \left\|\sum_{j=1}^{M} \delta 1\{X \in C_j\}\right\| \left\|\sum_{\ell=1}^{M} \delta 1\{Y \in D_\ell\}\right\| + R_2(\varepsilon, \alpha, M),$$

where $\lim_{\varepsilon \to 0} \limsup_{\alpha \to 0} R_2(\varepsilon, \alpha, M) = 0$. By the previous computations, we then derive that there exists a positive r (not depending on (M, ε, α)) such that

$$\begin{aligned}
&\mathrm{cov}\left(1\{X \in H^{\varepsilon}_C\}, 1\{Y \in H^{\varepsilon}_D\}\right) \\
&\leq r\left\|\sum_{j=1}^{M} \delta 1\{X \in C_j\}\right\| \left\|\sum_{\ell=1}^{M} \delta 1\{Y \in D_\ell\}\right\| + R_1(\varepsilon, \alpha, M) + rR_2(\varepsilon, \alpha, M).
\end{aligned}$$

This ends the proof of (8.27) and therefore of step 1.

Step 2. We show here that the proof of Lemma 8.18 can be reduced to prove Lemma 8.19. We start by showing that it is enough to prove that (8.25) holds for any $f \in \mathcal{ND}_k$ and any $g \in \mathcal{ND}_d$ such that $0 \leq f, g \leq 1$. With this aim, let M be a positive real and denote by φ_M the function from $\mathbb{R}$ to $[-M, M]$ defined by $\varphi_M(x) = (x \wedge M) \vee (-M)$. Let $f_M = \varphi_M \circ f$ and $g_M = \varphi_M \circ g$. Note that

$$\begin{aligned}\operatorname{cov}\left(f(X), g(Y)\right) &= \operatorname{cov}\left(f_M(X), g_M(Y)\right) + \operatorname{cov}\left((f-f_M)(X), g(Y)\right) \\ &\quad + \operatorname{cov}\left((g-g_M)(Y), f_M(X)\right).\end{aligned}$$

Since $\|\xi\| \leq \|\xi\|_2$ and $\|f_M(X)\|_2 \leq \|f\|_2$, we have

$$\begin{aligned}&\operatorname{cov}\left((f-f_M)(X), g(Y)\right) + \operatorname{cov}\left((g-g_M)(Y), f_M(X)\right) \\ &\leq \|(f-f_M)(X)\|_2 \|g(Y)\|_2 + \|(g-g_M)(X)\|_2 \|f(X)\|_2.\end{aligned}$$

Therefore

$$\operatorname{cov}\left(f(X), g(Y)\right) \leq \operatorname{cov}\left(f_M(X), g_M(Y)\right) + \eta_1(M),$$

where $\eta_1(M) \to 0$, as $M \to \infty$. Indeed, Lebesgue's dominated convergence theorem implies

$$\lim_{M\to\infty} \|(f-f_M)(X)\|_2 = 0 \text{ and } \lim_{M\to\infty} \|(g-g_M)(Y)\|_2 = 0. \tag{8.29}$$

Define now

$$f_M^* = \frac{f_M + M}{2M} \text{ and } g_M^* = \frac{g_M + M}{2M}.$$

They are coordinatewise non-decreasing functions such that $0 \leq f, g \leq 1$, and note that

$$\operatorname{cov}\left(f_M(X), g_M(Y)\right) = 4M^2 \operatorname{cov}\left(f_M^*(X), g_M^*(Y)\right).$$

It follows from the previous considerations that if (8.25) holds for any $f \in \mathcal{ND}_k$, $g \in \mathcal{ND}_d$ and such that $0 \leq f, g \leq 1$, there exists a positive r (not depending on M) such that

$$\operatorname{cov}\left(f(X), g(Y)\right) \leq 4M^2 r \|f_M^*(X)\| \|g_M^*(Y)\| + \eta_1(M).$$

But note that

$$\|f_M^*(X)\| = \sqrt{\operatorname{Var}(f_M^*(X))} = (2M)^{-1}\sqrt{\operatorname{Var}(f_M(X))},$$

and

$$\|f_M(X)\| \leq \|f(X)\| + \eta_2(M),$$

where $\eta_2(M) := \|(f-f_M)(X)\|_2 + \sqrt{2}\left(\|f(X)\|_2 \|(f-f_M)(X)\|_2\right)^{1/2} \to 0$, as $M \to \infty$, by (8.29). Similarly, we have

$$\|g_M^*(Y)\| \leq (2M)^{-1}\|g_M(Y)\| \text{ and } \|g_M(Y)\| \leq \|g(Y)\| + \eta_3(M)$$

where $\eta_3(M) \to 0$, as $M \to \infty$. So, overall, we get

$$\operatorname{cov}\left(f(X), g(Y)\right) \leq r\|f(X)\|\|g(Y)\| + \eta(M), \tag{8.30}$$

where $\eta(M) := \eta_1(M) + r\{\eta_3(M)\|f(X)\|_2 + \eta_2(M)\|g(Y)\|_2 + \eta_2(M)\eta_3(M)\}$. Since $\eta(M) \to 0$, as $M \to \infty$, letting M tend to infinity in (8.30) allows us to conclude that we can restrict the proof of Lemma 8.18 to functions which are coordinatewise non-decreasing functions and with values in $[0, 1]$. We then assume that below we are working with this class of functions.

We prove now that we can assume without loss of generality that X has values in $[0, 1]^k$ and Y has values in $[0, 1]^d$. With this aim, fix a positive real M. Denote by $X_M = (\varphi_M(X_1), ..., \varphi_M(X_k))$ where the X_i's are the coordinates of X and $\varphi_M(x) = (x \wedge M) \vee (-M)$. Similarly, let $Y_M = (\varphi_M(Y_1), ..., \varphi_M(Y_d))$. Note that, since $0 \leq f, g \leq 1$,

$$\begin{aligned}\operatorname{cov}\left(f(X), g(Y)\right) &= \operatorname{cov}\left(f(X_M), g(Y_M)\right) \\ &\quad + \operatorname{cov}\left(f(X) - f(X_M), g(Y)\right) + \operatorname{cov}\left(g(Y) - g(Y_M), f(X_M)\right) \\ &\leq \operatorname{cov}\left(f(X_M), g(Y_M)\right) + \|f(X) - f(X_M)\|_1^{1/2} + \|g(X) - g(Y_M)\|_1^{1/2}\end{aligned}$$

and

$$\mathbb{E}|f(X) - f(X_M)| \leq \mathbb{P}(X \neq X_M) \leq \sum_{i=1}^{k} \mathbb{P}(|X_i| > M) \text{ and also}$$

$$\mathbb{E}|g(Y) - g(Y_M)| \leq \sum_{i=1}^{d} \mathbb{P}(|Y_i| > M).$$

Using Lebesgue's dominated convergence theorem, it follows that

$$\operatorname{cov}\left(f(X), g(Y)\right) \leq \operatorname{cov}\left(f(X_M), g(Y_M)\right) + \eta_1(M),$$

where $\eta_1(M) \to 0$, as $M \to \infty$. Let h_M be the function from $[-M, M]$ to $[0, 1]$ defined by

$$h_M(x) = \frac{x + M}{2M},$$

and denoting by $b_M = h_M \circ \varphi_M$, set

$$X_M^* = (b_M(X_1), \ldots, b_M(X_k)) \text{ and } Y_M^* = (b_M(Y_1), \ldots, b_M(Y_d)).$$

Then the two vectors X_M^* and Y_M^* have values in $[0, 1]^k$ and $[0, 1]^d$ respectively. Note now that $f(X_M) = f_1(X_M^*)$ and $g(Y_M) = g_1(Y_M^*)$ where

$$\begin{aligned}f_1(x_1, \ldots x_k) &= f(2Mx_1 - M, \ldots, 2Mx_k - M) \text{ and} \\ g_1(x_1, \ldots x_d) &= g(2Mx_1 - M, \ldots, 2Mx_d - M).\end{aligned}$$

The functions f_1 and f_2 are still coordinatewise non-decreasing and with values in $[0, 1]$. It follows that if (8.25) holds for X and Y two vectors with values in $[0, 1]^k$ and $[0, 1]^d$ respectively, and when we consider the class of functions that are coordinatewise non-decreasing functions and with values in $[0, 1]$, then there exists a positive r (not depending on M) such that

$$\begin{aligned}\operatorname{cov}\left(f(X_M), g(Y_M)\right) &= \operatorname{cov}\left(f_1(X_M^*), g_1(Y_M^*)\right) \\ &\leq r\|f_1(X_M^*)\|\|f_1(X_M^*)\| = r\|f(X_M)\|\|f(Y_M)\|.\end{aligned}$$

But

$$\|f(X_M)\| \leq \|f(X)\| + \eta_2(M),$$

where $\eta_2(M) := \|f(X) - f(X_M)\|_1^{1/2} + \sqrt{2}\|f(X) - f(X_M)\|_1^{1/4} \to 0$, as $M \to \infty$. Similarly, we have $\|g(Y_M)\| \leq \|g(Y)\| + \eta_3(M)$, where $\eta_3(M) \to 0$, as $M \to \infty$. All the previous computations then prove that if (8.25) holds for X and Y, two vectors with values in $[0, 1]^k$ and $[0, 1]^d$ respectively, and when we consider the class of functions that are coordinatewise non-decreasing functions and with values in $[0, 1]$, then there exists a positive r (not depending on M) such that

$$\operatorname{cov}\left(f(X), g(Y)\right) \leq r\|f(X)\|\|f(Y)\| + \eta(M),$$

with $\eta(M) \to 0$, as $M \to \infty$. This ends the proof of step 2 and therefore of the lemma. ■

9

Gaussian Approximation under Asymptotic Negative Dependence

The aim of this chapter is to provide weakly negatively dependence conditions for the convergence of the partial sums process to either to a Gaussian process with independent increments or to a diffusion process with deterministic time varying volatility.

9.1 General Construction and Tightness

We are going to prove the invariance principle for triangular arrays, satisfying condition (A) defined in section 9.1.2. Apart from a traditional Lindeberg-type assumption for triangular array, the main restriction is a "weak dependence" assumption in the form of an asymptotic *rND* condition.

Throughout the section, we work with triangular arrays $X^{(n)} = (X_{1n}, \ldots, X_{nn})$, $n \geq 1$, of multivariate centered random variables with finite second order moments. As often, it is convenient to enlarge the array by adding $X_{jn} = 0$ for $j > n$.

9.1.1 Univariate Construction and Assumptions

We first assume that the triangular array is univariate and as a main dependence-type condition, we consider one of the following two conditions:

Asymptotic *rND* condition The triangular array $(X^{(n)}, n \geq 1)$ will be said to satisfy the asymptotic *rND* condition if there exists a non-increasing sequence $(r_j)_{j\geq 1}$ and such that for all n and all sets Q and Q^* with distance $d(Q, Q^*) \geq j$, and any two coordinatewise non-decreasing functions v and w

$$\mathrm{corr}(v(X_Q), w(X_{Q^*})) \leq r_j \to r < 1 \quad (j \to \infty), \tag{9.1}$$

where for the given n, $X_Q = (Y_1, Y_2, \ldots, Y_n)$ with $Y_i = X_{in}$ for $i \in Q$ and $Y_i = 0$ for $i \notin Q$.

Functional Gaussian Approximation for Dependent Structures. Florence Merlevède, Magda Peligrad and Sergey Utev. Oxford University Press (2019). © Florence Merlevède, Magda Peligrad and Sergey Utev 2019.
DOI: 10.1093/oso/9780198826941.001.0001

Asymptotic negative dependence condition The triangular array $(X^{(n)}, n \geq 1)$ will be called asymptotically negatively dependent if there exists a non-increasing sequence $(r_j)_{j\geq 1}$ such that for all n and all sets Q and Q^* with distance $d(Q, Q^*) \geq j$, and any two coordinatewise non-decreasing functions v and w

$$\operatorname{corr}(v(X_Q), w(X_{Q^*})) \leq r_j \to 0 \quad (j \to \infty). \tag{9.2}$$

9.1.2 *Condition (A)*

(a) The triangular array $(X^{(n)}, n \geq 1)$ satisfies the asymptotic *rND* **condition** (9.1)

(b) (Lindeberg's condition) Moreover, assume that

$$\sup_n \sum_{i=1}^{n} \mathbb{E}[X_{in}^2] = 1 \text{ and for any } \varepsilon > 0, \quad \sum_{i=1}^{n} \mathbb{E}[X_{in}^2 I_{(|X_{in}|>\varepsilon)}] \to 0 \tag{9.3}$$

as $n \to \infty$.

(c) (Construction of $W_n(t)$) Let $S_0 = S_{0n} = 0$, $t_0 = t_{0n} = 0$ and for $k = 1, \ldots, n$ define

$$S_k = S_{kn} = \sum_{i=1}^{k} X_{in} \text{ and } t_{kn} = \sum_{i=1}^{k} \mathbb{E}[X_{in}^2]$$

and define the normalized partial sums process by letting $W_n(t) = S_k$ for $t_{(k-1)n} < t \leq t_{kn}$, $k = 1, \ldots, n$ and $W_n(t) = S_n$ for $t \geq t_{nn}$. Also it is convenient to take $W_n(t) = W_n(1)$ for all $t \geq 1$.

Remark 9.1 *Condition (A)-(a) implies Condition (A)-(a′) which reads as follows: There exist a fixed real non-negative $r < 1$ and a positive integer m, such that, for all integers n and $j \in [1, m]$, the vector $X^{(n, m, j)} = \{X_{(j+km)n}, k = 0, \ldots, n\}$ satisfies the rND condition* (8.22) *with the same r since the distance between two disjoint sets of indices will be at least m.*

9.1.3 *Convention*

Although the partial sums process W_n has sample paths in $D[0, 1]$, the Lindeberg condition (9.3) implies that W_n has the same stochastic behavior as its continuous linear spline version, i.e. when points (t_{kn}, S_{kn}) and $(t_{(k+1)n}, S_{(k+1)n})$ are linearly connected, $k = 0, \ldots, n-1$. A sequence ξ_n of elements in $D[0, 1]$ is said to be tight in $C[0, 1]$ or with the uniform topology if there exists a tight sequence ζ_n in $C[0, 1]$ such that $\|\xi_n - \zeta_n\|_\infty \to 0$ as $n \to \infty$.

9.1.4 *Multivariate Construction and Multivariate Condition (A)*

Now, assume that $(X^{(n)}, n \geq 1)$, with $X(n) = (X_{1n}, \ldots, X_{nn})$, is a triangular array of centered random variables valued in the Euclidean vector space $\mathbb{R}^k$ with inner product $< x, y > = x_1 y_1 + \ldots + x_k y_k$.

We construct the partial sum process, as follows: For any $t \in [0, 1]$,

$$W_n(t) = (W_n^{(1)}(t), \dots, W_n^{(k)}(t)),$$

where for each $i = 1, \dots, k$, $W_n^{(i)}(t)$ is the univariate partial sum process as previously defined, constructed from $(X_{1n}^{(i)}, \dots, X_{nn}^{(i)})$ where $X_{kn}^{(i)}$ is the i-th coordinate of the vector X_{kn}.

Concerning the conditions, the main difference is in the *rND* condition. In what follows, we define $X^{(n)}(h) = (< X_{1n}, h >, \dots, < X_{nn}, h >)$ for any vector $h \in \mathbb{R}^k$.

(Multivariate) *rND* condition Now, we require that for each vector $h = (h_1, \dots, h_k)$ with non-negative coordinates, the univariate *rND* condition (8.22) to be satisfied for the univariate triangular array $(X^{(n)}(h), n \geq 1)$.

(Multivariate) Asymptotic *rND* condition We say that the triangular array $\{X^{(n)}, n \geq 1\}$ of multivariate random variables is satisfying the asymptotic *rND* condition if for each vector $h = (h_1, \dots, h_k)$, with non-negative coordinates, there exists a non-increasing sequence $(r_j(h))_{j\geq 1}$ such that the univariate array $(X^{(n)}(h), n \geq 1)$ satisfies the asymptotic *rND* condition (9.1).

(Multivariate) Asymptotic negative dependence condition Similarly, we say that the triangular array $\{X^{(n)}, n \geq 1\}$ of multivariate random variables is satisfying the asymptotic negative dependence condition if for each vector $h = (h_1, \dots, h_k)$, with non-negative coordinates, there exists a non-increasing sequence $(r_j(h))_{j\geq 1}$ such that the univariate array $(X^{(n)}(h), n \geq 1)$ satisfies the asymptotic negative dependence condition (9.2).

9.1.5 *Tightness*

First, we prove the tightness for triangular arrays satisfying Condition (A). Since the tightness condition in the finite-dimensional space holds, provided it holds for each coordinate, we only consider the univariate case.

Lemma 9.2 *Assume condition (A) is satisfied. Then*

$$\limsup_{n\to\infty} \mathbb{E}\left(\sup_{|s-t|\leq\delta} [W_n(s) - W_n(t)]^2 \right) \to 0 \quad as \quad \delta \to 0,$$

and thus, $\{W_n(t), t \in [0, 1]\}$ is tight in $D([0, 1])$ with the uniform topology. In addition $\{\sup_{t\in[0,1]} W_n^2(t)\}_{n\geq 1}$ is an uniformly integrable family.

Proof We start by making the additional assumption that the variables have moments of order $q > 2$; this restriction will be removed later in the proof. Let us work with the modulus of continuity defined as: For any $u \in [0, 1]$,

$$\delta_u(x) = \sup\{|x(s) - x(t)| \ ; \ |s - t| \leq u, \ s, t \in [0, 1]\} \ \text{ and set } \ L_n := \sum_{i=1}^{n} \mathbb{E}|X_{in}|^q.$$

According to Remark 9.1, condition (A)-(a) implies condition (A)-(a'). Let then m be a positive integer such that condition (A)-(a') is satisfied. We shall establish first that there exists a positive constant $c > 0$ which depends only on q and m such that, for all $0 \le u \le 1$,

$$\mathbb{E}[\delta_u(W_n)^q] \le c(L_n + u^{q/2-1}). \tag{9.4}$$

To see this, we first notice that by the construction of the process $W_n(t)$ for any s, with $t_{(k-1)n} < s \le t_{kn}$ and any t with $t_{(\ell-1)n} < t \le t_{\ell n}$ and $k \le \ell$, we have

$$\mathbb{E}\left(\sup_{s\le w\le t} |W_n(w) - W_n(s)|^q\right) \le \mathbb{E}\left(\max_{k+1\le j\le \ell}\left|\sum_{i=k+1}^{j} X_{in}\right|^q\right).$$

But, for any $j \in [k+1, \ell]$,

$$\sum_{i=k+1}^{j} X_{in} = \sum_{i=1}^{m}\sum_{r=[k/m]+1}^{[j/m]-1} X_{(i+rm)n} + \sum_{i=m[j/m]+1}^{j} X_{in} + \sum_{i=k+1}^{\ell\wedge([k/m]+1)m} X_{in},$$

giving

$$\begin{aligned}&\mathbb{E}\left(\sup_{s\le w\le t} |W_n(w) - W_n(s)|^q\right)\\ &\le (2m)^{q-1}\sum_{i=1}^{m}\mathbb{E}\left(\max_{k+1\le j\le \ell}\left|\sum_{r=[k/m]+1}^{[j/m]-1} X_{(i+rm)n}\right|^q\right) + 2^{q-1}(2m)^q\mathbb{E}\left(\max_{k+1\le i\le \ell}|X_{in}|^q\right).\end{aligned}$$

Applying Theorem 8.8 we infer that there is a constant C depending only on q and on m such that

$$\mathbb{E}\left(\sup_{s\le w\le t} |W_n(w) - W_n(s)|^q\right) \le C\left(\left(\sum_{i=k+1}^{\ell}\mathbb{E}(X_{in}^2)\right)^{q/2} + \sum_{i=k+1}^{\ell}\mathbb{E}|X_{in}|^q\right).$$

But, by construction, for $\ell > k$, $\sum_{i=k+1}^{\ell}\mathbb{E}(X_{in}^2) = \mathbb{E}(X_{\ell n}^2) + \sum_{i=k+1}^{\ell-1}\mathbb{E}(X_{in}^2) = \mathbb{E}(X_{\ell n}^2) + t_{(\ell-1)n} - t_{kn}$. Hence

$$\sum_{i=k+1}^{\ell}\mathbb{E}(X_{in}^2) \le \|X_{\ell n}\|_q^2 + t - s.$$

So overall, there is a constant c depending only on q and on m such that for any s, with $t_{(k-1)n} < s \le t_{kn}$ and any t with $t_{(\ell-1)n} < t \le t_{\ell n}$ and $k \le \ell$, we have

$$\mathbb{E}\left(\sup_{s\le w\le t} |W_n(w) - W_n(s)|^q\right) \le c\left(|t-s|^{q/2} + \sum_{i=k+1}^{\ell}\mathbb{E}|X_{in}|^q\right). \tag{9.5}$$

Further, we partition the interval $[0, 1]$ as $0 \le u \le 2u \le \cdots \le Nu \le 1$, where N is a positive integer such that $N \le 1/u$. Therefore

$$\delta_u(W_n) \le 3 \max_{0\le k\le N} \sup_{ku\le w\le (ku+u)\wedge 1} |W_n(w) - W_n(ku)| := 3I_n.$$

By applying the inequality (9.5) to bound the terms in I_n, we derive

$$\mathbb{E}[I_n^q] \le \sum_{k=0}^{N} \mathbb{E}\left(\sup_{ku\le w\le (ku+u)\wedge 1} |W_n(w) - W_n(ku)|^q \right) \le c \sum_{k=0}^{N} \left(u^{q/2} + \sum_{i=j_{k-1}+1}^{j_k} \mathbb{E}|X_{in}|^q \right)$$
$$\le 2c(u^{q/2}(1 + 1/u) + L_n),$$

where j_k is such that $t_{j_{k-1},n} < (k+1)u \le t_{j_k,n}$ for $k \ge 1$ and $j_{-1} = 0$. This proves the bound (9.4).

Now, let us drop the assumption that the variables have moments of order $q > 2$. So, in order to prove the tightness under the Lindeberg condition, we first find a positive sequence $\varepsilon_n \to 0$ such that

$$\Delta_n = \sum_{i=1}^{n} \mathbb{E}[X_{in}^2 I_{(|X_{in}|>\varepsilon_n)}] \to 0, \text{as} \quad n \to \infty$$

and define the class of truncation functions by $f_a(t) = t$ for $|t| \le a$, $f_a(t) = a$ for $t > a$ and $f_a(t) = -a$ for $t < -a$. Next, we define the truncated version of $W_n(t)$ by

$$S'_k = 0,\ S'_k = \sum_{i=1}^{k} \{f_{\varepsilon_n}(X_{in}) - \mathbb{E}[f_{\varepsilon_n}(X_{in})]\},\ k = 1, \dots, n.$$

We define then $\{W'_n(t), t \in [0, 1]\}$ by setting $W'_n(t) = S'_k$ for $t_{(k-1)n} < t \le t_{kn}$, $k = 1, \dots, n$ and $W'_n(t) = S'_n$ for $t \ge t_{nn}$. Notice that $||W_n - W'_n||_\infty = \max_{1\le k\le n} |S_k - S'_k|$. So, by taking into account the *rND* condition (A)-(a'), we infer that the maximal inequality of Theorem 8.8 implies that there exists a positive constant c depending on m such that

$$\mathbb{E}[||W_n - W'_n||_\infty^2] = \mathbb{E} \max_{1\le k\le n} |S_k - S'_k|^2 \le c\Delta_n \to 0 \text{ as } n \to \infty. \tag{9.6}$$

Now, the tightness property of the sequence of truncated processes $\{W'_n; n = 1, 2, \dots\}$ directly follows from (9.4) by taking into account that

$$\sum_{i=1}^{n} \mathbb{E}|f_{\varepsilon_n}(X_{in}) - \mathbb{E}[f_{\varepsilon_n}(X_{in})]|^q \le 2(2\varepsilon_n)^{q-2} \sum_{i=1}^{n} \mathbb{E}(X_{in}^2) = 2(2\varepsilon_n)^{q-2}.$$

The proof of the tightness is complete.

To prove that $\{\sup_{t\in[0,1]} W_n^2(t)\}_{n\ge1}$ is a uniformly integrable family, we first notice that, by taking into account (9.6), the uniform integrability of $\{\sup_{t\in[0,1]} W_n^2(t)\}_{n\ge1}$

follows from the uniform integrability of the family $\{\sup_{t\in[0,1]} W_n'^2(t)\}_{n\geq 1}$. This follows from (9.4) which gives

$$\sup_{n\to\infty} \mathbb{E}\left(\sup_{t\in[0,1]} W_n'^4(t)\right) \leq C,$$

where C is a positive constant depending on m. ■

9.1.6 *Some Properties of rND Processes*

Let us first recall the definition of a sequence of processes having *asymptotically independent increments*.

Definition 9.3 *We say that the sequence of processes $\{X_n(t),\ t \in [0, 1]\}$ has asymptotically independent increments if, for any positive integer m, any reals $u_1, \ldots, u_m$ and any reals $0 \leq t_0 < t_1 < \ldots < t_m \leq 1$,*

$$\mathbb{E}\exp\left(\sum_{k=1}^{m} iu_k(X_n(t_k) - X_n(t_{k-1}))\right) - \prod_{k=1}^{m} \mathbb{E}\exp(iu_k(X_n(t_k) - X_n(t_{k-1})) \to 0 \tag{9.7}$$

as $n \to \infty$.

In the case of stochastically continuous processes, the following equivalent modification of (9.7) to define a sequence of processes having *asymptotically independent increments* is useful in tackling mixing structures: We require that for any positive integer m, any reals $u_1, \ldots, u_m$, and any reals $0 \leq s_1 \leq t_1 < s_2 \leq t_2 < \cdots < s_m \leq t_m \leq 1$,

$$\mathbb{E}\exp\left(\sum_{k=1}^{m} iu_k(X_n(t_k) - X_n(s_k))\right) - \prod_{k=1}^{m} \mathbb{E}\exp(iu_k(X_n(t_k) - X_n(s_k)) \to 0,$$

$$\text{as } n \to \infty. \tag{9.8}$$

The intervals $[s_k, t_k]$, $k = 1, \ldots, m$, are disjoint and thus (9.8) automatically holds under strong mixing condition.

We now introduce the notion of processes with *rND* increments.

Definition 9.4 *We say that a stochastic process $\{X(t),\ t \in [0, 1]\}$ has rND increments (r-negatively dependent increments), if for any positive integer m and any reals $0 \leq s_1 \leq t_1 < s_2 \leq t_2 < \cdots < s_m \leq t_m \leq 1$, the vector $(X(t_k) - X(s_k),\ k = 1, \ldots, m)$ is r-negatively dependent (i.e. it satisfies condition (8.12) or more generally condition (8.22) in case of a process depending on n).*

Remark 9.5 *Note that if a stochastic process $\{X(t),\ t \in [0, 1]\}$ has continuous sample paths then according to Lemma 8.18, Definition 9.4 is equivalent to the following: For any positive integer m and any reals $0 \leq t_0 < t_1 < t_2 < \cdots < t_m \leq 1$, the vector $(X(t_k) - X(t_{k-1}),\ k = 1, \ldots, m)$ is r-negatively dependent (i.e. it satisfies condition (8.12) or more generally condition (8.22) in case of a process depending on n).*

We start with a proposition that will be useful in applications, on the stability of the property of having *rND* increments for independent summation of processes with *rND* increments. So, in particular, the sum of two independent processes with *rND* increments has *rND* increments. Its proof is very similar to the proof of Lemma 7.5 and is left as an exercise.

Proposition 9.6 *Suppose that* $X = (X_1, ..., X_n)$ *and* $Y = (Y_1, ..., Y_m)$ *are independent vectors satisfying* (8.22) *with the same* $r < 1$. *Then, so does the vector* $(X,Y) = (X_1, ..., X_n, Y_1, ..., Y_m)$.

Suppose that $\{X(t); t \in [0, 1]\}$ *and* $\{Y(t); t \in [0, 1]\}$ *are independent processes with rND increments; then, so are processes* $\{(X(t), Y(t)); t \in [0, 1]\}$ *and* $\{X(t) + Y(t); t \in [0, 1]\}$.

We shall establish now a lemma on the characterization of the limit of a process satisfying condition (A).

Lemma 9.7 *Let* $\{(X_{1n}, ..., X_{nn})\ ;\ n = 1, 2, ...\}$ *be a triangular array of variables satisfying condition* (A). *Assume that* $W_n \to^d W$ *in* $D[0, 1]$.

(i) *Then, W has continuous sample paths and rND increments.*

(ii) *for any* $q > 0$, *there exists a positive constant* c_q *with*

$$\mathbb{E}|W(t) - W(s)|^q \leq c_q |t - s|^{q/2}\ ,\ 0 \leq s < t \leq 1. \tag{9.9}$$

(iii) *For any partition* $0 = t_0 < t_1 < ... < t_n = 1$ *such that* $\max_j |t_j - t_{j-1}| \to 0$ *and for any* $p \geq 1$,

$$\mathbb{E}\left|\sum_{j=1}^{n}([W(t_j) - W(t_{j-1})]^2 - \mathbb{E}[W(t_j) - W(t_{j-1})]^2)\right|^p \to 0,\ \textit{as}\ n \to \infty. \tag{9.10}$$

(iv) *The process W is Gaussian with independent increments if and only if the sequence* $\{W_n\}$ *has asymptotically independent increments.*

(v) *If we assume in addition that* $\{(X_{1n}, ..., X_{nn})\ ;\ n = 1, 2, ...\}$ *is asymptotically negatively dependent, i.e. it satisfies condition* (9.2), *then W has negatively dependent increments (so its increments process satisfies* (8.12) *with* $r = 0$).

Proof The continuity of the sample paths follows from Lemma 9.2. On another hand, let m be a fixed positive integer, take reals $0 \leq u_1 \leq v_1 < u_2 \leq v_2 < \cdots < u_m \leq v_m \leq 1$ and define $Y_{\ell,n} = W_n(v_\ell) - W_n(u_\ell)$. Let $0 < \delta = \min_{1 \leq \ell \leq m-1}(u_{j+1} - v_j)$. Define also i_0 such that $u_{i_0+1} - v_{i_0} = \delta$ and, for any $t \in [0, 1]$, let $\ell_n(t)$ be the smallest integer k such that $\sum_{i=1}^{k} \mathbb{E}(X_{in}^2) := t_{kn} \geq t$. With this notation, $Y_k = \sum_{i=\ell_n(u_k)+1}^{\ell_n(v_k)} X_{in}$. Define

$$d_n = \ell_n(u_{i_0+1}) - \ell_n(v_{i_0}).$$

Let us prove that $d_n \to \infty$, as $n \to \infty$. With this aim, observe that by the second part of the Lindeberg condition, $\max_{1 \leq i \leq n} \mathbb{E}(X_{in}^2) \to 0$, as $n \to \infty$. Therefore, for any positive integer t, $\ell_n(t) \to \infty$ and $\sum_{i=1}^{\ell_n(t)} \mathbb{E}(X_{in}^2) := t_{\ell_n(t),n} \to t$, as $n \to \infty$. It follows that

$$\sum_{i=1}^{\ell_n(u_{i_0+1})} \mathbb{E}(X_{in}^2) - \sum_{i=1}^{\ell_n(v_{i_0})} \mathbb{E}(X_{in}^2) \to u_{i_0+1} - v_{i_0} = \delta > 0, \text{ as } n \to \infty.$$

Now if d_n does not converge to ∞, since $\max_{1\le i\le n} \mathbb{E}(X_{in}^2) \to 0$, we would have

$$\sum_{i=1}^{\ell_n(u_{i_0+1})} \mathbb{E}(X_{in}^2) - \sum_{i=1}^{\ell_n(v_{i_0})} \mathbb{E}(X_{in}^2) \le d_n \max_{1\le i\le n} \mathbb{E}(X_{in}^2) \to 0, \text{ as } n \to \infty,$$

which is a contradiction. This proves that $d_n \to \infty$, as $n \to \infty$.

Now, note that since $\{(X_{1n}, \ldots, X_{nn})\ ;\ n = 1, 2, \ldots\}$ satisfies the asymptotic *rND* condition (9.1) and $d_n \to \infty$, the vector $(Y_{1,n}, Y_{2,n}, \ldots, Y_{m,n})$ is r-negatively dependent (i.e. it satisfies condition (8.12)) for n large enough. By the continuous mapping theorem, it follows that the vector of increments $\{W(v_\ell) - W(u_\ell), \ell = 1, \ldots, m)$ satisfies (8.12) for any two functions v and w continuous, bounded and coordinatewise non-decreasing. By using Lemma 8.18, it follows that it satisfies also (8.12) for all coordinatewise non-decreasing functions. This ends the proof of (i).

To prove (ii) we notice that since the weak limits of $W_n(t)$ and of $W_n'(t)$ constructed by using the truncated random variables, as in the proof of Lemma 9.2, are the same, the moment inequality (9.9) follows from (9.4) applied with the r.v.'s $f_{\varepsilon_n}(X_{in}) - \mathbb{E}[f_{\varepsilon_n}(X_{in})]$.

To prove (iii), we first observe that the arrays $Y = ([W(t_i) - W(t_{i-1})]_+^2, i = 1, \ldots, n)$ and $Z = ([-(W(t_i) - W(t_{i-1}))]_+^2, i = 1, \ldots, n)$ satisfy condition (8.12) since W has *rND* increments by Item (i) and has continuous sample paths (see Remark 9.5). Hence by applying Lemma 8.7, we get that for any $p \ge 2$, there exists a positive constant c such that

$$\mathbb{E}\left|\sum_{j=1}^{n}([W(t_j) - W(t_{j-1})]_+^2 - \mathbb{E}[W(t_j) - W(t_{j-1})]_+^2)\right|^p$$
$$\le c\sum_{j=1}^{n} \mathbb{E}|W(t_j) - W(t_{j-1})|^{2p} + c\left(\sum_{j=1}^{n} \mathbb{E}|W(t_j) - W(t_{j-1})|^4\right)^{p/2}.$$

Hence by Item (ii), there exists a positive constant C such that

$$\mathbb{E}\left|\sum_{j=1}^{n}([W(t_j) - W(t_{j-1})]_+^2 - \mathbb{E}[W(t_j) - W(t_{j-1})]_+^2)\right|^p$$
$$\le C\max_j |t_j - t_{j-1}|^{p-1} + C\max_j |t_j - t_{j-1}|^{p/2} \to 0,$$

since $\max_j |t_j - t_{j-1}| \to 0$. Similarly, we get the same convergence for the quantity $\mathbb{E}\left|\sum_{j=1}^{n}([-(W(t_j) - W(t_{j-1}))]_+^2 - \mathbb{E}[-(W(t_j) - W(t_{j-1}))]_+^2)\right|^p$ which ends the proof of Item (iii).

Statement (iv) follows from the fact that among processes with independent increments the Gaussian process is the only one which is equivalent in distribution with a process with continuous sample paths.

Statement (v) is proved by using the same arguments as in Item (i). ■

Remark 9.8 *If the process W has negatively dependent increments (meaning that $r = 0$ in Definition 9.4 or equivalently in Remark 9.5 since W has continuous sample paths), then the inequality (9.9) can be strengthened, namely the increments of W are subgaussian, i.e. there exists a finite positive constant τ such that for any real h*

$$\mathbb{E}\exp(h[W(t) - W(s)]) \leq \exp(\tau h^2[t - s]), \ 0 \leq s < t \leq 1.$$

To prove it, we consider the triangular array $Y = (Y_{1n}, ..., Y_{nn})$ where

$$Y_{in} = W(s + (t - s)i/n) - W(s + (t - s)(i - 1)/n), \ i = 1, \dots, n.$$

Notice that by (9.9), for any $q > 0$, $\mathbb{E}(|Y_{in}|^q) < \infty$. By Shao's comparison theorem (2000), for any $q \geq 2$, we then get

$$\mathbb{E}[|W(t) - W(s)|^q] = \mathbb{E}[|Y_{1n} + ... + Y_{nn}|^q] \leq \mathbb{E}[|Y^*_{1n} + ... + Y^*_{nn}|^q],$$

where Y^*_{in} are mutually independent and such that Y_{in} and Y^*_{in} have the same distribution for each $i = 1, ..., n$. By the central limit theorem applied to triangular arrays of independent variables satisfying the Lindeberg condition, and by the uniform boundness of all absolute moments in (9.9), we then derive

$$\begin{aligned}\mathbb{E}[|W(t) - W(s)|^q] &\leq \lim_{n\to\infty} \mathbb{E}[|Y^*_{1n} + ... + Y^*_{nn}|^q] = \left(\mathbb{E}[|W(t) - W(s)|^2]\right)^{q/2} \mathbb{E}[|\eta|^q] \\ &\leq c_2^{q/2}|t - s|^{q/2}\mathbb{E}[|\eta|^q],\end{aligned}$$

for any $q \geq 2$, where η is a standard normal variable. Then the desired inequality follows via the comparison

$$\mathbb{E}\exp(h|W(t) - W(s)|^2) \leq \mathbb{E}\exp([c_2 h|t - s||\eta|^2) < \infty, \quad \text{for } |c_2 h(t - s)| < 1/2,$$

proving the subgaussian bound. ■

9.2 Convergence to a Gaussian Process with Independent Increments

9.2.1 Univariate Invariance Principle

First, we present necessary and sufficient conditions for the convergence to the Gaussian process with independent increments for the process W_n constructed in Section 9.1. The invariance principle for the partial sums process associated with a *rND*–triangular array is contained in the next theorem.

Theorem 9.9 *Let $\{(X_{1n}, ..., X_{nn}); n = 1, 2, ...\}$ be a triangular array of centered real-valued random variables satisfying condition (A). Then $W_n \to^d W$ where W is a centered Gaussian process with independent increments and such that $\mathrm{Var}(W(t)) = \sigma_t^2$ for all $t \in [0, 1]$, if and only if the following two conditions are satisfied:*

(i) The sequence $\{W_n\}$ has asymptotically independent increments.

(ii) $\mathrm{Var}(W_n(t)) \to \sigma_t^2$ for all $t \in [0, 1]$.

Proof The proof of this theorem directly follows from Lemma 9.2 and Lemma 9.7. ■

We note that Theorem 9.9 can be immediately applied to triangular arrays of strongly mixing sequences:

Definition 9.10 *We call a triangular array $\{(X_{1n}, ..., X_{nn}); n = 1, 2, ...\}$ strongly mixing if there exists a sequence $\alpha_m \to 0$, as $m \to \infty$, such that for all positive integers j,n and all sets $A \in \sigma(X_{1n}, \dots, X_{jn})$ and $B \in \sigma(X_{kn}, k \geq j + m)$,*

$$|\mathbb{P}(A \cap B) - \mathbb{P}(A)\mathbb{P}(B)| \leq \alpha_m \to 0 \quad (m \to \infty). \tag{9.11}$$

It is easy to see by recurrence that $W_n(t)$ constructed by using a strongly mixing triangular array has asymptotically independent increments. As a consequence we obtain the following corollary of Theorem 9.9.

Corollary 9.11 *Let $\{(X_{1n}, ..., X_{nn})\ ;\ n = 1, 2, ...\}$ be a strongly mixing triangular array of centered real-valued random variables satisfying condition (A). Then $W_n \to^d W$ where W is a centered Gaussian process with independent increments and $\mathrm{Var}(W(t)) = \sigma_t^2$ if and only if $\mathrm{Var}(W_n(t)) \to \sigma_t^2$ for all $t \in [0, 1]$.*

Remark 9.12 *By a non-random change of time $u \to v_n(u) = \min\{k > 0 : \mathrm{Var}(S_k) \geq u\}$, it follows that Theorem 9.9 implies the invariance principle proved in Utev and Peligrad (2003) (see their Theorem 2.2 and its corollaries). As a matter of fact, the asymptotic rND condition used in Corollary 9.11 is weaker than the interlaced mixing condition $\rho_n^* \to 0$. This condition is different from (9.1) in the sense that its left-hand side has an absolute value. In other words condition (9.1) is a one-sided variant of $\rho_n^* \to 0$.*

To treat the asymptotic negative dependent situation, we apply the following standard property:

Property 9.13 *Assume that $Y_1, ..., Y_k$ are negatively dependent random variables with finite second moments. Then, $Y_1, ..., Y_k$ are independent if and only if $Y_1, ..., Y_k$ are uncorrelated, i.e. $corr(Y_i, Y_j) = 0$ for all $i \neq j$.*

Proof We first notice that since $Y_1, ..., Y_k$ are negatively dependent then for all non-negative $u_1, ..., u_k$, we have $r^*(\sum_{i\in Q} u_iY_i, \sum_{j\in Q^*} u_jY_j) = 0$ for all sets of integers $Q \subset \{1, \dots, k\}$ and $Q^* = \{1, \dots, k\}\backslash Q$. Hence, according to Corollary 7.2, for any non-negative $u_1, ..., u_k$,

$$\mathbb{E}\exp\left(\sum_{j=1}^{k} iu_jY_j\right) = \prod_{k=1}^{m} \mathbb{E}\exp(iu_jY_j) \tag{9.12}$$

as soon as $\text{cov}(Y_i, Y_j) = 0$ for all $i \neq j$. Since the Laplace transform is analytic, (9.12) also holds for any reals $u_1, ..., u_k$ proving that $Y_1, ..., Y_k$ are independent if $Y_1, ..., Y_k$ are uncorrelated. ■

Our next result is particularly useful when applied to near stationary asymptotically negatively dependent variables.

Corollary 9.14 *Let $\{(X_{1n}, ..., X_{nn})\ ;\ n = 1, 2, ...\}$ be a triangular array of centered variables satisfying condition (9.3) and the asymptotic negative dependence condition (9.2). Then $W_n \to^d W$ where W is a centered Gaussian process with independent increments if and only if for any $0 \le s_1 \le t_1 < s_2 \le t_2 \le 1$,*

$$\text{cov}(W_n(t_1) - W_n(s_1), W_n(t_2) - W_n(s_2)) \to 0, \text{ as } n \to \infty. \tag{9.13}$$

Proof By Lemma 9.2 $\{W_n(t),\ t \in [0, 1]\}$ is tight in $D[0, 1]$ with the uniform topology. Therefore, from any subsequence of W_n we can extract a convergent subsequence to a continuous process X. We only need to show that X has independent increments. Indeed since X has continuous sample paths, it will be a Gaussian process and since by Lemma 9.2, for any $t \in [0, 1]$, $W_n^2(t)$ is a uniformly integrable family, we will have $\mathbb{E}(X(t)) = 0$ and $\mathbb{E}(X^2(t)) = \lim_{n\to\infty} \text{Var}(W_n(t))$ proving then that the limit in distribution of W_n is unique.

To prove that X has independent increments, we notice that by Item (v) of Lemma 9.7, X has negatively dependent increments. Using Property 9.13 and noticing that for any $t \in [0, 1]$, $X^2(t)$ is in $\mathbb{L}^1$ (since $W_n^2(t)$ is in $\mathbb{L}^1$ and is uniformly integrable), it follows that X will have independent increments if and only if its increments are uncorrelated, meaning that for any $0 \le s_1 \le t_1 < s_2 \le t_2 \le 1$,

$$\text{cov}(X(t_1) - X(s_1), X(t_2) - X(s_2)) = 0. \tag{9.14}$$

Let $0 \le s_1 \le t_1 < s_2 \le t_2 \le 1$. To prove (9.14), it suffices to notice that by the continuous mapping theorem $Z_n := (W_n(t_1) - W_n(s_1))(W_n(t_2) - W_n(s_2)) \to^d (X(t_1) - X(s_1))(X(t_2) - X(s_2))$ and since Z_n is uniformly integrable (since $W_n^2(t)$ is for all $t \in [0, 1]$), by the convergence of moments theorem, we have

$$\text{cov}(W_n(t_1) - W_n(s_1), W_n(t_2) - W_n(s_2)) \to \text{cov}(X(t_1) - X(s_1), X(t_2) - X(s_2)),$$

as $n \to \infty$.

Therefore (9.14) follows from (9.13). ■

Since, we are not using a particular ordering of the real line, the invariance principle admits a straightforward generalization to random fields, which then would relax conditions in Zhang (2000) and Yuan, Su and Hu (2003).

Remark 9.15 *Assume that* $\{(X_{1n}, \ldots, X_{nn})\,;\, n = 1, 2, \ldots\}$ *is a triangular array as in Corollary 9.14. If we can prove that there exists* σ^2 *such that* $\mathrm{Var}(W_n(b) - W_n(a)) \to \sigma^2(b-a)$ *for all* $0 \le a < b \le 1$, *then* $W_n \to^d \sigma W$ *where* W *is a standard Wiener process.*

Notice that for stationary sequences, the convergence of variances in the remark above follows from Proposition 7.10. This leads to:

Corollary 9.16 *(Invariance principle for stationary sequences satisfying asymptotic negative dependence) Let* $\{X_i\,;\, i = 1, 2, \ldots\}$ *be a stationary sequence with* $\mathbb{E}(X_1) = 0$, $\mathbb{E}(X_1^2) < \infty$ *and satisfying the asymptotic negative dependence condition* (9.2). *Then,*

$$\lim_{n\to\infty} \frac{\mathrm{Var}(S_n)}{n} = \sigma^2 \quad \text{and} \quad \left\{\frac{S_{[nt]}}{\sqrt{n}}, t \in [0,1]\right\} \to^d \sigma W \text{ as } n \to \infty,$$

where $S_k = \sum_{i=1}^{k} X_i$ *and* $W = \{W(t), t \in [0,1]\}$ *is a standard Brownian motion.*

Comment 9.17 *A careful analysis of the proofs of Theorem 9.9 and Corollary 9.14 reveals that they are also valid for the partial sums process associated with an infinite or a double infinite sum provided that condition* (9.3) *is replaced by*

$$\sup_n \sum_{i\in\mathbb{Z}} \mathbb{E}[X_{in}^2] = 1 \text{ and for any } \varepsilon > 0, \quad \sum_{i\in\mathbb{Z}} \mathbb{E}[X_{in}^2 I_{(|X_{in}|>\varepsilon)}] \to 0 \quad \text{as} \quad n \to \infty,$$

and the process $\{W_n(t), t \in [0,1]\}$ *is defined by* $W_n(0) = 0$ *and* $W_n(t) = \sum_{i=-\infty}^{k_n(t)} X_{in}$ *where*

$$k_n(t) = \min\{k \in \mathbb{Z} : \sum_{i=-\infty}^{k} \mathbb{E}[X_{in}^2] \ge t\}.$$

9.2.2 Multivariate Invariance Principle

Theorem 9.18 *(Multivariate IP for stationary sequences satisfying asymptotic negative dependence) Let* $X = (X_i)_{i\ge1} = (X_i^{(1)}, \ldots, X_i^{(k)})_{i\ge1}$ *be a stationary sequence with values in the Euclidean vector space* $\mathbb{R}^k$ *with inner product* $< x, y > = x_1y_1 + \ldots + x_ky_k$. *Assume that* $\mathbb{E}(X_1^{(i)}) = 0$ *and* $\mathbb{E}(X_1^{(i)})^2 < \infty$, *for all* $i = 1, \ldots, k$. *Suppose that for each vector* $h = (h_1, \ldots, h_k)$ *with non-negative coordinates, the triangular array* $(<X_i, h>)_{1\le i\le n,\, n\ge1}$ *satisfies the asymptotic negative dependence condition* (9.2). *Let* $S_n = X_1 + \ldots + X_n$. *Then, for any real* $u \in \mathbb{R}^k$, *there exists a limit*

$$\lim_{n\to\infty} \frac{\mathrm{Var}(< S_n, u >)}{n} = \sigma_u^2. \tag{9.15}$$

Moreover,

$$\left\{\frac{S_{[nt]}}{\sqrt{n}}, t \in [0,1]\right\} \to^d \{B(t), t \in [0,1]\} \text{ as } n \to \infty, \tag{9.16}$$

where $\{B(t)\ ;\ t \geq 0\}$ is the k-dimensional Brownian motion with independent increments, mean zero and variances $\mathrm{Var}(< B(t), u >) = t\sigma_u^2$, *for any vector* $u \in \mathbb{R}^k$ *and any* $t \in [0, 1]$.

Proof The proof is divided in three steps. We start by proving (9.15), then the multivariate central limit theorem and finally the multivariate invariance principle.

Step 1. From Corollary 9.16 it follows that there is σ_u^2 such that

$$\lim_{n\to\infty} \frac{\mathrm{Var}(< S_n, u >)}{n} = \sigma_u^2 \tag{9.17}$$

for $u \in \mathbb{R}_+^k = \{(u_1, \dots, u_k) : u_1 \geq 0, \dots, u_k \geq 0\}$.

Now, let $u, v \in \mathbb{R}_+^k$. Then from (9.17) it follows that

$$\lim_{n\to\infty} \frac{\mathbb{E}(< S_n, u >< S_n, v >)}{n} = \frac{1}{2}(\sigma_{u+v}^2 - \sigma_v^2 - \sigma_u^2). \tag{9.18}$$

Now if $u \in \mathbb{R}^k$ we write u as a difference of two vectors in $\mathbb{R}_+^k$ and we easily deduce from (9.18) that (9.17) also holds for any $u \in \mathbb{R}^k$.

Step 2. We prove now the multivariate CLT, that is:

$$\frac{S_n}{\sqrt{n}} \to^d Z \text{ as } n \to \infty, \tag{9.19}$$

where Z is a centered Gaussian vector of $\mathbb{R}^k$ with $\mathrm{Var}(< Z, u >) = \sigma_u^2$ for any $u \in \mathbb{R}^k$.

Its proof will be based on the following characterization property of the Gaussian vectors, whose proof will be left to the reader as an exercise.

Property 9.19 *Assume that $(X_1, \dots, X_n)$ is such that $a_1X_1 + \dots + a_nX_n$ has a Gaussian distribution for all $a_1, \dots, a_n \geq 0$. Then $(X_1, \dots, X_n)$ has a joint Gaussian distribution.*

To prove (9.19), we note that Corollary 9.16 also implies that for any $h \in \mathbb{R}_+^k$, the univariate sequence $\{n^{-1/2} <S_n, h>\}_{n\geq 1}$ is convergent to a normal distribution with variance σ_h^2, and in addition the second moments are convergent. We then notice that the process $S_n/\sqrt{n}$ is tight.

To characterize a partial limit $Z = (Z_1, \dots, Z_k)$ we observe that each marginal is normal and also $< Z, h >$ has a normal distribution for any $h \in \mathbb{R}_+^k$. By Property 9.19, this is enough to conclude that Z is a Gaussian vector with the covariance structure

$$\mathrm{cov}(Z_i, Z_j) = \lim_{n\to\infty} \frac{\mathrm{cov}(S_n^{(i)}, S_n^{(j)})}{n} \text{ as } n \to \infty, \tag{9.20}$$

where $S_n^{(i)} = \sum_{j=1}^n X_j^{(i)}$. This ends the proof of the second step.

Step 3. We prove now the multivariate invariance principle (9.16). We mention first that from Corollary 9.16 it also follows that, for any $h \in \mathbb{R}^k_+$, the sequence of univariate processes $\{n^{-1/2} <S_{[nt]}, h>, t \in [0, 1]\}_{n \geq 1}$ is convergent to W_h which is a univariate Brownian motion such that $\mathrm{Var}(W_h(t)) = t\sigma_h^2$. Then we notice that the sequence of processes $\{S_{[nt]}/\sqrt{n}, t \in [0, 1]\}_{n \geq 1}$ is tight.

Again, to characterize the limit let us consider any sublimit $B(t) = (B_1(t), \ldots, B_k(t))$. From the multivariate central limit theorem (9.15), for each $t \in [0, 1]$, fixed, $n^{-1/2} S_{[nt]}$ is convergent to a Gaussian vector Z_t with the covariance structure given by (9.19). This reduces the proof to showing that $B(t)$ has independent increments.

To simplify the exposition we shall consider only the case $k = 2$. Obviously the individual components $B_1(t)$ and $B_2(t)$ have independent increments. Because of the Gaussian structure we have then just to show that for any $0 \leq s < t < u < v$, we have

$$\mathrm{cov}(B_1(t) - B_1(s), B_2(v) - B_2(u)) = 0,$$

and, because we have the convergence of second moments we have only to show that

$$\mathrm{cov}(S_n(s, t), T_n(u, v)) \to 0 \quad \text{as} \quad n \to \infty, \tag{9.21}$$

where we used the notation

$$S_n(x, y) = \frac{1}{\sqrt{n}} \sum_{i=[nx]+1}^{[ny]} X_i^{(1)}, \; T_n(x, y) = \frac{1}{\sqrt{n}} \sum_{i=[nx]+1}^{[ny]} X_i^{(2)}.$$

To prove (9.21), we observe that since $\{aX_i^{(1)} + bX_i^{(2)}\}_{i \geq 1}$ with $a, b \geq 0$ is a stationary sequence satisfying the conditions of Corollary 9.16, the limiting process has independent increments and also the second moments converge and we have for all $0 \leq s < t < u < v$,

$$\mathrm{cov}(aS_n(s, t) + bT_n(s, t), aS_n(u, v) + bT_n(u, v)) \to 0 \quad \text{as} \quad n \to \infty.$$

In particular,

$$\mathrm{cov}(S_n(s, t), S_n(u, v)) \to 0\,, \; \mathrm{cov}(T_n(s, t), T_n(u, v)) \to 0 \quad \text{as} \quad n \to \infty.$$

By combining these considerations, it follows

$$\mathrm{cov}(S_n(s, t), T_n(u, v)) + \mathrm{cov}(T_n(s, t), S_n(u, v)) \to 0 \quad \text{as} \quad n \to \infty. \tag{9.22}$$

Now we use the fact that the sequences $\{a_i X_i^{(1)} + b_i X_i^{(2)}\}_{i \geq 1}$ satisfy the condition of weak linear negative dependence implying that

$$\limsup_{n\to\infty} \operatorname{cov}(S_n(x,y), T_n(u,v)) \leq 0 \text{ and also } \limsup_{n\to\infty} \operatorname{cov}(T_n(x,y), S_n(u,v)) \leq 0,$$

because $[nu] - [nt] + 1 \geq n(u-t) \to \infty$, as $n \to \infty$. The limit (9.21) then follows from (9.22). ■

By inspecting the proof, we establish a similar result for triangular arrays.

Corollary 9.20 *Let $\{(X_{1n}, ..., X_{nn}) ; n = 1, 2, ...\}$ be a triangular array of centered k-dimensional vectors of random variables satisfying the multivariate condition (A)(b). Suppose that for all vectors $h = (h_1, ..., h_k)$ with non-negative coordinates the triangular array of processes $\{(< X_{1n}, h >, ..., < X_{nn}, h >) ; n = 1, 2, ...\}$ satisfy the asymptotic negative dependence condition (9.2). Then $W_n \to^d W$ where W is a centered process with independent increments if and only if*

$$\operatorname{cov}(< W_n(t_1) - W_n(s_1), h >, < W_n(t_2) - W_n(s_2), h >) \to 0, \text{ as } n \to \infty$$

for all $0 \leq s_1 \leq t_1 < s_2 \leq t_2 \leq 1$ and all vectors $h \in \mathbb{R}^k_+$.

9.3 Convergence to a Diffusion Process with Deterministic Time Varying Volatility

Let W_n be the process of $D([0, 1])$ constructed in Section 9.1 from a triangular array of real-valued random variables $\{(X_{1n}, ..., X_{nn}) ; n = 1, 2, ...\}$. As we have seen in Theorem 9.9, if the triangular array satisfies condition (A) and if $\{W_n\}$ has asymptotically independent increments then a sufficient condition for W_n to converge to a Wiener process with parameter σ^2 is that $\operatorname{Var}(W_n(t)) \to \sigma^2 t$ for all $t \in [0, 1]$. A natural question is to find sufficient conditions ensuring $\{W_n\}$ to converge to other processes than Wiener processes. Theorem 9.21 below is in this direction and is in the spirit of Theorem 19.4 in Billingsley (1968). Note that if the function a: $[0, T] \times \mathbb{R} \to \mathbb{R}$ in condition (9.23) is not equal to zero then $\{W_n\}$ does not have asymptotically independent increments.

Theorem 9.21 *Let $\{(X_{1n}, ..., X_{nn}) ; n = 1, 2, ...\}$ be a triangular array of centered real-valued random variables satisfying condition (A) of Section 9.1.2. Assume there exist a function $a : [0, 1] \times \mathbb{R} \to \mathbb{R}$ and a positive constant L such that for all $t \in [0, 1]$ and all $x, y \in \mathbb{R}$,*

$$|a(t,x) - a(t,y)| \leq L|x-y| \text{ and } |a(t,x)| \leq L(1+|x|),$$

and, for any partition $0 = t_0 < t_1 < ... < t_k < t_{k+1} \leq 1$ with $t_{k+1} = t$, we have

$$\lim_{h\downarrow 0} h^{-1} \limsup_{n\to\infty} \mathbb{E}\left|\mathbb{E}[W_n(t+h) - W_n(t)|W_n(t_1), \dots, W_n(t_k), W_n(t)] - ha(t, W_n(t))\right| = 0. \tag{9.23}$$

Assume in addition that there exists a deterministic non-negative function $\sigma(\cdot)$ on $[0,1]$ such that $\sigma^2(\cdot)$ is Lebesgue integrable and, for any partition $0 = t_0 < t_1 < \ldots < t_k < t_{k+1} \le 1$ with $t_{k+1} = t$ and $\sup_{1\le j\le k+1} |t_j - t_{j-1}| = \eta \to 0$, we have

$$\lim_{\eta\to 0}\lim_{n\to\infty}\sum_{j=1}^{k+1}\mathrm{Var}[W_n(t_j) - W_n(t_{j-1})] = \int_0^t \sigma^2(y)dy. \tag{9.24}$$

Then, $\{W_n(t), t\in[0,1]\} \to^d X$ in $C[0,1]$ where X is the solution of the stochastic integral equation

$$X(t) = \int_0^t a(u, X(u))du + \int_0^t \sigma(u)dW(u) \text{ a.s., } t\in[0,1], \tag{9.25}$$

where $\{W(t); t\in[0,1]\}$ is a standard Brownian motion.

Corollary 9.22 *Under the assumptions of Theorem 9.21 and if $a(t,x) = a(t)x$, $\{W_n(t), t \in [0,1]\} \to^d X$ in $C([0,1])$ where X is a Gaussian process satisfying the stochastic integral equation (9.25).*

(See for instance Section 5.6 in Karatzas and Shreve (1991)).

Proof of Theorem 9.21 *Step 1.* By Lemma 9.2, we know that from any subsequence of W_n we can extract a convergent subsequence to a continuous process X. We then show that X must satisfy the integral equation (9.25). Notice that the weak solution to (9.25) is unique, which follows by considering a standard iterative procedure, sometimes referred to as the Picard–Lindelöf iterations,

$$X^{(0)}(t) = 0 \text{ and } X^{(n+1)}(t) = \int_0^t \sigma(u)dW(u) + \int_0^t a(u, X^{(n)}(u))du,\ n\ge 0,$$

and then showing that the fixed point exists and is unique.

Step 2. Now, we need to show that (9.25) holds. Recall that by Lemma 9.7, the process X has almost surely continuous sample paths and is in $\mathbb{L}^p([0,1])$ for all $p\ge 1$. Let $\mathcal{F}_t$ be the σ–field generated by $X(s), s\le t$. By the continuity of $X(t)$, condition (9.23), the continuity mapping theorem and the convergence of moments theorem, we infer that

$$\lim_{h\downarrow 0} h^{-1}\mathbb{E}\,|\mathbb{E}[X(t+h) - X(t)|\mathcal{F}_t] - ha(t, X(t))| = 0.$$

Fix $T\in[0,1]$. Then, for $\delta > 0$ and $t\in[0,T]$ we find $r = r(t,\delta) \le \delta^2$ such that, for all $0 < h \le r(t,\delta)$,

$$\mathbb{E}\,|\mathbb{E}[X(t+h) - X(t)|\mathcal{F}_t] - ha(t, X(t))| \le \delta h.$$

The class of open intervals $(t, t + r(t, \delta))$ covers the compact set $[0, T]$ so there exist an integer n_δ and a finite partition $0 = t_0 < t_1 < \ldots < t_{n_\delta} = T$ with $\max_j |t_j - t_{j-1}| \leq \delta^2$ such that

$$\mathbb{E}\left|\mathbb{E}[X(t_{k+1}) - X(t_k)|\mathcal{F}_{t_k}] - (t_{k+1} - t_k)a(t_k, X(t_k))\right| \leq \delta(t_{k+1} - t_k). \tag{9.26}$$

For the sequence $\delta_j = 1/j$, we first construct the associated partition $0 = t_{0,j} < t_{1,j} < \ldots < t_{n_j,j} < t_{n_{j+1},j} = T$ with $\max_k |t_{k+1,j} - t_{k,j}| \leq \delta_j^2$ and then we construct the triangular array of variables

$$Y_{kj} = X(t_{k+1,j}) - X(t_{k,j}) \, , \; k = 0, \ldots, n_j.$$

By Lemma 9.7, the triangular array $\{Y_{1j}, \ldots, Y_{n_j j}\}_{j\geq 1}$ has the following properties: There exists a positive constant c such that

$$\mathbb{E}(Y_{kj}^4) \leq c(t_{k+1,j} - t_{k,j})^2 \text{ for all } k = 0, \ldots, n_j, \tag{9.27}$$

and for any sequence of positive integers ℓ_j such that $\ell_j \leq n_j$,

$$\sum_{i=0}^{\ell_j}([Y_{ij}]^2 - \mathbb{E}[Y_{ij}^2]) \to^{\mathbb{L}^1} 0 \text{ as } j \to \infty. \tag{9.28}$$

Now we construct the martingale difference

$$\xi_{kj} = Y_{kj} - U_{kj} \text{ where } U_{kj} = \mathbb{E}[Y_{kj}|Y_{1j}, \ldots, Y_{(k-1)j}].$$

Let $t \in [0, 1]$ and

$$\ell_j(t) = \max\{k = 0, \ldots, n_j \; : \; t_{k,j} \leq t\} \text{ and define the martingale}$$

$$M_t^{(j)} = \sum_{i=0}^{\ell_j(t)} \xi_{ij} \, .$$

Note that by condition (9.24), the continuity mapping theorem and the convergence of moments theorem, the following convergence holds:

$$\lim_{j\to\infty} \sum_{i=0}^{\ell_j(t)} \mathbb{E}[Y_{ij}^2] = \int_0^t \sigma^2(u)du. \tag{9.29}$$

Assume now that one can prove that for any $t \in [0, T]$,

$$\sum_{i=0}^{\ell_j(t)} \xi_{ij}^2 \to^{\mathbb{P}} \int_0^t \sigma^2(u)du \text{ as } j \to \infty, \tag{9.30}$$

and

$$\mathbb{E}\left(\max_{0\le k\le n_j}|\xi_{kj}|\right)\to 0 \text{ as } j\to\infty. \tag{9.31}$$

Then, by the functional central limit theorem for triangular arrays of martingale differences as stated in Theorem 2.39, it will follow that

$$\left\{M_t^{(j)}, t\in[0,1]\right\}\to^d \left\{\int_0^t \sigma(u)dW(u), t\in[0,1]\right\} \text{ as } j\to\infty, \tag{9.32}$$

where W is a standard Brownian motion on $[0,1]$. Since $X(0)=0$ and $t_{0,j}=0$, note that for any $t\in[0,T]$,

$$\begin{aligned}
X(t)-\int_0^t a(u,X(u))du - M_t^{(j)} &= X(t)-\int_0^t a(u,X(u))du-\sum_{i=0}^{\ell_j(t)}\xi_{ij}\\
= X(t)-X(t_{\ell_j(t)+1,j}) &+ \sum_{k=0}^{\ell_j(t)}\mathbb{E}\left(X(t_{k+1,j})-X(t_{k,j})|\mathcal{F}_{t_{k,j}}\right)-\sum_{k=0}^{\ell_j(t)}(t_{k+1,j}-t_{k,j})a(t_{k,j},X(t_{k,j}))\\
&+\sum_{k=0}^{\ell_j(t)}(t_{k+1,j}-t_{k,j})a(t_{k,j},X(t_{k,j}))-\int_0^t a(u,X(u))du\\
= I_j'(t)+I_j''(t)+I_j'''(t), &\text{ say.}
\end{aligned}$$

Note that the first term $I_j'(t)\to 0$ as $j\to\infty$ in $\mathbb{L}^1$ uniformly on $t\in[0,1]$, by the continuity of $X(t)$.

The second term $I_j''(t)\to 0$ as $j\to\infty$ in $\mathbb{L}^1$ uniformly on $t\in[0,1]$ by (9.26), since

$$\begin{aligned}
\left\|\sup\nolimits_{t\le T}|I_j''(t)|\right\|_1 &\le \sum\nolimits_{k=0}^{n_j}\mathbb{E}\left|\mathbb{E}\left(X(t_{k+1,j})-X(t_{k,j})|\mathcal{F}_{t_{k,j}}\right)-(t_{k+1,j}-t_{k,j})a(t_{kj},X(t_{k,j}))\right|\\
&\le \delta_j\sum\nolimits_{k=0}^{n_j}(t_{k+1,j}-t_{k,j})\le 1/j.
\end{aligned}$$

For the third term, note that again by the continuity of $X(t)$ and by the conditions on the function a,

$$\sup_{t\le T}|I_j'''(t)|\le\sup_{t\le T}\left|\sum_{k=0}^{\ell_j(t)}(t_{k+1,j}-t_{k,j})a(t_{k,j},X(t_{k,j}))-\int_0^t a(u,X(u))du\right|\to^{a.s.} 0 \text{ as } j\to\infty.$$

Combining together the above bounds, we have

$$\sup_{t\le T}\left|X(t)-\int_0^t a(u,X(u))du-M_t^{(j)}\right|\to^{\mathbb{P}}_{j\to\infty} 0.$$

Thus, the martingale $M_t^{(j)}$ is convergent in probability. Denote by M_t its limit. Since $M_t^{(j)}$ are uniformly integrable martingales, it follows that M_t is a martingale. On the other hand, by the functional central limit theorem (9.32), the martingale M_t is a Gaussian process with independent increments, therefore one can construct a standard Wiener process $W = \{W(t), t \in [0, 1]\}$ by a non-random time transform of the martingale process, such that

$$M_t = \int_0^t \sigma(u)dW(u) \text{ a.s.}$$

This proves that the process X_t is a solution to an integral equation

$$X(t) - \int_0^t a(u, X(u))du = M_t = \int_0^t \sigma(u)dW(u) \text{ a.s.}$$

To end the proof of the theorem, it remains to show that (9.30) and (9.31) are satisfied. We start by proving (9.31). We have

$$\left(\mathbb{E}\left(\max_{0\le k\le n_j} |\xi_{kj}|\right)\right)^4 \le \mathbb{E}\left(\max_{0\le k\le n_j} |\xi_{kj}|^4\right) \le \sum_{k=0}^{n_j} \mathbb{E}(|\xi_{kj}|^4) \le 2^3 \sum_{k=0}^{n_j} \mathbb{E}(|Y_{kj}|^4 + |U_{kj}|^4).$$

Hence, by using Jensen's inequality for conditional expectation together with (9.27), we derive

$$\left(\mathbb{E}\left(\max_{0\le k\le n_j} |\xi_{kj}|\right)\right)^4 \le 2^4 c \sum_{k=0}^{n_j} (t_{k+1,j} - t_{k,j})^2 \le 2^4 c\delta_j^2 \sum_{k=0}^{n_j} (t_{k+1,j} - t_{k,j}) \le 2^4 c\delta_j^2,$$

which proves (9.31) since $\delta_j \to 0$ as $j \to \infty$.

We turn now to the proof of (9.30) and we notice that by (9.28) and (9.29) this is reduced to show that for any $t \in [0, T]$,

$$\sum_{i=0}^{\ell_j(t)} ([Y_{ij}]^2 - [\xi_{ij}]^2) \to^{\mathbb{P}} 0 \qquad \text{as } j \to \infty. \tag{9.33}$$

Note that $([Y_{ij}]^2 - [\xi_{ij}]^2) = U_{ij}(2Y_{ij} - U_{ij})$. So, for any $\varepsilon > 0$,

$$\sum_{i=0}^{\ell_j(t)} |[Y_{ij}]^2 - [\xi_{ij}]^2| \le \varepsilon^2 \sum_{i=0}^{\ell_j(t)} Y_{ij}^2 + (1 + \varepsilon^{-2}) \sum_{i=0}^{\ell_j(t)} U_{ij}^2.$$

Taking into account (9.29), it follows that (9.33) will hold if one can prove that

$$\sum_{i=0}^{n_j} U_{ij}^2 \to^{\mathbb{P}} 0 \qquad \text{as } j \to \infty. \tag{9.34}$$

Let $V_j^{(a)} := |U_{0,j}|^a + ... + |U_{n_j,j}|^a$, $a = 1, 2$. Notice that, by (9.27),

$$\mathbb{E}[U_{kj}^4] \leq \mathbb{E}[(Y_{kj})^4] \leq c(t_{k+1,j} - t_{k,j})^2 \leq c\delta_j^2(t_{k+1,j} - t_{k,j}),$$

implying that, for any $y > 0$,

$$\mathbb{P}\left(\max_{0\leq k\leq n_j} |U_{kj}| > y\right) \leq y^{-4}\sum_{k=0}^{n_j} \mathbb{E}[U_{kj}^4] \leq y^{-4}c\delta_j^2 \sum_{k=0}^{n_j}(t_{k+1,j} - t_{k,j}) \leq y^{-4}c\delta_j^2 \to 0,$$

as $j \to \infty$. We also notice that by (9.26), the fact that $|a(t, x)| \leq L(1 + |x|)$ for all $t \in [0, T]$ and $x \in \mathbb{R}$, and Item (ii) of Lemma 9.7,

$$\mathbb{E}V_j^{(1)} \leq \delta_j + L(1 + \sup_{0\leq t\leq T} \mathbb{E}|X(t)|) < \infty \text{ for all } j.$$

Hence, by the above considerations,

$$V_j^{(2)} \leq (\max_k |U_{kj}|)V_j^{(1)} \to^{\mathbb{P}} 0 \text{ as } j \to \infty,$$

since $\max_k |U_{kj}| \to 0$ in probability and $\{V_j^{(1)}\}$ is stochastically bounded. This proves (9.34) and then ends the proof of the theorem. ■

9.4 CLT and IP for Stationary Sequences of Asymptotic Negatively or Positively Dependent Variables

For stationary sequences, assumptions on the negative or positive dependence can be relaxed by considering the sequence of coefficients $(r_j^*)_{j\geq 1}$ or $(a_j^*)_{j\geq 1}$ defined by (7.17).

We begin by proving a central limit theorem for stationary sequences.

Theorem 9.23 *Let $\{X_i\ ;\ i = 1, 2, ...\}$ be a strictly stationary sequence of real-valued r.v.'s with $\mathbb{E}(X_1) = 0$ and $\mathbb{E}(X_1^2) < \infty$. Suppose that one of the following conditions is satisfied:*

(i) *Asymptotic negative dependence: $r_j^* \to 0$ as $j \to \infty$.*

(ii) *Asymptotic positive dependence: $a_j^* \to 0$ as $j \to \infty$ and in addition assume that there is a constant K such that for every finite subset A of $\mathbb{N}^*$,*

$$\mathbb{E}\left(\sum_{k\in A} X_k\right)^2 \leq K\sum_{k\in A} \mathbb{E}(X_k)^2 \tag{9.35}$$

Then under (i) or (ii),

$$\lim_{n\to\infty} \frac{\text{Var}(S_n)}{n} = \sigma^2 \quad \text{as } n \to \infty \tag{9.36}$$

and

$$\frac{S_n}{\sqrt{n}} \to^d \sigma N \text{ as } n \to \infty, \tag{9.37}$$

where N is a standard normal random variable.

Remark 9.24 *The conclusion of Theorem 9.23 still holds if instead of assuming the sequence $\{X_i\,; i = 1, 2, \ldots\}$ to be strictly stationary, we assume $\mathbb{L}^2$-stationarity and uniform integrability in $\mathbb{L}^2$.*

Proof of Theorem 9.23 Denote $\sigma_n^2 = \text{Var}(S_n)$. We are going to prove the case (i) only, the other being proved similarly. So we assume in the rest of the proof that $r_j^* \to 0$. Note first that (9.36) has already been proved in Proposition 7.10.

To prove the convergence in law of the normalized partial sums, observe that without restricting the generality, we may assume that the sequence is symmetric. Indeed, if not we construct a sequence $\{X_i'\ ;\ i = 1, 2, \ldots\}$ independent of $\{X_i\,; i = 1, 2, \ldots\}$ and identically distributed. Then for any positive integer k, we define $\tilde{X}_k = X_k - X_k'$. Using Lemma 7.5 and since $r_j^* \to 0$, we infer that the constructed sequence $\{\tilde{X}_i\ ;\ i = 1, 2, \ldots\}$ is symmetric, satisfies the condition (i) of the theorem and, therefore,

$$\lim_{n\to\infty} \frac{\text{Var}\left(\sum_{k=1}^n \tilde{X}_k\right)}{n} = 2\sigma^2 \quad \text{as } n \to \infty.$$

Let us show now that if

$$\frac{\sum_{k=1}^n \tilde{X}_k}{\sqrt{2n}} \to^d N(0, \sigma^2) \text{ as } n \to \infty, \tag{9.38}$$

then (9.37) holds. Denote $Z_n = \sum_{k=1}^n X_k/\sqrt{n}$ and $Z_n' = \sum_{k=1}^n X_k'/\sqrt{n}$ and notice they are identically distributed and independent. Since, by stationarity and Lemma 7.9, $\text{Var}(Z_n) < \infty$, it follows that both (Z_n) and (Z_n') are tight on the line and in addition uniformly integrable. From any subsequence (n') we can extract another one (n'') such that both $(Z_{n''})$ and $(Z_{n''}')$ are convergent is distribution. Denote their limits respectively by L and L' and note that L and L' are i.i.d. By (9.38),

$$L - L' = N(0, 2\sigma^2).$$

Now, Cramer's decomposition theorem states that if X_1 and X_2 are independent random variables such that $X_1 + X_2$ has a normal distribution, then each of the variables X_1, X_2 is normal. Therefore each of the variables L and L' are normal. Since L and L' are i.i.d., clearly $\text{Var}(L) = \sigma^2$. By the uniform integrability of (Z_n) and the fact that its mean is 0, by the convergence of moments Theorem (see Theorem 1.12), it follows that the mean of L is also 0. Hence L has a standard normal distribution. Since all the convergent subsequences have the same limiting distribution, it follows that Z_n is weakly convergent to a standard normal variable.

This ends the proof that if (9.38) holds so does (9.37). So, without restricting the generality, we assume from now on that the sequence $\{X_i\,;\,i=1,2,\ldots\}$ is symmetric.

To prove (9.37), we use the same blocking device as in the proof of Proposition 7.10. Let p be a fixed positive integer, $q = [\sqrt{p}]$ and $k = k_n = [n/(p+q)]$. Then, we partition the n variables into blocks. We start by a block of size p followed by a block of size q and so on. We are going to have $2k+1$ blocks, the last block containing at most $p+q$ variables. Denote the sums in the blocks of size p by $\{Y_i\}_{1\le i\le k}$ and the sums in the blocks of size q by $\{Z_i\}_{1\le i\le k}$. The last block is denoted by Z_{k+1} and we have a simple decomposition

$$S_n = Y_1 + Z_1 + \ldots + Y_k + Z_k + Z_{k+1}, \tag{9.39}$$

where $Y_i = S_{[p+q](i-1)+p} - S_{[p+q](i-1)}$ and $Z_i = S_{[p+q]i} - S_{[p+q](i-1)+p}$. Notice that the variables $\{Y_i\}_{1\le i\le k}$ have a gap of size q between two consecutive variables, whereas the variables $\{Z_i\}_{1\le i\le k+1}$ have a gap of size p between two consecutive ones. This last fact combined with Lemma 7.9 and stationarity implies that, for p large enough,

$$\begin{aligned}\mathbb{E}\left(\sum_{j=1}^{k+1} Z_j\right)^2 &\le \frac{1+r'_p}{1-r'_p}\left(\sum_{j=1}^{k-1}\mathbb{E}(Z_j^2) + \mathbb{E}(Z_k + Z_{k+1})^2\right)\\ &\le \frac{1+r'_p}{1-r'_p}\left(k\sigma_q^2 + (p+2q)^2\mathbb{E}(X_1^2)\right).\end{aligned}$$

It follows that, if p is large enough,

$$\limsup_{n\to\infty} n^{-1}\mathbb{E}(S_n - S_k^{(Y)})^2 \le C\sigma_q^2/p,$$

where $S_k^{(Y)} = Y_1 + \ldots + Y_k$. Using (9.36) and the fact that $q = [\sqrt{p}]$, we get

$$\lim_{p\to\infty}\limsup_{n\to\infty} n^{-1}\mathbb{E}\left(S_n - S_k^{(Y)}\right)^2 = 0. \tag{9.40}$$

Furthermore, observe that by (9.40) and (9.36)

$$\lim_{p\to\infty}\limsup_{k\to\infty} |n^{-1}\mathrm{Var}(S_k^{(Y)}) - \sigma^2| = 0. \tag{9.41}$$

Now we consider the class of variables/distributions $\mathcal{M}$ with zero means and finite second moments. For $X, Y \in \mathcal{M}$ define $d(X,Y) = \sup\{|\mathbb{E}[f(X)] - \mathbb{E}[f(Y)]|;$

$\sup_x |f''(x)| \leq 1\}$. Clearly, d is a metric on $\mathcal{M}$. By the first approximation result in Lemma 7.13, we obtain

$$d(n^{-1/2}S_k^{(Y)}, n^{-1/2}S_k^{(Y\varepsilon)})$$
$$\leq C_f\left[-\frac{1}{2}\left(\frac{\mathrm{Var}(S_k^{(Y)})}{n} - \frac{k\mathrm{Var}(S_p)}{n}\right) + c(r_q^*)\left(\frac{\mathrm{Var}(S_n)}{n} + \frac{k\mathrm{Var}(S_p)}{n}\right)\right],$$

where $c(x)$ is a non-negative function tending to zero as x goes to zero, the variables $\{\varepsilon_i\}$ are mutually independent and independent of the sequence $(Y_i)_{i\geq 1}$ and such that we have $\mathbb{P}(\varepsilon_i = \pm 1) = 1/2$, and $S_k^{(Y\varepsilon)} = \sum_{i=1}^k Y_i\varepsilon_i$.

As a consequence, by (9.36) and (9.41) it follows

$$\lim_{p\to\infty} \limsup_{n\to\infty} d(n^{-1/2}S_k^{(Y)}, n^{-1/2}S_k^{(Y\varepsilon)}) = 0. \tag{9.42}$$

By (9.40), (9.41) and (9.42), we can easily see that it remains to show that for any fixed p,

$$n^{-1/2}S_k^{(Y\varepsilon)} \to^d a_p Z \quad \text{as } n \to \infty \text{ and } a_p^2 \to \sigma^2 \text{ as } p \to \infty,$$

where Z has a standard normal distribution.

Note that the sequence $(M_k)_{k\geq 1}$ defined by $M_k = \sum_{i=1}^k \varepsilon_i Y_i$ is a martingale with square integrable and stationary increments with respect to the sigma algebra generated by (ε_i, Y_i) for $i \leq k$. Therefore, if one can prove that there exists a positive finite constant η_p such that

$$k^{-1}(Y_1^2 + \ldots + Y_k^2) \to^{\mathbb{P}} \eta_p \text{ as } n \to \infty, \tag{9.43}$$

then, by Theorem 2.29 (since for p fixed, it is clear that $k^{-1/2}\mathbb{E}(\max_{1\leq i\leq k} |\varepsilon_i Y_i|) \to 0$, as $n \to \infty$), we will have

$$k^{-1/2}S_k^{(Y\varepsilon)} \to^d \sqrt{\eta_p} Z \text{ as } n \to \infty.$$

The convergence (9.43) follows by Item (i) of Lemma 8.10 applied to $Y_i = S_{[p+q](i-1)+p} - S_{[p+q](i-1)}$. Moreover we have $\eta_p = \mathbb{E}(\eta_p) = \mathbb{E}(S_p^2)$ which implies that

$$n^{-1/2}S_k^{(Y\varepsilon)} \to^d a_p Z \text{ as } n \to \infty,$$

with $a_p^2 = \mathrm{Var}(S_p)/(p+q)$. It remains only to notice that $a_p^2 \to \sigma^2$ as $p \to \infty$ since $q/p \to 0$, completing the proof of point (i) of the theorem. The second part (ii) of this theorem follows by the same type of arguments, replacing only the lemmas involving the coefficients r^* with the corresponding lemmas involving the coefficients a^*. ■

Remark 9.25 *If $\{X_i\ ;\ i = 1, 2, ...\}$ is assumed to be negatively dependent (so $r'_j = 0$ for any $j \geq 1$, where r'_j is defined in (7.16)), then the convergence (9.36) holds with $\sigma^2 = \mathbb{E}(X_0^2) + 2\sum_{k\geq 1} \mathbb{E}(X_0 X_k)$. Indeed, the following lemma is valid (see Lemma 8 in Newman (1984)):*

Lemma 9.26 *Let $\{X_i\ ;\ i = 1, 2, ...\}$ be a $\mathbb{L}^2$-stationary sequence with $\text{cov}(X_i, X_j) \leq 0$ for $i \neq j$, then $\sum_{k\geq 1} \text{cov}(X_0, X_k)$ is absolutely convergent and*

$$\sigma^2 = \text{Var}(X_1) + 2\sum_{k\geq 1} \text{cov}(X_0, X_k) \in [0, \text{Var}(X_1)].$$

The central limit theorem above (Theorem 9.23) can be upgraded to an invariance principle provided we require more moments.

Theorem 9.27 *(Univariate IP for stationary sequences satisfying weak approximate negative dependence) Let $\{X_i\ ;\ i = 1, 2, ...\}$ be a stationary sequence. Assume that $\mathbb{E}(X_1) = 0$ and $\mathbb{E}|X_1|^t < \infty$ for some $t > 2$ and $r_j^* \to 0$ as $j \to \infty$. Then,*

$$\lim_{n\to\infty} \frac{\text{Var}(S_n)}{n} = \sigma^2 \quad \textit{and} \quad \left\{\frac{S_{[nt]}}{\sqrt{n}}, t \in [0, 1]\right\} \to^d \sigma W \textit{ as } n \to \infty,$$

where W is a standard Brownian motion.

In the multivariate setting, this gives:

Theorem 9.28 *(Multivariate IP for stationary sequences satisfying weak approximate negative dependence) Let $X = (X_i)_{i\geq 1} = (X_i^{(1)}, \dots, X_i^{(k)})_{i\geq 1}$ be a stationary sequence valued in the Euclidean vector space $\mathbb{R}^k$ with inner product $< x, y >= x_1 y_1 + ... + x_k y_k$. Assume that $\mathbb{E}(X_1^{(i)}) = 0$ and $\mathbb{E}|X_1^{(i)}|^t < \infty$, for some $t > 2$, for all $i = 1, ..., k$. Suppose that for each vector $h = (h_1, ..., h_k)$ with non-negative coordinates, $r_j^*(h) \to 0$ as $j \to \infty$, where $r_j^*(h)$ is the coefficient of weak linear negative dependence defined in (7.17) and associated with the process $< X, h >$.*

Let $S_n = X_1 + ... + X_n$. Then, for any vector u of $\mathbb{R}^k$, there exists a limit

$$\lim_{n\to\infty} \frac{\text{Var}(< S_n, u >)}{n} = \sigma_u^2 .$$

Moreover,

$$\left\{\frac{S_{[nt]}}{\sqrt{n}}, t \in [0, 1]\right\} \to^d B \textit{ as } n \to \infty,$$

where $B := \{B(t)\ ;\ t \in [0, 1]\}$ is the k-dimensional Brownian motion with independent increments, mean zero and variances $\text{Var}(<B(t), u>) = t\sigma_u^2$, for any vector u of $\mathbb{R}^k$ and any $t \geq 0$.

Proof They are almost the same as for the asymptotic *rND* case. The only difference is in the proof of tightness where now we use the Khintchine–Marcinkiewicz–Zygmund-type inequality given in Proposition 8.3 which is then upgraded to a maximal inequality by Lemma 1.20. Let us give the details in the univariate case.

Let m be a positive integer. By convexity,

$$\mathbb{E}(|S_n|^t) \leq 2^{t-1}\mathbb{E}\left|\sum_{i=1}^{m}\sum_{j=0}^{[(n-i)/m]} X_{i+jm}\right|^t + 2^{t-1}m^t\mathbb{E}|X_1|^t$$

$$\leq (2m)^{t-1}\sum_{i=1}^{m}\mathbb{E}\left|\sum_{j=0}^{[(n-i)/m]} X_{i+jm}\right|^t + 2^{t-1}m^t\mathbb{E}|X_1|^t.$$

Take now m large enough such that $r_m < 1$. Proposition (8.3) then implies that there exists a positive constant c depending on t, m and r_m such that

$$\mathbb{E}|S_n|^t \leq c\mathbb{E}\left(\sum_{i=1}^{n} X_i^2\right)^{t/2} \leq c\, n^{t/2}\mathbb{E}|X_1|^t,$$

which combined with Lemma 1.20 gives: For any $\lambda > 0$,

$$\mathbb{E}(\max_{1\leq j\leq n}|S_j|)^t \leq c' n^{t/2} \text{ implies } \lambda^2\mathbb{P}\left(\max_{1\leq j\leq n}|S_j| \geq \lambda\sqrt{n}\right) \leq c'\lambda^{2-t} \to 0, \text{ as } \lambda \to \infty.$$

This gives the tightness by Lemma 1.16. ∎

PART II

Examples

10

Examples of Stationary Sequences with Approximate Negative Dependence

In this chapter, we treat several examples of stationary processes which are asymptotically negatively dependent and for which the results of the previous chapter apply. Let us start by recalling the assumption of asymptotic negative dependence (9.2) of a triangular array:

A triangular array $(X^{(n)}, n \geq 1)$ with $X^{(n)} = (X_{1n}, ..., X_{nn})$ will be called asymptotically negatively dependent if there exists a decreasing sequence $(r_j)_{j\geq 1}$ such that for all n and all sets Q and Q^* with distance $d(Q, Q^*) \geq j$, and any two coordinatewise non-decreasing functions v and w, the condition (9.2) holds, namely:

$$\text{corr}(v(X_Q),\ w(X_{Q^*})) \leq r_j \ \to 0 \quad (j \to \infty),$$

where for the given n, $X_Q = (Y_1, Y_2, \dots, Y_n)$ with $Y_i = X_{in}$ for $i \in Q$ and $Y_i = 0$ for $i \notin Q$.

Recall also that if $r_1 < 1$, the process is said to satisfy the *rND* condition, whereas if $r_1 = 0$ it will be said to be negatively dependent (or negatively associated).

10.1 Determinantal Point Processes and Their Perturbations

Let E be a finite or countable set and $Q = (q_{ij})$ be a complex valued $E \times E$ matrix. For $A \subseteq E$, by $Q|A$ we denote the submatrix of Q whose rows and columns are indexed by A. Following Lyons (2003), we say that the random vector $X = \{X_i;\ i \in E\}$ of 0-1 valued random variables is the determinantal point process with kernel Q if there is a matrix Q as above such that

$$\mathbb{P}(X_i = 1, \quad \text{for all} \ \ i \in A) = \det(Q|A). \tag{10.1}$$

Let $\ell_2(E)$ denote the Hilbert space of complex-valued sequences with the inner product $(x, y) = \sum x_i \bar{y}_i$.

Functional Gaussian Approximation for Dependent Structures. Florence Merlevède, Magda Peligrad and Sergey Utev.
Oxford University Press (2019).
DOI: 10.1093/oso/9780198826941.001.0001

Subclass defined when Q is a positive contraction

An important particular case is when Q is a positive contraction, i.e. a self adjoint operator on $\ell_2(E)$ such that $0 \leq (Qu, u) \leq (u, u)$ for all $u \in \ell_2(E)$. Lyons (2003, Theorem 8.1) proved that when Q is a positive contraction the determinantal point process is uniquely defined. Moreover Lyons (2003) stated the following result (similar results are stated in Pemantle (2000)).

Fact 10.1 *Let Q be a positive contraction. Then $(X_n)_{n\in\mathbb{Z}}$ is negatively dependent.*

From this fact, Lyons then argued that various limit theorems for determinantal processes should follow from the general results for negatively dependent variables.

In all the rest of this section, unless otherwise stated, Q is supposed to be a positive contraction.

Stationary determinantal point processes and their perturbations

Various examples of stationary determinantal processes are suggested in Lyons and Steif (2003). As a particular case, following Soshnikov (2000), Lyons (2003) and Lyons and Steif (2003), let Q be the $\mathbb{Z} \times \mathbb{Z}$ Toeplitz matrix defined by

$$q_{ij} = z_a a^{|i-j|}, \quad \text{where} \quad z_a = (1-a)/(1+a),\ 0 < a < 1.$$

As is proved in Soshnikov (2000, Theorem 6, p.949), it is the only **renewal process** with stationary increments among determinantal point processes.

In this case, $\mathbb{P}(X_1 = 1) = z_a$ and $\mathbb{P}(X_0 = X_j = 1) = z_a^2(1 - a^{2j})$, and so, it is easy to see that

$$\mathrm{Var}(S_m) = mz_a(1-z_a) - 2mz_a^2 a^2/(1-a^2),$$

where $S_m = X_1 + \ldots + X_m$.

Because of the negative association, the invariance principle for this particular stationary determinantal point process follows from Shao's (2000) invariance principle which is based on the Newman (1980, 1984) central limit theorem. We can also consider the situation when a is not a constant. If $a = a(n) \to A < 1$ then, for any $0 \leq s \leq t \leq 1$,

$$n^{-1}\mathrm{Var}(S_{[nt]} - S_{[ns]}) \to (t-s)g_A^2,$$

where $g_A^2 = 2(1-A)A(1+A)^{-3}$, and from Corollary 9.14, we derive that $\{n^{-1/2}S_{[nt]}, t \in [0,1]\} \to^d g_A \cdot W$ where W is a standard Wiener process.

A partial representation via independent variables

Hough *et al.* (2006) proved that the statistic $S_{[nt]}$ is identically distributed as the sum of (non-identically distributed) independent Bernoulli variables.

Perturbed stationary determinantal point processes

Suppose that the data consists of n points such that at time i we have the point $Z_i = h(X_i, Y_i)$, where $(X_i)_{i\in\mathbb{Z}}$ is the determinantal point process such as considered above and the error $(Y_i)_{i\in\mathbb{Z}}$ is a non-degenerate stationary Gaussian sequence with positive continuous spectral density $f(x)$. By Theorem 9.8 in Bradley (2007), note that the following fact is valid:

Fact 10.2 *We have*

$$\rho_1^*(Y) < 1, \textit{ and } \lim_{k\to\infty} \rho_k^*(Y) = 0,$$

with

$$\rho_k^*(Y) = \sup_{Q,Q^*} \rho\left(\sigma(Y_i,\ i \in Q), \sigma(Y_i,\ i \in Q^*)\right), \tag{10.2}$$

where $\sup_{Q,Q^*}$ *means the supremum taken over non-empty subsets* Q *and* Q^* *of* $\mathbb{Z}$ *such that* $dist(Q, Q^*) \geq k$ *and the measure of dependence* $\rho(\mathcal{B}, \mathcal{C})$ *between two* σ*-algebras has been defined in Section* 5.1.1.

As a consequence of the fact above, the sequence $(Y_i)_{i\in\mathbb{Z}}$ satisfies the asymptotic negative dependence assumption i.e. $\lim_{k\to\infty} r_k(Y) = 0$.

Proceeding as in the proof of Lemma 7.5 (see also Remark 7.6), we then infer that:

Fact 10.3 *If* $(X_i)_{i\in\mathbb{Z}}$ *and* $(Y_i)_{i\in\mathbb{Z}}$ *are independent, the process* $(X_i, Y_i)_{i\in\mathbb{Z}}$ *is asymptotically negatively dependent.*

The following corollary gives the limiting behavior of partial sums associated with the sequence $(Z_i)_{i\in\mathbb{Z}}$. Notice that the function $h(x, y)$ is not assumed to be monotonic in any of its coordinates.

Corollary 10.4 *Let* $X = (X_i; i\in\mathbb{Z})$ *be a stationary determinantal point process and let* $Y = (Y_i; i \in \mathbb{Z})$ *be a non-degenerate stationary Gaussian sequence with positive continuous spectral density. Assume that X and Y are independent. Consider a function* $h(x, y)$ *defined on* $\{0, 1\} \times \mathbb{R}$ *and such that* $\mathbb{E}[h^2(X_1, Y_1)]<\infty$. *Define the stationary sequence by* $Z_i = h(X_i, Y_i) - \mathbb{E}[h(X_i, Y_i)]$, $i \in \mathbb{Z}$. *Then, for* $(Z_i)_{i\in\mathbb{Z}}$, *the conclusions of Corollary 9.16 hold, namely:* $\lim_{n\to\infty} n^{-1}\mathrm{Var}(\sum_{i=1}^n Z_i) = \sigma^2$ *and* $\left\{n^{-1/2} \sum_{i=1}^{[nt]} Z_i, t \in [0, 1]\right\} \to^d \sigma W$, *as* $n \to \infty$, *where* $W = \{W(t), t \in [0, 1]\}$ *is a standard Brownian motion.*

Proof We observe that

$$h(X_i, Y_i) = X_i h(1, Y_i) + (1 - X_i)h(0, Y_i) = X_i\xi_i - X_i\psi_i + \zeta_i, \tag{10.3}$$

where $\xi_i = (h(1,\ Y_i) - h(0,\ Y_i))_+$, $\psi_i = (-h(1,\ Y_i) + h(0,\ Y_i))_+$ and $\zeta_i = h(0,\ Y_i)$. Consider the stationary sequence of three-dimensional random vectors $(X_i\xi_i, X_i\psi_i, \zeta_i)_{i\in\mathbb{Z}}$.

Let $u = (u_1, u_2, u_3)$ be a set of non–negative numbers and let

$$f_u(x, y) = u_1 x(h(1, y) - h(0, y))_+ + u_2 x(-h(1, y) + h(0, y))_+ + u_3 h(0, y).$$

These functions are non-decreasing in x. Let $U_u := (U_i(u))_{i\in\mathbb{Z}}$ where $U_i(u) = f_u(X_i, Y_i) - \mathbb{E}(f_u(X_i, Y_i))$ for any $i \in \mathbb{Z}$. Proceeding as in the proof of Lemma 7.5, we infer that for any $u \in \mathbb{R}_+^3$,

$$r_k(U_u) \leq \max(r_k(X), \rho_k^*(Y)).$$

Hence, by Facts 10.1 and 10.2,

$$\lim_{k\to\infty} r_k(U_u) = 0. \tag{10.4}$$

Thus the conclusion of Corollary 10.4 follows from Theorem 9.18 applied to the three-dimensional stationary sequence vector $(X_i\xi_i - \mathbb{E}[X_i\xi_i], X_i\psi_i - \mathbb{E}[X_i\psi_i], \zeta_i - \mathbb{E}[\zeta_i])_{i\in\mathbb{Z}}$. ■

Comments

1) Our approach is useful for studying various statistics associated with (Z_i). Consider a simple moment statistic $\bar{X} = n^{-1}[Z_1 + ... + Z_n]$. Then, the estimate $\bar{X}$ is consistent and asymptotically normal as follows from Corollary 10.4.
2) One can also treat consistency and asymptotic normality of other statistics such as $S^2 = n^{-1}[(Z_1)^2 + ... + (Z_n)^2]$ by applying Corollary 10.4 again. Moreover, we can treat long-range dependent linear statistics associated with (Z_n) (see Section 12.5.2).

Approximation by sums of independent variables

In addition, the weak convergence results to other infinitely divisible laws, such as a compound distribution, are also available. To give an example of this kind, we assume that the spectral density f of $Y = (Y_i; i \in \mathbb{Z})$ (assumed to be a centered non-degenerate stationary Gaussian sequence) is such that $f \geq 1/K$ where $0 < K < \infty$ and admits the following interpolation by the trigonometric polynomial P_n of degree $n-1$,

$$\sup_{t\in[-\pi,\pi]} |f(t) - P_n(t)| \leq c(\ln n)^{-a}, \text{ for some } a > 1. \tag{10.5}$$

For example, f can be Hölder of any positive order.

Note that, for a centered stationary Gaussian process Y, we have

$$\rho_n^*(Y) = \sup \left| \mathrm{cov}\left(\sum_{i\in Q} a_i Y_i, \sum_{j\in S} a_j Y_j \right) \right| \Big/ \left\| \sum_{k\in Q} a_k Y_k \right\|_2 \left\| \sum_{k\in S} a_k Y_k \right\|_2,$$

where the supremum is taken over all pairs of non-empty, finite, disjoint sets $Q, S \in \mathbb{Z}$ satisfying $d(Q, S) = \inf\{|i-j|; i \in Q, j \in S\} \geq n$ and all choices of real numbers a_k, $k \in Q \cup S$ (see relation (7.1) in Bradley (2005)). Define $\xi_C = \sum_{k\in C} a_k Y_k$ and $e_A(t) = \sum_{k\in A} a_k \exp(ikt)$. Considering two sets A and B of integers with $\mathrm{d}(A, B) \geq n$, using (1.10), we have

$$\mathrm{cov}(\xi_A, \xi_B) = \int_{-\pi}^{\pi} e_A(t)\overline{e_B(t)}(f(t) - P_n(t))dt$$

$$
\begin{aligned}
&\leq c(\ln(n))^{-a}\left|\int_{-\pi}^{\pi}|e_A(t)|^2dt\right|^{1/2}\left|\int_{-\pi}^{\pi}|e_B(t)|^2dt\right|^{1/2}\\
&\leq c(\ln(n))^{-a}K\left|\int_{-\pi}^{\pi}|e_A(t)|^2f(t)dt\right|^{1/2}\left|\int_{-\pi}^{\pi}|e_B(t)|^2f(t)dt\right|^{1/2}\\
&= c(\ln(n))^{-a}K||\xi_A||_2||\xi_B||_2.
\end{aligned}
$$

Let now $X = (X_i; i \in \mathbb{Z})$ be a strictly stationary determinantal point process and let $Y = (Y_i; i \in \mathbb{Z})$ be a centered stationary Gaussian sequence with continuous spectral density f satisfying (10.5) and $f \geq 1/K$ where $0 < K < \infty$. Assume that X and Y are independent. Consider a function $f(x, y)$ defined on $\{0, 1\} \times \mathbb{R}$ and such that $f(1, y) \geq f(0, y)$ for all y. Let $Z_i = f(X_i, Y_i)$. Then, the strictly stationary sequence (Z_i) satisfies $r_j \leq c(\ln(j))^{-a}$, $j \geq 2$. To weaken the dependence we divide the variables in consecutive blocks of fixed size q and consider the sum of variables in each block. Obviously Lemma 7.14 is applicable and it is easy to see that the limiting behavior of partial sums associated with $\{Z_i\}$ can be then derived from the limiting behavior of independent sums of variables in blocks of fixed size q, and letting at the end $q \to \infty$.

10.2 Displaced Point Processes

Motivated by Gabrielli (2004), we consider point processes in a random time environment. More exactly, let $\{k, \xi_k\}$ be a point process and $k \to k + \tau_k$ be the random change of time. Then, assuming that we can only observe the presence of the particle and not the number of them, we then define the point process

$$\xi_j = I\left(\sum_k I_{j=k+\tau_k} X_k \geq 1\right) \tag{10.6}$$

that is $\xi_j = 1$ if there is a point k such that both $X_k = 1$ and $k + \tau_k = j$ hold.

Let (X_k, τ_k) be a strictly stationary sequence. Then, for any integer m,

$$
\begin{aligned}
(\xi_j)_{j\in\mathbb{Z}} &=^d \left(I\left(\sum_k I_{j=k+\tau_{k+m}} X_{k+m} \geq 1\right)\right)_{j\in\mathbb{Z}} = \left(I\left(\sum_{k'} I_{j+m=k'+\tau_{k'}} X_{k'} \geq 1\right)\right)_{j\in\mathbb{Z}}\\
&= (\xi_{j+m})_{j\in\mathbb{Z}}
\end{aligned}
$$

and so the sequence $(\xi_j)_{j\in\mathbb{Z}}$ is strictly stationary.

For asymptotically negatively dependent stationary sequences, we then derive:

Corollary 10.5 *Let $X = (X_i; i \in \mathbb{Z})$ be a strictly stationary point process which is asymptotically negatively dependent, i.e. satisfying (9.2). Let $Y = (Y_i; i \in \mathbb{Z})$ be a stationary Gaussian sequence with positive continuous spectral density. Assume that X and Y are independent.*

Let $f : \mathbb{R} \to \{-m, -m+1, \dots, 0, 1, \dots, m\}$ and assume that $\tau_k = f(Y_k)$. Define the displaced point process $(\xi_i)_{i\in\mathbb{Z}}$ by (10.6). Then, the stationary process $(\xi_i)_{i\in\mathbb{Z}}$ is also asymptotically negatively dependent and the conclusions of Corollary 9.16 hold with $S_n = \sum_{i=1}^{n}(\xi_i - \mathbb{E}(\xi_i))$.

Let us consider now asymptotically weakly positive dependent stationary sequences, i.e. $a_j^* \to 0$, as $j \to \infty$, where the sequence of coefficients $(a_j^*)_{j\geq 1}$ is defined by (7.17). In order to apply Theorem 9.23 we need to check the extra condition (9.35) about the behavior of variances of partial sums.

We say that the sequence $(X_i)_{i\in\mathbb{Z}}$ satisfies the good variance upper bound for class $\mathcal{F}$ if there exists a finite constant $C_{X,\mathcal{F}}$ such that, for all n and all functions $f_1, \dots, f_n \in \mathcal{F}$ with $\mathbb{E}[f_i(X_i)^2] < \infty$,

$$\mathrm{Var}\left(\sum_{i=1}^{n} f_i(X_i)\right) \leq C_{X,\mathcal{F}} \sum_{i=1}^{n} \mathrm{Var}[f_i(X_i)]. \tag{10.7}$$

Lemma 10.6 *Let $X = (X_i)_{i\in\mathbb{Z}}$ and $Y = (Y_i)_{i\in\mathbb{Z}}$ be two independent sequences satisfying (10.7) for classes $\mathcal{F}_X$ and $\mathcal{F}_Y$ respectively. Assume that $\{g_i(x,y), i \in \mathbb{Z}\}$ are functions such that for all fixed a, $\{g_i(x,a), i \in \mathbb{Z}\} \in \mathcal{F}_X$ and $\{\mathbb{E}(g_i(X_i,y)), i \in \mathbb{Z}\} \in \mathcal{F}_Y$. Then there exists a finite constant C such that, for any finite set A,*

$$\mathrm{Var}\left(\sum_{i\in A} g_i(X_i, Y_i)\right) \leq C \sum_{i\in A} \mathrm{Var}(g_i(X_i, Y_i)). \tag{10.8}$$

Proof Let

$$S_A = \sum_{i\in A} g_i(X_i, Y_i).$$

We observe that

$$\begin{aligned}\mathrm{Var}\left(\sum_{i\in A} g_i(X_i, Y_i)\right) &= \mathbb{E}\left(\mathbb{E}((S_A - \mathbb{E}(S_A))^2|Y)\right)\\ &\leq 2\mathbb{E}\left(\mathbb{E}((S_A - \mathbb{E}(S_A|Y))^2|Y)\right) + 2\mathbb{E}\left((\mathbb{E}(S_A|Y) - \mathbb{E}(S_A))^2\right).\end{aligned}$$

By independence of X and Y,

$$\mathbb{E}[(S_A - \mathbb{E}(S_A|Y))^2|Y] \leq C_{X,\mathcal{F}_X} \sum_{i\in A} \{\mathbb{E}[g_i^2(X_i, Y_i)|Y] - f_i^2(Y_i)\},$$

where $f_i(y_i) = \mathbb{E}[g_i(X_i, y_i)]$. On the other hand, by using again the independence of X and Y, we get

$$\mathbb{E}\left((\mathbb{E}(S_A|Y) - \mathbb{E}(S_A))^2\right) \leq C_{Y,\mathcal{F}_Y} \sum_{i\in A} \{\mathbb{E}(f_i^2(Y_i)) - [\mathbb{E}(f_i(Y_i))]^2\}.$$

Since $\mathbb{E}(f_i(Y_i)) = \mathbb{E}[g_i(X_i, Y_i)]$, we get overall that

$$\operatorname{Var}\left(\sum_{i\in A} g_i(X_i, Y_i)\right) \leq 2\max(C_{X,\mathcal{F}_Y}, C_{Y,\mathcal{F}_Y})\sum_{i\in A}\operatorname{Var}(g_i(X_i, Y_i)),$$

giving the lemma. ■

For asymptotically weakly positive dependent stationary sequences, we then derive:

Corollary 10.7 *Let $X = (X_i, i \in \mathbb{Z})$ be a strictly stationary point process which is asymptotically weakly positive dependent, i.e. $a_j^* \to 0$, as $j \to \infty$, and assume that it satisfies condition (10.7) for a class of non-decreasing functions. Let $Y = (Y_i, i \in \mathbb{Z})$ be a stationary Gaussian sequence with positive continuous spectral density. Assume that X and Y are independent. Let $f : \mathbb{R} \to \{-m, -m+1, \dots, 0, 1, \dots, m\}$ and assume that $\tau_k = f(Y_k)$. Define the displaced point process $(\xi_i)_{i\in\mathbb{Z}}$ by (10.6). Then, the strictly stationary process $(\xi_i)_{i\in\mathbb{Z}}$ is also asymptotically weakly positive dependent and the central limit theorem as stated in Theorem 9.23 holds with $S_n = \sum_{i=1}^{n}(\xi_i - \mathbb{E}(\xi_i))$.*

Proof To check condition (9.35) we apply Lemma 10.6 for the functions $g(x, y)$ which are monotone on x for any given y. ■

Comments

1) Aperiodic irreducible finite Markov chains $\{\tau_k\}$ are another reasonable choice for the random time environment. As it follows from Bradley (1997-b), the sequence $\{\tau_k\}_k$ is also interlaced mixing (even with exponential rates) and so the results remain valid for this case too.
2) If $\{\tau_k\}_k$ is a sequence of i.i.d. r.v.'s, the maximal inequality for the positively dependent variables follows easily from Newman and Wright (1981) and so the invariance principle holds in this case.

10.3 Exchangeable Processes via the *rND* Property

Let us consider a triangular array $\{X_{in}, i=1, \dots, n\}_{n\geq 1}$ of row exchangeable random variables meaning that for any $m \geq 1$ and any positive integers $k_1, \dots, k_m$, the random variables $X_{k_1 n}, \dots, X_{k_m n}$ are exchangeable in the sense that each permutation of them has the same distribution. Consider the partial sum process

$$X_n(t) = \sum_{i=1}^{[nt]} X_{in} \text{ for } t \in [0, 1].$$

(We set $X_n(t) = 0$ for $0 \leq t < 1/n$ and it can be convenient to extend $X_n(t)$ when $t \geq 1$ by setting $X_n(t) = X_n(1)$ for any $t \geq 1$.)

The aim of this section is to investigate when the Gaussian approximation holds. We certainly need an extra condition since otherwise any limit is possible as the following example shows.

Example: *For a random variable X, define* $X_{in} = X/n$, $i = 1,\ldots,n$. *Then* $\{X_{in}\}$ *are row exchangeable random variables and* $\{X_n(t), t \in [0,1]\} \to^d \{tX, t \in [0,1]\}$ *which is Gaussian if and only if X is.*

We shall assume, in addition to exchangeability, that the triangular array $\{X_{in}, i = 1, \ldots, n\}_{n\geq 1}$ satisfies the *rND* condition (8.22).

Permutation In the classical case, the triangular array $X_{1n}, \ldots, X_{nn}$ is a random permutation of real numbers $x_{1n}, \ldots, x_{nn}$, each of the $n!$ permutations having probability $1/n!$. Then, $X_{1n}, \ldots, X_{nn}$ are negatively dependent, so $r = 0$ in (8.22), as it was shown in Joag-Dev and Proshan (1983).

The next lemma is Theorem 24.1 in Billingsley (1968). As we will see, it is a consequence of Corollary 9.22.

Lemma 10.8 *Let* $X_{1n}, \ldots, X_{nn}$ *be a random permutation of real numbers* $x_{1n}, \ldots, x_{nn}$ *satisfying*

$$x_{1n} + \ldots + x_{nn} = 0, \; x_{1n}^2 + \ldots + x_{nn}^2 = 1 \; \text{and} \; \max_{1\leq i\leq n} |x_{in}| \to 0 \tag{10.9}$$

as $n \to \infty$. *Then,* $\{X_n(t), t \in [0,1]\} \to^d B$, *where B is a standard Brownian bridge.*

Proof Using exchangeability and condition (10.9), it is easy to see that the triangular array $\{X_{in}, i = 1, \ldots, n\}_{n\geq 1}$ is centered and such that $\mathbb{E}(X_{in}^2) = 1/n$ for all $i = 1, \ldots, n$. In addition, $\sum_{i=1}^n \mathbb{E}(X_{in}^4) = \sum_{i=1}^n x_{in}^4 \leq \max_{1\leq k\leq n} |x_{kn}|^2 \sum_{i=1}^n x_{in}^2$ which converges to zero by condition (10.9). This proves that $\{X_{in}, i = 1, \ldots, n\}_{n\geq 1}$ satisfies Lindeberg's condition. In addition, since $X_{1n}, \ldots, X_{nn}$ are negatively dependent, it follows that $\{X_{in}, i=1, \ldots, n\}_{n\geq 1}$ satisfies condition (*A*) of Section 9.1.2.

Observe now that for positive integers $n \geq b + k > b \geq u \geq 0$

$$\mathbb{E}\left[\sum_{i=b+1}^{b+k} X_{in} | X_{1n}, \ldots, X_{un}\right] = -\frac{k}{n-u} \sum_{i=1}^{u} X_{in} \tag{10.10}$$

(see Section 24 in Billingsley (1968) for more details). This proves that condition (9.23) holds on $[0, T]$, for any $0 < T < 1$, with $a(t, x) = -x/(1-t)$. Next, we note that by exchangeability, for any $0 \leq a < b \leq n$,

$$\text{Var}\left(\sum_{i=a+1}^{b} X_{in}\right) = (b-a)\mathbb{E}(X_{1n}^2) + 2(b-a)(b-a-1)\mathbb{E}(X_{1n}X_{2n}).$$

But, by (10.10) applied with $b = 1$, $k = 1$ and $u = 1$,

$$\mathbb{E}(X_{1n}X_{2n}) = \mathbb{E}(X_{1n}\mathbb{E}(X_{2n}|X_{1n})) = -\frac{1}{n-1}\mathbb{E}(X_{1n}^2) = -\frac{1}{n(n-1)}.$$

Therefore for each partition $\{t_j\}$ such that $0 = t_0 < t_1 < ... < t_k < t_{k+1} = t$,

$$\sum_{j=1}^{k+1} \mathrm{Var}\left(\sum_{i=[nt_{j-1}]+1}^{[nt_j]} X_{in} \right) = \frac{[nt]}{n} - \frac{2}{n(n-1)} \sum_{j=1}^{k+1} \left([nt_j] - [nt_{j-1}]\right)\left([nt_j] - [nt_{j-1}] - 1\right),$$

which proves that condition (9.24) holds with $\sigma^2(y) \equiv 1$.

Therefore, the conclusion of Theorem 9.21 is satisfied on $[0, T]$ for any $0 < T < 1$. By continuity of the sample paths, it follows that $\{X_n(t), t \in [0, 1]\} \to^d \{B(t), t \in [0, 1]\}$, where B is the solution of the stochastic integral equation

$$B(t) = -\int_0^t \frac{B(u)}{1-u} du + \int_0^t dW(u) \text{ a.s.}, \ t \in [0, 1].$$

Note that such a defined process B is a standard Brownian Bridge. ■

Now, we apply the following fact about conditional convergence.

Fact 10.9 *Assume that conditionally on the σ-fields $\mathcal{F}_n$,*

$$\mathbb{E}[f(X_n)|\mathcal{F}_n] \to^{\mathbb{P}} \mathbb{E}[f(X)] \text{ for all bounded and continuous function } f.$$

Then, the unconditional convergence also holds, i.e. $X_n \to^d X$.

We then derive:

Corollary 10.10 *Consider a triangular array $\xi_{1n}, ..., \xi_{nn}, n = 1, 2, ...,$ of exchangeable random variables such that*

$$\xi_{1n} + ... + \xi_{nn} \to^{\mathbb{P}} 0, \ \xi_{1n}^2 + ... + \xi_{nn}^2 \to^{\mathbb{P}} 1 \text{ and } \max_{1 \le i \le n} |\xi_{in}| \to^{\mathbb{P}} 0 \tag{10.11}$$

as $n \to \infty$. Then $\left\{\sum_{i=1}^{[nt]} \xi_{in}, t \in [0, 1]\right\} \to^d B$ where B is a standard Brownian bridge.

In the case of exchangeable arrays satisfying the *rND* condition (8.22), the following result holds.

Lemma 10.11 *Assume that $\{(X_{1n}, ..., X_{nn})\ ; n = 1, 2, ...\}$ is a triangular array of exchangeable random variables satisfying the rND condition (8.22). Suppose in addition that $\mathbb{E}[X_{1n}] = 0, \mathbb{E}[X_{1n}^2] = 1/n$, and for any $\varepsilon > 0$, $n\mathbb{E}[X_{1n}^2 I_{(|X_{1n}|>\varepsilon)}] \to 0$ as $n \to \infty$. Assume that $X_n(1) \to^d \eta$. Then, $\{X_n(t) - tX_n(1), t \in [0, 1]\} \to^d B$ where B is a standard Brownian bridge.*

Proof of Lemma 10.11 Define a new triangular array $\{(\xi_{1n}, ..., \xi_{nn}); n = 1, 2, ...\}$ by

$$\xi_{in} = X_{in} - S_n/n, n = 1, 2, \dots, \text{ and } i = 1 \dots, n$$

with $S_n = \sum_{i=1}^n X_{in}$. Clearly, $\{(\xi_{1n}, ..., \xi_{nn}); n = 1, 2, ...\}$ is an array of exchangeable centered variables such that $\sum_{i=1}^n \xi_{in} = 0$. Moreover, by taking into account condition (8.22), Lemma 7.9 implies that there exists a positive constant C, not depending on n, such that $\mathbb{E}(S_n^2) \leq C\sum_{i=1}^n \mathbb{E}[X_{in}^2] = C$. On the other hand, Lemma 8.11 implies that $\sum_{i=1}^n X_{in}^2 \to^{\mathbb{P}} 1$ as $n \to \infty$. Therefore,

$$\sum_{i=1}^n \xi_{in}^2 = \sum_{i=1}^n X_{in}^2 - \frac{S_n^2}{n} \to^{\mathbb{P}} 1, \text{ as } n \to \infty.$$

Finally, by using the Lindeberg condition: $n\mathbb{E}[X_{1n}^2 1_{(|X_{1n}|>\varepsilon)}] \to 0$ for any $\varepsilon > 0$, it follows that

$$\max_{1\leq i\leq n} |\xi_{in}| \leq |S_n|/n + \max_{1\leq i\leq n} |X_{in}| \to^{\mathbb{P}} 0,$$

as $n \to \infty$. So, overall, $\{(\xi_{1n}, ..., \xi_{nn})\,;\, n = 1, 2, ...\}$ is an array of exchangeable centered variables satisfying condition (10.11).

Define the partial sums process by $\xi_n(t) = \sum_{i=1}^{[nt]} \xi_{in}$, $t \in [0, 1]$. Then, by Corollary 10.10, $\{\xi_n(t), t \in [0, 1]\} \to^d B$. To end the proof, note that by construction, for any $t \in [0, 1]$,

$$X_n(t) - tX_n(1) = \xi_n(t) + \left(\frac{[nt]}{n} - t\right) S_n.$$

Since the second term in the right-hand side is going in probability to zero as $n \to \infty$, it follows, by Theorem 1.9, that $\{X_n(t) - tX_n(1), t \in [0, 1]\}$ and $\{\xi_n(t), t \in [0, 1]\}$ have the same asymptotic behavior which ends the proof of the lemma. ■

This approach provides another proof of the central limit theorem for random permutation established in Chernoff and Teicher (1958).

10.3.1 *Weighted Empirical Processes*

Notice that if U is uniformly distributed on $[0, 1]$, then the process $X = \{X(t) = I_{(U\leq t)}, t \in [0, 1]\}$ has negatively dependent (*ND*) increments. By Proposition 9.6, the sums of independent processes with *ND* increments has *ND* increments. Now, let $\{U, U_1, U_2, ...\}$ be an i.i.d. sequence such that U has a uniform distribution on the interval $[0, 1]$. Then, the empirical distribution function and more generally, finite bridges defined by

$$U_n(t) = \sum_{i=1}^n a_i(I_{(U_i>t)} - (1 - t)) \quad \text{with } a_i \geq 0 \text{ and } 0 \leq t \leq 1,$$

have *ND* increments.

Denoting by $G_n(t) = U_n([nt]/n)$, it can be proved that $\{U_n(t), t \in [0, 1]\}$ and $\{G_n(t), t \in [0, 1]\}$ have the same asymptotic behavior. Now, note that

$$G_n(t) = \sum_{i=1}^{[nt]} (U_n(i/n) - U_n((i-1)/n)) := \sum_{i=1}^{[nt]} Y_{i,n}.$$

Note also that $(Y_{i,n}, i{=}1, \dots, n)_{n\geq 1}$ is a triangular array of exchangeable centered random variables which are negatively dependent. In addition,

$$\mathbb{E}(Y_{i,n}^2) = \sum_{k=1}^{n} a_k^2 \mathbb{E}\left(I_{((i-1)/n \leq U_k < i/n)} - \frac{1}{n}\right)^2 = \frac{n-1}{n^2} \sum_{k=1}^{n} a_k^2,$$

and

$$\mathbb{E}(Y_{i,n}^4) = \sum_{k=1}^{n} a_k^4 \mathbb{E}\left(I_{((i-1)/n \leq U_k < i/n)} - \frac{1}{n}\right)^4 + 3\sum_{k=1}^{n} \sum_{j=1, j\neq k}^{n} \mathbb{E}\left(I_{((i-1)/n \leq U_k < i/n)} - \frac{1}{n}\right)^2$$
$$\cdot \mathbb{E}\left(I_{((i-1)/n \leq U_j < i/n)} - \frac{1}{n}\right)^2 \leq \frac{3}{n} \sum_{k=1}^{n} a_k^4 + \frac{1}{n^2}\left(\sum_{k=1}^{n} a_k^2\right)^2.$$

Therefore, setting

$$X_{in} = \frac{Y_{i,n}}{b_n} \text{ where } b_n^2 = \frac{n-1}{n} \sum_{k=1}^{n} a_k^2,$$

and assuming that $\sum_{k=1}^{n} a_k^4 / \left(\sum_{k=1}^{n} a_k^2\right)^2 \to 0$, as $n \to \infty$, we get overall that the triangular array of exchangeable variables $(X_{in}, i{=}1, \dots, n)_{n\geq 1}$ is negatively dependent and such that

$$\mathbb{E}[X_{1n}] = 0, \mathbb{E}[X_{1n}^2] = 1/n, \text{ and for any } \varepsilon > 0, n\mathbb{E}[X_{1n}^2 I_{(|X_{1n}|>\varepsilon)}] \to 0 \text{ as } n \to \infty.$$

Moreover $\sum_{k=1}^{n} X_{in} = b_n^{-1} U_n(1) = 0$ a.s. Hence applying Lemma 10.11, the following result holds:

Corollary 10.12 *If the following condition holds:* $\sum_{k=1}^{n} a_k^4 / (\sum_{k=1}^{n} a_k^2)^2 \to 0$, *as* $n \to \infty$, *then* $\{c_n^{-1} U_n(t), t \in [0,1]\} \to^d B$ *where* B *is a standard Brownian Bridge and* $c_n^2 = \sum_{k=1}^{n} a_k^2$.

Of course, in such a case, an application of the martingale CLT rather than our Lemma 10.11 is well-known (see for example Jacod and Shiryaev (2002, p.99)).

10.3.2 An Example with Convergence to a Non-Gaussian Process for Exchangeable Negatively Dependent Variables

We show next that indeed, the limit can be non-Gaussian even in the case of exchangeable arrays of negatively dependent variables. We base our proof on an interesting result by Pemantle (2000) and follow his notation. Let

$$n = 2m, \ b_k = \mathbb{P}(\text{Bin}(n, 1/2) = k) = \binom{n}{k} 2^{-2m}.$$

Fix $a > 0$ and define for $n \geq 4a^2$, $1/K_n = \mathbb{P}(|\text{Bin}(2m, 1/2) - (n/2)| \leq a\sqrt{n})$ and

$$a_k = K_n \begin{cases} b_k & \text{if } |k - (n/2)| \leq a\sqrt{n} \\ 0 & \text{otherwise.} \end{cases}$$

Finally let

$$A_k = a_k \Big/ \binom{n}{k} \text{ and } B_k = b_k \Big/ \binom{n}{k} = 2^{-2m}.$$

Clearly, $B_k^2 = B_{k-1}B_{k+1}$ for $k = 1, ..., n-1$ proving that the sequence $\{B_k; k = 0, ..., n\}$ is log-concave. Also the sequence $\{A_k;\, k = 0, ..., n\}$ is log-concave where it is non-zero, i.e. on the interval $\{k : |k - (n/2)| \leq a\sqrt{n}\}$.

Now, the sequence $\{a_k;\, k = 0, ..., n\}$ defines an exchangeable array of zero–one Bernoulli variables $\{(\varepsilon_{1n}, ..., \varepsilon_{nn})\}_{n\geq 1}$ by letting

$$\mathbb{P}(\varepsilon_{1n} + ... + \varepsilon_{nn} = k) = a_k.$$

The array $\{\varepsilon_{1n}, ..., \varepsilon_{nn}\}_{n\geq 1}$ is negatively dependent by the Pemantle criteria (2000, Theorem 2.7), more exactly because the sequence a_k is log-concave, namely:

$$a_k^2 \geq a_{k+1}a_{k-1}, \text{ for all } k = 1, \dots, n-1. \tag{10.12}$$

Moreover, we have: $\mathbb{P}(\varepsilon_{in} = 0) = \mathbb{P}(\varepsilon_{in} = 1) = 1/2$ for any $i = 1, ..., n$. Finally, we define

$$X_{in} = n^{-1/2}(2\varepsilon_{in} - 1), \; i = 1, \dots, n.$$

Then, the array $\{(X_{1n}, ..., X_{nn})\}_{n\geq 1}$ consists of normalized symmetric Rademacher variables, it is exchangeable and negatively dependent. Moreover,

$$\sum_{i=1}^{n} X_{in} \to^d \eta_a,$$

where η_a has a non-degenerate truncated Gaussian distribution. More precisely, the density of η_a is given by $f_a(x) = C_a^{-1}(2\pi)^{-1/2}e^{-x^2/2}\mathbf{1}_{\{-2a\leq x\leq 2a\}}$ where $C_a = \int_{-2a}^{2a}(2\pi)^{-1/2}e^{-x^2/2}dx$. Notice that an infinitely divisible variable X is bounded if and only if X is constant almost surely. Hence, in this example, the limiting law is not infinitely divisible. So we have proved that:

Proposition 10.13 *There exists an array $\{(X_{1n}, ..., X_{nn})\}_{n\geq 1}$ of negatively dependent exchangeable variables with $\mathbb{P}(\sqrt{n}X_{in} \in \{-1, 1\}) = 1$ for all $i = 1, ..., n$, such that $\sum_{i=1}^{n} X_{in} \to^d \eta_a$ where η_a has a truncated Gaussian law (that is not infinitely divisible and not a mixture of Gaussian distributions). In addition, if we define $X_n(t) = \sum_{k=1}^{[nt]} X_{kn}$ for $t \in [0, 1]$, then the process $\{X_n(t) - tX_n(1), t \in [0, 1]\} \to^d B$ where B is a standard Brownian bridge.*

10.3.3 Exchangeable Determinantal Point Processes

Here we shall continue the study of the determinantal point process. Let Q be a positive contraction. By Theorem 8.1 of Lyons (2003), recall that the process $X = \{X_i\}_{i \in \mathbb{Z}}$ defined in Section 10.1 is negatively dependent. Then, if we assume that the sequence is exchangeable, by de Finetti theorem X is the mixture of sequences of i.i.d. Bernoulli variables. This fact combined with the negative dependence implies that X_i are i.i.d. 0–1 random variables. For finitely exchangeable variables, by using the representation of S_n via sums of non-identically distributed independent Bernoulli variables due to Hough *et al.* (2006) we then immediately derive the following proposition:

Proposition 10.14 *Let* $X^{(n)} = \{X_{1n}, ..., X_{nn}\}$ *be a triangular array of exchangeable determinantal point processes defined by positive contraction operators* Q_n, $n \geq 1$. *Let* $S_n = \sum_{i=1}^{n} X_{in}$ *and assume that* $\mathrm{Var}(S_n) \to \infty$. *Then,* $(S_n - \mathbb{E}(S_n))/\sqrt{\mathrm{Var}(S_n)} \to^d N(0, 1)$.

In a contrast to this, we proved in Proposition 10.13 that for general triangular arrays of exchangeable negatively dependent variables, the non-Gaussian limits are possible. Moreover, below we show that the mixture of two Bernoulli processes is a determinantal point process, which is positively dependent, non-ergodic with partial sums approximated by mixtures of Gaussian distributions.

Example: For a positive integer n, let $E = \{1, ..., n\}$, where $n = \infty$ is also allowed. For real numbers a, x, y, we define the $n \times n$ matrix Q_n by

$$(Q_n)_{ii} = a, \ (Q_n)_{ij} = x, \ (Q_n)_{ji} = y, i < j, \ i, j = 1, \dots, n.$$

So for $n = 1$, $Q = [a]$ is the 1×1 matrix, etc. Further let $\psi_n(a, x, y) = \det(Q_n)$. It is then easy to compute by induction that, for $x \neq y$,

$$\psi_n(a, x, y) = c(x, y)(a - x)^n + (1 - c(x, y))(a - y)^n \quad \text{where} \quad c(x, y) = y/(y - x).$$

By letting $y \to x$ we obtain $\psi_n(a, x, x) = (a + (n - 1)x)(a - x)^{n-1}$.

Notice that, since the determinant does not change under transposition, we have $\psi_n(a, x, y) = \psi_n(a, y, x)$. There are two cases to consider (i) $a > y > 0 > x$ and (ii) $a > y \geq x > 0$.

Let $X(p) = (X_i(p))_{i \in \mathbb{Z}}$ denote the sequence of i.i.d. 0–1 random variables with $\mathbb{P}(X_1(p) = 1) = p$. Using the inclusion–exclusion argument, the function $\psi_n(a, x, y)$ determines a measure μ on $(\{0, 1\}^{\mathbb{Z}}, \mathcal{B})$ whose finite-dimensional distributions μ_I, for any finite subset I of $\mathbb{Z}$, satisfy

$$\mu_I(B) = c(x, y)\mathbb{P}((X_i(a - x))_{i \in I} \in B) + (1 - c(x, y))\mathbb{P}((X_i(a - y))_{i \in I} \in B),$$

where $B = \{i_k, 1 \leq k \leq |I|\}$ with $i_k \in \{0, 1\}$. For $a > y > 0 > x$, the measure μ is the distribution of an infinitely exchangeable sequence $X = (X_i)_{i \in \mathbb{Z}}$ and it is the mixture of two sequences of i.i.d. 0–1 random variables. Therefore the measure μ is positively dependent and, for any finite subset I of $\mathbb{Z}$, $\mu_I(B) = \mathbb{P}((X_i)_{i \in I} \in B)$. From this fact it follows that:

(i) the measure μ is not ergodic in contrast with the positive contraction case considered by Lyons (2003) and Shirai & Takahashi (2003);

(ii) partial sums satisfy, for any positive integer n and any $k = 0, \ldots, n$,

$$\mathbb{P}\left(\sum_{i=1}^{n} X_i = k\right) = c(x,y)\mathbb{P}\left(\sum_{i=1}^{n} X_i(a-x) = k\right) + (1-c(x,y))\mathbb{P}\left(\sum_{i=1}^{n} X_i(a-y) = k\right)$$

and therefore they are approximated by mixtures of two Gaussian laws;

(iii) it is easy to check that for any $1 \geq p_1 > p_2 > 0$ and $0 < \alpha < 1$, there exist $a > y > 0 > x$ such that $c(x, y) = \alpha$, $a - x = p_1$ and $a - y = p_2$, implying by the above considerations that the mixture of two Bernoulli processes is a determinantal point process.

For $a > y > x > 0$, by the Kendall condition (1967), the measure $\mu(A)$ represents the exchangeable sequence if and only if $(-1)^j \Delta^j(q(k)) \geq 0$, where $q(j) = \psi_j(a, x, y)$ and Δ is the forward difference operator. Using straightforward calculations, one can check that a such defined measure determines a probability distribution of a finite vector $X = (X_i, i = 1, \ldots, n)$ if and only if $y[1-a+x]^n \geq x[1-a+y]^n$. Moreover, it follows from the Pemantle criteria (10.12), that the measure μ is (conditionally) negatively dependent, in particular the finite exchangeable vector X is negatively dependent.

For $y = x$, the matrix Q is a positive contraction provided that $0 \leq (a + (n-1)x) \leq 1$ and $x \leq a$. Now, the negative association follows both from the treated case $y > x > 0$ and from Theorem 8.1 of Lyons (2003).

Proposition 10.15 *Let $a \geq y > 0$, $y \geq x$ and let Q be a $\mathbb{Z} \times \mathbb{Z}$ matrix defined by*

$$(Q)_{ii} = a, \ (Q)_{ij} = x, (Q)_{ji} = y, i < j.$$

If $x < 0$, then the matrix defines an infinitely exchangeable determinantal point process $X = \{X_i \ ; \ i \in \mathbb{Z}\}$ by setting, for any finite subset A of $\mathbb{Z}$, $\mathbb{P}(X_i = 1, \text{ for all } i \in A) = \det(Q|A)$, which is non-ergodic and positively dependent. Moreover, the mixture of two Bernoulli processes is a determinantal point process.

If $y \geq x > 0$, then the matrix defines the finitely exchangeable negatively dependent determinantal point process.

11

Stationary Sequences in a Random Time Scenery

In this chapter, we analyze the asymptotic behavior of the partial sums process associated with several examples of stationary sequences in a random time scenery. The examples considered are stationary sequences sampled by shifted renewal Markov chains and random walks in a strictly stationary scenery. The asymptotic behavior of the partial sums process is investigated with the help of the invariance principles stated in Chapter 4.

11.1 Sampling by a Shifted Markov Chain

11.1.1 Introduction

Here we are interested in investigating the limiting behavior of the partial sums associated with the process defined by

$$X_k = \zeta_{k+\phi_k},$$

where

$$\{\zeta_j\}_{j\in\mathbb{Z}} \text{ is a stationary sequence (observables/random scenery), and}$$
$$\{\phi_k\}_{k\geq 0} \text{ is a Markov chain (random time).}$$

We introduce two chains below. All our calculations are for the so-called renewal-type chain, which is the time-reversed version of the Chung chain.

11.1.2 The Chung Chain, the Renewal Chain and the Model

a) The Chung chain

One of the frequently used examples is the well-known Markov chain (originated in Chung (1960)) $\mathbf{Y} = \{Y_n, n = 0, 1, \ldots\}$, whose one step transition probabilities are given by: $p_{k,k+1} = t_k$ and $p_{k,0} = \bar{t}_k = 1 - t_k$, i.e. with the transition matrix

Functional Gaussian Approximation for Dependent Structures. Florence Merlevède, Magda Peligrad and Sergey Utev. Oxford University Press (2019).
DOI: 10.1093/oso/9780198826941.001.0001

$$Q = \begin{bmatrix} \bar{t}_0 & t_0 & 0 & 0 & 0 & \dots \\ \bar{t}_1 & 0 & t_1 & 0 & 0 & \dots \\ \bar{t}_2 & 0 & 0 & t_2 & 0 & \dots \\ & & & \dots & & \end{bmatrix}$$

If $a = \sum_{k=1}^{\infty} \prod_{j=0}^{k-1} t_j < \infty$, there is an invariant probability distribution μ with

$$\mu_k = \left(\prod_{j=0}^{k-1} t_j \right) (1 + a)^{-1}, \; k = 0, 1, \dots \tag{11.1}$$

Remark 11.1 *Let* $\sup_k t_k < 1$. *Then the Markov chain* $\mathbf{Y}$ *with one step transition probabilities given above is* ϕ*-mixing (see Rosenblatt (1971, page 214)) and thus exponentially* ϕ*-mixing. Indeed, it follows from Davydov (1973, Theorem 4) that if* $\mathbf{X} = \{X_n, n = 0, 1, \dots\}$ *is a Markov chain (not necessarily stationary) with state space, let say* $\mathbb{R}$, *and such that its* ϕ*-mixing coefficients satisfy* $\phi(n) < 1/2$ *for some* $n \geq 1$, *then* $\phi(n) \to 0$ *at least exponentially fast as* $n \to \infty$.

b) The Renewal chain

Let us consider now the time reversed Chung Markov chain, i.e. the process $\mathbf{X} = \{X_n, n = 0, 1, \dots\}$ such that for any non-negative integer m, $(X_0, X_1, \dots, X_m) =^d (Y_m, Y_{m-1}, \dots, Y_0)$ where $\mathbf{Y}$ is the Chung chain defined above with $a < \infty$. Then $\mathbf{X}$ is a Markov chain whose transition matrix is given by:

$$P = \begin{bmatrix} p_0 & p_1 & p_2 & p_3 & p_4 & \dots \\ 1 & 0 & 0 & 0 & 0 & \dots \\ 0 & 1 & 0 & 0 & 0 & \dots \\ & & & \dots & & \end{bmatrix} \quad \text{e.g. } p_k = (1 - t_k) \prod_{\ell=0}^{k-1} t_\ell.$$

Indeed we must have: $P_{i,j} = \dfrac{\mu_j}{\mu_i} Q_{j,i}$, for any $i, j = 0, 1, \dots$

This time reversed Chung Markov chain is often referred to as a renewal chain, and the times between successive visits to zero are independent random variables.

Example: *Take* $t_k \equiv t$ *for all* $k = 0, 1, \dots$ *with* $0 < t < 1$. *The Markov chain* $\mathbf{X}$ *has invariant probability distribution* π *given by* $\pi_k = (1-t)t^k$ *for all* $k = 0, 1, \dots$. *As quoted in Rosenblatt (1971), the chain is not* ϕ*-mixing and thus the original Chung chain is NOT* ψ*-mixing, unless* $t_k = 0$ *for* $k \geq k_0$. *Indeed, if we denote by* T *the transition operator on* $\mathbb{L}^1(\mu)$ *induced by the one step transition probabilities* $Q(\cdot, \cdot)$ *of the Chung chain* $\mathbf{Y}$, *then* $\sup_{f \perp 1} \dfrac{\|T^n f\|_{1,\mu}}{\|f\|_{1,\mu}}$ *does not converge to zero, as* $n \to \infty$. *Here* $\| \cdot \|_{1,\mu}$ *is the* $\mathbb{L}^1(\mu)$ *norm and* $f \perp 1$ *means* $\mu(f) = 0$.

c) The model: a stationary sequence sampled by shifted renewal chain

We shall consider the partial sums of a centered stationary sequence of random variables $\{\zeta_j\}_{j \in \mathbb{Z}}$, called the *random scenery*, sampled by the process $\{Y_k\}_{k \geq 0}$ which we define as

$$Y_k = k + \phi_k, \quad k \geq 0,$$

where $\{\phi_n\}_{n\geq 0}$ is a homogeneous "renewal"-type Markov chain introduced above, so its one step transition probabilities are given by $p_{0,i} = p_i := \mathbb{P}(\tau = i)$ for all $i \geq 0$ (with $\sum_{i\geq 0} p_i = 1$) and $p_{i,i-1} = 1$ for any $i \geq 1$. We then define

$$X_k = \zeta_{Y_k}, \ k \geq 0. \tag{11.2}$$

11.1.3 Notations and Properties of the "Renewal" Chain and of the Sampled Random Scenery

Let us assume in what follows that the process starts at the origin, i.e. $\phi_0 = 0$.

a) Analysis of the "renewal" chain

Let us introduce certain notations associated with the process ϕ_n. We denote the jumps of the chain by the i.i.d. sequence $T_0, T_1, \dots$ where $\mathbb{P}(T_0 = k) = p_k$ for $k \geq 0$, with the first jump taking place at time $n = 1$. Then with each time step, ϕ_n drops by one until it reaches 0 again,

$$\phi_1 = T_0, \ \phi_2 = T_0 - 1, \ \phi_3 = T_0 - 2, \ \dots, \ \phi_{T_0} = 1, \ \phi_{T_0+1} = 0.$$

We also define the sequence of stopping times $\nu_1, \nu_2, \dots$ as the times of returns to the origin,

$$\nu_1 \stackrel{\text{def}}{=} \inf\{k > 0 : \phi_k = 0\}, \quad \nu_{n+1} \stackrel{\text{def}}{=} \inf\{k > \nu_n : \phi_k = 0\}, \quad n \geq 1.$$

Finally we define the random times $\tau_1, \tau_2, \dots$ as the inter-arrival times to the origin, i.e. the times between successive returns to the origin,

$$\tau_1 = \nu_1 \text{ and } \tau_{n+1} = \nu_{n+1} - \nu_n, \quad n \geq 1.$$

Note that by definition of the process ϕ_n and basic properties of Markov chains, $(\tau_k)_{k\geq 1}$ forms a sequence of independent random variables each distributed as τ. Furthermore, a careful look at the paths of the process is enough to convince one that for $k \geq 1$,

$$\tau_k = T_{k-1} + 1 \text{ and } \nu_k = \tau_1 + \cdots + \tau_k.$$

b) The sampling process $Y_k = k + \phi_k$.

Let us have a look at how $Y_k = k + \phi_k$ behaves. We are only interested in $k \geq 1$. At time $k = 1$ the Markov chain jumps to T_0 and thus

$$Y_1 = 1 + \phi_1 = 1 + T_0.$$

Assume for the moment that the first jump is not zero, $T_0 \neq 0$. Then, at the next step $k = 2$, we have $Y_2 = 2 + \phi_2 = T_0 + 1$, and thus Y_k will stay at $T_0 + 1$ up to and including the time when ϕ_k returns to the origin; or in the notation of the previous subsection,

$$Y_1 = Y_2 = \cdots = Y_{\tau_1} = \tau_1 = T_0 + 1, \quad Y_{\tau_1+1} = \tau_1 + \tau_2.$$

Overall we have $Y_j = \nu_r = \tau_1 + \cdots + \tau_r$, for $\nu_{r-1} < j \leq \nu_r$. By now, it should be obvious that the process Y_j is non-decreasing, and constant on the intervals $(\nu_r, \nu_{r+1}]$. Also, since

$$Y_{\nu_j+1} = \nu_j + 1 + \phi_{\nu_j+1} = \nu_j + 1 + T_j > \nu_j,$$

we have that at times $\nu_j + 1$, the process $\{Y_k\}_{k\geq 0}$ makes a strictly positive jump.

c) Properties of the sampled random scenery

We investigate now the properties of the process $(X_k)_{k\geq 0}$ defined by (11.2), that is the properties of the time scenery $\{\zeta_j\}_{j\in\mathbb{Z}}$ sampled by the process Y_k, $k \geq 1$ introduced in the previous section. Note first that

$$X_j = \zeta_{\nu_1}, \quad 0 < j \leq \nu_1,$$
$$X_j = \zeta_{\nu_r}, \quad \nu_{r-1} < j \leq \nu_r, \ r \geq 2.$$

Long-range dependence. We use the following definition of long-range dependence: We call a stationary sequence of r.v.'s $\{U_j\}_{j\geq 0}$, in $\mathbb{L}^2$, long-range dependent if

$$\sum_{j\geq 1} |\mathrm{cov}(U_0, U_j)| = \infty.$$

Assume, in what follows, that the time scenery is i.i.d., with mean zero and finite variance $\sigma^2 > 0$. Then, as previously mentioned, for $0 < j \leq \nu_1, X_j = \zeta_{\nu_1}$, and for $\nu_1 < j \leq \nu_2$ we have $X_j = \zeta_{\nu_2}$ which is independent of ζ_{ν_1}. Let us then define

$$\mathcal{A} = \sigma(Y_k, k \geq 1),$$

and write

$$Z_j = \zeta_{\nu_j}, \ j \geq 1,$$

for the distinct consecutive variables $\zeta_\cdot$ sampled by Y_k. Then, since, as we mentioned earlier, at time $\nu_j + 1$, Y_k makes strictly positive jumps, conditionally on $\mathcal{A}$, the sequence $Z_1, Z_2, \ldots$ is i.i.d.

Overall we have that

$$\mathbb{E}(X_1 X_j | \mathcal{A}) = 0 \ \text{ or } \ \mathbb{E}(X_1 X_j | \mathcal{A}) = \sigma^2 \ \text{a.s.}$$

More precisely, note that

$$X_j = Z_1 \mathbf{1}\{j \leq \tau_1\} + \sum_{i=2}^{\infty} Z_i \mathbf{1}\{\nu_{i-1} < j \leq \nu_i\}.$$

Then we have that

$$\mathbb{E}(X_1X_j|\mathcal{A}) = \sigma^2\mathbf{1}\{j \leq \tau_1\} \text{ a.s. and thus } \mathbb{E}(X_1X_j) = \sigma^2\mathbb{P}(\tau_1 \geq j). \tag{11.3}$$

It follows that the sampled process $\{X_j\}_{j\geq 1}$ has long-range dependence if

$$\sum_{j\geq 1} |\mathbb{E}(X_1X_j)| = \infty, \quad \text{so that if and only if} \quad \sum_{j\geq 1} \mathbb{P}(\tau_1 \geq j) = \infty.$$

This last condition holds if and only if $\mathbb{E}(\tau_1) = \infty$, that is if and only if $\mathbb{E}(T_0) = \infty$.

Assume now that π is the stationary distribution of the Markov chain $\{\phi_n, n \geq 0\}$, that is $\pi P = \pi$, which translates to

$$\pi_1 = \pi_0(1 - p_0) = \pi_0\mathbb{P}(T_0 \geq 1),\ \pi_2 = \pi_0(1 - p_0 - p_1) = \pi_0\mathbb{P}(T_0 \geq 2) \text{ and so on.}$$

So

$$\pi_k = \pi_0\mathbb{P}(T_0 \geq k),\ k \geq 1,$$

implying that

$$\sum_{i=1}^{\infty} \pi_i = \pi_0 \sum_{i=1}^{\infty} \mathbb{P}(T_0 \geq i).$$

If this sum is infinite then the stationary distribution does not exist. Now it should be obvious that the sequence has long-range dependence if and only if the Markov chain does not have a stationary distribution.

Analysis of the stationarity of the sampled process. Still assuming that the time scenery is i.i.d. and that the chain starts at zero, we next compute $\mathbb{E}(X_2X_3)$ to see if the sequence is stationary. As it turns out, it is not equal to $\mathbb{E}(X_1X_2)$ and hence stationarity is ruled out immediately.

Before we actually show this, we introduce some further notations. We define the renewal process

$$N_k := \max\{j \geq 1 : \nu_j \leq k\} + 1, k \geq 0,$$

with the convention that $\max \emptyset = 0$. Note that N_k basically counts the number of visits to the origin up to time k, including the renewal at 0. Under this notation, we can now write

$$X_1 = Z_1 = Z_{N_0}, \quad X_2 = Z_{N_1}, \quad X_j = Z_{N_{j-1}}, j \geq 2,$$

where we recall that $Z_j = \zeta_{\nu_j}$. Also we can easily deduce that $1 \leq N_1 \leq 2$. More precisely, if the first jump is at least 1, then $N_1 = 1$. If the first jump is 0, then $N_1 = 2$. Thus we have

$$\mathbb{E}(X_2X_3|\mathcal{A}) = \sigma^2\mathbf{1}\{N_1 = N_2\} = \sigma^2\mathbf{1}\{N_2 = N_1 = 1\} + \sigma^2\mathbf{1}\{N_2 = N_1 = 2\} \text{ a.s.}$$

And taking expectations once more we have

$$\mathbb{E}(X_2X_3) = \sigma^2\mathbb{P}(N_1 = N_2 = 1) + \sigma^2\mathbb{P}(N_1 = N_2 = 2).$$

The event $\{N_1 = N_2 = 1\}$ occurs when the first jump T_0 is strictly greater than 1, because if it is 0, $N_1 = 2$, or if $T_0 = 1$ then $N_1 = 2 > N_1 = 1$. So it must be that $T_0 > 1$. On the other hand, $N_1 = 2$ means that $T_0 = 0$, and then if $N_2 = 2$ the second jump must be at least 1, otherwise $N_2 = 3$. Hence $\{N_1 = N_2 = 2\} = \{T_0 = 0, T_1 \geq 1\}$.

Thus we have

$$\begin{aligned}\mathbb{E}(X_2X_3) &= \sigma^2\mathbb{P}(T_0 > 1) + \sigma^2\mathbb{P}(T_0 = 0)\mathbb{P}(T_1 > 0)\\ &= \sigma^2\left(\mathbb{P}(\tau_1 > 2) + \mathbb{P}(\tau_1 = 1)\mathbb{P}(\tau_1 \geq 2)\right),\end{aligned}$$

where we have used the equality $\tau_1 = T_0 + 1$. Now recall that, by (11.3), we had

$$\mathbb{E}(X_1X_2) = \sigma^2\mathbb{P}(\tau_1 \geq 2),$$

which is clearly not equal to $\mathbb{E}(X_2X_3)$. So stationarity of the sampled time scenery is clearly ruled out.

11.1.4 Martingale Difference Scenery

From now on we shall assume that the chain starts at the origin and that $\mathbb{E}(\tau) < \infty$ which entails that the "renewal"-type Markov chain $(\phi_k)_{k\geq 0}$ has a stationary distribution. In this section we shall try to deduce sufficient conditions on the sampling process (ϕ_n) so that the sampled time scenery satisfies "projective"-type or "Gordin"-type criteria. In what follows we assume that (ζ_j) is a sequence of centered i.i.d. r.v.'s, or a sequence of martingale differences, i.e.

$$\mathbb{E}\left(\zeta_j \,|\mathcal{G}_{j-1}\right) = 0 \text{ a.s.}, \quad \text{where } \mathcal{G}_k \overset{\text{def}}{=} \sigma(\zeta_j; j \leq k),$$

such that $\mathbb{E}\zeta_j^2 = \sigma^2$, for all j.

Let us now introduce an appropriate adapted filtration for the sampled sequence $(X_k)_{k\geq 0}$. We have already defined the filtration $\mathcal{A}$ encoding all the history of the Markov sampling chain $\{\phi_n\}_{n\geq 0}$. We now define

$$\mathcal{F}_k = \sigma(\mathcal{A}; X_j, 1 \leq j \leq k),\ k \in \mathbb{Z}. \tag{11.4}$$

This filtration encodes all the history of the Markov chain, and all the history up to time k of the sampled time scenery

$$\zeta_{Y_1}, \zeta_{Y_2}, \ldots, \zeta_{Y_k}.$$

Note that by the above definition we have

$$\mathcal{F}_j = \mathcal{A}, \quad \text{for all } j \leq 0.$$

Then let us consider the quantity $\mathbb{E}(X_k|\mathcal{F}_{k-m})$. If $k \geq m$ we have

$$\begin{aligned}\mathbb{E}(X_k|\mathcal{F}_{k-m}) &= \mathbb{E}(\zeta_{Y_k}|\mathcal{A}, \zeta_{Y_1}, \dots, \zeta_{Y_{k-m}}) \\ &= \mathbb{E}\left(\zeta_{Y_k}\mathbf{1}_{\{Y_k=Y_{k-m}\}}|\mathcal{A}, \zeta_{Y_1}, \dots, \zeta_{Y_{k-m}}\right) \\ &\quad + \mathbb{E}\left(\zeta_{Y_k}\mathbf{1}_{\{Y_k\neq Y_{k-m}\}}|\mathcal{A}, \zeta_{Y_1}, \dots, \zeta_{Y_{k-m}}\right) \\ &= \zeta_{Y_{k-m}}\mathbf{1}_{\{Y_k=Y_{k-m}\}} \\ &\quad + \mathbf{1}_{\{Y_k\neq Y_{k-m}\}}\mathbb{E}\left(\zeta_{Y_k}|\mathcal{A}, \zeta_{Y_1}, \dots, \zeta_{Y_{k-m}}\right) \text{ a.s.}\end{aligned}$$

Therefore

$$\mathbb{E}(X_k|\mathcal{F}_{k-m}) = \zeta_{Y_{k-m}}\mathbf{1}_{\{N_{k-1}=N_{k-m-1}\}} + \mathbf{1}_{\{Y_k\neq Y_{k-m}\}}\mathbb{E}\left(\zeta_{Y_k}|\mathcal{A}, \zeta_{Y_1}, \dots, \zeta_{Y_{k-m}}\right) \text{ a.s.} \tag{11.5}$$

Recall now that we have assumed that (ζ_j) is either a sequence of i.i.d. centered r.v.'s, or a sequence of martingale differences. Then, note that on the event $\{Y_k \neq Y_{k-m}\}$ we must have

$$\mathbb{E}\left(\zeta_{Y_k}|\mathcal{A}, \zeta_{Y_1}, \dots, \zeta_{Y_{k-m}}\right) = 0 \text{ a.s.}$$

Thus we have for $k \geq m$, and (ζ_j) a sequence of i.i.d. centered r.v.'s, or a sequence of martingale differences,

$$\mathbb{E}(X_k|\mathcal{F}_{k-m}) = \zeta_{Y_{k-m}}\mathbf{1}_{\{N_{k-1}=N_{k-m-1}\}} \text{ a.s.} \tag{11.6}$$

Note that if $k < m$ then by definition of our filtration $\mathcal{F}_{k-m} = \mathcal{A}$ and thus by independence of the time scenery and the Markov process, and the fact that the time scenery is centered we have

$$\mathbb{E}(X_k|\mathcal{F}_{k-m}) = \mathbb{E}(\zeta_{Y_k}|\mathcal{A}) = 0 \text{ a.s.}$$

a) Gordin-type condition

Next we would like to better understand the quantity $\mathbb{P}(N_k = N_{k-m})$ which, in view of (11.6), will allow to state sufficient conditions for "Gordin"-type criteria as:

$$\sum_{m=1}^{\infty} \sup_{k\geq 0} ||\mathbb{E}(X_k|\mathcal{F}_{k-m})||_2 < \infty. \tag{11.7}$$

Since $\mathcal{F}_{k-m} = \mathcal{A}$ for $k \leq m$ and $\mathbb{E}(X_k|\mathcal{A}) = 0$, assume then that $k \geq m$, and let $\tilde{\nu}$ be the last renewal before time $k-m$. Then we have

$$\begin{aligned}\mathbb{P}(N_{k-m} = N_k) &= \sum_{r=0}^{k-m} \mathbb{P}(\tilde{\nu} = r, \tau_{\tilde{\nu}+1} > k-r) = \sum_{r=0}^{k-m} \mathbb{P}(\tilde{\nu} = r, \tau_{r+1} > k-r) \\ &\leq \sum_{r=0}^{k-m} \mathbb{P}(\tilde{\nu} = r, \tau_{r+1} > m) \leq \mathbb{P}(\tau > m).\end{aligned} \tag{11.8}$$

In this way, we have

$$\begin{aligned}||\mathbb{E}(X_k|\mathcal{F}_{k-m})||_2 &\le \mathbb{E}\left(\mathbb{E}(X_k|\mathcal{F}_{k-m})^2\right)^{1/2} \le \mathbb{E}\left(X_{k-m}^2 \mathbf{1}\{N_{k-1} = N_{k-1-m}\}\right)^{1/2} \\ &\le \sigma\sqrt{\mathbb{P}(N_{k-1} = N_{k-1-m})} \\ &\le \sigma\sqrt{\mathbb{P}(\tau > m)},\end{aligned}$$

and so

$$\begin{aligned}&||\mathbb{E}(X_k|\mathcal{F}_{k-m})||_2 \le \sigma\sqrt{\mathbb{P}(\tau > m)} \\ &\qquad \text{and } \sum_{m=1}^{\infty} \sup_{k\ge 0} ||\mathbb{E}(X_k|\mathcal{F}_{k-m})||_2 \le \sigma \sum_{m=1}^{\infty} \sqrt{\mathbb{P}(\tau > m)}. \qquad (11.9)\end{aligned}$$

Thus, if we require that $\sum_{m\ge 1} \sqrt{\mathbb{P}(\tau > m)} < \infty$, then the Gordin-type criteria (11.7) is satisfied.

b) Projective-type condition

We shall attempt to weaken the above condition by aiming for a projective condition. In other words, rather than finding sufficient conditions for (11.7) to be satisfied, we shall now attempt to find a sufficient condition so that the following "projective"-type criteria is satisfied:

$$\sup_{k\ge 0} ||\mathbf{P}_{k-m}(X_k)||_2 \le p_m \text{ and } \sum_{m=1}^{\infty} p_m < \infty,$$

where

$$\mathbf{P}_{k-m}(X_k) \overset{\text{def}}{=} \mathbb{E}(X_k|\mathcal{F}_{k-m}) - \mathbb{E}(X_k|\mathcal{F}_{k-m-1}).$$

Starting from (11.6), we get that, for $k > m + 1$

$$\begin{aligned}\mathbf{P}_{k-m}(X_k) &= \mathbb{E}(X_k|\mathcal{F}_{k-m}) - \mathbb{E}(X_k|\mathcal{F}_{k-m-1}) \\ &= X_k\left[\mathbf{1}\{N_{k-1} = N_{k-m-1}\} - \mathbf{1}\{N_{k-m-2} = N_{k-1}\}\right] \\ &= X_k \mathbf{1}\{N_{k-m-2} < N_{k-m-1} = N_{k-1}\}.\end{aligned}$$

Therefore, for any $k > m + 1$,

$$||\mathbf{P}_{k-m}(X_k)||_2 = \sigma\sqrt{\mathbb{P}(N_{k-m-2} < N_{k-m-1} = N_{k-1})}.$$

When $k = m + 1$ we have $||\mathbf{P}_{k-m}(X_k)||_2 = ||\mathbb{E}(X_{m+1}|\mathcal{F}_1)||_2 = \sigma\sqrt{\mathbb{P}(N_m = N_0)}$, and for $k \le m$ we have $||\mathbf{P}_{k-m}(X_k)||_2 = 0$, since $\mathcal{F}_j = \mathcal{A}$ for all $j \le 0$.

Notice then that on the event $\{N_{k-m-2} < N_{k-m-1} = N_{k-1}\}$, there must be a renewal at time $k - m - 1$, since $N_{k-m-1} > N_{k-m-2}$, and no renewal between times $k - m - 1$ and $k - 1$. Thus we have,

$$\mathbb{P}(N_{k-m-2} < N_{k-m-1} = N_{k-1}) = \sum_{r=0}^{k-m-2} \mathbb{P}(\tilde{\nu} = r, \tau_{\tilde{\nu}+1} = k - m - 1 - r, \tau_{\tilde{\nu}+2} > m)$$

$$= \sum_{r=0}^{k-m-2} \mathbb{P}(\tilde{\nu} = r, \tau_{r+1} = k - m - 1 - r)\mathbb{P}(\tau > m) \sim C\mathbb{P}(\tau > m),$$

where $\tilde{\nu}$ denotes the last renewal before time $k - m - 2$.

Therefore, in this situation too, to have $\sum_{m\geq 1} \sup_k \|\mathbf{P}_{k-m}(X_k)\|_2 < \infty$, we need $\sum_{m=1}^{\infty} \sqrt{\mathbb{P}(\tau > m)} < \infty$. Hence the projective criterion is not weaker than the Gordin-type condition we investigated earlier.

11.1.5 Stationary Time Scenery

The aim of this section is to analyze the quantities $\|\mathbb{E}(X_k|\mathcal{F}_{k-m})\|_2$ when $\mathcal{F}_k$ is still defined by (11.4) but the random scenery $\{\zeta_j\}$ is, this time, a general stationary sequence, dropping then the i.i.d. or martingale difference requirement. The main aim is then to investigate under what additional conditions a Gordin-type criterion on the sequence $\{\zeta_j\}$ can be preserved under sampling by the renewal Markov chain $\{\phi_n\}$, when we still consider the chain starting from 0.

So, in what follows, we still assume that the chain starts from the origin, we suppose that $\{\zeta_j\}_{j\in\mathbb{Z}}$ is a strictly stationary sequence of centered r.v.'s in $\mathbb{L}^2$, and we consider the stationary non-decreasing filtration $\{\mathcal{G}_j\}_{j\in\mathbb{Z}} = \sigma(\zeta_k : k \leq j)$. Denote by

$$a(k) = \|\mathbb{E}(\zeta_k|\mathcal{G}_0)\|_2,\ k \geq 0.$$

Recall that $Y_j = \nu_r$, for $\nu_{r-1} < j \leq \nu_r$, and by definition of N_k, we have that $N_k - 1 = \max\{j \geq 1 : \nu_j \leq k\}$, the last time the increasing random walk $\{\nu_j : j \geq 0\}$ is below the level k. Thus N_k is the hitting time of the set (k, ∞), or the time of first entry in the interval $[k + 1, \infty)$ of the random walk $\{\nu_j : j \geq 0\}$. Thus $\nu_{N_k - 1} < k + 1$ and $\nu_{N_k} \geq k + 1$. This fact with k replaced by $k - 1$ gives

$$\nu_{N_{k-1}-1} < k \leq \nu_{N_{k-1}}.$$

Since $Y_j = \nu_r$ for $\nu_{r-1} < j \leq \nu_r$, we can conclude that $Y_k = \nu_{N_{k-1}}$ for all $k \geq 1$. Then once again writing $\mathcal{A}$ for $\sigma(\phi_j, j \in \mathbb{Z})$, $\mathcal{F}_k = \sigma(\mathcal{A}, X_j, 1 \leq j \leq k)$, and $\mathbb{E}_{\mathcal{A}}(\cdot)$ for $\mathbb{E}(\cdot|\mathcal{A})$, we have

$$\begin{aligned}\mathbb{E}\left(X_k\,|\mathcal{F}_{k-m}\right) &= \mathbb{E}_{\mathcal{A}}\left(\zeta_{Y_k}\,|\zeta_{Y_1}, \dots, \zeta_{Y_{k-m}}\right)\\ &= \mathbb{E}_{\mathcal{A}}\left(\zeta_{\nu_{N_{k-1}}}\,|\zeta_{\nu_1}, \dots, \zeta_{\nu_{N_{k-m-1}}}\right) \text{ a.s.}\end{aligned}$$

Before we continue we need to recall the following fact:

Fact 11.2 *If $\mathcal{B} \subset \mathcal{C}$ are two sigma fields, and X an arbitrary integrable r.v. then* $\|\mathbb{E}(X|\mathcal{B})\|_2 \leq \|\mathbb{E}(X|\mathcal{C})\|_2$.

Going back to our calculation, writing k instead of $k-1$ to simplify notation, we have

$$\left\|\mathbb{E}_{\mathcal{A}}\left(\zeta_{\nu_{N_k}} \middle| \zeta_{\nu_1}, \dots, \zeta_{\nu_{N_{k-m}}}\right)\right\|_2 \leq \left\|\mathbb{E}_{\mathcal{A}}\left(\zeta_{\nu_{N_k}} \middle| \mathcal{G}_{\nu_{N_{k-m}}}\right)\right\|_2.$$

We defined earlier $\mathbb{E}\left(\mathbb{E}(\zeta_m|\mathcal{G}_0)^2\right) = \|\mathbb{E}(\zeta_m|\mathcal{G}_0)\|_2^2 = a^2(m)$, and thus we have

$$\left\|\mathbb{E}_{\mathcal{A}}\left(\zeta_{\nu_{N_k}} \middle| \mathcal{G}_{\nu_{N_{k-m}}}\right)\right\|_2^2 \leq \mathbb{E}\left[a^2(\nu_{N_k} - \nu_{N_{k-m}})\right].$$

Let us continue the computations. Note that

$$\nu_{N_k} - \nu_{N_{k-m}} = \sum_{i=N_{k-m}+1}^{N_k} \tau_i.$$

If the sum is started from N_{k-m} then we would be done, since by definition

$$\nu_{N_{k-m}-1} \leq k-m \text{ and } \nu_{N_k} > k,$$

implying that $\nu_{N_k} - \nu_{N_{k-m}-1} > m$, and then $\left\|\mathbb{E}_{\mathcal{A}}\left(\zeta_{\nu_{N_k}} \middle| \mathcal{G}_{\nu_{N_{k-m}}}\right)\right\|_2 \leq a(m)$ by monotonicity of $(a(k))_k$.

However, the object we are dealing with is more complicated: ν_{N_k} and $\nu_{N_{k-m}}$ are the points of first entry in the intervals (k, ∞), $(k-m, \infty)$ respectively, and thus their difference cannot be bounded below, since, for example, the two hitting times could coincide if a huge jump is taken. Obviously this possibility happens with probability $\mathbb{P}(\tau > m)$.

Recall that N_k is the first time the positive random walk ν_j exceeds the value k. Thus

$$\nu_{N_k} - \nu_{N_{k-m}} = \nu_{\sigma_k^+} - \nu_{\sigma_{k-m}^+} > k - (\nu_{\sigma_{k-m}^+} - (k-m)) - (k-m) = m - (\nu_{\sigma_{k-m}^+} - (k-m)),$$

where $\sigma_k^+ = \inf\{j \geq 0 : \nu_j > k\}$, and thus $\nu_{\sigma_{k-m}^+} - (k-m)$ is the overshoot.

A quick calculation gives for $l = [m/2]$:

$$\begin{aligned}\mathbb{E}\left[a^2(\nu_{N_k} - \nu_{N_{k-m}})\right] &= \mathbb{E}\left[a^2(\nu_{N_k} - \nu_{N_{k-m}})\mathbf{1}\{N_{k-m} \leq N_{k-l} - 1\}\right] \\ &\quad + \mathbb{E}\left[a^2(\nu_{N_k} - \nu_{N_{k-m}})\mathbf{1}\{N_{k-m} = N_{k-l}\}\right] \\ &= I_1 + I_2,\end{aligned}$$

where the first event corresponds to the fact that there is at least one renewal between times $k-m$ and $k-l$, while the second event corresponds to no renewals.

Then we have

$$I_1 = \mathbb{E}\left[a^2\left(\nu_{N_k} - \nu_{N_{k-l}-1} + \nu_{N_{k-l}-1} - \nu_{N_{k-m}}\right)\mathbf{1}\{N_{k-m} \le N_{k-l} - 1\}\right]$$
$$\le a^2([m/2])\mathbb{P}(N_{k-l} > N_{k-m}) \le a^2([m/2]).$$

Indeed by previous considerations we have that $\nu_{N_k} - \nu_{N_{k-l}-1} > l$, while on the event $\{N_{k-m} \le N_{k-l} - 1\}$, clearly $\nu_{N_{k-l}-1} - \nu_{N_{k-m}} \ge 0$ and thus

$$\nu_{N_k} - \nu_{N_{k-l}-1} + \nu_{N_{k-l}-1} - \nu_{N_{k-m}} \ge l.$$

Therefore, since $a(\cdot)$ is non-increasing,

$$a^2\left(\nu_{N_k} - \nu_{N_{k-l}-1} + \nu_{N_{k-l}-1} - \nu_{N_{k-m}}\right) \le a^2(l).$$

For the second term we have $I_2 \le a^2(0)\mathbb{P}(N_{k-m} = N_{k-l})$. So, according to (11.8), we get that for $k \ge m$,

$$I_2 \le a^2(0)\mathbb{P}(\tau \ge m - l) \le a^2(0)\mathbb{P}(\tau \ge [m/2]).$$

With similar arguments, we can investigate quantities as $\sup_{k,\ell\ge 0} \|\mathbb{E}(X_{k+m}X_{k+m+\ell}|\mathcal{F}_k) - \mathbb{E}_{\mathcal{A}}(X_{k+m}X_{k+m+\ell})\|_1$. We summarize the above computations in the fact below.

Fact 11.3 *For any non-negative integer k, let*

$$a(k) = \|\mathbb{E}(\zeta_k|\mathcal{G}_0)\|_2 \text{ and } c(k) = \sup_{j\ge i\ge k} \|\mathbb{E}(\zeta_i\zeta_j|\mathcal{G}_0) - \mathbb{E}(\zeta_i\zeta_j)\|_1.$$

Then, for any non-negative integer m,

$$\sup_{k\ge 0} \|\mathbb{E}(X_{k+m}|\mathcal{F}_k)\|_2^2 \le a^2([m/2]) + a^2(0)\mathbb{P}(\tau > [m/2])$$

and

$$\sup_{k,\ell\ge 0} \|\mathbb{E}(X_{k+m}X_{k+m+\ell}|\mathcal{F}_k) - \mathbb{E}_{\mathcal{A}}(X_{k+m}X_{k+m+\ell})\|_1 \le c([m/2]) + c(0)\mathbb{P}(\tau > [m/2]).$$

11.1.6 Invariance Principle for a Process Sampling by the Shifted Renewal Markov Chain

When the random scenery is a martingale differences sequence, an application of Theorem 2.29 allows to derive the following CLT for the partial sums associated with $(\zeta_{k+\phi_k})$.

Theorem 11.4 *Let $\{\zeta_j\}_{j\ge 0}$ be a sequence of centered random variables, uniformly integrable in $\mathbb{L}^2$, independent of $(\phi_k)_{k\ge 0}$ and such that*

$$\mathbb{E}(\zeta_i|\mathcal{G}_{i-1}) = 0 \text{ and } \mathbb{E}(\zeta_i^2|\mathcal{G}_{i-1}) = \sigma^2 \text{ for any } i \ge 1, \quad (11.10)$$

where $\mathcal{G}_i = \sigma(\zeta_k, 1 \leq k \leq i)$ *for* $i \geq 1$ *and* $\mathcal{G}_0 = \{\emptyset, \Omega\}$*. In addition, assume that* $\mathbb{E}(\tau^2) < \infty$*. Let* $\{X_k\}_{k\geq 0}$ *be defined by* (11.2) *and* $S_n = \sum_{k=1}^n X_k$*. Then, if the chain starts from zero,* $n^{-1/2}S_n$ *converges in distribution to* cN *where* $c^2 = \sigma^2\mathbb{E}(\tau^2)/\mathbb{E}(\tau)$ *and* N *is a standard Gaussian.*

Remark 11.5 *Since* $\mathbb{E}(\tau^2) = \mathbb{E}((T_0+1)^2)$*,* $\pi_0 = 1/\mathbb{E}(\tau)$ *and* $\pi_k = \pi_0\mathbb{P}(T_0 \geq k)$*, we easily derive that* $c^2 = \sigma^2\left(1 + 2\sum_{i=1}^{\infty} i\pi_i\right)$.

Proof We assume throughout the proof that the chain starts from zero and use the notations introduced in Section 11.1.3. By independence of the random scenery and the Markov chain and the fact that the ζ_i's are centered, $\mathbb{E}(S_n|\mathcal{A}) = 0$. Therefore, the direct martingale differences projective-type decomposition gives

$$\frac{1}{\sqrt{n}}S_n = \sum_{j=1}^{n}\frac{1}{\sqrt{n}}\sum_{k=j}^{n}\mathbf{P}_j(X_k) =: \sum_{j=1}^{n} d_{j,n},$$

where $\mathbf{P}_j(X_k) = \mathbb{E}(X_k|\mathcal{F}_j) - \mathbb{E}(X_k|\mathcal{F}_{j-1})$ with $\mathcal{F}_j = \sigma(\mathcal{A}, X_\ell, 1 \leq \ell \leq j)$ for $j \geq 1$ and $\mathcal{F}_j = \mathcal{A}$ for $j \leq 0$.

The theorem will then follow directly from Theorem 2.29 if we can prove

$$\sum_{j=1}^{n} d_{j,n}^2 \to^{\mathbb{P}} c^2 \text{ as } n \to \infty, \tag{11.11}$$

and that condition (2.42) holds, that is $\mathbb{E}(\max_{1\leq j\leq n}|d_{j,n}|) \to 0$, as $n \to \infty$. Notice that, for any $j \geq 1$ and any $k \geq j$, by (11.6),

$$\mathbf{P}_j(X_k) = X_j\mathbf{1}_{\{N_{j-2}<N_{j-1}=N_{k-1}\}}\mathbf{1}_{\{j\geq 2\}} + X_1\mathbf{1}_{\{N_0=N_{k-1}\}}\mathbf{1}_{\{j=1\}}.$$

Hence, for any $j \geq 0$,

$$d_{j+1,n} = \frac{1}{\sqrt{n}}X_{j+1}\delta_{j,n}\ , \ \text{ where } \delta_{j,n} = \sum_{k=j}^{n-1}\mathbf{1}_{\{N_{j-1}<N_j=N_k\}},$$

with the convention that $N_{-1} = 0$. Observe that, for $j \geq 1$, $\delta_{j,n}$ is not zero only if the renewal occurs at time j, that is $j \in \{\nu_1, \nu_2, ...\}$. Let $j \geq 1$. Given the renewal occurs at $j = \nu_u \leq n-1$, we then have $N_j = N_{\nu_u} = u+1$,

$$X_{j+1} = \zeta_{\nu_{N_j}} = \zeta_{\nu_{u+1}} \text{ and } \delta_{j,n} = \tau_{u+1}\mathbf{1}_{\{\nu_{u+1}\leq n\}} + (n-\nu_u)\mathbf{1}_{\{\nu_{u+1}>n\}} = \tau_{u+1}\wedge(n-\nu_u).$$

In addition,

$$d_{1,n} = n^{-1/2}\zeta_{\nu_1}(\tau_1 \wedge n).$$

Let T be a positive real and define

$$\eta_n = \max\{k \geq 1 \ ; \ \nu_k \leq n-1\}, \ v(T) = \sup_{i\geq 1} \mathbb{E}(\zeta_i^2 I(|\zeta_i| > T)), w(T) = \mathbb{E}(\tau^2 I(\tau > T)).$$

Note that

$$I_n = \max_{1\leq j\leq n} |d_{j,n}| \leq \frac{1}{\sqrt{n}} \max_{0\leq u\leq \eta_n} |\zeta_{\nu_{u+1}}|\tau_{u+1}.$$

Taking into account the algebraic inequality (4.84), stating that for any reals a and b: $|ab|1_{\{|ab|>T\}} \leq a^2 1_{\{a^2>T\}} + b^2 1_{\{b^2>T\}}$, we get that for any positive integer u,

$$\mathbb{E}\left\{|\zeta_{\nu_u}\tau_u|^2 I(|\zeta_{\nu_u}\tau_u| > T)\right\} \leq \mathbb{E}[\zeta_{\nu_u}^2 I(|\zeta_{\nu_u}| > \sqrt{T})] + w(\sqrt{T}) \leq v(\sqrt{T}) + w(\sqrt{T}),$$

where we have used the independence between the time scenery and the Markov chain for the last inequality. Since $v(\sqrt{T}) + w(\sqrt{T}) \to 0$, as $T \to \infty$, $\{|\zeta_{\nu_u}|\tau_u\}_{u\geq 1}$ is a uniformly square integrable family and (2.42) follows because $\eta_n \leq n$.

On the other hand, to verify (11.11), we note that, for any $j \geq 0$,

$$\delta_{j,n} = \sum_{k=j}^{\infty} \mathbf{1}_{\{N_{j-1}<N_j=N_k\}} - \sum_{k=n}^{\infty} \mathbf{1}_{\{N_{j-1}<N_j=N_k\}} := \delta_j' + \delta_{j,n}''.$$

By independence between the time scenery and the Markov chain, we have

$$\begin{aligned}\frac{1}{n}\sum_{j=0}^{n-1} \mathbb{E}((X_{j+1}\delta_{j,n}'')^2) &\leq 2\sigma^2 \frac{1}{n} \sum_{j=0}^{n-1}\sum_{k=n}^{\infty}\sum_{k'=n}^{k} \mathbb{E}\left(1_{\{N_j=N_k\}}1_{\{N_j=N_{k'}\}}\right) \\ &\leq 2\sigma^2 \frac{1}{n}\sum_{j=0}^{n-1}\sum_{k=n}^{\infty}(k-n+1)\mathbb{P}\left(N_j = N_k\right) \leq 2\sigma^2\frac{1}{n}\sum_{j=0}^{n-1}\sum_{k=n}^{\infty}(k-n+1)\mathbb{P}(\tau > k-j),\end{aligned}$$

where we used (11.8) for the last inequality. Therefore,

$$\begin{aligned}\frac{1}{n}\sum_{j=0}^{n-1} \mathbb{E}((X_{j+1}\delta_{j,n}'')^2) &\leq 2\sigma^2 \frac{1}{n}\sum_{\ell=0}^{n-1}\sum_{k=n}^{\infty}(k-n+1)\mathbb{P}(\tau > k-n+1+\ell) \\ &\leq 2\sigma^2\frac{1}{n}\sum_{\ell=0}^{n-1}\sum_{k=\ell+1}^{\infty} k\mathbb{P}(\tau > k),\end{aligned}$$

which converges to zero as $n \to \infty$ since by assumption $\sum_{k\geq 1} k\mathbb{P}(\tau > k) < \infty$ (indeed $\mathbb{E}(\tau^2) < \infty$).

Therefore, we infer that (11.11) will be satisfied if one can prove that

$$\frac{1}{n}\sum_{j=0}^{n-1}(X_{j+1}\delta_j')^2 \to^{\mathbb{P}} c^2 \text{ as } n \to \infty. \tag{11.12}$$

But, as before, for $j \geq 1$, δ_j' is not zero only if the renewal occurs at time j. Given that the renewal occurs at $j = \nu_u$ we then have $\delta_j' = \tau_{u+1}$. Hence,

$$\begin{aligned}\frac{1}{n}\sum_{j=0}^{n-1}(X_{j+1}\delta_j')^2 &= \frac{1}{n}\sum_{u=0}^{\eta_n}\zeta_{\nu_{u+1}}^2\tau_{u+1}^2\\ &= \sigma^2\frac{\eta_n+1}{n}\frac{1}{\eta_n+1}\sum_{u=0}^{\eta_n}\tau_{u+1}^2 + \frac{1}{n}\sum_{u=1}^{\eta_n+1}(\zeta_{\nu_u}^2-\sigma^2)\tau_u^2\\ &= I_n + J_n, \text{ say.}\end{aligned}$$

The first term $I_n \to \sigma^2\mathbb{E}(\tau^2)/\mathbb{E}(\tau)$ a.s. by the law of large numbers. To show that $J_n \to^{\mathbb{P}} 0$, we use standard arguments involving first the truncation of τ_i at the level K ($K \to \infty$), second the truncation of the $\zeta_{\nu_u}^2$ at a level T, that is the decomposition:

$$\zeta_{\nu_u}^2-\sigma^2 = \zeta_{\nu_u}^2\mathbf{1}_{\{|\zeta_{\nu_u}|>T\}}-\mathbb{E}\left(\zeta_{\nu_u}^2\mathbf{1}_{\{|\zeta_{\nu_u}|>T\}}|\mathcal{F}_{\nu_{u-1}}\right)+\left(\zeta_{\nu_u}^2\mathbf{1}_{\{|\zeta_{\nu_u}|\leq T\}} - \mathbb{E}\left(\zeta_{\nu_u}^2\mathbf{1}_{\{|\zeta_{\nu_u}|\leq T\}}|\mathcal{F}_{\nu_{u-1}}\right)\right).$$

Invoking the independence between the time scenery and the Markov chain, proving $J_n \to^{\mathbb{P}} 0$ is reduced to show that $n^{-1}\sum_{u=1}^{\eta_n+1} D_{u,T}\tau_u^2 1_{\{|\tau_u|\leq K\}} \to^{\mathbb{P}} 0$, as $n \to \infty$, where $D_{u,T} = \zeta_{\nu_u}^2 1_{\{|\zeta_{\nu_u}|\leq T\}} - \mathbb{E}\left(\zeta_{\nu_u}^2 1_{\{|\zeta_{\nu_u}|\leq T\}}|\mathcal{F}_{\nu_{u-1}}\right)$. This can be proved easily by using the weak law of large numbers for uniformly integrable martingale differences. ■

As we shall see next, we can relax the assumption that the random scenery is a martingale differences sequence, and in Theorem 11.6 below we shall make the following assumption on the random time scenery:

Condition (*H*) $\{\zeta_j\}_{j\geq 0}$ is a strictly stationary sequence of centered random variables in $\mathbb{L}^2(\mathbb{P})$, independent of $(\phi_k)_{k\geq 0}$ and such that

$$\sum_{k\geq 1}\frac{\|\mathbb{E}(\zeta_k|\mathcal{G}_0)\|_2}{\sqrt{k}} < \infty \text{ and } \lim_{n\to\infty}\sup_{j\geq i\geq n}\|\mathbb{E}(\zeta_i\zeta_j|\mathcal{G}_0) - \mathbb{E}(\zeta_i\zeta_j)\|_1 = 0, \tag{11.13}$$

where $\mathcal{G}_i = \sigma(\zeta_k, k \leq i)$.

Applying Theorem 4.24, we can prove the following result concerning the asymptotic behavior of $\{n^{-1/2}S_{[nt]}, t \in [0,1]\}$ when the chain starts from zero.

Theorem 11.6 *Let $\{\zeta_j\}_{j\geq 0}$ be a sequence of real-valued random variables satisfying condition (H). Assume that $\mathbb{E}(\tau^2) < \infty$. Let $\{X_k\}_{k\geq 0}$ be defined by (11.2), $S_0 = 0$ and*

$S_n = \sum_{k=1}^{n} X_k$ *for any* $n \geq 1$. *Then, if the chain starts from 0,* $\{n^{-1/2}S_{[nt]},\ t \in [0, 1]\}$ *converges in distribution in* $D([0, 1])$ *to a Brownian motion with parameter* c^2 *defined by*

$$c^2 = \mathbb{E}(\zeta_0^2)\left(1 + 2\sum_{i\geq 1} i\pi_i\right) + 2\sum_{m\geq 1}\mathbb{E}(\zeta_0\zeta_m)\sum_{j=1}^{m}(P^j)_{0,m-j}. \tag{11.14}$$

Note that $\mathbb{E}(\tau^2) < \infty$ is equivalent to $\sum_{i\geq 1} i\,\mathbb{P}(\tau > i) < \infty$ and therefore to the finiteness of the series $\sum_{i\geq 1} i\pi_i$.

Comment 11.7 *The conclusion of Theorem* 11.6 *still holds if instead of requiring the strict stationarity in condition (H), we require stationarity in* $\mathbb{L}^2$, $\sum_{m\geq 0}|\mathbb{E}(\zeta_0\zeta_m)| < \infty$,

$$\lim_{T\to\infty}\sup_{i\geq 1}\mathbb{E}(\zeta_i^2 I(|\zeta_i| > T)) = 0$$

and if (11.15) *is replaced by*

$$\sum_{k\geq 1}\frac{\max_{m\geq 0}\|\mathbb{E}(\zeta_{k+m}|\mathcal{G}_m)\|_2}{\sqrt{k}} < \infty, \text{ and}$$

$$\lim_{n\to\infty}\sup_{j\geq i\geq n, m\geq 0}\|\mathbb{E}(\zeta_{i+m}\zeta_{j+m}|\mathcal{G}_m) - \mathbb{E}(\zeta_{i+m}\zeta_{j+m})\|_1 = 0. \tag{11.15}$$

The proof is left to the reader.

Proof of Theorem 11.6 First note that c^2 is well defined. Indeed, the following series $\sum_{m\geq 1}\mathbb{E}(\zeta_0\zeta_m)\sum_{j=1}^{m}(P^j)_{0,m-j}$ is absolutely convergent. To see this, we note that since π is the stationary distribution, we have that for any non-negative integers m and i, $\pi_m = \sum_{j\geq 0}\pi_j(P^i)_{j,m}$. Therefore $(P^j)_{0,m-j} \leq \pi_0^{-1}\pi_{m-j}$ which entails that

$$\begin{aligned}\sum_{m\geq 1}|\mathbb{E}(\zeta_0\zeta_m)|\sum_{j=1}^{m}(P^j)_{0,m-j} &\leq \pi_0^{-1}\sum_{m\geq 1}|\mathbb{E}(\zeta_0\zeta_m)|\sum_{j=1}^{m}\pi_{m-j}\\ &\leq \sum_{m\geq 1}|\mathbb{E}(\zeta_0\zeta_m)|\sum_{j=0}^{m-1}\mathbb{P}(T_0 \geq j) \leq \mathbb{E}(\tau)\sum_{m\geq 1}|\mathbb{E}(\zeta_0\zeta_m)|.\end{aligned}$$

So the absolute convergence of the series $\sum_{m\geq 1}\mathbb{E}(\zeta_0\zeta_m)\sum_{j=1}^{m}(P^j)_{0,m-j}$ follows from the finiteness of the series $\sum_{m\geq 1}|\mathbb{E}(\zeta_0\zeta_m)|$, which holds under the first part of condition (11.13) (see Remark 4.13 and the last conclusion of Theorem 4.10).

The invariance principle will be proved by an application of Theorem 4.24. The conditions (4.53) (and then also (4.44)) and $\sup_{n\geq 1} n^{-1}\sum_{j=1}^{n}\mathbb{E}(X_j^2) \leq C < \infty$ hold by using the independence between the time scenery and the random walk and by taking into account the stationarity of the scenery. On another hand, using Fact 11.3

together with the first part of condition (11.13), we deduce that condition (4.51) (and then (4.45)) will be satisfied as soon as

$$\sum_{m\geq 1} \frac{1}{\sqrt{m}}\sqrt{\mathbb{P}(\tau > [m/2])} < \infty, \tag{11.16}$$

which obviously holds since $\mathbb{E}(\tau^2) < \infty$ is equivalent to $\sum_{i\geq 1} i\,\mathbb{P}(\tau > i) < \infty$. To apply Theorem 4.24, it remains to verify (4.50). With this aim, we shall apply Proposition 4.26. Condition (4.52) obviously holds by taking into account Fact 11.3 together with the second part of condition (11.13) and the fact that $\mathbb{E}(\tau) < \infty$. Since, as already mentioned, condition (4.53) holds, to end the proof of the theorem, it remains to verify condition (4.54), which holds if, for any $\varepsilon > 0$ and any $t \in [0, 1]$,

$$\lim_{m\to\infty} \limsup_{n\to\infty} \mathbb{P}\left(\frac{1}{m}\sum_{K=1}^{m-1}\left|\frac{1}{n}\sum_{k=1}^{[nt]}\left\{\mathbb{E}_{\mathcal{A}}(X_k^2) + 2\sum_{\ell=1}^{K}\mathbb{E}_{\mathcal{A}}(X_kX_{k+\ell})\right\} - tc^2\right| > \varepsilon\right) = 0. \tag{11.17}$$

To see this we note that for any integer ℓ in $[0, K]$ where K is a fixed integer, because of the independence between the time scenery and the Markov chain, by the stationarity of the random time scenery,

$$\begin{aligned}\mathbb{E}_{\mathcal{A}}(X_kX_{k+\ell}) &= \sum_{m\geq\ell} \mathbf{1}_{\phi_k=m}\mathbb{E}(\zeta_{k+m}^2) + \sum_{i=1}^{\ell}\sum_{m\geq 0}\mathbf{1}_{\phi_k=\ell-i}\mathbf{1}_{\phi_{k+\ell}=m}\mathbb{E}(\zeta_{k+\ell-i}\zeta_{k+\ell+m}) \\ &= \mathbb{E}(\zeta_0^2)\mathbf{1}_{\phi_k\geq\ell} + \sum_{i=1}^{\ell}\sum_{m\geq 0}\mathbf{1}_{\phi_k=\ell-i}\mathbf{1}_{\phi_{k+\ell}=m}\mathbb{E}(\zeta_0\zeta_{m+i}).\end{aligned}$$

It follows that

$$\begin{aligned}&\left|\frac{1}{n}\sum_{k=1}^{[nt]}\left(\mathbb{E}_{\mathcal{A}}(X_k^2) + 2\sum_{\ell=1}^{K}\mathbb{E}_{\mathcal{A}}(X_kX_{k+\ell})\right) - c^2t\right| \leq \sigma^2\left|\frac{[nt]}{n} - t\right| \\ &\quad + 2\mathbb{E}(\zeta_0^2)\left|\frac{1}{n}\sum_{k=1}^{[nt]}\sum_{\ell=1}^{K}\mathbf{1}_{\phi_k\geq\ell} - t\sum_{i\geq 1} i\pi_i\right| \\ &\quad + 2\left|\frac{1}{n}\sum_{k=1}^{[nt]}\sum_{\ell=1}^{K}\sum_{i=1}^{\ell}\sum_{r\geq 0}U(k,\ell,i,r) - \sum_{r\geq 1}\mathbb{E}(\zeta_0\zeta_r)\sum_{j=1}^{r}(P^j)_{0,r-j}\right|,\end{aligned} \tag{11.18}$$

where

$$U(k,\ell,i,r) := \mathbf{1}_{\phi_k=\ell-i}\mathbf{1}_{\phi_{k+\ell}=r}\mathbb{E}(\zeta_0\zeta_{r+i}). \tag{11.19}$$

Note that since the Markov chain is irreducible and positive recurrent, by the law of large numbers for Markov chains, for any $t \in [0, 1]$,

$$\lim_{n\to\infty}\frac{1}{n}\sum_{k=1}^{[nt]}\sum_{\ell\geq 1}\mathbf{1}_{\phi_k\geq \ell}=\lim_{n\to\infty}\frac{1}{n}\sum_{k=1}^{[nt]}\phi_k=t\sum_{i\geq 1}i\pi_i \text{ a.s.,}$$

whatever the initial law is. In addition, by (11.8),

$$\frac{1}{n}\sum_{k=1}^{[nt]}\sum_{\ell\geq K+1}\mathbb{E}_{\phi_0=0}(\mathbf{1}_{\phi_k\geq \ell})=\frac{1}{n}\sum_{k=1}^{[nt]}\sum_{\ell\geq K+1}\mathbb{P}_{\phi_0=0}(N_{k-1}=N_{k+\ell-1})\leq \sum_{\ell\geq K+1}\mathbb{P}(\tau>\ell),$$

where here and below $\mathbb{E}_{\phi_0=0}$ is the expectation under $\mathbb{P}_{\phi_0=0}$, the probability when the chain starts at 0.

Hence,

$$\frac{1}{m}\sum_{K=1}^{m-1}\frac{1}{n}\sum_{k=1}^{[nt]}\sum_{\ell\geq K+1}\mathbb{E}_{\phi_0=0}(\mathbf{1}_{\phi_k\geq \ell})\leq \frac{1}{m}\sum_{K=1}^{m-1}\sum_{\ell\geq K+1}\mathbb{P}(\tau>\ell),$$

which converges to zero as $m\to\infty$. So, we have proved that, for any $\varepsilon>0$ and any $t\in[0,1]$,

$$\lim_{m\to\infty}\limsup_{n\to\infty}\mathbb{P}\left(\frac{1}{m}\sum_{K=1}^{m-1}\left|\frac{1}{n}\sum_{k=1}^{[nt]}\sum_{\ell=1}^{K}\mathbf{1}_{\phi_k\geq \ell}-t\sum_{i\geq 1}i\pi_i\right|\geq\varepsilon\right)=0. \tag{11.20}$$

On the other hand, let ℓ be a positive fixed integer, by the ergodic Theorem applied to the irreducible and positive recurrent Markov chain $Z_k=(\phi_k,\phi_{k+\ell})$, we have, for $\ell\geq i$,

$$\lim_{n\to\infty}\frac{1}{n}\sum_{k=1}^{[nt]}\mathbf{1}_{\phi_k=\ell-i}\mathbf{1}_{\phi_{k+\ell}=r}=t\pi_{\ell-i}(P^\ell)_{\ell-i,r} \text{ a.s. ,}$$

whatever the initial law is. But, for any $\ell\geq i$, by homogeneity of the Markov chain,

$$\begin{aligned}(P^\ell)_{\ell-i,r}&=\mathbb{P}(\phi_\ell=r|\phi_0=\ell-i)=\mathbb{P}(\phi_\ell=r,\phi_{\ell-i}=0|\phi_0=\ell-i)\\&=\mathbb{P}(\phi_\ell=r|\phi_{\ell-i}=0)=(P^i)_{0,r}.\end{aligned}$$

Hence, for any K and L fixed,

$$\begin{aligned}&\lim_{n\to\infty}\frac{1}{n}\sum_{k=1}^{[nt]}\sum_{\ell=1}^{K}\sum_{i=1}^{\ell}\sum_{r=0}^{L}\mathbf{1}_{\phi_k=\ell-i}\mathbf{1}_{\phi_{k+\ell}=r}\mathbb{E}(\zeta_0\zeta_{r+i})\\&=t\sum_{\ell=1}^{K}\sum_{i=1}^{\ell}\pi_{\ell-i}\sum_{r=0}^{L}(P^i)_{0,r}\mathbb{E}(\zeta_0\zeta_{r+i}) \text{ a.s.,}\end{aligned} \tag{11.21}$$

whatever the initial law is. On the other hand,

$$\mathbb{E}_{\phi_0=0}\left|\frac{1}{n}\sum_{k=1}^{[nt]}\sum_{\ell=1}^{K}\sum_{i=1}^{\ell}\sum_{r\geq L+1}\mathbf{1}_{\phi_k=\ell-i}\mathbf{1}_{\phi_{k+\ell}=r}\mathbb{E}(\zeta_0\zeta_{r+i})\right|$$
$$\leq \frac{1}{n}\sum_{k=1}^{n}\sum_{\ell=1}^{K}\sum_{i=1}^{\ell}\sum_{r\geq L+1}|\mathbb{E}(\zeta_0\zeta_{r+i})|\mathbb{P}_{\phi_0=0}(\phi_k=\ell-i,\phi_{k+\ell}=r)$$
$$\leq \frac{1}{n}\sum_{k=1}^{n}\sum_{\ell=1}^{K}\sum_{i=1}^{\ell}\sum_{r\geq L+1}|\mathbb{E}(\zeta_0\zeta_{r+i})|(P^{\ell})_{\ell-i,r}(P^k)_{0,\ell-i}.$$

But, since π is the stationary distribution, we have that for any non-negative integers m and i, $\pi_m=\sum_{j\geq 0}\pi_j(P^i)_{j,m}$. Therefore $\pi_0(P^k)_{0,\ell-i}\leq\pi_{\ell-i}$, which implies that

$$\mathbb{E}_{\phi_0=0}\left|\frac{1}{n}\sum_{k=1}^{[nt]}\sum_{\ell=1}^{K}\sum_{i=1}^{\ell}\sum_{r\geq L+1}\mathbf{1}_{\phi_k=\ell-i}\mathbf{1}_{\phi_{k+\ell}=r}\mathbb{E}(\zeta_0\zeta_{r+i})\right|$$
$$\leq \pi_0^{-1}\sum_{\ell=1}^{K}\sum_{i=1}^{\ell}\sum_{r\geq L+1}|\mathbb{E}(\zeta_0\zeta_{r+i})|\pi_{\ell-i}\leq K\pi_0^{-1}\sum_{r\geq L+2}|\mathbb{E}(\zeta_0\zeta_r)|.$$

But, as already mentioned, $\sum_{m\geq 0}|\mathbb{E}(\zeta_0\zeta_m)|<\infty$ under the first part of condition (11.13). Hence, for any positive integer K fixed,

$$\lim_{L\to\infty}\limsup_{n\to\infty}\mathbb{E}_{\phi_0=0}\left|\frac{1}{n}\sum_{k=1}^{[nt]}\sum_{\ell=1}^{K}\sum_{i=1}^{\ell}\sum_{r\geq L+1}\mathbf{1}_{\phi_k=\ell-i}\mathbf{1}_{\phi_{k+\ell}=r}\mathbb{E}(\zeta_0\zeta_{r+i})\right|=0. \tag{11.22}$$

Next, we show that, for any positive integer K fixed,

$$\lim_{L\to\infty}\sum_{\ell=1}^{K}\sum_{i=1}^{\ell}\pi_{\ell-i}\sum_{r=L+1}^{\infty}(P^i)_{0,r}\mathbb{E}(\zeta_0\zeta_{r+i})=0 \tag{11.23}$$

and

$$\lim_{K\to\infty}\sum_{\ell=K+1}^{\infty}\sum_{i=1}^{\ell}\pi_{\ell-i}\sum_{r\geq 0}(P^i)_{0,r}\mathbb{E}(\zeta_0\zeta_{r+i})=0. \tag{11.24}$$

Noting that

$$\sum_{\ell=1}^{\infty}\sum_{i=1}^{\ell}\pi_{\ell-i}\sum_{r=0}^{\infty}(P^i)_{0,r}\mathbb{E}(\zeta_0\zeta_{r+i})=\sum_{i=1}^{\infty}\sum_{r=0}^{\infty}(P^i)_{0,r}\mathbb{E}(\zeta_0\zeta_{r+i})=\sum_{r\geq 1}\mathbb{E}(\zeta_0\zeta_r)\sum_{j=1}^{r}(P^j)_{0,r-j},$$

and considering (11.21), (11.22), (11.23) and (11.24), we will get that, for any $\varepsilon>0$ and any $t\in[0,1]$,

$$\lim_{m\to\infty} \limsup_{n\to\infty} \mathbb{P}\left(\frac{1}{m}\sum_{K=1}^{m-1} \left| \frac{1}{n}\sum_{k=1}^{[nt]}\sum_{\ell=1}^{K}\sum_{i=1}^{\ell}\sum_{r\geq 0} U(k,\ell,i,r) - \sum_{r\geq 1}\mathbb{E}(\zeta_0\zeta_r)\sum_{j=1}^{r}(P^j)_{0,r-j} \right| \geq \varepsilon \right) = 0, \tag{11.25}$$

where $U(k, \ell, i, r)$ is defined in (11.19).

Starting from (11.18) and considering (11.20) and (11.25), the convergence in (11.17) follows. Hence to end the proof of the theorem, it remains to show (11.23) and (11.24). We start by proving (11.23). We observe that

$$\sum_{\ell=1}^{K}\sum_{i=1}^{\ell}\pi_{\ell-i}\sum_{r=L+1}^{\infty}(P^i)_{0,r}|\mathbb{E}(\zeta_0\zeta_{r+i})| \leq \sum_{i=1}^{K}\sum_{\ell=i}^{K}\pi_{\ell-i}\sum_{r=L+1+i}^{\infty}\mathbb{E}(\zeta_0\zeta_r)$$
$$\leq K\sum_{r=L+2}^{\infty}|\mathbb{E}(\zeta_0\zeta_r)|,$$

which converges to zero as $L \to \infty$ since, as mentioned before, $\sum_{r\geq 0}|\mathbb{E}(\zeta_0\zeta_r)| < \infty$ under the first part of condition (11.13). This proves (11.23).

To prove (11.24), we first write the following decomposition

$$\sum_{\ell=K+1}^{\infty}\sum_{i=1}^{\ell}\pi_{\ell-i}\sum_{r=0}^{\infty}(P^i)_{0,r}|\mathbb{E}(\zeta_0\zeta_{r+i})| \leq \sum_{\ell=K+1}^{\infty}\sum_{i=1}^{[\ell/2]}\pi_{\ell-i}\sum_{r=0}^{\infty}(P^i)_{0,r}|\mathbb{E}(\zeta_0\zeta_{r+i})|$$
$$+\sum_{\ell=K+1}^{\infty}\sum_{i=[\ell/2]+1}^{\ell}\pi_{\ell-i}\sum_{r=0}^{\infty}(P^i)_{0,r}|\mathbb{E}(\zeta_0\zeta_{r+i})| := I_K + J_K.$$

Now, since π is the stationary distribution, $(P^i)_{0,r} \leq \pi_0^{-1}\pi_r$ for any positive integer i. Therefore,

$$I_K \leq \pi_0^{-1}\sum_{\ell=K+1}^{\infty}\sum_{i=1}^{[\ell/2]}\pi_{\ell-i}\sum_{r=0}^{\infty}\pi_r|\mathbb{E}(\zeta_0\zeta_{r+i})|$$
$$\leq \pi_0^{-1}\sum_{i=1}^{[(K+1)/2]}\sum_{\ell=K+1}^{\infty}\pi_{\ell-i}\sum_{r=0}^{\infty}\pi_r|\mathbb{E}(\zeta_0\zeta_{r+i})|$$
$$+\pi_0^{-1}\sum_{i\geq[(K+1)/2]}\sum_{\ell\geq 2i}\pi_{\ell-i}\sum_{r=0}^{\infty}\pi_r|\mathbb{E}(\zeta_0\zeta_{r+i})|$$
$$\leq \pi_0^{-1}\sum_{\ell=[(K+1)/2]}^{\infty}\pi_\ell\sum_{r=0}^{\infty}\pi_r\sum_{k\geq 1}|\mathbb{E}(\zeta_0\zeta_k)|$$
$$+\pi_0^{-1}\sum_{\ell\geq[(K+1)/2]}\pi_\ell\sum_{r=0}^{\infty}\pi_r\sum_{j\geq[(K+1)/2]}|\mathbb{E}(\zeta_0\zeta_j)|.$$

Hence,

$$I_K \leq 2\pi_0^{-1} \sum_{\ell \geq [(K+1)/2]} \pi(\ell) \sum_{j \geq 1} |\mathbb{E}(\zeta_0 \zeta_j)|,$$

which converges to zero as $K \to \infty$ using again that $\sum_{r\geq 0} |\mathbb{E}(\zeta_0\zeta_r)| < \infty$. On the other hand, since $(P^i)_{0,r} \leq \pi_0^{-1}\pi_r$ for any positive integer i,

$$\begin{aligned} J_K &\leq \pi_0^{-1} \sum_{i\geq[(K+1)/2]+1} \sum_{\ell=i}^{2i} \pi_{\ell-i} \sum_{r=0}^{\infty} \pi_r |\mathbb{E}(\zeta_0\zeta_{r+i})| \\ &\leq \pi_0^{-1} \sum_{i\geq[(K+1)/2]+1} \sum_{r=0}^{\infty} \pi_r |\mathbb{E}(\zeta_0\zeta_{r+i})| \leq \pi_0^{-1} \sum_{j\geq[(K+1)/2]+1} |\mathbb{E}(\zeta_0\zeta_j)|, \end{aligned}$$

which converges to zero as $K \to \infty$ using once again that $\sum_{r\geq 0} |\mathbb{E}(\zeta_0\zeta_r)| < \infty$. This ends the proof of (11.24) and then of the theorem. ■

11.2 Projective Approach for RWRS

In this section we consider random walks in a strictly stationary random scenery and we shall investigate conditions in terms of projective criteria for the associated partial sums suitably normalized to converge to a Brownian motion.

Let us describe more precisely the framework. We assume that the random scenery $\{\zeta_i, \mathcal{G}_i\}_{i\in\mathbb{Z}}$ is strictly stationary with associated strictly stationary filtration, i.e. for any integers k, i, m, h and any bounded measurable function $f: \mathbb{R}^m \to \mathbb{R}$

$$\mathbb{E}[f(\zeta_k, \dots, \zeta_{k+m-1})|\mathcal{G}_i] = \mathbb{E}[f(\zeta_{k+h}, \dots, \zeta_{k+m-1+h})|\mathcal{G}_{i+h}].$$

Now, let $\{Y, Y_i\}$ be an i.i.d. integer-valued sequence independent of $\{\zeta_i, \mathcal{G}_i\}$. Define $S_0 = 0$ and introduce the random walk

$$S_n = Y_1 + \dots + Y_n \ , \ n = 1, 2, \dots .$$

We continue the random walk to the left, for $n = -1, -2, \dots$ by reflecting it. In other words, for any $n \geq 1$, $S_{-n} = -\sum_{k=1}^{n} Y_{-k}$ and then $(S_{-n})_{n\geq 0} \stackrel{d}{=} (-S_n)_{n\geq 0}$. The main statistic is the scenery sampled by the random walk

$$X_n = \zeta_{S_n}, \ n \in \mathbb{Z}, \tag{11.26}$$

and we shall investigate conditions in terms of projective criteria on the random scenery in order for the normalized partial sums process $\{n^{-1/2} \sum_{i=1}^{[nt]} X_i, t \in [0, 1]\}$ to converge to a Brownian motion.

In this section, we shall only consider random walks such that $\mathbb{E}(Y) > 0$. Other models of random walks in random scenery will be treated in Chapter 13.

Remark 11.8 *Since we are sampling the strictly stationary random scenery by a random walk with strictly stationary increments independent of it, stationarity is preserved, and as a stationary filtration we define the natural filtration $\mathcal{F}_i = \sigma(X_k, k \leq i)$, i.e. the σ-field generated by X_k with indices $k \leq i$.*

The main result is the following invariance principle.

Lemma 11.9 *Assume that $(\zeta_n)_{n\in\mathbb{Z}}$ is a strictly stationary sequence with finite second moment, zero mean and such that*

$$\sum_{n=1}^{\infty} \frac{1}{\sqrt{n}} \|\mathbb{E}(\zeta_n|\mathcal{G}_0)\|_2 < \infty. \tag{11.27}$$

Assume that $\{Y, Y_k\}$ is an i.i.d. integer-valued sequence, independent of $\{\zeta_i, \mathcal{G}_i\}$, such that $\mathbb{E}[|Y|^p] < \infty$ for some $p > 2$ and $\mathbb{E}(Y) > 0$. Then

$$\sum_{n=1}^{\infty} \frac{1}{\sqrt{n}} \|\mathbb{E}(X_n|\mathcal{F}_0)\|_2 < \infty.$$

This lemma combined with Theorem 4.17 and Remark 4.13 gives the following result concerning the limiting behavior of the partial sums process associated with the random scenery sampled by the random walk:

Corollary 11.10 *Assume that $(\zeta_n)_{n\in\mathbb{Z}}$ and $\{Y, Y_n\}_{n\in\mathbb{Z}}$ are as in Lemma 11.9. Suppose in addition that $(\zeta_n)_{n\in\mathbb{Z}}$ is ergodic. Then*

$$\left\{ n^{-1/2} \sum_{i=1}^{[nt]} X_i, t \in [0,1] \right\} \to^d \sigma W \text{ in } D([0,1]),$$

where W is a standard Brownian motion on $[0, 1]$ and $\sigma^2 = \sum_{k\in\mathbb{Z}} \mathbb{E}(X_0 X_k)$ where the series is absolutely convergent.

Proof of Lemma 11.9 To highlight the arguments, the proof is divided in two parts. In the first one we treat the simplest case where $0 \leq Y \leq M$ a.s. with $\mathbb{E}[Y] > 0$, and in the second part, we consider more general random walks.

Part I. (Increasing random walks with bounded jumps.) In this part, we prove Lemma 11.9 when $0 \leq Y \leq M$ a.s. with $\mathbb{E}[Y] > 0$. Once again we write $\mathcal{A}$ for the σ-field generated by the random walk, that is $\mathcal{A} = \sigma(Y_i, i \in \mathbb{Z})$. The following notation will be also used:

$$b(n) := \|\mathbb{E}(\zeta_n|\mathcal{G}_0)\|_2, \quad \text{for any } n \in \mathbb{N}. \tag{11.28}$$

To prove the lemma, we start by noticing that by definition, for any $n \in \mathbb{N}$ and any $m \in \mathbb{Z}$,

$$\mathbb{E}(X_{n+m}|\mathcal{F}_m) = \mathbb{E}(\zeta_{S_{n+m}}|\zeta_{S_i}, i \leq m).$$

Using the fact that, for σ-fields such that $\mathcal{B} \subseteq \mathcal{C}$, $\|\mathbb{E}(X|\mathcal{B})\|_2 \leq \|\mathbb{E}(X|\mathcal{C})\|_2$, we get

$$\|\mathbb{E}(X_{n+m}|\mathcal{F}_m)\|_2 \leq \|\mathbb{E}(\zeta_{S_{n+m}}|\mathcal{A}, \mathcal{G}_{S_m})\|_2.$$

Using stationarity and the independence between the random scenery and the random walk, we then infer that, for any $n \in \mathbb{N}$ and any $m \in \mathbb{Z}$,

$$\|\mathbb{E}(X_{n+m}|\mathcal{F}_m)\|_2^2 \leq \|\mathbb{E}(\zeta_{S_{n+m}-S_m}|\mathcal{A}, \mathcal{G}_0)\|_2^2 \leq \mathbb{E}\left(b^2(S_n)\right). \tag{11.29}$$

Note now that $\mathcal{F}_0 \subset \mathcal{F}_1$ and thus, by using stationarity, we have, for any $j \geq 1$,

$$b(j) = \|\mathbb{E}(\zeta_j|\mathcal{F}_0)\|_2 \leq \|\mathbb{E}(\zeta_j|\mathcal{F}_1)\|_2 = b(j-1),$$

showing that $(b(j))_{j\geq 0}$ is non-increasing. Therefore, for any $n \geq 1$ and any positive real t, we get

$$\mathbb{E}\left(b^2(S_n)\right) \leq b^2([nt]) + b^2(0)\mathbb{P}(S_n \leq [nt]).$$

Taking into account (11.29), this implies that

$$\sum_{n\geq 1} \frac{1}{\sqrt{n}}\|\mathbb{E}(X_n|\mathcal{F}_0)\|_2 \leq \sum_{n\geq 1} \frac{1}{\sqrt{n}} b([nt]) + b(0) \sum_{n\geq 1} \frac{1}{\sqrt{n}} \sqrt{\mathbb{P}(S_n \leq [nt])}. \tag{11.30}$$

The first sum is finite. Indeed,

$$\begin{aligned}\sum_{n\geq 1} \frac{1}{\sqrt{n}} b([nt]) &\leq \sqrt{t} \sum_{n\geq 1} \frac{1}{\sqrt{[nt]}} b([nt]) = \sqrt{t} \sum_{n\geq 1}\sum_{k\geq 1} \mathbf{1}_{\{nt-1<k\leq nt\}} \frac{b(k)}{\sqrt{k}} \\ &\leq \sqrt{t}(1+1/t) \sum_{k\geq 1} \frac{b(k)}{\sqrt{k}},\end{aligned}$$

which is finite by condition (11.27).

To deal with the second sum in (11.30), we shall use Hoeffding's inequality as stated in Theorem 2.18. Let $\mu = \mathbb{E}(Y)$ and choose $t \in]0, \mu[$. Notice that

$$\mathbb{P}(S_n \leq nt) = \mathbb{P}(S_n - n\mu \leq nt - n\mu) = \mathbb{P}\left(\sum_{i=1}^{n} (\mu - Y_i) \geq n(\mu - t)\right).$$

Since $0 \leq Y_i \leq M$ a.s., Theorem 2.18 gives that for any $t \in]0, \mu[$,

$$\mathbb{P}\left(\sum_{i=1}^{n} (\mu - Y_i) \geq n(\mu - t)\right) \leq \exp\left(-\frac{2n(\mu - t)^2}{M^2}\right).$$

So, overall, starting from (11.30) and taking into account the above considerations, we have proved that, for any $t \in]0, \mu[$,

$$\sum_{n\geq 1} \frac{1}{\sqrt{n}} \|\mathbb{E}(X_n|\mathcal{F}_0)\|_2 \leq \sqrt{t}(1 + 1/t) \sum_{k\geq 1} \frac{b(k)}{\sqrt{k}} + b(0) \sum_{n\geq 1} \frac{1}{\sqrt{n}} \exp\left(-\frac{n(\mu - t)^2}{M^2}\right),$$

which ends the proof of the lemma for the simple case.

Part II. (General random walks.) In this part, we prove Lemma 11.9 in the general case, that is when the random walks may have negative jumps and are not necessarily bounded, although we still require that $\mathbb{E}[Y] > 0$. The situation is more delicate and the quantities we shall first control are $\|\mathbb{E}(X_j|\widetilde{\mathcal{F}}_i)\|_2$ for $j \geq i > 0$ (rather than directly $\|\mathbb{E}(X_j|\mathcal{F}_i)\|_2$) where the following notation is used:

$$\widetilde{\mathcal{F}}_i = \sigma(X_k,\ 0 \leq k \leq i) \text{ for } i \geq 1 \text{ and } \widetilde{\mathcal{F}}_i = \{\emptyset, \Omega\} \text{ for } i \leq 0. \tag{11.31}$$

Note that the filtration above is not stationary. With the notation (11.28), we have, by setting

$$S_m^* := \max_{0\leq i\leq m} S_i,$$

and by taking into account the arguments leading to (11.29), that, for any integers $k \geq 0$ and $m \geq 1$,

$$\begin{aligned}
\|\mathbb{E}(X_{k+m}|\widetilde{\mathcal{F}}_m)\|_2 &= \|\mathbb{E}(\zeta_{S_{k+m}}|\sigma(\zeta_{S_i}, 0 \leq i \leq m))\|_2 \\
&= \|\mathbb{E}(\zeta_{S_{k+m}-S_m^*}|\sigma(\zeta_{S_i-S_m^*}, 0 \leq i \leq m))\|_2 \\
&\leq \|\mathbb{E}(\zeta_{S_{k+m}-S_m^*}|\mathcal{A}, \mathcal{G}_0)\|_2 \leq \|b((S_{k+m} - S_m^*)_+)\|_2.
\end{aligned}$$

Let $S[a, b] = S_b - S_a$. Then

$$\begin{aligned}
S_{k+m} - \max_{0\leq i\leq m} S_i &= S[m, k + m] - \max_{0\leq i\leq m} (S_i - S_m) \\
&= S[m, k + m] - \max_{0\leq i\leq m} (-S[i, m]).
\end{aligned}$$

Now observe that

$$\max_{0\leq i\leq m} (-S[i, m]) = \max_{0\leq i\leq m} (S_i - S_m) =^d \max_{0\leq i\leq m} (-S_i),$$

by i.i.d. increments, time reversal and symmetry. Note now that

$$S[m, k + m] \text{ is independent of } \max_{0\leq i\leq m} (-S[i, m])$$

and then

$$S[m,k+m] - \max_{0\le i\le m}(-S[i,m]) =^d S[m,k+m] + \inf_{0\le i\le m} \tilde{S}_i,$$

where $\tilde{S}_i$ is an independent copy of the random walk.

So, overall we get that, for any positive integers k and m,

$$\|\mathbb{E}(X_{k+m}|\tilde{\mathcal{F}}_m)\|_2 \le \|b((S[m,k+m] + \tilde{I}_m)_+)\|_2,$$

where we introduced the notation

$$\tilde{I}_m := \inf_{0\le i\le m} \tilde{S}_i.$$

As in the Part I of the proof, we split the expectation of $b^2((S[m,k+m] + \tilde{I}_m)_+)$ according to whether $S[m,k+m] + \tilde{I}_m$ is larger or smaller than $[tk]$ for some positive real t. Using the fact that $b(\cdot)$ is decreasing we obtain the bound

$$\|b((S[m,k+m] + \tilde{I}_m)_+)\|_2^2 \le b^2([tk]) + b^2(0)\mathbb{P}(S[m,k+m] + \tilde{I}_m \le tk).$$

Observe that

$$\{X \le 2tk\}^c \cap \{Y < -tk\}^c \subset \{X + Y \le tk\}^c,$$

and thus, for any positive real t,

$$\|b(S[m,k+m] + \tilde{I}_m)\|_2 \le b([tk]) + b(0)\left(\sqrt{\mathbb{P}(S[m,k+m] \le 2tk)} + \sqrt{\mathbb{P}(\tilde{I}_m < -tk)}\right). \quad (11.32)$$

To deal with the second term, we write $\mu = \mathbb{E}(Y)$ and consider a positive real t such that $t < \mu/2$. By using stationarity, we then write

$$\mathbb{P}(S[m,k+m] \le 2tk) = \mathbb{P}(S_k \le 2tk) = \mathbb{P}\left(\sum_{i=1}^{k}(\mu - Y_i) \ge k(\mu - 2t)\right) \le \mathbb{P}\left(\left|\sum_{i=1}^{k}(\mu - Y_i)\right| \ge k(\mu - 2t)\right).$$

Since $\mu_p = \mathbb{E}(|Y|^p) < \infty$ for $p \ge 2$, Markov's inequality together with the Rosenthal inequality as stated in Theorem 2.17 give that for any non-negative integers k, m and any $t \in]0, \mu/2[$,

$$\mathbb{P}(S[m,k+m] \le 2tk) \le c_p(\mu - 2t)^{-p}k^{-p}\left((k\mu_2)^{p/2} + k\mu_p\right),$$

where c_p is a positive universal constant depending only on p.

We deal now with the last term on the right-hand side of (11.32). With this aim, we shall first recall the following result which is due to Kiefer and Wolfowitz (1956, Theorems 5 and 6).

Fact 11.11 *Let $Z_1, Z_2, \ldots$ be i.i.d. real-valued random variables. Set $S_0 = 0$, $S_n = \sum_{i=1}^n Z_i$ for any $n \geq 1$ and $M_\infty = \sup_{n\geq 0} S_n$. Denote $x^+ = \max(0, x)$ for all real x. If $-\infty \leq \mathbb{E}(Z_1) < 0$ and $\mathbb{E}(Z_1^+)^{p+1} < \infty$ for $p > 0$ then $\mathbb{E}(M_\infty)^p < \infty$ and if $-\infty < \mathbb{E}(Z_1) < 0$ and $\mathbb{E}(Z_1^+)^{p+1} = \infty$ for $p > 0$ then $\mathbb{E}(M_\infty)^p = \infty$.*

Let $I_m = \inf_{0\leq i\leq m} S_i$ be the infimum of the random walk up to time m. First of all, note that stochastically $0 \leq -I_m \leq -\inf_{i\geq 0} S_i \equiv -I$. Since $-I = \sup_{n\geq 0} \sum_{k=1}^n -Y_k$ with $\mathbb{E}(-Y_1) < 0$, by Fact 11.11, we derive that if $\mathbb{E}(|Y_1|^p) < \infty$, for some $p \geq 2$, then $\mathbb{E}(|I|^{p-1}) < \infty$ and then by using Markov's inequality,

$$\mathbb{P}(I_m < -tk) \leq \mathbb{P}(-I > tk) \leq C(tk)^{-(p-1)}.$$

So overall, since $p > 2$, we have shown that for any $k, m \geq 1$,

$$\mathbb{E}[(\mathbb{E}(X_{k+m}|X_0, X_1, \ldots, X_m))^2] \leq a_k^2 \text{ where } \sum_{k=1}^{\infty} \frac{a_k}{\sqrt{k}} < \infty.$$

To end the proof, we make the following observation. For any positive integer k, define

$$Z_m = \mathbb{E}(X_k|X_{-m}, \ldots, X_0) \text{ and } \mathcal{Z}_m = \sigma(X_{-m}, \ldots, X_0) \text{ for } m \geq 0.$$

Observe that $(Z_m)_{m\geq 0}$ is a martingale with respect to the non-decreasing sequence $(\mathcal{Z}_m)_{m\geq 0}$ of σ-algebras.

Moreover, $\sup_{m\geq 0} \|Z_m\|_2^2 < \infty$. Hence by the convergence martingale theorem (see for instance Corollary 2.2 in Hall and Heyde (1980)), $\mathbb{E}(X_k|X_{-m}, \ldots, X_0)$ converges, as $m \to \infty$, in $\mathbb{L}^2$ and with probability 1, to $\mathbb{E}(X_k|\mathcal{F}_0)$. Therefore, using in addition stationary,

$$\begin{aligned}\|\mathbb{E}(X_k|\mathcal{F}_0)\|_2 &= \lim_{m\to\infty} \|\mathbb{E}(X_k|X_{-m}, \ldots, X_0)\|_2 \\ &= \lim_{m\to\infty} \|\mathbb{E}(X_{k+m}|X_0, \ldots, X_m)\|_2 \leq a_k,\end{aligned}$$

and thus

$$\sum_{k\geq 1} k^{-1/2} \|\mathbb{E}(X_k|\mathcal{F}_0)\|_2 \leq \sum_{k\geq 1} \frac{a_k}{\sqrt{k}} < \infty.$$

The proof of Lemma 11.9 is complete. ■

Remark 11.12 *Consider a random walk $S_k = Y_1 + \ldots + Y_k$ with i.i.d. increments $\{Y, Y_i\}_i$ with finite second moment $\mathbb{E}(Y^2) < \infty$ and non-zero mean $\mathbb{E}(Y) \neq 0$, so the walk is transient. Notice that the following sums are absolutely convergent and bounded*

$$\sup_x \sum_{k\geq 0} \mathbb{P}(S_k = x) \leq \sum_{k\geq 0} \mathbb{P}(S_k = 0) < \infty.$$

The first inequality follows by conditioning on the first visit, say at time η_x, to the state x and using the strong Markov property, i.e. $S_{k+\eta_x} - S_{\eta_x}$ is a copy of the original random walk. The boundness follows by using the Fourier transform representation

$$\sum_{k\geq 0} \mathbb{P}(S_k = 0) = \frac{1}{2\pi} \int_{-\pi}^{\pi} \frac{1}{1 - \mathbb{E}[e^{itY}]} dt.$$

12

Linear Processes

In this chapter, we apply different methods to establish the Gaussian approximation to linear statistics of a stationary sequence $\{\xi_i\}$, called the innovation sequence. Define

$$X_k = \sum_i a_{ik}\xi_i \quad \text{and associated sums} \quad S_n = \sum_{k=1}^{n} X_k = \sum_i \xi_i \sum_{k=1}^{n} a_{ik}.$$

(i) We first treat stationary linear processes when $a_{ij} = a_{i-j}$.

(ii) We then move to near stationary processes when the sums $b_{in} = \sum_{k=1}^{n} a_{ik}$ do not vary too much with respect to i (more precisely when they satisfy the near linearity property (12.9)).

(iii) Finally, we treat special sums; namely, discrete Fourier transforms that are of the form

$$S_n(\theta) = \sum_{j=1}^{n} X_j e^{ij\theta}.$$

12.1 Linear Processes with Short Memory

In this section we shall discuss when limiting properties of stationary processes are preserved under infinite linear transforms.

Let $(\xi_i)_{i\in\mathbb{Z}}$ be a strictly stationary sequence with $\mathbb{E}(|\xi_0|) < \infty$ and $\mathbb{E}(\xi_0) = 0$ and let $\mathcal{I}$ be its invariant σ–field. Define

$$X_k = \sum_{j=-\infty}^{\infty} a_j \xi_{k-j} \quad \text{and assume} \quad \sum_{i=-\infty}^{\infty} |a_i| < \infty. \tag{12.1}$$

Functional Gaussian Approximation for Dependent Structures. Florence Merlevède, Magda Peligrad and Sergey Utev. Oxford University Press (2019).
DOI: 10.1093/oso/9780198826941.001.0001

The following notation is useful in various parts of the section:

$$S_k^{(\xi)} = \sum_{j=1}^{k} \xi_j.$$

We begin with a representation of S_n in terms of partial sums and illustrate its applications to the CLT and FCLT.

We first notice that by interchanging finite and infinite sum

$$S_n = S_n^{(X)} = \sum_{k=1}^{n} X_k = \sum_{k=1}^{n} \sum_{j=-\infty}^{\infty} a_j \xi_{k-j} = \sum_j a_j \sum_{k=1}^{n} \xi_{k-j} =: \sum_j a_j S_{n,j}^{(\xi)},$$

where

$$S_{n,j}^{(\xi)} = \xi_{1-j} + \ldots + \xi_{n-j} \text{ and } S_{n,0}^{(\xi)} = \xi_1 + \ldots + \xi_n = S_n^{(\xi)}.$$

Since for each finite j, $S_{n,j}^{(\xi)}$ and $S_n^{(\xi)}$ are identically distributed, it suggests the following intuitive approximation

$$S_n^{(X)} \approx \sum_i a_i S_n^{(\xi)} = A S_n^{(\xi)}, \text{ where } \quad A = \sum_i a_i.$$

To make the argument rigorous we shall use the following simple argument:

Lemma 12.1 *Let $\sum_{i\in\mathbb{Z}} |a_i| < \infty$. Assume that for a sequence of positive constants b_n the following conditions hold:*

$$a_{n,i} := \frac{\mathbb{E}|S_{n,i}^{(\xi)} - S_n^{(\xi)}|}{b_n} \to 0 \quad \textit{for each } i, \textit{ and} \quad C := \sup_n \frac{\mathbb{E}|S_n^{(\xi)}|}{b_n} < \infty.$$

Then,

$$\frac{\mathbb{E}|S_n^{(X)} - A S_n^{(\xi)}|}{b_n} \to 0.$$

Remark 12.2 *This lemma shows that if we have a weak limit theorem for the innovations, so if $S_n^{(\xi)}/b_n \to^d L$, then a related result holds for partial sums of linear processes with the same normalization, i.e. $S_n^{(X)}/b_n \to^d AL$.*

Proof Notice that

$$S_n^{(X)} - A S_n^{(\xi)} = \sum_{i\in\mathbb{Z}} a_i (S_{n,i}^{(\xi)} - S_n^{(\xi)})$$

and thus by the triangular inequality

$$\begin{aligned}\mathbb{E}|S_n^{(X)} - AS_n^{(\xi)}| &\leq \sum_{i\in\mathbb{Z}} |a_i| \mathbb{E}|S_{n,i}^{(\xi)} - S_n^{(\xi)}| \\ &\leq \sum_{|i|>M} |a_i| \sup_n \sup_{i\in\mathbb{Z}} \mathbb{E}|S_{n,i}^{(\xi)} - S_n^{(\xi)}| + \sum_{i\in\mathbb{Z}} |a_i| \max_{|i|\leq M} \mathbb{E}|S_{n,i}^{(\xi)} - S_n^{(\xi)}|.\end{aligned}$$

It remains to notice that by stationarity

$$\mathbb{E}|S_n^{(X)} - AS_n^{(\xi)}|/b_n \leq 2C \sum_{|i|>M} |a_i| + \sum_{i\in\mathbb{Z}} |a_i| \max_{|i|\leq M} a_{n,i}.$$

The result follows by letting $n \to \infty$ followed by $M \to \infty$. ■

By using the same approach, our next proposition allows to compare the maximum of partial sums of the innovations to the maximum of partial sums of the linear process with short memory (i.e. $\sum_i |a_i| < \infty$).

Proposition 12.3 *Let $\sum_{i\in\mathbb{Z}} |a_i| < \infty$. Assume the representation (12.1) is satisfied and in addition, there is a constant $C > 0$ and a sequence of positive reals $b_n \to \infty$ such that for all n,*

$$\mathbb{E}\left(\max_{1\leq j\leq n} |S_j^{(\xi)}|\right) \leq Cb_n \tag{12.2}$$

and

$$b_n^{-1} \max_{1\leq j\leq n} |\xi_j| \to^{\mathbb{P}} 0 \quad \text{as } n \to \infty. \tag{12.3}$$

Then,

$$\frac{1}{b_n}(\max_{1\leq j\leq n} |S_j^{(X)} - AS_j^{(\xi)}|) \to^{\mathbb{P}} 0 \quad \text{as } n \to \infty. \tag{12.4}$$

If the innovations are assumed in $\mathbb{L}^p$, $p \geq 1$, (12.2) is replaced by $\mathbb{E}\left(\max_{1\leq j\leq n} |S_j^{(\xi)}|^p\right) \leq Cb_n^p$ and the convergence in (12.3) holds in $\mathbb{L}^p$, then

$$\frac{1}{b_n^p}\mathbb{E}(\max_{1\leq j\leq n} |S_j^{(X)} - AS_j^{(\xi)}|^p) \to 0 \quad \text{as } n \to \infty.$$

Proof The proof is similar. Note that, for any positive integer M less than n,

$$\max_{1\leq m\leq n} |S_m^{(X)} - AS_m^{(\xi)}| \leq \sum_{i\in\mathbb{Z}} |a_i| \left(\max_{1\leq m\leq n} |S_{m,i}^{(\xi)} - S_m^{(\xi)}|\right)$$

$$\leq \sum_{|i|>M} |a_i|(\max_{1\leq m\leq n} |S_{m,i}^{(\xi)} - S_m^{(\xi)}|) + \sum_{i\in\mathbb{Z}} |a_i| \max_{|i|\leq M} \max_{1\leq m\leq n} |S_{m,i}^{(\xi)} - S_m^{(\xi)}|$$

$$\leq \sum_{|i|>M} |a_i|(\max_{1\leq m\leq n} |S_{m,i}^{(\xi)} - S_m^{(\xi)}|) + 2M\sum_{i\in\mathbb{Z}} |a_i|(\max_{-M\leq i\leq n+M} |\xi_i|)$$

$$:= I + II.$$

For M fixed the second term II divided by b_n converges to 0 in probability as $n \to \infty$ by condition (12.3).

Next, by Markov inequality and stationarity, for any $\varepsilon > 0$,

$$\mathbb{P}(I > \varepsilon b_n) \leq 2(\varepsilon b_n)^{-1} \sum_{|i|>M} |a_i| \mathbb{E}(\max_{1\leq m\leq n} |S_m^{(\xi)}|) \leq C\varepsilon^{-1} \sum_{|i|>M} |a_i|,$$

which converges to 0 when $M \to \infty$. So the first part of the proposition follows by letting first $n \to \infty$ followed by $M \to \infty$. The proof of the second part of the proposition is done by using similar arguments. ■

Theorem 12.4 *Let $\sum_{i\in\mathbb{Z}} |a_i| < \infty$. Assume that representation (12.1) and condition (12.2) are satisfied. Moreover, assume that the innovations satisfy the invariance principle $\{b_n^{-1}S_{[nt]}^{(\xi)}, t \in [0,1]\} \to^d \eta W$ in $D([0,1])$ as $n \to \infty$, where η is $\mathcal{I}$-measurable and W is a standard Brownian motion on $[0,1]$ independent on $\mathcal{I}$. Then the linear process also satisfies the invariance principle, i.e. $\{b_n^{-1}S_{[nt]}^{(X)}, t \in [0,1]\} \to^d \eta AW$ in $D([0,1])$ as $n \to \infty$.*

Proof Notice that the convergence in probability in (12.3) follows from the invariance principle $\{b_n^{-1}S_{[nt]}^{(\xi)}, t \in [0,1]\} \to^d \eta W$ in $D([0,1])$ as $n \to \infty$, since the modulus of continuity is convergent to 0 in probability. All the conditions in Proposition 12.3 are then satisfied which imply the conclusion of the theorem. ■

From Theorem 12.4 we easily derive the following useful consequence.

Corollary 12.5 *($\mathbb{L}^p$–invariance principle.) Let $\sum_{i\in\mathbb{Z}} |a_i| < \infty$. Assume the representation (12.1) holds and $p \geq 1$. Then,*

$$\lim_{n\to\infty} \mathbb{E}(\sup_{0\leq t\leq 1} |b_n^{-1}S_{[nt]}^{(\xi)} - \eta W(t)|^p) = 0 \Rightarrow \quad \lim_{n\to\infty} \mathbb{E}(\sup_{0\leq t\leq 1} |b_n^{-1}S_{[nt]} - \eta AW(t)|^p) = 0.$$

Discussion

Theorem 12.4 and its Corollary 12.5 work for many dependent structures such as surveyed in Peligrad (1986), Doukhan (1994), Bradley (2007) and Merlevède, Peligrad and Utev (2006). Various invariance principles can be extended from the original sequence to the linear process with short memory. Here we mention some traditional and also some recently developed dependence conditions for innovations whose partial sums satisfy both a maximal inequality and the invariance principle and therefore Theorem 12.4 and its Corollary 12.5 apply.

For instance, let us assume that $(\xi_i)_{i\in\mathbb{Z}}$ is a stationary ergodic sequence with $\mathbb{E}(\xi_0^2) < \infty$ and $\mathbb{E}(\xi_0) = 0$ and let $\mathcal{F}_k = \sigma(\xi_i, i \le k)$. For all the structures below the family $(\max_{1\le k\le n}(S_k^{(\xi)})^2/n)_{n\ge 1}$ is uniformly integrable and the conclusion of Corollary 12.5 holds with $b_n = \sqrt{n}$ and with $p = 2$. Moreover, since $(\xi_i)_{i\in\mathbb{Z}}$ is assumed to be ergodic then there is a non-negative constant σ such that $\eta = \sigma$.

(i) Hannan (1979) (see also its extension to Hilbert space in Dedecker and Merlevède (2003)):

$$\sum_{n=1}^{\infty} \|\mathbf{P}_0(\xi_n)\|_2 < \infty \text{ and } \mathbb{E}(\xi_0|\mathcal{F}_{-\infty}) = 0 \; a.s.$$

where $\mathbf{P}_k(X) = \mathbb{E}(X|\mathcal{F}_k) - \mathbb{E}(X|\mathcal{F}_{k-1})$ is the projection operator (see Theorem 4.17).

(ii) Newman and Wright (1981): $(\xi_i)_{i\in\mathbb{Z}}$ is an associated sequence (i.e. such that any two coordinatewise non-decreasing functions of any finite subcollection of the ξ_i's (of finite variance) are non-negatively correlated) which satisfies in addition

$$\sum_{k=1}^{\infty} \operatorname{cov}(\xi_k, \xi_0) < \infty.$$

(iii) Doukhan, Massart and Rio (1994):

$$\sum_{k=0}^{\infty}(k+1)\int_0^{\alpha(k)} Q^2(u)du < \infty,$$

where Q denotes the càdlàg inverse of the function $t \to \mathbb{P}(|\xi_0| > t)$ and $(\alpha(k))_{k\ge 0}$ is the sequence of strong mixing coefficients associated with $(\xi_i)_{i\in\mathbb{Z}}$ (see Theorem 6.39).

(iv) Dedecker and Rio (2000):

$$\mathbb{E}(\xi_0 S_n^{(\xi)}|\mathcal{F}_0) \text{ converges in } \mathbb{L}^1.$$

(See Theorem 4.18).

(v) Peligrad and Utev (2005), by developing Maxwell and Woodroofe (2000),

$$\sum_{n=1}^{\infty} \frac{\|\mathbb{E}(S_n^{(\xi)}|\mathcal{F}_0)\|_2}{n^{3/2}} < \infty.$$

(See Theorem 4.16).

(vi) Peligrad, Utev and Wu (2007), which guarantees the $\mathbb{L}^p$-invariance principle, $p \ge 2$.

$$\sum_{n=1}^{\infty} \frac{\|\mathbb{E}(S_n^{(\xi)}|\mathcal{F}_0)\|_p}{n^{3/2}} < \infty.$$

Comments

(a) If $\mathbb{E}(\xi_0^2) < \infty$ and $b_n \geq \sqrt{n}$ then condition (12.3) automatically holds.

(b) The set of indexes $\mathbb{Z}$ can be replaced by $\mathbb{Z}^d$ where d is a positive integer allowing for the treatment of random fields.

(c) A natural extension is to consider innovations with values in functional spaces that also facilitate the study of estimation and forecasting problems for several classes of continuous time processes (see Bosq (2000)). The linear processes are still defined by the formula (12.1) with the difference that now the innovations $(\xi_k)_{k\in\mathbb{Z}}$ take values in a separable Hilbert space H and the sequence of constants is replaced by the sequence of bounded linear operators $\{a_k\}_{k\in\mathbb{Z}}$ from H to H. Merlevède, Peligrad and Utev (1997) treated the problem of the central limit theorem for this case under the summability condition

$$\sum_{j=-\infty}^{\infty} \|a_j\|_{L(H)} < \infty,$$

where $\|a_j\|_{L(H)}$ denotes the usual operators norm. It was discovered that, if this condition is not satisfied, then the central limit theorem fails even for the case of independent innovations. The approach developed in that paper shows that the central limit theorem results stated can be strengthened to the invariance principle (some results in this direction for strongly mixing sequences are established in Merlevède (2003)).

(d) In all the examples (i)–(iv) the variance of partial sums is linear in n.

12.2 Functional CLT using Coboundary Decomposition

Here we shall establish a functional CLT by using the coboundary decomposition.

Lemma 12.6 *Assume that the sequence of innovations $(\xi_j)_{j\in\mathbb{Z}}$ is strictly stationary, centered with finite second moment and has a bounded spectral density. Assume in addition that the sequence of constants $(a_k)_{k\geq 0}$ satisfies the following two conditions*

$$A = \sum_{k=0}^{\infty} a_k \text{ is convergent and } \sum_{j\geq 0}\left(\sum_{k=j+1}^{\infty} a_k\right)^2 < \infty.$$

For any $k \in \mathbb{Z}$, define the causal linear process

$$X_k = \sum_{j\geq 0} a_j \xi_{k-j}.$$

Then there is a stationary sequence $(\nu_k)_k$ of square integrable real-valued random variables such that, for any $\ell \in \mathbb{Z}$, the following coboundary decomposition holds:

$$X_\ell = A\xi_\ell + \nu_{\ell-1} - \nu_\ell. \tag{12.5}$$

Proof Let us introduce the stationary sequence

$$\nu_\ell = \sum_{j\geq 0}\left(\sum_{k=j+1}^{\infty} a_k\right)\xi_{\ell-j}, \quad \ell \in \mathbb{Z}.$$

This sequence is well defined in $\mathbb{L}^2$ under the conditions of this theorem. With this definition of ν_ℓ, it is immediate that (12.5) is satisfied. ■

As an immediate consequence we obtain:

Theorem 12.7 *Assume that the sequence of innovations and of constants satisfy the conditions of Lemma 12.6. Moreover assume that the innovations satisfy the invariance principle $\{b_n^{-1}S_{[nt]}^{(\xi)}, t \in [0,1]\} \to^d \eta W$ in $D([0,1])$ as $n \to \infty$, where η is $\mathcal{I}$-measurable and W is a standard Brownian motion on $[0,1]$ independent on $\mathcal{I}$. Suppose also that*

$$\frac{\max_{1\leq \ell\leq n}|\nu_\ell|}{b_n} \to^{\mathbb{P}} 0.$$

Then the linear process also satisfies the invariance principle, i.e. $\{b_n^{-1}S_{[nt]}^{(X)}, t \in [0,1]\} \to^d \eta AW$ in $D([0,1])$ as $n \to \infty$.

Proof The functional CLT follows from the corresponding one for the innovations by noticing that, for any $t \in [0,1]$,

$$S_{[nt]}^{(X)} = AS_{[nt]}^{(\xi)} + \nu_0 - \nu_{[nt]}. \quad ■$$

12.3 Toward Linear Processes with Long Memory

We consider in this section linear processes $X_j = \sum_{k\in\mathbb{Z}} a_{j-k}\xi_k$. In many situations the process $\{X_j, j \in \mathbb{Z}\}$ is well defined under weaker conditions that (12.1). For instance, when the innovations form an i.i.d. sequence of centered random variables which are also square integrable the necessary and sufficient condition for the existence of X_0 in the almost sure sense is

$$\sum_{i\in\mathbb{Z}} a_i^2 < \infty \tag{12.6}$$

(implied by $\sum |a_i| < \infty$). So, whenever X_0 is well defined we can represent the partial sums as a linear statistic

$$S_n = \sum_k \xi_k \sum_{i=1}^{n} a_{i-k} := \sum_{k\in\mathbb{Z}} \xi_k b_{k,n},$$

where $b_{k,n} := a_{1-k} + ... + a_{n-k}$.

It should be noted that in this case the variance of partial sums might not be asymptotically linear in n. As a matter of fact, if the innovations are i.i.d. centered with finite second moment the variance of S_n can range from a constant to close of n^2. When the variance is not asymptotically linear in n we shall refer to that case as long memory.

We shall obtain first the central limit theorem for the linear statistic for various dependence structures that will allow us to obtain the CLT for the classes of linear processes satisfying condition (12.6). The invariance principle is much more delicate in the long memory case. The discussion of the invariance principle will follow the section on the central limit theorem.

12.3.1 *CLT for Linear Statistics with Dependent Innovations via Martingale Approximation*

Let $(\xi_i)_{i\in\mathbb{Z}}$ be a strictly stationary sequence of centered real-valued random variables with finite second moment. The aim of this section is to study the asymptotic behavior of linear statistics of the type

$$S_n = \sum_{k\in\mathbb{Z}} b_{k,n}\xi_k, \tag{12.7}$$

where $\{b_{k,n}, k \in \mathbb{Z}\}$ is a triangular array of numerical constants satisfying $\sum_{k\in\mathbb{Z}} b_{k,n}^2 < \infty$ for any n. Note that the linear statistic S_n is properly defined if and only if the stationary sequence of innovations $(\xi_i)_{i\in\mathbb{Z}}$ has a bounded spectral density.

The main result of this section is the following:

Theorem 12.8 *Let $(\xi_i)_{i\in\mathbb{Z}}$ be a strictly stationary and ergodic sequence of real-valued random variables with finite second moment, centered at expectations, adapted to a stationary filtration $(\mathcal{F}_i)_{i\in\mathbb{Z}}$ and satisfying*

$$\Gamma_j = \sum_{k=0}^{\infty} |\mathbb{E}[\xi_k \mathbb{E}(\xi_0|\mathcal{F}_{-j})]| < \infty \quad \textit{for all } j \in \mathbb{Z} \textit{ and } \frac{1}{p}\sum_{j=1}^{p}\Gamma_j \to 0 \quad \textit{as } p \to \infty. \tag{12.8}$$

Let $\{b_{k,n}, k \in \mathbb{Z}\}$ be a triangular array of numerical constants such that

$$b_n^2 = \sum_{k\in\mathbb{Z}} b_{k,n}^2 < \infty \textit{ and } \lim_{n\to\infty} \frac{1}{b_n^2}\sum_{k\in\mathbb{Z}}(b_{k,n} - b_{k-1,n})^2 = 0. \tag{12.9}$$

Let S_n be defined by (12.7). Then, $(\xi_i)_{i\in\mathbb{Z}}$ has a continuous spectral density f,

$$\lim_{n\to\infty} \frac{\mathrm{Var}(S_n)}{b_n^2} = 2\pi f(0) \text{ and } \frac{S_n}{b_n} \to^d \sqrt{2\pi f(0)}N \text{ as } n \to \infty, \tag{12.10}$$

where N is a standard normal variable.

Comment 12.9 *This result is due to Peligrad and Utev (2006). We note that if we do not assume the sequence $(\xi_i)_{i\in\mathbb{Z}}$ to be ergodic then it can be proven that there is a non-negative random variable η^2 measurable with respect to $\mathcal{I}$ (the invariant σ-field) such that $n^{-1}\mathbb{E}((\sum_{k=1}^n \xi_k)^2|\mathcal{F}_0) \to \eta^2$ in $\mathbb{L}^1$ as $n \to \infty$ and $\mathbb{E}(\eta^2) = 2\pi f(0)$. In this situation, the second part of (12.10) has to be modified as follows: $b_n^{-1}S_n \to^d \eta N$ as $n \to \infty$, where N is a standard normal variable independent of η. For a complete proof of this comment, we refer the reader to the proof of Theorem 1 in Peligrad and Utev (2006).*

As a corollary of Theorem 12.8, we can derive the following central limit theorem for the partial sums associated with a linear process $\{X_k, k \in \mathbb{Z}\}$ defined by

$$X_k = \sum_{j\in\mathbb{Z}} a_{k-j}\xi_j,\ k \in \mathbb{Z}, \tag{12.11}$$

when $\sum_{j\in\mathbb{Z}} a_j^2 < \infty$ and $(\xi_i)_{i\in\mathbb{Z}}$ satisfies the conditions of Theorem 12.8.

Corollary 12.10 *Let $(\xi_i)_{i\in\mathbb{Z}}$ be as in Theorem 12.8 and $(a_k)_{k\in\mathbb{Z}}$ be a sequence of real numbers such that $\sum_{j\in\mathbb{Z}} a_j^2 < \infty$. Let*

$$b_n^2 = \sum_{k\in\mathbb{Z}} b_{k,n}^2 := \sum_{k\in\mathbb{Z}} (a_{1-k} + \dots + a_{n-k})^2. \tag{12.12}$$

and assume that $b_n^2 \to \infty$ as $n \to \infty$. Let $\{X_k\}_{k\in\mathbb{Z}}$ be defined by (12.11). Then

$$b_n^{-1}\sum_{k=1}^n X_k \to^d \sqrt{2\pi f(0)}N \text{ as } n \to \infty,$$

where N is a standard normal variable and f is the spectral density of $(\xi_i)_{i\in\mathbb{Z}}$.

Indeed, recall the representation

$$S_n = \sum_{k\in\mathbb{Z}} \xi_k \sum_{i=1}^n a_{i-k} =: \sum_{k\in\mathbb{Z}} \xi_k b_{k,n},$$

where $b_{k,n} := a_{1-k} + \dots + a_{n-k}$. Observe that, by the Cauchy inequality,

$$b_{k,n}^2 = (a_{1-k} + \dots + a_{n-k})^2 \le n(a_{1-k}^2 + \dots + a_{n-k}^2)$$

and so the series

$$\sum_{k\in\mathbb{Z}} b_{k,n}^2 \le n \sum_{i\in\mathbb{Z}} (a_{1-i}^2 + \ldots + a_{n-i}^2) = n^2 \sum_{i\in\mathbb{Z}} a_i^2,$$

which is certainly finite if $\sum_{i\in\mathbb{Z}} a_i^2 < \infty$. To end the proof of the corollary, observe that

$$\sum_{j\in\mathbb{Z}} (b_{j,n} - b_{j-1,n})^2 \le 4 \sum_{k\in\mathbb{Z}} a_k^2.$$ ■

Comment 12.11 *Condition (12.8) is satisfied under various different dependent conditions. We just mention in what follows that it holds if $(\xi_i)_{i\in\mathbb{Z}}$ satisfies Hannan's condition (4.29), namely:*

$$\mathbb{E}(\xi_0|\mathcal{F}_{-\infty}) = 0 \text{ a.s. and } \sum_{i\geq 0} \|\mathbf{P}_0(\xi_i)\|_2 < \infty,$$

where we recall that $\mathbf{P}_0(\cdot) = \mathbb{E}(\cdot|\mathcal{F}_0) - \mathbb{E}(\cdot|\mathcal{F}_{-1})$. Indeed, since $\mathbb{E}(\xi_0|\mathcal{F}_{-\infty}) = 0$ a.s., we have the following representation:

$$\xi_k = \sum_{i=-\infty}^{k} \mathbf{P}_i(\xi_k) \quad \text{and} \quad \mathbb{E}(\xi_0|\mathcal{F}_{-j}) = \sum_{i=-\infty}^{-j} \mathbf{P}_i(\xi_0).$$

By stationarity $\|\mathbf{P}_{-n}(\xi_0)\|_2 = \|\mathbf{P}_{-n+k}(\xi_k)\|_2$ for any $k \in \mathbb{Z}$. Next, $\mathbf{P}_i(\xi_0)$ and $\mathbf{P}_j(\xi_k)$ are uncorrelated for $i \neq j$, implying that

$$\mathbb{E}[\xi_k \mathbb{E}(\xi_0|\mathcal{F}_{-j})] = \sum_{i=-\infty}^{\min(k,-j)} \mathbb{E}\,(\mathbf{P}_i(\xi_k)\mathbf{P}_i(\xi_0)).$$

As a consequence

$$|\mathbb{E}[\xi_k \mathbb{E}(\xi_0|\mathcal{F}_{-j})]| \le \sum_{i=-\infty}^{-j} \|\mathbf{P}_i(\xi_k)\|_2 \|\mathbf{P}_i(\xi_0)\|_2 = \sum_{i=-\infty}^{-j} \|\mathbf{P}_{i-k}(\xi_0)\|_2 \|\mathbf{P}_i(\xi_0)\|_2.$$

Therefore,

$$\begin{aligned}\Gamma_j &= \sum_{k=0}^{\infty} |\mathbb{E}[\xi_k \mathbb{E}(\xi_0|\mathcal{F}_{-j})]| \le \sum_{i=-\infty}^{-j} \|\mathbf{P}_i(\xi_0)\|_2 \sum_{k\geq 0} \|\mathbf{P}_{-k}(\xi_0)\|_2 \\ &= \sum_{i\geq j} \|\mathbf{P}_0(\xi_i)\|_2 \sum_{k\geq 0} \|\mathbf{P}_0(\xi_k)\|_2.\end{aligned}$$

Whence, under Hannan's condition, we derive that $\lim_{j\to\infty} \Gamma_j = 0$, proving the validity of condition (12.8). We refer to the paper by Peligrad and Utev (2006) for other dependence conditions implying condition (12.8).

Proof of Theorem 12.8 Since by assumption, $\sum_{k=0}^{\infty} |\mathbb{E}[\xi_k \xi_0]| = \Gamma_0 < \infty$, by Lemma 1.5, $(\xi_i)_{i\in\mathbb{Z}}$ has a continuous spectral density and the first part of (12.10) holds. The proof of the second part of (12.10) will be divided in two steps. In step 1, we prove that, under the additional assumption that $(\xi_i)_{i\in\mathbb{Z}}$ forms a sequence of martingale differences then the conclusion of Theorem 12.8 is satisfied. In the second step, and with the help of step 1, we use a martingale approximation to show that Theorem 12.8 holds in its full generality under condition (12.8).

Step 1. In this step, we shall assume that $(\xi_i)_{i\in\mathbb{Z}}$ is a strictly stationary and ergodic sequence of martingale differences with finite second moment, and prove that, in this case $S_n \to^d \mathcal{N}(0, \mathbb{E}(\xi_0^2))$.

We shall apply Corollary 2.32 to the triangular array of martingale differences $(d_{n,j})_{j\in\mathbb{Z},n\geq 1}$ where, for any $j \in \mathbb{Z}$ and any $n \geq 1$,

$$d_{n,j} = b_n^{-1}(b_{j,n}\xi_j).$$

Note first that, for any positive integer n, $b_n^{-2}\sum_{j\in\mathbb{Z}} \mathbb{E}(b_{j,n}\xi_j)^2 = \mathbb{E}(\xi_0^2)$. The conclusion of the theorem will then follow from Corollary 2.32 if one can prove that the triangular array $\{d_{n,j}, j \in \mathbb{Z}\}_{n\geq 1}$ satisfies the conditions (2.44) and (2.45).

To verify (2.44), that is: $b_n^{-1}\mathbb{E}(\sup_{k\in\mathbb{Z}} |b_{k,n}\xi_k|) \to 0$ as $n \to \infty$, it is enough to check the Lindeberg condition: For any $\varepsilon > 0$,

$$b_n^{-2}\sum_{i\in\mathbb{Z}} \mathbb{E}[b_{i,n}^2 \xi_i^2 \mathbf{1}_{\{b_n^{-1}|b_{i,n}\xi_i|>\varepsilon\}}] \to 0, \text{ as } n \to \infty. \tag{12.13}$$

But, by stationarity, for any $\varepsilon > 0$,

$$\begin{aligned} b_n^{-2}\sum_{i\in\mathbb{Z}} \mathbb{E}[b_{i,n}^2 \xi_i^2 1_{\{b_n^{-1}|b_{i,n}\xi_i|>\varepsilon\}}] &\leq \max_{i\in\mathbb{Z}} \mathbb{E}\left(\xi_1^2 1_{\{b_n^{-1}|b_{i,n}\xi_1|>\varepsilon\}}\right) \\ &\leq \mathbb{E}\left(\xi_1^2 1_{\{b_n^{-1}\max_{i\in\mathbb{Z}}|b_{i,n}||\xi_1|>\varepsilon\}}\right). \end{aligned}$$

We then conclude that (12.13) will hold if $b_n^{-1}\max_{j\in\mathbb{Z}} |b_{j,n}| \to 0$. This latter condition holds under condition (12.9) as stated in the following lemma about sequences, whose proof will be given at the end of the proof of the theorem.

Lemma 12.12 *Under* (12.9), $\lim_{n\to\infty} b_n^{-1}\max_{k\in\mathbb{Z}} |b_{k,n}| = 0$.

It remains to show that the triangular array $\{d_{n,j}, j \in \mathbb{Z}\}_{n\geq 1}$ satisfies condition (2.45). So we shall show in what follows that the following convergence holds:

$$J_n := \frac{1}{b_n^2}\sum_{j=-\infty}^{\infty} b_{j,n}^2 \xi_j^2 \to^{\mathbb{P}} \mathbb{E}(\xi_0^2) \text{ as } n \to \infty. \tag{12.14}$$

The approach to this proof is to make blocks of variables and replace each variable in a block by the average of variables forming the block. The following lemma, with the Hilbert space type language, is convenient.

Lemma 12.13 *Let ℓ_2 be a Hilbert space of double sequences $x = \{x_j\}_{j\in\mathbb{Z}}$ with the norm $\|x\|_2^2 = \sum_j |x_j|^2$. Set also $\|x\|_1 = \sum_j |x_j|$. Let the translation operator be denoted by $Tx(j) = x_{j+1}$. Given a fixed positive integer p, by I_k denote the set of integers $I_k = \{(p(k-1)+1, ..., kp\}$ and associate the sequence $\{(A_p x)_j\}_{j\in\mathbb{Z}}$ based on the average of the terms in a block:*

$$(A_p x)_j = \frac{1}{p}\sum_{i\in I_k} x_i, \quad for\ j \in I_k.$$

In addition, define $(x^2)_k = (x_k^2)$. Then, for any positive integers j and p,

$$\|x - T^j x\|_2 \le j\|x - Tx\|_2, \ \|x - A_p x\|_2 \le p\|x - Tx\|_2$$

and

$$\|x^2 - A_p x^2\|_1 \le 2p\|x - Tx\|_1, \ \|x^2 - y^2\|_1 \le \|x - y\|_2^2 + 2\|x - y\|_2\|x\|_2.$$

In particular, assume that we have a sequence of elements $x_{(n)} \in \ell_2$ such that $\|x_{(n)}\|_2 = 1$ and $\|x_{(n)} - Tx_{(n)}\|_2 \to 0$ as $n \to \infty$. Then, for any positive integers j and p,

$$\|x_{(n)}^2 - T^j x_{(n)}^2\|_1 \to 0, \ \|x_{(n)}^2 - (A_p x_{(n)})^2\|_1 \to 0 \ and \ \|x_{(n)}^2 - A_p(x_{(n)}^2)\|_1 \to 0,$$

as $n \to \infty$.

Proof The proof requires easy algebraic manipulations and is left to the reader. ■

With the help of this lemma, let us prove (12.14). Fix a positive integer p and make small blocks of normalized sums of consecutive random variables. Define

$$(x_{(n)})_i = x_{i,n} = b_{i,n}/b_n, \ t_{k,n} = (A_p(x_{(n)}^2))_j = p^{-1}\sum_{i\in I_k} x_{i,n}^2 \text{ for } j \in I_k,$$

and decompose the sum in (12.14) in the following way

$$J_n = \sum_{j=-\infty}^{\infty} x_{j,n}^2\xi_j^2 = \sum_{k=-\infty}^{\infty} pt_{k,n}\left(\frac{1}{p}\sum_{j=p(k-1)+1}^{pk}\xi_j^2\right) + \sum_{k=-\infty}^{\infty}\sum_{j=p(k-1)+1}^{pk}[x_{j,n}^2 - t_{k,n}]\xi_j^2$$

$$:= J_{n,1} + J_{n,2}.$$

Notice first that $\Sigma_k pt_{k,n} = b_n^2$ and as a consequence, by stationarity and the $\mathbb{L}_1$ ergodic theorem (see Theorem 1.3), the following convergence holds uniformly in n

$$\mathbb{E}|J_{n,1} - \mathbb{E}(\xi_0^2)| \le \mathbb{E}\left|\left(\frac{1}{p}\sum_{j=p(k-1)+1}^{pk}\xi_j^2\right) - \mathbb{E}(\xi_0^2)\right| \to 0 \text{ as } p \to \infty.$$

On the other hand,

$$\begin{aligned}
\mathbb{E}|J_{n,2}| &\le \mathbb{E}[\xi_0^2] \sum_{k=-\infty}^{\infty} \sum_{j=p(k-1)+1}^{pk} |x_{j,n}^2 - t_{k,n}| \\
&= \mathbb{E}[\xi_0^2]\, ||x_{(n)}^2 - A_p(x_{(n)}^2)||_1 \to 0 \text{ as } n \to \infty,
\end{aligned}$$

by Lemma 12.13. This ends the proof of (12.14) and then of Step 1.

Step 2. We show now that Theorem 12.8 holds in its full generality. So, we assume from now on that $(\xi_i)_{i\in\mathbb{Z}}$ is a strictly stationary and ergodic sequence of centered real-valued random variables with finite second moment that satisfies condition (12.8).

This step is based on a blocking procedure and then on an approximation of the sums of variables in blocks by martingale differences. As before, let p be a fixed positive integer and denote by $I_k = \{(k-1)p+1, ..., kp\}$. So I_k's are blocks of consecutive integers of size p and $\mathbb{Z} = \cup_{k=-\infty}^{\infty} I_k$. Let

$$x_{j,n} = \frac{b_{j,n}}{b_n},\ c_{k,n} = \frac{1}{p}\sum_{i\in I_k} x_{i,n}.$$

We start with the following decomposition

$$\begin{aligned}
W_n &= \sum_{j=-\infty}^{\infty} x_{j,n}\xi_j = \sum_{k=-\infty}^{\infty} \sum_{j\in I_k} x_{j,n}\xi_j \\
&= \sum_{k=-\infty}^{\infty} \sqrt{p}c_{k,n}\left(\frac{1}{\sqrt{p}}\sum_{j\in I_k}\xi_j\right) + \sum_{k=-\infty}^{\infty}\sum_{j\in I_k}[x_{j,n} - c_{k,n}]\xi_j \ := B_{n,1} + B_{n,2}.
\end{aligned}$$

We shall show first that $B_{n,2}$ is negligible for the convergence in distribution. As noticed at the beginning of the proof, $(\xi_i)_{i\in\mathbb{Z}}$ has a continuous spectral density and by the second inequality in part (i) of Lemma 1.5, the variance of $B_{n,2}$ is bounded by

$$\mathbb{E}(B_{n,2}^2) \le \left(\mathbb{E}[\xi_0^2] + 2\sum_{k=1}^{\infty}|\mathbb{E}(\xi_0\xi_k)|\right) \sum_{k=-\infty}^{\infty}\sum_{j\in I_k}(x_{j,n} - c_{k,n})^2.$$

On the other hand, notice that, by Lemma 12.13,

$$\sum_{k=-\infty}^{\infty}\sum_{j\in I_k}(x_{j,n} - c_{k,n})^2 = ||x_{(n)} - A_p x_{(n)}||_2^2 \to 0 \text{ as } n \to \infty,$$

which proves that

$$\mathbb{E}(B_{n,2}^2) \to 0 \quad \text{as } n \to \infty.$$

To analyze $B_{n,1}$ we denote the weighted sum in a block of size p by

$$Y_k^{(p)} = \frac{1}{\sqrt{p}} \sum_{j \in I_k} \xi_j, \ k \in \mathbb{Z} \quad \text{and} \quad \mathcal{G}_k = \mathcal{F}_{kp}.$$

Then, $Y_k^{(p)}$ is $\mathcal{G}_k$-measurable and define

$$Z_k^{(p)} = \mathbb{E}(Y_k^{(p)} | \mathcal{G}_{k-1}) \text{ and } V_k^{(p)} = Y_k^{(p)} - Z_k^{(p)}.$$

Obviously $V_k^{(p)}$ is a stationary sequence of martingale differences and $Y_k^{(p)} = Z_k^{(p)} + V_k^{(p)}$. It follows that $B_{n,1}$ can be decomposed into a linear process with stationary martingale differences innovations and another one involving $Z_k^{(p)}$.

We shall show first that the term involving $Z_k^{(p)}$ is negligible for the convergence in distribution in the sense that

$$\lim_{p\to\infty} \limsup_{n\to\infty} \left\| \sum_{k=-\infty}^{\infty} \sqrt{p} c_{k,n} (Z_k^{(p)}) \right\|_2^2 = 0. \tag{12.15}$$

Let $\delta_{n,k} = \sqrt{p} c_{k,n}$. Observe that $\delta_n^2 := \sum_{k\in\mathbb{Z}} \delta_{n,k}^2 \leq 1$. In addition,

$$\sum_{k\in\mathbb{Z}} \delta_{n,k}^2 = \sum_{k=-\infty}^{\infty} p c_{k,n}^2 = 1 + \sum_{k=-\infty}^{\infty} \sum_{j\in I_k} (c_{k,n}^2 - x_{j,n}^2),$$

and, by Lemma 12.13 and the construction,

$$\sum_{k=-\infty}^{\infty} \sum_{j\in I_k} |x_{j,n}^2 - c_{k,n}^2| = \|x_{(n)}^2 - (A_p x_{(n)})^2\|_1 \to 0 \text{ as } n \to \infty.$$

Hence,

$$\sum_{k\in\mathbb{Z}} \delta_{n,k}^2 \to 1 \text{ as } n \to \infty.$$

Moreover,

$$\sum_{k\in\mathbb{Z}} |\delta_{n,k} - \delta_{n,k-1}|^2 \leq \frac{p^4}{b_n^2} \sum_{j\in\mathbb{Z}} |b_{j,n} - b_{j-1,n}|^2.$$

Hence, since (12.9) is assumed, the sequence $\{\delta_{n,k}\}$ satisfies condition (1.11). Therefore, according to Lemma 1.5, part (iii), we deduce that

$$\lim_{n\to\infty} \left\| \sum_{k=-\infty}^{\infty} \sqrt{p} c_{k,n} (Z_k^{(p)}) \right\|_2^2 = 2\pi f_{(p)}(0),$$

where $f_{(p)}(x)$ denotes the spectral density of the process $\{Z_k^{(p)}\}_{k\in\mathbb{Z}}$. On the other hand, since

$$f_{(p)}(0) = \lim_{n\to\infty} \frac{1}{n} \left\| \sum_{k=1}^{n} (Z_k^{(p)}) \right\|_2^2 ,$$

in order to establish (12.15) it is enough to show that

$$\lim_{p\to\infty} \sum_{k=1}^{\infty} |\mathbb{E}(Z_1^{(p)} Z_k^{(p)})| = 0.$$

First, we observe that, for any $k \geq 1$,

$$|\mathbb{E}(Z_1^{(p)} Z_k^{(p)})| = \frac{1}{p} \left| \mathbb{E}\left[\mathbb{E}\left(\sum_{i=1}^{p} \xi_i \,|\mathcal{F}_0 \right) \sum_{j=(k-1)p+1}^{kp} \xi_j \right] \right| .$$

By the triangle inequality and Condition (12.8) obviously

$$\sum_{k=1}^{\infty} |\mathbb{E}(Z_1^{(p)} Z_k^{(p)})| \leq 2\frac{1}{p} \sum_{i=1}^{p} \sum_{n=i}^{\infty} |\mathbb{E}[\mathbb{E}(\xi_i|\mathcal{F}_0)\xi_n]| \leq 2\frac{1}{p} \sum_{i=1}^{p} \Gamma_i \to 0 \quad \text{as } p \to \infty.$$

To complete the prove we have to show that the remaining linear process involving the martingale differences satisfies the desired CLT. We shall denote by

$$X_k^{(p)} = \sum_{j=-\infty}^{\infty} \sqrt{p} c_{k,n} V_j^{(p)} \text{ and } S_n^{(p)} = \sum_{k=1}^{n} X_k^{(p)}.$$

Since the coefficients $\delta_{n,k} = \sqrt{p} c_{k,n}$ satisfy (1.11), by step 1, it follows that for any p fixed,

$$S_n^{(p)} \to \sqrt{\mathbb{E}([V_1^{(p)}]^2)} N \text{ as } n \to \infty.$$

In order to complete the proof, by Theorem 1.10 we have only to establish that

$$\mathbb{E}([V_1^{(p)}]^2) \to 2\pi f(0) \quad \text{as} \quad p \to \infty.$$

With this aim, we notice that, by stationarity,

$$\begin{aligned}
\mathbb{E}([V_1^{(p)}]^2) &= p^{-1}\mathbb{E}\left(\sum_{i=1}^{p} (\xi_i - \mathbb{E}(\xi_i|\mathcal{F}_0)) \right)^2 \\
&= \mathbb{E}(\xi_0^2) + 2p^{-1} \left(\sum_{i=1}^{p-1} (p-i)\mathbb{E}(\xi_0\xi_i) \right) - p^{-1} \sum_{i,j=1}^{p} \mathbb{E}\left(\xi_i \mathbb{E}(\xi_j|\mathcal{F}_0) \right).
\end{aligned}$$

Since $\Gamma_0 < \infty$,

$$\lim_{p\to\infty}\left\{\mathbb{E}(\xi_0^2) + 2p^{-1}\left(\sum_{i=1}^{p-1}(p-i)\mathbb{E}(\xi_0\xi_i)\right)\right\} = 2\pi f(0).$$

Therefore, the proof will be complete if one can prove that

$$\lim_{p\to\infty} p^{-1}\left|\sum_{i,j=1}^{p}\mathbb{E}\left(\xi_i\mathbb{E}(\xi_j|\mathcal{F}_0)\right)\right| = 0. \tag{12.16}$$

But, by stationarity,

$$\frac{1}{p}\left|\sum_{i,j=1}^{p}\mathbb{E}\left(\xi_i\mathbb{E}(\xi_j|\mathcal{F}_0)\right)\right| \le \frac{2}{p}\sum_{i=1}^{p}\sum_{k\ge 0}\left|\mathbb{E}\left(\xi_k\mathbb{E}(\xi_0|\mathcal{F}_{-i})\right)\right| := \frac{2}{p}\sum_{i=1}^{p}\Gamma_i,$$

proving (12.16) by taking into account condition (12.8). This ends the proof of step 2. To end the proof of Theorem 12.8, it remains to prove Lemma 12.12.

Proof of Lemma 12.12 To prove it, we proceed as follows. Let m be a fixed positive integer. Note first that if i_0 is such that $|b_{i_0,n}| = \max_{k\in\mathbb{Z}}|b_{k,n}|$ then, since $i_0 = j_0\ (\mathrm{mod}(m))$ where $j_0 \in \{0, ..., m-1\}$, it follows that

$$\max_{k\in\mathbb{Z}}|b_{k,n}| \le \max_{j\in\{0,\dots,m-1\}}\max_{k\in\mathbb{Z}}|b_{km+j,n}|.$$

It follows that

$$\max_{k\in\mathbb{Z}} b_{k,n}^4 \le \sum_{j=0}^{m-1}\sum_{k\in\mathbb{Z}} b_{km+j,n}^4 \le \sum_{j=0}^{m-1}\left(\sum_{k\in\mathbb{Z}} b_{km+j,n}^2\right)^2.$$

Therefore, the lemma will follow if we can prove that

$$\lim_{m\to\infty}\limsup_{n\to\infty}\frac{1}{b_n^4}\sum_{j=0}^{m-1}\left(\sum_{k\in\mathbb{Z}} b_{km+j,n}^2\right)^2 = 0. \tag{12.17}$$

With this aim, observe that, for any $j = 0, ..., m-1$,

$$\begin{aligned}\left|\sum_{k\in\mathbb{Z}} b_{km+j,n}^2 - \frac{1}{m}\sum_{k\in\mathbb{Z}} b_{k,n}^2\right| &= \left|\sum_{k\in\mathbb{Z}} b_{km+j,n}^2 - \frac{1}{m}\sum_{\ell=0}^{m-1}\sum_{k\in\mathbb{Z}} b_{km+\ell,n}^2\right| \\ \le \frac{1}{m}\sum_{\ell=0}^{m-1}\sum_{k\in\mathbb{Z}}\left|b_{km+j,n}^2 - b_{km+\ell,n}^2\right| &\le \frac{1}{m}\sum_{\ell=0}^{j-1}\sum_{k\in\mathbb{Z}}\left|\sum_{u=\ell+1}^{j} b_{km+u,n}^2 - b_{km+u-1,n}^2\right|\end{aligned}$$

$$+\frac{1}{m}\sum_{\ell=j}^{m-1}\sum_{k\in\mathbb{Z}}\left|\sum_{u=j+1}^{\ell} b^2_{km+u,n}-b^2_{km+u-1,n}\right|,$$

implying that

$$\left|\sum_{k\in\mathbb{Z}} b^2_{km+j,n}-\frac{1}{m}\sum_{k\in\mathbb{Z}} b^2_{k,n}\right|\le\sum_{u=1}^{m-1}\sum_{k\in\mathbb{Z}}\left|b^2_{km+u,n}-b^2_{km+u-1,n}\right|\le\sum_{i\in\mathbb{Z}}\left|b^2_{i,n}-b^2_{i-1,n}\right|.$$

Therefore,

$$\begin{aligned}\frac{1}{b_n^4}\sum_{j=0}^{m-1}\left(\sum_{k\in\mathbb{Z}} b^2_{km+j,n}\right)^2 &\le \frac{2m}{b_n^4}\left(\sum_{i\in\mathbb{Z}}\left|b^2_{i,n}-b^2_{i-1,n}\right|\right)^2+\frac{2}{mb_n^4}\left(\sum_{k\in\mathbb{Z}} b^2_{k,n}\right)^2\\ &\le \frac{8m}{b_n^2}\sum_{i\in\mathbb{Z}}\left|b_{i,n}-b_{i-1,n}\right|^2+\frac{2}{m},\end{aligned}$$

which proves (12.17) by taking into account condition (12.9). ■

12.4 Invariance Principle for Linear Processes

We move now to explore the invariance principle for the partial sums associated with a causal linear process defined by

$$X_k=\sum_{j=0}^{\infty} a_j\xi_{k-j},\ k\in\mathbb{Z}. \tag{12.18}$$

Assume that X_0 exists, is in $\mathbb{L}^2$ and is defined as before. Let us associate with the partial sums the following process in $D([0,1])$

$$W_n(t)=\sum_{j=1}^{[nt]} X_j/\sigma_n,\ t\in[0,1],\ \text{ where } \sigma_n^2=\operatorname{Var}(S_n).$$

We would like to find the limiting distribution of this process when a CLT is available. It should be noted that if $\{W_n(t), t\in[0,1]\}$ converges weakly to a standard Brownian motion, then necessarily $\sigma_n^2=nh(n)$ where $h(n)$ is a slowly varying function (i.e. a regularly varying function with exponent 1). This is so since for $t\in[0,1]$ fixed we have $S_{[nt]}/\sigma_n\to^d N(0,t)$ and, in addition, taking $t=1$ we have S_n^2/σ_n^2 is uniformly integrable (by the convergence of moments theorem), implying $\sigma^2_{[nt]}/\sigma_n^2\to t$.

A natural conjecture is that if we assume that the innovations (ξ_k) are i.i.d. centered with finite second moment $\mathbb{E}(\xi_0^2)>0$, $\sum_{j=0}^{\infty} a_j^2<\infty$ and $\sigma_n^2=nh(n)$ with $h(n)$ a slowly varying function, then $\{W_n(t), t\in[0,1]\}$ converges weakly to a standard Brownian motion. This conjecture however has a negative answer.

12.4.1 Construction of the Counterexample

Let us first give a characterization for the variance of the partial sums of a linear process with i.i.d. innovations to be linearly varying with exponent 1 (see Definition 12.17).

Lemma 12.14 *Let $(\xi_i, i \in \mathbb{Z})$ be a sequence of i.i.d. centered real-valued r.v.'s with finite second moment $\mathbb{E}[\xi_1^2] = \sigma^2 > 0$. In addition, let $(a_i,\ i \geq 0)$ be a sequence of real numbers such that $\sum_{i\geq 0} a_i^2 < \infty$. Then we consider the causal linear process defined by (12.18). Let $b_n = a_0 + \cdots + a_n$ and assume that*

$$\sum_{k=0}^{n} b_k^2 \to \infty, \text{ as } n \to \infty, \tag{12.19}$$

and

$$\sum_{j=0}^{\infty} (b_{n+j} - b_j)^2 = o\left(\sum_{k=0}^{n-1} b_k^2\right). \tag{12.20}$$

Then, $\left(\sum_{k=0}^{n-1} b_k^2\right)^{-1} \sigma_n^2 \to \sigma^2$ and $\sigma_n^2 = nh(n)$ where $h(n)$ is a slowly varying function.

Remark 12.15 *Notice that Wu and Woodroofe (2004) pointed out that (12.20) is a necessary and sufficient condition in order for (4.86) to hold for the sequence $(X_k)_{k\in\mathbb{Z}}$.*

Proof Notice that

$$S_n = \sum_{j=1}^{n} b_{n-j}\xi_j + \sum_{j=0}^{\infty} (b_{n+j} - b_j)\xi_{-j}$$

and then

$$\sigma_n^2 = \sigma^2 \left\{ \sum_{k=0}^{n-1} b_k^2 + \sum_{j=0}^{\infty} (b_{n+j} - b_j)^2 \right\}. \tag{12.21}$$

This shows that $\left(\sum_{k=0}^{n-1} b_k^2\right)^{-1} \sigma_n^2 \to \sigma^2$. Let now $\mathcal{F}_k = \sigma(\xi_\ell, \ell \leq k)$ and note that

$$\mathbb{E}(S_n|\mathcal{F}_0) = \sum_{j=0}^{\infty} (b_{n+j} - b_j)\xi_{-j}.$$

Then $\mathbb{E}(\mathbb{E}(S_n|\mathcal{F}_0))^2 = \sigma^2 \sum_{j=0}^{\infty} (b_{n+j} - b_j)^2$. Notice that our conditions imply

$$\sigma_n^2 \to \infty \text{ and } \sigma_n^{-1} \|\mathbb{E}(S_n|\mathcal{F}_0)\|_2 \to 0, \text{ as } n \to \infty.$$

The result follows by applying Proposition 4.30. ∎

The next example shows that conditions (12.19) and (12.20) are not sufficient to ensure that linear processes $(X_k)_{k\in\mathbb{Z}}$, as defined above, satisfy the weak invariance principle.

Proposition 12.16 *There exist a sequence of i.i.d. innovations $\{\xi_i\}$, centered with positive and finite second moments and a sequence $(a_i, i \geq 0)$ of real numbers, satisfying $\sum_{i\geq 0} a_i^2 < \infty$, such that the linear process $(X_k)_{k\in\mathbb{Z}}$ defined by (12.18) satisfies $\sigma_n^2 = nh(n)$ with $h(n)$ slowly varying and such that the weak invariance principle does not hold.*

Proof of Proposition 12.16 Our example is inspired by the construction of examples as in Herrndorf (1983), and also by the paper of Wu and Woodroofe (2004). Let us define two sequences $\{a_n, n \geq 0\}$ and $\{a'_n, n \geq 0\}$ as follows:

$$a_0 = 0, a_1 = \frac{1}{\log 2} \text{ and } a_n = \frac{1}{\log(n+1)} - \frac{1}{\log n}, \text{ for } n \geq 2$$

and

$$a'_0 = 1, a'_1 = -1, a'_2 = 0 \text{ and } a'_n = a_{n-2}, \text{ for } n \geq 3.$$

Let now $(\xi_i, i \in \mathbb{Z})$ be a sequence of independent, identically distributed and symmetric random variables such that

$$\mathbb{P}(\xi_0 > x) \sim \frac{1}{x^2 \log^{3/2} x}, \text{ for } x \text{ large enough.} \tag{12.22}$$

Define now two linear processes:

$$X_k = \sum_{j\geq 0} a_j \xi_{k-j} \text{ and } X'_k = \sum_{j\geq 0} a'_j \xi_{k-j}.$$

Denote $\sigma_n^2 = \mathbb{E}(\sum_{k=1}^n X_k)^2$ and $\sigma_n'^2 = \mathbb{E}(\sum_{k=1}^n X'_k)^2$. Let $b_n = a_0 + \cdots + a_n$. Since

$$\sum_{j=0}^{\infty} (b_{n+j} - b_j)^2 = O\left(\frac{n}{\log^3 n}\right) \text{ and } \sum_{k=0}^{n-1} b_k^2 \sim \frac{n}{(\log n)^2},$$

it follows that both of the linear processes $\{X_k, k \in \mathbb{Z}\}$ and $\{X'_k, k \in \mathbb{Z}\}$ satisfy the conditions of Lemma 12.14.

Now observe that

$$X'_{k+2} = X_k + \xi_{k+2} - \xi_{k+1},$$

then: $\sum_{k=1}^n X'_{k+2} = \sum_{k=1}^n X_k + \xi_{n+2} - \xi_2$. It follows that $\sigma_n'^2 \sim \sigma_n^2$. In addition, according to (12.22), classical computations yield that for every $\varepsilon > 0$,

$$\limsup_{n\to\infty} \mathbb{P}(\max_{1\leq k\leq n} |\xi_{k+2} - \xi_{k+1}| \geq \varepsilon\sqrt{n}/\log n) = 1. \tag{12.23}$$

As a consequence, it follows that the sequences of processes $\{\{\sigma_n^{-1}\sum_{i=1}^{[nt]} X_i, t \in [0,1]\}$ and $\{(\sigma_n')^{-1}\sum_{i=1}^{[nt]} X_i', t \in [0,1]\}$ cannot satisfy the weak invariance principle at the same time. Indeed, if for instance the sequence $\{\sigma_n^{-1}\sum_{i=1}^{[nt]} X_i, t \in [0,1]\}$ satisfies the weak invariance principle, then necessarily for every $\varepsilon \geq 0$, $\mathbb{P}(\max_{1\leq i\leq n}|X_i| \geq \varepsilon\sigma_n) \to 0$, as $n \to \infty$, and consequently, from (12.23),

$$\limsup_{n\to\infty} \mathbb{P}(\max_{1\leq i\leq n}|X_i'| \geq \varepsilon\sigma_n') \geq \limsup_{n\to\infty} \mathbb{P}(\max_{1\leq k\leq n}|\xi_{k+2}-\xi_{k+1}| \geq \varepsilon\sqrt{n}/\log n) = 1.$$

Then, for linear processes, the weak invariance principle cannot hold without additional assumptions to the conditions of Lemma 12.14. ∎

12.4.2 *Finite-dimensional Distributions*

In this subsection we shall describe the behavior of the finite-dimensional distributions of the process

$$W_n(t) = S_{[nt]}/b_n,\ 0 \leq t \leq 1$$

where the sequence of innovations, constants and notations are as in Corollary 12.10.

Definition 12.17 *We say that a positive sequence $(v_n^2)_{n\geq 1}$ is regularly varying with exponent $\beta > 0$ if, for any $t \in [0,1]$,*

$$\frac{v_{[nt]}^2}{v_n^2} \to t^\beta, \text{ as } n \to \infty. \tag{12.24}$$

Definition 12.18 *A Gaussian process is called a standard fractional Brownian motion on $[0,1]$ with Hurst index $\alpha \in (0,1)$ if it has the following covariance structure: For any $s,t \in [0,1]$,*

$$\operatorname{cov}(W_H(s), W_H(t)) = \frac{1}{2}(s^{2\alpha} + t^{2\alpha} - |t-s|^{2\alpha}).$$

Theorem 12.19 *Assume the innovations and the constants satisfy the conditions of Corollary 12.10. Let $\beta \in]0,2]$ and assume that b_n^2 is regularly varying with exponent β. Then the finite-dimensional distributions of $\{b_n^{-1}S_{[nt]}, t \in [0,1]\}$ converge to the corresponding ones of $\sqrt{2\pi f(0)}W_H$, where W_H is the standard fractional Brownian motion with Hurst index $H = \beta/2$.*

Example 1 *Let us consider the linear process X_k defined by*

$$X_k = (1-B)^{-d}\xi_k = \sum_{i\geq 0} a_i\xi_{k-i} \text{ with } a_i = \frac{\Gamma(i+d)}{\Gamma(d)\Gamma(i)}, \tag{12.25}$$

where $0 < d < 1/2$, B is the lag operator, and $(\xi_i)_{i\in\mathbb{Z}}$ is a strictly stationary sequence satisfying the condition of Theorem 12.8. Then Theorem 12.19 applies with $\beta = 2d+1$, since $a_k \sim \kappa_d k^{d-1}$ for some $\kappa_d > 0$.

Example 2 *Now, if we consider the following choice of* $(a_k)_{k\geq 0}$: $a_0 = 1$ *and* $a_i = (i+1)^{-\alpha} - i^{-\alpha}$ *for* $i \geq 1$ *with* $\alpha \in]0, 1/2[$, *then the theorem also applies. Indeed for this choice,* $b_n^2 \sim \kappa_\alpha n^{1-2\alpha}$, *where* κ_α *is a positive constant depending on* α.

Example 3 *For the selection* $a_i \sim i^{-\alpha}\ell(i)$ *where* ℓ *is a slowly varying function at infinity and* $1/2 < \alpha < 1$ *then,* $b_n^2 \sim \kappa_\alpha n^{3-2\alpha}\ell^2(n)$ *(see for instance Relations (12) in Wang et al. (2001)), where* κ_α *is a positive constant depending on* α.

Example 4 *Finally, if* $a_i \sim i^{-1/2}(\log i)^{-\alpha}$ *for some* $\alpha > 1/2$, *then* $b_n^2 \sim n^2(\log n)^{1-2\alpha}/(2\alpha - 1)$ *(see again Relations (12) in Wang et al. (2001)). Hence (12.24) is satisfied with* $\beta = 2$.

Proof of Theorem 12.19 To prove the convergence of the finite-dimensional distributions, we shall apply the Cramér–Wold device. Let m be a positive integer. Let $0 < t_1 < t_2 < \cdots < t_m \leq 1$ and set $n_\ell = [nt_\ell]$. For $\lambda_1, \ldots, \lambda_m \in \mathbb{R}$, notice that

$$\frac{\sum_{\ell=1}^m \lambda_\ell S_{n_\ell}}{b_n} = \frac{1}{b_n}\sum_{j\in\mathbb{Z}}\left(\sum_{\ell=1}^m \lambda_\ell b_{j,n_\ell}\right)\xi_j, \tag{12.26}$$

where $b_{j,n} = a_{1-j} + \cdots + a_{n-j}$ for all $j \in \mathbb{Z}$, and $b_n^2 = \sum_{j\in\mathbb{Z}} b_{j,n}^2$.

We shall apply Theorem 12.8 to the linear process $\sum_{j\in\mathbb{Z}} B_{j,n}\xi_j$ where

$$B_{j,n} = \sum_{\ell=1}^m \lambda_\ell b_{j,n_\ell}. \tag{12.27}$$

As a first step we calculate the limit over n of the following quantity

$$\frac{1}{b_n^2}\sum_{j\in\mathbb{Z}} B_{j,n}^2 = \frac{1}{b_n^2}\sum_{j\in\mathbb{Z}}\sum_{\ell=1}^m\sum_{k=1}^m \lambda_\ell\lambda_k b_{j,n_\ell}b_{j,n_k}.$$

For any $1 \leq \ell \leq k \leq m$, by using the fact that for any two real numbers A and B we have $A(A+B) = 2^{-1}(A^2 + (A+B)^2 - B^2)$, we get that

$$\begin{aligned}\frac{1}{b_n^2}\sum_{j\in\mathbb{Z}} b_{j,n_\ell}b_{j,n_k} &= \frac{1}{2b_n^2}\sum_{j\in\mathbb{Z}}\left(b_{j,n_\ell}^2 + b_{j,n_k}^2 - \left(b_{j,n_\ell} - b_{j,n_k}\right)^2\right)\\ &= \frac{1}{2b_n^2}\sum_{j\in\mathbb{Z}}\left(b_{j,n_\ell}^2 + b_{j,n_k}^2 - b_{j,n_k-n_\ell}^2\right).\end{aligned}$$

Now, by using condition (12.24), we derive that, for any $1 \leq \ell \leq k \leq m$,

$$\frac{\sum_{j\in\mathbb{Z}} b_{j,n_\ell}b_{j,n_k}}{b_n^2} \to \frac{1}{2}\left(t_\ell^\beta + t_k^\beta - (t_k - t_\ell)^\beta\right). \tag{12.28}$$

It follows from (12.28) that

$$\lim_{n\to\infty}\frac{1}{b_n^2}\sum_{j\in\mathbb{Z}}B_{j,n}^2=\frac{1}{2}\sum_{\ell=1}^{m}\sum_{k=1}^{m}\lambda_\ell\lambda_k\left(t_\ell^\beta+t_k^\beta-|t_k-t_\ell|^\beta\right). \tag{12.29}$$

Moreover,

$$\frac{1}{b_n^2}\sum_{j\in\mathbb{Z}}(B_{j,n}-B_{j-1,n})^2\le\left(\sum_{\ell=1}^{m}\lambda_\ell^2\right)\sum_{k=1}^{m}\frac{1}{b_n^2}\sum_{j\in\mathbb{Z}}(b_{j,n_k}-b_{j-1,n_k})^2\to 0,\ \text{ as } n\to\infty,$$

since $\sum_{j\in\mathbb{Z}}(b_{j,n_k}-b_{j-1,n_k})^2\le 4\sum_{i\in\mathbb{Z}}a_i^2$ and $b_n\to\infty$. Therefore, the conditions of Theorem 12.8 being satisfied, using (12.29), we can deduce that

$$\frac{\sum_{\ell=1}^{m}\lambda_\ell S_{n_\ell}}{b_n}\ \text{converges in distribution to}\ \Lambda_{m,\beta}\sqrt{2\pi f(0)}N,$$

where

$$\Lambda_{m,\beta}=\frac{1}{2}\sum_{\ell=1}^{m}\sum_{k=1}^{m}\lambda_\ell\lambda_k\left(t_\ell^\beta+t_k^\beta-|t_k-t_\ell|^\beta\right)$$

ending the proof of the convergence of the finite-dimensional distributions. ■

12.4.3 *Tightness*

We shall comment about the functional CLT only for i.i.d. innovations.

Theorem 12.20 *Let $(a_i)_{i\in\mathbb{Z}}$ be in ℓ^2 and let $(\xi_i)_{i\in\mathbb{Z}}$ be i.i.d. centered real-valued random variables with $\|\xi_0\|_2=1$. Let $\beta\in]0,2]$ and assume that σ_n^2 (the variance of S_n) is regularly varying with exponent β. If $\beta\in]1,2]$ then the process $\{\sigma_n^{-1}S_{[nt]},t\in[0,1]\}$ converges in distribution in $D([0,1])$ to W_H where W_H is a standard fractional Brownian motion with Hurst index $H=\beta/2$. If $\beta\in]0,1]$ and we assume in addition that, for a real $q>2/\beta$, we have $\|\xi_0\|_q<\infty$, then the process $\{\sigma_n{}^{-1}S_{[nt]},t\in[0,1]\}$ converges in distribution in $D([0,1])$ to W_H.*

Proof By Theorem 12.19, it is enough to prove tightness. With this aim, we apply Corollary 1.22. By the Khintchine–Burkholder inequality applied to linear statistics of i.i.d. variables $\{\xi_i\}_{i\in\mathbb{Z}}$, we get that, if $\|\xi_0\|_q<\infty$ for some $q\ge 2$, then

$$\mathbb{E}|S_n|^q=\mathbb{E}\left|\sum_{j\in\mathbb{Z}}b_{j,n}\xi_j\right|^q\le c\mathbb{E}|\xi_1|^q\left(\sum_{j\in\mathbb{Z}}b_{j,n}^2\right)^{q/2}=c\mathbb{E}|\xi_1|^q\sigma_n^q,$$

where $b_{j,n}=a_{1-j}+\ldots+a_{n-j}$. Recall that $\sigma_n^2=n^\beta h(n)$ where h is slowly varying. Therefore by Corollary 1.22, the tightness follows by considering either $q>2/\beta$ if $\beta\in]0,1]$ or $q=2$ if $\beta\in]1,2]$. ■

12.5 IP for Linear Statistics with Weakly Associated Innovations

12.5.1 The Case of Asymptotically Negative Dependent Innovations

From Corollary 9.14, we can prove the following result:

Corollary 12.21 *Let $\xi = (\xi_k)_{k\in\mathbb{Z}}$ be a $\mathbb{L}^2$-stationary sequence of real-valued centered random variables with a continuous spectral density function f on $(-\pi, \pi]$, satisfying the asymptotic negative dependence condition (9.2) and such that $\{\xi_k^2\}$ is an uniformly integrable family. Assume in addition that $\|\xi_0\|_2 = 1$. Consider a triangular array of non-negative numerical constants $\{b_{k,n}, k\in\mathbb{Z}\}$ satisfying the condition (12.9). Define the triangular array $\{X_{k,n}, k\in\mathbb{Z}\}_{n\geq 1}$ by*

$$X_{k,n} := b_{k,n}\xi_k/b_n.$$

For $0\leq t\leq 1$ we set

$$k_n(t) = \min\left\{k\in\mathbb{Z}: \sum_{i=-\infty}^{k} \mathbb{E}[X_{i,n}^2] \geq t\right\}$$

and $W_n(0) = 0$, $W_n(t) = \sum_{i=-\infty}^{k_n(t)} X_{i,n}$. Then $\{W_n(t), t\in[0,1]\} \to^d V$ in $D([0,1])$ where $V = \sqrt{2\pi f(0)}W$ with W the standard Brownian motion. In particular

$$\frac{1}{b_n}\sum_{k\in\mathbb{Z}} b_{k,n}\xi_k \to^d \sqrt{2\pi f(0)}N,$$

where $N\sim\mathcal{N}(0,1)$.

Proof of Corollary 12.21 To prove the corollary, we shall take into account Comment 9.17 and apply Corollary 9.14 to the triangular array $\{X_{k,n}, k\in\mathbb{Z}\}_{n\geq 1}$. Since the sequence $(\xi_k)_{k\in\mathbb{Z}}$ is asymptotically negative dependent and the coefficients $b_{k,n}$ are non-negative then the triangular array $\{X_{k,n}, k\in\mathbb{Z}\}_{n\geq 1}$ is also asymptotically negative dependent. Next, for any $\varepsilon > 0$,

$$\sum_{k\in\mathbb{Z}} \mathbb{E}(X_{k,n}^2 1_{\{|X_{n,k}|>\varepsilon\}}) \leq \sup_{k\in\mathbb{Z}} \mathbb{E}(\xi_k^2 1_{\{|\xi_k|>\delta_n^{-1}\varepsilon\}}),$$

where $\delta_n = b_n^{-1}\max_{k\in\mathbb{Z}}|b_{k,n}|$. But by Lemma 12.12, since (12.9) is assumed, $\lim_{n\to\infty}\delta_n = 0$. This convergence, together with the uniform integrability of $\{\xi_k^2\}$, entail that $\{X_{k,n}, k\in\mathbb{Z}\}_{n\geq 1}$ satisfies the Lindeberg condition.

To apply Corollary 9.14 (see also Remark 9.15), it remains to show that for all $0\leq a<b$

$$\mathrm{Var}(W_n(b) - W_n(a)) \to 2\pi f(0)(b-a). \tag{12.30}$$

Note that

$$\mathrm{Var}(W_n(b) - W_n(a)) = \mathbb{E}\left(\sum_{i=k_n(a)+1}^{k_n(b)} X_{i,n}\right)^2 = b_n^{-2}\mathbb{E}\left(\sum_{k=k_n(a)+1}^{k_n(b)} b_{k,n}\xi_k\right)^2.$$

Using the fact that $(\xi_k)_{k\in\mathbb{Z}}$ is $\mathbb{L}^2$-stationary with a continuous spectral density function f on $(-\pi, \pi]$ and proceeding as in the proof of point (iii) of Lemma 1.5, we infer that (12.30) will hold if one can prove that, for any $i \in \mathbb{Z}$,

$$\lim_{n\to\infty} \left|A_{n,i} - (b-a)\right| = 0, \tag{12.31}$$

where

$$A_{n,i} = b_n^{-2} \sum_{k=k_n(a)+1}^{k_n(b)} b_{k,n} b_{k+i,n}.$$

To prove the convergence (12.31), note that for any $i \in \mathbb{Z}$,

$$A_{n,i} = b_n^{-2} \sum_{k=k_n(a)+1}^{k_n(b)} b_{k,n}^2 + b_n^{-2} \sum_{k=k_n(a)+1}^{k_n(b)} b_{k,n}(b_{k+i,n} - b_{k,n}).$$

Hence,

$$\left|A_{n,i} - (b-a)\right| \le \left|b_n^{-2} \sum_{k=k_n(a)+1}^{k_n(b)} b_{k,n}^2 - (b-a)\right| + b_n^{-2}\left(\sum_{k=k_n(a)+1}^{k_n(b)} b_{k,n}^2\right)^{1/2} \times \left(\sum_{k=k_n(a)+1}^{k_n(b)} (b_{k+i,n} - b_{k,n})^2\right)^{1/2}.$$

The last term in the right-hand side is going to zero as $n \to \infty$, by taking into account condition (12.9). To show that the first term is going to zero as $n \to \infty$, we observe that since we have proved that the triangular array $(X_{j,n})_{j\in\mathbb{Z}}$ satisfies the Lindeberg condition, we have $\sup_{i\in\mathbb{Z}} \mathbb{E}(X_{i,n}^2) \to 0$. This implies that for any $t \in [0,1]$, $\sum_{i\le k_n(t)} \mathbb{E}[X_{i,n}^2] = b_n^{-2}\sum_{k=-\infty}^{k_n(t)} b_{k,n}^2 \to t$. This proves that the first term in the right-hand side is going to zero as $n \to \infty$. This ends the proof of (12.31) and then of (12.30). ■

12.5.2 *The Case of Long-Range Dependent Statistics of Stationary Perturbed Determinantal Point Processes*

We start by proving the asymptotic normality of linear statistics when the innovation process is a stationary perturbed determinantal point process.

Corollary 12.22 *Let $X = (X_i, i \in \mathbb{Z})$ be a stationary determinantal point process and let $Y = (Y_i, i \in \mathbb{Z})$ be a non-degenerate stationary Gaussian sequence with positive continuous spectral density. Assume that X and Y are independent. Also, consider a measurable function $h(x, y)$ defined on $\{0, 1\} \times \mathbb{R}$ and such that $\mathbb{E}[h^2(X_1, Y_1)] < \infty$. Define the stationary sequence $\xi = (\xi_i, i \in \mathbb{Z})$ by $\xi_i = h(X_i, Y_i) - \mathbb{E}[h(X_i, Y_i)]$, $i \in \mathbb{Z}$. Let $\{b_{k,n}, k \in \mathbb{Z}\}$ be a triangular array of non-negative numerical constants satisfying (12.9). Then ξ has a continuous spectral density f on $(-\pi, \pi]$ and*

$$b_n^{-1} \sum_{k \in \mathbb{Z}} b_{k,n} \xi_k \to^d \sqrt{2\pi f(0)} N, \tag{12.32}$$

where $b_n^2 = \sum_{k \in \mathbb{Z}} b_{k,n}^2$ and $N \sim \mathcal{N}(0, 1)$.

Proof of Corollary 12.22 By (10.3), recall that

$$\xi_i = X_i v_i - X_i \psi_i + \zeta_i,$$

where $v_i = (h(1, Y_i) - h(0, Y_i))_+$, $\psi_i = (-h(1, Y_i) + h(0, Y_i))_+$ and $\zeta_i = h(0, Y_i)$. For any $u = (u_1, u_2, u_3) \in \mathbb{R}^3$, let

$$f_u(x, y) = u_1 x(h(1, y) - h(0, y))_+ + u_2 x(-h(1, y) + h(0, y))_+ + u_3 h(0, y)$$

and define the process $U_u := (U_i(u))_{i \in \mathbb{Z}}$ by

$$U_i(u) = f_u(X_i, Y_i) - \mathbb{E}(f_u(X_i, Y_i)) \text{ for any } i \in \mathbb{Z}. \tag{12.33}$$

The fact that ξ has a continuous spectral density f on $(-\pi, \pi]$ comes from Lemma 12.23 (the proof of which follows) by applying it with $u = (1, -1, 1)$.

Lemma 12.23 *Let $X = (X_i, i \in \mathbb{Z})$, $Y = (Y_i, i \in \mathbb{Z})$ and $h(\cdot, \cdot)$ be as in Corollary 12.22. Then, for $u = (u_1, u_2, u_3) \in \mathbb{R}^3$, the stationary process $U = (U_i(u), i \in \mathbb{Z})$ defined by (12.33) has a continuous spectral density f_u on $(-\pi, \pi]$.*

Let us prove now the convergence in distribution of $b_n^{-1} \sum_{k \in \mathbb{Z}} b_{k,n} \xi_k$. With this aim, we first recall that, for any $u \in \mathbb{R}_+^3$, $\lim_{n \to 0} r_n(U_u) = 0$ (see (10.4)). This convergence together with Lemma 12.23 and Corollary 12.21 prove that, for any $u \in \mathbb{R}_+^3$,

$$b_n^{-1} \sum_{k \in \mathbb{Z}} b_{k,n} U_k(u) \to^d \sqrt{2\pi f_u(0)} N, \tag{12.34}$$

where f_u is the spectral density of U_u. Now, by using the same arguments as those developed in the proof of Theorem 9.18 (and in particular Property 9.19), it follows that the convergence (12.34) also holds for any $u \in \mathbb{R}^3$. Therefore taking $u = (1, -1, 1)$, the convergence (12.32) follows. To end the proof of the corollary, it remains to prove Lemma 12.23.

Proof of Lemma 12.23 The lemma will follow from Theorem 1.7 if one can prove that, for any $u \in \mathbb{R}^3$,

$$\lim_{n\to\infty} \zeta_n(U) = 0 \text{ where } \zeta_n(U) := \sup \left| \mathbb{E}\left(\sum_{i\in Q} U_i(u) \sum_{j\in S} U_j(u) \right) \right| / \mathrm{Card}(Q \cup S), \quad (12.35)$$

the supremum being taken over all pairs of non-empty, finite, disjoint sets Q, $S \in \mathbb{Z}$ satisfying $d(Q, S) = \min_{q\in Q, s\in S} |q - s| \geq n$.

With this aim, we first notice that since $X = (X_i)_{i\in\mathbb{Z}}$ and $Y = (Y_i)_{i\in\mathbb{Z}}$ are independent,

$$\mathbb{E}\left(\sum_{i\in Q} U_i(u) \sum_{j\in S} U_j(u) \right) = \mathbb{E}\left(\mathrm{cov}_Y\left(\sum_{i\in Q} f_u(X_i, Y_i), \sum_{j\in S} f_u(X_j, Y_j) \right) \right)$$
$$+ \mathrm{cov}\left(\mathbb{E}_Y\left(\sum_{i\in Q} f_u(X_i, Y_i) \right), \mathbb{E}_Y\left(\sum_{j\in S} f_u(X_j, Y_j) \right) \right). \quad (12.36)$$

But, setting $y = (y_i)_{i\in Q\cup S}$,

$$\mathrm{cov}_{Y=y}\left(\sum_{i\in Q} f_u(X_i, Y_i), \sum_{j\in S} f_u(X_j, Y_j) \right) = \mathrm{cov}\left(\sum_{i\in Q} b(y_i)X_i, \sum_{j\in S} b(y_j)X_j \right),$$

where $b(y_i) = u_1(h(1, y_i) - h(0, y_i))_+ + u_2(-h(1, y_i) + h(0, y_i))_+$. With definition (1.17), we derive that, for any pairs (Q, S) of non-empty, finite, disjoint sets in $\mathbb{Z}$, satisfying $d(Q, S) \geq n$,

$$\left| \mathrm{cov}_{Y=y}\left(\sum_{i\in Q} f_u(X_i, Y_i), \sum_{j\in S} f_u(X_j, Y_j) \right) \right| \leq \kappa_n(X) \left(\left(\sum_{k\in Q} |b(y_k)|^2 \right)^{1/2} \left(\sum_{k\in S} |b(y_k)|^2 \right)^{1/2} \right),$$

implying that

$$\left| \mathbb{E}\left(\mathrm{cov}_Y\left(\sum_{i\in Q} f_u(X_i, Y_i), \sum_{j\in S} f_u(X_j, Y_j) \right) \right) \right| \leq \kappa_n(X) \left(\left(\sum_{k\in Q} \mathbb{E}|b(Y_k)|^2 \right)^{1/2} \left(\sum_{k\in S} \mathbb{E}|b(Y_k)|^2 \right)^{1/2} \right).$$

Therefore, by the stationarity of Y,

$$\left| \mathbb{E}\left(\mathrm{cov}_Y\left(\sum_{i\in Q} f_u(X_i, Y_i), \sum_{j\in S} f_u(X_j, Y_j) \right) \right) \right| \leq \kappa_n(X)\mathbb{E}|b(Y_0)|^2 \mathrm{Card}(Q \cup S). \quad (12.37)$$

On the other hand, note that, by the stationarity of X,

$$\mathbb{E}(f_u(X_i, y)) = u_1 \mathbb{E}(X_0)(h(1,y) - h(0,y))_+ + u_2 \mathbb{E}(X_0)(-h(1,y) + h(0,y))_+ \\ + u_3 h(0,y) := v(y).$$

Therefore

$$\operatorname{cov}\left(\mathbb{E}_Y\left(\sum_{i\in Q} f_u(X_i, Y_i)\right), \mathbb{E}_Y\left(\sum_{j\in S} f_u(X_j, Y_j)\right)\right) \\ = \operatorname{cov}\left(\sum_{i\in Q} v(Y_i), \sum_{j\in S} v(Y_j)\right),$$

implying that, for any pairs (Q, S) of non-empty, finite, disjoint sets of $\mathbb{Z}$ satisfying $d(Q, S) \geq n$,

$$\left|\operatorname{cov}\left(\mathbb{E}_Y\left(\sum\nolimits_{i\in Q} f_u(X_i, Y_i)\right), \mathbb{E}_Y\left(\sum\nolimits_{j\in S} f_u(X_j, Y_j)\right)\right)\right| \\ \leq \rho_n^*(Y) \left\|\sum\nolimits_{i\in Q} v(Y_i)\right\|_2 \left\|\sum\nolimits_{j\in S} v(Y_j)\right\|_2 .$$

Since Y has a positive and continuous spectral density, according to Fact 10.2, $\rho_1^*(Y) < 1$ and then $r_1'(Y) < 1$ which implies by Lemma 7.9 and stationarity that

$$\left\|\sum_{i\in Q} v(Y_i)\right\|_2^2 \leq c \operatorname{Card}(Q) \|v(Y_0)\|_2^2,$$

where $c = \frac{1+r_1'(Y)}{1-r_1'(Y)} < \infty$. Therefore,

$$\left|\operatorname{cov}\left(\mathbb{E}_Y\left(\sum_{i\in Q} f_u(X_i, Y_i)\right), \mathbb{E}_Y\left(\sum_{j\in S} f_u(X_j, Y_j)\right)\right)\right| \\ \leq c\rho_n^*(Y)\mathbb{E}|v(Y_0)|^2 \operatorname{Card}(Q \cup S). \tag{12.38}$$

So, overall, starting from (12.36) and taking into account (12.37) and (12.38), it follows that there exists a positive finite constant K such that, for any $n \geq 1$,

$$\zeta_n(U) \leq K(\kappa_n(X) + \rho_n^*(Y)). \tag{12.39}$$

By Fact 10.1, X is negatively dependent. Hence, by taking into account Remark 9.25, it follows that

$$\sum_{k\geq 1} |\operatorname{cov}(X_0, X_k)| = -\sum_{k\geq 1} \operatorname{cov}(X_0, X_k) < \infty,$$

and then X has a continuous spectral density. Therefore, by Theorem 1.7, $\kappa_n(X) \to 0$ as $n \to \infty$. On the other hand, since Y is a Gaussian process with positive continuous spectral density, by Fact 10.2, $\rho_n^*(Y) \to 0$ as $n \to \infty$. Taking into account these considerations in (12.39), it follows that (12.35) holds. This ends the proof of the lemma (and then the proof of Corollary 12.22). ■

Corollary 12.22 gives the following central limit theorem for linear processes constructed with non-negative numerical constants and generated by asymptotically negative dependent innovations.

Corollary 12.24 *Suppose $X = (X_k)_{k\in\mathbb{Z}}$ and $Y = (Y_k)_{k\in\mathbb{Z}}$ satisfy the condition of Corollary 12.22. Let $h(x, y)$ be a measurable function defined on $\{0,1\} \times \mathbb{R}$ and such that $\mathbb{E}[h^2(X_1, Y_1)] < \infty$. Define the stationary sequence $\xi = (\xi_i, \in \mathbb{Z})$ by $\xi_i = h(X_i, Y_i) - \mathbb{E}[h(X_i, Y_i)]$, $i \in \mathbb{Z}$. Let $\{a_j, j \in \mathbb{Z}\}$ be a non-negative sequence such that $\sum_{j\in\mathbb{Z}} a_j^2 < \infty$. Let*

$$U_k = \sum_{j\in\mathbb{Z}} a_j \xi_{k-j},\ S_n = \sum_{k=1}^{n} U_k,\ b_{k,n} = \sum_{i=1}^{n} a_{k+i} \text{ and } b_n^2 = \sum_{k\in\mathbb{Z}} b_{k,n}^2.$$

If $b_n \to \infty$ then

$$b_n^{-1} S_n \to^d \sqrt{2\pi f(0)} N.$$

where $N \sim \mathcal{N}(0,1)$ and f is the spectral density of Z.

Proof of Corollary 12.24 It suffices to notice that $S_n = \sum_{k\in\mathbb{Z}} b_{k,n}\xi_k$ and to apply Corollary 12.22. Indeed since $\sum_{j\in\mathbb{Z}} a_j^2 < \infty$, $b_n^2 < \infty$. On another hand $\sum_{k\in\mathbb{Z}}(b_{k,n} - b_{k-1,n})^2 \leq \sum_{j\in\mathbb{Z}} a_j^2$ which shows that the second part of condition (12.9) holds since $b_n^2 \to \infty$ and $\sum_{j\in\mathbb{Z}} a_j^2 < \infty$. ■

Comment *The central limit theorem for triangular arrays of linear statistics of the following form*

$$\sum_{j=1}^{k} a_j \sum_{i\in L_j} X_i,$$

where $L_j := L_j^{(n)}$ are sets and $(X_i)_{i\in\mathbb{Z}}$ is a determinantal point process, was established in Soshnikov (2002). It is an open question whether the central limit theorem holds for long-range dependent statistics of perturbed determinantal processes for arbitrary sequences (a_j) with $\sum a_j^2 < \infty$.

12.6 Discrete Fourier Transform and Periodogram

An important application of martingale theory is in the analysis of periodogram. Given a stochastic process $(X_j)_{j\in\mathbb{Z}}$, the periodogram is defined as

$$I_n(\theta) = \frac{1}{2\pi n}\left|\sum_{j=1}^{n} X_j e^{\mathrm{i}j\theta}\right|^2, \qquad \theta \in \mathbb{R},$$

where $\mathrm{i} = \sqrt{-1}$ is the imaginary unit. The periodogram is related to the discrete Fourier transform

$$S_n(\theta) = \sum_{j=1}^{n} X_j e^{\mathrm{i}j\theta}.$$

In this section we consider a strictly stationary ergodic sequence $(X_k)_{k\in\mathbb{Z}}$ of centered real-valued random variables with finite second moments, adapted to a non-decreasing filtration $(\mathcal{F}_k)_{k\in\mathbb{Z}}$, and we shall impose the following regularity condition:

$$\mathbb{E}(X_0|\mathcal{F}_{-\infty}) = 0 \ \mathbb{P}\text{-almost surely,} \tag{12.40}$$

where $\mathcal{F}_{-\infty} = \bigcap_{n\in\mathbb{Z}}\mathcal{F}_n$ is the tail sigma field.

12.6.1 A CLT for Almost All Frequencies

We present below a central limit theorem for almost all frequencies that has been obtained by Peligrad and Wu in 2010. In Theorem 12.25, we let the parameter θ be in the space $[0, 2\pi]$, endowed with Borel sigma algebra and Lebesgue measure λ.

Theorem 12.25 *Let $(X_k)_{k\in\mathbb{Z}}$ be a strictly stationary ergodic sequence of centered real-valued random variables with finite second moments, adapted to a non-decreasing filtration $(\mathcal{F}_k)_{k\in\mathbb{Z}}$, and such that (12.40) is satisfied. Then, for almost all $\theta \in [0, 2\pi]$, the following convergence holds:*

$$\lim_{n\to\infty} \frac{\mathbb{E}|S_n(\theta)|^2}{n} = 2\pi f(\theta), \tag{12.41}$$

where f is the spectral density of $(X_k)_{k\in\mathbb{Z}}$. Furthermore,

$$\frac{1}{\sqrt{n}}[\mathrm{Re}(S_n(\theta)), \mathrm{Im}(S_n(\theta))] \Rightarrow [N_1(\theta), N_2(\theta)] \ \textit{under}\ \mathbb{P}, \tag{12.42}$$

where $N_1(\theta)$ and $N_2(\theta)$ are independent identically distributed normal random variables with mean 0 and variance $\pi f(\theta)$.

An implication of this results is that we obtain the limiting distribution of the periodogram as follows:

Remark 12.26 *As a consequence of Theorem 12.25, for sequences satisfying (12.40), the periodogram $n^{-1}|S_n(\theta)|^2$ is asymptotically distributed as $f(\theta)\chi^2(2)$ for almost all $\theta \in [0, 2\pi]$, where $\chi^2(2)$ is the chi-squared distribution with 2 degrees of freedom.*

The proof of Theorem 12.25 is a combination of martingale techniques with results from ergodic theory and harmonic analysis.

Some preparatory materials. From harmonic analysis we shall use the first three facts below. The last one comes from ergodic theory.

Fact 1 *Carleson Theorem (Carleson 1966): If $(a_k)_{k\geq 0}$ are real numbers such that $\sum_{l=0}^{\infty} a_l^2 < \infty$, then $\sum_{j=1}^{n} a_j e^{ij\theta}$ converges λ-almost surely on $[0, 2\pi]$.*

Fact 2 *Hunt and Young (1974): There is a constant C such that*

$$\int_0^{2\pi} \max_{1\leq k\leq n} \left| \sum_{\ell=1}^{k} e^{i\ell\theta} a_\ell \right|^2 d\theta \leq C \sum_{\ell=1}^{n} a_\ell^2.$$

Fact 3 *Fejér–Lebesgue Theorem (cf Bary, 1964, p. 139 or Theorem 15.7 in Champeney, 1989). If $g(\theta)$ is integrable on $[0, 2\pi]$, with Fourier coefficients*

$$b_\ell = \int_0^{2\pi} e^{i\ell\theta} g(\theta) d\theta$$

then, denoting

$$h_m(\theta) = \frac{1}{2\pi} \sum_{\ell=-m}^{m} e^{-i\ell\theta} b_\ell,$$

we have

$$g(\theta) = \lim_{n\to\infty} \frac{1}{n} \sum_{m=-n}^{n} h_m(\theta) \ \lambda\text{-almost surely and in } \mathbb{L}^1[0, 2\pi]. \tag{12.43}$$

Fact 4 *Next proposition gives the strong law of large numbers for discrete Fourier transform. We notice that we need only the variables to be identically distributed to get an almost sure result. We refer the reader to Proposition 30 and Lemma 32 in Cuny, Merlevède and Peligrad (2013) and to Zhang (2017) for more general results on this type.*

In the sequel λ denotes the Lebesgue measure on $[0, 2\pi]$.

Proposition 12.27 *Let $(X_k)_{k\geq 0}$ denote a sequence of identically distributed random variables on $(\Omega, \mathcal{K}, \mathbb{P})$, with finite first moment. Then, for λ-almost all θ in $[0, 2\pi]$,*

$$\lim_{n\to\infty} \frac{1}{n} S_n(\theta) = 0 \ \mathbb{P}\text{-a.s.}$$

where $S_n(\theta) = \sum_{k=1}^{n} e^{ik\theta} X_k$.

Proof We shall use a truncation argument. Define

$$Y_k = X_k I(|X_k| \leq k) \text{ and } S_n^*(\theta) = \sum_{k=1}^{n} e^{ik\theta} Y_k.$$

Then,

$$\sum_{k=1}^{\infty} \mathbb{P}(X_k \neq Y_k) = \sum_{k=1}^{\infty} \mathbb{P}(|X_k| > k) \leq \mathbb{E}(|X_0|) < \infty.$$

By the Borel–Cantelli lemma, applied on Ω, $\mathbb{P}(X_n \neq Y_n \text{ i.o.}) = 0$. Therefore, for all θ in $[0, 2\pi]$,

$$\frac{1}{n} S_n(\theta) \text{ and } \frac{1}{n} S_n^*(\theta) \text{ have the same } \mathbb{P}\text{-a.s. limit.}$$

The proposition is then reduced to show that, for almost all θ in $[0, 2\pi]$, we have that $S_n^*(\theta)/n \to 0$, $\mathbb{P}$ -a.s.

In order to prove it, note that, by using the Kronecker lemma, it is enough to show that, for almost all θ in $[0, 2\pi]$,

$$\sum_{k=1}^{\infty} \frac{e^{ik\theta} Y_k}{k} \text{ converges } \mathbb{P}\text{-a.s.} \tag{12.44}$$

To prove it, we use Carleson's (1966) theorem. Clearly,

$$\sum_{k=1}^{\infty} \frac{\mathbb{E}\left(Y_k^2\right)}{k^2} = \mathbb{E}\left(X_0^2 \sum_{k=1}^{\infty} \frac{I(|X_0| \leq k)}{k^2}\right) = \mathbb{E}\left(X_0^2 \sum_{k \geq |X_0|}^{\infty} \frac{1}{k^2}\right) \leq c\mathbb{E}(|X_0|),$$

for some positive constant c. The latter relation implies that there is $\Omega' \subset \Omega$ with $\mathbb{P}(\Omega') = 1$, such that, for all $\omega \in \Omega'$,

$$\sum_{k=1}^{\infty} \frac{Y_k^2(\omega)}{k^2} < \infty.$$

By Carleson's (1966) theorem, for such ω,

$$\sum_{k=1}^{\infty} \frac{e^{ik\theta} Y_k(\omega)}{k} \text{ converges for } \lambda\text{-almost all } \theta.$$

Denote now

$$A = \left\{(\theta, \omega) \in [0, 2\pi] \times \Omega;\ \text{where } \sum_{k=1}^{\infty} \frac{e^{ik\theta} Y_k(\omega)}{k} \text{ is convergent.}\right\}$$

By Fubini's theorem, in the product space $[0, 2\pi] \times \Omega$, we have $\lambda \times \mathbb{P}(A) = 1$. Again by Fubini's theorem, (12.44) follows. The proof is complete. ■

Some preparatory lemmas. For $k \in \mathbb{Z}$ we define the projection operator by

$$\mathbf{P}_k(\cdot) = \mathbb{E}(\cdot|\mathcal{F}_k) - \mathbb{E}(\cdot|\mathcal{F}_{k-1}). \tag{12.45}$$

Lemma 12.28 *Let*

$$T_n(\theta) = \sum_{j=0}^{n} X_j e^{ij\theta} = X_0 + S_n(\theta).$$

Under (12.40), *for λ-almost all θ, we have*

$$\mathbf{P}_0(T_n(\theta)) \to \sum_{l=0}^{\infty} e^{il\theta}\mathbf{P}_0(X_l) =: D_0(\theta) \ \ \mathbb{P}\text{-almost surely and in } \mathbb{L}^2. \tag{12.46}$$

Proof By (12.40) we have $\sum_{k\in\mathbb{Z}} \|\mathbf{P}_k(X_0)\|_2^2 = \|X_0\|_2^2 < \infty$, whence $\sum_{k\in\mathbb{Z}} |\mathbf{P}_0(X_k)|^2 < \infty$ $\mathbb{P}$-almost surely. Therefore, by Carleson's (1966) theorem, for almost all ω, $\sum_{1\le k\le n} e^{ik\theta}\mathbf{P}_0(X_k)$ converges λ-almost surely. Denote the limit by $D_0 = D_0(\theta)$. We now consider the set

$$A = \{(\theta, \omega) \subset [0, 2\pi] \times \Omega, \text{ where } \{\mathbf{P}_0(S_n(\theta))\}_n \text{ does not converge}\}$$

and notice that almost all sections for ω fixed have Lebesgue measure 0. So by Fubini's theorem the set A has measure 0 in the product space and therefore, again by Fubini's theorem, almost all sections for θ fixed have probability 0. It follows that for almost all θ, $\mathbf{P}_0(S_n(\theta)) \to D_0$ almost surely under $\mathbb{P}$. Next, by the maximal inequality in Hunt and Young (1974), there is a constant C such that

$$\int_0^{2\pi} \left[\sup_n |\mathbf{P}_0(S_n(\theta))|^2\right] \lambda(d\theta) \le C \sum_k |\mathbf{P}_0(X_k)|_2^2$$

and then we integrate

$$\int_0^{2\pi} \mathbb{E}\left[\sup_n |\mathbf{P}_0(S_n(\theta))|^2\right] \lambda(d\theta) \le C\|X_0\|_2^2 < \infty.$$

Therefore,

$$\mathbb{E}\left[\sup_n |\mathbf{P}_0(S_n(\theta))|^2\right] < \infty \ \text{ for almost all } \theta.$$

Since $|\mathbf{P}_0(S_n(\theta))| < \sup_n |\mathbf{P}_0(S_n(\theta))|$, and the last one is integrable for almost all θ, by the Lebesgue dominated convergence we have that $\mathbf{P}_0(S_n(\theta))$ converges in $\mathbb{L}^2$. ■

Lemma 12.29 *Let $g(\theta) = (2\pi)^{-1}\mathbb{E}|D_0(\theta)|^2$. For all $j \in \mathbb{Z}$, we have*

$$\int_0^{2\pi} g(\theta)e^{ij\theta}\,d\theta = c_j, \tag{12.47}$$

where $c_j = \mathrm{cov}(X_0, X_j)$ for all $j \in \mathbb{Z}$. So $(c_j)_{j\in\mathbb{Z}}$ are the Fourier coefficients of g implying that $g = f$. Additionally, for almost all θ,

$$\lim_{n\to\infty} \frac{\mathbb{E}|S_n(\theta)|^2}{n} = 2\pi f(\theta). \tag{12.48}$$

Proof As before, let

$$\mathbf{P}_0(T_n(\theta)) = \sum_{l=0}^{n} e^{il\theta}\mathbf{P}_0(X_l).$$

It follows that

$$\frac{1}{2\pi}\int_0^{2\pi} |\mathbf{P}_0(T_n(\theta))|^2 e^{ij\theta}\,d\theta = \frac{1}{2\pi}\sum_{u=0}^{n}\sum_{v=0}^{n}\int_0^{2\pi} \mathbf{P}_0(X_u)\mathbf{P}_0(X_v)e^{i(u-v+j)\theta}\,d\theta.$$

Since

$$\frac{1}{2\pi}\int_0^{2\pi} e^{ij\theta}\,d\theta = 0 \text{ if } j \neq 0,$$

for $0 \leq j \leq n$,

$$\frac{1}{2\pi}\int_0^{2\pi} |\mathbf{P}_0(T_n(\theta))|^2 e^{ij\theta}\,d\theta = \sum_{l=0}^{n-j} \mathbf{P}_0(X_l)\mathbf{P}_0(X_{l+j}), \tag{12.49}$$

and 0 in rest (i.e. if $j < 0$ or $j > n$). Since $X_j = \sum_{l\in\mathbb{Z}} \mathbf{P}_l(X_j)$, by orthogonality of martingale differences and stationarity we have that

$$\begin{aligned} c_j &= \lim_{N\to\infty} \mathbb{E}\left[\left(\sum_{l=-N}^{0} \mathbf{P}_l(X_0)\right)\left(\sum_{l=-N}^{0} \mathbf{P}_l(X_j)\right)\right] \\ &= \lim_{N\to\infty} \sum_{l=-N}^{0} \mathbb{E}[\mathbf{P}_l(X_0)\mathbf{P}_l(X_j)] = \lim_{n\to\infty} \sum_{l=0}^{n-j} \mathbb{E}[(\mathbf{P}_0X_l)(\mathbf{P}_0X_{l+j})]. \end{aligned}$$

Combining the latter equality with (12.49) we obtain

$$c_j = \lim_{n\to\infty} \frac{1}{2\pi}\int_0^{2\pi} \mathbb{E}|\mathbf{P}_0(T_n(\theta))|^2 e^{ij\theta}\,d\theta.$$

We know that, by (12.46), for λ-almost all θ,

$$\mathbf{P}_0(T_n(\theta)) \to D_0(\theta) \ \ \mathbb{P}\text{-almost surely and in } \mathbb{L}^2.$$

By the Lebesgue dominated convergence theorem, as in the proof of Lemma 12.28, (12.47) follows in view of the Hunt and Young maximal inequality since $\sup_n |\mathbf{P}_0(T_n(\theta))|$ is integrable.

Now we prove (12.48). By stationarity, we have

$$\begin{aligned}\frac{1}{n}\mathbb{E}|S_n(\theta)|^2 &= \frac{1}{n}\sum_{j=1}^{n}\sum_{l=1}^{n}\mathbb{E}(X_jX_l)\mathrm{e}^{\mathrm{i}j\theta}\,\mathrm{e}^{-\mathrm{i}l\theta}\\ &= \frac{1}{n}\sum_{j=1}^{n-1}\sum_{l=1}^{n} c_{j-l}\mathrm{e}^{\mathrm{i}(j-l)\theta} = \sum_{j=-(n-1)}^{n-1}\left(1-\frac{|j|}{n}\right)c_j\mathrm{e}^{\mathrm{i}j\theta}.\end{aligned}$$

Namely $\mathbb{E}|S_n(\theta)|^2/n$ is the Cesàro average of the sum $\sum_{j=-l}^{l} c_j\mathrm{e}^{\mathrm{i}j\theta}$. Note that $\|D_0(\theta)\|_2^2$ is integrable over $[0, 2\pi]$. Therefore by the Fejér–Lebesgue Theorem (Fact 3 above), relation (12.48) holds for λ-almost all $\theta \in [0, 2\pi]$. ■

Proof of Theorem 12.25 The first assertion of Theorem 12.25 is just Lemma 12.29. We now prove (12.42).

Step 1. The construction of a martingale differences sequence. Define the projector operator by (12.45). Then we construct

$$\mathbf{P}_1(S_n(\theta)) = \mathbb{E}(S_n(\theta)|\mathcal{F}_1) - \mathbb{E}(S_n(\theta)|\mathcal{F}_0) = \sum_{k=1}^{n}\mathrm{e}^{\mathrm{i}k\theta}\mathbf{P}_1(X_k).$$

Then, by Lemma 12.28 for λ-almost all θ,

$$\mathbf{P}_1(S_n(\theta)) \to D_1(\theta) \text{ in } \mathbb{L}^2. \tag{12.50}$$

Note that

$$\mathbb{E}(\mathbf{P}_1(S_n(\theta))|\mathcal{F}_0) = 0 \text{ almost surely under } \mathbb{P}.$$

Therefore, by the contractive property of the conditional expectation,

$$0 = \mathbb{E}(\mathbf{P}_1(S_n(\theta))|\mathcal{F}_0) \to \mathbb{E}(D_1(\theta)|\mathcal{F}_0) \text{ in } \mathbb{L}^2.$$

For λ-almost all θ, we then construct a sequence of stationary martingale differences $(D_k(\theta))_{k\geq 1}$, given by

$$\mathbf{P}_k(S_{n+k}(\theta) - S_k(\theta)) \to D_k(\theta) \text{ in } \mathbb{L}^2.$$

Step 2. Martingale approximation. By step 1 we know that there is a set $\Gamma_1 \subset [0, 2\pi]$, such that $\lambda(\Gamma_1) = 2\pi$ and for all $\theta \in \Gamma_1$ the martingale

$$M_n(\theta) = \sum_{1 \le k \le n} e^{ik\theta} D_k(\theta)$$

is well defined in $\mathbb{L}^2$. We show now that, for almost all θ,

$$\frac{\mathbb{E}|S_n(\theta) - M_n(\theta)|^2}{n} \to 0. \tag{12.51}$$

To this end, note that $S_n(\theta) - \mathbb{E}(S_n(\theta)|\mathcal{F}_0)$ and $\mathbb{E}(S_n(\theta)|\mathcal{F}_0)$ are orthogonal, and we have

$$||S_n(\theta)||_2^2 = ||S_n(\theta) - \mathbb{E}(S_n(\theta)|\mathcal{F}_0)||_2^2 + ||\mathbb{E}(S_n(\theta)|\mathcal{F}_0)||_2^2. \tag{12.52}$$

Next, for $\theta \in \Gamma_1$, by the orthogonality of martingale differences, the stationarity and Lemma 12.28, we have that

$$\begin{aligned}||S_n(\theta) - \mathbb{E}(S_n(\theta)|\mathcal{F}_0) - M_n(\theta)||_2^2 &= \sum_{k=1}^{n} ||\mathbf{P}_k(S_n(\theta) - M_n(\theta))||_2^2 \\ = \sum_{k=1}^{n} ||\mathbf{P}_k S_n(\theta) - e^{ik\theta} D_k(\theta)||_2^2 &= \sum_{k=1}^{n} ||\mathbf{P}_0(T_{n-k}(\theta)) - D_0(\theta)||_2^2 = o(n).\end{aligned} \tag{12.53}$$

Hence,

$$\lim_{n\to\infty} \frac{|||S_n(\theta) - \mathbb{E}(S_n(\theta)|\mathcal{F}_0)||_2 - ||M_n(\theta)||_2|}{\sqrt{n}} = 0. \tag{12.54}$$

But $||M_n(\theta)||_2^2 = n||D_0(\theta)||_2^2$ and by Lemma 12.37, for almost all θ, $\lim_{n\to\infty} n^{-1}||S_n(\theta)||_2^2 = ||D_0(\theta)||_2^2$. This consideration together with (12.52) and (12.54) entail that, for almost all θ,

$$\limsup_{n\to\infty} \frac{||\mathbb{E}(S_n(\theta)|\mathcal{F}_0)||_2^2}{n} = 0.$$

This implies (12.51) by taking into account (12.53) and the decomposition

$$||S_n(\theta) - M_n(\theta)||_2^2 = ||S_n(\theta) - \mathbb{E}(S_n(\theta)|\mathcal{F}_0) - M_n(\theta)||_2^2 + ||\mathbb{E}(S_n(\theta)|\mathcal{F}_0)||_2^2.$$

Step 3. The CLT for the approximating martingale. Here we shall consider θ in a set of measure 2π where all the functions make sense.

By Theorem 1.9, it remains just to prove the central limit theorem for complex valued martingale:

$$\frac{1}{\sqrt{n}}\sum_{1\le k\le n} \mathrm{e}^{ik\theta} D_k(\theta).$$

As a matter of fact we shall provide a central limit theorem for the real part and imaginary part and show that in the limit they are independent. By the Cramér–Wold device, we then only have to show that, for any reals a and b,

$$\frac{1}{\sqrt{n}}\sum_{1\le k\le n} \left(a\,\mathrm{Re}(\mathrm{e}^{ik\theta}D_k(\theta)) + b\,\mathrm{Im}(\mathrm{e}^{ik\theta}D_k(\theta))\right)$$
$$\to^d N \sim \mathcal{N}(0, \pi(a^2+b^2)f(\theta)), \text{ as } n \to \infty.$$

With this aim, we shall apply Theorem 2.29. Clearly, since $D_0(\theta)$ is square integrable and $(D_k(\theta))_{k\in\mathbb{Z}}$ is stationary,

$$\frac{1}{\sqrt{n}}\mathbb{E}(\max_{1\le k\le n}|a\,\mathrm{Re}(\mathrm{e}^{ik\theta}D_k(\theta)) + b\,\mathrm{Im}(\mathrm{e}^{ik\theta}D_k(\theta))|) \le \frac{|a|+|b|}{\sqrt{n}}\mathbb{E}(\max_{1\le k\le n}|D_k(\theta)|) \to 0.$$

It remains to verify

$$\frac{1}{n}\sum_{1\le k\le n}\left(a\,\mathrm{Re}(\mathrm{e}^{ik\theta}D_k(\theta)) + b\,\mathrm{Im}(\mathrm{e}^{ik\theta}D_k(\theta))\right)^2 \to^{\mathbb{P}} \frac{(a^2+b^2)}{2}\mathbb{E}|D_0(\theta)|^2,$$

or equivalently if $a^2+b^2=1$,

$$\frac{1}{n}\sum_{1\le k\le n}\left(a\,\mathrm{Re}(\mathrm{e}^{ik\theta}D_k(\theta)) + b\,\mathrm{Im}(\mathrm{e}^{ik\theta}D_k(\theta))\right)^2 \to^{\mathbb{P}} \frac{1}{2}\mathbb{E}|D_0(\theta)|^2. \tag{12.55}$$

For convenience we write $D_k = D_k(\theta)$ and $D_k = A_k + iB_k$. So,

$$\begin{aligned}(a\,\mathrm{Re}(\mathrm{e}^{ik\theta}D_k(\theta)) &+ b\,\mathrm{Im}(\mathrm{e}^{ik\theta}D_k(\theta)))^2 \\ &= (a\cos(k\theta)+b\sin(k\theta))^2A_k^2 + (b\cos(k\theta)-a\sin(k\theta))^2B_k^2 \\ &\quad + 2(a\cos(k\theta)+b\sin(k\theta))(b\cos(k\theta)-a\sin(k\theta))A_kB_k.\end{aligned}$$

By using basic trigonometric formulas, it follows that, if $a^2+b^2=1$,

$$\begin{aligned}(a\,\mathrm{Re}(\mathrm{e}^{ik\theta}D_k(\theta)) + b\,\mathrm{Im}(\mathrm{e}^{ik\theta}D_k(\theta)))^2 &= \frac{A_k^2+B_k^2}{2} + \frac{(a^2-b^2)\cos(2k\theta)}{2}(A_k^2-B_k^2) \\ + ab\sin(2k\theta)(A_k^2-B_k^2) &+ ab(\cos(2k\theta)+\sin(2k\theta))A_kB_k \\ + (b^2-a^2)\sin(2k\theta)A_kB_k.\end{aligned}$$

By stationarity and the ergodic theorem,

$$\frac{1}{n}\sum_{1\le k\le n}\frac{A_k^2+B_k^2}{2}=\frac{1}{2n}\sum_{1\le k\le n}|D_k(\theta)|^2\to\frac{1}{2}\mathbb{E}|D_0(\theta)|^2\ \mathbb{P}\text{-almost surely}$$

On another hand, by Proposition 12.27, for almost all θ,

$$(a\,\mathrm{Re}(\mathrm{e}^{\mathrm{i}k\theta}D_k(\theta))+b\,\mathrm{Im}(\mathrm{e}^{\mathrm{i}k\theta}D_k(\theta)))^2-\frac{A_k^2+B_k^2}{2}\to 0\ \mathbb{P}\text{-almost surely.}$$

By these arguments, (12.55) follows and it remains to apply Lemma 12.29 to obtain the result. ■

12.6.2 *Examples*

We present several examples of processes for which the conclusions of Theorem 12.29 hold.

Clearly condition (12.40) is satisfied if the left tail sigma field $\mathcal{F}_{-\infty}$ is trivial. These processes are called regular (see Chapter 2, vol. 1 in Bradley, 2007).

Example 1 *(Mixing sequences.) Assume that $(X_k)_{k\in\mathbb{Z}}$ is a strictly stationary sequence of real-valued r.v.'s. If the sequence is additionally strong mixing (so $\lim_{n\to\infty}\alpha(n)=0$ where the strong mixing coefficients $\alpha(n)$ have been defined in Section 5.1), the tail sigma field is trivial; see Claim 2.17a in Bradley (2007). Examples of this type include Harris recurrent Markov chains. Now if we denote by $(\rho(n))_{n\ge1}$ the sequence of maximal correlation coefficients associated with $(X_k)_{k\in\mathbb{Z}}$ (see again Section 5.1 for the definition), and if $\lim_{n\to\infty}\rho(n)<1$, then the tail sigma field is also trivial; see Proposition 5.6 in Bradley (2007).*

Example 2 *(Functions of i.i.d. random variables.) Let $(\varepsilon_k)_{k\in\mathbb{Z}}$ be a sequence of i.i.d. and consider $X_n=f(\varepsilon_k,k\le n)$. These are regular processes and therefore Theorem 12.25 applies. Examples include linear processes, functions of linear processes and iterated random functions (Wu and Woodroofe, 2004) among others. For example let $X_n=\sum_{j=0}^{\infty}a_j\varepsilon_{n-j}$, where the ε_j's are i.i.d. with mean 0 and variance 1 and a_j are real coefficients with $\sum_{j=1}^{\infty}a_j^2<\infty$. In this case X_n is well-defined and, by Lemmas 12.28 and 12.29, the spectral density is*

$$f(\theta)=\frac{1}{2\pi}\left|\sum_{j=0}^{\infty}a_j\mathrm{e}^{\mathrm{i}j\theta}\right|^2.$$

Example 3 *(Reversible Markov chains.) As in Chapter 14, we consider $(X_j)_{j\in\mathbb{Z}}$ a strictly stationary, centered, ergodic and reversible Markov chain with values in a measurable space. By computation (14.4),*

$$\|\mathbb{E}(X_0|\mathcal{F}_{-n})\|_2^2=\mathrm{cov}(X_0,X_{2n}):=c_{2n}.$$

If we assume $c_n\to0$ then the conclusion of Theorem 12.25 holds.

13

Random Walk in Random Scenery

In this chapter we investigate the central limit behavior (and its functional form) of the partial sums associated with a stationary sequence sampled by a recurrent one-dimensional strongly aperiodic random walk.

13.1 Introduction and Main Results

We shall consider the partial sums of a centered $\mathbb{L}^2$-stationary sequence $X = (X_k)_{k\in\mathbb{Z}}$ sampled by a recurrent one-dimensional aperiodic random walk $(S_k)_{1\le k\le n}$ defined as follows:

Definition 13.1 *Let $(S_n)_{n\ge 0}$ be a random walk associated with a sequence $\xi = (\xi_i)_{i\ge 1}$ of i.i.d. $\mathbb{Z}$-valued random variables and independent of the sequence X. Hence $S_0 = 0$ and $S_n = \sum_{i=1}^n \xi_i$ for $n \ge 1$. We shall say that $(S_n)_{n\ge 0}$ satisfies condition A_0 if it is strongly aperiodic in the sense that there is no proper subgroup $L \subset \mathbb{Z}$ such that $\mathbb{P}(\xi_i - x \in L) = 1$ for some $x \in \mathbb{Z}$, and if the characteristic function of the ξ_i's satisfies*

$$\phi(t) = \mathbb{E}[e^{it\xi_1}] = 1 - \gamma|t| + R(t), \quad t \in [-\pi, \pi). \tag{13.1}$$

where $\gamma > 0$ and $R(t) = o(|t|)$ as $t \to 0$.

Note that the strong aperiodicity assumption implies that $\phi(t) = 1$ if and only if $t = 0$ (see Lemma 21.6 in Bhattacharya and Rao (1976)).

The aim of this section is to study the convergence in distribution of

$$Z_n = \sum_{k=0}^{n} X_{S_k},\ n \ge 0, \tag{13.2}$$

suitably normalized, or of its associated partial sums process. The suitably normalizing constants sequence will be b_n defined by

Functional Gaussian Approximation for Dependent Structures. Florence Merlevède, Magda Peligrad and Sergey Utev. Oxford University Press (2019).
DOI: 10.1093/oso/9780198826941.001.0001

$$b_n^2 = \frac{2n \log n}{\pi \gamma}. \tag{13.3}$$

A useful representation for the partial sums Z_n is given by the local times of the random walk. If we write $N_{k,n}$ for the local time of the random walk at point $k \in \mathbb{Z}$ up to time n, i.e.

$$N_{k,n} = \sum_{i=0}^{n} 1_{\{S_i = k\}}, \tag{13.4}$$

then we have for $n \geq 0$

$$Z_n = \sum_{i=0}^{n} X_{S_i} = \sum_{k \in \mathbb{Z}} N_{k,n} X_k.$$

In this section, we consider three kinds of conditions for the sequence $X = (X_k)_{k\in\mathbb{Z}}$ usually called the *random scenery*.

Condition A_1 *$X = (X_k)_{k\in\mathbb{Z}}$ is a $\mathbb{L}^2$-stationary sequence of real-valued centered random variables with a continuous spectral density function on $(-\pi, \pi]$, satisfying the asymptotic negative dependence condition (9.2) and such that $\{X_k^2\}$ is an uniformly integrable family.*

Condition A_2 *For any $k \in \mathbb{Z}$, $X_k = h(\phi_i, \psi_i) - \mathbb{E}h(\phi_i, \psi_i)$ where $(\phi_i; i \in \mathbb{Z})$ is a stationary determinantal point process, $(\psi_i; i \in \mathbb{Z})$ is a non-degenerate stationary Gaussian sequence with positive continuous spectral density and independent of $(\phi_i; i \in \mathbb{Z})$, and $h(\cdot, \cdot)$ is a measurable function from $\{0, 1\} \times \mathbb{R}$ to $\mathbb{R}$ and such that $\mathbb{E}[h^2(X_1, Y_1)] < \infty$.*

Condition A_3 *$(X_i)_{i\in\mathbb{Z}}$ is a strictly stationary and ergodic sequence of centered real-valued random variables with finite second moment, adapted to a stationary filtration $(\mathcal{F}_i)_{i\in\mathbb{Z}}$ and satisfying*

$$\Gamma_j = \sum_{k=0}^{\infty} |\mathbb{E}[X_k \mathbb{E}(X_0|\mathcal{F}_{-j})]| < \infty \quad \textit{for all } j \in \mathbb{Z} \textit{ and } \frac{1}{p}\sum_{j=1}^{p} \Gamma_j \to 0 \quad \textit{as } p \to \infty. \tag{13.5}$$

Recall that under conditions A_2 and A_3, X has a continuous spectral density function f on $(-\pi, \pi]$ (see Corollary 12.26 and Theorem 12.8).

13.1.1 On the Central Limit Behavior of Z_n

The main result of this section is the following:

Theorem 13.2 *Assume that $X = (X_k)_{k\in\mathbb{Z}}$ satisfies either condition A_1, or condition A_2, or condition A_3. Let $(S_n)_{n\geq 0}$ be as in Definition 13.1 and satisfying condition A_0. Define Z_n by (13.2) and b_n^2 by (13.3). Then*

$$b_n^{-1} Z_n \to^d \sqrt{2\pi f(0)} N,$$

where $N \sim \mathcal{N}(0, 1)$ and f is the spectral density of X.

Proof Let

$$\Omega_1 = \left\{ w \in \Omega \ : \ \frac{1}{b_n^2} \sum_{k \in \mathbb{Z}} N_{k,n}^2(\omega) \to 1, \ \text{as } n \to \infty \right\},$$

and

$$\Omega_2 = \left\{ w \in \Omega \ : \ \frac{1}{b_n^2} \sum_{k \in \mathbb{Z}} \left(N_{k,n}(\omega) - N_{k-1,n}(\omega) \right)^2 \to 0, \ \text{as } n \to \infty \right\}.$$

Denote now by $\mathbb{P}_\xi$ the probability $\mathbb{P}$ given all the process $\xi = (\xi_k)$. Denote by $\mathbb{E}_\xi$ (resp. by Var_ξ) the expectation (resp. the variance) under $\mathbb{P}_\xi$. Working under $\mathbb{P}_\xi$ is like working with deterministic $N_{k,n}$'s. Since the random variables $N_{k,n}$ are non-negative, by applying either Corollary 12.21 if condition A_1 holds, Corollary 12.22 if condition A_2 does or Theorem 12.8 under condition A_3, we get that, for any continuous and bounded function g, the following convergence holds on the set $\Omega_1 \cap \Omega_2$:

$$\mathbb{E}_\xi \left(g \left(\sum_{k \in \mathbb{Z}} \alpha_{kn} X_k \right) \right) \to \mathbb{E}_\xi (g(\sigma N)),$$

where $\sigma^2 = 2\pi f(0)$ and N is a standard normal random variable. Therefore if we can prove that

$$\frac{1}{b_n^2} \sum_{k \in \mathbb{Z}} N_{k,n}^2 \to 1 \text{ a.s. and } \frac{1}{b_n^2} \sum_{k \in \mathbb{Z}} \left(N_{k,n} - N_{k-1,n} \right)^2 \to 0 \text{ a.s.} \tag{13.6}$$

as $n \to \infty$, then Theorem 13.2 will follow. Observe that

$$\sum_{k \in \mathbb{Z}} N_{k,n}^2 = \sum_{k \in \mathbb{Z}} \left(\sum_{i=0}^{n} 1_{\{S_i = k\}} \right)^2 = \sum_{i,j=0}^{n} 1_{\{S_i = S_j\}} := V_n(0).$$

Therefore Lemma 13.11 of Section 13.2 implies the first part of (13.6). The second part of (13.6) holds by Corollary 13.12 of Section 13.2. This ends the proof of the theorem. ■

13.1.2 *On the Functional form of the Central Limit Theorem*

The aim of this section is to derive the limiting behavior of the partial sums process associated with Z_n and defined as follows:

$$W_n := \left\{ b_n^{-1} \sum_{i=0}^{[nt]} X_{S_i}, \ t \in [0, 1] \right\}. \tag{13.7}$$

Some comments in the case of an i.i.d. scenery. When $(X_i)_{i\in\mathbb{Z}}$ is a sequence of i.i.d. real-valued centered random variables with finite second moment σ^2, then it has been proved in Deligiannidis and Utev (2011), that

$$W_n \to^d \sigma W \text{ in } D([0,1]),$$

where W is a standard Brownian motion. To prove such a convergence result, it suffices, as usual, to prove

(i) the convergence of finite-dimensional distributions, namely: For any $m \geq 1$, any $a_1, \ldots, a_m$ in $\mathbb{R}$ and any $0 = t_0 < t_1 < \ldots < t_m \leq 1$,

$$\sum_{i=1}^{m} a_i(W_n(t_i) - W_n(t_{i-1})) \to^d \mathcal{N}\left(0, \sigma^2 \sum_{i=1}^{m} a_i^2(t_i - t_{i-1})\right). \tag{13.8}$$

(ii) the tightness of the sequence of processes $\{W_n(t), t \in [0,1]\}_{n\geq 1}$.

Item (i) can be proved by using similar arguments as used to show the central limit theorem. Let us now comment on the tightness condition. In the case when $(X_i)_{i\in\mathbb{Z}}$ is a strictly stationary sequence so is the sequence $(X_{S_i})_{i\in\mathbb{Z}}$. Therefore, according to Lemma 1.16, to prove that the sequence (W_n) is tight, it suffices to prove that

$$\lim_{\lambda\to\infty} \limsup_{n\to\infty} \lambda^2 \mathbb{P}\left(\max_{0\leq k\leq n}\left|\sum_{i=0}^{k} X_{S_i}\right| > \lambda b_n\right) = 0. \tag{13.9}$$

In the case where the X_k's are in addition independent, the tightness criterion (13.9) can be proved via the maximal inequality as stated in Theorem 2 in Newman and Wright (1981) (see Bolthausen (1989), pages 114 and 115 for more details). When the X_k's are not assumed to be independent, the Newman–Wright maximal inequality cannot be used anymore, but, as we shall see, if we assume that the X_k's are in $\mathbb{L}^{2+\delta}$, tightness can be proved if conditions A_1 or A_2 are satisfied.

Theorem 13.3 *Let $X = (X_k)_{k\in\mathbb{Z}}$ be a $\mathbb{L}^2$-stationary sequence of real-valued centered random variables such that $\sup_{k\in\mathbb{Z}} \mathbb{E}(|X_k|^{2+\delta}) < \infty$ for some $\delta > 0$. Suppose in addition that X satisfies either condition A_1 or condition A_2. Let $(S_n)_{n\geq 0}$ be as in Definition 13.1 and satisfying condition A_0. Define W_n by (13.7). Then*

$$W_n \to^d \sqrt{2\pi f(0)}W \text{ in } D([0,1]),$$

where W is a standard Brownian motion.

Proof We first prove the convergence of finite-dimensional distributions and then the tightness of W_n.

Convergence of finite-dimensional distributions. For any $m \geq 1$ and any $0 = t_0 < t_1 < \ldots < t_m \leq 1$, let

$$W_n^{(m)} = (W_n(t_i) - W_n(t_{i-1}))_{1\le i\le m}\,.$$

Let now $a = (a_1, \dots, a_m) \in \mathbb{R}_+^m$ and note that

$$< W_n^{(m)}, a >= \frac{1}{b_n} \sum_{k\in\mathbb{Z}} \sum_{i=1}^{m} a_i \left(N_{k,[nt_i]} - N_{k,[nt_{i-1}]}\right) X_k,$$

where $N_{k,n}$ is defined in (13.4). Since the a_i's have been assumed to be non-negative, it follows that for any $k \in \mathbb{Z}$, the random variables $\sum_{i=1}^{m} a_i \left(N_{k,[nt_i]} - N_{k,[nt_{i-1}]}\right)$ are non-negative. Therefore, as done in the proof of Theorem 13.2, an application of Corollary 12.21 if condition A_1 holds and of Corollary 12.22 if condition A2 does, will show that

$$< W_n^{(m)}, a >\to^d \mathcal{N}(0, \sigma_a^2) \text{ where } \sigma_a^2 := 2\pi f(0) \sum_{i=1}^{m} a_i^2 (t_i - t_{i-1}), \tag{13.10}$$

if one can prove that, as $n \to \infty$,

$$\begin{aligned} &\frac{1}{b_n^2} \sum_{k\in\mathbb{Z}} \tilde{N}_{k,n}^2 \to \sum_{i=1}^{m} a_i^2 (t_i - t_{i-1}) \text{ a.s. and} \\ &\frac{1}{b_n^2} \sum_{k\in\mathbb{Z}} \left(\sum_{i=1}^{m} a_i (N_{k,[nt_i]} - N_{k-1,[nt_i]}) \right)^2 \to 0 \text{ a.s.} \end{aligned} \tag{13.11}$$

with

$$\tilde{N}_{k,n} = \sum_{i=1}^{m} a_i \left(N_{k,[nt_i]} - N_{k,[nt_{i-1}]}\right).$$

The second part of (13.11) follows from Corollary 13.12 since

$$\left(\sum_{i=1}^{m} a_i (N_{k,[nt_i]} - N_{k-1,[nt_i]}) \right)^2 \le \sum_{\ell=1}^{m} a_\ell^2 \sum_{i=1}^{m} \left(N_{k,[nt_i]} - N_{k-1,[nt_i]}\right)^2.$$

To prove the first part of (13.11), note that

$$\begin{aligned} \tilde{N}_{k,n}^2 &= \sum_{i=1}^{m} a_i^2 \left(N_{k,[nt_i]}^2 - N_{k,[nt_{i-1}]}^2 - 2N_{k,[nt_{i-1}]} \left(N_{k,[nt_i]} - N_{k,[nt_{i-1}]}\right)\right) \\ &\quad + 2 \sum_{i=1}^{m-1} \sum_{j=i+1}^{m} a_i a_j \left(N_{k,[nt_i]} - N_{k,[nt_{i-1}]}\right) \left(N_{k,[nt_j]} - N_{k,[nt_{j-1}]}\right). \end{aligned} \tag{13.12}$$

This proves the first part of (13.11) by taking into account Lemmas 13.11 and 13.13. This ends the proof of (13.10).

In addition, by using Lemma 1.5 and the independence between the scenery and the random walk, we infer that

$$\mathbb{E}\left(< W_n^{(m)}, a >^2\right) \to \sigma_a^2,$$

provided that, as $n \to \infty$,

$$\frac{1}{b_n^2} \sum_{k\in\mathbb{Z}} \mathbb{E}(\tilde{N}_{k,n}^2) \to \sum_{i=1}^{m} a_i^2(t_i - t_{i-1}) \text{ and } \frac{1}{b_n^2} \sum_{k\in\mathbb{Z}} \mathbb{E}\left((N_{k,[nt_i]} - N_{k-1,[nt_i]})^2\right) \to 0.$$

The second part comes from Lemma 13.5, whereas to prove the first part it suffices to consider the equality (13.12) and to take into account Lemmas 13.5 and 13.13.

From all these considerations, we then conclude that the process $W_n^{(m)}$ is tight. In addition each partial limit $N = (N_1, ..., N_m)$ is such that each of its marginals is Gaussian and $< N, a >$ is a centered Gaussian random variable with the following variance: $2\pi f(0) \sum_{i=1}^{m} a_i^2(t_i - t_{i-1})$ for any $a = (a_1, \dots, a_m) \in \mathbb{R}_+^m$. By Property 9.19, this is enough to conclude that N is a centered Gaussian vector such that for any $a \in \mathbb{R}^m$, $< N, a >\sim \mathcal{N}(0, \sigma_a^2)$ which ends the proof of the convergence of finite-dimensional distributions.

Tightness of W_n. Tightness will follow from Lemma 1.16 if we can prove that (1.24) holds. With this aim, it is enough to prove that, for $0 < \delta \le 2$ such that $\sup_k \mathbb{E}(|X_k|^{2+\delta}) < \infty$,

$$\lim_{m\to\infty} \limsup_{n\to\infty} \sum_{i=1}^{m} \mathbb{E}\left(\sup_{(i-1)m^{-1}<t\le im^{-1}} \left|W_n(t) - W_n((i-1)m^{-1})\right|^{2+\delta} \right) = 0. \tag{13.13}$$

Let

$$S_{j,n}^{(i)} = S_{j+[n(i-1)m^{-1}]} \text{ and } k_{n,m} = [1 + n/m],$$

and note that

$$\sup_{(i-1)m^{-1}<t\le im^{-1}} |W_n(t) - W_n((i-1)m^{-1})| \le b_n^{-1} \max_{0\le k\le k_{n,m}} \left| \sum_{j=0}^{k} X_{S_{j,n}^{(i)}} \right| . \tag{13.14}$$

To give an upper bound of $\mathbb{E}\left(\max_{0\le k\le k_{n,m}} \left|\sum_{j=0}^{k} X_{S_{j,n}^{(i)}}\right|^{2+\delta} \right)$ we shall use the chaining Lemma 1.21. Therefore, if we can prove that there exists a positive constant C_1 not depending on (n, i, m) such that for any positive integer ℓ,

$$\mathbb{E}\left| \sum_{j=2^k(\ell-1)+1}^{2^k\ell} X_{S_{j,n}^{(i)}} \right|^{2+\delta} \le C_1\, b_{2^k}^{2+\delta} \tag{13.15}$$

then, by using the fact that $b_n^2 = \dfrac{2n\log n}{\pi\gamma}$, we will get that there exists a positive constant C_2 not depending on (n, i, m) such that

$$\left\| \max_{0\le k\le k_{n,m}} \left| \sum_{j=0}^{k} X_{S_{j,n}^{(i)}} \right| \right\|_{2+\delta}^{2+\delta} \le C_2 \frac{b_n^{2+\delta}}{m^{1+\delta/2}}.$$

Combining this upper bound with (13.14) leads to (13.13) and will end the proof of the tightness of W_n.

Therefore, the proof of the theorem will be complete if one can prove (13.15). With this aim, we write

$$\sum_{j=2^k(\ell-1)+1}^{2^k\ell} X_{S_{j,n}^{(i)}} = \sum_{x\in\mathbb{Z}} \tilde{\alpha}_{x,n}^{(k,\ell)} X_x \text{ where } \tilde{\alpha}_{x,n}^{(k,\ell)} := \sum_{j=1}^{2^k} \mathbf{1}_{S_{j+2^k(\ell-1),n}^{(i)}=x}.$$

Moreover, for any positive integer u and any $j = 0, \ldots, u-1$, we write

$$j(u) := \{x \in \mathbb{Z} \,:\, x = j \bmod(u)\},$$

which gives

$$\sum_{j=2^k(\ell-1)+1}^{2^k\ell} X_{S_{j,n}^{(i)}} = \sum_{j=0}^{u-1} \sum_{x\in j(u)} \tilde{\alpha}_{x,n}^{(k,\ell)} X_x.$$

Therefore, for any positive integer u,

$$\begin{aligned}\mathbb{E}\left| \sum_{j=2^k(\ell-1)+1}^{2^k\ell} X_{S_{j,n}^{(i)}} \right|^{2+\delta} &\le u^{1+\delta} \sum_{j=0}^{u-1} \mathbb{E}\left| \sum_{x\in j(u)} \tilde{\alpha}_{x,n}^{(k,\ell)} X_x \right|^{2+\delta} \\ &= u^{1+\delta} \sum_{j=0}^{u-1} \mathbb{E}\left(\mathbb{E}_\xi \left(\left| \sum_{x\in j(u)} \tilde{\alpha}_{x,n}^{(k,\ell)} X_x \right|^{2+\delta} \right)\right),\end{aligned}$$

where $\mathbb{E}_\xi$ is the expectation under $\mathbb{P}_\xi$ (the probability $\mathbb{P}$ given all the process $\xi = (\xi_k)$).

Let us assume, in what follows, that in addition to satisfy $\sup_k \mathbb{E}(|X_k|^{2+\delta}) < \infty$, the process X satisfies condition A_1. We then have $\lim_{n\to 0} r_n = 0$. This implies that for u large enough, the coefficient r_1' associated with the sequence $\{X_i, i \in j(u)\}$ and defined by (7.16) satisfies $r_1' < 1$. Hence, using Fatou's Lemma and applying Proposition 8.3 (note that the random variables $\tilde{\alpha}_{x,n}^{(k,\ell)}$ are non-negative), it follows that for any positive integer u fixed but large enough, there exists a constant c depending only on δ and r_1', such that

$$\mathbb{E}\left(\mathbb{E}_\xi\left(\left|\sum_{x\in j(u)}\tilde{\alpha}_{x,n}^{(k,\ell)}X_x\right|^{2+\delta}\right)\right)\le c\,\mathbb{E}\left(\mathbb{E}_\xi\left(\sum_{x\in j(u)}\left(\tilde{\alpha}_{x,n}^{(k,\ell)}\right)^2X_x^2\right)^{(2+\delta)/2}\right)$$
$$\le c\sup_k\mathbb{E}(|X_k|^{2+\delta})\mathbb{E}\left(\sum_{x\in j(u)}\left(\tilde{\alpha}_{x,n}^{(k,\ell)}\right)^2\right)^{(2+\delta)/2}.$$

So, overall, for any positive integer u fixed but large enough,

$$\mathbb{E}\left|\sum_{j=2^k(\ell-1)+1}^{2^k\ell}X_{S_{j,n}^{(i)}}\right|^{2+\delta}\le c\,u^{2+\delta}\sup_k\mathbb{E}(|X_k|^{2+\delta})\mathbb{E}\left(\sum_{x\in\mathbb{Z}}\left(\tilde{\alpha}_{x,n}^{(k,\ell)}\right)^2\right)^{(2+\delta)/2},\tag{13.16}$$

which will prove (13.15) (and then the tightness of W_n) if one can show that there exists a positive constant c_2 such that

$$\mathbb{E}\left(\sum_{x\in\mathbb{Z}}\left(\tilde{\alpha}_{x,n}^{(k,\ell)}\right)^2\right)^{(2+\delta)/2}\le c_2\,b_{2^k}^{2+\delta}.\tag{13.17}$$

With this aim, note that since $\{\xi_i, i\in\mathbb{Z}\}$ is a strictly stationary sequence,

$$\mathbb{E}\left(\sum_{x\in\mathbb{Z}}\left(\tilde{\alpha}_{x,n}^{(k,\ell)}\right)^2\right)^{(2+\delta)/2}=\mathbb{E}\left(\sum_{i,j=1}^{2^k}1_{\{S_i=S_j\}}\right)^{(2+\delta)/2}\le\mathbb{E}\left(V_{2^k}(0)\right)^{(2+\delta)/2}.$$

(See Definition (13.18)). Since $0<\delta\le 2$, by convexity, we have

$$\mathbb{E}\left(V_{2^k}(0)\right)^{(2+\delta)/2}\le 2^{\delta/2}\left(\mathbb{E}(V_{2^k}(0))\right)^{(2+\delta)/2}+2^{\delta/2}\|V_{2^k}(0)-\mathbb{E}(V_{2^k}(0))\|_2^{(2+\delta)/2}.$$

Hence, by Lemmas 13.5 and 13.8, we derive that there exists a positive constant C not depending on k such that

$$\mathbb{E}\left(V_{2^k}(0)\right)^{(2+\delta)/2}\le C\left(b_{2^k}^{2+\delta}+2^{k(2+\delta)/2}\right)\le 2C\left(b_{2^k}^{2+\delta}\right),$$

which ends the proof of (13.17) and then of the tightness of W_n if X satisfies condition A_1 and $\sup_k\mathbb{E}(|X_k|^{2+\delta})<\infty$.

Now, if X satisfies condition A_2 rather than condition A_1, we use the decomposition (10.3) to write

$$X_k=\left(X_k^{(1)}-\mathbb{E}\left(X_k^{(1)}\right)\right)-\left(X_k^{(2)}-\mathbb{E}\left(X_k^{(2)}\right)\right)+\left(X_k^{(3)}\right)-\mathbb{E}\left(X_k^{(3)}\right),$$

where $X_k^{(1)}=\phi_k(h(1,\psi_k)-h(0,\psi_k))_+$, $X_k^{(2)}=\phi_k(-h(1,\psi_k)+h(0,\psi_k))_+$ and $X_k^{(3)}=h(0,\psi_k)$. It has been proved, in the proof of Corollary 10.4, that for each $i=1,2,3$, the process $X^{(i)} = \{X_k^{(i)}, k \in \mathbb{Z}\}$ satisfies $\lim_{n\to 0} r_n(X^{(i)}) = 0$ and therefore condition A_1. Hence each of the processes $W_n^{(i)} = \left\{b_n^{-1} \sum_{k=0}^{[nt]} (X_{S_k}^{(i)} - \mathbb{E}(X_{S_k}^{(i)})), t \in [0,1]\right\}$, for $i = 1, 2, 3$, is tight which obviously implies the tightness of W_n. ■

Comment 13.4 *Analyzing the proof of Theorem* 13.3, *we can formulate other dependence conditions on the random scenery for its conclusion to hold. For instance if* $(X_i)_{i\in\mathbb{Z}}$ *is a strictly stationary and ergodic sequence of centered real-valued random variables adapted to a non-decreasing filtration* $(\mathcal{F}_i)_{i\in\mathbb{Z}}$ *and satisfying the following reinforced Hannan's condition:*

$$\mathbb{E}(X_0|\mathcal{F}_{-\infty}) = 0 \text{ a.s. and } \sum_{i\geq 0} \|\mathbf{P}_0(\xi_i)\|_{2+\delta} < \infty \text{ for some } \delta > 0,$$

then the conclusion of Theorem 13.3 *holds. Indeed, due to Comment* 12.11, *the condition above implies condition* A3 *and then the convergence of the finite-dimensional distributions. To prove the tightness of* W_n, *it suffices to use Theorem* 3.10 *instead of Proposition* 8.3 *to get* (13.16).

13.2 Properties of the Random Walk

Let $(S_n)_{n\geq 0}$ be defined in Definition 13.1 and satisfying condition A_0. The aim of this section is to collect some properties of the random walk $(S_n)_{n\geq 0}$ that are useful to prove Theorem 13.2. Let us introduce the following notation: For any $a \in \mathbb{Z}$,

$$V_n(a) = \sum_{k\in\mathbb{Z}} N_{k,n} N_{k+a,n}, \tag{13.18}$$

where $N_{k,n}$ is the local time of the random walk at point $k \in \mathbb{Z}$ up to time n (see (13.4)). Note that $V_n(a)$ can be rewritten as

$$V_n(a) = \sum_{i,j=0}^{n} \mathbf{1}_{\{S_j = S_i + a\}}. \tag{13.19}$$

Lemma 13.5 *For any* $a \in \mathbb{Z}$,

$$\mathbb{E}(V_n(a)) = \frac{2n \log n}{\pi\gamma} + O(n) \text{ as } n \to \infty.$$

Proof For any $a \in \mathbb{Z}$, note that

$$\begin{aligned}
\mathbb{E}(V_n(a)) &= \sum_{i=0}^{n-1} \sum_{j=i+1}^{n} \mathbb{P}(S_j - S_i = a) + \sum_{i=1}^{n} \sum_{j=0}^{i-1} \mathbb{P}(S_i - S_j = -a) + (n+1)\mathbf{1}_{a=0} \\
&= \sum_{i=1}^{n} \sum_{j=1}^{i} \big(\mathbb{P}(S_j = a) + \mathbb{P}(S_j = -a)\big) + (n+1)\mathbf{1}_{a=0}.
\end{aligned} \tag{13.20}$$

As in the proof of Lemma 2.3 in Bolthausen (1989), to evaluate $\sum_{j=1}^{i}\left(\mathbb{P}(S_j = a) + \mathbb{P}(S_j = -a)\right)$ for i large enough, we shall use the Tauberian Theorem for power series recalled in Fact 13.6 (see Theorem 5 p. 447 in Feller (1971)):

Fact 13.6 *Let $q_n \geq 0$ and suppose that*

$$Q(\lambda) = \sum_{n\geq 0} q_n \lambda^n \text{ converges for } 0 \leq \lambda < 1.$$

If L varies slowly at infinity and $0 \leq \rho < \infty$ then each of the two relations below implies the other:

(1) $Q(\lambda) \sim \frac{1}{(1-\lambda)^{\rho}} L\left(\frac{1}{1-\lambda}\right), \lambda \to 1^-$

and

(2) $q_0 + q_1 + \ldots + q_{n-1} \sim \frac{1}{\Gamma(\rho+1)} n^{\rho} L(n)$, as $n \to \infty$,

where Γ is the Euler gamma function. Furthermore if the sequence $\{q_n\}$ is monotonic and $0 < \rho < \infty$, then the relation (1) is equivalent to

$$q_n \sim \frac{1}{\Gamma(\rho)} n^{\rho-1} L(n), \; n \to \infty.$$

Therefore, for any $0 \leq \lambda < 1$, we set

$$Q(\lambda) = \sum_{i=1}^{\infty} \lambda^i \left(\mathbb{P}(S_i = a) + \mathbb{P}(S_i = -a)\right),$$

and our aim is to evaluate $Q(\lambda)$ as $\lambda \to 1^-$. By the fact that the ξ_i's are i.i.d., the inverse Fourier formula gives

$$\mathbb{P}(S_i = a) + \mathbb{P}(S_i = -a) = \frac{1}{\pi}\int_{-\pi}^{\pi} \cos(at)\phi^i(t)\mathrm{d}t.$$

Hence,

$$Q(\lambda) = \frac{1}{\pi}\int_{-\pi}^{\pi} \frac{\lambda\cos(at)\phi(t)}{1-\lambda\phi(t)}\mathrm{d}t.$$

By taking into account the relation (13.20) and Fact 13.6, the lemma will follow if we can prove:

Lemma 13.7 *For any $a \in \mathbb{Z}$, as $\lambda \to 1^-$,*

$$\frac{1}{\pi}\int_{-\pi}^{\pi} \frac{\lambda\cos(at)\phi(t)}{1-\lambda\phi(t)}\mathrm{d}t \sim \frac{2}{\pi\gamma}\log\left(\frac{1}{1-\lambda}\right). \tag{13.21}$$

To end the proof, it remains to prove Lemma 13.7 above. With this aim, let us consider an arbitrary small $\delta > 0$ and note that

$$\frac{1}{\pi}\int_{t=-\pi}^{\pi} \frac{\lambda\cos(at)\phi(t)}{1-\lambda\phi(t)}\mathrm{d}t = \frac{1}{\pi}\int_{|t|\le\delta} \frac{\lambda\cos(at)\phi(t)}{1-\lambda\phi(t)}\mathrm{d}t + \frac{1}{\pi}\int_{\pi\ge|t|>\delta} \frac{\lambda\cos(at)\phi(t)}{1-\lambda\phi(t)}\mathrm{d}t.$$

By strong aperiodicity, for small enough $\delta > 0$ there exists a small positive constant $D(\delta)$ such that $|\varphi(t)| < 1 - D(\delta)$ when $|t| > \delta$. Thus,

$$\left|\frac{1}{\pi}\int_{\pi\ge|t|>\delta} \frac{\lambda\cos(at)\phi(t)}{1-\lambda\phi(t)}\mathrm{d}t\right| \le CD(\delta)^{-1}, \tag{13.22}$$

for all $0 \le \lambda < 1$. On the other hand, for δ small enough,

$$\frac{1}{\pi}\int_{|t|\le\delta} \frac{\lambda\cos(at)\phi(t)}{1-\lambda\phi(t)}\mathrm{d}t = \frac{1}{\pi}\int_{|t|\le\delta} \frac{\lambda\cos(at)\phi(t)}{1-\lambda(1-\gamma|t|)}\mathrm{d}t + I(\lambda,\delta), \tag{13.23}$$

where, for any $0 \le \lambda < 1$,

$$\begin{aligned}|I(\lambda,\delta)| &\le \frac{1}{\pi}\int_{|t|\le\delta}\left|\frac{\lambda\phi(t)}{1-\lambda\phi(t)} - \frac{\lambda\phi(t)}{1-\lambda(1-\gamma|t|)}\right|\mathrm{d}t\\ &\le \frac{1}{\pi}\int_{|t|\le\delta}\frac{|\phi(t)-1+\gamma|t||}{|1-\lambda\phi(t)|\times|1-\lambda(1-\gamma|t|)|}\mathrm{d}t.\end{aligned}$$

Now, by condition (13.1), for any $\delta > 0$, there exists a positive constant $C(\delta) = o_\delta(1)$ such that for $|t| < \delta$, $|\varphi(t) - 1 - \gamma|t|| \le C(\delta)|t|$. In addition for $\delta > 0$ small enough and $|t| < \delta$,

$$|1-\lambda\phi(t)| \ge 1 - |\phi(t)| \ge \frac{\gamma}{2}|t|,$$

for any $0 \le \lambda < 1$. Therefore, for $\delta > 0$ small enough,

$$\begin{aligned}|I(\lambda,\delta)| &\le C(\delta)\frac{1}{\pi}\int_{|t|\le\delta}\frac{|t|}{|1-\lambda\phi(t)|\times|1-\lambda(1-\gamma|t|)|}\mathrm{d}t\\ &\le C(\delta)\frac{4}{\gamma\pi}\int_{t=0}^{\delta}\frac{1}{1-\lambda+\lambda\gamma t}\mathrm{d}t\\ &= O\left(C(\delta)\frac{1}{\lambda}\log\left(\frac{1}{1-\lambda}\right)\right),\end{aligned} \tag{13.24}$$

for any $0 \le \lambda < 1$ (we interpret $0/0$ as 0). In addition, as $\lambda \to 1^-$, we have, for any $a \in \mathbb{Z}$,

$$\frac{1}{\pi}\int_{|t|\le\delta}\frac{\lambda\cos(at)\phi(t)}{1-\lambda(1-\gamma|t|)}dt \sim \frac{2}{\pi}\int_{t=0}^{\delta}\frac{1}{1-\lambda+\lambda\gamma t}dt + o_\delta(1)$$

$$\sim \frac{2}{\pi\gamma}\int_{t=0}^{\lambda\gamma\delta/(1-\lambda)}\frac{dt}{1+t} + o_\delta(1) \sim \frac{2}{\pi\gamma}\log\left(\frac{1}{1-\lambda}\right) + o_\delta(1). \tag{13.25}$$

Therefore, by taking into account (13.24) and (13.25) in (13.23), we get, as $\lambda \to 1^-$, that

$$\frac{1}{\pi}\int_{|t|\le\delta}\frac{\lambda\cos(at)\phi(t)}{1-\lambda\phi(t)}dt \sim \frac{2}{\pi\gamma}\log\left(\frac{1}{1-\lambda}\right)(1+O(C(\delta)))+o_\delta(1)$$
$$\sim \frac{2}{\pi\gamma}\log\left(\frac{1}{1-\lambda}\right),$$

since δ is arbitrarily small and $C(\delta)\to 0$ and $o_\delta(1)\to 0$ as $\delta\to 0$. This estimation when $\lambda\to 1^-$ combined with (13.22) proves the lemma. ∎

Lemma 13.8 *For any $a\in\mathbb{Z}$,*

$$\mathrm{Var}(V_n(a)) = O(n^2) \text{ as } n\to\infty.$$

Remark 13.9 *Note that a more precise computation for $Var(V_n(0))$ is given in Deligiannidis and Utev (2011). Their proof is based on a Darboux–Wiener type result (see their Lemma 3.1).*

Proof To prove the variance bound we follow closely the approach in Deligiannidis and Utev (2011).

$$\begin{aligned}\mathrm{Var}(V_n(a)) &= \sum_{x,y}\left[\mathbb{E}(N_{x,n}N_{x+a,n}N_{y,n}N_{y+a,n}) - \mathbb{E}(N_{x,n}N_{x+a,n})\mathbb{E}(N_{y,n}N_{y+a,n})\right]\\ &= \sum_{x,y}\sum_{i,j,k,l}\big[\mathbb{P}(S_i=x,S_j=x+a,S_k=y,S_l=y+a)\\ &\quad -\mathbb{P}(S_i=x,S_j=x+a)\mathbb{P}(S_k=y,S_l=y+a)\big].\end{aligned}$$

Notice that if $i=j$ or $k=l$ both terms of the difference are zero for any $a\in\mathbb{Z}^*$. In addition, if $a=0$ and if $i=j$ or $k=l$, summing over x and y, the difference gives also zero.

We shall only treat the case $i<j$ and $k<l$, other cases being similar. There are three possible configurations and for each one we define a set of quadruples:

1. $I_1=\{(i,j,k,l): 0\le i<j\le k<l\le n\}$,
2. $I_2=\{(i,j,k,l): 0\le i\le k<j\le l\le n\}$,
3. $I_3=\{(i,j,k,l): 0\le i\le k<l\le j\le n\}$.

We treat the sum over each of these sets one by one. The sum over I_1 can be easily shown to vanish, using the fact that the ξ_i's are independent.

Sum over I_2.

For the sum over I_2, we define the set M_n of 5-tuples $(m_1, \ldots, m_5)$ such that

$$i = m_1,\ k = m_1 + m_2,\ j = m_1 + m_2 + m_3,\ l = m_1 + \cdots + m_4,\ n = m_1 + \cdots + m_5,$$

where $m_1, m_2, m_4, m_5 \geq 0$ and $m_3 > 0$. Then the term under consideration is given by

$$\begin{aligned} a_1(n) = \sum_{x,y} \sum_{I_2} \Big[&\mathbb{P}(S_i = x, S_j = x + a, S_k = y, S_l = y + a) \\ &- \mathbb{P}(S_i = x, S_j = x + a)\mathbb{P}(S_k = y, S_l = y + a)\Big] \\ \leq \sum_{x,y} \sum_{M_n} &\mathbb{P}(S_{m_1} = x, S_{m_1+m_2} = y, S_{m_1+m_2+m_3} = x + a, S_{m_1+\cdots+m_4} = y + a), \end{aligned}$$

where we obtained the upper bound by dropping the negative term. Let $J = (-\pi, \pi]$. After changing variables $y = x + z$ and using Fourier inversion formula we obtain,

$$\begin{aligned} a_1'(n) &:= \sum_{x,y} \sum_{M_n} \mathbb{P}(S_{m_1} = x, S_{m_1+m_2} = y, S_{m_1+m_2+m_3} = x + a, S_{m_1+\cdots+m_4} = y + a) \\ &= \sum_{x,z} \sum_{M_n} \mathbb{P}(S_{m_1} = x)\mathbb{P}(S_{m_2} = z)\mathbb{P}(S_{m_3} = a - z)\mathbb{P}(S_{m_4} = z) \\ &= \sum_{z} \sum_{M_n} \mathbb{P}(S_{m_2} = z)\mathbb{P}(S_{m_3} = a - z)\mathbb{P}(S_{m_4} = z) \\ &= \sum_{z} \sum_{M_n} \mathbb{P}(S_{m_3} = a - z)\frac{1}{(2\pi)^2} \int_{J^2} \mathrm{e}^{-\mathrm{i}(t_1+t_2)z} \phi^{m_2}(t_1)\phi^{m_4}(t_2)\mathrm{d}t_1\mathrm{d}t_2 \\ &= \frac{1}{(2\pi)^2} \sum_{M_n} \int_{J^2} \mathrm{e}^{-\mathrm{i}(t_1+t_2)a} \phi^{m_2}(t_1)\phi^{m_4}(t_2) \sum_{z} \mathbb{P}(S_{m_3} = a - z)\mathrm{e}^{\mathrm{i}(t_1+t_2)(a-z)}\mathrm{d}t_1\mathrm{d}t_2 \\ &= \frac{1}{(2\pi)^2} \sum_{M_n} \int_{J^2} \mathrm{e}^{-\mathrm{i}(t_1+t_2)a} \phi^{m_2}(t_1)\phi^{m_4}(t_2)\phi^{m_3}(t_1 + t_2)\mathrm{d}t_1\mathrm{d}t_2. \end{aligned}$$

Next for $\lambda \in [0, 1)$ we define $f_1(\lambda) := \sum_{n\geq 0} \lambda^n a_1'(n)$. Note that, by simple algebra,

$$\begin{aligned} &\sum_{n=0}^{\infty} \sum_{M_n} \lambda^n \phi^{m_2}(t_1)\phi^{m_4}(t_2)\phi^{m_3}(t_1 + t_2) \\ &\qquad = (1 - \lambda)^{-2} \frac{\lambda\phi(t_1 + t_2)}{(1 - \lambda\phi(t_1))(1 - \lambda\phi(t_2))(1 - \lambda\phi(t_1 + t_2))}. \end{aligned}$$

Therefore, for any $\lambda \in [0, 1)$,

$$f_1(\lambda) = \frac{(1-\lambda)^{-2}}{(2\pi)^2} \int_{J^2} e^{-i(t_1+t_2)a} \frac{\lambda\phi(t_1+t_2)}{(1-\lambda\phi(t_1))(1-\lambda\phi(t_2))(1-\lambda\phi(t_1+t_2))} dt_1 dt_2.$$

Using the technique introduced in Bolthausen (1989) and strong aperiodicity, one can see that the important region is $U_\epsilon = [-\epsilon, \epsilon]^2$, since

$$f_1(\lambda) = \frac{(1-\lambda)^{-2}}{(2\pi)^2} \times \int_{U_\epsilon} e^{-i(t_1+t_2)a} \frac{\lambda\phi(t_1+t_2)}{(1-\lambda\phi(t_1))(1-\lambda\phi(t_2))(1-\lambda\phi(t_1+t_2))} dt_1 dt_2 + O\left((1-\lambda)^{-2}\right).$$

First notice that for $|s| \leq 2\epsilon$, $|R(s)| \leq C(\epsilon)|s|$, where $C(\epsilon) \to 0$ as $\epsilon \to 0$. Thus, by using the triangle inequality,

$$\begin{aligned} |1-\lambda\phi(s)| &= |1-\lambda(1-\gamma|s|+R(s))| \\ &\geq |1-\lambda+\lambda\gamma|s|| - |R(s)| \\ &= 1-\lambda+\lambda\gamma|s| - |R(s)| \\ &\geq 1-\lambda+\frac{\gamma}{2}|s|, \end{aligned}$$

for small enough ϵ and λ close to 1 (at this stage we only take limits wrt to λ).

Therefore, as $\lambda \to 1^-$, we have

$$\begin{aligned} |f_1(\lambda)| &\leq C(1-\lambda)^{-2} \int_{U_\epsilon} \frac{dtds}{(1-\lambda+\frac{\gamma}{2}|t|)(1-\lambda+\frac{\gamma}{2}|s|)(1-\lambda+\frac{\gamma}{2}|t+s|)} \\ &\quad + O\left((1-\lambda)^{-2}\right) \leq C(1-\lambda)^{-3} \int_{-\infty}^{\infty}\int_{-\infty}^{\infty} \frac{drds}{(1+\gamma|r|)(1+\gamma|s|)(1+\gamma|r+s|)} \\ &\leq C(1-\lambda)^{-3}. \end{aligned}$$

A simple application of the Tauberian Theorem for power series (see Fact 13.6) gives that there exists a positive constant C not depending on (a, n) such that

$$a_1(n) \leq a_1'(n) \leq Cn^2. \tag{13.26}$$

Third sum.

We now consider the final sum over I_3. We define the set M_n of 5-tuples $(m_1, \ldots, m_5)$ such that

$$i = m_1,\ k = m_1 + m_2,\ l = m_1 + m_2 + m_3,\ j = m_1 + \cdots + m_4,\ n = m_1 + \cdots + m_5,$$

where $m_1, m_2, m_4, m_5 \geq 0$ and $m_3 > 0$. Hence we write

$$a_2(n) = \sum_{x,y} \sum_{I_2} \left[\mathbb{P}(S_i = x, S_j = x + a, S_k = y, S_l = y + a) \right.$$
$$\left. -\mathbb{P}(S_i = x, S_j = x + a)\mathbb{P}(S_k = y, S_l = y + a) \right]$$
$$= \sum_{x,y} \sum_{M_n} \left[\mathbb{P}(S_{m_1} = x, S_{m_1+m_2} = y, S_{m_1+m_2+m_3} = y + a, S_{m_1+\cdots+m_4} = x + a) \right.$$
$$\left. -\mathbb{P}(S_{m_1} = x, S_{m_1+\cdots+m_4} = x + a)\mathbb{P}(S_{m_1+m_2} = y, S_{m_1+m_2+m_3} = y + a) \right].$$

Therefore, using the independence of the ξ_i's and the fact that they are identically distributed, we get, setting $z = y - x$,

$$a_2(n) = \sum_{M_n} \left[\sum_z \mathbb{P}(S_{m_2} = z)\mathbb{P}(S_{m_3} = a)\mathbb{P}(S_{m_4} = -z) \right.$$
$$\left. - \mathbb{P}(S_{m_2+m_3+m_4} = a)\mathbb{P}(S_{m_3} = a) \right],$$

and

$$\sum_z \mathbb{P}(S_{m_2} = z)\mathbb{P}(S_{m_4} = -z) = \sum_z \mathbb{P}(S_{m_2} = z)\mathbb{P}(S_{m_4+m_2} - S_{m_2} = -z)$$
$$= \sum_z \mathbb{P}(S_{m_2} = z, S_{m_4+m_2} - S_{m_2} = -z) = \sum_z \mathbb{P}(S_{m_2} = z, S_{m_4+m_2} = 0)$$
$$= \mathbb{P}(S_{m_4+m_2} = 0).$$

Therefore,

$$a_2(n) = \sum_{M_n} \mathbb{P}(S_{m_3} = a) \left(\mathbb{P}(S_{m_2+m_4} = 0) - \mathbb{P}(S_{m_2+m_3+m_4} = a) \right).$$

Hence, by change of variables,

$$a_2(n) = \sum_{m_1=0}^{n} \sum_{m_3=1}^{n-m_1} \mathbb{P}(S_{m_3} = a) \sum_{m_2=0}^{n-m_1-m_3} \sum_{m_4=0}^{n-m_1-m_3-m_2} \left[\mathbb{P}(S_{m_2+m_4} = 0) - \mathbb{P}(S_{m_2+m_3+m_4} = a) \right]$$
$$= \sum_{j=1}^{n} \sum_{k=1}^{j} \mathbb{P}(S_k = a) \sum_{\ell=0}^{j-k} (\ell + 1) \left[\mathbb{P}(S_\ell = 0) - \mathbb{P}(S_{\ell+k} = a) \right].$$

We write now

$$a_2(n) = a_2'(n) + a_2''(n),$$

where

$$a_2'(n) = \sum_{j=1}^{n}\sum_{k=1}^{j} \mathbb{P}(S_k = a) \sum_{\ell=0}^{j-k} (\ell+1)\,[\mathbb{P}(S_\ell = 0) - \mathbb{P}(S_{\ell+k} = 0)]$$

and

$$a_2''(n) = \sum_{j=1}^{n}\sum_{k=1}^{j} \mathbb{P}(S_k = a) \sum_{\ell=0}^{j-k} (\ell+1)\,[\mathbb{P}(S_{\ell+k} = 0) - \mathbb{P}(S_{\ell+k} = a)]\,. \tag{13.27}$$

Note that

$$\begin{aligned}
&1_{j\geq 2k} \sum_{\ell=0}^{j-k} (\ell+1)\,[\mathbb{P}(S_\ell = 0) - \mathbb{P}(S_{\ell+k} = 0)] \\
&= 1_{j\geq 2k} \sum_{\ell=0}^{j-k} ((\ell+1)\mathbb{P}(S_\ell = 0) - (\ell+k+1)\mathbb{P}(S_{\ell+k} = 0)) + k1_{j\geq 2k} \sum_{\ell=0}^{j-k} \mathbb{P}(S_{\ell+k} = 0) \\
&= 1_{j\geq 2k} \sum_{\ell=0}^{k-1} (\ell+1)\mathbb{P}(S_\ell = 0) - 1_{j\geq 2k} \sum_{\ell=j-k+1}^{j} (\ell+1)\mathbb{P}(S_\ell = 0) \\
&\quad + k1_{j\geq 2k} \sum_{\ell=0}^{j-k} \mathbb{P}(S_{\ell+k} = 0),
\end{aligned}$$

and, on the other hand,

$$1_{j<2k} \sum_{\ell=0}^{j-k} (\ell+1)\,[\mathbb{P}(S_\ell = 0) - \mathbb{P}(S_{\ell+k} = 0)] \leq 1_{j<2k} \sum_{\ell=0}^{k-1} (\ell+1)\mathbb{P}(S_\ell = 0).$$

Therefore, for any $j \geq k$,

$$\sum_{\ell=0}^{j-k} (\ell+1)\,[\mathbb{P}(S_\ell = 0) - \mathbb{P}(S_{\ell+k} = 0)] \leq \sum_{\ell=0}^{k-1} (\ell+1)\mathbb{P}(S_\ell = 0) + k1_{j\geq 2k} \sum_{\ell=0}^{j-k} \mathbb{P}(S_{\ell+k} = 0),$$

implying that

$$\begin{aligned}
a_2(n) \leq |a_2''(n)| &+ \sum_{j=1}^{n}\sum_{k=1}^{j} \mathbb{P}(S_k = a) \sum_{\ell=0}^{k-1} (\ell+1)\mathbb{P}(S_\ell = 0) \\
&+ \sum_{j=1}^{n}\sum_{k=1}^{j} k1_{j\geq 2k}\mathbb{P}(S_k = a) \sum_{\ell=0}^{j-k} \mathbb{P}(S_{\ell+k} = 0).
\end{aligned} \tag{13.28}$$

We use now the following lemma whose proof is given later in this section:

Lemma 13.10 *There exist positive constants C_1 and C_2 such that for any positive integer m and any $w \in \mathbb{Z}$,*

$$\sup_{w\in\mathbb{Z}} \mathbb{P}[S_m = w] \le \frac{C_1}{m} \text{ and } \Big|\mathbb{P}[S_m = w] - \mathbb{P}[S_m = 0]\Big| \le C_2 \frac{|w|}{m^2}.$$

This gives

$$\sum_{j=1}^{n}\sum_{k=1}^{j} \mathbb{P}(S_k = a) \sum_{\ell=0}^{k-1} (\ell+1)\mathbb{P}(S_\ell = 0) \le \sum_{j=1}^{n}\sum_{k=1}^{j} \mathbb{P}(S_k = a)$$
$$+ 2\sum_{j=1}^{n}\sum_{k=2}^{j} \mathbb{P}(S_k = a) \sum_{\ell=1}^{k-1} \ell\, \mathbb{P}(S_\ell = 0) \le c_1 n \log n + c_2 n^2,$$

where c_1 and c_2 are two positive constants not depending on (a, n). Moreover,

$$\sum_{j=1}^{n}\sum_{k=1}^{j} k 1_{j\ge 2k} \mathbb{P}(S_k = a) \sum_{\ell=0}^{j-k} \mathbb{P}(S_{\ell+k} = 0) \le c_3 \sum_{k=1}^{n}\sum_{j=2k}^{n} \log\left(\frac{j+1}{k}\right)$$
$$\le c_3 n \sum_{k=1}^{n} \log\left(\frac{n+k}{k}\right) \le c_3 n \int_0^n \log\left(1+\frac{n}{x}\right) dx = c_3 n^2 \int_1^\infty \frac{\log(1+y)}{y^2} dy \le c_4 n^2,$$

where c_3 and c_4 are positive constants not depending on (a, n). So, overall, starting from (13.28) and taking into account the above computations, it follows that there exists a positive constant c not depending on (a, n) such that

$$a_2(n) \le |a_2''(n)| + cn^2. \tag{13.29}$$

We give now an upper bound for the quantity $|a_2''(n)|$ where $a_2''(n)$ is defined by (13.27). Using Lemma 13.10 again, it follows that there exists a positive constant c not depending on a such that

$$|a_2''(n)| \le c|a| \sum_{j=1}^{n}\sum_{k=1}^{j} \frac{1}{k} \sum_{\ell=0}^{j-k} \frac{1}{\ell+k} \le 4c|a| n (\log(n+1))^2.$$

Taking into account this last upper bound in (13.29) gives the following: There exists a positive constant C not depending on (a, n) such that

$$a_2(n) \le Cn\left(n + |a|(\log n)^2\right). \tag{13.30}$$

Lemma 13.8 then follows by taking into account the upper bounds (13.26) and (13.30). ■

To complete the proof it remains to prove Lemma 13.10.

Proof of Lemma 13.10 We only prove the second inequality, the first one being simpler. For any $\epsilon > 0$, by strong aperiodicity we have that for all $|t| > \epsilon$, $|\varphi(t)| < C(\epsilon) < 1$. Therefore,

$$\begin{aligned}|\mathbb{P}[S_m = 0] - \mathbb{P}[S_m = w]| &\leq \int_{-\pi}^{\pi} |1 - e^{itw}||\phi(t)|^m dt \\ &\leq \int_{|t|<\epsilon} |1 - e^{itw}||\phi(t)|^m dt + 4\pi C(\epsilon)^m,\end{aligned}$$

where the second term decays exponentially fast. For the first term, we note that by (13.1), there exists a constant $D(\epsilon)$ which vanishes as $\epsilon \to 0$ and such that $|\varphi(t) - 1 + \gamma|t|| \leq D(\epsilon)|t|$. Therefore, for ϵ small enough,

$$|\phi(t)| \leq |1 - \gamma|t|| + D(\epsilon)|t| \leq 1 - \frac{\gamma}{2}|t|,$$

implying that

$$\begin{aligned}\int_{|t|<\epsilon} |1 - e^{itw}||\phi(t)|^m dt &\leq 2\int_{|t|<\epsilon} |t||w| \left(1 - \frac{\gamma}{2}|t|\right)^m dt \\ &= 4|w| \int_{t=0}^{\epsilon} t\left(1 - \frac{\gamma t}{2}\right)^m dt \\ &\leq 4|w| \int_{t=0}^{\epsilon} t \exp\left(-\frac{m\gamma t}{2}\right) dt \leq C\frac{|w|}{m^2},\end{aligned}$$

and then the lemma follows. ■

Lemma 13.11 *For any $a \in \mathbb{Z}$,*

$$V_n(a) \sim b_n^2 \quad \text{a.s. as } n \to \infty.$$

Proof Let $\tau_n = [\alpha^n]$ for a fixed $\alpha > 1$, where brackets $[\cdot]$ denote the integer part. Let us then show that

$$\frac{V_{\tau_n}(a)}{b_{\tau_n}^2} \to 1 \quad \text{a.s. as } n \to \infty. \tag{13.31}$$

By using Lemma 13.5, this is equivalent to showing that

$$\frac{V_{\tau_n}(a) - \mathbb{E}(V_{\tau_n}(a))}{b_{\tau_n}^2} \to 0 \quad \text{a.s. as } n \to \infty. \tag{13.32}$$

But, by Markov inequality and Lemma 13.8, for any $\varepsilon > 0$,

$$\mathbb{P}\left(\left|V_{\tau_n}(a) - \mathbb{E}V_{\tau_n}(a)\right| \geq \varepsilon b_{\tau_n}^2\right) \leq \frac{\text{Var}(V_{\tau_n}(a))}{\varepsilon^2 b_{\tau_n}^2} \leq C\frac{\tau_n^2}{\varepsilon^2 \tau_n^2 (\log \tau_n)^2}.$$

Therefore (13.32) (and then (13.31)) follows by the Borel–Cantelli lemma. The rest of the proof is standard but we give it for completeness. Since

$$V_n(a) = \sum_{i,j=0}^{n} 1_{\{S_j = S_i + a\}},$$

$(V_n(a))_{n\geq 0}$ is a non-decreasing sequence as well as the normalizing sequence $(b_n^2)_{n\geq 1}$. Therefore, if $1 < \tau_n \leq k \leq \tau_{n+1}$,

$$\frac{b_{\tau_n}^2}{b_{\tau_{n+1}}^2}\frac{V_{\tau_n}(a)}{b_{\tau_n}^2} \leq \frac{V_k(a)}{b_k^2} \leq \frac{b_{\tau_{n+1}}^2}{b_{\tau_n}^2}\frac{V_{\tau_{n+1}}(a)}{b_{\tau_{n+1}}^2}.$$

But $\dfrac{b_{\tau_{n+1}}^2}{b_{\tau_n}^2} \to \alpha$ as $n \to \infty$. Therefore, by using (13.31),

$$\frac{1}{\alpha} \leq \liminf_{k\to\infty} \frac{V_k(a)}{b_k^2} \leq \limsup_{k\to\infty} \frac{V_k(a)}{b_k^2} \leq \alpha.$$

Taking the intersection of all such events for rationals $\alpha > 1$, the lemma follows. ■

From Lemma 13.11, we have the following easy (but useful) consequence.

Corollary 13.12 *For any $a \in \mathbb{Z}$,*

$$\frac{1}{b_n^2}\sum_{k\in\mathbb{Z}} \left(N_{k,n} - N_{k+a,n}\right)^2 \to 0 \text{ a.s. as } n \to \infty.$$

The following lemma is also valid:

Lemma 13.13 *For any $0 < A < B$, as $n \to \infty$,*

$$\sum_{i=1}^{[An]} \sum_{j=[An]+1}^{[Bn]} \mathbb{P}(S_i = S_j) = o(b_n^2) \text{ and } \sum_{i=1}^{[An]} \sum_{j=[An]+1}^{[Bn]} 1_{\{S_i = S_j\}} = o(b_n^2) \text{ a.s.}$$

Proof The proof of the first part of the lemma follows from simple computations by taking into account the first part of Lemma 13.10. To prove the second part, we consider a fixed $\alpha > 1$ and set $\tau_n = [\alpha^n]$. If $1 < \tau_n \leq k \leq \tau_{n+1}$,

$$b_k^{-2}\sum_{i=1}^{[Ak]} \sum_{j=[Ak]+1}^{[Bk]} 1_{\{S_i = S_j\}} \leq b_{\tau_n}^{-2} \sum_{i=1}^{[A\tau_{n+1}]} \sum_{j=[A\tau_n]+1}^{[B\tau_{n+1}]} 1_{\{S_i = S_j\}}, \tag{13.33}$$

where we have used the fact that $(b_n^2)_{n\geq 1}$ is an increasing sequence. Note now that

$$\sum_{i=1}^{[A\tau_{n+1}]}\sum_{j=[A\tau_n]+1}^{[B\tau_{n+1}]} \mathbb{E}\left(1_{\{S_i=S_j\}}\right) = \sum_{i=[A\tau_n]+1}^{[A\tau_{n+1}]}\sum_{j=[A\tau_n]+1}^{[B\tau_{n+1}]} \mathbb{P}(S_i = S_j) + \sum_{i=1}^{[A\tau_n]}\sum_{j=[A\tau_n]+1}^{[B\tau_{n+1}]} \mathbb{P}(S_i = S_j)$$

$$= \sum_{i,j=[A\tau_n]+1}^{[A\tau_{n+1}]} \mathbb{P}(S_i = S_j) + \sum_{i=[A\tau_n]+1}^{[A\tau_{n+1}]}\sum_{j=[A\tau_{n+1}]+1}^{[B\tau_{n+1}]} \mathbb{P}(S_i = S_j) + \sum_{i=1}^{[A\tau_n]}\sum_{j=[A\tau_n]+1}^{[B\tau_{n+1}]} \mathbb{P}(S_i = S_j)$$

$$:= I_n^{(1)} + I_n^{(2)} + I_n^{(3)}.$$

Taking into account the first part of the lemma, we have

$$b_{\tau_n}^{-2}\left(I_n^{(2)} + I_n^{(3)}\right) \to 0, \text{ as } n \to \infty.$$

Moreover, setting

$$\ell_n = [A\tau_{n+1}] - [A\tau_n] - 1,$$

we have, by strict stationarity of the sequence $(\xi_i)_{i\in\mathbb{Z}}$,

$$I_n^{(1)} = \mathbb{E}\left(V_{\ell_n}(0)\right).$$

But, by Lemma 13.5, $\mathbb{E}\left(V_{\ell_n}(0)\right) \sim b_{\ell_n}^2$. Since $b_{\tau_n}^{-2}b_{\ell_n}^2 \sim A(\alpha-1)$, it follows that

$$b_{\tau_n}^{-2}I_n^{(1)} \to A(\alpha-1), \text{ as } n \to \infty.$$

So, overall,

$$b_{\tau_n}^{-2}\sum_{i=1}^{[A\tau_{n+1}]}\sum_{j=[A\tau_n]+1}^{[B\tau_{n+1}]} \mathbb{P}(S_i = S_j) \to A(\alpha-1), \text{ as } n \to \infty. \tag{13.34}$$

On the other hand, by analyzing the proof of Lemma 13.8, we infer that

$$\mathrm{Var}\left(\sum_{i=1}^{[A\tau_{n+1}]}\sum_{j=[A\tau_n]+1}^{[B\tau_{n+1}]} 1_{\{S_i=S_j\}}\right) = O(\tau_n^2),$$

showing, by using Markov inequality together with the Borel–Cantelli lemma, that

$$b_{\tau_n}^{-2}\left(\sum_{i=1}^{[A\tau_{n+1}]}\sum_{j=[A\tau_n]+1}^{[B\tau_{n+1}]} \left(1_{\{S_i=S_j\}} - \mathbb{P}(S_i = S_j)\right)\right) \to 0 \ a.s., \text{ as } n \to \infty. \tag{13.35}$$

Combining (13.34) and (13.35), we derive that

$$b_{\tau_n}^{-2} \sum_{i=1}^{[A\tau_{n+1}]} \sum_{j=[A\tau_n]+1}^{[B\tau_{n+1}]} 1_{\{S_i=S_j\}} \to A(\alpha-1)\ a.s., \text{ as } n \to \infty. \tag{13.36}$$

Therefore, starting from (13.33) and taking into account (13.36), we have

$$0 \le \liminf_{k\to\infty} b_k^{-2} \sum_{i=1}^{[Ak]} \sum_{j=[Ak]+1}^{[Bk]} 1_{\{S_i=S_j\}} \le \limsup_{k\to\infty} b_k^{-2} \sum_{i=1}^{[Ak]} \sum_{j=[Ak]+1}^{[Bk]} 1_{\{S_i=S_j\}} \le A(\alpha-1).$$

Taking the intersection of all such events for rationals $\alpha > 1$, the lemma follows. ■

13.3 CLT for Random Walk Self-intersections—Non-linear Normalizer

In this section we prove a central limit theorem for the self-intersection local time of random walks in $d \ge 3$. This is a quite remarkable example of a CLT for stationary sequence with non-linear normalization $\sqrt{n \log n}$.

The stationary sequence. Let S_n be a $\mathbb{Z}^d$-valued random walk,

$$S_n = \xi_1 + \cdots + \xi_n,$$

where $\xi_1, \xi_2, \ldots$ are i.i.d., strongly aperiodic, symmetric $\mathbb{Z}^d$-valued r.v.'s, with second moments and covariance matrix Γ. It is a well-known fact that this random walk is transient—this is important for the sequence to be stationary. Let also $S_0 = 0$.

Define the stationary sequence as follows:

$$Z_j := \sum_{k=j+1}^{\infty} I(S_k = S_j), \quad j \ge 0,$$

Notice that we can make this two-sided by continuing the random walk on the left.

First of all notice that

$$\sum_{j=1}^{n-1} Z_j = \sum_{j=1}^{n-1} \sum_{k=j+1}^{n} I(S_j = S_k) + \sum_{j=1}^{n-1} \sum_{k=n+1}^{\infty} I(S_j = S_k)$$

$$= Q_n + \sum_{j=1}^{n-1} \sum_{k=n+1}^{\infty} I(S_j = S_k) := Q_n + \sum_{j=1}^{n-1} W_j^n,$$

where $Q_n := \sum_{1 \le i < j \le n} I(S_i = S_j)$ is a modified version of the self-intersection local time V_n that we introduced earlier (so Q_n measures intersections off the diagonal).

The following result is an adaptation of Theorem 5.5.3(1) in Chen (2010).

Theorem 13.14 *With Z_j as above,*

$$\frac{1}{\sqrt{n\log n}}\sum_{j=1}^{n-1}\left(Z_j - \mathbb{E}Z_j\right) \to^d \mathcal{N}\left(0, \frac{1}{2\pi^2|\Gamma|}\right). \tag{13.37}$$

The proof can be found in Chen (2010), and it relies heavily on the fact that for $d = 3$, $\sup_z \mathbb{P}(S_n = z) \le Cn^{-3/2}$. This allows to prove that $\mathbb{E}(\sum_{j=1}^n W_j^n)^2 = O(n)$, so that under the normalization $\sqrt{n\log n}$, the result follows from the CLT for Q_n.

To prove the CLT for Q_n, Chen uses a blocking technique, with γ_n independent blocks, where $\gamma_n \to \infty$ and $\gamma_n = o(\sqrt{\log n})$. More precisely,

$$Q_n = \sum_{i=1}^{\gamma_n}\sum_{n_{i-1}<j<k\le n_i} I(S_j = S_k) + R_n,$$

where $0 = n_0 < n_1 < \ldots < n_{\gamma_n} = n$ is an integer partition of $[0, n]$ such that for each $1 \le i \le \gamma_n$, $n_i - n_{i-1} = [n\gamma_n^{-1}]$ or $[n\gamma_n^{-1}] + 1$, and the remainder R_n is such that $\mathbb{E}(R_n^2) = o(n\log n)$. The random variables $\sum_{n_{i-1}<j<k\le n_i} I(S_j = S_k)$, $i = 1, \ldots, \gamma_n$, are independent and such that

$$\sum_{n_{i-1}<j<k\le n_i} I(S_j = S_k) =^d Q_{n_i - n_{i-1}}. \tag{13.38}$$

Moreover, by Lemma 5.5.2 in Chen (2010), we have

$$\mathrm{Var}(Q_n) \sim n\log n/(2\pi^2|\Gamma|), \quad \text{as } n \to \infty.$$

Therefore, since $\mathbb{E}(R_n^2) = o(n\log n)$,

$$\mathrm{Var}(Q_n - R_n) \sim n\log n/(2\pi^2|\Gamma|), \quad \text{as } n \to \infty,$$

implying, by independence, that

$$\sum_{i=1}^{\gamma_n}\mathrm{Var}\left(\sum_{n_{i-1}<j<k\le n_i} I(S_j = S_k)\right) \sim n\log n/(2\pi^2|\Gamma|), \quad \text{as } n \to \infty.$$

The CLT then follows provided the Lindeberg condition is satisfied. Using (13.38), it remains then to show that, for any $\varepsilon > 0$,

$$\lim_{n\to\infty}\frac{1}{n\log n}\sum_{i=1}^{\gamma_n}\mathbb{E}\left(\left|Q_{n_i-n_{i-1}} - \mathbb{E}(Q_{n_i-n_{i-1}})\right|^2 \mathbb{1}_{\{|Q_{n_i-n_{i-1}} - \mathbb{E}(Q_{n_i-n_{i-1}})| > \varepsilon\sqrt{n\log n}\}}\right) = 0.$$

This follows from the relation (6.4.3) in Chen (2010) which states that, for any $m \geq 1$,

$$\mathbb{E}(|Q_n - \mathbb{E}(Q_n)|^m) = O((n \log n)^{m/2}), \quad \text{as } n \to \infty. \tag{13.39}$$

Indeed, for any $\varepsilon > 0$ and any $m > 2$, by taking into account (13.39), the choice of the partition of γ_n, we get

$$\frac{1}{n \log n} \sum_{i=1}^{\gamma_n} \mathbb{E}\left(|Q_{n_i - n_{i-1}} - \mathbb{E}(Q_{n_i - n_{i-1}})|^2 \, 1_{\{|Q_{n_i-n_{i-1}} - \mathbb{E}(Q_{n_i-n_{i-1}})| > \varepsilon\sqrt{n \log n}\}} \right)$$
$$\leq \frac{1}{\varepsilon^{m-2}(n \log n)^{m/2}} \sum_{i=1}^{\gamma_n} \mathbb{E}\left(|Q_{n_i - n_{i-1}} - \mathbb{E}(Q_{n_i - n_{i-1}})|^m \right) = O\left(\frac{\gamma_n}{\varepsilon^{m-2} \gamma_n^{m/2}} \right),$$

which converges to zero as $n \to \infty$.

14

Reversible Markov Chains

This chapter is devoted to the study of the weak invariance principle for the partial sums process associated with reversible Markov chains, both under the standard normalization $\sqrt{n}$ and under more general normalization.

14.1 Introduction

Reversible Markov chains play a very important role in applications to infinite particle systems, random walks, processes in random media and Metropolis–Hastings algorithms. For instance, Kipnis and Varadhan (1986) and Kipnis and Landim (1999) considered applications to interacting particle systems, while Tierney (1994), Zhao, Woodroofe and Volný (2010), and Longla, Peligrad and Peligrad (2012) discussed applications to Monte Carlo Markov chains.

We assume that $(\xi_n)_{n\in\mathbb{Z}}$ is a stationary Markov chain defined on a probability space $(\Omega, \mathcal{F}, \mathbb{P})$ with values in a general state space $(S, \mathcal{A})$. The marginal distribution is denoted by $\pi(A) = \mathbb{P}(\xi_0 \in A)$. Assume that there is a regular conditional distribution for ξ_1 given ξ_0 denoted by $Q(x, A) = \mathbb{P}(\xi_1 \in A|\, \xi_0 = x)$. Let Q also denote the Markov operator acting via $(Qf)(x) = \int_S f(s)Q(x, ds)$. Next, let $\mathbb{L}_0^2(\pi)$ be the set of measurable functions on S such that $\int f^2 d\pi < \infty$ and $\int f d\pi = 0$. For some function $g \in \mathbb{L}_0^2(\pi)$, let

$$X_i = g(\xi_i),\ S_n = \sum_{i=1}^{n} X_i, \sigma_n = (\mathbb{E}S_n^2)^{1/2}, \tag{14.1}$$

and

$$W_n = \{W_n(t), t \in [0, 1]\} \text{ where } W_n(t) = n^{-1/2} S_{[nt]}.$$

For any integrable random variable X we denote $\mathbb{E}_{\xi_n}(X) = \mathbb{E}(X|\xi_n)$. The symbol $\to^d$ denotes, as usual, convergence in distribution.

The Markov chain is called reversible if $Q = Q^*$, where Q^* is the adjoint operator of Q. The condition of reversibility is equivalent to requiring that (ξ_0, ξ_1) and (ξ_1, ξ_0) have the same distribution or

Functional Gaussian Approximation for Dependent Structures. Florence Merlevède, Magda Peligrad and Sergey Utev. Oxford University Press (2019).
DOI: 10.1093/oso/9780198826941.001.0001

$$\int_A Q(\omega, B)\pi(d\omega) = \int_B Q(\omega, A)\pi(d\omega),$$

for all Borel sets $A, B \in \mathcal{A}$.

14.2 Functional Central Limit Theorem for Reversible Markov Chains under Normalization $\sqrt{n}$

In 1986, Kipnis and Varadhan proved the following striking theorem. Denote by $c_k = \text{cov}(X_0, X_k)$.

Theorem 14.1 *Assume that the Markov chain is reversible, stationary and ergodic, and that*

$$\sum_{k=1}^{n} c_k \to c, \quad as\, n \to \infty. \tag{14.2}$$

Then conditions (4.7) and (4.11) are satisfied. In addition W_n converges in distribution in $D([0,1])$ equipped with the uniform norm to $\sigma_g W$ where W is a standard Brownian motion and $\sigma_g^2 = \|X_0\|_2^2 + 2c$.

Proof We first establish a martingale approximation via Theorem 4.5. We verify first condition (4.7) with $\mathcal{F}_k = \sigma(\xi_j, j \le k)$. This will follow if we can prove that $(\mathbf{P}_1(S_n))$ is a Cauchy sequence in $\mathbb{L}^2$. By the properties of the conditional expectation and stationarity, we have

$$\begin{aligned}\mathbb{E}(\mathbf{P}_1(S_n - S_m))^2 &= \mathbb{E}(\mathbb{E}_1(S_n - S_m))^2 - \mathbb{E}(\mathbb{E}_0(S_n - S_m))^2\\ &= \mathbb{E}(\mathbb{E}_0(S_{n-1} - S_{m-1}))^2 - \mathbb{E}(\mathbb{E}_0(S_n - S_m))^2\\ &= \mathbb{E}\left((\mathbb{E}_0(X_m) - \mathbb{E}_0(X_n))(\mathbb{E}_0(S_{n-1} - S_{m-1}) + \mathbb{E}_0(S_n - S_m))\right). \end{aligned} \tag{14.3}$$

To continue this computation, let us look at the terms of the type $\mathbb{E}(\mathbb{E}_0(X_k)\mathbb{E}_0(X_j))$. The property of reversibility transforms these terms in covariances. Indeed, notice that

$$\begin{aligned}\mathbb{E}(\mathbb{E}_0(X_k)\mathbb{E}_0(X_j)) &= \mathbb{E}(\mathbb{E}_0(X_{-k})\mathbb{E}_0(X_j)) = \mathbb{E}(X_{-k}\mathbb{E}_0(X_j))\\ &= \mathbb{E}(\mathbb{E}_0(X_{-k}X_j)) = \mathbb{E}(X_{-k}X_j) = \mathbb{E}(X_0X_{k+j}) = c_{k+j}.\end{aligned} \tag{14.4}$$

So,

$$\begin{aligned}\mathbb{E}(\mathbf{P}_1(S_n - S_m))^2 &= (c_{2m} + \dots + c_{n+m-1}) + (c_{2m+1} + \dots + c_{n+m})\\ &- (c_{n+m} + \dots + c_{2n-1}) - (c_{n+m+1} + \dots + c_{2n}).\end{aligned}$$

Because of our condition (14.2) clearly

$$\lim_{n>m\to\infty} \mathbb{E}(\mathbf{P}_1(S_n - S_m))^2 = 0,$$

implying that $(\mathbf{P}_1(S_n))$ is a Cauchy sequence in $\mathbb{L}^2$. Therefore, by stationarity, we can then define a stationary sequence $(D_k)_{k\in\mathbb{Z}}$ of martingale differences in $\mathbb{L}^2$ by setting, for any $k \in \mathbb{Z}$,

$$D_k = \lim_{n\to\infty} D_k^n \text{ in } \mathbb{L}^2 \text{ where } D_k^n := \mathbf{P}_k(X_k + X_{k+1} + \dots + X_{k+n-1}). \tag{14.5}$$

Note that since the Markov chain is ergodic so is the sequence $(D_k)_{k\in\mathbb{Z}}$.

We shall verify now condition (4.11). By using (14.4), note that

$$\frac{1}{n}\|\mathbb{E}(S_n|\mathcal{F}_0)\|_2^2 = \frac{1}{n}\sum_{1\le j,k\le n} \mathbb{E}(\mathbb{E}_0(X_k)\mathbb{E}_0(X_j)) = \frac{1}{n}\sum_{k=1}^{n}\sum_{u=k+1}^{n+k} c_u,$$

which converges to 0 as $n \to \infty$ by (14.2). Hence (4.11) is proved.

Note that, since conditions (4.7) and (4.11) hold, Theorem 4.5 implies that

$$\frac{1}{n}\|S_n - M_n\|_2^2 \to 0, \tag{14.6}$$

where $M_n = \sum_{i=1}^n D_i$ and the D_i's are defined in (14.5). The convergence (14.6), together with the usual CLT for a stationary and ergodic sequence of martingale differences (see Theorem 2.37), implies that $n^{-1/2}S_n$ converges in distribution to a centered Gaussian random variable with variance

$$\mathbb{E}(D_1^2) = \lim_{n\to\infty}\frac{1}{n}\mathbb{E}(S_n^2) = \|X_0\|_2^2 + 2c,$$

where we have used condition (14.2) for the last convergence. We turn now to the proof of the functional form of the central limit theorem. We first notice that the convergence (14.6) also implies that for any positive integer m, any reals $a_1, \dots, a_m$ and any $t_0 = 0 < t_1 < t_2 < \dots < t_m \le 1$,

$$\frac{1}{n}\Big\|\sum_{\ell=1}^{m} a_\ell\left(S_{[nt_\ell]} - M_{[nt_\ell]}\right)\Big\|_2^2 \to 0.$$

The convergence above together with the Cramér–Wold device and Theorem 2.37 implies that the finite-dimensional distributions of W_n converge to those of $\sigma_g W$. To complete the proof, it remains to show that the process W_n is tight. According to Lemma 1.16, since $(X_k)_{k\in\mathbb{Z}}$ is a stationary sequence, it is then enough to prove (1.25), which obviously will hold if one can show that

$$\frac{\max_{1\le k\le n} S_k^2}{n} \text{ is uniformly integrable.} \tag{14.7}$$

With this aim, we shall prove a forward–backward martingale decomposition. It combines ideas from Wu and Woodroofe (2004) with forward–backward martingale approximations by Meyer and Zheng (1984) and Lyons and Zheng (1988).

It is convenient to obtain a decomposition for the term $X_k + X_{k+1}$ first. We shall show that

$$X_k + X_{k+1} = D_{k+1} + \tilde{D}_k, \tag{14.8}$$

where $(D_k)_{k\in\mathbb{Z}}$ is the stationary and ergodic sequence of martingale differences in $\mathbb{L}^2$ defined in (14.5) and adapted to $\mathcal{F}_k = \sigma(\xi_j, j \le k)$, and $(\tilde{D}_k)_{k\in\mathbb{Z}}$ is a stationary and ergodic sequence of reverse martingale differences in $\mathbb{L}^2$ adapted to $\mathcal{G}_k := \sigma(\xi_j, j \ge k)$. To construct the sequence $(\tilde{D}_k)_{k\in\mathbb{Z}}$ we introduce the notation $\tilde{\mathbb{E}}_k(X_j) = \mathbb{E}(X_j|\mathcal{G}_k) = \mathbb{E}(X_j|\xi_k)$ and we set, for any $k \in \mathbb{Z}$,

$$\tilde{D}_k^n = \tilde{\mathbb{E}}_k(X_{k-n} + \ldots + X_k) - \tilde{\mathbb{E}}_{k+1}(X_{k-n} + \ldots + X_k).$$

Therefore

$$X_k = \tilde{D}_k^n + \tilde{\eta}_{k+1}^n - \tilde{\eta}_k^n + \tilde{\mathbb{E}}_{k+1}(X_{k-n}) \text{ where } \tilde{\eta}_k^n = \tilde{\mathbb{E}}_k(X_{k-n} + \cdots + X_{k-1}). \tag{14.9}$$

On the other hand, using the triangular martingale representation (4.25) of Remark 4.7, we have for any $k \in \mathbb{Z}$,

$$X_{k+1} = D_{k+1}^n + \eta_k^n - \eta_{k+1}^n + \mathbb{E}_{k+1}(X_{k+n+1}) \text{ where } \eta_k^n = \mathbb{E}_k(X_{k+1} + \ldots + X_{k+n}), \tag{14.10}$$

and D_{k+1}^n is defined in (14.5). Note now that, by reversibility, $\tilde{\mathbb{E}}_{k+1}(X_{k-n}) = \mathbb{E}_{k+1}(X_{k+n+2})$ and that $\tilde{\eta}_k^n = \eta_k^n$. Hence, adding relations (14.10) and (14.9) leads to

$$X_{k+1} + X_k = D_{k+1}^n + \tilde{D}_k^n + \mathbb{E}_{k+1}(X_{k+n+2}) + \mathbb{E}_{k+1}(X_{k+n+1}). \tag{14.11}$$

Note that by stationarity, (14.4) and (14.2),

$$||\mathbb{E}_{k+1}(X_{k+n+2})||_2^2 = ||\mathbb{E}_0(X_{n+1})||_2^2 = c_{2(n+1)} \to 0, \text{ as } n \to \infty.$$

Similarly $||\mathbb{E}_{k+1}(X_{k+n+1})||_2 \to 0$ as n tend to infinity. So, overall, starting from (14.11) and taking into account that D_{k+1}^n converges in $\mathbb{L}^2$, we get that, for any $k \in \mathbb{Z}$,

$$\tilde{D}_k^n \text{ converges in } \mathbb{L}^2 \text{ to } \tilde{D}_k$$

and (14.8) holds. It follows that for any positive integer k,

$$2S_k + (X_0 - X_k) = \sum_{i=1}^{k}(D_i + \tilde{D}_{i-1}),$$

and so

$$S_k = \frac{1}{2}\sum_{i=1}^{k}(D_i + \tilde{D}_{i-1}) - \frac{1}{2}(X_0 - X_k). \tag{14.12}$$

Since $(X_k)_{k\in\mathbb{Z}}$ is a square integrable stationary sequence, we have

$$\max_{1\le k\le n} \frac{|X_k|}{\sqrt{n}} \to 0, \text{ a.s. and in } \mathbb{L}^2.$$

Therefore starting from (14.12) and setting $M_k = \sum_{i=1}^{k} D_i$ and $\tilde{M}_k = \sum_{i=1}^{k} \tilde{D}_{i-1}$, it follows that, in order to prove (14.7) (and then the tightness of W_n), it is enough to prove that both $(n^{-1}\max_{1\le k\le n} M_k^2)$ and $(n^{-1}\max_{1\le k\le n} \tilde{M}_k^2)$ are uniformly integrable families. By Corollary 2.10 and Inequality (2.3), it is enough to prove that $(n^{-1}M_n^2)$ and $(n^{-1}\tilde{M}_n^2)$ are uniformly integrable. Note that the usual CLT for a stationary and ergodic sequence of martingale differences (see Theorem 2.37) combined with the moment convergence theorem (see Theorem 1.12) implies the uniform integrability of both $(n^{-1}M_n^2)$ and $(n^{-1}\tilde{M}_n^2)$. This ends the proof of (14.7). ■

Remark 14.2 *For reversible Markov chains, actually* (14.2) *is equivalent to*

$$\lim_{n\to\infty} \frac{1}{n}\mathbb{E}(S_n^2) = \sigma_g^2.$$

This result is easy to prove by spectral calculus whose details can be found in Kipnis and Varadhan (1986) or in Borodin and Ibragimov (1995, Chapter 4).

Let us give a sketch of the proof for people familiar with operator spectral analysis. We start by noticing that

$$\text{cov}(X_0, X_k) = \int_{-1}^{1} t^k \rho_g(dt), \tag{14.13}$$

where ρ_g is the spectral operator measure. This observation together with (14.3) and (14.4) imply that, for $n \ge m$,

$$\begin{aligned}
\mathbb{E}(\mathbf{P}_1(S_n - S_m))^2 &= \int_{-1}^{1}\left[\left(\sum_{k=m}^{n-1} t^k\right)^2 - \left(\sum_{k=m+1}^{n} t^k\right)^2\right]\rho_g(dt) \\
&= \int_{-1}^{1}\left[\left(\sum_{k=m}^{n-1} t^k\right)^2 - t^2\left(\sum_{k=m}^{n-1} t^k\right)^2\right]\rho_g(dt) \\
&= \int_{-1}^{1}\left[\frac{t^{2m}(1+t)(1-t^{n-m})^2}{(1-t)}\right]\rho_g(dt).
\end{aligned}$$

It follows that

$$||S_n||_2^2 = \sum_{j=1}^{n} ||\mathbf{P}_1(S_j)||_2^2 + ||\mathbb{E}(S_n|\mathcal{F}_0)||_2^2 = \int_{-1}^{1} \frac{(1+t)}{(1-t)} \sum_{j=1}^{n} (1-t^j)^2 \rho_g(dt)$$
$$+ \int_{-1}^{1} \left(\sum_{k=1}^{n} t^k \right)^2 \rho_g(dt).$$

This implies that if $\lim_{n\to\infty} n^{-1}\mathbb{E}(S_n^2) = \sigma_g^2$, then for any $\varepsilon > 0$,

$$\sigma_g^2 \geq \int_0^{1-\varepsilon} \frac{1}{1-t} \rho_g(dt),$$

implying the following condition:

$$\int_{-1}^{1} \frac{1}{1-t} \rho_g(dt) < \infty. \tag{14.14}$$

This last convergence obviously implies (14.2) by taking into account (14.13).

Remark 14.3 *As a matter of fact, condition (14.2) is equivalent to condition (14.14), which is a condition on the function g in relation with the transition operator Q of the Markov chain. It is known that the spectrum of Q for a reversible Markov chain is concentrated on $[-1, 1]$. If there is a positive number $a < 1$, such that the spectrum Q is concentrated on $[-1, a)$ then condition (14.14) is satisfied for every function $g \in \mathbb{L}_0^2(\pi)$. Therefore only the top of the spectrum is responsible for the main condition in the central limit theorem and its invariance principle. This was the starting point of several results by Roberts and Rosenthal (2008). They showed that there is a positive number $a < 1$ such that the spectrum of Q is concentrated on $[-1, a)$ if and only if there is a positive constant K such that for every $g \in \mathbb{L}_0^2(\pi)$ the asymptotic variance $\lim_{n\to\infty} n^{-1}\mathbb{E}\left(\sum_{k=1}^{n} g(\xi_k)\right)^2$ exists and the following inequality holds:*

$$\sigma_g^2 = \lim_{n\to\infty} n^{-1}\mathbb{E}\left(\sum_{k=1}^{n} g(\xi_k)\right)^2 \leq K\mathbb{E}(g(\xi_1))^2,$$

for some positive finite constant K. Roberts and Rosenthal (2008) called this property "variance bounding". They additionally showed that, at least for reversible Markov chains, this "variance bounding" condition is equivalent to the fact that the central limit theorem holds for $\sum_{k=1}^{n} g(\xi_k)/\sqrt{n}$ for every $g \in \mathbb{L}_0^2(\pi)$. This condition was further studied by Derriennic and Lin (2011), who showed that for any stationary and ergodic Markov chain (not necessarily reversible) the "variance bounding" condition implies the central limit theorem for any $g \in \mathbb{L}_0^2(\pi)$.

Remark 14.4 *In the context of Markov chains, a very interesting problem is to show that the central limit theorem and its invariance principle still hold when the Markov chain is not started from equilibrium, but rather at a point. These types of results are also known under the name of quenched limit theorems. There is a vast literature in this field. Kipnis and Varadhan (1986) conjectured that under the conditions of Theorem 14.1, the central limit theorem and its functional form also hold when the process is started at a point. Steps for solving this conjecture can be found in papers by Derriennic and Lin (2001, 2003) and Cuny and Peligrad (2012).*

14.3 Maximal Inequalities

As in the Doob maximal inequalities for martingales, we shall compare moments and tail distributions of the maximum of partial sums with those of the corresponding partial sums.

Proposition 14.5 *Let $(X_i)_{i\in\mathbb{Z}}$ be defined by (14.1) and $Q = Q^*$. Let $p > 1$ and let $q > 1$ such that $1/p + 1/q = 1$. Then for all $n \geq 1$,*

$$\| \max_{1\leq i\leq n} |S_i| \|_p \leq \| \max_{1\leq i\leq n} |X_i| \,\|_p + (4q+3) \max_{1\leq i\leq n} \|S_i\|_p. \tag{14.15}$$

Remark 14.6 *Let $p = 2$. Since $(X_i)_{i\in\mathbb{Z}}$ is stationary, it is well known that*

$$\| \max_{1\leq i\leq n} |X_i| \|_2 = o(n^{1/2}) \text{ as } n \to \infty.$$

If we assume in addition $\liminf_n \sigma_n^2/n > 0$ *we deduce that there exists $C > 0$ not depending on n, such that*

$$\| \max_{1\leq i\leq n} |S_i| \|_2 \leq C \max_{1\leq i\leq n} \|S_i\|_2.$$

Proof of Proposition 14.5 For this proof, it is convenient to use the triangular difference of martingale decomposition given by (4.19), namely: For any $k \geq 0$,

$$X_k = D^n_{k+1} + \theta^n_k - \theta^n_{k+1} + \frac{1}{n}\mathbb{E}_k(S_{k+n} - S_k), \tag{14.16}$$

where

$$\theta^n_k = \frac{1}{n}\sum_{i=0}^{n-1} \mathbb{E}_k(X_k + \ldots + X_{k+i}) \text{ and } D^n_k = \theta^n_k - \mathbb{E}_{k-1}(\theta^n_k).$$

We construct now a triangular martingale approximation for the reversed process adapted to the filtration $\mathcal{G}_n = \sigma(\xi_i, i \geq n)$. We introduce the notation $\tilde{\mathbb{E}}_1(X_0) = \mathbb{E}(X_0|\mathcal{G}_1) = \mathbb{E}(X_0|\xi_1) = (Q^*f)(\xi_1)$.

Now, let

$$\tilde{\theta}_k^n = \frac{1}{n}\sum_{i=0}^{n-1} \tilde{\mathbb{E}}_k(X_{k-i} + \ldots + X_k).$$

With this notation

$$X_{k+1} = \tilde{D}_k^n + \tilde{\theta}_{k+1}^n - \tilde{\theta}_k^n + \frac{1}{n}\tilde{\mathbb{E}}_{k+1}(X_{-n+k+1} + \ldots + X_k), \tag{14.17}$$

where $\tilde{D}_k^n$ are martingale differences with respect to the decreasing filtration $\mathcal{G}_k = \sigma(\xi_i, k \geq n)$, $\tilde{D}_k^n = \tilde{\theta}_k^n - \mathbb{E}_{k+1}\tilde{\theta}_k^n$.

If we assume that $Q = Q^*$, we have $\tilde{\mathbb{E}}_1(X_0) = \mathbb{E}(X_2|\xi_1) = (Qf)(\xi_1)$. Therefore, $\tilde{\theta}_k^n = \theta_k^n, \tilde{\theta}_{k+1}^n = \theta_{k+1}^n$ and $\tilde{\mathbb{E}}_{k+1}(X_{-n+k+1} + \ldots + X_k) = \mathbb{E}_{k+1}(X_{k+2} + \ldots + X_{k+n+1})$. Adding relations (14.16) and (14.17) leads to

$$X_k + X_{k+1} = D_{k+1}^n + \tilde{D}_k^n + \frac{1}{n}\mathbb{E}_k(S_n - S_k) + \frac{1}{n}\mathbb{E}_{k+1}(S_{k+n+1} - S_{k+1}).$$

Summing these relations we obtain the representation

$$\sum_{i=0}^{k-1}(X_i + X_{i+1}) = \sum_{i=1}^{k}\left[(D_i^n + \tilde{D}_{i-1}^n) + \frac{1}{n}\mathbb{E}_{i-1}(S_{n+i-1} - S_{i-1}) + \frac{1}{n}\mathbb{E}_i(S_{n+i} - S_i)\right].$$

So, $\qquad 2S_k + (X_0 - X_k) = \sum_{i=1}^{k}(D_i^n + \tilde{D}_{i-1}^n) + \bar{R}_k^n,$

where $\quad \bar{R}_k^n = n^{-1}\sum_{i=1}^{k}[\mathbb{E}_{i-1}(S_{n+i-1} - S_{i-1}) + \mathbb{E}_i(S_{n+i} - S_i)].$

Therefore, in the reversible case, we get the following backward–forward martingale representation

$$S_k = \frac{1}{2}[(X_k - X_0) + (M_k^n + \tilde{M}_k^n) + \bar{R}_k^n], \tag{14.18}$$

where $M_k^n = \sum_{i=1}^{k} D_i^n$ is a forward martingale adapted to the non-decreasing filtration $(\mathcal{F}_k)_{k\in\mathbb{Z}}$ and $\tilde{M}_k^n = \sum_{i=0}^{k-1}\tilde{D}_i^n$ is a backward martingale adapted to the decreasing filtration $(\mathcal{G}_k)_{k\in\mathbb{Z}}$.

From (4.21), we also have the following decomposition: For any positive integers n and m,

$$S_m = M_m^n + R_m^n, \tag{14.19}$$

where $R_m^n = \bar{\theta}_0^n - \bar{\theta}_m^n + \frac{1}{n}\sum_{k=0}^{m-1}\mathbb{E}_k(S_{k+n} - S_k)$ with $\bar{\theta}_k^n = \frac{1}{n}\sum_{i=1}^{n-1}\mathbb{E}_k(S_{k+i} - S_k)$.

We are now in position to end the proof of Proposition 14.5. We start from (14.18) and take the maximum on both sides. We easily obtain

$$\max_{1\le i\le n} |S_i| \le \frac{1}{2}\Big\{|X_0| + \max_{1\le i\le n} |X_i| + \max_{1\le i\le n} |M_i^n + \tilde{M}_i^n| + \max_{1\le i\le n} |\bar{R}_i^n|\Big\}. \tag{14.20}$$

Note that

$$\max_{1\le i\le n} |\bar{R}_i^n| \le \frac{1}{n}\sum_{i=1}^{n}(|\mathbb{E}_{i-1}(S_{n+i-1} - S_{i-1})| + |\mathbb{E}_i(S_{n+i} - S_i)|),$$

whence, by Minkowski's inequality and stationarity, for any $p \ge 1$

$$\|\max_{1\le i\le n} |\bar{R}_i^n|\,\|_p \le \frac{2}{n}\sum_{i=1}^{n} \|\mathbb{E}_i(S_{n+i} - S_i)\|_p = 2\|\mathbb{E}_0(S_n)\|_p. \tag{14.21}$$

Taking into account that $\max_{1\le k\le n} |\tilde{M}_k^n| \le |\tilde{M}_n^n| + \max_{1\le k\le n} |\tilde{M}_n^n - \tilde{M}_k^n|$, we deduce

$$\|\max_{1\le k\le n} |M_k^n + \tilde{M}_k^n|\,\|_p \le \|\max_{1\le k\le n} |M_k^n|\,\|_p + \|\max_{1\le k\le n} |\tilde{M}_n^n - \tilde{M}_k^n|\,\|_p + \|\tilde{M}_n^n\|_p,$$

whence, by Doob's maximal inequality applied twice, stationarity and reversibility, we get

$$\|\max_{1\le k\le n} |M_k^n + \tilde{M}_k^n|\,\|_p \le q\|M_n^n\|_p + (q+1)\|\tilde{M}_n^n\|_p = (2q+1)\|M_n^n\|_p,$$

where q is the conjugate of p.

From (14.19) we have $M_n^n = S_n - R_n^n$, and from Minkowski's inequality we deduce that

$$\|M_n^n\|_p \le \|S_n\|_p + \frac{2}{n}\sum_{i=0}^{n-1} \|\mathbb{E}_0(S_i)\|_p + \|\mathbb{E}_0(S_n)\|_p$$

$$\text{whence,}\quad \|M_n^n\|_p \le \|S_n\|_p + 3\max_{1\le i\le n} \|\mathbb{E}_0(S_i)\|_p. \tag{14.22}$$

From (14.20), (14.21) and (14.22) we deduce the following extension of the Doob maximal inequality for reversible processes:

$$\|\max_{1\le i\le n} |S_i|\,\|_p \le \frac{1}{2}(\|X_0\|_p + \|\max_{1\le i\le n} |X_i|\,\|_p + (2q+1)[\|S_n\|_p + 3\max_{1\le i\le n} \|\mathbb{E}_0(S_i)\|_p] + 2\|\mathbb{E}_0(S_n)\|_p).$$

Taking now into account that $\|\mathbb{E}_0(S_i)\|_p \le \|S_i\|_p$, Proposition 14.5 is established. ■

Proposition 14.7 *Let $(X_i)_{i\in\mathbb{Z}}$ be defined by (14.1) and $Q = Q^*$. Then, for every $x > 0$ and $n \geq 1$,*

$$\mathbb{P}\left(\max_{1\leq i\leq n}|S_i| > x\right) \leq \frac{2}{x}\left\{18\,\mathbb{E}\left(|S_n|\mathbf{1}_{\{|S_n|>x/12\}}\right) + 55\max_{1\leq i\leq n}\|\mathbb{E}_0(S_i)\|_1 + \|\max_{1\leq i\leq n}|X_i|\|_1\right\}. \tag{14.23}$$

The proposition above allows sufficient conditions for tightness, that will ensure continuity of every limiting process.

Proposition 14.8 *Assume (X_i) is defined by (14.1), conditions (14.31) and (14.34) are satisfied and (S_n/σ_n) is convergent in distribution. Then, $\{\sigma_n^{-1}S_{[nt]}, t \in [0,1]\}$ is tight in $D([0,1])$ endowed with uniform topology and any limiting process is continuous.*

Proof of Proposition 14.7 For every $x \geq 0$, using (14.20), we obtain

$$\mathbb{P}\left(\max_{1\leq i\leq n}|S_i| > x\right) \leq \mathbb{P}\left(\max_{1\leq i\leq n}|M_i^n + \tilde{M}_i^n| > x\right) + \mathbb{P}\left(|X_0| + \max_{1\leq i\leq n}|X_i| + \max_{1\leq i\leq n}|\bar{R}_i^n| > x\right). \tag{14.24}$$

Applying the Markov inequality, then the triangle inequality followed by (14.21) with $p = 1$, leads to

$$\mathbb{P}\left(|X_0| + \max_{1\leq i\leq n}|X_i| + \max_{1\leq i\leq n}|\bar{R}_i^n| > x\right) \leq \frac{2}{x}\left(\|\max_{1\leq i\leq n}|X_i|\|_1 + \|\mathbb{E}_0(S_n)\|_1\right). \tag{14.25}$$

By the triangle inequality and reversibility,

$$\begin{aligned}\mathbb{P}\left(\max_{1\leq i\leq n}|M_i^n + \tilde{M}_i^n| > x\right) &\leq \mathbb{P}\left(\max_{1\leq i\leq n}|M_i^n| > x/3\right) \\ &\quad + \mathbb{P}\left(\max_{1\leq i\leq n}|\sum_{k=i}^{n}\tilde{D}_i^n| > x/3\right) + \mathbb{P}\left(|\tilde{M}_n^n| > x/3\right) \\ &\leq 3\,\mathbb{P}\left(\max_{1\leq i\leq n}|M_i^n| > x/3\right).\end{aligned}$$

Using Proposition 2.9, it follows that

$$\mathbb{P}\left(\max_{1\leq i\leq n}|M_i^n + \tilde{M}_i^n| > x\right) \leq \frac{18}{x}\mathbb{E}\left(|M_n^n|1_{\{|M_n^n|>x/6\}}\right).$$

Let us express the right-hand side in terms of S_n. By (14.19) we have $M_n^n = S_n - R_n^n$ and using the fact that for all positive real numbers x, y, a, we have $(x+y)\mathbf{1}\{x+y>a\} \leq 2x\mathbf{1}\{x>a/2\} + 2y\mathbf{1}\{y>a/2\} \leq 2x\mathbf{1}\{x>a/2\} + 2y$, we obtain

$$\mathbb{E}\left(|M_n^n|\mathbf{1}_{\{|M_n^n|>x/6\}}\right) \le 2\mathbb{E}\left(|S_n|\mathbf{1}_{\{|S_n|>x/12\}}\right) + 2\|R_n^n\|_1$$
$$\le 2\mathbb{E}\left(|S_n|\mathbf{1}_{\{|S_n|>x/12\}}\right) + 6\max_{1\le i\le n}\|\mathbb{E}_0(S_i)\|_1.$$

Therefore,

$$\mathbb{P}\left(\max_{1\le i\le n}|M_i^n + \tilde{M}_i^n| > x\right) \le \frac{18}{x}\left\{2\,\mathbb{E}\left(|S_n|\mathbf{1}_{\{|S_n|>x/12\}}\right) + 6\max_{1\le i\le n}\|\mathbb{E}_0(S_i)\|_1\right\}. \tag{14.26}$$

Starting from (14.24) and considering the bounds (14.25) and (14.26), inequality (14.23) follows. ■

Proof of Proposition 14.8 Tightness will follow if we can prove that, for any $\varepsilon > 0$,

$$\lim_{\delta\to 0^+}\limsup_{n\to\infty}\frac{1}{\delta}\mathbb{P}\left(\max_{1\le k\le[n\delta]}|S_k| > \varepsilon\sigma_n\right) = 0. \tag{14.27}$$

Assume that $S_n/\sigma_n \to^d L$. By the continuous mapping theorem (see Theorem 1.13) and Theorem 1.11, we notice that the limit has finite second moment, namely

$$\mathbb{E}(L^2) \le \liminf_{n\to\infty}\|S_n\|_2^2/\sigma_n^2 = 1. \tag{14.28}$$

Furthermore, note that by (14.31), for any $\varepsilon > 0$,

$$\limsup_{n\to\infty}\frac{1}{\sigma_n}\mathbb{E}\left(|S_{[n\delta]}|1_{\{|S_{[n\delta]}|>\varepsilon\sigma_n/12\}}\right)$$
$$\le \sqrt{\delta}\limsup_{n\to\infty}\frac{1}{\sigma_{[n\delta]}}\mathbb{E}\left(|S_{[n\delta]}|1_{\{|S_{[n\delta]}|>\varepsilon\sigma_{[n\delta]}/(24\sqrt{\delta})\}}\right)$$
$$\le 2\sqrt{\delta}\limsup_{n\to\infty}\mathbb{E}\left(\frac{|S_{[n\delta]}|}{\sigma_{[n\delta]}} - \frac{\varepsilon}{48\sqrt{\delta}}\right)_+.$$

Since $(|S_n|/\sigma_n)_{n\ge 1}$ is uniformly integrable (because $\mathbb{E}(S_n^2)/\sigma_n^2 = 1$), by the continuous mapping theorem and the convergence of moments theorem (see Theorem 1.12), it follows that

$$\limsup_{n\to\infty}\frac{1}{\sigma_n}\mathbb{E}\left(|S_{[n\delta]}|\mathbf{1}_{\{|S_{[n\delta]}|>\varepsilon\sigma_n/12\}}\right) \le 2\sqrt{\delta}\mathbb{E}\left(L - \frac{\varepsilon}{48\sqrt{\delta}}\right)_+$$
$$\le 96\varepsilon^{-1}\delta\,\mathbb{E}\left(L^2\mathbf{1}_{\{|L|>\varepsilon/(48\sqrt{\delta})\}}\right). \tag{14.29}$$

We now remind that we work under the assumption that $\liminf_n \sigma_n^2/n > 0$. Note that, by stationarity and the fact that X_0 is square integrable, it follows that $\max_{1\le i\le n}|X_i|/\sqrt{n} \to 0$ *a.s.* and in $\mathbb{L}^2$ and therefore

$$\max_{1\le i\le n}|X_i|/\sigma_n \to 0\ \textit{a.s.}\text{ and in }\mathbb{L}^2. \tag{14.30}$$

Also, since (14.31) is assumed, the convergence (14.33) holds. Therefore, by passing to the limit in relation (14.23) and using (14.29), (14.30) and (14.33), we obtain that, for any $\delta > 0$,

$$\limsup_{n\to\infty} \frac{1}{\delta}\mathbb{P}\left(\max_{1\leq k\leq [n\delta]} |S_k| > \varepsilon\sigma_n\right) \leq \frac{36\times 96}{\varepsilon}\mathbb{E}\left(L^2 I\left(|L| > \frac{\varepsilon}{48\sqrt{\delta}}\right)\right).$$

Finally, taking into account (14.28), the convergence (14.27) follows by letting $\delta \to 0^+$. ■

14.4 Invariance Principle for Reversible Markov Chains under General Normalization

14.4.1 Introduction and Result

Zhao, Woodroofe and Volný (2010) addressed the central limit theorem question for reversible Markov chains under the weaker condition,

$$\sigma_n^2 = nh(n), \tag{14.31}$$

where h is a slowly varying function (i.e. $\lim_{n\to\infty} h(nu)/h(n) = 1$ for all $u > 0$). First, they proved that if $\liminf_n h(n) = 0$, then necessarily

$$2S_n = (1 + (-1)^{n-1})X_1 \;\; a.s. \tag{14.32}$$

and if $\liminf_{n\to\infty} h(n) = c^2 \neq 0$ then $\sigma_n^2/n \to c^2$. Then, in their Proposition 2, they showed that the representation (14.31) implies

$$||\mathbb{E}_0(S_n)||_2 = o(\sigma_n). \tag{14.33}$$

Note then that, by Proposition 4.30, if $\sigma_n^2 \to \infty$, for a stationary reversible Markov chain $(X_n)_{n\in\mathbb{Z}}$ defined by (14.1), the relations (14.31) and (14.33) are equivalent.

In the following, to avoid the trivial case, we shall assume that

$$\liminf_{n\to\infty} \sigma_n^2/n > 0. \tag{14.34}$$

Zhao, Woodroofe and Volný (2010) also showed, by a class of examples, the surprising result that the distribution of S_n/σ_n needs not to converge to the standard normal distribution under (14.31). Their example satisfies the central limit theorem, in the sense that

$$\frac{S_n}{\sigma_n} \to^d |c|Z, \tag{14.35}$$

where Z is a standard normal variable and $|c| \neq 0$, 1. A large class of examples satisfying (14.35) is given in Deligiannidis, Peligrad and Utev (2015). This paper also contains necessary and sufficient conditions for (14.31) in terms of the operator spectral measure.

A natural question is whether, in the context of reversible Markov chains, the central limit theorem in (14.35) implies the invariance principle namely

$$\left\{\frac{S_{[ns]}}{\sigma_n}, s \in [0, 1]\right\} \to^d |c|W, \tag{14.36}$$

where W is the standard Brownian motion on $[0, 1]$. This question is interesting in itself, especially in the light of recent examples of stationary sequences which satisfy CLT but not its functional form (Giraudo and Volný, 2014).

A step in this direction is Theorem 2 in Longla, Peligrad and Peligrad (2012) which shows that conditional CLT implies the functional CLT in (14.36) for functions of reversible Markov chains. By the conditional CLT we understand that for some $c > 0$ and for all t,

$$\mathbb{E}_{\xi_0}\left(\exp it\frac{S_n}{\sigma_n}\right) \to \exp\left(-\frac{t^2}{2c^2}\right) \text{ in probability.} \tag{14.37}$$

This form of the conditional CLT was essentially used in Longla, Peligrad and Peligrad (2012), in order to establish the convergence of finite-dimensional distributions.

In this section we show that actually (14.35) and (14.31) imply (14.36) for functionals of reversible Markov chains. More precisely we shall establish the following theorem, which is a combination of the results obtained in Longla, Peligrad and Peligrad (2012) and in Peligrad and Utev (2016).

Theorem 14.9 *Assume that (X_n) is defined by (14.1) and the Markov chain is stationary and reversible. Then, the following statements are equivalent:*

(a) *The functional CLT in (14.36) holds.*

(b) *The CLT in (14.35) and the representation (14.31) hold.*

(c) *The conditional CLT in (14.37) and the representation (14.31) hold.*

When we assume irreducibility and aperiodicity, the following corollary holds:

Corollary 14.10 *Assume $(\xi_n)_{n\in\mathbb{Z}}$ is a stationary, reversible, irreducible and aperiodic Markov chain such that (14.31) is satisfied. Then $S_n/\sigma_n \to^d L$ implies that $\{\sigma_n^{-1}S_{[nt]}, t \in [0, 1]\} \to^d cW$, where W is a standard Brownian motion and c is the standard deviation of L.*

14.4.2 Proof of Theorem 14.9

To soothe the notation, we set throughout the proof $W_n(t) = \sigma_n^{-1}S_{[nt]}$, for any $t \in [0, 1]$.

The fact that (a) implies (b) follows by standard arguments in the following way. Clearly, since the partial sum is just a finite-dimensional distribution for $s = 1$, the functional CLT in (14.36) implies the CLT in (14.35). Then, by (14.35), we have that

$S_{[ns]}/\sigma_{[ns]} \to^d |c|Z$, for every $s > 0$. On the other hand, the convergence in (14.36) implies $S_{[ns]}/\sigma_n \to^d |c|sZ$ for all $s > 0$. By the theorem of types (see Theorem 1.14), it follows that $\sigma^2_{[ns]}/\sigma^2_n \to s$.

We prove now that (c) implies (a). Because conditional convergence in distribution implies weak convergence, it follows that $S_n/\sigma_n \to^d L$. Then, by the second part of Proposition 14.8, $W_n(t)$ is tight in $C([0, 1])$ endowed with uniform topology with all possible limits in $C([0, 1])$. Now, let us consider a convergent subsequence, say $W_{n'} \to^d X$. Then X is continuous and since S_n/σ_n is conditionally convergent in distribution, X has independent increments (by the next lemma in this subsection applied on subsequences). It is well known [see, for instance, Doob (1953), Ch. VIII)] that the process X has the representation $X(t) = at + bW(t)$ for some constants a and b, where W is the standard Brownian motion. Without restricting the generality, by symmetry, we can assume $b > 0$. To identify the constants, we use the convergence of moments in the limit theorem, namely Theorem 1.12. Notice that $(S_n/\sigma_n)_{n\geq 1}$ is uniformly integrable in $\mathbb{L}^1$ since it is bounded in $\mathbb{L}^2$. We use this remark to obtain

$$\mathbb{E}L = \lim_{n\to\infty} \mathbb{E}S_n/\sigma_n = 0 = \lim_{n'\to\infty} \mathbb{E}W_{n'}(1) = \mathbb{E}X(1) = a + b\mathbb{E}W(1) = a,$$

so $a = 0$. Finally, by the same argument, it follows that

$$\lim_{n\to\infty} \mathbb{E}|S_n|/\sigma_n = \mathbb{E}|L| = \lim_{n'\to\infty} \mathbb{E}|W_{n'}(1)| = \mathbb{E}|X(1)| = b\mathbb{E}|W(1)| = b\sqrt{2/\pi},$$

and so $b = \mathbb{E}|L|\sqrt{\pi/2}$. It follows that $X(t) = \left(\mathbb{E}|L|\sqrt{\pi/2}\right) W(t)$. In particular, it follows that L has normal distribution and therefore $\mathbb{E}|L|\sqrt{\pi/2}$ is the standard deviation of L.

Lemma 14.11 *If (c) holds and $\{W_n(t), t \in [0, 1]\} \to^d X$, then X has independent increments.*

Proof Without loss of generality, for simplicity we consider only two increments. For any $0 \leq s < t \leq 1$, we shall show that

$$(W_n(s), W_n(t) - W_n(s)) \to^d (X(s), X(t) - X(s)),$$

where $X(s)$ and $X(t) - X(s)$ are independent. By the Cramér–Wold device, it is enough to show that for any two real numbers a and b,

$$A = \mathbb{E}\exp[iaW_n(s) + ib(W_n(t) - W_n(s))] - \mathbb{E}\exp[iaX(s)]\mathbb{E}\exp[ib(X(t) - X(s))] \to 0.$$

To see this, notice that

$$\begin{aligned}&\mathbb{E}\exp[iaW_n(s) + ib(W_n(t) - W_n(s))]\\ &\quad= \mathbb{E}\exp[iaW_n(s)]\mathbb{E}_{[ns]}\exp[ib(W_n(t) - W_n(s))].\end{aligned}$$

By adding and subtracting $\mathbb{E}\exp[iaW_n(s)]\mathbb{E}\exp[ib(X(t) - X(s))]$ to A, we easily obtain

$$\begin{aligned}|A| \leq \mathbb{E}|\mathbb{E}_{[ns]}\exp[ib(W_n(t) - W_n(s)] - \mathbb{E}\exp[ib(X(t) - X(s))]| +\\ + |\mathbb{E}\exp[iaW_n(s)] - \mathbb{E}\exp[iaX(s)]| = I + II.\end{aligned}$$

Since we assume in particular that $W_n(s) \to^d X(s)$, it follows that $II \to 0$. Furthermore, by (14.31), $X(s)$ and $s^{1/2}L$ are identically distributed.

To treat the term I, notice that by stationarity and the definition of $W_n(t)$ we have that

$$I = \mathbb{E}|\mathbb{E}_0 \exp[ib(S_{[nt]-[ns]}/\sigma_n)] - \mathbb{E}\exp[ib(X(t) - X(s))]|. \qquad (14.38)$$

Because we assume that $\sigma_n \to \infty$ we have

$$\frac{1}{\sigma_n}\mathbb{E}|S_{[nt]-[ns]} - S_{[n(t-s)]}| \to 0, \qquad (14.39)$$

which easily implies that for all b,

$$\mathbb{E}|\mathbb{E}_0 \exp[ib(S_{[nt]-[ns]}/\sigma_n)] - \mathbb{E}_0 \exp[ib(S_{[n(t-s)]}/\sigma_n)]| \to 0. \qquad (14.40)$$

Now, since $S_{[n(t-s)]}/\sigma_n \to^d X(t-s)$ and $(S_{[nt]} - S_{[ns]})/\sigma_n \to^d X(t) - X(s)$, we deduce from (14.39) and stationarity that $X(t-s)$ and $X(t) - X(s)$ have the same distribution. Furthermore, by (14.31), we deduce that $S_{[n(t-s)]}/\sigma_n$ is also conditionally convergent in distribution; so, we also have $X(t-s)$ is distributed as $(t-s)^{1/2}L$. Whence, by taking also into account (14.38) and (14.40) it follows that

$$\limsup_{n\to\infty} I = \limsup_{n\to\infty} \mathbb{E}|\mathbb{E}_0 \exp[ib(S_{[n(t-s)]}/\sigma_n)] - \mathbb{E}\exp[ib(X(t-s))]| = 0,$$

leading to the conclusion.

It remains to prove that (b) implies (c). The idea of the proof is to show that from any subsequence of $\mathbb{E}_{\xi_0}(\exp it(S_n/\sigma_n))$ we can extract one converging to $\exp(-t^2/(2c^2))$ in probability. To achieve this goal, we need a technical lemma concerning conditional convergence.

Lemma 14.12 *Assume (V_n, η) is convergent in distribution to (V, Y). Then, we can construct on the same probability space a sequence (V_n', η'), where each vector is distributed as (V_n, η) and a vector (V', η') distributed as (V, Y) such that, for all t,*

$$\mathbb{E}_{\eta'} \exp(itV_n') \to \mathbb{E}_{\eta'} \exp(itV') \textit{ in probability.}$$

Proof of Lemma 14.12 By the Skorohod theorem (see Skorohod, 1956), we can construct on the same probability space a sequence (V_n', η'), where each vector is distributed as (V_n, η), and a vector (V', Y') distributed as (V, Y) such that $(V_n', \eta')(\omega) \to (V', Y')(\omega)$ for all ω. Clearly $\eta' = Y'$ and $V_n'(\omega) \to V'(\omega)$ for all ω. Then, by the mean value theorem, for any t and $\delta > 0$,

$$|\mathbb{E}_{\eta'} \exp(itV_n') - \mathbb{E}_{\eta'} \exp(itV')| \leq$$
$$\mathbb{E}_{\eta'}(|\exp(itV_n') - \exp(itV')| I(|V_n' - V'| \leq \delta)) + 2\mathbb{P}_{\eta'}(|V_n' - V'| > \delta)$$
$$\leq |t|\delta + 2\mathbb{P}_{\eta'}(|V_n' - V'| > \delta) \text{ a.s.}$$

By taking the expectation

$$\mathbb{E}|\mathbb{E}_{\eta'}\exp(itV_n') - \mathbb{E}_{\eta'}\exp(itV')| \le |t|\delta + 2\mathbb{P}(|V_n' - V'| > \delta),$$

which tends to 0 as $n \to \infty$ and then $\delta \to 0$. ■

We continue to prove Theorem 14.9 by proving that (b) implies (c). We remind that we work under the assumption that $\liminf \sigma_n^2/n > 0$ and therefore the convergence (14.30) holds, namely: $\max_{1\le i\le n}|X_i|/\sigma_n \to 0$ *a.s.* and in $\mathbb{L}^2$. This property will allow us to adjust the sums by a few variables without changing the limiting distribution. We shall use some notations: $\bar{S}_n = X_{n+1} + \dots + X_{2n}$; $\mathcal{P}_n = \sigma(\xi_i, i \le n)$ is the past sigma field.

Step 1. As a preliminary computation, we show that for all t,

$$\mathbb{E}\left(\mathbb{E}_{\xi_0}\left(\exp\frac{itS_n}{\sigma_n}\right)\right)^2 \to \exp\left(-\frac{t^2}{c^2}\right) \quad \text{as } n \to \infty. \tag{14.41}$$

By (14.30), the properties of conditional expectation and the Markov property,

$$\begin{aligned}\mathbb{E}\left(\exp\frac{itS_{2n}}{\sigma_n}\right) &= \mathbb{E}\left(\exp\frac{it(S_n+\bar{S}_n)}{\sigma_n}\right) = \mathbb{E}\left(\left(\exp\frac{itS_n}{\sigma_n}\right)\mathbb{E}\left(\exp\frac{it\bar{S}_n}{\sigma_n}\Big|\mathcal{P}_n\right)\right)\\ &= \mathbb{E}\left(\left(\exp\frac{itS_n}{\sigma_n}\right)\mathbb{E}_{\xi_n}\left(\exp\frac{it\bar{S}_n}{\sigma_n}\right)\right).\end{aligned}$$

We write now $S_n = (X_0 + \dots + X_{n-1}) - X_0 + X_n = S_{[0,n-1]} + (X_n - X_0)$.

Clearly, by simple computations and (14.30),

$$\begin{aligned}&\left|\mathbb{E}\left(\exp\frac{itS_n}{\sigma_n}\mathbb{E}_{\xi_n}\left(\exp\frac{it\bar{S}_n}{\sigma_n}\right)\right) - \mathbb{E}\left(\exp\frac{itS_{[0,n-1]}}{\sigma_n}\mathbb{E}_{\xi_n}\left(\exp\frac{it\bar{S}_n}{\sigma_n}\right)\right)\right|\\ &\quad\le \mathbb{E}\left|\exp\frac{it(X_n - X_0)}{\sigma_n} - 1\right| \to 0 \text{ as } n \to \infty.\end{aligned}$$

Therefore, by combining these facts we obtain

$$\mathbb{E}\left(\exp\frac{itS_{2n}}{\sigma_n}\right) - \mathbb{E}\left(\exp\frac{itS_{[0,n-1]}}{\sigma_n}\mathbb{E}_{\xi_n}\left(\exp\frac{it\bar{S}_n}{\sigma_n}\right)\right) \to 0. \tag{14.42}$$

Note now that, by the properties of conditional expectation,

$$\mathbb{E}\left(\exp\frac{itS_{[0,n-1]}}{\sigma_n}\mathbb{E}_{\xi_n}\left(\exp\frac{it\bar{S}_n}{\sigma_n}\right)\right) = \mathbb{E}\left(\mathbb{E}_{\xi_n}\left(\exp\frac{itS_{[0,n-1]}}{\sigma_n}\right)\mathbb{E}_{\xi_n}\left(\exp\frac{it\bar{S}_n}{\sigma_n}\right)\right).$$

By taking into account the reversibility of the process, we obtain

$$\mathbb{E}_{\xi_n}\left(\exp\frac{itS_{[0,n-1]}}{\sigma_n}\right) = \mathbb{E}_{\xi_n}\left(\exp\frac{it(X_{n+1}+\dots+X_{2n})}{\sigma_n}\right) = \mathbb{E}_{\xi_n}\left(\exp\frac{it\bar{S}_n}{\sigma_n}\right).$$

Therefore, by combining this identity with (14.42), it follows that

$$\mathbb{E}\left(\exp\frac{itS_{2n}}{\sigma_n}\right)-\mathbb{E}\left(\mathbb{E}_{\xi_n}\left(\exp\frac{it\bar{S}_n}{\sigma_n}\right)\right)^2\to 0.$$

So, by using now the stationary, the representation of the variance in (14.31) and the central limit theorem in (14.35), the convergence given in relation (14.41) follows.

Step 2. We show now that from any subsequence (n') we can extract one $(n'')\subset(n')$ for which there is a pair of random variables (V,ξ), with ξ distributed as ξ_0 and V normally distributed with mean 0 and variance c^2 (i.e. $N(0,c^2)$), such that

$$\mathbb{E}_\xi\exp(itV'_{n''})\to\mathbb{E}_\xi\exp(itV)\text{ in probability,}\tag{14.43}$$

where $(V'_{n''},\xi)$ is distributed as $(S_{n''}/\sigma_{n''},\xi_0)$. In addition,

$$\mathbb{E}\left(\mathbb{E}_{\xi_0}\left(\exp\frac{itS_{n''}}{\sigma_{n''}}\right)\right)^2\to\mathbb{E}(\mathbb{E}_\xi\exp(itV))^2.\tag{14.44}$$

Indeed, since $(S_n/\sigma_n,\xi_0)$ is tight, from any subsequence (n') we can extract one $(n'')\subset(n')$ such that $(S_{n''}/\sigma_{n''},\xi_0)$ is convergent in distribution to (V,Y) where, by (14.35), V is $N(0,c^2)$ and Y is distributed as ξ_0. By Lemma 14.12, applied to $V_{n''}=S_{n''}/\sigma_{n''}$, there exist two pairs of variables: $(V'_{n''},\xi)$ distributed as $(S_{n''}/\sigma_{n''},\xi_0)$ and (V,ξ), with ξ distributed as ξ_0 and V centered normal with variance c^2, such that convergence in (14.43) holds. Now by starting from (14.43) and applying the Lebesgue dominated convergence theorem, (14.44) follows.

Step 3. In order to finish the proof of the theorem we shall show that the limit in (14.43) does not depend on the subsequence. As a matter of fact we are going to establish the following characterization.

Lemma 14.13 *Let V be $N(0,c^2)$ such that*

$$\mathbb{E}(\mathbb{E}_\xi\exp(itV))^2=\exp(-\frac{t^2}{c^2}).\tag{14.45}$$

Then,

$$\mathbb{E}_\xi\exp(itV)=\exp(-\frac{t^2}{2c^2})\text{ a.s.}\tag{14.46}$$

Proof of Lemma 14.13 Note that, in order to establish (14.46), it is enough to show that, for all integers $m\geq 0$,

$$\mathbb{E}_\xi(V^m)=\mathbb{E}(V^m)\text{ a.s.}\tag{14.47}$$

With this aim, we redefine V on a larger probability space together with two independent variables uniformly distributed U and $\tilde{U}$ which are also independent on ξ and such that $V = f(\xi, U)$ and $\tilde{V} = f(\xi, \tilde{U})$. Note that V and $\tilde{V}$ are conditionally independent given ξ and (V, ξ) has the same distribution as $(\tilde{V}, \xi)$. In addition, let $N, \tilde{N}$ be i.i.d. random variables $N(0, c^2)$. Then, by (14.45),

$$\mathbb{E}(\mathbb{E}_\xi \exp(itV))^2 = \mathbb{E}(\exp(it(V + \tilde{V})))$$
$$= \sum_{m=0}^{\infty} \mathbb{E}(V + \tilde{V})^m \frac{(it)^m}{m!} = \sum_{m=0}^{\infty} \mathbb{E}(N + \tilde{N})^m \frac{(it)^m}{m!}.$$

Hence, for all $n = 0, 1, 2, \ldots$,

$$\mathbb{E}(V + \tilde{V})^n = \mathbb{E}(N + \tilde{N})^n. \tag{14.48}$$

Further, we proceed by induction to prove (14.47). Note that (14.47) obviously holds for $m = 0$. Assume (14.47) holds for $m \leq k$. To prove it for $k + 1$ we use (14.48) with $n = 2k + 2$ and develop the binomial

$$\sum_{\ell=0}^{2k+2} C_{2k+2}^{\ell} \mathbb{E} V^\ell \tilde{V}^{2k+2-\ell} = \sum_{\ell=0}^{2k+2} C_{2k+2}^{\ell} \mathbb{E} N^\ell \mathbb{E} N^{2k+2-\ell}. \tag{14.49}$$

By the induction hypothesis, conditional independence of V and $\tilde{V}$ and the properties of conditional expectation we obtain, for $0 \leq \ell \leq k$, and every integer m,

$$\mathbb{E} V^\ell \tilde{V}^m = \mathbb{E}(\mathbb{E}_\xi(V^\ell \tilde{V}^m)) = \mathbb{E}(\mathbb{E}_\xi V^\ell \mathbb{E}_\xi V^m) = \mathbb{E}(V^\ell)\mathbb{E}(V^m) = \mathbb{E}(N^\ell)\mathbb{E}(N^m)$$

and similarly,

$$\mathbb{E} \tilde{V}^\ell V^m = \mathbb{E}(N^\ell)\mathbb{E}(N^m).$$

By using both these estimates in (14.49), we observe that the terms in the sum from $0 \leq \ell \leq k$ and $k + 2 \leq \ell \leq 2k + 2$ all cancel, and it follows that

$$\mathbb{E} V^{k+1} \tilde{V}^{k+1} = (\mathbb{E} N^{k+1})^2 = (\mathbb{E} V^{k+1})^2.$$

Taking into account that

$$\mathbb{E} V^{k+1} \tilde{V}^{k+1} = \mathbb{E}(\mathbb{E}_\xi V^{k+1} \mathbb{E}_\xi \tilde{V}^{k+1}) = \mathbb{E}(\mathbb{E}_\xi V^{k+1})^2,$$

we obtain by the above arguments that

$$\mathbb{E}(\mathbb{E}_\xi V^{k+1})^2 = (\mathbb{E} V^{k+1})^2.$$

It follows that

$$\mathbb{E}(\mathbb{E}_\xi V^{k+1} - \mathbb{E} V^{k+1})^2 = 0,$$

which completes the proof of (14.47) and therefore of (14.46). ■

Combining (14.46) with (14.43) we get

$$\mathbb{E}_{\xi} \exp\left(\frac{itS'_{n''}}{\sigma_{n''}}\right) \to \exp\left(-\frac{t^2}{2c^2}\right) \text{ in probability.}$$

Consequently,

$$\mathbb{E}_{\xi_0} \exp\left(\frac{itS_{n''}}{\sigma_{n''}}\right) \to \exp\left(-\frac{t^2}{2c^2}\right) \text{ in probability,}$$

completing the proof of the theorem.

14.4.3 *Proof of Corollary 14.10*

Once again to soothe the notation, we set, throughout the proof, $W_n(t) = \sigma_n^{-1} S_{[nt]}$, for any $t \in [0, 1]$. The proof of this corollary follows the lines of Theorem 14.9 with the exception that we replace Lemma 14.11 by the following lemma.

Lemma 14.14 *Under the assumptions of Corollary* 14.10, *if* $\{W_n(t), t \in [0, 1]\} \to^d X$ *then* X *has independent increments.*

Proof We mention that, by the fact that the Markov chain is stationary, irreducible and aperiodic, it follows that it is absolutely regular in the sense that its β-mixing coefficients defined in Section 5.1.1 are decreasing to zero (see Example 3 given in Section 5.1.2). Recall now that $2\alpha(n) \leq \beta(n)$ where

$$\alpha(n) = \sup \mathbb{P}(A \cap B) - \mathbb{P}(A)\mathbb{P}(B);$$

here the supremum is taken over all $A \in \sigma(\xi_i, i \leq 0)$ and $B \in \sigma(\xi_i, i \geq n)$. Because we know from the proof of Theorem 14.9 that the process X is continuous, it is enough to show that for all positive integer k and $0 < s_1 < t_1 < s_2 < t_2 < \ldots < s_k < t_k < 1$, the increments $(X(t_i) - X(s_i))_{1 \leq i \leq k}$ are independent. Now, using the definitions of $\alpha(n)$ and $W_n(t)$, we get by recurrence

$$|\mathbb{P}(\cap_{i=1}^k \{W_n(t_i) - W_n(s_i) \in A_i\}) - \Pi_{i=1}^k \mathbb{P}(W_n(t_i) - W(s_i) \in A_i)|$$
$$\leq \min_{1 \leq i \leq k-1} \alpha([n(s_{i+1} - t_i)]) \to 0 \text{ as } n \to \infty,$$

for any Borelians $A_1, \ldots, A_k$. The conclusion follows by passing to the limit with n. ∎

14.5 Application to a Metropolis-Hastings Algorithm

In this section we analyze a standardized example of a stationary, irreducible and aperiodic Metropolis–Hastings algorithm with uniform marginal distribution. This type of Markov chain is interesting since it can easily be transformed into Markov chains with different marginal distributions. We point out a central limit theorem under a normalization other than the variance of partial sums. Markov chains of this type are often studied in the literature from different points of view, as in Doukhan, Massart and Rio (1994), Rio (2000 and 2009) and Merlevède and Peligrad (2010). The idea of considering Metropolis–Hastings algorithms in this context comes from Zhao, Woodroofe and Volný. (2010).

Let $E = [-1, 1]$. As in Example 2 given in Section 5.1.2, we define now the transition probabilities of a Markov chain by

$$Q(x, A) = (1 - |x|)\delta_x(A) + |x|\upsilon(A),$$

where δ_x denotes the Dirac measure and υ on $[-1, 1]$ satisfies

$$\upsilon(dx) = |x|dx. \tag{14.50}$$

Then, there is a unique invariant measure, the uniform distribution on $[-1, 1]$,

$$\pi(dx) = dx/2,$$

and the stationary Markov chain $(\xi_i)_i$ with values in E and transition probability $Q(x, A)$ is reversible and positively recurrent. Moreover, for any odd function f we have

$$Q^k(f)(\xi_0) = \mathbb{E}(f(\xi_k)|\xi_0) = (1 - |\xi_0|)^k f(\xi_0) \text{ a.s.} \tag{14.51}$$

For the odd function $f(x) = sign\ x$, define $X_i = sign\ \xi_i$. In this context we shall show:

Example: *Let $(\xi_i)_{i\geq 1}$ be a stationary Markov chain with transition kernel described above and let $X_i = sign\ \xi_i$. Let $S_n = \sum_{k=1}^n X_k$ and $\sigma_n^2 = \mathrm{Var}(S_n)$. Then $\sigma_n^2/(2n\log n) \to 1$ and*

$$\{W_n(t), t \in [0, 1]\} \to^d \frac{1}{2^{1/2}}W,$$

where $W_n(t) = \sigma_n^{-1}\sum_{j=1}^{[nt]} X_j$ and W is the standard Brownian motion.

Proof For any $m \geq 0$ we have

$$\mathbb{E}(X_0X_m) = \mathbb{E}(f(\xi_0)Q^m(f)(\xi_0)) = \int_E (1 - |x|)^m\pi(dx) = 1/(m + 1).$$

Therefore, by simple computations, we obtain

$$\sigma_n^2 \sim 2n\log n \text{ as } n \to \infty.$$

Now, to find the limiting distribution of S_n properly normalized, we study the regeneration process. Let

$$T_0 = \inf\{i > 0 : \xi_i \neq \xi_0\}$$

and, for $k \geq 0$,

$$T_{k+1} = \inf\{i > T_k : \xi_i \neq \xi_{i-1}\},\ \tau_k = T_{k+1} - T_k.$$

It is well known that $(\xi_{T_k}, \tau_k)_{k\geq 0}$ are i.i.d. random variables with ξ_{T_k} having the distribution υ. Furthermore the conditional distribution of τ_k given $\xi_{T_k} = x$ is the geometric distribution $\mathcal{G}(1-x)$, namely,

$$\mathbb{P}(\tau_1 > n|\xi_{T_1} = x) = (1-|x|)^n.$$

Then, it follows that

$$\mathbb{E}(\tau_1|\xi_{T_1} = x) = \frac{1}{|x|} \text{ and } \mathbb{E}(\tau_1) = 2.$$

So, by the law of large numbers $T_n/n \to 2$ a.s.

On the other hand, since

$$\mathbb{P}(\tau_1|X_{T_1}| > y|\xi_{T_1} = x) = \mathbb{P}(\tau_1 > y|\xi_{T_1} = x) = (1-|x|)^y,$$

by integration we obtain

$$\mathbb{P}(\tau_1 > y) = \int_{-1}^{1} (1-|x|)^y |x| dx = 2\int_0^1 (1-x)^y x dx \sim 2y^{-2} \text{ as } y \to \infty.$$

Moreover, $\mathbb{E}(\tau_k X_{T_k}) = 0$ by symmetry. Also

$$H(y) = \mathbb{E}(\tau_1^2 I(\tau_1 \leq y)) \sim 4\ln y \text{ as } y \to \infty.$$

Define a normalization satisfying $b_n^2 \sim nH(b_n)$. In our case, $b_n^2 \sim 4n\ln b_n$, implying that $b_n^2 \sim 2n\ln n$.

Note now that for any positive integer k, we have

$$\sum_{\ell=1}^{T_k-1} X_k = (T_0 - 1)X_0 + \sum_{\ell=0}^{k-1} \tau_k X_{T_k}.$$

For each n, let then m_n be such that $T_{m_n} \leq n+1 < T_{m_n+1}$. Denoting by $Y_k = \tau_k X_{T_k}$, we then have the following representation:

$$\sum_{k=1}^{n} X_k - \sum_{k=0}^{[n/2]-1} Y_k = (T_0 - 1)X_0 + \left(\sum_{k=0}^{m_n-1} \tau_k X_{T_k} - \sum_{k=0}^{[n/2]-1} \tau_k X_{T_k}\right) + \sum_{k=T_{m_n}}^{n} X_k. \quad (14.52)$$

Since $(Y_k)_{k\geq 0}$ is a centered i.i.d. sequence in the domain of attraction of a normal law, by the limit theorem for i.i.d. variables in the domain of attraction of a stable law (see pages 576–578 in Feller (1971)) we obtain,

$$\frac{\sum_{k=0}^{[n/2]-1} Y_k}{b_{[n/2]}} \to^d N(0,1). \tag{14.53}$$

By Theorem 1.9, the CLT for $(\sum_{k=1}^{n} X_k)/b_{[n/2]}$ will follow from (14.52) and (14.53) provided we show that the normalized quantity in the right-hand side of (14.52) converges in probability to 0. Clearly, because $b_{[n/2]} \to \infty$, we have

$$\frac{(T_0-1)X_0}{b_{[n/2]}} \to^{\mathbb{P}} 0.$$

Also,

$$\frac{\left\|\sum_{k=T_{m_n}}^{n} X_k\right\|_1}{b_{[n/2]}} \leq \frac{\mathbb{E}|\tau_{m_n}|}{b_{[n/2]}} = \frac{2}{b_{[n/2]}} \to 0.$$

Therefore it remains to study the middle term. Let $\delta \in]0, 1/2[$,

$$\begin{aligned}\mathbb{P}\left(\left|\sum_{k=0}^{m_n-1} Y_k - \sum_{k=0}^{[n/2]-1} Y_k\right| > \varepsilon b_{[n/2]}\right) &\leq \mathbb{P}\left(\left|\frac{m_n}{n} - \frac{1}{2}\right| \geq \delta\right) \\ &+ \mathbb{P}\left(\max_{n/2-\delta n<l<n/2+\delta n}\left|\sum_{k=0}^{l-1} Y_k - \sum_{k=0}^{[n/2]-1} Y_k\right| > \varepsilon b_{[n/2]}\right) = I + II.\end{aligned}$$

Then, by the definition of m_n and the law of large numbers for the i.i.d. sequence $(\tau_i)_{i\geq 0}$ we know that $m_n/n \to 1/\mathbb{E}(\tau_1) = 1/2$ a.s. Therefore the first term converges to 0 for every δ fixed as $n \to \infty$. To deal with the second term, we first note that, by stationarity,

$$II \leq 2\,\mathbb{P}\left(\max_{1\leq l\leq [\delta n]+1}\left|\sum_{k=0}^{l-1} Y_k\right| > \varepsilon b_{[n/2]}/2\right).$$

Next since the Y_k's are i.i.d., Theorem 1.1.5 in de la Peña and Giné (1999) gives

$$II \leq 18\,\mathbb{P}\left(\left|\sum_{k=0}^{[n\delta]} Y_k\right| > \varepsilon b_{[n/2]}/60\right).$$

Then, by the central limit theorem in (14.53) and the fact that $b_n^2 \sim 2n \ln n$, we have

$$\limsup_{n\to\infty} \mathbb{P}\left(\left|\sum_{k=0}^{[n\delta]} Y_k\right| > \varepsilon b_{[n/2]}/60\right) = \limsup_{n\to\infty} \mathbb{P}\left(\left|\sum_{k=0}^{[\delta n]} Y_k\right| / b_{[\delta n]} > \varepsilon b_{[n/2]}/60 b_{[\delta n]}\right)$$
$$\leq \mathbb{P}\left(N(0,1) > \varepsilon \delta^{-1/2}/120\right),$$

which converges to 0 as $\delta \to 0$.

It follows that

$$\frac{S_n}{b_{[n/2]}} \to^d N(0,1).$$

We recall that $\sigma_n^2 = 2n \log n = b_n^2$, implying that

$$\frac{S_n}{\sigma_n} \to^d N\left(0, \frac{1}{2}\right).$$

Consequently, because the chain is irreducible and aperiodic, by Corollary 14.10, $\{W_n(t), t \in [0,1]\} \to^d 2^{-1/2} W$. ■

15

Functional Central Limit Theorem for Empirical Processes

In this chapter we discuss the Gaussian approximation for the empirical process under different kinds of dependence assumptions.

Let $(X_n)_{n\in\mathbb{Z}}$ be a sequence of random variables with common distribution F, and define its associated empirical process by

$$\mathcal{E}_{n,\mathbf{X}}(t) = \frac{1}{\sqrt{n}} \sum_{k=1}^{n} \left(1_{\{X_k \le t\}} - F(t)\right), \; t \in \mathbb{R}. \tag{15.1}$$

Let F^{-1} be the generalized inverse of F. Recall that if F is continuous then $F(F^{-1}(t)) = t$. Hence $U_i = F(X_i)$ for all $i \in \mathbb{Z}$ are uniformly distributed over $[0, 1]$. We note also that the mixing coefficients associated with $(U_i)_{i\in\mathbb{Z}}$ are the same as those associated with $(X_i)_{i\in\mathbb{Z}}$. In addition, concerning weaker coefficients, it is easy to see that $\alpha_{k,\mathbf{U}}(n)$ and $\alpha_{k,\mathbf{X}}(n)$ are of the same order of magnitude and the same is true for $\beta_{k,\mathbf{U}}(n)$ and $\beta_{k,\mathbf{X}}(n)$. So, overall, provided that F is continuous, we have

$$\left\{\mathcal{E}_{n,\mathbf{X}}(F^{-1}(t)), 0 \le t \le 1\right\} = \left\{\mathcal{E}_{n,\mathbf{U}}(t), 0 \le t \le 1\right\}, n = 1, 2, \ldots$$

Therefore, all the theorems proved for $\mathcal{E}_{n,\mathbf{U}}$ will hold for $\mathcal{E}_{n,\mathbf{X}}$. So, in the situation when F is continuous (which we will always assume throughout the chapter, unless specified otherwise), we can restrict ourselves to the case when the distribution of the X_i's is uniform over $[0, 1]$. To soothe the notation, we shall redenote $\mathcal{E}_{n,\mathbf{X}}$ by $U_n(t)$. Therefore we will mainly be concerned with the asymptotic behavior of

$$U_n(t) = \frac{1}{\sqrt{n}} \sum_{j=1}^{n} (I_{\{U_j \le t\}} - t), \; t \in [0, 1], \tag{15.2}$$

where $(U_i)_{i\in\mathbb{Z}}$ is a stationary sequence of uniformly distributed random variables on $[0, 1]$.

Functional Gaussian Approximation for Dependent Structures. Florence Merlevède, Magda Peligrad and Sergey Utev. Oxford University Press (2019).
DOI: 10.1093/oso/9780198826941.001.0001

We shall apply the central limit theorems developed in previous chapters to describe the behavior of the finite-dimensional distributions. To prove tightness, our higher moment inequalities for sums and their maximum will come into play. Let us comment first about tightness, that is: For any $\varepsilon > 0$,

$$\lim_{\delta\to 0}\limsup_{n\to\infty}\sum_{i=0}^{[\delta^{-1}]-1}\mathbb{P}(\sup_{i\delta<s\leq(i+1)\delta}|U_n(s)-U_n(i\delta)|\geq\varepsilon)=0,$$

which obviously holds if: For all $t\geq 0$,

$$\lim_{\delta\to 0}\limsup_{n\to\infty}\frac{1}{\delta}\mathbb{P}(\sup_{t<s\leq t+\delta}|U_n(s)-U_n(t)|\geq\varepsilon)=0, \tag{15.3}$$

As always, the simplest idea to obtain such convergences is to establish inequalities of the type $\mathbb{E}|U_n(s)-U_n(t)|\leq|t-s|^{1+\alpha}$ for some $\alpha>0$. However, this inequality does not hold even in the independent case.

15.1 Tightness Criteria

Let us first state a result mentioned by Shao and Yu (1996) in their Remark 2.1, which is inspired by Theorem 22.1 in Billingsley (1968).

Theorem 15.1 *Assume that for some $p>2$, there exist $\alpha>0$, $\beta\in[0,1]$ and $r>(1-\beta)/2$ such that for any $0\leq s\leq t\leq 1$, we have*

$$\mathbb{E}|U_n(t)-U_n(s)|^p\leq K((t-s)^{1+\alpha}+n^{-r}(t-s)^{\beta}), \tag{15.4}$$

where K is a positive constant not depending on (s,t,n). Then (15.3) *holds.*

Proof Fix $\varepsilon\in]0,1[$. Let $q=q_n=\varepsilon/\sqrt{n}$ and $m=m_n(\delta)=[\delta/q]+1$. Taking advantage of the process structure, from Billingsley (1968, relation (22.18)) we easily derive that

$$\mathbb{P}\left(\sup_{s\leq t\leq s+\delta}|U_n(t)-U_n(s)|\geq 5\varepsilon\right)\leq\mathbb{P}\left(\max_{i\leq m}|U_n(s+iq)-U_n(s)|\geq\varepsilon\right). \tag{15.5}$$

By (15.5), in order to prove (15.3), it is enough to show that

$$\lim_{\delta\to 0}\limsup_{n\to\infty}\frac{1}{\delta}\mathbb{P}\left(\max_{i\leq m_n(\delta)}|U_n(s+iq)-U_n(s)|\geq\varepsilon\right)=0.$$

Note that

$$\max_{i\leq m}|U_n(s+iq)-U_n(s)|=\max_{i\leq m}|S_{i,n}|,$$

where

$$S_{i,n} = \sum_{j=1}^{i} Z_{j,n} \text{ with } Z_{j,n} = U_n(s + jq) - U_n(s + (j-1)q).$$

We shall then apply the following maximal inequality (see (1.28)): Let u be the non-negative integer such that $2^u - 1 < m \leq 2^u$ and let $v > 1$. Then, by setting $S_{0,n} = 0$, we have

$$\| \max_{i \leq m} |S_{i,n}| \|_v \leq \sum_{l=0}^{u} \left(\sum_{k=0}^{2^{u-l}-1} \|S_{(k+1)2^l,n} - S_{k2^l,n})\|_v^v \right)^{1/v}. \tag{15.6}$$

For u such that $2^u - 1 < m \leq 2^u$ and for any integers $l \in [0, u]$ and $k \in [0, 2^{u-l} - 1]$, note that

$$S_{(k+1)2^l,n} - S_{k2^l,n} = U_n(s + (k+1)2^l q) - U_n(s + k2^l q).$$

By condition (15.4), it follows that for $p > 2$,

$$\|S_{(k+1)2^l,n} - S_{k2^l,n})\|_p^p \leq K\left((2^l q)^{1+\alpha} + n^{-r}(2^l q)^{\beta}\right). \tag{15.7}$$

Taking into account (15.7) in (15.6), we get

$$\begin{aligned}
\| \max_{i \leq m} |S_{i,n}| \|_p &\leq C \sum_{l=0}^{u} \left(2^{u-l}(2^l q)^{1+\alpha} + n^{-r} 2^{u-l}(2^l q)^{\beta}\right)^{1/p} \\
&\leq C' \left\{(mq)^{(1+\alpha)/p} + (n^{-r} m (\log m) q^{\beta})^{1/p}\right\} \\
&\leq C'' \left\{\delta^{1+\alpha} + (\varepsilon n^{-1/2})^{1+\alpha} + n^{-r+(1-\beta)/2}(\delta + \varepsilon n^{-1/2})\varepsilon^{\beta-1}(\log(1 + \varepsilon^{-1}\delta\sqrt{n}))\right\}^{1/p},
\end{aligned}$$

where C, C' and C'' are positive constants not depending on $(s, n, \varepsilon, \delta)$. The logarithm terms appear in the formula only if $\beta = 1$. Then, since $r > (1 - \beta)/2$ and $\alpha > 0$,

$$\limsup_{n \to \infty} \frac{1}{\delta} \mathbb{P}\left(\sup_{s \leq t \leq s+\delta} |U_n(t) - U_n(s)|) \geq 5\varepsilon \right) \leq C'' \varepsilon^{-p} \delta^{\alpha},$$

and the result follows by letting $\delta \to 0^+$ since we imposed $\alpha > 0$. ■

15.2 Applications to Dependent Sequences

Inequality (15.4) (and then tightness of the empirical process) can be verified for many different structures of dependence. In all this section, $(U_k)_{k \in \mathbb{Z}}$ will be a stationary sequence of real-valued random variables with uniform distribution over $[0, 1]$.

15.2.1 ρ-mixing Sequences

Let $(\rho(n))_{n\geq 0}$ be the maximal correlation coefficients sequence associated with $(U_k)_{k\in\mathbb{Z}}$ (recall definition (5.3)). The following theorem is valid.

Theorem 15.2 *Assume that $A := \sum_{n\geq 0}\rho(2^n) < \infty$. Then there exists a centered Gaussian process $G = \{G(t), t \in [0, 1]\}$ such that $\{U_n(t), t \in [0, 1]\}$ converges in distribution in $D([0, 1])$ to G. The covariances of G are defined via*

$$\mathrm{Var}(G(t) - G(s)) = \lim_{n\to\infty} \mathrm{Var}(U_n(t) - U_n(s)) \tag{15.8}$$

and satisfy $\mathrm{Var}(G(t) - G(s)) \leq c|t - s|$ with finite positive c depending only on A.

Remark 15.3 *Notice that,*

$$\mathrm{cov}(G(s), G(t)) = \sum_{k\in\mathbb{Z}} \mathrm{cov}\left(1_{\{U_0\leq s\}}, 1_{\{U_k\leq t\}}\right), \ (s, t) \in [0, 1]^2, \tag{15.9}$$

provided that the series above converges absolutely.

Proof First, we show that (15.4) is satisfied. Let $0 \leq s < t \leq 1$. For any $k \in \mathbb{Z}$, let $X_k = 1_{\{s<U_k\leq t\}} - (t - s)$. $(X_k)_{k\in\mathbb{Z}}$ is then a stationary sequence of centered real-valued random variables such that $\rho_{\mathbf{X}}(n) \leq \rho(n)$ where $(\rho_{\mathbf{X}}(n))_{n\geq 0}$ is the maximal correlation coefficients sequence associated with $(X_k)_{k\in\mathbb{Z}}$. Using the Rosenthal inequality for ρ-mixing sequence, as stated in Theorem 6.13, and noticing that $\|X_1\|_p^p \leq t - s$ for any $p \geq 2$, it follows that for any $p \geq 2$,

$$\begin{aligned}\mathbb{E}|U_n(t) - U_n(s)|^p \leq C(t - s)^{p/2} \exp\left(C \sum_{i=0}^{[\log_2 n]} \rho(2^i)\right) \\ + Cn^{1-p/2}(t - s) \exp\left(C \sum_{i=0}^{[\log_2 n]} \rho^{2/p}(2^i)\right),\end{aligned}$$

where C is a positive constant depending only on p and $\rho(\cdot)$. Since $\sum_{n\geq 0}\rho(2^n) < \infty$, $\exp\left(C\sum_{i=0}^{[\log_2 n]} \rho^{2/p}(2^i)\right)$ is a slowly varying function, and then choosing $p > 2$, there exists ε with $0 < \varepsilon < (p - 2)/2$ such that $\exp\left(C\sum_{i=0}^{[\log_2 n]} \rho^{2/p}(2^i)\right) \leq \kappa n^{\varepsilon}$. So, overall, for any $p > 2$, there exists a positive constant K not depending on n and on s, t such that

$$\mathbb{E}|U_n(t) - U_n(s)|^p \leq K\left((t - s)^{p/2} + n^{1+\varepsilon-p/2}(t - s)\right),$$

with $0 < \varepsilon < (p - 2)/2$. This proves (15.4) with $1 + \alpha = p/2$, $\beta = 1$ and $r = p/2 - 1 - \varepsilon$.

Applying now the central limit theorem given in Theorem 6.50, it is easy to see that, under condition $\sum_{n\geq 0}\rho(2^n) < \infty$, the finite-dimensional marginals of $\{U_n(t), t \in [0,1]\}$ converge to those of a centered Gaussian process G whose covariance is defined via (15.8). Moreover, by inequality (6.9),

$$\mathrm{Var}(U_n(t) - U_n(s)) \leq c\mathrm{Var}(X_1) \leq c|t-s|,$$

where one can take $c = c_1 \exp(3A)$ with $c_1 = 2(\sqrt{2}+1)^2\left(1 + \frac{1}{\sqrt{2}(2^{1/6}-1)}\right)^2$. ■

15.2.2 α-dependent Sequences

Let $(\alpha_{2,\mathbf{U}}(n))_{n\geq 0}$ be the α-dependence coefficients sequence of order 2 associated with $(U_k)_{k\in\mathbb{Z}}$ (recall definition (5.13)). The following proposition is valid.

Proposition 15.4 *Assume that for any $n \geq 1$, $\alpha_{2,\mathbf{U}}(n) = O(n^{-a})$ for some $a > \frac{5+\sqrt{17}}{4}$. Then (15.4) is satisfied.*

Proof Let $0 \leq s < t \leq 1$ and set again $X_k = \mathbf{1}_{\{s<U_k\leq t\}} - (t-s)$, for any $k \in \mathbb{Z}$. Denote by $\mathcal{F}_0 = \sigma(U_k, k \leq 0)$ and $\mathbb{E}_0(\cdot) = \mathbb{E}(\cdot|\mathcal{F}_0)$. Let $p > 2$. We apply this time the Rosenthal inequality given in Theorem 3.17 together with the estimates given in Lemma 3.18. Let $\gamma > 1$ and note that since $\|X_\ell\|_\infty \leq 1$,

$$\begin{aligned}\|\mathbb{E}_0(X_\ell)\|_p^p &\leq \|\mathbb{E}_0(X_\ell)\|_2^2 \leq \mathbb{E}\left(\mathbf{1}_{s<U_\ell\leq t}|\mathbb{E}_0(X_\ell)|\right)\\ &\leq (t-s)^\gamma\|\mathbb{E}_0(X_\ell)\|_{\gamma/(\gamma-1)} \leq 2(t-s)^{1/\gamma}\left(\alpha_{1,\mathbf{U}}(\ell)\right)^{1-1/\gamma}.\end{aligned}$$

Similarly, since $p/2 \geq 1$,

$$\|X_0\mathbb{E}_0(X_\ell)\|_{p/2}^{p/2} \leq \|X_0\mathbb{E}_0(X_\ell)\|_1 \leq 4(t-s)^{1/\gamma}\left(\alpha_{1,\mathbf{U}}(\ell)\right)^{1-1/\gamma},$$

and, for any $j \geq i$,

$$\begin{aligned}\|\mathbb{E}_0(X_iX_j) - \mathbb{E}(X_iX_j)\|_{p/2}^{p/2} &\leq 2^{(p-2)/2}\|\mathbb{E}_0(X_iX_j) - \mathbb{E}(X_iX_j)\|_1\\ &\leq 2^{(p-2)/2}\|\mathbb{E}_0(X_iX_j) - \mathbb{E}(X_iX_j)\|_\gamma\|\mathbb{E}_0(X_iX_j) - \mathbb{E}(X_iX_j)\|_{\gamma/(\gamma-1)}\\ &\leq 2^{p/2}\|X_i\|_\gamma\|\mathbb{E}_0(X_iX_j) - \mathbb{E}(X_iX_j)\|_{\gamma/(\gamma-1)}\\ &\leq 2^{(p+2)/2}(t-s)^{1/\gamma}\left(4\times 2^{1/(\gamma-1)}\alpha_{2,\mathbf{U}}(i)\right)^{1-1/\gamma}.\end{aligned}$$

On the other hand, for any $\beta > 1$, we have

$$|\mathbb{E}(X_0X_j)| \leq \|X_0\mathbb{E}_0(X_j)\|_1 \leq 4(t-s)^{1/\beta}\left(\alpha_{1,\mathbf{U}}(j)\right)^{1-1/\beta}.$$

Taking into account the fact that for any $k \geq 1$, $\alpha_{1,\mathbf{U}}(k) \leq \alpha_{2,\mathbf{U}}(k) = O(k^{-a})$ and the estimates given in Lemma 3.18, we infer that, for any $\beta > 1$ such that $a(1 - 1/\beta) > 1$ and any $\gamma > 1$,

$$\mathbb{E}|U_n(t) - U_n(s)|^p \\ \leq K\left\{(t-s)^{p/(2\beta)} + (t-s)^{1/\gamma}\left(\frac{n}{n^{p/2}}(\log n)^{p\max(1,(2\delta)^{-1})} + \frac{n^{p/2}}{n^{a(1-1/\gamma)}}\right)\right\}, \quad (15.10)$$

where $\delta = \min\left(1, 1/(p-2)\right)$ and K is a constant not depending on (n, s, t). This upper bound proves that (15.4) holds if we can select p, β and γ as follows:

$$\gamma > 1,\ p/2 - 1 > \frac{\gamma - 1}{2\gamma},\ 1 < \beta < p/2,\ a(1 - 1/\beta) > 1 \text{ and } a(1 - 1/\gamma) - p/2 > \frac{\gamma - 1}{2\gamma}.$$

All these conditions are compatible if we select γ large enough and a such that $a > \frac{a}{a-1} + \frac{1}{2}$ which holds if $a > \frac{5+\sqrt{17}}{4}$. ■

Applying the central limit theorem given in Theorem 4.18 together with its comment below, it is easy to see that the finite-dimensional marginals of $\{U_n(t), t \in [0, 1]\}$ converge to those of a centered Gaussian process G whose covariance is given by (15.9), as soon as, for any $t \in [0, 1]$,

$$\sum_{k\geq 1} \|\mathbb{E}(I_{\{U_k \leq t\}}|\mathcal{F}_0) - t\|_1 < \infty, \quad (15.11)$$

where $\mathcal{F}_0 = \sigma(U_k, k \leq 0)$, and provided that the process $(U_k)_{k\in\mathbb{Z}}$ is 2-ergodic. Obviously, $\sum_{k\geq 1} \alpha_{1,\mathbf{U}}(k) < \infty$ implies condition (15.11), and $\lim_{n\to\infty} \alpha_{2,\mathbf{U}}(n) = 0$ entails the 2-ergodicity. Together with Proposition 15.4, this gives the following theorem:

Theorem 15.5 *Under the assumption of Proposition 15.4, $\{U_n(t), t \in [0, 1]\}$ converges in distribution in $D([0, 1])$ to a centered Gaussian process G whose covariance is given by (15.9).*

Comment 15.6 *The condition on the α-dependent coefficients, namely: $\alpha_{2,\mathbf{U}}(n) = O(n^{-a})$ for some $a > \frac{5+\sqrt{17}}{4}$, is weaker than the one required in Theorem 10.2 of Dedecker et al. (2007) (more precisely they ask $a > 7/3$). For strongly mixing sequences, Rio (2000, Theorem 7.3) obtained the condition $\alpha_n = O(n^{-a})$, $a > 1$. Its proof relies heavily on the properties of the strong mixing coefficients. It is an open question if, for α-dependent sequences, the Rio condition is sufficient for the validity of the functional CLT for the empirical process.*

15.2.3 β-dependent Sequences

Let $(\beta_{2,\mathbf{U}}(n))_{n\geq 0}$ be the β-dependence coefficients sequence of order 2 associated with $(U_k)_{k\in\mathbb{Z}}$ (recall the definition (5.15)). The following proposition is valid.

Proposition 15.7 *Assume that for any $n \geq 1$, $\beta_{2,\mathbf{U}}(n) = O(n^{-a})$ for some $a > 1$. Then* (15.3) *is satisfied.*

Proof To use the strength of the β-dependence coefficients, we start by introducing a new probability measure on $\mathbb{R}$ as in Dedecker (2010). Let

$$b(U_0, j) = \sup_{t \in [0,1]} \left| P_{U_j|U_0}(]-\infty, t]) - t \right|,$$

and note that $\mathbb{E}(b(U_0, j)) \leq \beta_{1,\mathbf{U}}(j) \leq \beta_{2,\mathbf{U}}(j)$. Hence, since we assumed that $\beta_{2,\mathbf{U}}(n) = O(n^{-a})$ for some $a > 1$, $\sum_{j=1}^{\infty} \mathbb{E}(b(U_0, j)) < \infty$. Recall now that here $\lambda_{[0,1]}$ is the law of U_0. Let, then, Q be the probability law on $\mathbb{R}$ whose density with respect to $\lambda_{[0,1]}$ is

$$C^{-1}\left(1 + 4\sum_{j=1}^{\infty} b(x, j)\right) \text{ with } C = 1 + 4\sum_{j=1}^{\infty} \mathbb{E}(b(U_0, j)). \tag{15.12}$$

Let F_Q be the distribution function of Q. Note that F_Q is continuous. Denote by $Y_k = F_Q(U_k)$. Clearly $\mathcal{E}_{n,\mathbf{U}} = \mathcal{E}_{n,\mathbf{Y}} \circ F_Q$ almost surely. Hence the tightness criteria (15.3) will hold if one can prove that, for any $\varepsilon > 0$ and all $t \in [0, 1]$,

$$\lim_{\delta \to 0} \limsup_{n \to \infty} \frac{1}{\delta} \mathbb{P}\big(\sup_{t < s \leq t + \delta} |\mathcal{E}_{n,\mathbf{Y}}(s) - \mathcal{E}_{n,\mathbf{Y}}(t)| \geq \varepsilon\big) = 0. \tag{15.13}$$

Note also since since F_Q is non-decreasing, the β-dependence coefficients of $\mathbf{Y} = (Y_i)_{i \in \mathbb{Z}}$ are smaller than those of $\mathbf{U} = (U_i)_{i \in \mathbb{Z}}$.

To give an insight into the interest of having introduced the probability law Q and of looking at the empirical process $\mathcal{E}_{n,\mathbf{Y}}$ instead at $\mathcal{E}_{n,\mathbf{U}}$, let us compute the covariances. For any $0 \leq s \leq t \leq 1$,

$$\begin{aligned} \left|\mathrm{cov}(1_{\{s \leq Y_0 \leq t\}}, 1_{\{s \leq Y_j \leq t\}})\right| &= \left|\mathbb{E}\left(1_{\{s \leq Y_0 \leq t\}}\left(\mathbb{E}(1_{\{s \leq Y_j \leq t\}}|U_0) - \mathbb{E}(1_{\{s \leq Y_j \leq t\}})\right)\right)\right| \\ &\leq 2\mathbb{E}\left(1_{\{s \leq Y_0 \leq t\}} b(U_0, j)\right). \end{aligned}$$

Hence, for any positive integer k,

$$\begin{aligned} \mathrm{Var}\left(\sum_{j=1}^{k} 1_{\{s < Y_j \leq t\}}\right) &\leq k \int_0^1 1_{\{s \leq F_Q(x) \leq t\}} \left(1 + 4\sum_{j=1}^{\infty} b(x, j)\right) dx \\ &= kC \int_0^1 1_{\{s \leq F_Q(x) \leq t\}} Q(dx). \end{aligned}$$

Note now that for any random variable V such that $\mathcal{L}(V) = Q$, $F_Q(V) \sim \mathcal{U}([0, 1])$. Hence,

$$\operatorname{Var}\left(\sum_{j=1}^{k} 1_{\{s<Y_j\le t\}}\right) \le kC(t-s), \tag{15.14}$$

which is a better bound that what we would obtain with the α-dependence coefficients.

To prove now that (15.13) holds, we follow the first lines of the proof of Theorem 15.1 and then infer that (15.13) will hold if we can prove that, for any $\varepsilon \in]0,1[$ and any $s \in [0,1]$,

$$\lim_{\delta\to 0} \limsup_{n\to\infty} \frac{1}{\delta}\mathbb{P}\left(\max_{i\le m_n(\delta)} |\mathcal{E}_{n,\mathbf{Y}}(s+iq) - \mathcal{E}_{n,\mathbf{Y}}(s)| \ge \varepsilon\right) = 0, \tag{15.15}$$

where $q = q_n = \varepsilon/\sqrt{n}$ and $m = m_n(\delta) = [\delta/q] + 1$. Note that

$$\max_{i\le m_n(\delta)} |\mathcal{E}_{n,\mathbf{Y}}(s+iq) - \mathcal{E}_{n,\mathbf{Y}}(s)| = \max_{i\le m} |S_{i,n}|,$$

where

$$S_{i,n} = \sum_{j=1}^{i} Z_{j,n} \text{ with } Z_{j,n} = \mathcal{E}_{n,\mathbf{Y}}(s+jq) - \mathcal{E}_{n,\mathbf{Y}}(s+(j-1)q).$$

To simplify the exposition we take $s = 0$. By using stationarity and Proposition 6.36, we get that, for any positive integer n, any non-negative integers k, l and any $p > 2$, there exists a non-negative $\mathcal{F}_0$-measurable random variable $A_0(n,p)$ satisfying $\mathbb{E}(A_0(n,p)) \le \sum_{k=1}^{n} k^{p-2}\beta_{2,\mathbf{U}}(k)$ and such that

$$\begin{aligned} \|S_{(k+1)2^l,n} - S_{k2^l,n}\|_p^p &= n^{-p/2}\left\|\sum_{i=1}^{n}\left(1_{\{k2^lq<Y_i\le(k+1)2^lq\}} - \mathbb{E}(1_{\{k2^lq<Y_i\le(k+1)2^lq\}})\right)\right\|_p^p \\ &\ll \left(\sum_{i=0}^{n-1} |\operatorname{cov}(1_{\{k2^lq<Y_0\le(k+1)2^lq\}}, 1_{\{k2^lq<Y_i\le(k+1)2^lq\}})|\right)^{p/2} \\ &+ n^{1-p/2}\mathbb{E}\left(1_{\{k2^lq<Y_0\le(k+1)2^lq\}}A_0(n,p)\right) \\ &+ n^{1-p/2}\mathbb{E}\left(1_{\{k2^lq<Y_0\le(k+1)2^lq\}}\right)\sum_{k=0}^{n}(k+1)^{p-2}\beta_{2,\mathbf{U}}(k). \end{aligned}$$

Arguing as we did to get (15.14), we infer that

$$\sum_{i=0}^{n-1} |\operatorname{cov}(1_{\{k2^lq<Y_0\le(k+1)2^lq\}}, 1_{\{k2^lq<Y_i\le(k+1)2^lq\}})| \le C2^lq.$$

Hence,

$$\sum_{k=0}^{2^{u-l}-1} \|S_{(k+1)2^l,n} - S_{k2^l,n})\|_p^p \ll 2^{u-l}(C2^l q)^{p/2} + n^{1-p/2}\mathbb{E}\left(\mathbf{1}_{\{0<Y_0\le 2^u q\}} A_0(n,p)\right)$$
$$+ n^{1-p/2}\mathbb{E}\left(\mathbf{1}_{\{0<Y_0\le 2^u q\}}\right)\sum_{k=0}^{n}(k+1)^{p-2}\beta_{2,\mathbf{U}}(k)$$
$$\ll 2^u \times q^{p/2}2^{l(p-2)/2} + n^{1-p/2}\sum_{k=0}^{n}(k+1)^{p-2}\beta_{2,\mathbf{U}}(k).$$

Therefore, using the fact that, for any $n \ge 1$, $\beta_{2,\mathbf{U}}(n) = O(n^{-a})$ for some $a > 1$, we get that, for any positive integer n, any non-negative integers a, l and any $p > 2$,

$$\sum_{k=0}^{2^{u-l}-1} \|S_{(k+1)2^l,n} - S_{k2^l,n})\|_p^p \ll 2^u \times q^{p/2}2^{l(p-2)/2} + n^{1-p/2}\log n + n^{p/2-a}.$$

Considering this upper bound in (15.6) with u being the non-negative integer such that $2^u - 1 < m \le 2^u$, we get that, for any $p > 2$,

$$\| \max_{i\le m} |S_{i,n}|\|_p^p \ll (mq)^{p/2} + (\log m)^p \left\{n^{1-p/2}\log n + n^{p/2-a}\right\},$$

where the constant involved in $\ll$ does not depend on (m, q, n, s). So, overall, for any $p > 2$, there exists a positive constant K not depending on (n, δ, ε), such that

$$\frac{1}{\delta}\mathbb{P}\left(\max_{i\le m_n(\delta)} |\mathcal{E}_{n,\mathbf{Y}}(s+iq) - \mathcal{E}_{n,\mathbf{Y}}(s)| \ge \varepsilon\right)$$
$$\ll K\varepsilon^{-p}\left(\delta^{(p-2)/2} + \delta^{-1}\varepsilon^{p/2}n^{-p/4} + \delta^{-1}(\log(\varepsilon^{-1}\delta\sqrt{n}+1))^p\left\{n^{1-p/2}\log n + n^{p/2-a}\right\}\right).$$

This implies (15.13) by taking $2 < p < 2a$ and by letting n tend to infinity and after δ tend to zero. ■

As we have seen before, the condition $\sum_{k\ge1}\alpha_{2,\mathbf{U}}(k) < \infty$ implies the convergence in distribution of the finite-dimensional marginals of $\{U_n(t), t \in [0,1]\}$ to those of a centered Gaussian process G whose covariance is given by (15.9). Since $\sum_{k\ge1}\alpha_{2,\mathbf{U}}(k) < \infty$ is implied by $\sum_{k\ge1}\beta_{2,\mathbf{U}}(k) < \infty$, we get:

Theorem 15.8 *Under the assumption of Proposition 15.7, $\{U_n(t),\ t \in [0,\ 1]\}$ converges in distribution in $D([0,\ 1])$ to a centered Gaussian process G whose covariance is given by (15.9).*

Comment 15.9 *Note that the sample paths of G are almost surely uniformly continuous with respect to the usual distance on* $[0, 1]$. *Indeed, using the notations of the proof of Proposition 15.7, we have, for any* $s, t \in [0, 1]$,

$$\mathrm{cov}(G(s), G(t)) = \sum_{k \in \mathbb{Z}} \mathrm{cov}\left(1_{\{Y_0 \leq F_Q(s)\}}, 1_{\{Y_k \leq F_Q(t)\}}\right).$$

By similar computations as the ones leading to (15.14), *we get that, for any* $0 \leq s \leq t \leq 1$,

$$\mathrm{Var}(G(t) - G(s)) \leq \left(1 + 4 \sum_{j=1}^{\infty} \beta_{1,U}(k)\right)(F_Q(t) - F_Q(s)).$$

Hence, the sample paths of the Gaussian process G are almost surely uniformly continuous with respect to the distance d_Q *defined by* $d_Q(x, y) = |F_Q(x) - F_Q(y)|$. *Note now that the uniform continuity with respect to* d_Q *is equivalent to the uniform continuity with respect to the usual distance on* $[0, 1]$ *since Q and* $\lambda_{[0,1]}$ *are equivalent.*

Remark 15.10 *The fact that the empirical central limit theorem holds under the condition* $\beta_{2,U}(n) = O(n^{-a})$ *for some* $a > 1$ *has been obtained in Dedecker (2010) using a different approach to prove tightness. For* β*-mixing sequences, the empirical central limit theorem given in Theorem 15.8 is satisfied as soon as* $\sum_{n>0} \beta_n < \infty$, *as shown by Rio (1998).*

16

Application to the Uniform Laws of Large Numbers for Dependent Processes

In this chapter, we shall see how the coupling results mentioned in Section 5.1.3 are very useful in proving asymptotic results and we focus here on the uniform laws of large numbers.

Uniform laws of averages extend the classical laws of large numbers from a single function to a collection of such functions. The uniform law addresses then the question of finding a class $\mathcal{F}$ of functions such that

$$\lim_{n\to\infty} \sup_{f\in\mathcal{F}} \frac{1}{n} \left| \sum_{i=1}^{n} f(X_i) - \mathbb{E}f(X_i) \right| = 0, \tag{16.1}$$

in some stochastic sense.

If the convergence in (16.1) holds with probability one, $\mathcal{F}$ is said to be a Glivenko–Cantelli class. Alternatively $\mathcal{F}$ is said to satisfy a uniform law of averages with respect to the sequence $(X_i)_{i\geq 1}$.

Vapnik and Cervonenkis (1981) found necessary and sufficient conditions under which a class of functions $\mathcal{F}$ is almost surely uniformly convergent with respect to an i.i.d. process $(X_i)_{i\geq 1}$. The conditions involve the asymptotic behavior of covering numbers associated with the class $\mathcal{F}$ and sample sequences of the process $(X_i)_{i\geq 1}$. For more details, we refer the reader to Vapnik and Cervonenkis (1981) or Pollard (1984).

Concerning mixing sequences, the uniform convergence has been studied by many authors. Among them Philipp (1982), Yukich (1986), Massart (1988), Yu (1991), Doukhan, Massart and Rio (1995). All these results used specific rates of convergence to zero of the mixing coefficients.

Below we shall present results about the uniform law of large numbers for mixing sequences. This chapter is divided into several parts. First we deal with absolutely regular sequences, then with $\tilde{\phi}$-mixing sequences and finally with strong mixing sequences. In these results no condition is imposed on the rates of convergence to zero of the mixing coefficients.

Since the proofs are mainly based on coupling results, we shall assume in what follows that the probability space is large enough to accommodate all the variables constructed.

Functional Gaussian Approximation for Dependent Structures. Florence Merlevède, Magda Peligrad and Sergey Utev. Oxford University Press (2019).
DOI: 10.1093/oso/9780198826941.001.0001

Let $\mathcal{F}$ be a class of real-valued measurable functions on a Polish space S. A measurable function $F : S \to \mathbb{R}$ is said to be an envelope for $\mathcal{F}$, if $|f(x)| \leq F(x)$, for every $f \in \mathcal{F}$. The class $\mathcal{F}$ is said to be uniformly bounded if for some $0 \leq K < \infty$, $F = K$ is an envelope for $\mathcal{F}$. If the class $\mathcal{F}$ is uncountable it must satisfy regularity conditions in order to insure that no problems of measurability will arise when considering quantities such as the supremum. Following Pollard, we use the term *permissible* to indicate that the class $\mathcal{F}$ satisfies such conditions (for more details see Dudley (1978) or Pollard (1984)).

16.1 The Case of Absolutely Regular Sequences

Nobel and Dembo (1993) obtained a uniform law for the class of absolutely regular sequences ($\beta_n \to 0$, as $n \to \infty$ where β_n is defined in Section 5.1.1) without imposing any condition on the rate of convergence to 0 of the dependence coefficients. In fact they obtained the almost sure uniform convergence for absolutely regular strictly stationary sequences as a direct consequence of uniform convergence for independent random variables without imposing any additional conditions on the class of functions. Their result is the following:

Theorem 16.1 *Let $(X_j)_{j\in\mathbb{Z}}$ be a strictly stationary sequence of S-valued random variables defined on $(\Omega, \mathcal{T}, \mathbb{P})$ and such that $\beta_n \to 0$ as $n \to \infty$. Let $(X_j^*)_{j\in\mathbb{Z}}$ be a sequence of S-valued independent random variables such that, for all $j \in \mathbb{Z}$, X_j^* and X_j have the same distribution. Let $\mathcal{F}$ be a permissible class of real-valued functions on S, with an envelope F such that $F \in \mathbb{L}^1(\mathbb{P})$ and such that*

$$\sup_{f\in\mathcal{F}} \left| \frac{1}{n} \sum_{i=1}^{n} (f(X_i^*) - \mathbb{E}f(X_i^*)) \right| \to 0 \quad a.s.$$

Then

$$\sup_{f\in\mathcal{F}} \left| \frac{1}{n} \sum_{i=1}^{n} (f(X_i) - \mathbb{E}f(X_i)) \right| \to 0 \quad a.s.$$

Let us notice that the coefficient of absolute regularity used by Nobel and Dembo involves all the future of the process which means that the process is ergodic and the proofs of their theorem benefitted by the results in ergodic theory.

16.2 The Case of $\tilde{\phi}$-Mixing Sequences

Concerning the case of $\tilde{\phi}$-mixing sequences, meaning that $\tilde{\phi}_n \to 0$, as $n \to \infty$, where $\tilde{\phi}_n$ is defined by (5.4), Peligrad (2001) has obtained a uniform strong law without imposing any rate of convergence to zero on $\tilde{\phi}_n$ and without requiring ergodicity. She has shown that the class of strictly stationary and $\tilde{\phi}$-mixing sequences satisfies a uniform law of averages if an independent sequence having the same marginals does. Therefore the results available for independent sequences translate to similar results for $\tilde{\phi}$-mixing sequences. Her result can be formulated as follows:

Theorem 16.2 *Let $\mathcal{F}$ be a permissible class of real-valued functions on S having an envelope $F \in \mathbb{L}^1(\mathbb{P})$. Then $\mathcal{F}$ satisfies a uniform strong law of averages with respect to a strictly stationary $\tilde{\phi}$-mixing sequence of S-valued random variables $(X_i)_{i\in\mathbb{Z}}$ if it satisfies the same law with respect to a sequence of S-valued i.i.d. random variables $(X_i^*)_{i\in\mathbb{Z}}$ such that $X_0^* =^d X_0$.*

The proof of the above theorem can be divided in two steps: The first one involves a class of functions $\mathcal{F}$ which is uniformly bounded and the second one involves a truncation argument. In fact, by analyzing these two steps we can give two additional results on the uniform law of convergence for $\tilde{\phi}$-mixing sequences without imposing stationarity. This is the aim of the next statements.

Proposition 16.3 *Let $(X_j)_{j\geq 1}$ be a sequence of S-valued variables such that $\tilde{\phi}_n \to 0$ as $n \to \infty$. Let $(X_j^*)_{j\geq 1}$ be a sequence of independent random variables such that, for all $j \geq 1$, X_j^* and X_j have the same distribution. Let $\mathcal{F}$ be a class of real-valued functions on S which is uniformly bounded and such that for a fixed $p \geq 1$ and each $1 \leq j \leq p$ we have*

$$\lim_{k\to\infty} \sup_{f\in\mathcal{F}} \left| \frac{1}{k} \sum_{i=0}^{k-1} \left(f(X_{ip+j}^*) - \mathbb{E}f(X_{ip+j}^*) \right) \right| = 0 \quad a.s. \tag{16.2}$$

Then

$$\lim_{n\to\infty} \sup_{f\in\mathcal{F}} \left| \frac{1}{n} \sum_{i=1}^{n} (f(X_i) - \mathbb{E}f(X_i)) \right| = 0 \quad a.s. \tag{16.3}$$

Proof of Proposition 16.3 For all $n \geq 1$, we shall denote by $f_n(X_n) := f(X_n) - \mathbb{E}f(X_n)$. However, for the sake of simplicity, we use the notation $f(X_n)$ instead of $f_n(X_n)$.

Let p be a natural number $p < n$. Denote by $k = [n/p] + 1$, and extend the vector $(X_1, ..., X_n)$ to the sequence $(X_1, ..., X_n, 0, 0, ...)$.

Define $Y_{i,j} = X_{ip+j}$ for $0 \leq i \leq k-1$, $1 \leq j \leq p$. For each $1 \leq j \leq p$, consider the vector $(Y_{0,j}^*, \dots, Y_{k-1,j}^*)$ constructed in Theorem 5.5, and notice that

$$\sup_{f\in\mathcal{F}} \frac{1}{n} \left| \sum_{i=1}^{n} f(X_i) \right| \leq \sup_{f\in\mathcal{F}} \frac{1}{n} \left| \sum_{i=1}^{n} f(X_i) - \sum_{i=0}^{k-1} \sum_{j=1}^{p} f(Y_{i,j}^*) \right| + \sup_{f\in\mathcal{F}} \frac{2}{kp} \sum_{j=1}^{p} \left| \sum_{i=0}^{k-1} f(Y_{i,j}^*) \right|. \tag{16.4}$$

Under condition (16.2), the second term in the right-hand side in the above inequality tends to zero almost surely by letting n tend to infinity and then p. Then it is enough to show that

$$\lim_{p\to\infty} \limsup_{n\to\infty} \sup_{f\in\mathcal{F}} \frac{1}{n} \left| \sum_{i=1}^{n} f(X_i) - \sum_{i=0}^{k-1} \sum_{j=1}^{p} f(Y_{i,j}^*) \right| = 0 \quad a.s. \tag{16.5}$$

By the triangle inequality we have:

$$\frac{1}{n}\left|\sum_{i=1}^{n} f(X_i) - \sum_{i=0}^{k}\sum_{j=1}^{p} f(Y^*_{i,j})\right| \leq \max_{1\leq j\leq p} \frac{2}{k}\left|\sum_{i=0}^{k-1}\left(f(Y_{i,j}) - f(Y^*_{i,j})\right)\right|. \tag{16.6}$$

Fix j, $1 \leq j \leq p$ and let $\eta > 0$ be such that $\tilde{\phi}_p < \dfrac{1}{2(1+\eta)}$. Let $\ell_k = \left[\dfrac{k\ln(2(1+\eta))}{|\ln(\tilde{\phi}_p)|}\right]$.

We write

$$\frac{1}{k}\left|\sum_{i=0}^{k-1}\left(f(Y_{i,j}) - f(Y^*_{i,j})\right)\right| =: I_1 + I_2, \tag{16.7}$$

where

$$I_1 = \frac{1}{k}\left|\sum_{i=0}^{k-1}\left(f(Y_{i,j}) - f(Y^*_{i,j})\right)\right| 1\left(\sum_{i=0}^{k-1} 1(Y_{i,j} \neq Y^*_{i,j}) < \ell_k\right)$$

and

$$I_2 = \frac{1}{k}\left|\sum_{i=0}^{k-1}\left(f(Y_{i,j}) - f(Y^*_{i,j})\right)\right| 1\left(\sum_{i=0}^{k-1} 1(Y_{i,j} \neq Y^*_{i,j}) \geq \ell_k\right).$$

Since $\mathcal{F}$ is bounded, we can find $C > 0$ such that, for all $f \in \mathcal{F}$,

$$I_2 \leq C1\left(\sum_{i=0}^{k-1} 1(Y_{i,j} \neq Y^*_{i,j}) \geq \ell_k\right) \quad a.s. \tag{16.8}$$

and also

$$I_1 \leq C\frac{\ell_k}{k} \leq C\ln(2(1+\eta)) \times |\ln \tilde{\phi}_p|^{-1}. \tag{16.9}$$

By combining (16.6)–(16.9) we can find a constant K such that

$$\frac{1}{n}\left|\sum_{i=1}^{n} f(X_i) - \sum_{i=0}^{k-1}\sum_{j=1}^{p} f(Y^*_{i,j})\right| \leq K\left\{1\left(\sum_{i=0}^{k-1} 1(Y_{i,j} \neq Y^*_{i,j}) \geq \ell_k\right)\right.$$
$$\left. + \ln(2(1+\eta)) \times |\ln \tilde{\phi}_p|^{-1}\right\} \quad a.s.$$

By applying now (5.6) of Corollary 5.7, it is easy to obtain via the Borel–Cantelli lemma that

$$\mathbb{P}\left(\sum_{i=0}^{k-1} 1(Y_{i,j} \neq Y^*_{i,j}) \geq \ell_k \ i.o.\right) = 0,$$

which yields

$$\limsup_{n\to\infty}\sup_{f\in\mathcal{F}}\frac{1}{n}\left|\sum_{i=1}^{n}f(X_i)-\sum_{i=0}^{k-1}\sum_{j=1}^{p}f(Y^*_{i,j})\right|\leq K\ln(2(1+\eta))\times|\ln\tilde{\phi}_p|^{-1}\quad a.s.$$

The result follows by letting $p\to\infty$. ■

Remark 16.4 *In view of the proof, Proposition* 16.3 *holds with convergence in probability replacing almost sure convergence in* (16.2) *and* (16.3).

In the next theorem we remove the condition of boundedness of the class of function $\mathcal{F}$ and replace it with an integrability condition imposed on the envelope of the class.

However, the result is now a convergence in probability uniform law instead of an almost sure result.

Theorem 16.5 *Assume that* $(X_j)_{j\geq 1}$ *and* $(X_j^*)_{j\geq 1}$ *are as in Proposition* 16.3. *Let* $\mathcal{F}$ *be a class of functions with an envelope* F *such that* $\{F(X_i)\}_{i\geq 1}$ *is an uniformly integrable family and*

$$\sup_{f\in\mathcal{F}}\left|\frac{1}{n}\sum_{i=1}^{n}(f(X_i^*)-\mathbb{E}f(X_i^*))\right|\overset{\mathbb{P}}{\to}0.\tag{16.10}$$

Then

$$\sup_{f\in\mathcal{F}}\left|\frac{1}{n}\sum_{i=1}^{n}(f(X_i)-\mathbb{E}f(X_i))\right|\overset{\mathbb{P}}{\to}0.\tag{16.11}$$

Proof of Theorem 16.5 Once again for all $n\geq 1$, we shall use the notation $f(X_n)$ to mean $f(X_n)-\mathbb{E}f(X_n)$.

Let C be a positive constant and let $f\in\mathcal{F}$. Denote by $f_1=f\mathbf{1}(F\leq C)$ and by $f_2=f\mathbf{1}(F>C)$. We obviously have:

$$\begin{aligned}\sup_{f\in\mathcal{F}}\frac{1}{n}\left|\sum_{i=1}^{n}f(X_i)\right| &= \sup_{f_1,f\in\mathcal{F}}\frac{1}{n}\left|\sum_{i=1}^{n}(f_1(X_i)-\mathbb{E}f_1(X_i))\right|\\ &\quad+\sup_{f_2,f\in\mathcal{F}}\frac{1}{n}\left|\sum_{i=1}^{n}(f_2(X_i)-\mathbb{E}f_2(X_i))\right|.\end{aligned}\tag{16.12}$$

In order to treat the last term in the right-hand side of equality (16.12), it suffices to notice that

$$\mathbb{E}\sup_{f_2,f\in\mathcal{F}}\frac{1}{n}\left|\sum_{i=1}^{n}(f_2(X_i)-\mathbb{E}f_2(X_i))\right|\leq\frac{2}{n}\sum_{i=1}^{n}\mathbb{E}|F(X_i)|\mathbf{1}(|F(X_i)|>C),$$

which converges to zero by letting $n \to \infty$ and then $C \to \infty$ and using the uniform integrability of $\{F(X_i)\}_{i\geq 1}$.

We handle now the first term in the right-hand side of equality (16.12). By Proposition 16.3 and Remark 16.4, since the family $\{f_1, f \in \mathcal{F}\}$ has a bounded envelope, we will obtain that the first term in the right-hand side of equality (16.12) converges to zero in probability as soon as condition (16.2) holds in probability. Note that due to the uniform integrability of the envelope, condition (16.10) entails that the convergence still holds in $\mathbb{L}^1$. This last consideration together with the following statement prove the desired convergence, i.e. condition (16.2) holds in probability, and then end the proof of Theorem 16.5.

Claim *Let $(Z_i)_{i\geq 1}$ be a sequence of independent S-valued random variables defined on $(\Omega, \mathcal{T}, \mathbb{P})$ and let $\mathcal{F}$ be a permissible class of real-valued functions on S such that*

$$\sup_{f\in\mathcal{F}} \left| \frac{1}{n} \sum_{i=1}^{n} (f(Z_i) - \mathbb{E}f(Z_i)) \right| \overset{\mathbb{L}^1}{\to} 0.$$

Then for each p, $p \geq 1$ and all $1 \leq i_1 < i_2 < \cdots < i_k \leq n$, where $k = [n/p] + 1$ we have

$$\sup_{f\in\mathcal{F}} \left| \frac{1}{k} \sum_{j=1}^{k} (f(Z_{i_j}) - \mathbb{E}f(Z_{i_j})) \right| \overset{\mathbb{L}^1}{\to} 0.$$

Proof of the claim The proof is based on the fact that if X and Y are two S-valued random variables independent such that for all $f \in \mathcal{F}$ we have $\mathbb{E}f(Y) = 0$, then

$$\mathbb{E}\left(\sup_{f\in\mathcal{F}} |f(X)|\right) \leq \mathbb{E}\left(\sup_{f\in\mathcal{F}} |f(X) + f(Y)|\right). \tag{16.13}$$

We give here a short proof of (16.13). Notice first that since $\mathbb{E}f(Y) = 0$ and X and Y are independent

$$f(X) = \mathbb{E}\left(f(X) + f(Y)|X\right).$$

Then

$$\begin{aligned}
\mathbb{E}\left(\sup_{f\in\mathcal{F}} |f(X)|\right) &= \mathbb{E}\left(\sup_{f\in\mathcal{F}} |\mathbb{E}\left(f(X) + f(Y)|X\right)|\right) \\
&\leq \mathbb{E}\left(\mathbb{E}\left(\sup_{f\in\mathcal{F}} |f(X) + f(Y)| \Big| X\right)\right) \\
&\leq \mathbb{E}\left(\sup_{f\in\mathcal{F}} |f(X) + f(Y)|\right).
\end{aligned}$$

■

16.3 The Case of Strong Mixing Sequences

Concerning the case of strong mixing sequences of real-valued random variables, Theorem 16.6 below has been obtained by Peligrad (2002). The proof is mainly based on the coupling result (c) of Theorem 5.9 and on a decomposition in "p-sequences" as in the proof of Proposition 16.3. The result is again obtained without imposing any rate of convergence to zero of the strong mixing coefficients and without requiring ergodicity.

Let us also notice that since, the proof is mainly based on the coupling result for strongly mixing sequences, due to Remark 5.6, results for strong mixing sequences taking their values in a Polish space cannot be derived by a similar approach.

For the sake of simplicity, we set $\overline{\alpha}_n = \overline{\alpha}(\mathcal{F}_1^{n-1}, X_n)$, where $\overline{\alpha}$ is defined in (5.8).

Theorem 16.6 *Assume that $\{X_n\}_{n\geq 1}$ is a sequence of identically distributed real-valued random variables in $\mathbb{L}^1$ such that $\overline{\alpha}_n \to 0$ as $n \to \infty$. Let $\{X_n^*\}_{n\geq 1}$ be a sequence of independent identically distributed real-valued random variables such that X_1^* has the same distribution as X_1. Let $\mathcal{F}$ be a class of Lipschitz functions with the Lipschitz constants bounded by C $(C > 0)$ such that*

$$\lim_{n\to\infty} \mathbb{E} \sup_{f\in\mathcal{F}} \frac{1}{n} \left| \sum_{i=1}^{n} (f(X_i^*) - \mathbb{E}f(X_i^*)) \right| = 0. \tag{16.14}$$

Then

$$\lim_{n\to\infty} \mathbb{E} \sup_{f\in\mathcal{F}} \frac{1}{n} \left| \sum_{i=1}^{n} (f(X_i) - \mathbb{E}f(X_i)) \right| = 0. \tag{16.15}$$

In the statement of Theorem 16.6, let us notice that the assumption that the variables have the same distribution can be relaxed by imposing some weaker forms of stationarity. Moving in this direction, we would like to mention the following result:

Proposition 16.7 *If we assume in Theorem 16.6 that $\{X_n\}_{n\geq 1}$ is uniformly integrable instead of being identically distributed then Theorem 16.6 still holds.*

Proof of Proposition 16.7 Once again for all $n \geq 1$, we shall use the notation $f(X_n)$ to mean $f(X_n) - \mathbb{E}f(X_n)$.

To prove the result, we consider the "p-sequences" as in the proof of Proposition 16.3. So, let p be a positive integer less than n, set $k = [n/p] + 1$, and, for each $1 \leq j \leq p$, let $(Y_{0,j}^*, \ldots, Y_{k-1,j}^*)$ be the vector constructed, this time, as in Theorem 5.9. We notice that by the triangle inequality, we have

$$\begin{aligned}\mathbb{E} \sup_{f\in\mathcal{F}} \frac{1}{n} \left| \sum_{i=1}^{n} f(X_i) \right| &\leq \mathbb{E} \sup_{f\in\mathcal{F}} \frac{1}{n} \sum_{j=1}^{p} \left| \sum_{i=0}^{k-1} \left(f(Y_{i,j}) - f(Y_{i,j}^*) \right) \right| \\ &\quad + 2\,\mathbb{E} \sup_{f\in\mathcal{F}} \frac{1}{kp} \sum_{j=1}^{p} \left| \sum_{i=0}^{k-1} f(Y_{i,j}^*) \right|. \end{aligned} \tag{16.16}$$

By using the claim stated in the proof of Theorem 16.5 together with condition (16.14), the last term in the right-hand side of inequality (16.16) tends to zero by letting n tend to infinity. Now, to treat the first term in the right-hand side, we first notice that since f is Lipschitz, by Theorem 5.9, we derive that

$$\mathbb{E}\sup_{f\in\mathcal{F}}\frac{1}{n}\sum_{j=1}^{p}\left|\sum_{i=0}^{k-1}\left(f(Y_{i,j})-f(Y_{i,j}^{*})\right)\right| \le \frac{4}{n}C\sum_{j=1}^{p}\sum_{i=0}^{k-1}\int_0^{\overline{\alpha}_p} Q_{|X_{ip+j}|}(u)du$$
$$\le 8C\sup_{1\le i\le n}\int_0^{\overline{\alpha}_p} Q_{|X_i|}(u)du.$$

To obtain the desired result we let first n tend to infinity and then we apply the following lemma, by letting p tend to infinity:

Lemma 16.8 *Suppose that $\{X_n\}_{n\ge1}$ is an uniformly integrable sequence of random variables, then*

$$\lim_{\varepsilon\to0}\sup_{i\ge1}\int_0^{\varepsilon} Q_{|X_i|}(u)du = 0. \tag{16.17}$$

Proof of Lemma 16.8 Let us suppose that $\operatorname{ess\,sup}_{i\ge1}|X_i| = +\infty$, otherwise (16.17) is trivial. Now notice that

$$\sup_{i\ge1}\int_0^{\varepsilon} Q_{|X_i|}(u)du \le \sup_{i\ge1}\mathbb{E}\{|X_i|1(|X_i|\ge Q_{|X_i|}(\varepsilon))\},$$

which converges to zero when $\varepsilon\to0$ by using the fact that $\lim_{\varepsilon\to0}Q_{|X_i|}(\varepsilon) = +\infty$ together with the uniform integrability condition. ∎

We can also formulate the following result:

Theorem 16.9 *Assume $\{X_n\}_{n\ge1}$ is a sequence of uniformly integrable, centered real-valued random variables and such that $\overline{\alpha}_n\to0$ as $n\to\infty$. Then*

$$\lim_{n\to\infty}\mathbb{E}\left(\max_{1\le j\le n}\frac{1}{n}\left|\sum_{i=1}^{j}X_i\right|\right) = 0.$$

Proof of Theorem 16.9 Let us first truncate the random variables in the following way: Take A to be a positive number and set

$$X_i' = X_i1(|X_i|\le A) - \mathbb{E}X_i1(|X_i|\le A)$$

and

$$X_i'' = X_i1(|X_i|>A) - \mathbb{E}X_i1(|X_i|>A).$$

Now we set $S_k = \sum_{i=1}^{k} X_i$ and $S'_k = \sum_{i=1}^{k} X'_i$ and we notice that

$$\frac{1}{n}\mathbb{E}\left(\max_{1\le j\le n}|S_j|\right) \le \frac{\sum_{i=1}^{n}\mathbb{E}(|X''_i|)}{n} + \frac{1}{n}\mathbb{E}\left(\max_{1\le j\le n}|S'_j|\right). \tag{16.18}$$

Under the condition of uniform integrability of the sequence $\{X_i\}_{i\ge 1}$, it is clear that the first term in the right-hand side of the above inequality tends to zero by letting n tend to infinity and then likewise A. Then it remains to treat the last term in inequality (16.18).

Let p be a positive integer such that $p < n$. Denote by $k = [n/p] + 1$, and extend the vector $(X'_1, \dots, X'_n)$ to the sequence $(X'_1, \dots, X'_n, 0, 0, \dots)$. Define $Y'_{i,j} = X'_{ip+j}$ for $0 \le i \le k-1$, $1 \le j \le p$. For each $1 \le j \le p$, consider the vector $(Y'^*_{0,j}, \dots, Y'^*_{k-1,j})$ constructed in Theorem 5.9, and notice that

$$\frac{1}{n}\mathbb{E}\left(\max_{1\le j\le n}|S'_j|\right) = \frac{1}{n}\mathbb{E}\max_{1\le m\le p}\max_{1\le \ell\le k}\left|\sum_{j=1}^{m}\sum_{i=0}^{\ell-1}X'_{ip+j}\right|$$

$$\le \frac{1}{n}\mathbb{E}\max_{1\le m\le p}\max_{1\le \ell\le k}\left|\sum_{j=1}^{m}\sum_{i=0}^{\ell-1}Y'^*_{i,j}\right| + \frac{1}{n}\mathbb{E}\max_{1\le m\le p}\max_{1\le \ell\le k}\left|\sum_{j=1}^{m}\sum_{i=0}^{\ell-1}\left(Y'_{i,j}-Y'^*_{i,j}\right)\right|. \tag{16.19}$$

First we treat the last term in the right-hand side of the above inequality. By using Theorem 5.9, we derive that

$$\frac{1}{n}\mathbb{E}\max_{1\le m\le p}\max_{1\le \ell\le k}\left|\sum_{j=1}^{m}\sum_{i=0}^{\ell-1}\left(Y'_{i,j}-Y'^*_{i,j}\right)\right| \le \frac{1}{n}\sum_{j=1}^{p}\sum_{i=0}^{k-1}\mathbb{E}\left|Y'_{i,j}-Y'^*_{i,j}\right|$$

$$\le 8\sup_{1\le i\le n}\int_0^{\overline{\alpha}_p} Q_{|X'_i|}(u)du \le 16A\overline{\alpha}_p,$$

which converges to zero by letting respectively n and then p tend to infinity. To treat now the first term in the right-hand side of inequality (16.19), first we write

$$\frac{1}{n}\mathbb{E}\max_{1\le m\le p}\max_{1\le \ell\le k}\left|\sum_{j=1}^{m}\sum_{i=0}^{\ell-1}Y'^*_{i,j}\right| \le \frac{1}{n}\sum_{j=1}^{p}\mathbb{E}\max_{1\le \ell\le k}\left|\sum_{i=0}^{\ell-1}Y'^*_{i,j}\right|.$$

By using Rosenthal's inequality for independent random variables (see for instance Theorem 6.17), it follows that there exists a constant C such that

$$\mathbb{E}\max_{1\le \ell\le k}\left|\sum_{i=0}^{\ell-1}Y'^*_{i,j}\right| \le C\left\{\left(\sum_{i=0}^{k-1}\mathbb{E}(Y'^*_{i,j})^2\right)^{1/2} + \mathbb{E}\max_{1\le i\le k-1}|Y'^*_{i,j}|\right\}.$$

Therefore,

$$\frac{1}{n}\mathbb{E}\max_{1\leq m\leq p}\max_{1\leq \ell\leq k}\left|\sum_{j=1}^{m}\sum_{i=0}^{\ell-1}Y'^{*}_{i,j}\right|\leq 2AC\left(\sqrt{\frac{2p}{n}}+\frac{p}{n}\right),$$

which converges to zero by letting n tend to infinity.

Gathering all these considerations, we obtain the desired result. ■

17

Examples and Counterexamples

In this chapter, we study the $\mathbb{L}^p$ $(p \geq 2)$ properties of a renewal Markov chain allowing us to show that some projective criteria given in Chapter 4 to get the functional central limit theorem are essentially optimal.

17.1 The Renewal-Type Markov Chain Example

We begin by reminding what is the *renewal chain* as previously defined in Section 11.1.2. Let $\{Y_k; k \geq 0\}$ be a discrete Markov chain with the state space $\mathbb{Z}^+$ and transition matrix $P = (p_{i,j})$ given by $p_{k,k-1} = 1$ for $k \geq 1$ and $p_j = p_{0,j-1} = \mathbb{P}(\tau = j), j = 1, 2, \ldots,$ (that is whenever the chain hits 0, $Y_t = 0$, it then regenerates with the probability p_j). The Markov chain $\{Y_k; k \geq 0\}$ has the transition matrix

$$\begin{bmatrix} p_1 & p_2 & p_3 & p_4 & p_5 & \ldots \\ 1 & 0 & 0 & 0 & 0 & \ldots \\ 0 & 1 & 0 & 0 & 0 & \ldots \\ & & & \ldots & & \end{bmatrix}.$$

Notice that if $p_1 > 0$, and in addition $p_{n_j} > 0$ along $n_j \to \infty$, then the chain $\{Y_k; k \geq 0\}$ is irreducible and aperiodic. Moreover, the stationary distribution exists if and only if $\mathbb{E}[\tau] < \infty$ and it is given by

$$\pi_j = \pi_0 \sum_{i=j+1}^{\infty} p_i \,, \; j = 1, 2 \ldots, \text{ and } \pi_0 = 1/\mathbb{E}[\tau].$$

In each of the following examples, $\{X_i\}_{i \geq 0}$ is the stationary sequence constructed by

$$X_i = g(Y_i), \; i \geq 0,$$

for a certain functional g. Moreover, $S_n = X_1 + \ldots + X_n$ and $S_0 = 0$.

Functional Gaussian Approximation for Dependent Structures. Florence Merlevède, Magda Peligrad and Sergey Utev. Oxford University Press (2019).
DOI: 10.1093/oso/9780198826941.001.0001

17.1.1 $\mathbb{L}^2$ Analysis for the Renewal Chain Example

In this section we are going to prove the following theorem which is due to Peligrad and Utev (2005).

Theorem 17.1 *For any non-negative sequence $a_n \to 0$, there exist a stationary ergodic discrete Markov chain $(Y_k)_{k\geq 0}$ and a functional g such that $X_i = g(Y_i)$; $i \geq 0$, $\mathbb{E}[X_1] = 0$, $\mathbb{E}[X_1^2] < \infty$ and*

$$\sum_{n=1}^{\infty} a_n \frac{\|\mathbb{E}(S_n|Y_0)\|_2}{n^{3/2}} < \infty \quad \text{but} \quad \frac{S_n}{\sqrt{n}} \quad \text{is not stochastically bounded.} \tag{17.1}$$

This shows that, in its generality, the Maxwell–Woodroofe condition (4.30) is optimal to get the central limit theorem.

Proof of theorem 17.1 The example is constructed on the renewal chain. Let us consider now an arbitrary non-negative sequence $a_n \to 0$ as in our theorem 17.1. Notice that, without loss of generality, it is enough to assume that $(a_n)_{n\geq 1}$ is a strictly decreasing sequence of real positive numbers.

The choice of p_j further depends on this arbitrary non-negative sequence $(a_n)_{n\geq 1}$ and on the sequence $\{u_k;\, k = 1, 2, ...\}$ of positive integers defined by

$$u_1 = 1\ ,\ u_2 = 2\ ,\ u_k^4 + 1 < u_{k+1} \ \text{ for } k \geq 3 \quad \text{and} \quad a_\ell \leq k^{-2} \ \text{ for } \ell \geq u_k. \tag{17.2}$$

Then, for $i \geq 1$, we take

$$p_i = \begin{cases} c/u_j^2 & \text{if} \quad i = u_j \quad \text{for some} \quad j \geq 1 \\ 0 & \text{if} \quad i \neq u_j \quad \text{for all} \quad j \geq 1, \end{cases}$$

that is, for each positive integer $j \geq 1$, $p_{u_j} = c/u_j^2$ and $p_i = 0$ for $u_j < i < u_{j+1}$.

Clearly,

$$\mathbb{E}[\tau] < \infty \quad \text{but} \quad \mathbb{E}[\tau^2] = \infty. \tag{17.3}$$

As a functional g we take $\mathbf{1}_{\{x=0\}} - \pi_0$, where $\pi_0 = \mathbb{P}_\pi(Y_0 = 0)$ under the stationary distribution denoted by $\mathbb{P}_\pi$ ($\mathbb{E}_\pi$ denotes the expectation for the process started with the stationary distribution). The stationary sequence is then defined by

$$X_j = 1_{\{Y_j=0\}} - \pi_0 \quad \text{so that} \quad S_n = \sum_{j=1}^{n} X_j = \sum_{j=1}^{n} 1_{\{Y_j=0\}} - n\pi_0. \tag{17.4}$$

By $\mathbb{P}_k$ and $\mathbb{E}_k$ we denote the probability and the expectation operator when the Markov chain is started at k, i.e. $\mathbb{P}(Y_0 = k) = 1$. Let

$$\nu = \min\{m \geq 1 \ :\ Y_m = 0\}\ ,\ A_n = \mathbb{E}_0[S_n]\ ,\ x \wedge y = \min(x, y). \tag{17.5}$$

We shall first check the first part of theorem 17.1. With this aim, we need the following fact.

Fact 17.2

$$V_n = \|\mathbb{E}(S_n|Y_0)\|_2 \le \|\nu \wedge n\|_2 \; + \max_{1\le i\le n} |A_i|$$

$$\equiv I_n + J_n,$$

where $\|x\|_2^2 = \sum_{k=0}^{\infty} x_k^2 \pi_k$.

Proof We first notice that $|S_n| \le n$ and $\mathbb{P}_k(\nu = k) = 1$, for any $k > 0$, and, conditionally on $Y_0 = k$ (with $0 < k \le n$),

$$\mathbb{E}_k(S_n) = \mathbb{E}_k(S_k) + \mathbb{E}_k(S_n - S_k),$$

so that the first term is bounded by k and the second term is equal to $\mathbb{E}_0(S_{n-k})$, since $Y_k = 0$. Thus,

$$|\mathbb{E}_k(S_n)| \le k \wedge n + |A_{n-k}|,$$

ending the proof of the fact above by using Minkowski's inequality. ■

By Fact 17.2, to prove that $\sum_{n\ge 1} a_n \|\mathbb{E}(S_n|Y_0)\|_2 n^{-3/2} < \infty$, it is enough to show that

$$\sum_{n=1}^{\infty} a_n I_n / n^{3/2} + \sum_{n=1}^{\infty} a_n J_n / n^{3/2} < \infty. \tag{17.6}$$

To see that the first sum is finite we note that

$$I_n^2 = \mathbb{E}_\pi(\nu \wedge n)^2 = \sum_{j=1}^{n} j^2 \pi_j + n^2 \sum_{j=n+1}^{\infty} \pi_j.$$

But, for $u_{\ell-1} \le j$,

$$\pi_j \le \pi_0 \sum_{i=u_{\ell-1}+1}^{\infty} p_i \le c\pi_0 \sum_{i\ge \ell} \frac{1}{u_i^2} \le \frac{c'}{u_\ell^2}.$$

Therefore, for $u_k < n \le u_{k+1}$, we write

$$I_n^2 \le \left[\sum_{\ell=1}^{k}\left(\sum_{j=u_{\ell-1}+1}^{u_\ell} j^2\pi_j\right)\right] + \left(\sum_{j=u_k+1}^{n} j^2\pi_j\right) + n^2 \sum_{\ell=k+1}^{\infty}\left(\sum_{j=u_{\ell-1}+1}^{u_\ell} \pi_j\right)$$

$$\le C(u_k \; + \; n^3/u_{k+1}^2 \; + \; n^2/u_{k+1}).$$

It follows that there exists a positive constant K such that

$$\sum_{n=2}^{\infty} \frac{a_n I_n}{n^{3/2}} = \sum_{k=1}^{\infty} \sum_{n=u_k+1}^{u_{k+1}} \frac{a_n I_n}{n^{3/2}} \leq \sum_{k=1}^{\infty} \frac{1}{k^2} \sum_{n=u_k+1}^{u_{k+1}} I_n n^{-3/2} \leq K \sum_{k\geq 1} \frac{1}{k^2},$$

proving that the first sum in (17.6) is finite.

To prove that the second sum in (17.6) is finite, we need to analyze A_n, which satisfies the renewal equation

$$A_n = \mathbb{E}_0[S_{n\wedge \nu}] + \sum_{j=1}^{n-1} A_{n-j} p_j.$$

We now use probabilistic arguments to analyze this renewal equation. With this aim, we define

$$T_0 = 0,\ T_k = \min\{t > T_{k-1} : Y_t = 0\},\ \tau_k = T_k - T_{k-1}\,,\ k = 1, 2, \ldots \tag{17.7}$$

Then, $\{\tau_j\}_{j\geq 1}$ are independent variables. For $j \geq 2$, they are equally distributed, and their common law is the law of τ_1 when the Markov chain starts from zero, that is the law of τ. Let $\xi_j = 1 - \pi_0 \tau_j$ and introduce the stopping time

$$\nu_n = \min\{j \geq 1 : T_j \geq n\}\,.$$

Clearly, $S_{T_k} = \sum_{j=1}^{k} \xi_j = k - \pi_0 T_k$, $\nu_n \leq n$ and since $\mathbb{E}(\tau) = 1/\pi_0$, $\mathbb{E}_0[\xi_1] = 0$. Thus, by the Wald identity

$$\mathbb{E}_0[S_{T_{\nu_n}}] = \mathbb{E}_0\left(\sum_{j=1}^{\nu_n} \xi_j\right) = 0.$$

Hence, since $|S_a - S_b| \leq |a - b|$, by the definition of A_n we obtain

$$|A_n| = \left|\mathbb{E}_0[S_{T_{\nu_n}} - S_n]\right| \leq \mathbb{E}_0[\tau_{\nu_n}] \leq \mathbb{E}_0[\max_{1\leq i\leq n} \tau_i].$$

Writing $M_n := \max_{1\leq i\leq n} \tau_i$, we then have

$$J_n = \max_{1\leq i\leq n} |A_i| \leq \mathbb{E}_0[M_n].$$

To analyze $\mathbb{E}_0[M_n]$, we notice that

$$\mathbb{E}_0[M_n] = \sum_{t=1}^{\infty} u_t \mathbb{P}_0(M_n = u_t)$$

$$\text{and} \quad \mathbb{P}_0(M_n = u_t) \leq \min(1, n\mathbb{P}(\tau = u_t)) \leq c_1 \min(1, n/u_t^2).$$

Fix n, $u_k < n \leq u_{k+1}$. Notice first that for $t \leq k-1$, we have $u_t \leq u_{k-1} \leq u_k^{1/4} \leq n^{1/4}$. Also, $\sum_{j=k+1}^{\infty} 1/u_j \leq c_2/u_{k+1}$ and thus, splitting the sum into three parts according to t: $t \leq k-1, t = k$ and $t \geq k+1$, we obtain the bound

$$\mathbb{E}_0[M_n] \leq c_3 \left(n^{1/4} + \frac{n}{u_{k+1}} + u_k \min(1, n/u_k^2) \right).$$

Using the construction of u_n and its relation to a_n, it can be easily derived that $\sum_{n=1}^{\infty} a_n J_n / n^{3/2} < \infty$. This completes the proof of (17.6).

To end the proof, it remains then to show the stochastic unboundedness of $S_n/\sqrt{n}$. We proceed by contradiction; that is we assume that

$$\left\{ S_n/\sqrt{n} \, ; \, n \geq 1 \right\} \quad \text{is stochastically bounded}$$

and show that $\mathbb{E}(\tau^2) < \infty$, which is in contradiction with (17.3).

Let $\{\tau_j\}$ be independent variables equally distributed as τ. Define

$$T_k = \tau_1 + ... + \tau_k \ , \ \eta_n = \max\{i \geq 1 \, : \, T_i \leq n\},$$

$$T(i, n] = T_n - T_i \ , \ \eta_n(\xi) = \max\{i \geq 1 \, : \, \xi + T(1, i] \leq n\}$$

(where $\max_{i \in \emptyset} a_i = 0$). With these notations, observe that

$$S_n = \eta_n(\nu) - na,$$

where $a = 1/\mathbb{E}[\tau_1] = \pi_0$ and ν is defined in (17.5).

The following proposition shows the stochastic unboundedness of $S_n/\sqrt{n}$, and then ends the proof of Theorem 17.1.

Proposition 17.3 *Assume that for a non-negative integer valued variable ξ*

$$\left\{ \frac{\eta_n(\xi) - an}{\sqrt{n}} \, ; \, n \geq 1 \right\} \quad \textit{is stochastically bounded.} \tag{17.8}$$

Then, $\mathbb{E}[\tau_1^2] < \infty$.

Proof First, let η_n' be a copy of the renewal process $\{\eta_n : n \geq 1\}$ which does not depend on ξ. Then, $\eta_n(\xi)$ is identically distributed as $\eta_{n-\xi}'$ and so any finite number of renewals does not affect the stochastic boundedness of the normalized renewal processes. As a consequence, condition (17.8) implies that, for any positive real M,

$$\mathbb{P}\left([an - \sqrt{n}M] \leq \eta_n < [an + \sqrt{n}M] \right) \geq 1 - \varepsilon_M,$$

where $\varepsilon_M \to 0$ as $M \to \infty$.

Next, we apply the standard relationship $\{\eta_n \geq k\} = \{T_k \leq n\}$, yielding

$$\begin{aligned}\mathbb{P}\left([an - \sqrt{n}M] \leq \eta_n < [an + \sqrt{n}M]\right) &= \mathbb{P}\left(T_{[an-\sqrt{n}M]} \leq n, T_{[an+\sqrt{n}M]} > n\right)\\ &= \mathbb{P}\left(T_L \leq n, T_R > n\right) =: I \geq 1 - \varepsilon_M,\end{aligned}$$

where

$$L = L[n, M] = [an - \sqrt{n}M], \quad R = R[n, M] = [an + \sqrt{n}M].$$

Now, we take $k = R - L$. Since $T(i, n] = T_n - T_i$ is identically distributed as T_{n-i}, we can write that, for any positive real N,

$$\begin{aligned}I &= \mathbb{P}\left(T_L \leq n, T_L + T(L, R] > n\right)\\ &= \mathbb{P}\left(T_L \leq n - kN, T_L + T(L, R] > n\right) + \mathbb{P}\left(n - kN < T_L \leq n, T_L + T(L, R] > n\right)\\ &\leq \mathbb{P}(T(L, R] > kN) + \mathbb{P}\left(n - kN < T_L \leq n\right) = \mathbb{P}\left(T_k > kN\right) + \mathbb{P}(n - kN < T_L \leq n).\end{aligned}$$

By the law of large numbers

$$\mathbb{P}(T_k/k > N) \leq \delta_N,$$

where $\delta_N \to 0$ as $N \to \infty$. Thus,

$$\mathbb{P}\left(n - kN < T_L \leq n\right) \geq 1 - \varepsilon_M - \delta_N.$$

Since $2\sqrt{n}M - 1 \leq k \leq 2\sqrt{n}M + 1$ we derive

$$\mathbb{P}\left(|T_L - n|/\sqrt{n} \leq (2M + 1)N\right) \geq 1 - \varepsilon_M - \delta_N.$$

Now we use the symmetrization argument. We consider an independent copy of $\{\tau_j\}$, namely $\{\tau'_j\}$, and denote by $T'_k = \tau'_1 + \ldots + \tau'_k$, $T^s_k = T_k - T'_k$. Clearly

$$\mathbb{P}\left(|T^s_L|/\sqrt{n} \geq 2(2M + 1)N\right) \leq \varepsilon_M + \delta_N.$$

Here

$$\lim_{n\to\infty} L(n, M)/n = a.$$

By standard arguments involving an application of the Lévy maximal inequality for sums of symmetric independent random variables we easily derive that the sequence $\{T^s_n/\sqrt{n}\}$ is stochastically bounded. By Theorem 3 in Esseen and Janson (1985), the fact that $\{T^s_n/\sqrt{n}\}$ is stochastically bounded implies that $\mathbb{E}(\tau_1 - \tau'_1)^2 < \infty$. Thus, $\mathbb{E}(\tau_1^2) < \infty$. ■

17.1.2 $\mathbb{L}^p$ Analysis for the Renewal Chain Example

Next result shows the minimality of the condition $\delta_{\infty,p} < \infty$ in Theorem 3.6, for $||S_n^*||_p = O(\sqrt{n})$ to hold, and is due to Peligrad, Utev and Wu (2007).

Theorem 17.4 *For any $p > 2$ and any non-negative sequence $a_n \to 0$ as $n \to \infty$, there exist a stationary ergodic discrete Markov chain $(Y_k)_{k\geq 0}$ and a functional g such that $X_i = g(Y_i)$; $i \geq 0$, $\mathbb{E}(X_1) = 0$, $\mathbb{E}(|X_1|^p) < \infty$ and*

$$\sum_{n=1}^{\infty} a_n n^{-3/2} || \mathbb{E}(S_n|Y_0)||_p < \infty \text{ but } \limsup_{n\to\infty} n^{-1/2}||S_n^*||_p = \infty. \tag{17.9}$$

Proof of Theorem 17.4 Let $(Y_k)_{k\geq 0}$ be a discrete Markov chain with the state space $\mathbb{Z}^+$ and transition matrix $P = (p_{i,j})$ given in Section 17.1. To specify the sequence $(p_j)_{j\geq 1}$ defining P, we consider a non-negative sequence $a_n \to 0$ satisfying (17.9), and define $p_i = cj/u_j^{1+p/2}$, when $i = u_j$ for some $j \geq 1$, and $p_i = 0$ elsewhere. Here, the sequence $\{u_k; k = 1, 2, ...\}$ of positive integers satisfies (17.2). Notice that

$$\mathbb{E}(\tau^2) < \infty \text{ but } \mathbb{E}(\tau^p) = \infty.$$

We start the chain with the stationary distribution $(\pi_j)_{j\geq 0}$. Further, we define the stationary sequence $(X_i)_{i\geq 0}$ by (17.4). With the same notations as before, similarly to Fact 17.2, we have

$$||\mathbb{E}(S_n|Y_0)||_p \leq ||\nu \wedge n||_p + \max_{1\leq i\leq n} |A_i| = I_n + J_n \text{ (say), where } ||x||_p^p = \sum_{k=0}^{\infty} |x_k|^p \pi_k.$$

The convergence part of the $\mathbb{L}^p$ example is similar as for the $\mathbb{L}^2$ case and is then left to the reader. And so we have

$$\sum_{n=1}^{\infty} a_n I_n/n^{3/2} < \infty \text{ and } \sum_{n=1}^{\infty} a_n J_n/n^{3/2} < \infty.$$

It remains to prove that

$$\limsup_{n\to\infty} n^{-p/2}||S_n^*||_p^p = \infty. \tag{17.10}$$

With this aim, we recall that setting $\xi_j = 1 - \pi_0\tau_j$, we have $S_{T_k} = \sum_{j=1}^k \xi_j$ where the T_k's and the τ_k's have been defined in (17.7). Notice that, for any positive integer K,

$$\max_{0\leq j\leq Kn} |S_j| \geq \max_{T_j\leq Kn} |S_{T_j}| = \pi_0 \max_{T_j\leq Kn} |T_j - \mathbb{E}_0(T_j)| \geq \pi_0|T_n - \mathbb{E}_0(T_n)| 1_{\{T_n\leq Kn\}},$$

since $\mathbb{E}_0(T_j) = j\mathbb{E}(\tau) = j/\pi_0$. Thus, in order to show (17.10), it is enough to prove that

$$\limsup_{n\to\infty} \frac{1}{n^{p/2}} \mathbb{E}_0\left[|T_n - \mathbb{E}_0(T_n)|^p 1_{\{T_n \le Kn\}}\right] = \infty, \tag{17.11}$$

where we recall that $\mathbb{E}_0$ denotes the expectation when the Markov chain starts at zero. Recall also that $T_n = \sum_{i=1}^n \tau_i$ where, when the Markov chain starts at zero, $(\tau_i)_{i\ge1}$ is a sequence of i.i.d. random variables and their common law is the law of τ. In what follows, we assume that the chain starts from zero. We shall use a truncation argument. Starting with the decomposition

$$\tau_i - \mathbb{E}(\tau) = \{\tau_i 1_{\{\tau_i \le \sqrt{n}\}} - \mathbb{E}[\tau 1_{\{\tau\le\sqrt{n}\}}]\} - \mathbb{E}[\tau 1_{\{\tau>\sqrt{n}\}}] + \tau_i 1_{\{\tau_i>\sqrt{n}\}},$$

and letting $X_{in} = n^{-1/2}\{\tau_i 1_{\{\tau_i\le\sqrt{n}\}} - \mathbb{E}[\tau 1_{\{\tau\le\sqrt{n}\}}]\}$, $Y_{in} = n^{-1/2}\tau_i 1_{\{\tau_i>\sqrt{n}\}} 1_{\{T_n\le Kn\}}$ and $W_n = X_{1n} + \ldots + X_{nn}$, we write

$$n^{-1/2}\|[T_n - \mathbb{E}_0(T_n)]1_{\{T_n\le Kn\}}\|_p \ge \left\|\sum_{i=1}^n Y_{in}\right\|_p - n^{1/2}\mathbb{E}[\tau 1_{\{\tau>\sqrt{n}\}}] - \|W_n\|_p,$$

where here and below $\|\xi\|_p^p = \mathbb{E}_0(|\xi|^p)$. Notice that $n^{1/2}\mathbb{E}[\tau 1_{\{\tau>\sqrt{n}\}}] \le \mathbb{E}(\tau^2) < \infty$, $|X_{in}| \le 1$ and $\mathrm{Var}(W_n) = n\mathrm{Var}(X_{1n}) \le \mathbb{E}(\tau^2) < \infty$. So, by independence and Rosenthal's inequality, $\|W_n\|_p = O(1)$. Thus, the divergence (17.11) will hold if

$$\limsup_{n\to\infty} n^{-1/2}\left\|\sum_{i=1}^n \tau_i 1_{\{\tau_i>\sqrt{n}\}} 1_{\{T_n\le Kn\}}\right\|_p = \infty.$$

Since the τ_i's are i.i.d. and $p > 2$, we can write

$$\begin{aligned}\mathbb{E}_0\left|\sum_{i=1}^n \tau_i 1_{\{\tau_i>\sqrt{n}\}} 1_{\{T_n\le Kn\}}\right|^p &\ge \sum_{i=1}^n \mathbb{E}_0[\tau_i^p 1_{\{\tau_i>\sqrt{n}\}} 1_{\{T_n\le Kn\}}] \\ &\ge n\mathbb{E}_0[\tau_n^p 1_{\{\sqrt{n}<\tau_n\le n\}} 1_{\{T_n\le Kn\}}].\end{aligned}$$

Since $\{\tau_n \le n\} \cap \{\tau_1 + \ldots + \tau_{n-1} \le (K-1)n\} \subset \{\tau_1 + \ldots + \tau_n \le Kn\}$, by using the independence of the τ_i's we obtain

$$\begin{aligned}&\mathbb{E}_0[\tau_n^p 1_{\{\sqrt{n}<\tau_n\le n\}} 1_{\{T_n\le Kn\}}] \ge \mathbb{E}_0\left[\tau_n^p 1_{\{\tau_n>\sqrt{n}\}} 1_{\{\tau_n\le n\}} 1_{\{\tau_1+\ldots+\tau_{n-1}\le(K-1)n\}}\right] \\ &\quad = \mathbb{E}_0[\tau_n^p 1_{\{\tau_n>\sqrt{n}\}} 1_{\{\tau_n\le n\}}]\mathbb{P}_0[\tau_1 + \ldots + \tau_{n-1} \le (K-1)n].\end{aligned}$$

By the law of large numbers we have $\mathbb{P}_0[\tau_1 + \ldots + \tau_{n-1} \le (K-1)n] \to 1$ as $n \to \infty$ when $K - 1 > \mathbb{E}(\tau)$.

Overall, for all n sufficiently large and $K > 2 + \mathbb{E}(\tau)$,

$$n\mathbb{E}_0\left[\tau_n^p 1_{\{\sqrt{n}<\tau_n\leq n\}}1_{\{T_n\leq Kn\}}\right] \geq \frac{1}{2}n\mathbb{E}\left[\tau^p 1_{\{\sqrt{n}<\tau\leq n\}}\right].$$

So, it remains to prove that

$$\limsup_{n\to\infty} \frac{n}{n^{p/2}}\mathbb{E}[\tau^p 1_{\{\sqrt{n}<\tau\leq n\}}] = \infty.$$

To prove the last statement, we observe that, along the subsequence $(u_j)_{j\geq 1}$ defined by (17.2) and letting $j \to \infty$,

$$\frac{u_j}{u_j^{p/2}}\mathbb{E}\left[\tau^p \mathbf{1}_{\{\sqrt{u_j}<\tau\leq u_j\}}\right] \geq u_j^{1-p/2}u_j^p\mathbb{P}(\tau = u_j) = u_j^{1-p/2}u_j^p cju_j^{-1-p/2} = cj \to \infty.$$

Therefore we have established the desired result. ∎

17.2 Telescopic Examples

In this section, we analyze several telescopic-type examples, so for any positive integer i, $X_i = Z_i - Z_{i-1}$ where $\{Z_i\}_{i\geq 0}$ is a stationary sequence. Notice that the partial sums $S_n = X_1 + ... + X_n$ then have a simple representation:

$$S_n = \sum_{i=1}^{n}(Z_i - Z_{i-1}) = Z_n - Z_0,$$

which implies that the sequence S_n is stochastically bounded. Hence the CLT holds without any extra moment assumptions and under any normalization $b_n \to \infty$, i.e.

$$\frac{S_n}{b_n} \to^d 0 = N(0, 0). \tag{17.12}$$

Consider now the normalized partial sum process

$$\{S_{[nt]}/b_n, t \in [0, 1]\} = \{b_n^{-1}(Z_{[nt]} - Z_0), t \in [0, 1]\}.$$

Here and after, we apply the following intuitive fact.

Fact 17.5 *Under the construction above, the invariance principle holds, i.e. the normalized partial sums process* $\{S_{[nt]}/b_n, t \in [0, 1]\} \to^d 0$ *in* $D([0, 1])$ *if and only if*

$$\frac{1}{b_n}\max_{1\leq k\leq n}|Z_k - Z_0| \to^{\mathbb{P}} 0.$$

17.2.1 CLT Does NOT Imply IP for Stationary Sequences under any Normalization

In this part we combine the renewal chain structure with the telescopic behavior to show that any normalization b_n for which the CLT holds, i.e. $S_n/b_n \to N(0, c^2)$, does NOT imply that the normalized partial sum process $\{S_{[nt]}/b_n, t \in [0,1]\}$ is stochastically bounded. More exactly, we show the following result.

Theorem 17.6 *Let $g_n \to \infty$ be a non-negative sequence and define $b_n = n/g_n$. Assume that $b_n \to \infty$. There exist a stationary ergodic discrete Markov chain $(Y_k)_{k\geq 0}$ and a function f such that $X_i = f(Y_i) - f(Y_{i-1})$; $i \geq 1$, $\mathbb{E}[X_1] = 0$, $\mathbb{E}[X_1^2] < \infty$ and*

$$\frac{S_n}{b_n} \to 0. \quad \textit{However,} \ \max_{1\leq k\leq n} \frac{|S_k|}{b_n} \quad \textit{is not stochastically bounded.} \tag{17.13}$$

Proof of Theorem 17.6 Let $(Y_i)_{i\geq 0}$ be a renewal Markov chain considered in Section 17.1. We need to identify the probabilities p_j and the function f. We first note that there exist an infinite subset S of $\mathbb{N}^*$ that we denote by $S := \{\ell_i, i \geq 1\}$ where $(\ell_i)_{i\geq 1}$ is an increasing sequence of integers, and a non-decreasing subsequence $(n_{\ell_i})_{i\geq 1}$ of (n) converging to infinity and such that the two following properties are satisfied:

$$\sum_{i\geq 1} \frac{2^{\ell_i}}{n_{\ell_i}} < \infty, \tag{17.14}$$

and there exist positive finite constants i_0 and κ such that for any $i \geq i_0$,

$$b_{n_{\ell_i}} = \frac{n_{\ell_i}}{g_{n_{\ell_i}}} \leq \kappa 2^{\ell_i}. \tag{17.15}$$

Note that most of the time, these properties are satisfied with $S = \mathbb{N}$. This is the case for instance when $g_n = n^r$ with $r \in]0,1[$.
Construction. Let r_0 be a positive integer to be chosen later and

$$\alpha = \frac{1}{n_{\ell_1}} + \sum_{r\geq r_0} \frac{1}{n_{\ell_r}}.$$

Define now the probabilities $(p_j)_{j\geq 1}$ as follows: Let

$$p_1 = \mathbb{P}(\tau = 1) = \frac{1}{\alpha n_{\ell_1}}, \ \mathbb{P}(\tau = 2^{\ell_r} + 1) = p_{2^{\ell_r}+1} = \frac{1}{\alpha n_{\ell_r}} \ \text{for } r \geq r_0 \tag{17.16}$$

and

$$p_j = \mathbb{P}(\tau = j) = 0 \ \text{for } 2 \leq j \leq 2^{\ell_{r_0}} \ \text{or } j \neq 2^{\ell_r} + 1 \ \text{for } r \geq r_0. \tag{17.17}$$

Note that τ is a positive random variable taking values in the set $E = \{1, (2^{\ell_r} + 1)_{r\geq r_0}\}$ and each element of E has a positive probability to occur. In addition, by (17.14),

$$\mathbb{E}(\tau) = p_1 + \sum_{r=r_0}^{\infty} (2^{\ell_r} + 1) p_{2^{\ell_r}+1} \leq \frac{2}{\alpha} \sum_{i \geq 1} \frac{2^{\ell_i}}{n_{\ell_i}} < \infty.$$

The idea now is to construct f such that it vanishes at big jumps, more precisely such that

$$\mathbb{P}(f(\tau - 1) = 0) = \mathbb{P}(f(\tau) = 0) = 1. \tag{17.18}$$

With this aim, we let $f(0) = 0$ and

$$f(j) = \sum_{i=1}^{j} a_i, \; j \geq 1, \tag{17.19}$$

where the sequence $(a_i)_{i \geq 1}$ is chosen such that

$$a_i = 0, \; 1 \leq i \leq 2^{\ell_{r_0}-1} \text{ and for } k \geq \ell_{r_0}, a_i = \begin{cases} 0, & i = 2^{k-1} + 1, 2^k, \\ 1, & 2^{k-1} + 2 \leq i \leq 2^{k-1} + 2^{k-5}, \\ 0, & 2^{k-1} + 2^{k-5} < i \leq 2^k - 2^{k-5}, \\ -1, & 2^k - 2^{k-5} + 1 \leq i \leq 2^k - 1, \end{cases} \tag{17.20}$$

where we choose r_0 such that $\ell_{r_0} > 6$ to guarantee that the sets in the definition of a_i are non-empty. By construction,

$$a_1 = 0, \sum_{i=1}^{2^{\ell_r}} a_i = \sum_{i=1}^{2^{\ell_r}+1} a_i = 0 \text{ for } r \geq r_0 \text{ and } a_k \in \{0, -1, 1\} \text{ for any } k \geq 1. \tag{17.21}$$

Notice that (17.16), (17.17), (17.19) and (17.21) imply (17.18).

The convergence $S_n/b_n \to^d 0$ is then straightforward since $S_n = \sum_{i=1}^n X_i = f(Y_n) - f(Y_0)$ is stochastically bounded as a telescopic sum process and $b_n \to \infty$.

Let us show that $\mathbb{E}[X_1^2] < \infty$ and $\mathbb{E}[X_1] = 0$. By construction, and since $Y_0 = j > 0 \Rightarrow Y_1 = j - 1$ and $Y_0 = 0 \Rightarrow Y_1 = \tau$ which is independent of Y_0,

$$\begin{aligned} \mathbb{E}[X_1^2] &= \mathbb{E}\{[f(Y_1) - f(Y_0)]^2\} \\ &= \sum_{j \geq 0} \mathbb{E}\{(f(Y_1) - f(j))^2 | Y_0 = j\} \mathbb{P}(Y_0 = j) \\ &= \sum_{j \geq 1} (f(j-1) - f(j))^2 \mathbb{P}(Y_0 = j) + \mathbb{E}\{[f(\tau) - f(0)]^2\} \mathbb{P}(Y_0 = 0). \end{aligned}$$

Now, since $f(0) = 0$ and $f(\tau) = 0$ a.s., by (17.18) and because $|a_j| \leq 1$ by construction, it follows that

$$\mathbb{E}[X_1^2] = \sum_{j \geq 1} a_j^2 \mathbb{P}(Y_0 = j) \leq 1.$$

Observe that the same arguments imply that

$$\begin{aligned}\mathbb{E}[X_1] &= -\sum_{j\geq 1} a_j \mathbb{P}(Y_0 = j) = -\sum_{j=1}^{\infty} a_j \pi_j \\ &= -\pi_0 \sum_{j=1}^{\infty} a_j \sum_{i\geq j+1} p_i = -\pi_0 \sum_{i=2}^{\infty} p_i \sum_{j=1}^{i-1} a_j \\ &= -\pi_0 \mathbb{E}[f(\tau - 1)] = 0.\end{aligned}$$

We prove now that $\max_{1\leq k\leq n} |S_k|/b_n$ is not stochastically bounded. Clearly, it is enough to prove it when the chain starts from zero, which we assume from now on. Notice that, from the construction (17.20) of the cyclic sequence a_j, for any $r \geq r_0$,

$$\max_{1\leq p\leq 2^{\ell_r}} \left| \sum_{i=1}^{p} a_i \right| \geq 2^{\ell_r - 5} - 1$$

and hence, since we choose r_0 such that $\ell_{r_0} > 6$,

$$A := \inf_{n\geq 2^{\ell_{r_0}}} \frac{1}{n} \max_{1\leq p\leq n} \left| \sum_{i=1}^{p} a_i \right| > 0. \tag{17.22}$$

Recall now the definition (17.7) for the sequence $(T_k)_{k\geq 1}$ of renewal times: For any positive integer k, $T_k = \tau_1 + \ldots + \tau_k$ where $\{\tau_j\}_{j\geq 1}$ are independent variables. For $j \geq 2$, they are identically distributed, and their common law is the law of τ. Consider also the last renewal defined by

$$\eta_n = \max\{k \geq 1 \, : \, T_k \leq n\}.$$

By Fact 17.5 and since $f(Y_0) = 0$ (recall that we assumed that the chain starts from zero and $f(0) = 0$), we need to show that

$$\frac{g_n}{n} \max_{1\leq k\leq n} |f(Y_k)| \not\to^{\mathbb{P}} 0. \tag{17.23}$$

Note that

$$\max_{1\leq k\leq n} |f(Y_k)| = \max_{1\leq k\leq n} \left| \sum_{i=1}^{Y_k} a_i \right| := J_n. \tag{17.24}$$

Note also that if $k = T_j$ then

$$Y_{k+1} = \tau_{j+1} - 1.$$

Then if $\tau_{j+1} - 1 = w > 0$, the variables $Y_{k+1}, Y_{k+2}, ..., Y_{k+w}$ have respective values w, $w-1, ..., 1$. Therefore,

$$J_n \geq \max_{1\leq \ell \leq \eta_n - 1} \max_{T_\ell < k \leq T_{\ell+1}} \left| \sum_{i=1}^{Y_k} a_i \right| = \max_{1\leq \ell \leq \eta_n - 1} \max_{T_\ell + 1 \leq k \leq T_{\ell+1} - 1} \left| \sum_{i=1}^{Y_k} a_i \right|$$

$$= \max_{1\leq \ell \leq \eta_n - 1} \max_{1\leq k \leq \tau_{\ell+1} - 1} \left| \sum_{i=1}^{Y_{T_\ell + k}} a_i \right|,$$

implying that

$$J_n \geq \max_{2\leq k \leq \eta_n} \max_{1\leq m \leq \tau_k - 1} \left| \sum_{i=1}^{m} a_i \right|.$$

Introduce a new characteristic by

$$I_n = \max_{2\leq k \leq [cn]} \max_{1\leq m \leq \tau_k - 1} \left| \sum_{i=1}^{m} a_i \right| \quad \text{where } c = \frac{1}{2\mathbb{E}(\tau)}. \tag{17.25}$$

Notice that by the renewal theorem the number of renewals in the interval $\{1, ..., n\}$ is approximately $n/\mathbb{E}(\tau)$. Therefore,

$$\mathbb{P}(J_n \geq I_n) \to 1, \text{ as } n \to \infty. \tag{17.26}$$

By combining (17.24)–(17.26) and taking into account (17.22), we see that (17.23) holds provided

$$\frac{1}{b_n} \max_{2\leq k \leq [cn]} |\tau_k - 1| \not\to^{\mathbb{P}} 0. \tag{17.27}$$

But, for $r \geq i_0$, by (17.15),

$$\mathbb{P}\left(\max_{2\leq k \leq [cn_{\ell_r}]} |\tau_k - 1| \geq b_{n_{\ell_r}}/\kappa \right) \geq \mathbb{P}\left(\max_{2\leq k \leq [cn_{\ell_r}]} |\tau_k - 1| \geq 2^{\ell_r} \right)$$
$$= 1 - \left(1 - \mathbb{P}(\tau \geq 2^{\ell_r} + 1)\right)^{[cn_{\ell_r}]-1}.$$

Hence, for $r \geq \max(i_0, r_0)$, by (17.16),

$$\mathbb{P}\left(\max_{2\leq k \leq [cn_{\ell_r}]} |\tau_k - 1| \geq b_{n_{\ell_r}}/\kappa \right) \geq 1 - \left(1 - \mathbb{P}(\tau = 2^{\ell_r} + 1)\right)^{[cn_{\ell_r}]-1}$$
$$= 1 - \left(1 - \frac{1}{\alpha\, n_{\ell_r}}\right)^{[cn_{\ell_r}]-1},$$

showing that

$$\liminf_{r\to\infty} \mathbb{P}\left(\max_{2\le k\le [cn_{\ell_r}]} |\tau_k - 1| \ge b_{n_{\ell_r}}/\kappa \right) \ge 1 - \exp(-c/\alpha),$$

and hence (17.27). This ends the proof of Theorem 17.6. ■

17.2.2 Linear Asymptotic Variance and CLT Do Not Imply IP

As we have seen in Chapter 4, the Heyde's condition (4.14), namely:

$$\mathbf{P}_1(S_n) \text{ is convergent in } \mathbb{L}^2 \text{ and } \frac{\|S_n\|_2^2}{n} \to \|D_1\|_2^2 ,$$

is sufficient to get the CLT of $S_n/\sqrt{n}$. In this part, we provide a dynamical system example showing that condition (4.14) does not imply, in general, the functional version of the central limit theorem. The ideas for the counterexample follow the ideas of Volný and Samek (2000) as developed in Dedecker, Merlevède and Volný (2007).

Proposition 17.7 *There exists $X_0 \in \mathbb{L}^2$ measurable with respect to a σ-algebra $\mathcal{F}_0$, and a bijective and bimeasurable transformation T preserving the probability $\mathbb{P}$ such that X_0 is regular, $\mathbb{P}$ is ergodic and with $X_i = X_0 \circ T^i$, $S_n = \sum_{i=1}^n X_i$,*

$$\sum_{i\ge 0} \mathbf{P}_0(X_i) \text{ converges to } 0 \text{ in } \mathbb{L}_2, \text{ and } \frac{\|S_n\|_2}{\sqrt{n}} \to 0 \text{ but } \max_{1\le i\le n} \frac{|S_i|}{\sqrt{n}} \text{ is not tight.} \tag{17.28}$$

Therefore the linear asymptotic behavior and the CLT hold but the IP does not hold for the normalization $s_n = \sqrt{n}$.

In the proposition above X_0 is regular, which means that $\lim_{n\to\infty} \mathbb{E}(X_0|T^n(\mathcal{F}_0)) = 0$ a.s.

Proof of Proposition 17.7 We present a telescopic-type example using the dynamic system language. Let $(\Omega, \mathcal{A}, \mathbb{P}) = ([0,1], \mathcal{B}([0,1]), \lambda)^{\mathbb{Z}}$, where λ is the Lebesgue measure, and let T be the left shift, i.e. $(T(\omega))_i = \omega_{i+1}$ for all $i \in \mathbb{Z}$. For all $i \in \mathbb{Z}$, let $\pi_i : \Omega \to [0,1]$ be the projections such that $\pi_i(\omega) = \omega_i$, and let $\mathcal{F}_k = \sigma(\pi_i, i \le k)$. For $k \ge 1$ and $1 \le j \le k$, let $\bar{A}_{k,j}$ be independent subsets of $[0,1]$ such that $\lambda(\bar{A}_{k,j}) = 1/(k4^k)$, and let $A_{k,j} = \pi_0^{-1}(\bar{A}_{k,j})$. Notice that for all $k \ge 1$, $1 \le j \le k$ and $i \in \mathbb{Z}$, the sets $A_{k,j} \circ T^i$ are independent, hence the random variables $\mathbf{1}_{A_{k,j}} \circ T^i$, $i \in \mathbb{Z}$ are mutually independent. With this notation, we then define

$$\tilde{e}_{k,j} = 2^{k-j}\mathbf{1}_{A_{k,j}} \text{ and } e_{k,j} = \tilde{e}_{k,j} - \mathbb{E}(\tilde{e}_{k,j}),\ n_j = \sum_{i=1}^{j-1} 2^i,\ m_k = \sum_{\ell=1}^{k-1} n_{\ell+1},$$

$$f_{k,j} = \left(\sum_{i=0}^{2^j-1} e_{k,j} \circ T^{-i} - \sum_{i=2^j}^{2^{j+1}-1} e_{k,j} \circ T^{-i} \right) \circ T^{-2(m_k+n_j)},\ f_k = \sum_{j=1}^{k} f_{k,j}$$

and finally $X_0 = \sum_{k\geq 1} f_{2^k},\ X_i = X_0 \circ T^i\ ,\ S_n = \sum_{i=1}^{n} X_i\ .$

Notice that $\sum_{k\geq 1}\sum_{j=1}^{2^k} \|f_{2^k,j}\|_2 < \infty$, so that X_0 is well defined in $\mathbb{L}^2$. Note now that each of the functions $f_{k,j}$ is a coboundary:

$$f_{k,j} = g_{k,j} - g_{k,j} \circ T^{-1},$$

where

$$\begin{aligned} g_{k,j} &= \left(\sum_{i=0}^{2^j-1} \sum_{\ell=0}^{2^j-1} \tilde{e}_{k,j} \circ T^{-(i+\ell)} \right) \circ T^{-2(m_k+n_j)} \\ &= \left(\sum_{i=0}^{2^j-1} (i+1)\tilde{e}_{k,j} \circ T^{-i} + \sum_{i=2^j}^{2^{j+1}-1} (2^{j+1} - i - 1)\tilde{e}_{k,j} \circ T^{-i} \right) \circ T^{-2(m_k+n_j)}. \end{aligned}$$

We then have that

$$\sum_{\ell=1}^{n} f_{k,j} \circ T^{\ell} = g_{k,j} \circ T^n - g_{k,j},\ n \geq 1. \tag{17.29}$$

Moreover, since

$$\|g_{k,j} - \mathbb{E}(g_{k,j})\|_2^2 \leq 2 \sum_{i=0}^{2^j-1} (i+1)^2 \|e_{k,j}\|_2^2 \leq \frac{2^{j+1}}{k},$$

we get that for all $n \geq 1$ (and then in particular for $n \geq 2^j$),

$$\left\| \sum_{\ell=1}^{n} f_{k,j} \circ T^{\ell} \right\|_2^2 \leq \frac{2^{j+3}}{k}. \tag{17.30}$$

Then, a tedious analysis built upon (17.30) and the construction of the functions $f_{k,j}$ gives the following result denoted by Step 1.

Step 1. $n^{-1}\|S_n\|_2^2 \to 0$.

We then show that

Step 2. $\sum_{\ell=0}^{N} \mathbf{P}_0(X_\ell)$ *converges in* $\mathbb{L}^2$ *to zero.*

To prove it, we first notice that

$$\begin{aligned}\mathbf{P}_{-\ell}(X_0) &= e_{2^k,j}\circ T^{-\ell},\ \text{for } \ell = 2(m_{2^k}+n_j)+i, 0\le i\le 2^j-1,\\ \mathbf{P}_{-\ell}(X_0) &= -e_{2^k,j}\circ T^{-\ell}\ ,\ \text{for } \ell = 2(m_{2^k}+n_j)+i, 2^j\le i\le 2^{j+1}-1.\end{aligned}$$

All other terms are zero. Thus, provided the series exists, the limit should be zero. Then Step 2 follows by observing that the series is convergent since $\mathbb{E}(2^j e_{2^k,j})^2 \le 2^{-k}$.

Finally, we show that:

Step 3. $\{n^{-1/2}S_{[nt]}\ ;\ 0\le t\le 1\}$ *is not tight.*

Note that

$$\sum_{m\ge 1}\sum_{j=1}^{2^m} g_{2^m,j} \ \text{ is almost surely finite}$$

and that X_0 is well defined and

$$X_0 = \sum_{m\ge 1}\sum_{j=1}^{2^m} g_{2^m,j} - \left(\sum_{m\ge 1}\sum_{j=1}^{2^m} g_{2^m,j}\right)\circ T^{-1}.$$

Moreover, X_0 is a coboundary implying that the sequence $X_i = X_0\circ T^i$ is telescopic.

Thus, to establish Step 3, we need to show that $n^{-1/2}\max_{1\le\ell\le n}|S_\ell|$ is NOT convergent to zero in probability.

Using (17.29) and the fact that the functions $g_{k,j}$ are non-negative, notice first that, for all $k\ge 1$,

$$\max_{1\le\ell\le n}\left|\sum_{i=1}^{\ell} X_i\right| \ge \max_{1\le\ell\le n}\sum_{j=1}^{2^k} g_{2^k,j}\circ T^{\ell} - \sum_{m\ge 1}\sum_{j=1}^{2^m} g_{2^m,j}$$

and the last term in the left-hand side is finite. Hence it suffices to prove that there exists a subsequence $n(k)$ such that

$$\frac{1}{\sqrt{n(k)}}\max_{1\le\ell\le n(k)}\sum_{j=1}^{2^k} g_{2^k,j}\circ T^{\ell} \text{ does not converge to 0 in probability.} \tag{17.31}$$

Take $n(k) = 4^{2^k}$ and notice that

$$\mathbb{P}\left(\max_{1\le\ell\le n(k)}\sum_{j=1}^{2^k} g_{2^k,j}\circ T^{\ell} \ge \frac{\sqrt{n(k)}}{2}\right) = \mathbb{P}\left(\bigcup_{\ell=1}^{4^{2^k}}\left\{\sum_{j=1}^{2^k} g_{2^k,j}\circ T^{\ell} \ge \frac{2^{2^k}}{2}\right\}\right).$$

Now let B_k be the sets defined by

$$B_k = \{\omega \in \Omega \text{ such that } \omega \text{ is in only one of the sets } A_{2^k,j} \circ T^{-(2^j-1)-2(m_{2^k}+n_j)+\ell}$$
$$\text{for } 1 \leq \ell \leq 4^{2^k} \text{ and } 1 \leq j \leq 2^{k-1}\}.$$

On B_k, one of the functions $g_{2^k,j} \circ T^\ell$ reaches the level $2^{2^k}/2$ and the others are non-negative. Hence

$$\mathbb{P}\left(\max_{1\leq \ell\leq n(k)} \sum_{j=1}^{2^k} g_{2^k,j} \circ T^\ell \geq \frac{\sqrt{n(k)}}{2}\right) \geq \mathbb{P}(B_k) > \left(1 - \frac{1}{2^k 4^{2^k}}\right)^{2^{k-1}4^{2^k}} \to e^{-1/2},$$

which proves (17.31). ■

REFERENCES

Alam, K. and Saxena, K.M.L. (1981). Positive dependence in multivariate distributions. *Comm. Statist. A—Theory Methods* **10**, 1183–1196.

Aldous, D. (1978). Stopping times and tightness. *Ann. Probab.* **6**, no. 2, 335–340.

Andrews, D.W.K. (1984). Nonstrong mixing autoregressive processes. *J. Appl. Probab.* **21**, no. 4, 930–934.

Arcones, M.A. (2003-a). The large deviation principle for stochastic processes I. *Theory Probab. Appl.* **47**, no. 4, 567–583.

Arcones, M.A. (2003-b). The large deviation principle for stochastic processes II. *Theory Probab. Appl.* **48**, no. 1, 19–44.

Azuma, K. (1967). Weighted sums of certain dependent random variables. *Tôhoku Math. J.* **19**, 357–367.

von Bahr, B. and Esseen, C.G. (1965). Inequalities for the r-th absolute moment of a sum of random variables, $1 \leq r \leq 2$. *Ann. Math. Statist.* **36**, 299–303.

Bary, N.K. (1964). *A Treatise on Trigonometric Series*. Macmillan, New York.

Bentkus, V. (2004). On Hoeffding's inequalities. *Ann. Probab.* **32**, no. 2, 1650–73.

Bennett, G. (1962). Probability inequalities for the sum of independent random variables. *Journal of the American Statistical Association* **57**, 33–45.

Berbee, H.P.C. (1979). Random walks with stationary increments and renewal theory, *Math. Center Tracts* **112**. Mathematisch Centrum, Amsterdam.

Bercu, B., Delyon, B. and Rio, E. (2015). *Concentration inequalities for sums and martingales*. SpringerBriefs in Mathematics. Springer, Cham.

Berger, E. (1982). *Fastsichere Approximation von Partialsummen unabhängiger und stationärer ergodischer Folgen von Zufallsvektoren*. Diss., Univ. Gottingen.

Berkes, I. and Philipp, W. (1977/78). An almost sure invariance principle for the empirical distribution function of mixing random variables. *Z. Wahr. verw. Gebiete* **41**, no. 2, 115–137.

Berkes, I. and Philipp, W. (1979). Approximation theorems for independent and weakly dependent random variables. *Ann. Probab.* **7**, no. 1, 29–54.

Berkes, I. and Philipp, W. (1998). Limit theorems for mixing sequences without rate assumptions. *Ann. Probab.* **26**, no. 2, 805–831.

Bhattacharya, R. N. and Rao, R. (1976). *Normal approximation and asymptotic expansions*. Wiley Series in Probability and Mathematical Statistics. John Wiley & Sons, New York-London-Sydney.

Billingsley, P. (1968). *Convergence of Probability Measures*. John Wiley & Sons, New York.

Billingsley, P. (1995). *Probability and Measure*. John Wiley & Sons, New York.

Billingsley, P. (1999) *Convergence of Probability Measures*. Second edition. Wiley Series in Probability and Statistics. John Wiley & Sons, Inc., New York.

Birkel, T. (1987). The invariance principle for associated processes. *Stochastic Process Appl.* **27**, no. 1, 57–71.

Birkel, T. (1993). A functional central limit theorem for positively dependent random variables. *J. Multivariate Anal.* **44**, no. 2, 314–320.

Bishop, E. (1966). An upcrossing inequality with applications. *Michigan Math. J.* **13**, 1–13.

Bolthausen, E. (1989). A central limit theorem for two-dimensional random walks in random sceneries. *Ann. Probab.* **17**, no. 1, 108–115.

Borodin, A.N. and Ibragimov, I.A. (1995). *Limit theorems for functionals of random walks.* Trudy Mat. Inst. Steklov. **195**. Transl. into English: Proc. Steklov Inst. Math.

Borovkova, S., Burton, R. and Dehling, H.G. (2001). Limit theorems for functionals of mixing processes with applications to U-Statistics and dimension estimation. *Trans. Amer. Math. Soc.* **353**, no. 11, 4261–4318.

Bosq, D. (2000). *Linear processes in function spaces. Theory and applications.* Lecture Notes in Statistics, **149**. Springer-Verlag, New York.

Bradley, R.C. (1983). Approximation theorems for strongly mixing random variables, *Michigan Math. J.* **30**, no. 1, 69–81.

Bradley, R.C. (1992). On the spectral density and asymptotic normality of weakly dependent random fields. *J. Theor. Probab.* **5**, no. 2, 355–373.

Bradley, R.C. (1997-a). On quantiles and the central limit question for strongly mixing sequences. Dedicated to Murray Rosenblatt. *J. Theoret. Probab.* **10**, no. 2, 507–555.

Bradley, R.C. (1997-b). Every "lower psi-mixing" Markov chain is "interlaced rho-mixing". *Stochastic Process. Appl.* **72**, 221–239.

Bradley, R.C. (2005). Basic properties of strong mixing conditions. A survey and some open questions. *Probab. Surv.* **2**, 107–144.

Bradley, R.C. (2007). *Introduction to strong mixing conditions.* Vol. 1,2,3. Kendrick Press, Heber City, UT.

Bradley, R.C. and Bryc, W. (1985). Multilinear forms and measures of dependence between random variables. *J. Multivariate Anal.* **16**, no. 3, 335–367.

Brockwell, P.J. and Davis, R.A. (1991). *Time series: Theory and methods.* Second edition. Springer Series in Statistics. Springer-Verlag, New York.

Brown, B.M. and Eagleson, G.K. (1971). Martingale convergence to infinitely divisible laws with finite variances. *Trans. Amer. Math. Soc.* **162**, 449–453.

Bryc, W. (1982). On the approximation theorem of Berkes and Philipp, *Demonstratio Matematica* **15**, no. 3, 807–816.

Bryc, W. and Smolenski, W. (1993). Moment conditions for almost sure convergence of weakly correlated random variables. *Proc. Amer. Math. Soc.* **119**, no. 2, 629–635.

Burkholder, D.L. (1973). Distribution function inequalities for martingales. *Ann. Probab.* **1**, no. 1, 19–42.

Burkholder, D.L. (1979). A sharp inequality for martingale transforms. *Ann. Probab.* 7, no. 5, 858–863.

Burkholder, D.L. (1988). Sharp inequalities for martingales and stochastic integrals. Colloque Paul Lévy sur les Processus Stochastiques (Palaiseau, 1987). *Astérisque* **157–158**, 75–94.

Burton, R.M., Dabrowski, A.R. and Dehling, H. (1986). An invariance principle for weakly associated random vectors. *Stochastic Process Appl.* **23**, no. 2, 301–306.

Carleson, L. (1966). On convergence and growth of partial sums of Fourier series. *Acta Math.* **116**, 135–157.

Champeney, D.C. (1989). *A Handbook of Fourier Theorems.* Cambridge Univ. Press, Cambridge.

Chen, X. (2010). *Random walk intersections. Large deviations and related topics.* Mathematical Surveys and Monographs, **157**. American Mathematical Society, Providence, RI.

Chernoff, H. and Teicher, H. (1958). A central limit theorem for sums of interchangeable random variables. *Ann. Math. Statist* **29**, no. 1, 118–130.

Chung, K.L. (1960). *Markov chains with stationary transition probabilities.* Die Grundlehren der mathematischen Wissenschaften, Bd. 104. Springer-Verlag, Berlin-Göttingen-Heidelberg.

Cox, J.Th. and Grimmett, G. (1984). Central limit theorem for associated random variables and the percolation model. *Ann.Probab.* **12**, no. 2, 514–528.

Cuny, C., Dedecker, J. and Jan, C. (2017). Limit theorems for the left random walk on $GL_d(\mathbb{R})$. *Ann. Inst. H. Poincaré Probab. Statist.* **123**, no. 4, 1839–1865.

Cuny, C., Merlevède, F. and Peligrad, M. (2013). Law of the iterated logarithm for the periodogram. *Stochastic Process. Appl.* **123**, no. 11, 4065–4089.

Cuny, C. and Peligrad, M. (2012). Central limit theorem started at a point for stationary processes and additive functional of reversible Markov Chains. *J. Theoret. Probab.* **25**, no. 1, 171–188.

Dabrowski, A.R. (1982). A note on a theorem of Berkes and Philipp for dependent sequences, *Statist. Probab. Lett.* **1**, no. 2, 53–55.

Davydov, Yu.A. (1968). Convergence of distributions generated by stationary stochastic processes. *Theory Probab. Appl.* **13**, no. 4, 691–696.

Davydov, Yu.A. (1973). Mixing conditions for Markov chains. *Theory Probab. Appl.* **18**, no. 2, 312–328.

Deddens, J., Peligrad, M. and Yang, T. (1987). On strong consistency of kernel estimators under dependence assumptions. *Proceedings of the Sixth Pannonian Symposium of Mathematical Statistics*, D. Reidel Publishing Company, 33–43.

Dedecker, J. (2010). An empirical central limit theorem for intermittent maps. *Probab. Theory Related Fields* **148**, no. 1–2, 177–195.

Dedecker, J., Dehling, H. and Taqqu, M. (2015). Weak convergence of the empirical process of intermittent maps in $\mathbb{L}^2$ under long-range dependence. *Stoch. Dyn.*, **15**, no. 2., 1550008, 29 pp.

Dedecker, J. and Doukhan, P. (2003). A new covariance inequality and applications. *Stochastic Process. Appl.* **106**, no. 1, 63–80.

Dedecker, J., Doukhan, P., Lang, G., León, R., Louhichi, S. and Prieur, C. (2007). *Weak dependence: With examples and applications*. Lecture Notes in Statistics, **190**. Springer, New York.

Dedecker, J., Gouëzel, S. and Merlevède, F. (2010). Some almost sure results for unbounded functions of intermittent maps and their associated Markov chains. *Ann. Inst. Henri Poincaré Probab. Stat.* **46**, no. 3, 796–821.

Dedecker, J. and Merlevède, F. (2002). Necessary and sufficient conditions for the conditional central limit theorem. *Ann. Probab.* **30**, no. 3, 1044–1081.

Dedecker, J. and Merlevède, F. (2003). The conditional central limit theorem in Hilbert spaces. *Stochastic Process. Appl.* **108**, no. 2, 229–262.

Dedecker, J. and Merlevède, F. (2017-a). A deviation bound for α-dependent sequences with applications to intermittent maps. *Stoch. Dyn.* **17**, no. 1, 1750005, 27 pp.

Dedecker, J. and Merlevède, F. (2017-b). Density estimation for $\tilde{\beta}$. -dependent sequences. *Electron. J. Stat.* **11**, no. 1, 981–1021.

Dedecker, J., Merlevède, F., Peligrad, M. and Utev, S. (2009). Moderate deviations for stationary sequences of bounded random variables. *Ann. Inst. Henri Poincaré Probab. Stat.* **45**, no. 2, 453–476.

Dedecker, J., Merlevède, F. and Pène, F. (2013). Empirical central limit theorems for ergodic automorphisms of the torus. *ALEA, Lat. Am. J. Probab. Math. Stat.*, **10**, no. 2, 731–766.

Dedecker, J., Merlevède, F. and Volný, D. (2007). On the weak invariance principle for non-adapted sequences under projective criteria. *J. Theoret. Probab.* **20**, no. 4, 971–1004.

Dedecker, J. and Prieur, C. (2005). New dependence coefficients. Examples and applications to statistics. *Probab. Theory Related Fields* **132**, no. 2, 203–236.

Dedecker, J. and Prieur, C. (2007). An empirical central limit theorem for dependent sequences. *Stochastic Process. Appl.* **117**, no. 1, 121–142.

Dedecker, J. and Rio, E. (2000). On the functional central limit theorem for stationary processes. *Ann. Inst. Henri Poincaré Probab. Stat.* **36**, no. 1, 1–34.

Dedecker, J. and Rio, E. (2008). On mean central limit theorems for stationary sequences. *Ann. Inst. Henri Poincaré Probab. Stat.* **44**, no. 4, 693–726.

Dedecker, J. and Saulière, G. (2017). The Mann-Whitney U-statistic for α-dependent sequences. *Math. Methods Statist.* **26**, no. 2, 111–133.

Dehling, H. (1983). A note on a theorem of Berkes and Philipp, *Z. Wahr. verw. Gebiete* **62**, no. 1, 39–42.

Dehling, H., Denker, M. and Philipp, W. (1986). Central limit theorems for mixing sequences of random variables under minimal conditions. *Ann. Probab.* **14**, no. 4, 1359–1370.

Dehling, H. and Philipp, W. (1982). Almost sure invariance principles for weakly dependent vector valued random variables, *Ann. Probab.* **10**, no. 3, 689–701.

Dehling, H. and Philipp, W. (2002). Empirical process techniques for dependent data, in: *Empirical process techniques for dependent data* (pp. 3-113). Birkhäuser, Boston.

de la Peña, V. and Giné, E. (1999). *Decoupling: From dependence to independence.* Probability and Its Applications. Springer-Verlag, New York.

Deligiannidis, G. and Utev, S. (2011). Computation of the asymptotics of the variance of the number of self-intersections of stable random walks using the Wiener-Darboux theory. (Russian) *Sibirsk. Mat. Zh.* **52**, no. 4, 809–22; translation in *Sib. Math. J.* **52** (2011), no. 4, 639–650.

Deligiannidis, G., Peligrad M. and Utev, S. (2015). Asymptotic variance of stationary reversible and normal Markov processes. *Electron. J. Probab.* **20**, no. 20, 26 pp.

Dembo, A. and Zeitouni, O. (1998). *Large Deviations Techniques and Applications.* Second edition. Springer, New York.

Denker, M. (1986). Uniform integrability and the central limit theorem for strongly mixing processes. in: *Dependence in Probability and Statistics* (Proc. Conf. Oberwolfach. 1985). Boston (pp. 269–289).

Derriennic, Y. and Lin, M. (2001). The central limit theorem for Markov chains with normal transition operators, started at a point. *Probab Theory Relat. Fields* **119**, no. 4, 508–528.

Derriennic, Y. and Lin, M. (2003). The central limit theorem for Markov chains started at a point. *Probab Theory Relat. Fields* **125**, no. 1, 73–76.

Derriennic, Y. and Lin, M. (2011). Variance bounding Markov chains, L_2 -uniform mean ergodicity and the CLT. *Stoch. Dyn.* **11**, no. 1, 81–94.

Diaconis, P. and Freedman, D. (1999). Iterated random functions. *SIAM Rev.* **41**, no. 1, 45–76.

Doeblin, W. (1937). Exposé de la théorie des chaînes simples constantes de Markov à un nombre fini d'états, *Rev. Math. de l'Union Interbalkanique* **2**, 77–105.

Doob, J. (1953). *Stochastic Processes.* Wiley, New York.

Doukhan, P. (1994). *Mixing. Properties and examples.* Lecture Notes in Statistics, 85. Springer-Verlag, New York.

Doukhan, P. and Louhichi, S. (1999). A new weak dependence condition and applications to moment inequalities. *Stochastic Process. Appl.* **84**, no. 2, 313–342.

Doukhan, P., Massart, P. and Rio, E. (1994). The functional central limit theorem for strongly mixing processes. *Ann. Inst. H. Poincaré Probab. Statist.* **30**, no. 1, 63–82.

Doukhan, P., Massart, P. and Rio, E. (1995). Invariance principles for absolutely regular empirical processes, *Ann. Inst. H. Poincaré Probab. Statist.* **31**, no. 2, 393–427.

Dubins, L. and Gilat, D. (1978). On the distribution of maxima of martingales. *Proc. Amer. Math. Soc.* **68**, no. 3, 337–338.

Dudley, R.M. (1968). Distances of probability measures and random variables, *Ann. Math. Stat.* **39**, no. 5, 1563–1572.

Dudley, R.M. (1978). Central limit theorems for empirical measures, *Ann. Probab.* **6**, no. 6, 899–929.

Dudley, R.M., and Philipp, W. (1983). Invariance principles for sums of Banach space valued random elements and empirical processes. *Z. Wahr. verw. Gebiete* **62**, no. 4, 509–552.

Eaton, M.L. (1974). A probability inequality for linear combinations of bounded random variables. *Ann. Statist.* **2**, no. 3, 609–614.

Eberlein, E. (1983). Strong approximation of very weak Bernoulli processes, *Z. Wahr. verw. Gebiete* **62**, no. 1, 17–37.

Esary, J. D., Proschan, F. and Walkup, D. W. (1967). Association of random variables, with applications. *Ann. Math. Statist.* **38**, no. 5, 1466–1474.
Esseen, C.-G. and Janson, S. (1985). On moment conditions for normed sums of independent variables and martingale differences. *Stochastic Process. Appl.* **19**, no. 1, 17–182.
Feller, W. (1971). *An introduction to probability theory and its applications.* Vol. II. Second edition John Wiley & Sons, Inc., New York-London-Sydney.
Freedman, D.A. (1975). On tail probabilities for martingales. *Ann. Probab.* **3**, 100–118.
Fuk, D.H. and Nagaev, S.V. (1971). Probabilistic inequalities for sums of independent random variables. *Theory Probab. Appl.* **16**, no. 4, 660–675.
Gabrielli, A. (2004). Point processes and stochastic displacement fields. *Phys. Rev. E* **70**, no. 6, 066131, 16 pp.
Gänssler, P. and Häusler, E. (1979). Remarks on the functional central limit theorem for martingales. *Z. Wahr. verw. Gebiete* **50**, no. 3, 237–243.
Garsia, A. M. (1965). A simple proof of E. Hopfs maximal ergodic theorem. *J. of Maths. and Mech.* **14**, 381–382.
Giraudo, D. and Volný, D. (2014). A strictly stationary β-mixing process satisfying the central limit theorem but not the weak invariance principle. *Stoch. Proc. Appl.* **124**, no. 11, 3769–3781.
Goldstein, S. (1979). Maximal coupling, *Z. Wahr. verw. Gebiete* **46**, no. 2, 193–204.
Gordin, M.I. (1969). The central limit theorem for stationary processes, *Soviet Math. Dokl.* **10**, 1174–1176.
Gordin, M.I. (1973). Abstracts of Communication, T.1: A-K, International Conference on Probability Theory, Vilnius.
Gordin, M.I. and Peligrad, M. (2011). On the functional central limit theorem via martingale approximation. *Bernoulli* **17**, no. 1, 424–440.
Griffeath, D. (1975). A maximal coupling for Markov chains, *Z. Wahr. verw. Gebiete* **31**, 95–106.
Hahn, M.G., Kuelbs, J. and Samur, J.D. (1987). Asymptotic normality of trimmed sums of ϕ-mixing random variables. *Ann. Probab.* **15**, no. 4, 1395–1418.
Hall, P. and Heyde, C.C. (1980). *Martingale limit theory and its application.* Academic Press, New York-London.
Hannan, E.J. (1979). The central limit theorem for time series regression. *Stoch. Processes Appl.* **9**, no. 3, 281–289.
Helland, I.S. (1982). Central limit theorems for martingales with discrete or continuous time. *Scand. J. Statist.* **9**, no. 2, 79–94.
Herrndorf, N. (1983). The invariance principle for ϕ-mixing sequences. *Z. Wahr. verw. Gebiete* **63**, no. 1, 97–108.
Heyde, C.C. (1974). On the central limit theorem for stationary processes. *Z. Wahr. verw. Gebiete* **30**, no. 4, 315–320.
Hoeffding, W. (1963). Probability inequalities for sums of bounded random variables. *J. Amer. Statist. Assoc.* **58**, 13–30.
Hough, J.B., Krishnapur, M., Peres, Y. and Virag, B. (2006). Determinantal processes and independence. *Probab. Surv.* **3**, 206–229.
Hunt, G.A. (1955). An inequality in probability theory. *Proc. Amer. Math. Soc.* **6**, no. 3, 506–510.
Hunt, G.A. and Young, W.S. (1974). A weighted norm inequality for Fourier series. *Bull. Amer. Math. Soc.* **80**, 274–277.
Ibragimov, I.A. (1961). Spectral functions of certain classes of stationary Gaussian processes. *Soviet Math. Dokl.* **2**, 403–405.
Ibragimov, I.A. (1962). Some limit theorems for stationary processes. *Theory Probab. Appl.* **7**, no. 4, 361–92.

Ibragimov, I.A. (1975). A note on the central limit theorem for dependent random variables. *Theory Probab. Appl.* **20**, no. 1, 134–140.

Ibragimov, I.A. and Linnik, Yu.V. (1971). *Independent and Stationary Sequences of Random Variables.* Wolters-Noorhoff, Groningen.

Ibragimov, I.A. and Rozanov, Yu.A. (1970). On the connection between two characteristics of dependence of Gaussian random vectors. *Theory Probab. Appl.* **15**, no. 2, 295–299.

Iosifescu, M. (1980). Recent advances in mixing sequences of random variables. Third International Summer School on Probability Theory and Mathematical Statistics (Varna, 1978) (Bulgarian), pp. 111–138, Bulgar. Akad. Nauk, Sofia, 1980.

Iosifescu, M. and Theodorescu, R. (1969). *Random processes and learning.* Die Grundlehren der mathematischen Wissenschaften, Band 150. Springer-Verlag, New York.

Ivanov, V.V. (1996). Oscillations of averages in the ergodic theorem. *Dokl. Akad. Nauk.* **347**, no. 6, 736–738.

Jacod, J. and Shiryaev, A.N. (2003). *Limit theorems for stochastic processes.* Second edition. Springer-Verlag, Berlin.

Jakubowski, A. (1986). Principle of conditioning in limit theorems for sums of random variables. *Ann. Probab.* **14**, no. 3, 902–915.

Joag-Dev, K. and Proschan, F. (1983). Negative association of random variables, with applications. *Ann. Statist.* **11**, no. 1, 286–295.

Karatzas, I. and Shreve, S. (1991). *Brownian motion and stochastic calculus.* Second edition. Springer-Verlag, New York.

Kendall, D.G. (1967). On finite and infinite sequences of exchangeable events. *Studia Sci. Math. Hungar.* **2**, 319–327.

Kiefer, J. and Wolfowitz, J. (1956). On the characteristics of the general queueing process, with applications to random walk. *Ann. Math. Statist.* **27**, no 1, 147–161.

Kipnis, C. and Landim, C. (1999). *Scaling Limits of Interacting Particle Systems.* Springer, New York.

Kipnis, C. and Varadhan, S.R.S. (1986). Central limit theorem for additive functionals of reversible Markov processes and applications to simple exclusions. *Comm. Math. Phys.* **104**, no. 1, 1–19.

Kolmogorov, A.N. and Rozanov, J.A. (1960). On strong mixing conditions for stationary Gaussian processes. *Theory Probab. Appl.* **5**, no. 4, 204–208.

Komlós, J. (1967). A generalization of a problem of Steinhaus. *Acta Math. Acad. Sci. Hungar.* **18**, no. 1–2, 217–229.

Kuelbs, J., and Philipp, W. (1980). Almost sure invariance principles for partial sums of mixing *B* valued random variables. *Ann. Probab.* **8**, no. 6, 1003–1036.

Lasota, A. and Yorke, J.A. (1973). On the existence of invariant measures for piecewise monotonic transformations. *Trans. Amer. Math. Soc.* **186**, 481–488.

Lehmann, E. (1966). Some concepts of dependence. *Ann. Math. Statist.* **37**, no. 5, 1137–1153.

Lin, M. and Sine, R. (1983). Ergodic theory and the functional equation $(IT)x=y$. *J. Operator Theory* **10**, no. 1, 153–166.

Liverani, C., Saussol, B. and Vaienti, S. (1999). A probabilistic approach to intermittency. *Ergodic Theory Dynam. Systems* **19**, no. 3, 671–685.

Longla, M., Peligrad, C. and Peligrad, M. (2012). On the functional CLT for reversible Markov Chains with nonlinear growth of the variance. *J. Appl. Probab.* **49**, no. 4, 1091–1105.

Longla, M. and Peligrad, M. (2012). Some aspects of modeling dependence in copula-based Markov chains. *J. Multivariate Anal.* **111**, 234–240.

Lyons, R. (2003). Determinantal probability measures. *Publ. Math. Inst. Hautes Études Sci.* **98**, no. 1, 167–212.

Lyons, R. and Steif, J. E. (2003). Stationary determinantal processes: Phase multiplicity, Bernoullicity, entropy, and domination. *Duke Math. J.* **120**, no. 3, 515–575.

Lyons, T.J. and Zheng, W.A. (1988). A crossing estimate for the canonical process on a Dirichlet space and a tightness result. Colloque Paul Lévy sur les processes stochastiques. *Asterisque* **157–158**, 249–272.

Major, P. (1978). On the invariance principle for sums of identically distributed random variables. *J. Multivariate Anal.* **8**, no. 4, 487–517.

Massart, P. (1988). *Invariance principles for empirical processes: The weakly dependent case.* PhD Thesis, Univ. of Paris (Paris).

Matula, P. (2003). On some inequalities for positively and negatively dependent random variables with applications. *Publ. Math. Debrecen* **63**, no. 4, 511–522.

Maxwell, M. and Woodroofe, M. (2000). Central limit theorem for additive fonctionals of Markov chains. *Ann. Probab.* **28**, no. 2, 713–724.

Merlevède, F. (2003). On the central limit theorem and its weak invariance principle for strongly mixing sequences with values in a Hilbert space via martingale approximation. *J. Theoret. Probab.* **16**, no. 3, 625–653.

Merlevède, F., Peligrad, C. and Peligrad, M. (2012). Almost sure invariance principles via martingale approximation. *Stochastic Process. Appl.* **122**, no. 1, 170–190.

Merlevède, F. and Peligrad, M. (2000). The functional central limit theorem under the strong mixing condition. *Ann. Probab.* **28**, no. 3, 1336–1352.

Merlevède, F. and Peligrad, M. (2002). On the coupling of dependent random variables and applications, in: *Empirical process techniques for dependent data* (pp. 171–193). Birkhäuser Boston, Boston, MA.

Merlevède, F. and Peligrad, M. (2006). On the weak invariance principle for stationary sequences under projective criteria. *J. Theoret. Probab.* **19**, no. 3, 647–689.

Merlevède, F. and Peligrad, M. (2013). Rosenthal-type inequalities for the maximum of partial sums of stationary processes and examples. *Ann. Probab.* **41**, no. 2, 914–960.

Merlevède, F., Peligrad, M. and Utev, S. (1997). Sharp conditions for the CLT of linear processes in a Hilbert space. *J. Theoret. Probab.* **10**, no. 3, 681–693.

Merlevède, F., Peligrad, M. and Utev, S. (2006). Recent advances in invariance principles for stationary sequences. *Probab. Surv.* **3**, 1–36.

Merlevède, F., Peligrad, M. and Utev, S. (2019). Functional CLT for martingale-like nonstationary dependent structures. To appear in *Bernoulli.*

Meyer, P.A. and Zheng, W.A. (1984). *Construction du processus de Nelson reversible.* Sém. Probab. XIX, Lect. Notes in Math. No 1123, 12–26.

Mokkadem, A. (1988). Mixing properties of ARMA processes. *Stochastic Process. Appl.* **29**, no. 2, 309–315.

Nachbin, L. (1965). *Topology and Order.* Van Nostrand, New York.

Newman, C.M. (1980). Normal fluctuations and the FKG inequalities. *Comm. Math. Phys.* **75**, no. 2, 119–128.

Newman, C.M. (1984). Asymptotic independence and limit theorems for positively and negatively dependent random variables. in: Y.L. Tong, ed., *Inequalities in Statistics and Probability* (pp. 127–140). Institute of Mathematical Statistics, Hayward, California.

Newman, C.M. and Wright, A.L. (1981). An invariance principle for certain dependent sequences. *Ann. Probab.* **9**, no. 4, 671–675.

Newman, C.M. and Wright, A.L. (1982) Associated random variables and martingale inequalities. *Z. Wahr. verw. Gebiete.* **59**, no. 3, 361–371.

Nobel, A. and Dembo, A. (1993). A note on uniform laws of averages for dependent processes, *Stat. and Prob. Letters* **17**, no. 3, 169–172.

Oliveira, P.E. (2012). *Asymptotics for associated random variables.* Springer, Heidelberg.

Orey, S. (1959). Recurrent Markov chains. *Pacific J. Math.* **9**, no. 3, 805–827.

Ornstein, D. (1974). *Ergodic theory, randomness, and dynamical systems.* Yale University Press, New Haven and London.

Peligrad, M. (1985). An invariance principle for ϕ-mixing sequences. *Ann. Probab.* **13**, no. 4, 1304–1313.

Peligrad, M. (1986). Recent advances in the central limit theorem and its weak invariance principle for mixing sequences of random variables (a survey). *Dependence in probability and statistics* (Oberwolfach, 1985), pp. 193–223, *Progr. Probab. Statist.*, **11**. Birkhäuser, Boston, MA.

Peligrad, M. (1990). On Ibragimov–Iosifescu conjecture for ϕ-mixing sequences. *Stochastic Process. Appl.* **35**, no. 2, 293–308.

Peligrad, M. (1993). Asymptotic results for ϕ-mixing sequences. *Doeblin and modern probability* (Blaubeuren, 1991), pp. 163–169, *Contemp. Math.*, **149**. Amer. Math. Soc., Providence, RI.

Peligrad, M. (2001). A note on the uniform laws for dependent processes via coupling. *J. Theoret. Probab.* **14**, no. 4, 979–988.

Peligrad, M. (2002). Some remarks on coupling of dependent random variables. *Statist. Probab. Lett.* **60**, no. 2, 201–209.

Peligrad, M. (2010). Conditional central limit theorem via martingale approximation, in: *Dependence in probability, analysis and number theory* (pp. 295–309). Kendrick Press, Heber City, UT.

Peligrad, M. and Utev, S. (1997). Central limit theorem for linear processes. *Ann. Probab.* **25**, no. 1, 443–456.

Peligrad, M. and Utev, S. (2005). A new maximal inequality and invariance principle for stationary sequences. *Ann. Probab.* **33**, no. 2, 798–815.

Peligrad, M. and Utev, S. (2006). Central limit theorem for stationary linear processes. *Ann. Probab.* **34**, no. 4, 1608–1622.

Peligrad, M. and Utev, S. (2016). On the invariance principle for reversible Markov chains. *J. Appl. Probab.* **53**, no. 2, 593–599.

Peligrad, M., Utev, S. and Wu, W.B. (2007). A maximal L_p-inequality for stationary sequences and its applications. *Proc. Amer. Math. Soc.* **135**, no. 2, 541–550.

Peligrad, M. and Wu, W.B. (2010). Central limit theorem for Fourier transforms of stationary processes. *Ann. Probab.* **38**, no. 5, 2009–2022.

Pemantle, R. (2000). Towards a theory of negative dependence. Probabilistic techniques in equilibrium and nonequilibrium statistical physics. *J. Math. Phys.* **41**, no. 3, 1371–1390.

Philipp, W. (1979). Almost sure invariance principles for sums of B-valued random variables, Probability in Banach spaces (Proc. Conf. Oberwolfach 1978), *Lecture Notes In Math.* **709**, 171–193. Springer, New- York- Heidelberg- Berlin.

Philipp, W. (1982). Invariance principles for sums of mixing random elements and the multivariate empirical process, in: *Colloq. MathSoc. János Bolyai No. 36. Limit Theorems in Probability and Statistics, Veszprém, Hungary* (North-Holland, Amsterdam), pp. 843–873.

Philipp, W. (1986). Invariance principles for independent and weakly dependent random variables, in: *Dependence in Probability and Statistics* (Proc. Conf. Oberwolfach. 1985), Boston, pp. 225–268.

Philipp, W. (1988). Limit theorems for sums of partial quotients of continued fractions. *Monatsh. Math.* **105**, no. 3, 195–206.

Pinelis, I. (1994). Optimum bounds for the distributions of martingales in Banach spaces. *Ann. Probab.* **22**, no 4, 1679–1706.

Pinelis, I. (2015). Best possible bounds of the von Bahr–Esseen type. *Ann. Funct. Anal.* **6**, no 4, 1–29.

Pollard, D. (1984). *Convergence of Stochastic Processes*. Springer, New York.

Puhalskii, A. (1994). Large deviations of semimartingales via convergence of the predictable characteristics. *Stoch. Stoch. Rep.* **49**, no. 1–2, 27–85.

Rio, E. (1993). Covariance inequalities for strongly mixing processes. *Ann. Inst. Henri Poincaré Probab. Stat.* **29**, no 4, 587–597.

Rio, E. (1995). The functional law of the iterated logarithm for stationary strongly mixing sequences. *Ann. Probab.* **23**, no. 3, 1188–1203.

Rio, E. (1998). Absolutely regular empirical processes and universal entropy. *Probab. Theory Related Fields* **111**, no. 4, 585–608.

Rio, E. (2000). *Théorie asymptotique des processus al éatoires faiblement dépendants*. Mathématiques et applications de la SMAI. **31**. Springer-Verlag, Berlin-Heidelberg.

Rio, E. (2009). Moment inequalities for sums of dependent random variables under projective conditions. *J. Theoret. Probab.* **22**, no. 1, 146–163.

Rio, E. (2017). About the constants in the Fuk–Nagaev inequalities. *Electron. Commun. Probab.* **22**, no. 28, 1–12.

Roberts, G.O. and Rosenthal, J.S. (2008). Variance bounding Markov chains. *Ann. Appl. Probab.* **18**, no. 3, 1201–1214.

Rootzén, H. (1980). On the functional central limit theorem for martingales. II. *Z. Wahrsch. Verw. Gebiete* **51**, no. 1, 79–93.

Rosenblatt, M. (1956). A central limit theorem and a strong mixing condition. *Proc. Nat. Acad. Sci. USA* **42**, 43–47.

Rosenblatt, M. (1971). *Markov Processes. Structure and Asymptotic Behavior.* Springer-Verlag, New York.

Rosenblatt, M. (1972). Central limit theorems for stationary processes, in: *Proceedings of the Sixth Berkeley Symposium on Probability and Statistics*, vol. **2**, pp. 551–561. University of California Press, Los Angeles.

Rosenblatt, M. (1974). *Random processes*. Second edition. Graduate Texts in Mathematics, No. 17. Springer-Verlag, New York-Heidelberg.

Rosenblatt, M. (1980). Linear processes and bispectra. *J. Appl. Probab.* **17**, no. 1, 265–270.

Rosenblatt, M. (1985). *Stationary Sequences and Random Fields*. Birkhäuser, Boston, MA.

Rosenthal, H.P. (1970). On the subspaces of L_p (p>2) spanned by sequences of independent random variables. *Israel J. Math.* **8**, no. 3, 273–303.

Samur, J.D. (1984). Convergence of sums of mixing triangular arrays of random vectors with stationary rows. *Ann. Probab.* **12**, no. 2, 390–426.

Schwarz, G. (1980). Finitely determined processes–an indiscrete approach. *J. Math. Anal. Appl.* **76**, no. 1, 146–158.

Seip, K. and Ulanovskii, A.M. (1997). The Beurling–Malliavin density of a random sequence. *Proc. Am. Math. Soc.* **125**, no. 6, 1745–1749.

Shao, Q. M. (1988). A remark on the invariance principle for ρ-mixing sequences of random variables. *Chinese Ann. Math. Ser. A* **9**, no. 4, 409–412.

Shao, Q. M. (1995). Maximal inequalities for partial sums of ρ-mixing sequences. *Ann. Probab.* **23**, no. 2, 948–965.

Shao, Q. M. (2000). A comparison theorem on moment inequalities between negatively associated and independent random variables. *J. Theoret. Probab.* **13**, no. 2, 343–356.

Shao, Q. M. and Yu H. (1996). Weak convergence for weighted empirical processes of dependent sequences. *Ann. Probab.* **24**, no. 4, 2098–2127.

Shields, P. (1973). *The Theory of Bernoulli Shifts*. University of Chicago Press, Chicago.

Shirai, T. and Takahashi, Y. (2003). Random point fields associated with certain Fredholm determinants. II. Fermion shifts and their ergodic and Gibbs properties. *Ann. Probab.* **31**, no. 3, 1533–1564.

Skorohod, A.V. (1956). Limit theorems for stochastic processes. *Theory Prob. Appl.* **1**, no. 1, 261–290.

Soshnikov, A. (2000). Determinantal random point fields. *Usp. Mat. Nauk.* **55**, no. 5, 107–160.

Soshnikov, A. (2002). Gaussian limit for determinantal random point fields. *Ann. Probab.* **30**, no. 1, 171–187.

Stein, C. (1972). A bound for the error in the normal approximation to the distribution of a sum of dependent random variables. *Proceedings of the Sixth Berkeley Symposium on Probability and Statistics*, vol. **2**, pp. 583–602. University of California Press, Los Angeles.

Strassen, V. (1965). The existence of probability measures with given marginals. *Ann. Math. Statist.* **36**, no. 2, 423–439.

Tierney, L. (1994). Markov chains for exploring posterior distribution (with discussion). *Ann. Statist.* **22**, no. 4, 1701–1762.

Utev, S. (1990). Central limit theorem for dependent random variables. *Probab. Theory Math. Statist.* Vol. II (Vilnius, 1989), 519–528, "Mokslas", Vilnius.

Utev, S. (1991). Sums of random variables with ϕ-mixing [translation of *Trudy Inst. Mat.* (Novosibirsk) 13 (1989), Asimptot. Analiz Raspred. Sluch. Protsess., 78–100]. *Siberian Adv. Math.* **1**, no. 3, 124–155.

Utev, S. and Peligrad, M. (2003). Maximal inequalities and an invariance principle for a class of weakly dependent random variables. *J. Theoret. Probab.* **16**, no. 1, 101–115.

Vapnik, V.N. and Cervonenkis, A.Ya. (1981). Necessary and sufficient conditions for the uniform convergence of means to their expectations. *Theory Probab. Appl.* **26**, no. 3, 532–553.

Vasershtein, L.N. (1969). Markov processes on countable product spaces describing large systems of automata. *Problemy Peredači Informacii* **5**, no. 3, 64–72.

Volkonskii, V.A. and Rozanov, Yu.A. (1959). Some limit theorems for random functions. I. *Theory Probab. Appl.* **4**, no. 2, 178–197.

Volný, D. (1993). Approximating martingales and the central limit theorem for strictly stationary processes. *Stoch. Processes Appl.* **44**, no. 1, 41–74.

Volný, D. and Samek, P. (2000). On the invariance principle and the law of iterated logarithm for stationary processes, in: *Mathematical Physics and Stochastic Analysis*, pp. 424–38. World Sci. Publishing, River Edge.

Wang, Q., Lin, X.-Y. and Gulati, C.M. (2001). Asymptotics for moving average processes with dependent innovations. *Statist. Probab. Lett.* **54**, no. 4, 347–356.

Wooldridge, J.M. and White, H. (1988). Some invariance principles and central limit theorems for dependent heterogeneous processes. *Econometric Theory* **4**, no. 2, 210–230.

Worms, J. (2000). *Principes de déviations modérées pour des martingales et applications statistiques*. PhD thesis, University of Marne-la-Vallée, France.

Wu, W.B. (2007). Strong invariance principles for dependent random variables. *Ann. Probab.* **35**, no. 6, 2294–2320.

Wu, W.B. and Woodroofe, M. (2004). Martingale approximations for sums of stationary processes. *Ann. Probab.* **32**, no. 2, 1674–1690.

Wu, W.B. and Shao, X. (2004). Limit theorems for iterated random functions. *J. Appl. Probab.* **41**, no. 2, 425–436.

Wu, W.B. and Zhao, Z. (2008). Moderate deviations for stationary processes. *Statistica Sinica* **18**, no. 2, 769–782.

Yu, B. (1991). Rates of convergence and central limit theorems for empirical processes of stationary mixing sequences, *Tech. Rept. 260*. Dept. of Statistics, University of California, Berkeley, CA.

Yuan, M., Su, C. and Hu, T. (2003). A central limit theorem for random fields of negatively associated processes. *J. Theoret. Probab.* **16**, no. 2, 309–323.

Yukich, J.E. (1986). Rates of convergence for classes of functions: The non-i.i.d. case, *J. Multivariate Anal.* **20**, no. 2, 175–189.

Zhang, L.X. (2000). A functional central limit theorem for asymptotically negatively dependent random variables. *Acta Math. Hungar.* **86**, no. 3, 237–259.

Zhang, Na. (2017). On the law of large numbers for discrete Fourier transform. *Statist. Probab. Lett.* **120**, 101–107.

Zhao, O., Woodroofe, M. and Volný, D. (2010). A central limit theorem for reversible processes with nonlinear growth of variance. *J. Appl. Probab.* **47**, no. 4, 1195–1202.

SUBJECT INDEX